The Tube Amplifier Schematic Bible Volume 1

A-F

Salvatore Gambino

ISBN: 1500447366
ISBN-13: 978-1500447366

INTRODUCTION

This book of amp schematics was assembled with service and repair in mind. I have always had a very deep respect for the design and performance that tube amps produce. Let's face it, guitar tube amps don't always get the respect that they deserve. Tube amplifiers have always worked hard and should be looked at as a major part of your sound as they inspire you to dig deep into your playing. If you feel somewhat the same way I do about tube amps, then you know each amplifier has their own characteristics and tone. I hope you can use this educational information to understand how tube amps are designed and how they work.
Look for my other book

Reading Schematics Made Easy.

CONTENTS

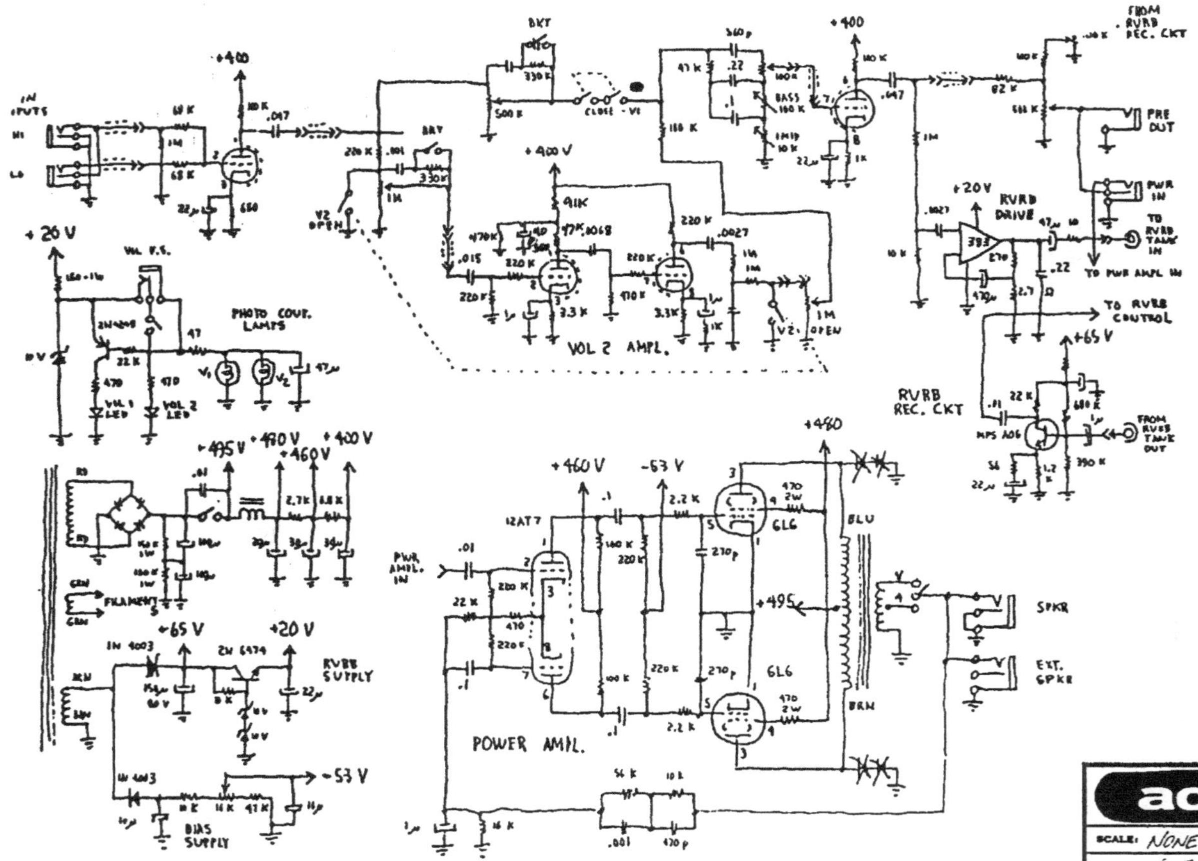

acoustic

SCALE: NONE	APPROVED BY:	DRAWN BY JEB
DATE: 06-30-81		REVISED
Model: G60T TUBE GUITAR AMPLIFIER		
		DRAWING NUMBER

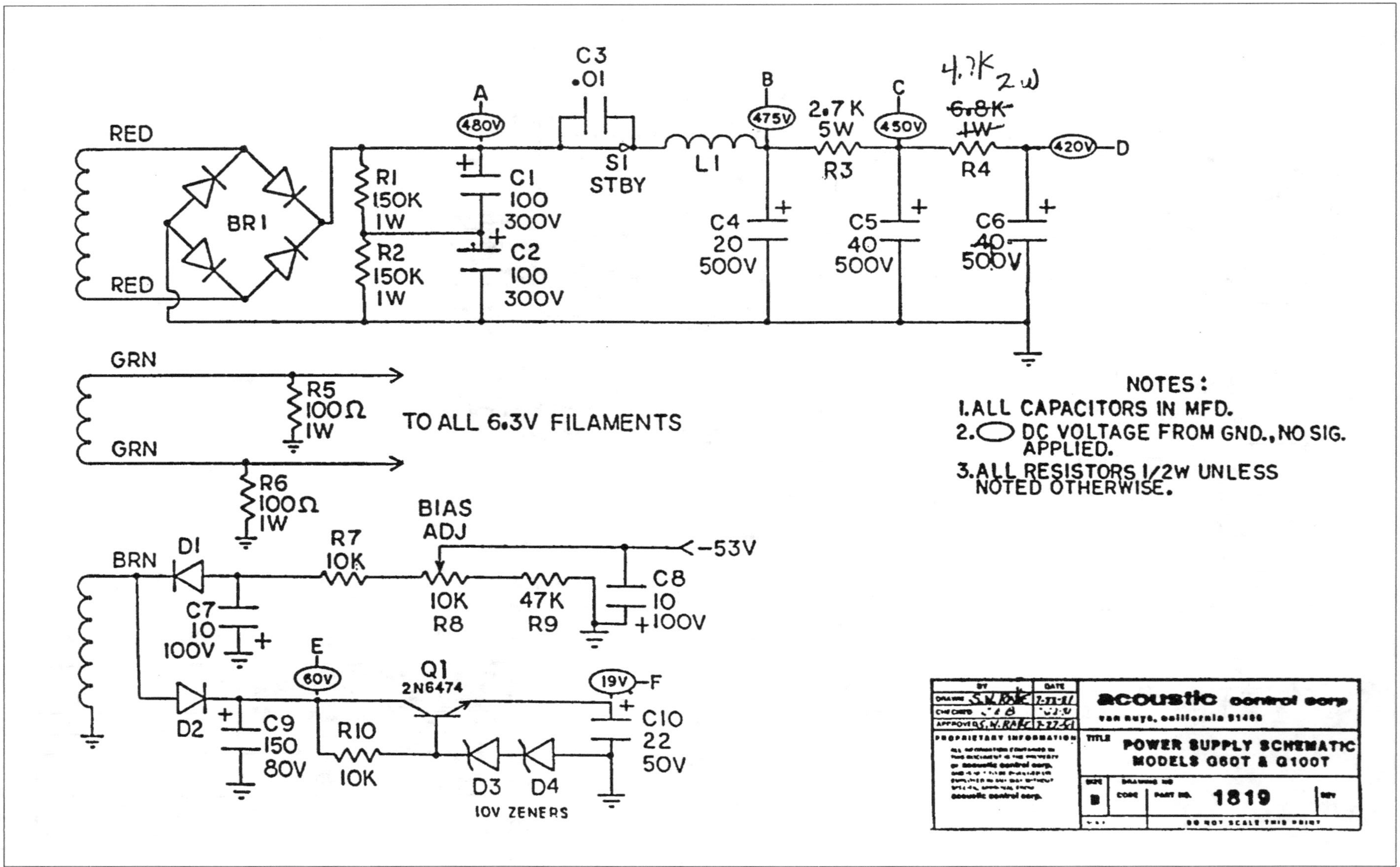
RED
RED
BR1
A
480V
R1
150K
1W
R2
150K
1W
C1
100
300V
C2
100
300V
C3
.01
S1
STBY
L1
B
475V
C4
20
500V
2.7K
5W
R3
C
450V
C5
40
500V
4.7K 2W
6.8K
1W
R4
C6
40
500V
420V
D
GRN
GRN
R5
100Ω
1W
R6
100Ω
1W
TO ALL 6.3V FILAMENTS
BRN
D1
C7
10
100V
R7
10K
BIAS
ADJ
10K
R8
47K
R9
-53V
C8
10
100V
D2
C9
150
80V
E
60V
R10
10K
Q1
2N6474
D3
D4
10V ZENERS
19V
F
C10
22
50V
NOTES:
1. ALL CAPACITORS IN MFD.
2. DC VOLTAGE FROM GND., NO SIG. APPLIED.
3. ALL RESISTORS 1/2W UNLESS NOTED OTHERWISE.
acoustic control corp
van nuys, california 91406
POWER SUPPLY SCHEMATIC
MODELS G60T & G100T
1819
DO NOT SCALE THIS PRINT

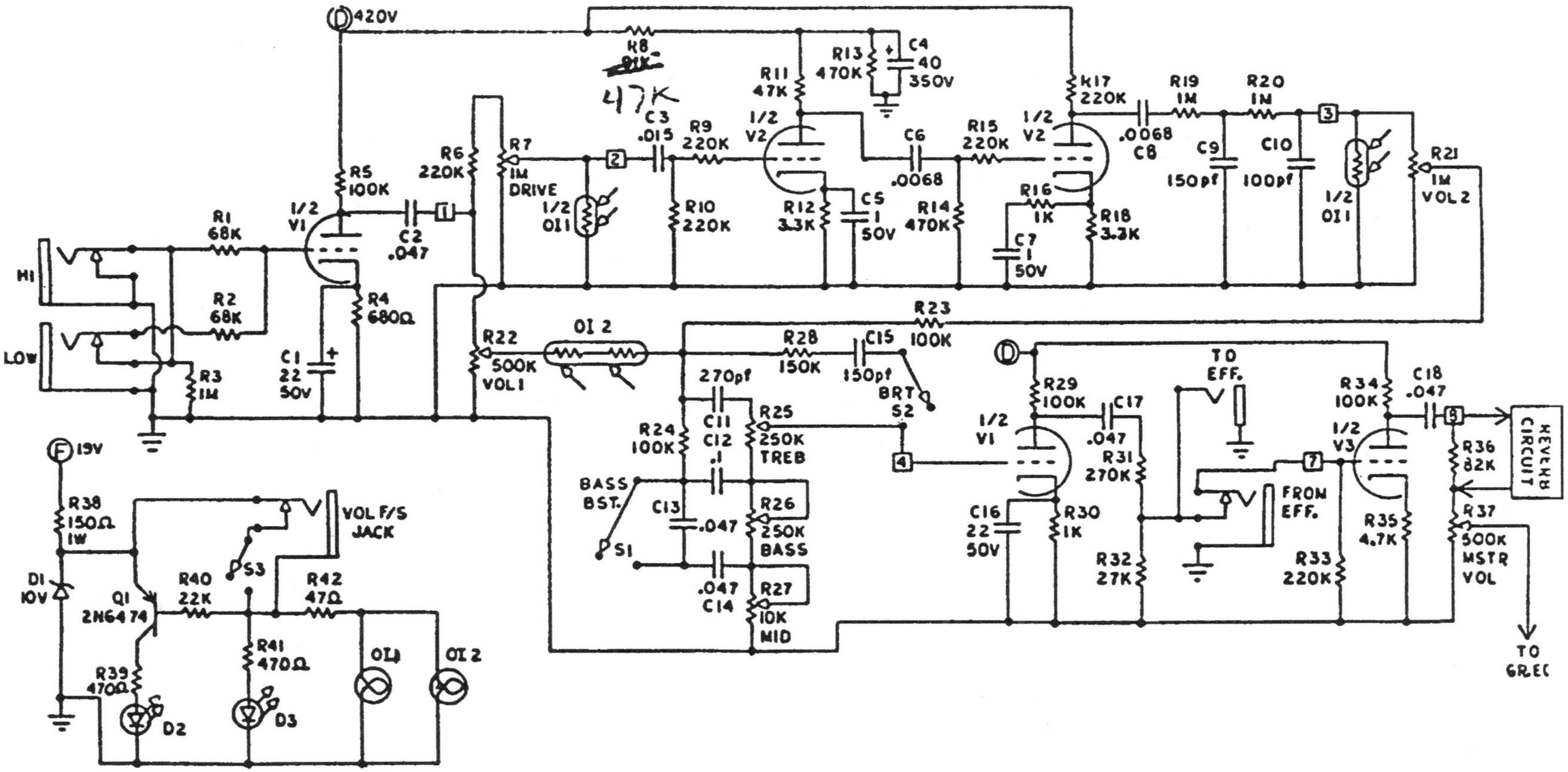

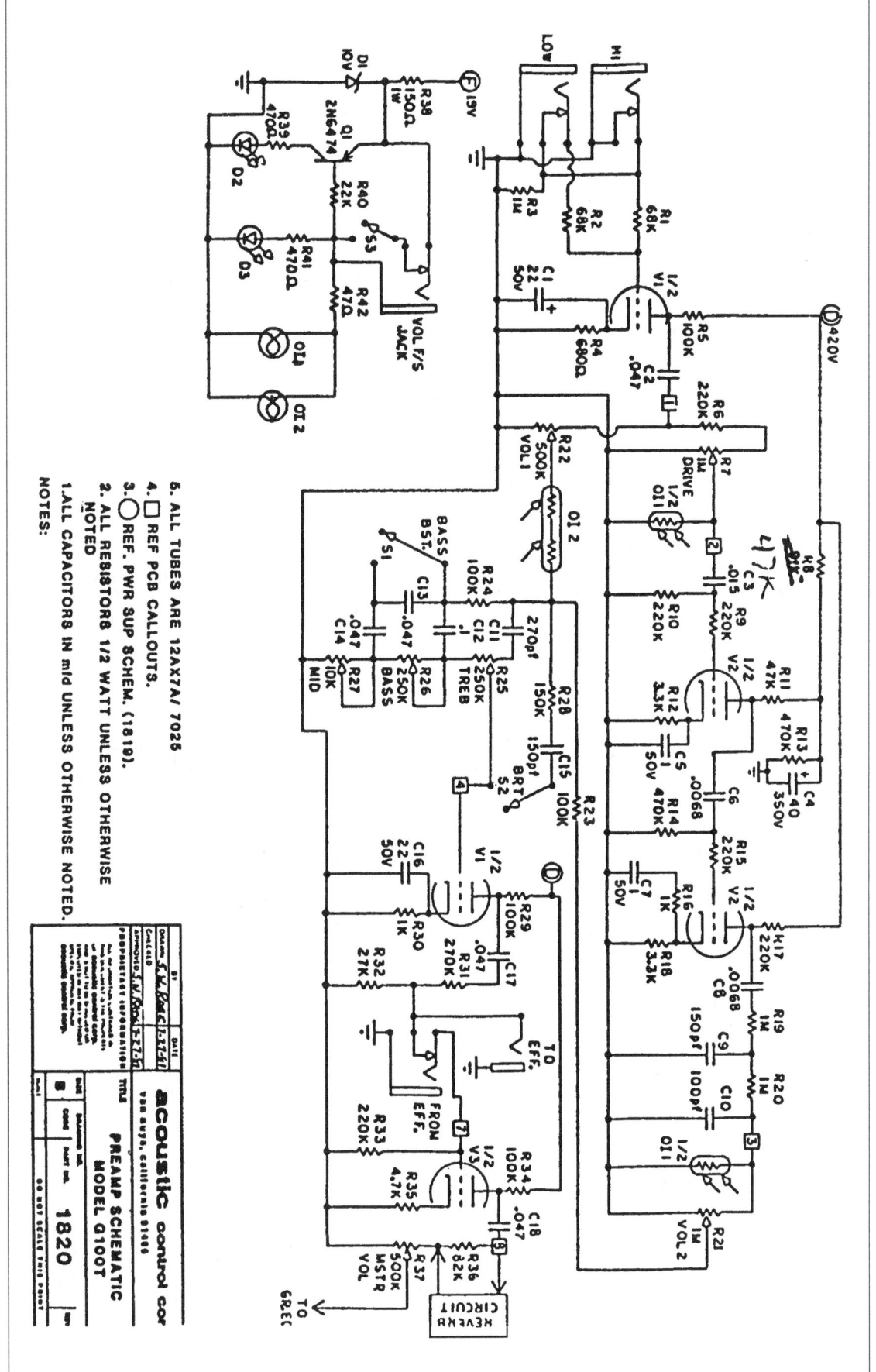

5. ALL TUBES ARE 12AX7A/ 7025
4. □ REF PCB CALLOUTS.
3. ○ REF. PWR SUP SCHEM. (1819).
2. ALL RESISTORS 1/2 WATT UNLESS OTHERWISE NOTED
1. ALL CAPACITORS IN mfd UNLESS OTHERWISE NOTED.
NOTES:

DRAWN S.W. Rose	7-27-81	**acoustic** control cor
CHECKED		van nuys, california 91406
APPROVED S.W. Rose	7-27-81	
PROPRIETARY INFORMATION		TITLE PREAMP SCHEMATIC MODEL G100T
		SIZE B · CODE · PART NO. 1820 · REV
		DO NOT SCALE THIS PRINT

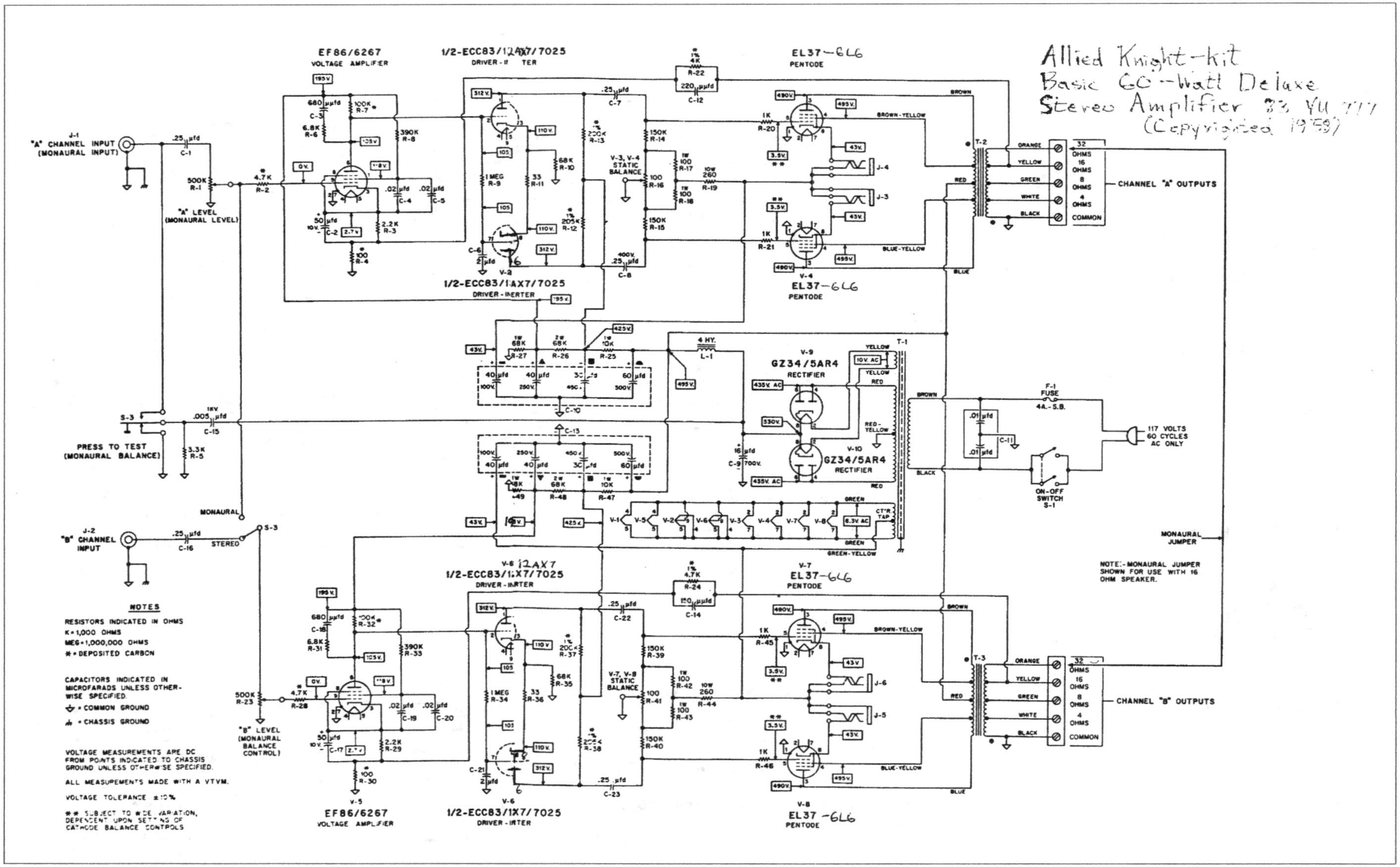
Allied Knight-kit
Basic 60-Watt Deluxe
Stereo Amplifier 83 YU 777
(Copyrighted 1959)
EF86/6267
VOLTAGE AMPLIFIER
1/2-ECC83/12AX7/7025
DRIVER-INVERTER
EL37-6L6
PENTODE
J-1
"A" CHANNEL INPUT
(MONAURAL INPUT)
"A" LEVEL
(MONAURAL LEVEL)
V-3, V-4
STATIC
BALANCE
CHANNEL "A" OUTPUTS
CHANNEL "B" OUTPUTS
GZ34/5AR4
RECTIFIER
F-1
FUSE
4A.-S.B.
117 VOLTS
60 CYCLES
AC ONLY
ON-OFF
SWITCH
S-1
PRESS TO TEST
(MONAURAL BALANCE)
MONAURAL
STEREO
J-2
"B" CHANNEL
INPUT
"B" LEVEL
(MONAURAL
BALANCE
CONTROL)
V-7, V-8
STATIC
BALANCE
MONAURAL
JUMPER
NOTE:-MONAURAL JUMPER
SHOWN FOR USE WITH 16
OHM SPEAKER.
NOTES
RESISTORS INDICATED IN OHMS
K=1,000 OHMS
MEG=1,000,000 OHMS
✱=DEPOSITED CARBON
CAPACITORS INDICATED IN
MICROFARADS UNLESS OTHER-
WISE SPECIFIED.
= COMMON GROUND
= CHASSIS GROUND
VOLTAGE MEASUREMENTS ARE DC
FROM POINTS INDICATED TO CHASSIS
GROUND UNLESS OTHERWISE SPECIFIED.
ALL MEASUREMENTS MADE WITH A VTVM.
VOLTAGE TOLERANCE ±10%
✱✱ SUBJECT TO WIDE VARIATION,
DEPENDENT UPON SETTING OF
CATHODE BALANCE CONTROLS

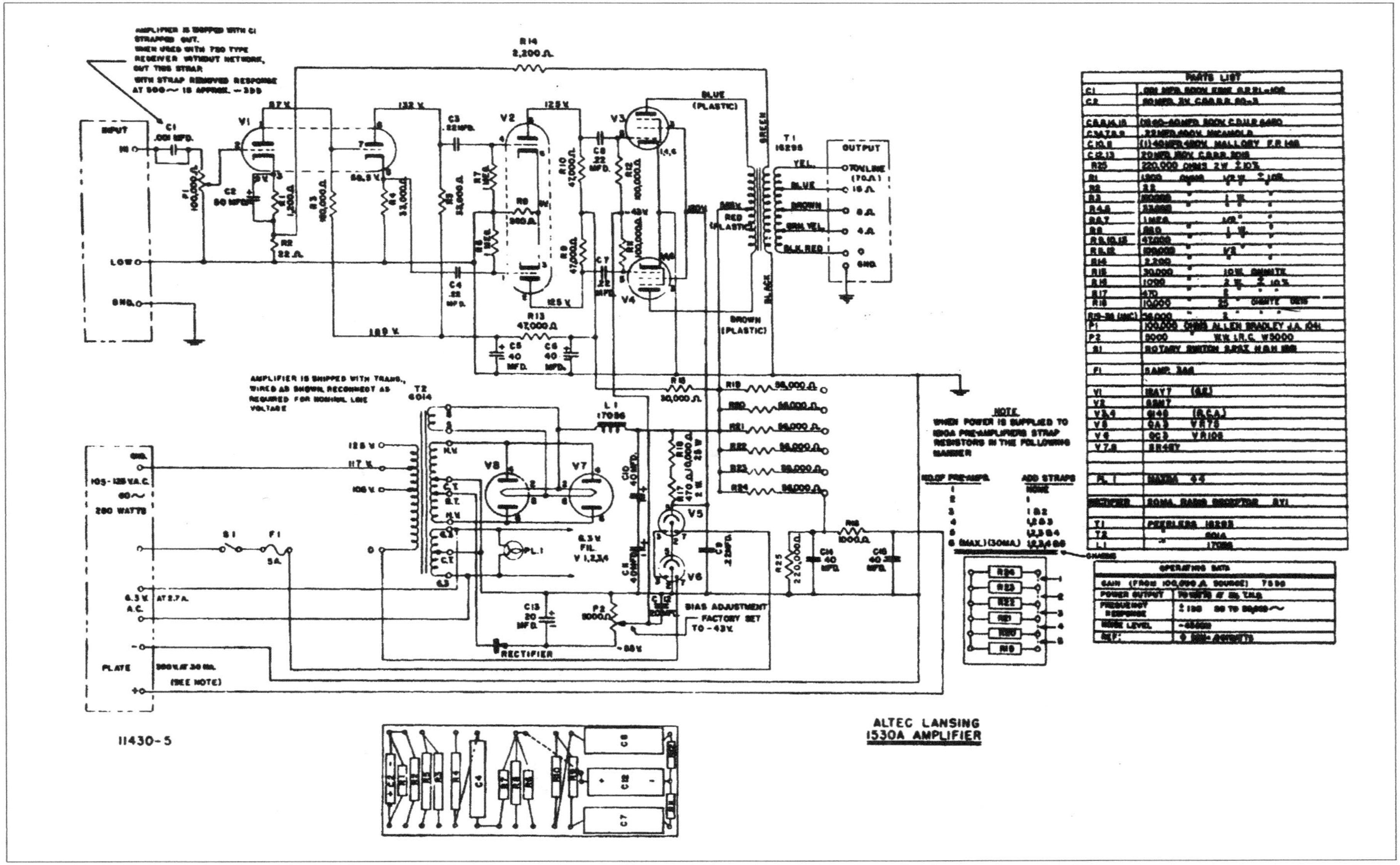
INPUT
HI
LOW
GND.
C1
.001 MFD.
P1
100,000 Ω
C2
R1
1,200 Ω
R2
22 Ω
R3
R4
33,000 Ω
R5
33,000 Ω
C3
.22 MFD.
C4
.22 MFD.
R14
2,200 Ω
V1
V2
V3
V4
V5
V6
V7
V8
R13
47,000 Ω
C5
40 MFD.
C6
40 MFD.
R8
47,000 Ω
R10
47,000 Ω
R12
100,000 Ω
BLUE
(PLASTIC)
RED
(PLASTIC)
BROWN
(PLASTIC)
GREEN
BLACK
T1
16295
OUTPUT
YEL.
BLUE
BROWN
GRN. YEL.
BLK. RED
16 Ω
8 Ω
4 Ω
0
GND.
R15
30,000 Ω
AMPLIFIER IS SHIPPED WITH TRANS., WIRED AS SHOWN. RECONNECT AS REQUIRED FOR NOMINAL LINE VOLTAGE
T2
6014
L1
17086
125 V.
117 V.
105 V.
GND.
105-125 V.A.C.
60~
280 WATTS
S1
F1
5A.
6.3 V. A.C.
AT 2.7 A.
PLATE
(SEE NOTE)
6.3 V. FIL.
V 1,2,3,4
PL.1
C13
20 MFD.
P2
5000 Ω
RECTIFIER
BIAS ADJUSTMENT
FACTORY SET
TO -43V.
C9
.22 MFD.
R25
220,000 Ω
C14
40 MFD.
R16
1000 Ω
NOTE
WHEN POWER IS SUPPLIED TO 1530A PRE-AMPLIFIERS STRAP RESISTORS IN THE FOLLOWING MANNER
NO. OF PRE-AMPS.
ADD STRAPS
NONE
PARTS LIST
OPERATING DATA
11430-5
ALTEC LANSING
1530A AMPLIFIER

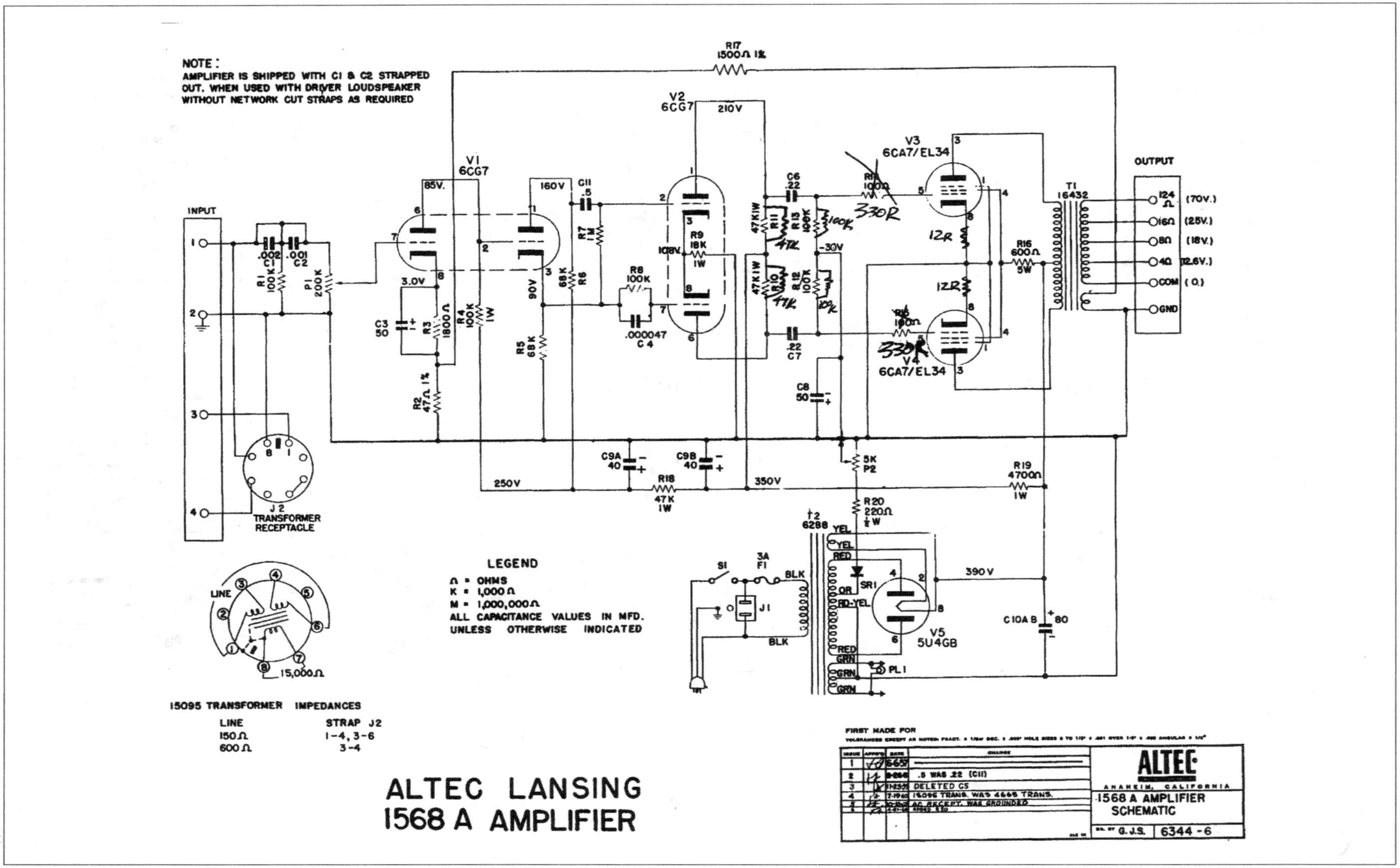
NOTE:
AMPLIFIER IS SHIPPED WITH C1 & C2 STRAPPED OUT. WHEN USED WITH DRIVER LOUDSPEAKER WITHOUT NETWORK CUT STRAPS AS REQUIRED
INPUT
OUTPUT
124 Ω (70V.)
16Ω (25V.)
8Ω (18V.)
4Ω (12.6V.)
COM (0.)
GND
V1 6CG7
V2 6CG7
V3 6CA7/EL34
V4 6CA7/EL34
V5 5U4GB
T1 16432
T2 6288
R17 1500Ω 1%
J2 TRANSFORMER RECEPTACLE
LEGEND
Ω = OHMS
K = 1,000Ω
M = 1,000,000Ω
ALL CAPACITANCE VALUES IN MFD. UNLESS OTHERWISE INDICATED
15095 TRANSFORMER IMPEDANCES
LINE 150Ω 600Ω
STRAP J2 1-4, 3-6 3-4
ALTEC LANSING
1568 A AMPLIFIER
FIRST MADE FOR
.5 WAS .22 (C11)
DELETED C5
15095 TRANS. WAS 4665 TRANS.
AC RECEPT. WAS GROUNDED
ALTEC
ANAHEIM, CALIFORNIA
1568 A AMPLIFIER SCHEMATIC
G.J.S.
6344-6

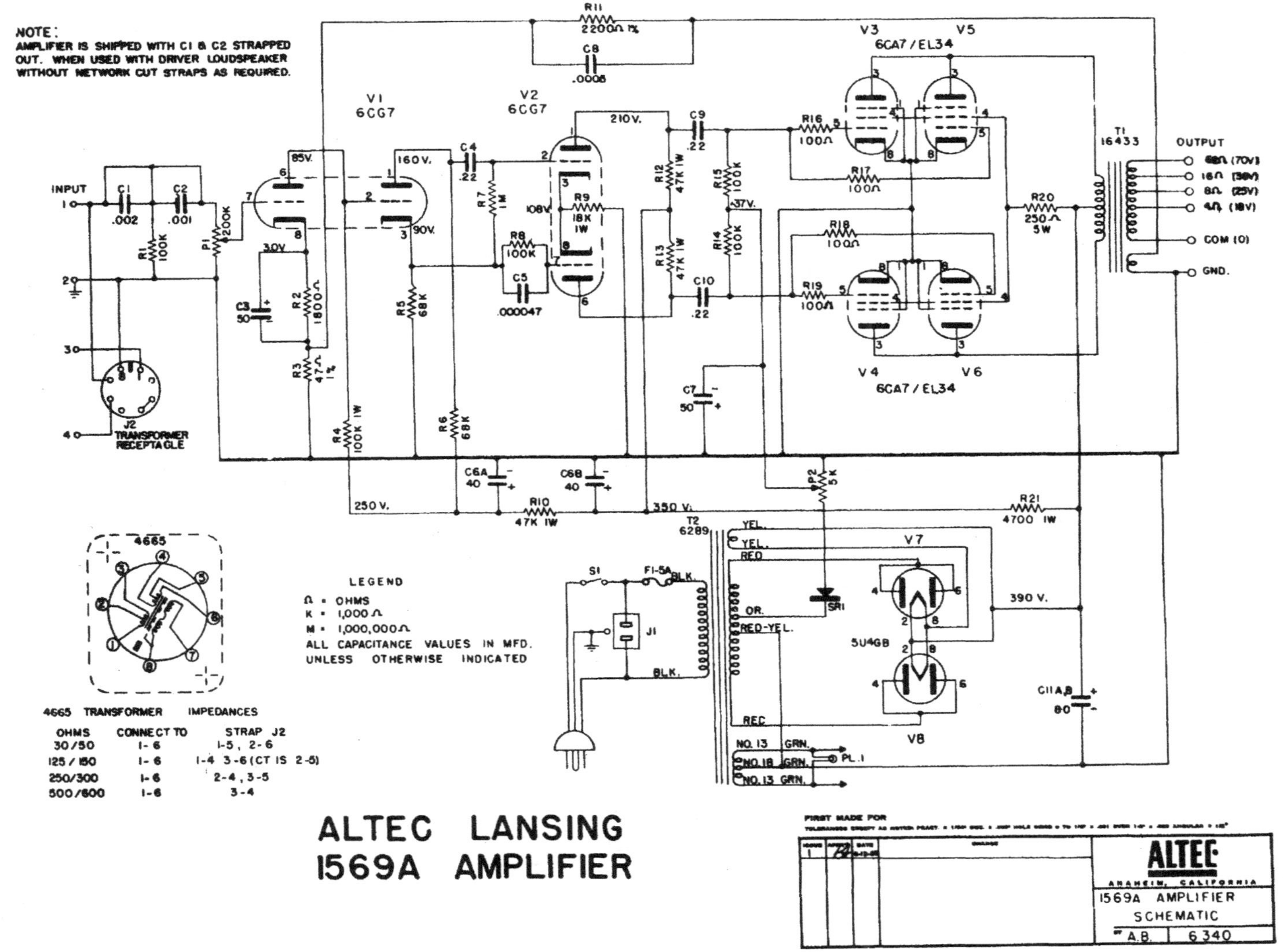
NOTE:
AMPLIFIER IS SHIPPED WITH C1 & C2 STRAPPED OUT. WHEN USED WITH DRIVER LOUDSPEAKER WITHOUT NETWORK CUT STRAPS AS REQUIRED.
INPUT
J2 TRANSFORMER RECEPTACLE
V1 6CG7
V2 6CG7
V3 6CA7/EL34
V5
V4 6CA7/EL34
V6
V7
V8
5U4GB
T1 16433
T2 6289
OUTPUT
8Ω (70V)
16Ω (35V)
8Ω (25V)
4Ω (18V)
COM (0)
GND.
R11 2200Ω 1%
C8 .0006
R1 100K
P1 200K
R2 1800Ω
R3 47Ω 1%
R4 100K 1W
R5 68K
R6 68K
R7 1M
R8 100K
R9 18K 1W
R10 47K 1W
R12 47K 1W
R13 47K 1W
R14 100K
R15 100K
R16 100Ω
R17 100Ω
R18 100Ω
R19 100Ω
R20 250Ω 5W
R21 4700 1W
P2 5K
C1 .002
C2 .001
C3 50
C4 .22
C5 .000047
C6A 40
C6B 40
C7 50
C9 .22
C10 .22
C11A,B 80
85V.
160V.
90V.
30V
108V
210V.
-37V.
250 V.
350 V.
390 V.
S1
F1-5A
J1
SR1
PL.1
BLK.
YEL.
RED
OR.
RED-YEL.
REC
NO. 13 GRN.
NO. 18 GRN.
4665
LEGEND
Ω = OHMS
K = 1,000Ω
M = 1,000,000Ω
ALL CAPACITANCE VALUES IN MFD. UNLESS OTHERWISE INDICATED
4665 TRANSFORMER IMPEDANCES
OHMS CONNECT TO STRAP J2
30/50 1-6 1-5, 2-6
125/150 1-6 1-4 3-6 (CT IS 2-5)
250/300 1-6 2-4, 3-5
500/600 1-6 3-4
ALTEC LANSING
1569A AMPLIFIER
FIRST MADE FOR
ALTEC
ANAHEIM, CALIFORNIA
1569A AMPLIFIER
SCHEMATIC
AB 6340

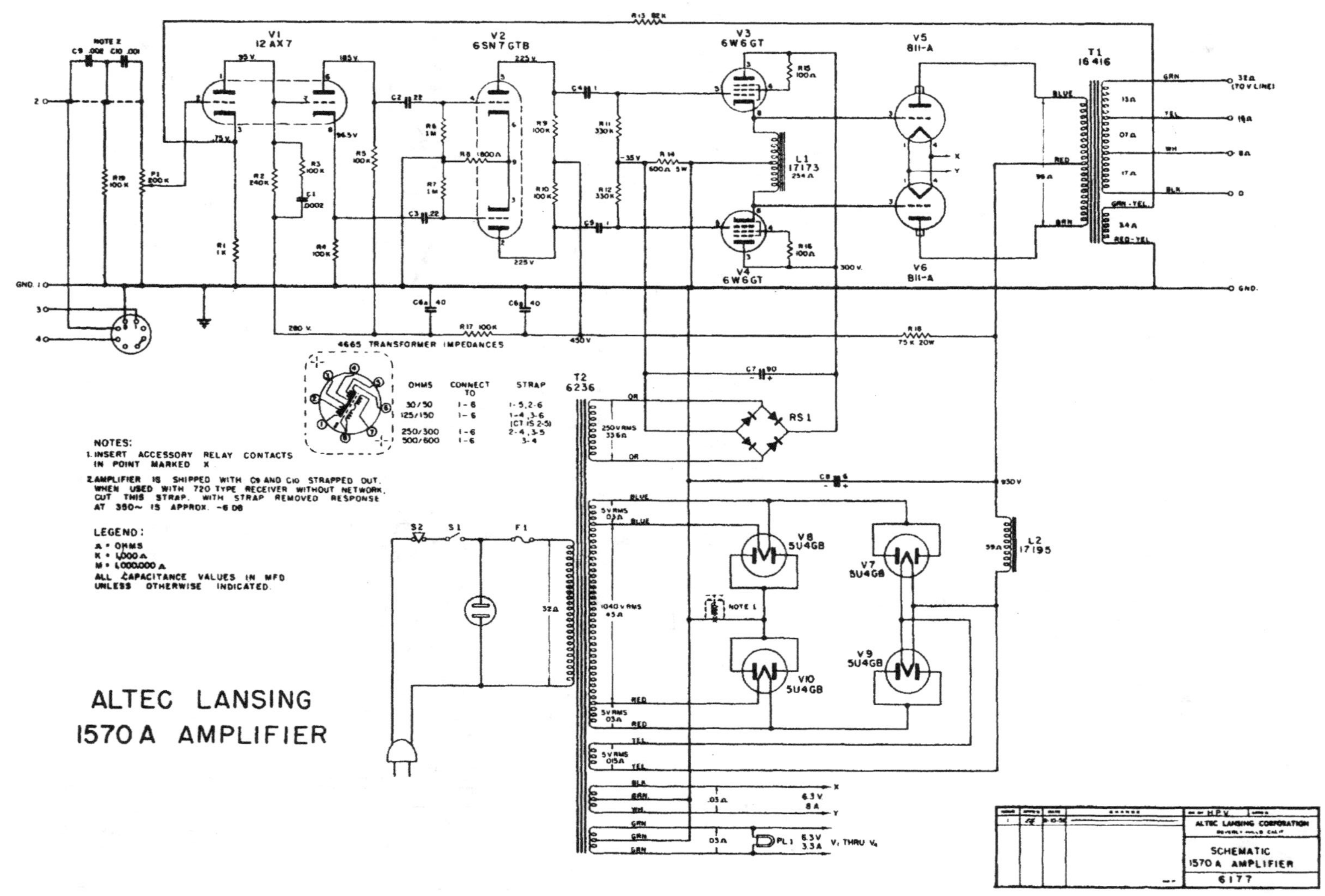
V1
12AX7
V2
6SN7GTB
V3
6W6GT
V4
6W6GT
V5
811-A
V6
811-A
T1
16416
L1
17173
L2
17195
T2
6236
RS1
V7
5U4GB
V8
5U4GB
V9
5U4GB
V10
5U4GB
4665 TRANSFORMER IMPEDANCES
OHMS CONNECT TO STRAP
30/50 1-8 1-5, 2-6
125/150 1-6 1-4, 3-6 (CT IS 2-5)
250/300 1-6 2-4, 3-5
500/600 1-6 3-4
NOTES:
1. INSERT ACCESSORY RELAY CONTACTS IN POINT MARKED X.
2. AMPLIFIER IS SHIPPED WITH C9 AND C10 STRAPPED OUT. WHEN USED WITH 720 TYPE RECEIVER WITHOUT NETWORK, CUT THIS STRAP. WITH STRAP REMOVED RESPONSE AT 350~ IS APPROX. -6 DB
LEGEND:
Ω = OHMS
K = 1,000 Ω
M = 1,000,000 Ω
ALL CAPACITANCE VALUES IN MFD UNLESS OTHERWISE INDICATED.
ALTEC LANSING
1570A AMPLIFIER
ALTEC LANSING CORPORATION
SCHEMATIC
1570A AMPLIFIER
6177

TROUBLESHOOTING MUSICAL INSTRUMENT AMPLIFIERS

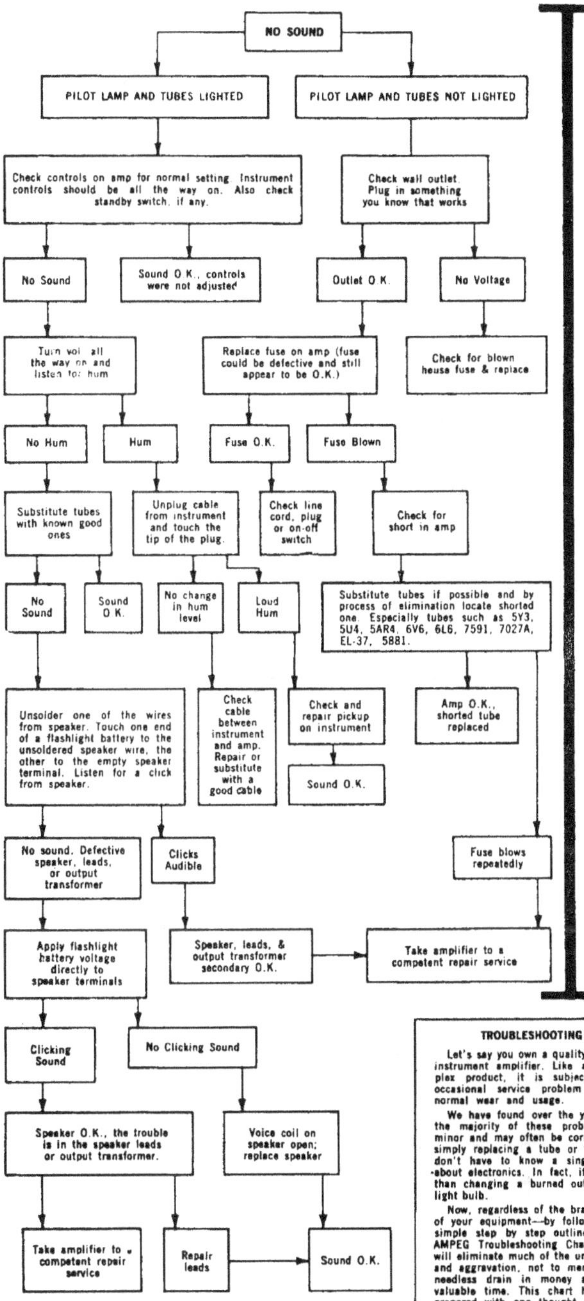

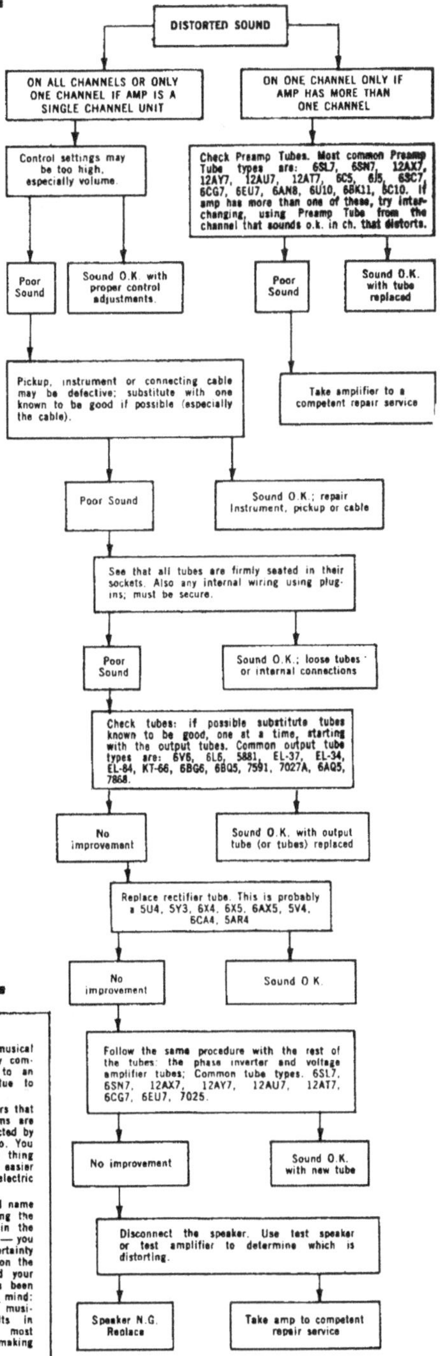

TROUBLESHOOTING

Let's say you own a quality musical instrument amplifier. Like any complex product, it is subject to an occasional service problem due to normal wear and usage.

We have found over the years that the majority of these problems are minor and may often be corrected by simply replacing a tube or two. You don't have to know a single thing about electronics. In fact, it's easier than changing a burned out electric light bulb.

Now, regardless of the brand name of your equipment—by following the simple step by step outline in the AMPEG Troubleshooting Chart — you will eliminate much of the uncertainty and aggravation, not to mention the needless drain in money and your valuable time. This chart has been prepared with one thought in mind: to help you—the professional musician—obtain maximum results in sound and performance and most importantly, to keep you making good music.

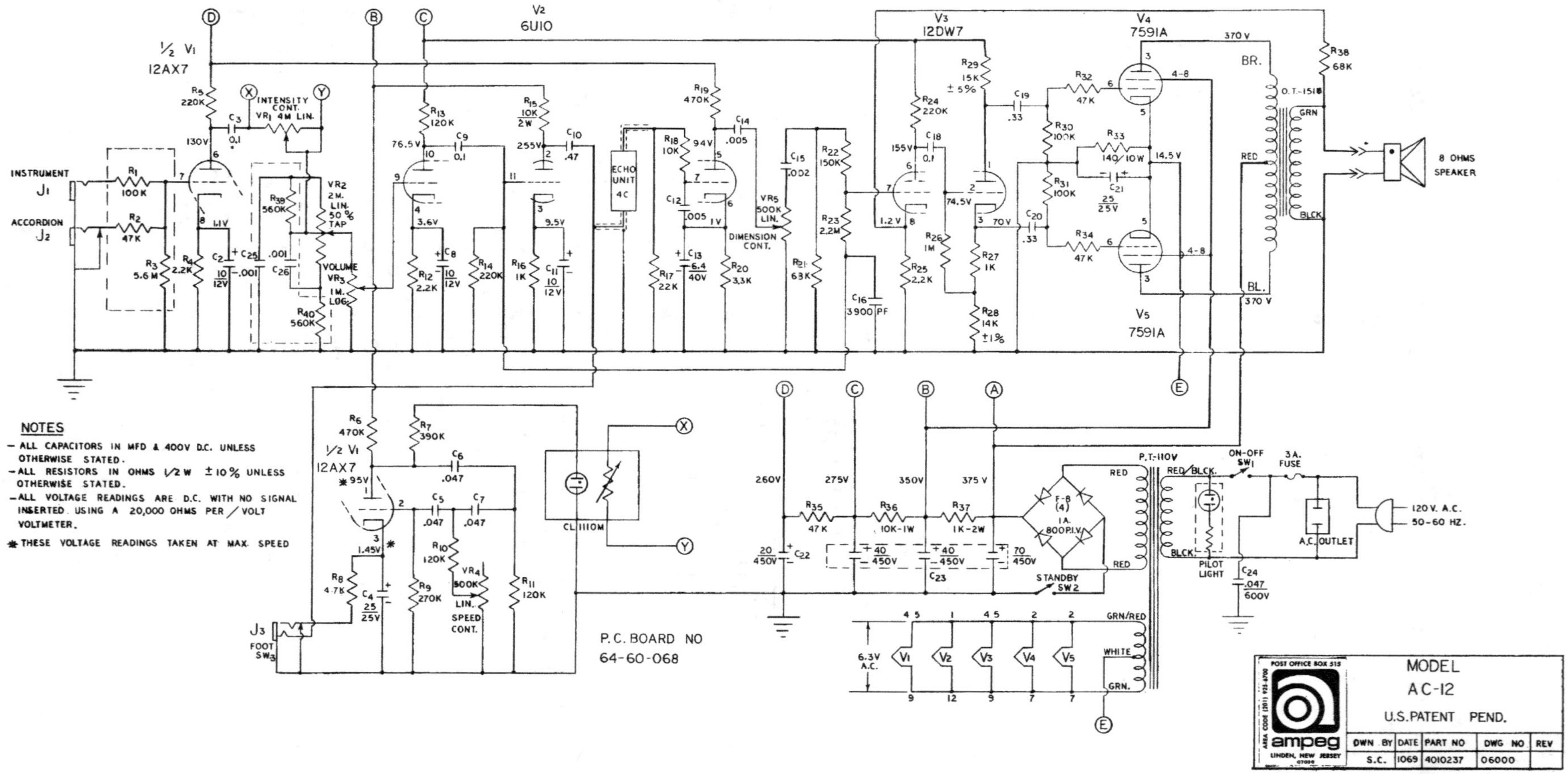
1/2 V1
12AX7
INSTRUMENT
J1
ACCORDION
J2
R1
100K
R2
47K
R3
5.6M
R4
2.2K
R5
220K
130V
1.1V
C2
10
12V
C3
0.1
C25
.001
C26
.001
R39
560K
R40
560K
INTENSITY CONT.
VR1 4M LIN.
VR2
2M.
LIN.
50 %
TAP
VOLUME
VR3
1M.
LOG.
V2
6U10
R12
2.2K
R13
120K
76.5V
3.6V
C8
10
12V
C9
0.1
R14
220K
R15
10K
2W
255V
C10
.47
R16
1K
9.5V
C11
10
12V
ECHO
UNIT
4C
R17
22K
R18
10K
94V
R19
470K
C12
.005
C13
6.4
40V
C14
.005
1V
R20
3.3K
VR5
500K
LIN.
DIMENSION
CONT.
C15
.002
R21
63K
R22
150K
R23
2.2M
C16
3900 PF
V3
12DW7
155V
C18
0.1
R24
220K
1.2 V
R25
2.2K
R26
1M
74.5V
R27
1K
R28
14K
±1%
R29
15K
± 5%
C19
.33
70V
C20
.33
V4
7591A
V5
7591A
R30
100K
R31
100K
R32
47K
R33
140/10W
C21
25
25V
14.5V
R34
47K
370 V
BR.
O.T.-1518
GRN
RED
BLCK
BL.
370 V
R38
68K
8 OHMS
SPEAKER
1/2 V1
12AX7
R6
470K
R7
390K
C6
.047
95V
C5
.047
C7
.047
1.45V
R8
4.7K
C4
25
25V
R9
270K
R10
120K
VR4
500K
LIN.
SPEED
CONT.
R11
120K
J3
FOOT
SW3
CL1110M
P.C. BOARD NO
64-60-068
260V
275V
350V
375 V
R35
47K
R36
10K-1W
R37
1K-2W
C22
20
450V
40
450V
C23
40
450V
70
450V
F-8
(4)
1A.
800P.I.V.
STANDBY
SW2
P.T.-110V
RED
RED/BLCK.
BLCK.
PILOT
LIGHT
ON-OFF
SW1
3A.
FUSE
A.C. OUTLET
120V. A.C.
50-60 HZ.
C24
.047
600V
GRN/RED
WHITE
GRN.
6.3V
A.C.
NOTES
- ALL CAPACITORS IN MFD & 400V D.C. UNLESS OTHERWISE STATED.
- ALL RESISTORS IN OHMS 1/2 W ± 10 % UNLESS OTHERWISE STATED.
- ALL VOLTAGE READINGS ARE D.C. WITH NO SIGNAL INSERTED. USING A 20,000 OHMS PER / VOLT VOLTMETER.
* THESE VOLTAGE READINGS TAKEN AT MAX. SPEED
POST OFFICE BOX 515
ampeg
LINDEN, NEW JERSEY
MODEL
AC-12
U.S. PATENT PEND.
DWN BY
DATE
PART NO
DWG NO
REV
S.C.
1069
4010237
06000

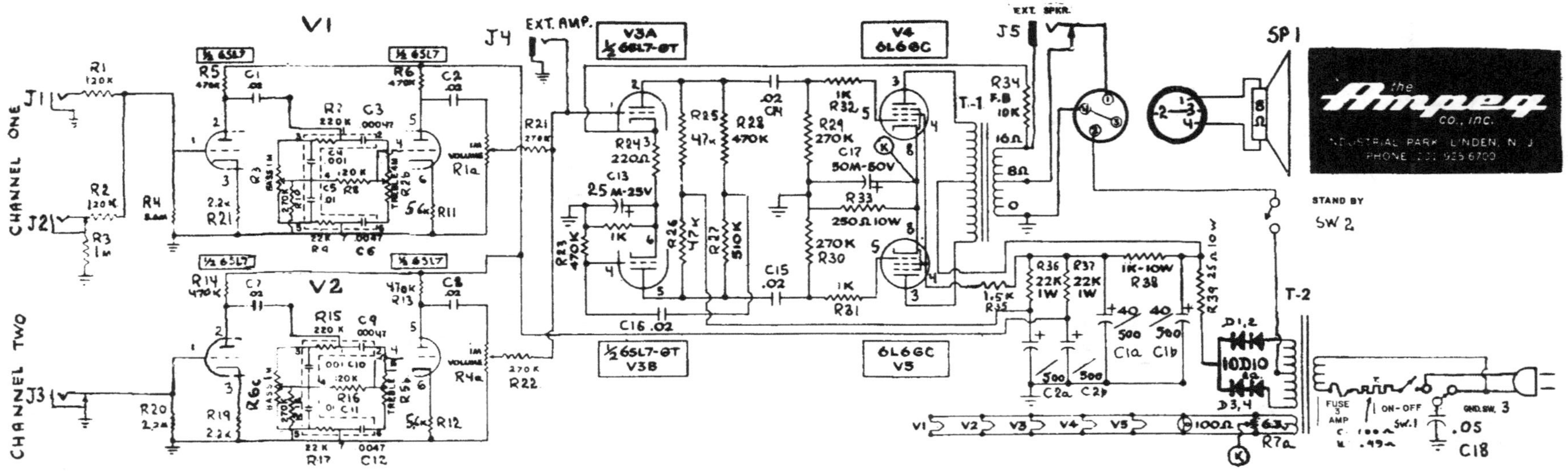

MODEL B-12NB, B-15NB

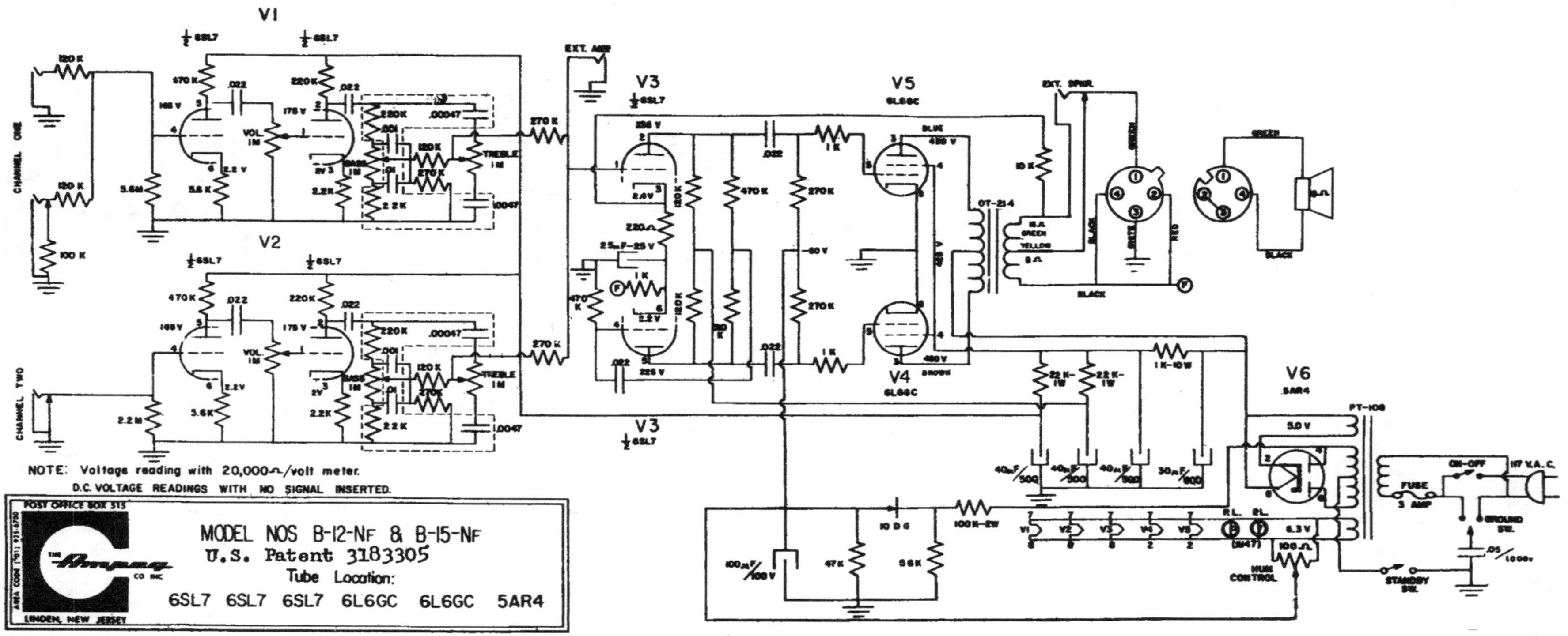
V1
V2
V3
V4
V5
V6
CHANNEL ONE
CHANNEL TWO
EXT. AMP
EXT. SPKR.
OT-214
PT-108
HUM CONTROL
STANDBY SW.
GROUND SW.
FUSE 3 AMP
ON-OFF
117 V.A.C.
NOTE: Voltage reading with 20,000-Ω/volt meter.
D.C. VOLTAGE READINGS WITH NO SIGNAL INSERTED.
POST OFFICE BOX 315
LINDEN, NEW JERSEY
MODEL NOS B-12-NF & B-15-NF
U.S. Patent 3183305
Tube Location:
6SL7 6SL7 6SL7 6L6GC 6L6GC 5AR4

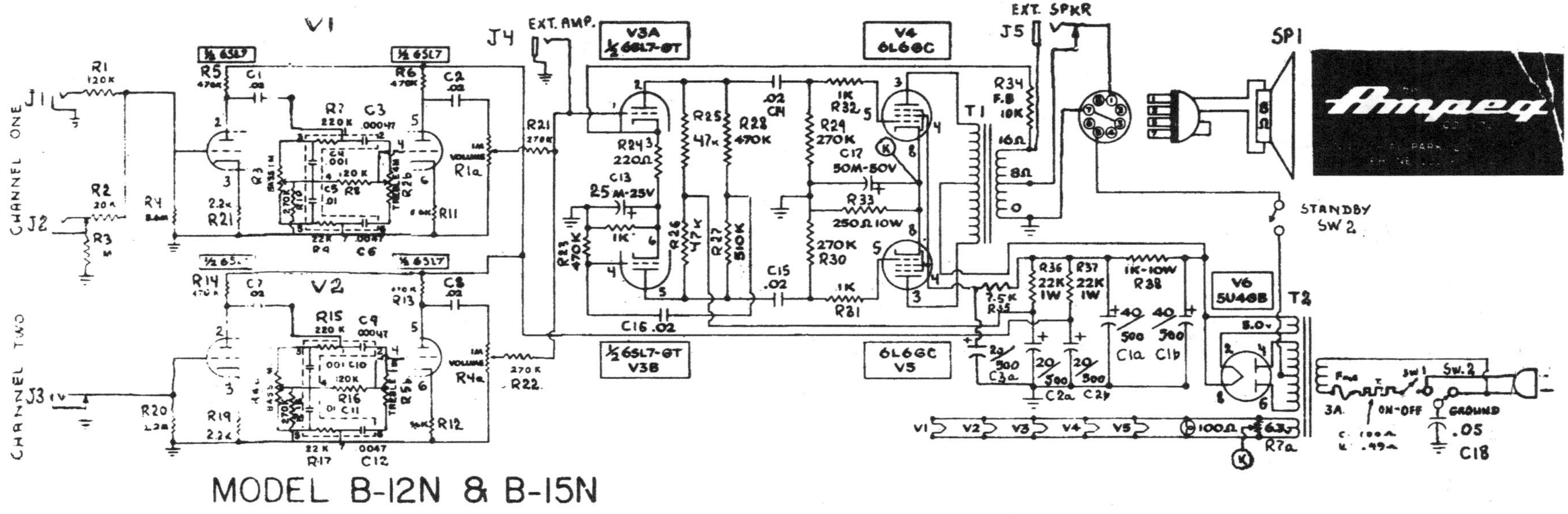
CHANNEL ONE
CHANNEL TWO
J1
J2
J3
J4
EXT. AMP.
J5
EXT. SPKR
V1
V2
½ 6SL7
V3A ½ 6SL7-GT
½ 6SL7-GT V3B
V4 6L6GC
6L6GC V5
V6 5U4GB
T1
T2
SP1
STANDBY SW 2
R21 270K
R22 270K
R23 470K
R24 220Ω
C13 25 M-25V
R25 47K
R26 47K
R27 510K
R28 470K
C16 .02
C15 .02
R32 1K
R24 270K
C17 50M-50V
R33 250Ω 10W
270K R30
1K R31
R34 F.B 10K
16Ω
8Ω
0
R36 22K 1W
R37 22K 1W
1K-10W R38
7.5K R35
C1a C1b
C2a C2b
C3a
3A
ON-OFF
GROUND
.05 C18
SW 1
SW. 2
R7a
MODEL B-12N & B-15N

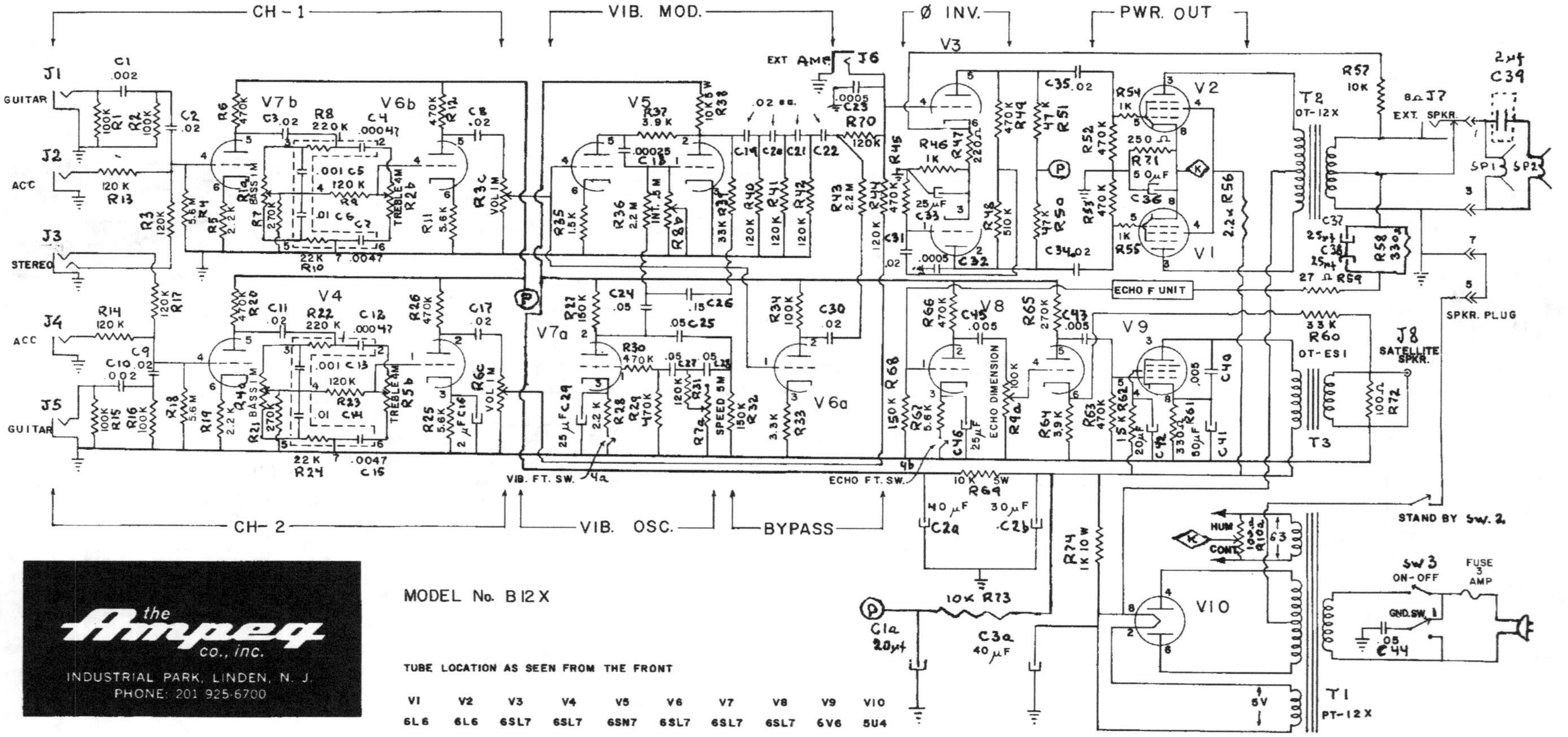

MODEL No. B12X

TUBE LOCATION AS SEEN FROM THE FRONT

V1	V2	V3	V4	V5	V6	V7	V8	V9	V10
6L6	6L6	6SL7	6SL7	6SN7	6SL7	6SL7	6SL7	6V6	5U4

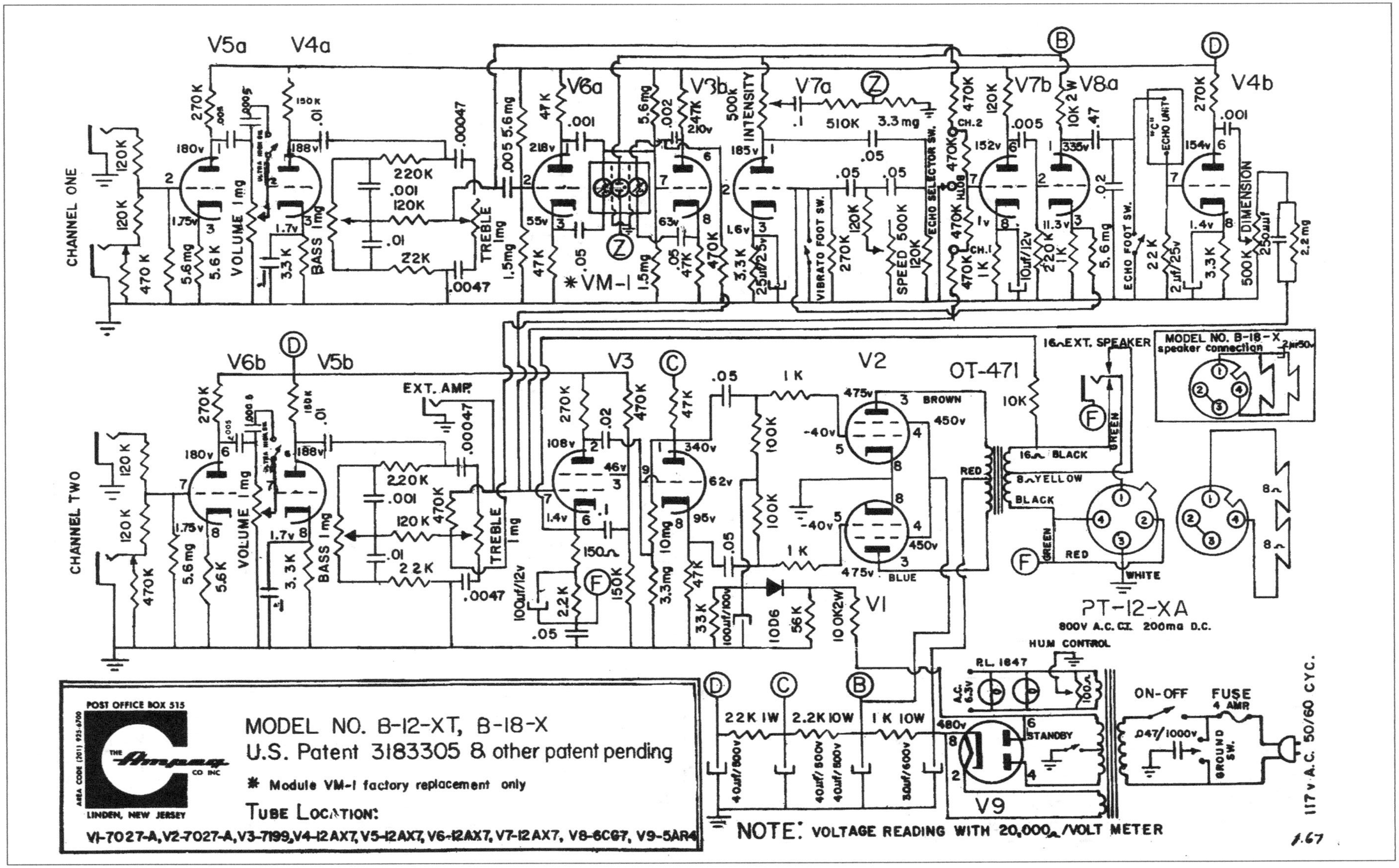
CHANNEL ONE
CHANNEL TWO
V5a
V4a
V6a
V3b
V7a
V7b
V8a
V4b
V6b
V5b
V3
V2
V1
V9
OT-471
PT-12-XA
800V A.C.C.T. 200ma D.C.
VOLUME 1mg
BASS 1mg
TREBLE 1mg
INTENSITY
DIMENSION
*VM-1
VIBRATO FOOT SW.
SPEED 500K
ECHO SELECTOR SW.
ECHO FOOT SW.
"C" ECHO UNIT
EXT. AMP.
16Ω EXT. SPEAKER
MODEL NO. B-18-X speaker connection
HUM CONTROL
STANDBY
ON-OFF
FUSE 4 AMP
GROUND SW.
117 v A.C. 50/60 CYC.
POST OFFICE BOX 515
AREA CODE (201) 925-6700
THE Ampeg CO INC
LINDEN, NEW JERSEY
MODEL NO. B-12-XT, B-18-X
U.S. Patent 3183305 & other patent pending
* Module VM-1 factory replacement only
TUBE LOCATION:
V1-7027-A, V2-7027-A, V3-7199, V4-12AX7, V5-12AX7, V6-12AX7, V7-12AX7, V8-6CG7, V9-5AR4
NOTE: VOLTAGE READING WITH 20,000Ω/VOLT METER
1.67

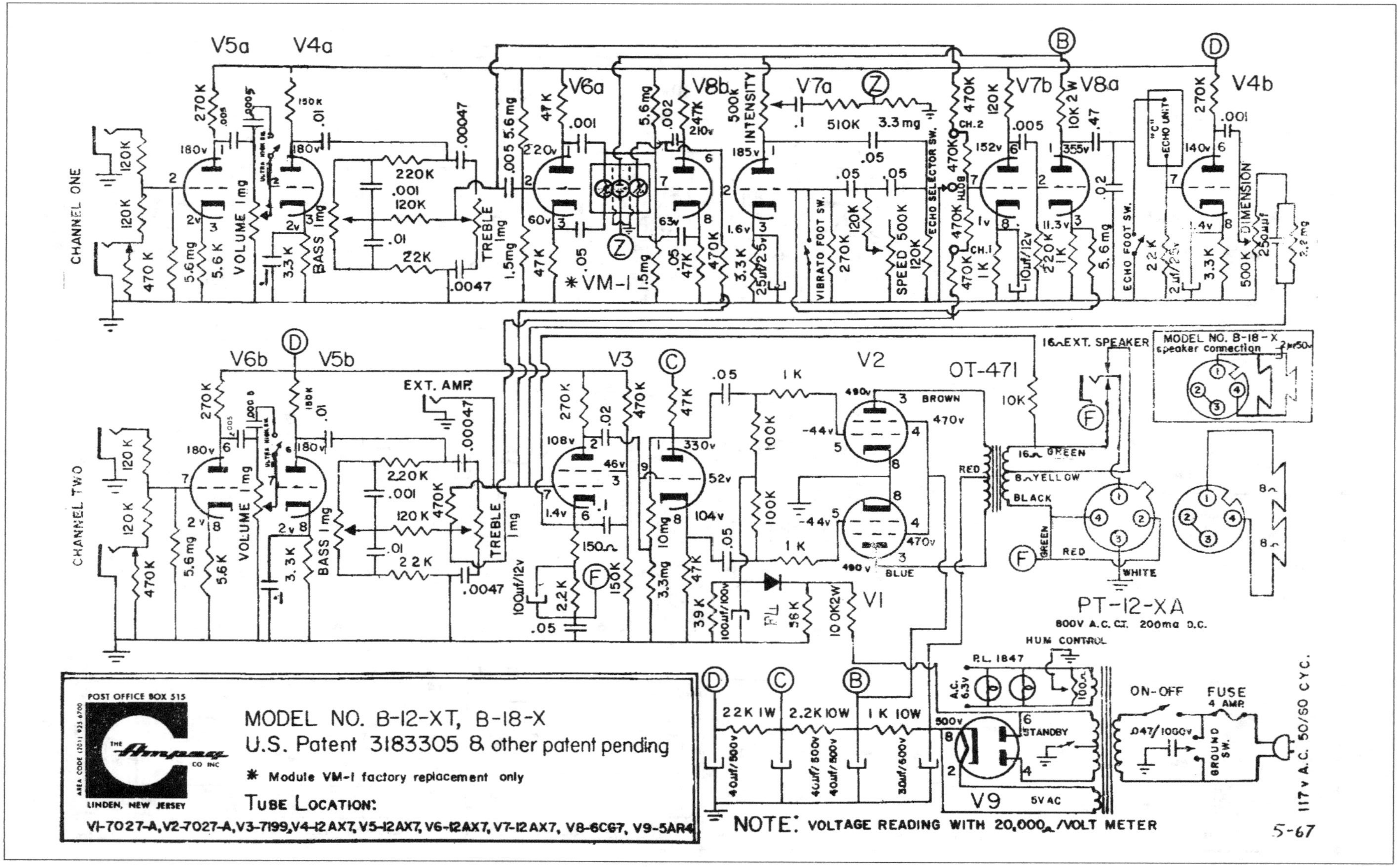
CHANNEL ONE
CHANNEL TWO
V5a
V4a
V6a
V8b
V7a
V7b
V8a
V4b
V6b
V5b
V3
V2
V1
V9
OT-471
PT-12-XA
800V A.C. C.T. 200ma D.C.
HUM CONTROL
P.L. 1847
ON-OFF
FUSE
4 AMP.
GROUND SW.
STANDBY
117v A.C. 50/60 CYC.
VOLUME 1mg
BASS 1mg
TREBLE 1mg
INTENSITY
SPEED 500K
DIMENSION
ECHO FOOT SW.
ECHO SELECTOR SW.
VIBRATO FOOT SW.
"C" ECHO UNIT
EXT. AMP.
16Ω EXT. SPEAKER
MODEL NO. B-18-X speaker connection
*VM-1
BROWN
BLUE
RED
GREEN
YELLOW
BLACK
WHITE
MODEL NO. B-12-XT, B-18-X
U.S. Patent 3183305 & other patent pending
* Module VM-1 factory replacement only
TUBE LOCATION:
V1-7027-A, V2-7027-A, V3-7199, V4-12AX7, V5-12AX7, V6-12AX7, V7-12AX7, V8-6CG7, V9-5AR4
POST OFFICE BOX 515
LINDEN, NEW JERSEY
NOTE: VOLTAGE READING WITH 20,000Ω/VOLT METER
5-67

MODEL No. B12XY
TUBE LOCATION AS SEEN FROM THE FRONT
V1 V2 V3 V4 V5 V6 V7 V8 V9 V10
7027A 7027A 7199 6SL7 6SN7 6SN7 6D10 6D10 7591 5AR4
the Ampeg co., inc.
INDUSTRIAL PARK, LINDEN, N.J.
PHONE 201 925 6700
EXT. AMP.
ECHO SPKR. SP 3,4
STAND by SW-2
117v A.C.
FUSE 4A
ECHO FOOT SW. 5
FOOT SW. 4 VIBRATO
"F" UNIT
EXT. SPKR.
OT 470
OT ES1
PT 12XA

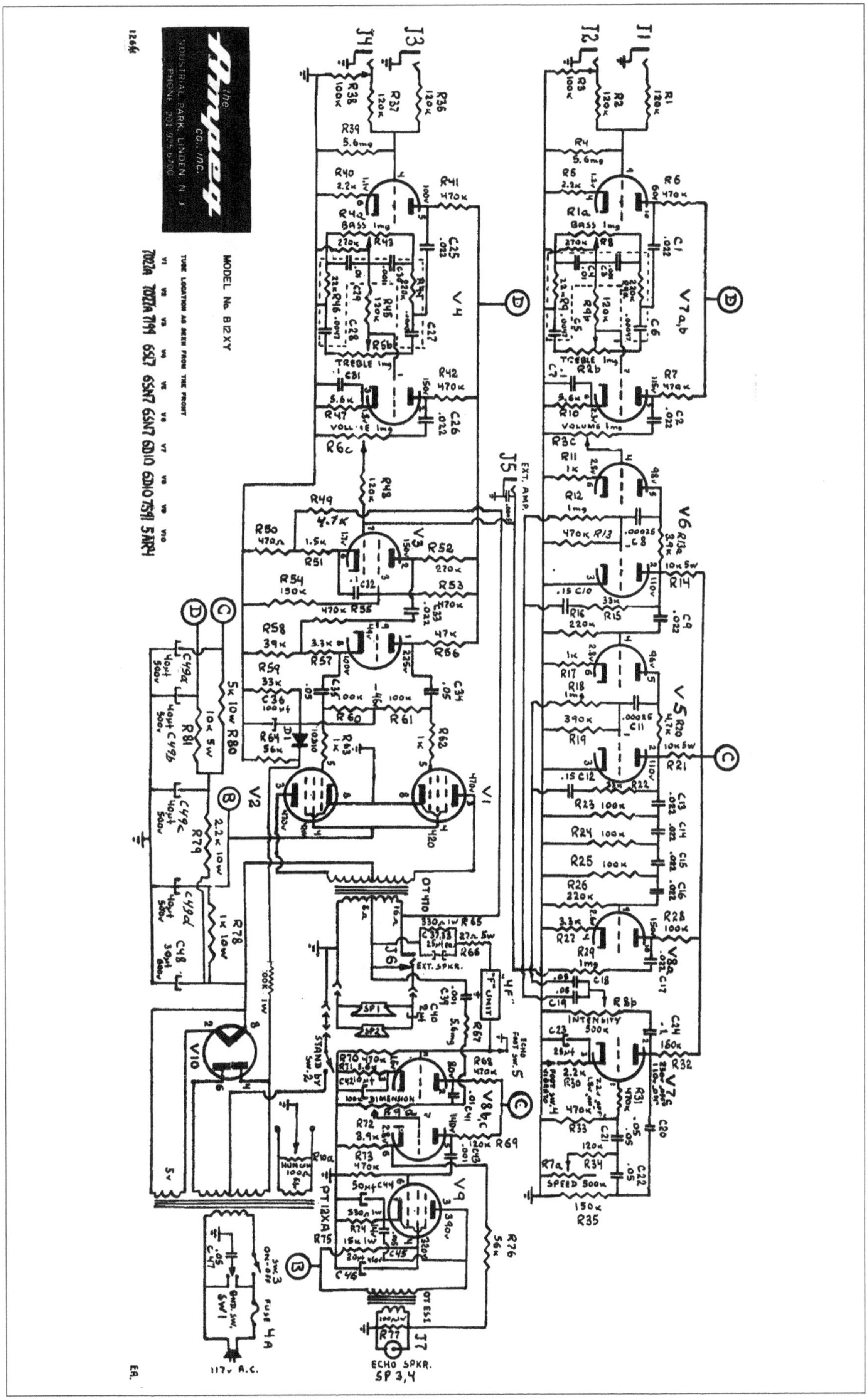

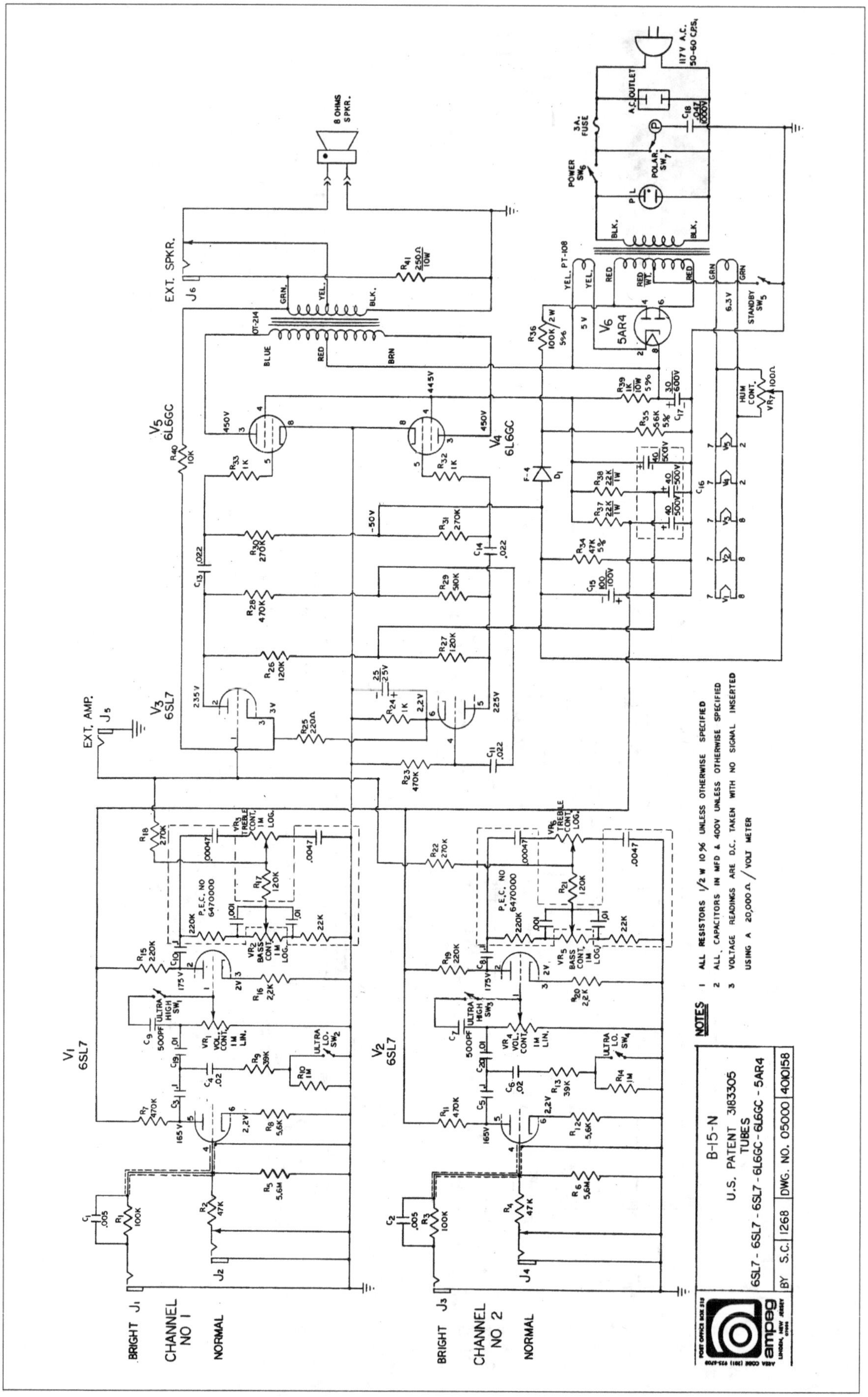
BRIGHT J1
CHANNEL NO 1
NORMAL
BRIGHT J3
CHANNEL NO 2
NORMAL
V1 6SL7
V2 6SL7
V3 6SL7
V4 6L6GC
V5 6L6GC
V6 5AR4
EXT. AMP. J5
EXT. SPKR. J6
8 OHMS SPKR.
117V A.C. 50-60 CPS.
NOTES
1 ALL RESISTORS 1/2 W 10% UNLESS OTHERWISE SPECIFIED
2 ALL CAPACITORS IN MFD & 400V UNLESS OTHERWISE SPECIFIED
3 VOLTAGE READINGS ARE D.C. TAKEN WITH NO SIGNAL INSERTED USING A 20,000 Ω / VOLT METER
B-15-N
U.S. PATENT 3183305
TUBES
6SL7 - 6SL7 - 6SL7 - 6L6GC - 6L6GC - 5AR4
BY S.C. 1268
DWG. NO. 05000 4010158

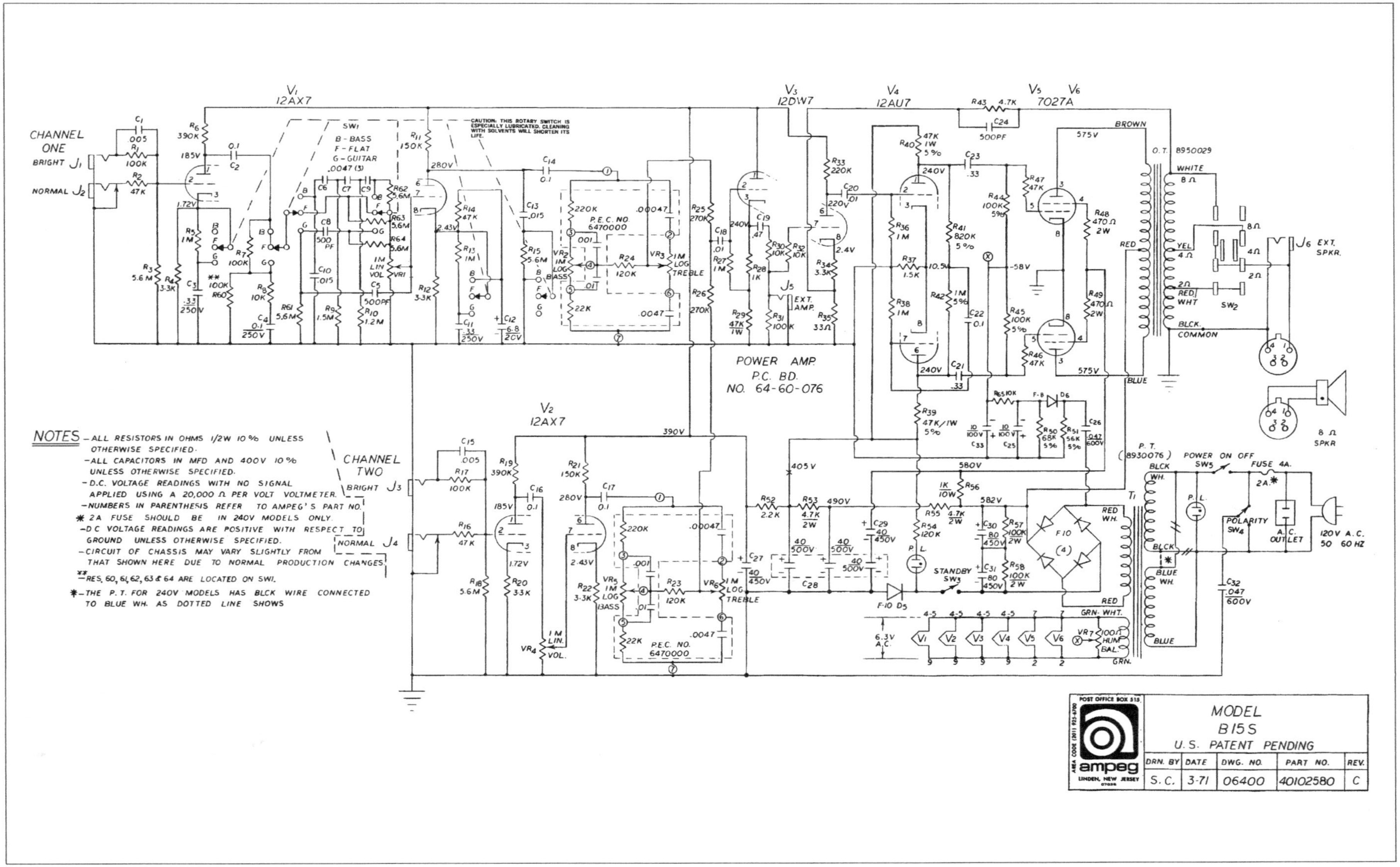
CHANNEL ONE
BRIGHT J1
NORMAL J2
V1 12AX7
V2 12AX7
V3 12DW7
V4 12AU7
V5 V6 7027A
CAUTION: THIS ROTARY SWITCH IS ESPECIALLY LUBRICATED. CLEANING WITH SOLVENTS WILL SHORTEN ITS LIFE.
CHANNEL TWO
BRIGHT J3
NORMAL J4
POWER AMP. P.C. BD. NO. 64-60-076
P.E.C. NO. 6470000
O.T. 8950029
P.T. (8930076)
POWER ON OFF SW5
FUSE 4A.
POLARITY SW4
A.C. OUTLET
120V A.C. 50 60 HZ
STANDBY SW3
J6 EXT. SPKR.
8 Ω SPKR.
NOTES – ALL RESISTORS IN OHMS 1/2W 10% UNLESS OTHERWISE SPECIFIED.
– ALL CAPACITORS IN MFD AND 400V 10% UNLESS OTHERWISE SPECIFIED.
– D.C. VOLTAGE READINGS WITH NO SIGNAL APPLIED USING A 20,000 Ω PER VOLT VOLTMETER.
– NUMBERS IN PARENTHESIS REFER TO AMPEG'S PART NO.
✱ 2A FUSE SHOULD BE IN 240V MODELS ONLY.
– D.C VOLTAGE READINGS ARE POSITIVE WITH RESPECT TO GROUND UNLESS OTHERWISE SPECIFIED.
– CIRCUIT OF CHASSIS MAY VARY SLIGHTLY FROM THAT SHOWN HERE DUE TO NORMAL PRODUCTION CHANGES.
✱✱ RES. 60, 61, 62, 63 & 64 ARE LOCATED ON SW1.
✱ THE P.T. FOR 240V MODELS HAS BLCK WIRE CONNECTED TO BLUE WH. AS DOTTED LINE SHOWS
POST OFFICE BOX 515
AREA CODE (201) 925-6700
ampeg
LINDEN, NEW JERSEY 07036
MODEL B15S
U.S. PATENT PENDING
DRN. BY S.C.
DATE 3-71
DWG. NO. 06400
PART NO. 40102580
REV. C

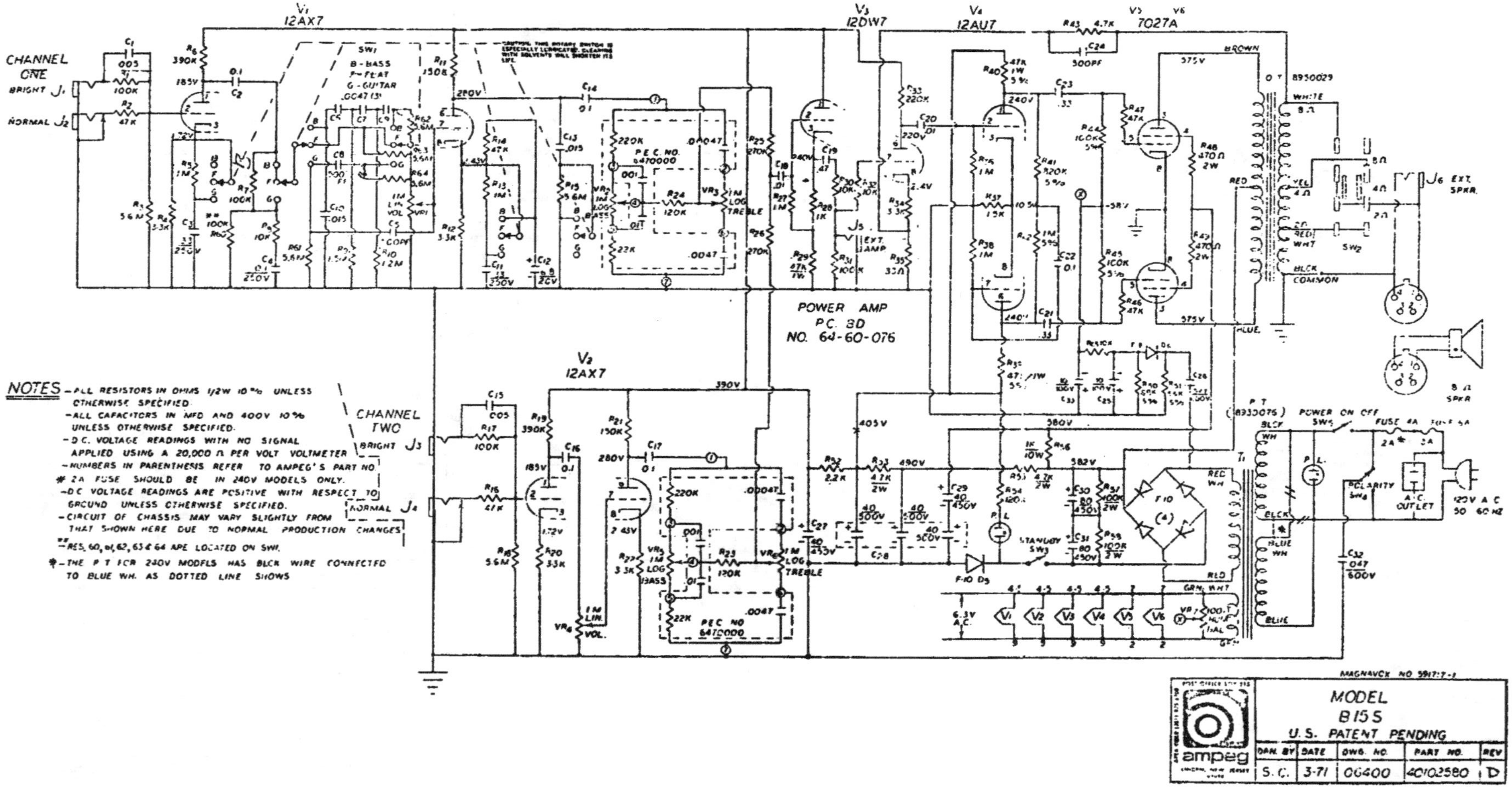
CHANNEL ONE
BRIGHT J1
NORMAL J2
CHANNEL TWO
BRIGHT J3
NORMAL J4
V1 12AX7
V2 12AX7
V3 12DW7
V4 12AU7
V5 V6 7027A
SW1
B - BASS
F - FLAT
G - GUITAR
P.E.C. NO. 6470000
POWER AMP
P.C. 3D
NO. 64-60-076
J5 EXT. AMP
J6 EXT. SPKR.
8 Ω SPKR
POWER ON OFF SW5
FUSE 4A
POLARITY SW4
A.C. OUTLET
120V A C 50 60 HZ
STANDBY SW3
6.3V A.C.
NOTES — ALL RESISTORS IN OHMS 1/2W 10% UNLESS OTHERWISE SPECIFIED.
— ALL CAPACITORS IN MFD AND 400V 10% UNLESS OTHERWISE SPECIFIED.
— D.C. VOLTAGE READINGS WITH NO SIGNAL APPLIED USING A 20,000 Ω PER VOLT VOLTMETER
— NUMBERS IN PARENTHESIS REFER TO AMPEG'S PART NO.
* 2A FUSE SHOULD BE IN 240V MODELS ONLY.
— D.C. VOLTAGE READINGS ARE POSITIVE WITH RESPECT TO GROUND UNLESS OTHERWISE SPECIFIED.
— CIRCUIT OF CHASSIS MAY VARY SLIGHTLY FROM THAT SHOWN HERE DUE TO NORMAL PRODUCTION CHANGES.
** RES. 60, 61, 62, 63 & 64 ARE LOCATED ON SW1.
* — THE P.T. FOR 240V MODELS HAS BLCK WIRE CONNECTED TO BLUE WH. AS DOTTED LINE SHOWS
MAGNAVOX NO 5917:7-1
ampeg
MODEL
B15S
U.S. PATENT PENDING
DRN. BY S.C.
DATE 3-71
DWG. NO. 06400
PART NO. 40102580
REV D

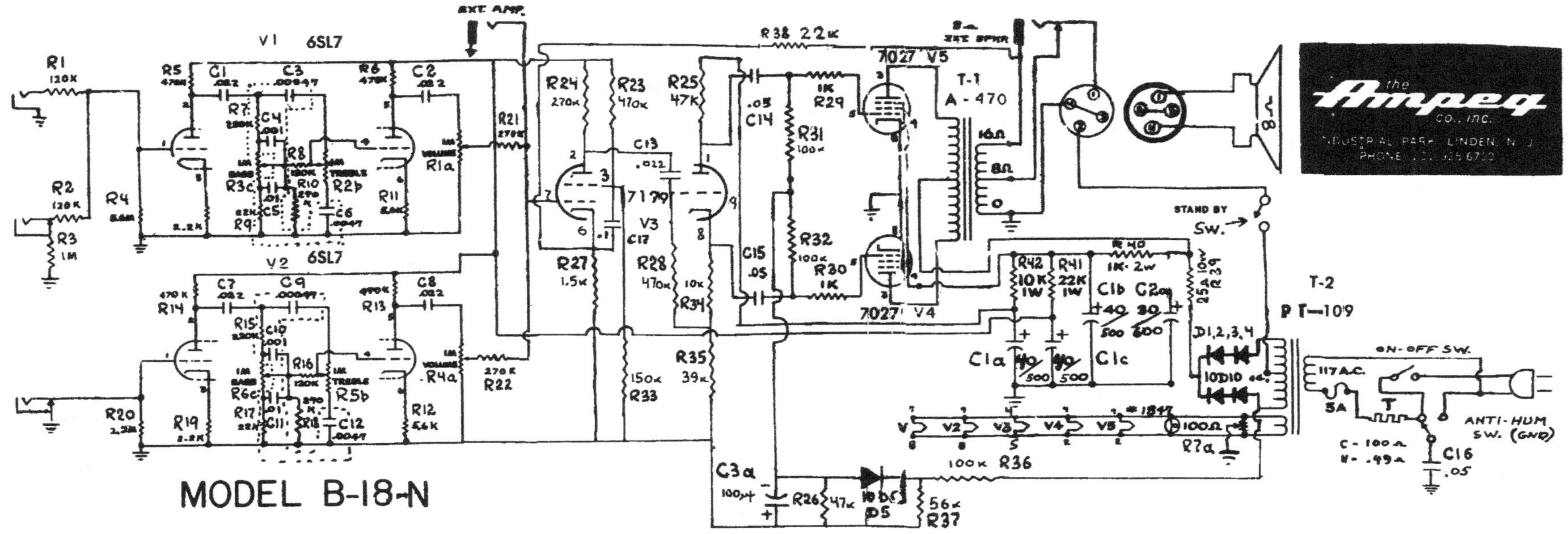
MODEL B-18-N
V1 6SL7
V2 6SL7
V3
7199
7027 V5
7027 V4
T-1
A-470
T-2
PT-109
EXT. AMP.
STAND BY SW.
ON-OFF SW.
ANTI-HUM SW. (GND)
117 A.C.
5A
R1 120K
R2 120K
R3 1M
R4
R20
R21 270K
R22 270K
R23 470K
R24 270K
R25 47K
R27 1.5K
R28 470K
R33 150K
R35 39K
R36 100K
R37 56K
R38 2.2K
R29 1K
R30 1K
R31 100K
R32 100K
R41 22K 1W
R42 10K 1W
R40 1K 2W
C1a
C1b
C1c
C2a
C3a
C13 .022
C14 .05
C15 .05
C16 .05
D1,2,3,4
D5
16Ω
8Ω
the Ampeg Co., Inc.

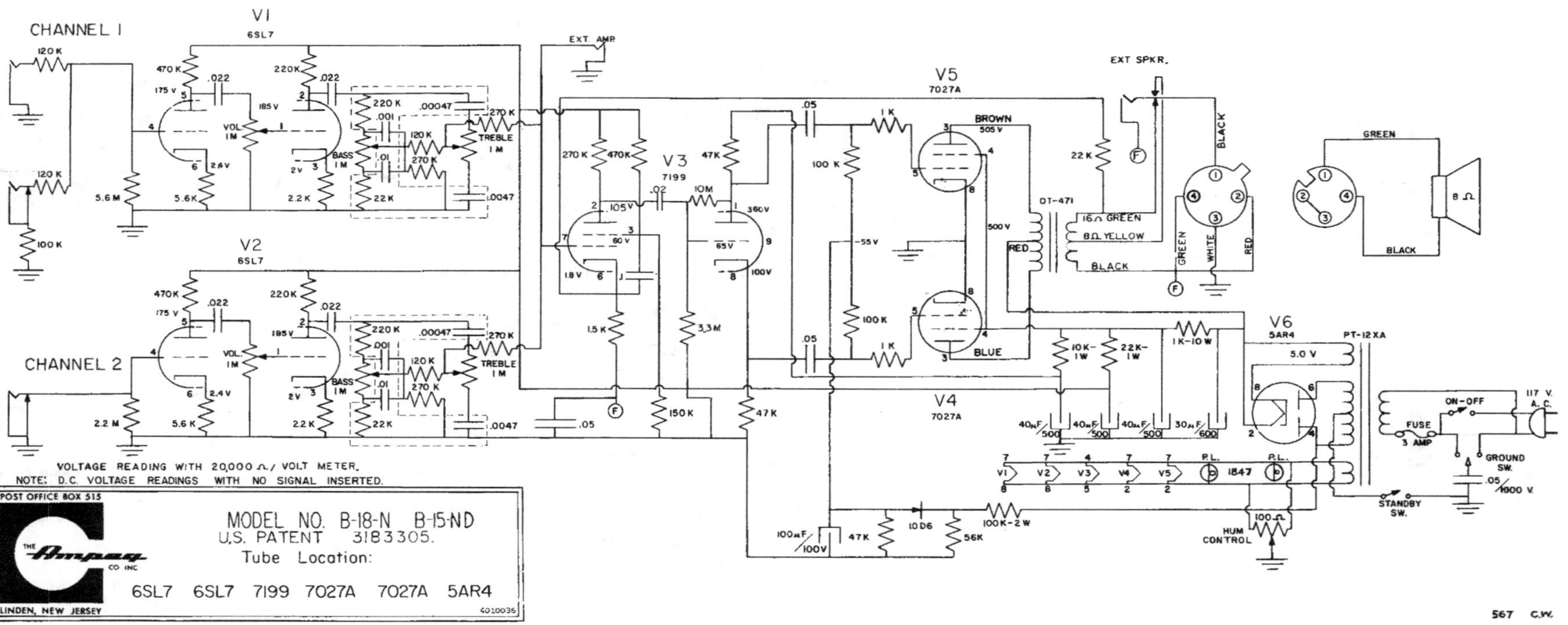
CHANNEL 1
CHANNEL 2
V1
6SL7
V2
6SL7
V3
7199
V4
7027A
V5
7027A
V6
5AR4
EXT. AMP
EXT SPKR.
BROWN
505 V
BLUE
OT-471
16Ω GREEN
8Ω YELLOW
BLACK
RED
GREEN
WHITE
8 Ω
PT-12XA
5.0 V
ON-OFF
FUSE
3 AMP
117 V.
A. C.
GROUND
SW.
STANDBY
SW.
HUM
CONTROL
TREBLE
1 M
BASS
1 M
VOL.
1 M
NOTE: D.C. VOLTAGE READINGS WITH NO SIGNAL INSERTED.
VOLTAGE READING WITH 20,000 Ω / VOLT METER.
POST OFFICE BOX 515
THE Ampeg CO INC
LINDEN, NEW JERSEY
AREA CODE (201) 925 6700
MODEL NO. B-18-N B-15-ND
U.S. PATENT 3183305.
Tube Location:
6SL7 6SL7 7199 7027A 7027A 5AR4
4010036
567 C.W.

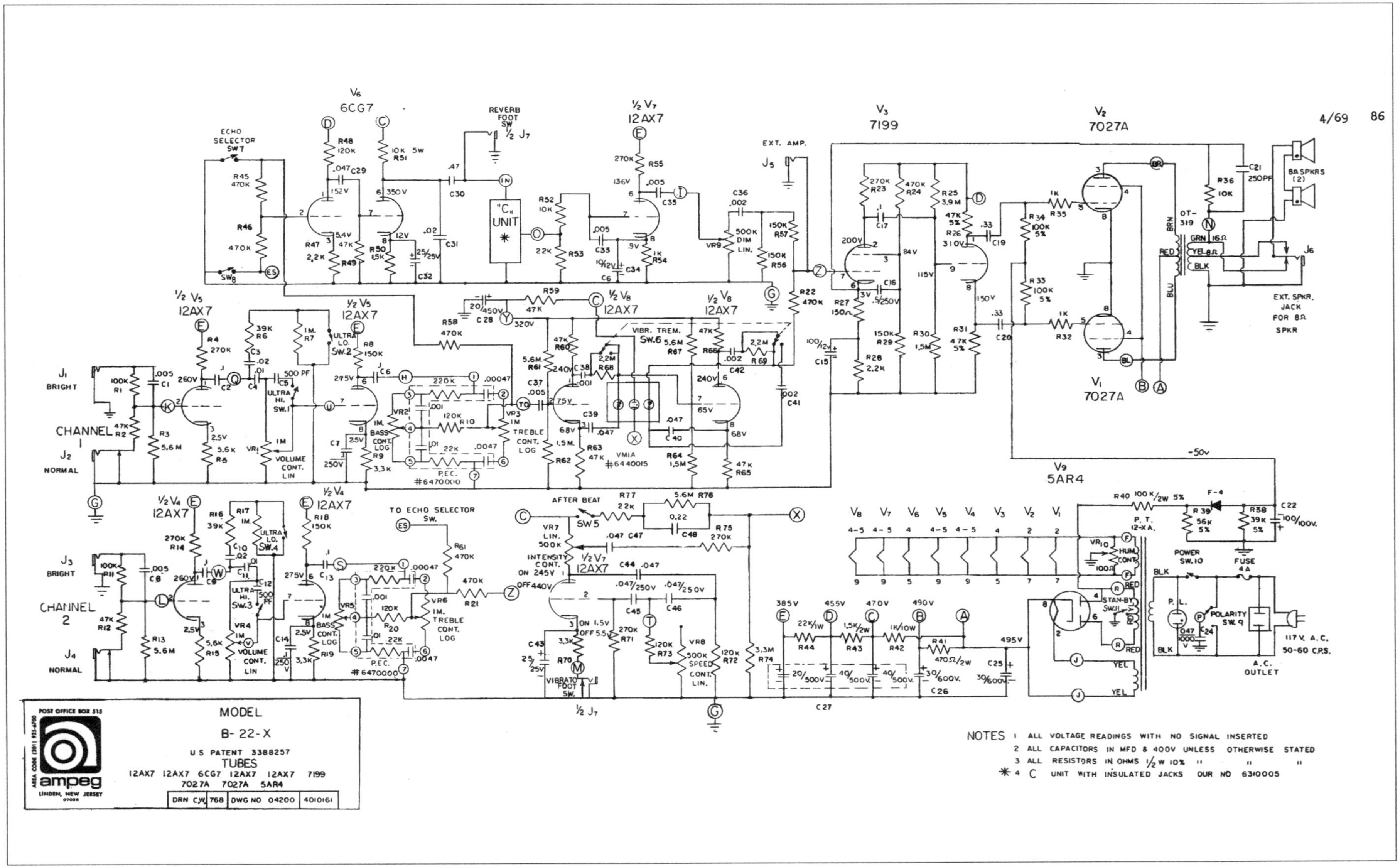
4/69 86
V6 6CG7
ECHO SELECTOR SW7
REVERB FOOT SW ½ J7
"C4" UNIT *
½ V7 12AX7
EXT. AMP. J5
V3 7199
V2 7027A
V1 7027A
8Ω SPKRS (2)
EXT. SPKR. JACK FOR 8Ω SPKR
J6
OT-319
½ V5 12AX7
J1 BRIGHT
CHANNEL 1
J2 NORMAL
½ V8 12AX7
VIBR. TREM. SW.6
VMIA #6440015
P.E.C. #6470000
TREBLE CONT. LOG
BASS CONT. LOG
VOLUME CONT. LIN
ULTRA HI. SW.1
ULTRA LO. SW.2
-50V
V9 5AR4
½ V4 12AX7
J3 BRIGHT
CHANNEL 2
J4 NORMAL
ULTRA HI. SW.3
ULTRA LO. SW.4
TO ECHO SELECTOR SW.
AFTER BEAT SW5
INTENSITY CONT. 500K LIN.
½ V7 12AX7
VIBRATO FOOT SW. ½ J7
SPEED CONT. LIN.
V8 V7 V6 V5 V4 V3 V2 V1
P.T. 12-X A.
POWER SW.10
FUSE 4A
POLARITY SW.9
STAN-BY SW.11
HUM CONT.
117V. A.C. 50-60 CPS.
A.C. OUTLET
POST OFFICE BOX 315
AREA CODE (201) 925-6700
ampeg
LINDEN, NEW JERSEY 07036
MODEL
B-22-X
U S PATENT 3388257
TUBES
12AX7 12AX7 6CG7 12AX7 12AX7 7199
7027A 7027A 5AR4
DRN C.W. 768 DWG NO 04200 4010161
NOTES 1 ALL VOLTAGE READINGS WITH NO SIGNAL INSERTED
2 ALL CAPACITORS IN MFD & 400V UNLESS OTHERWISE STATED
3 ALL RESISTORS IN OHMS ½W 10% " " "
* 4 C UNIT WITH INSULATED JACKS OUR NO 6310005

Ampeg

B-25

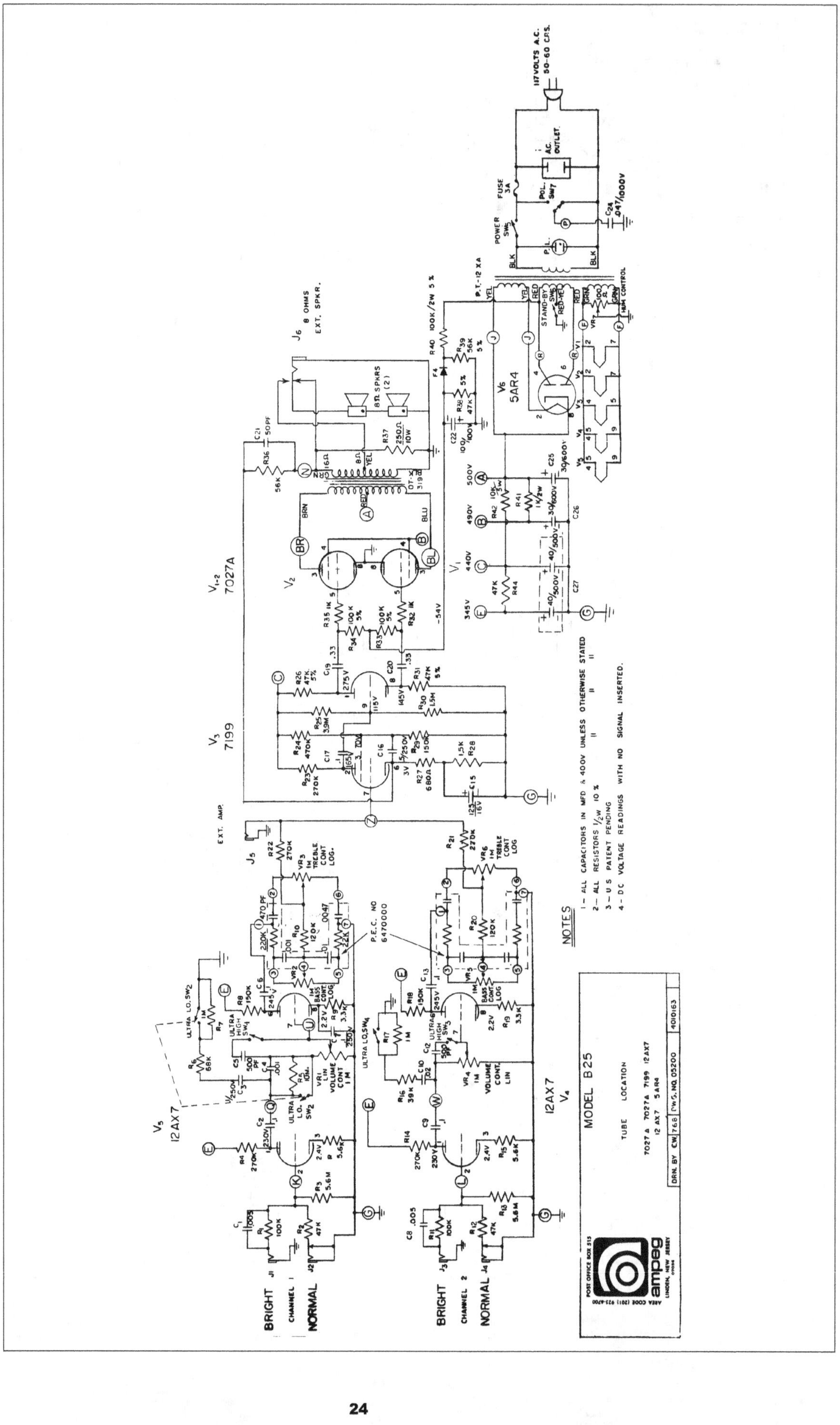

MODEL B25
TUBE LOCATION
7027A 7027A 7199 12AX7
12 AX7 5AR4
DRN. BY CW 7-68 DWG. NO. 05200 4010163
NOTES
1 - ALL CAPACITORS IN MFD & 400V UNLESS OTHERWISE STATED
2 - ALL RESISTORS 1/2W 10%
3 - U S PATENT PENDING
4 - DC VOLTAGE READINGS WITH NO SIGNAL INSERTED.
117 VOLTS A.C. 50-60 C.P.S.

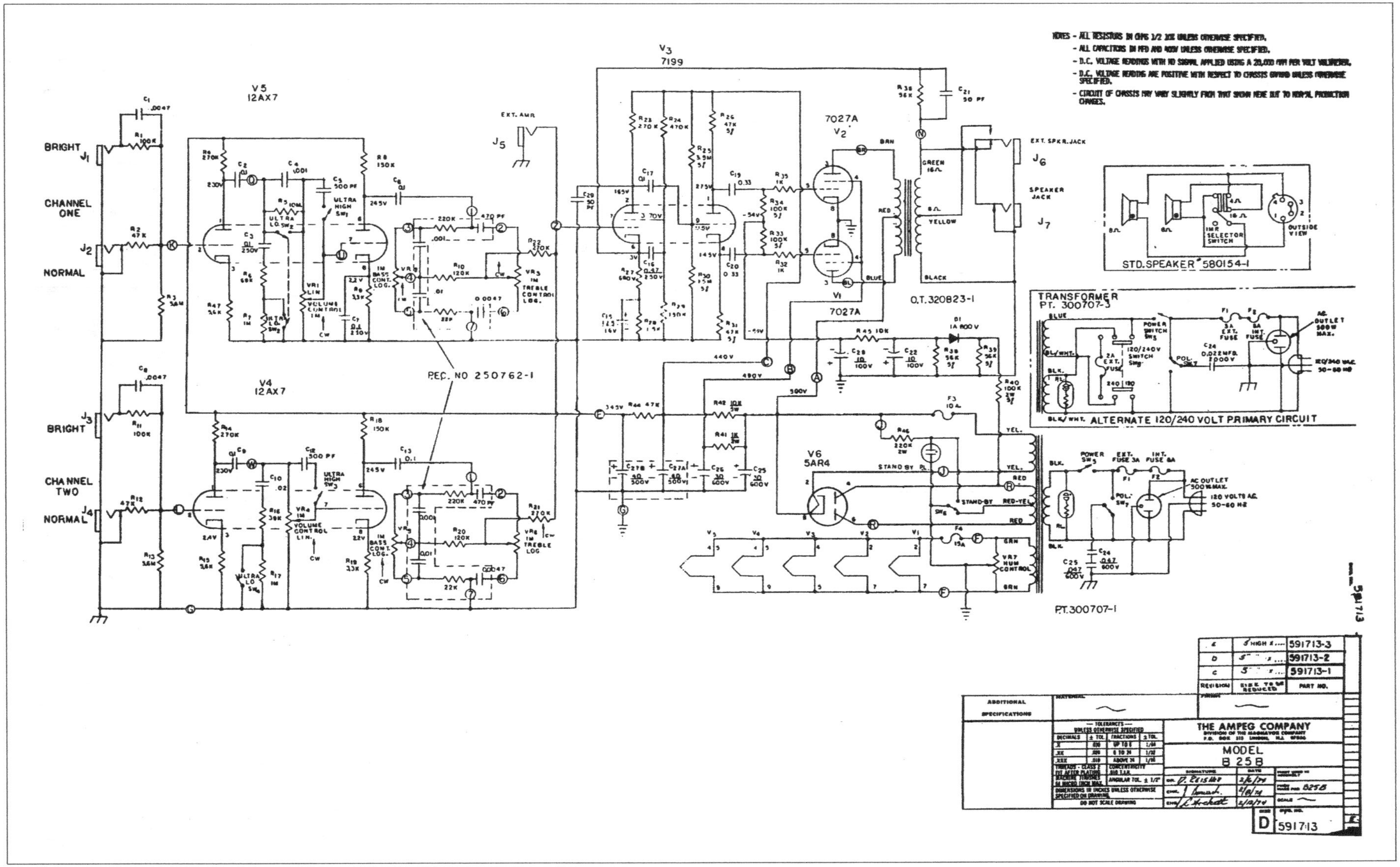
NOTES - ALL RESISTORS IN OHMS 1/2 W UNLESS OTHERWISE SPECIFIED.
- ALL CAPACITORS IN MFD AND 400V UNLESS OTHERWISE SPECIFIED.
- D.C. VOLTAGE READINGS WITH NO SIGNAL APPLIED USING A 20,000 OHM PER VOLT VOLTMETER.
- D.C. VOLTAGE READING ARE POSITIVE WITH RESPECT TO CHASSIS GROUND UNLESS OTHERWISE SPECIFIED.
- CIRCUIT OF CHASSIS MAY VARY SLIGHTLY FROM THAT SHOWN HERE DUE TO NORMAL PRODUCTION CHANGES.
BRIGHT J1
CHANNEL ONE
NORMAL J2
BRIGHT J3
CHANNEL TWO
NORMAL J4
V5 12AX7
V4 12AX7
V3 7199
7027A V2
7027A V1
V6 5AR4
EXT. AMP. J5
EXT. SPKR. JACK J6
SPEAKER JACK J7
PEC. NO 250762-1
O.T. 320823-1
STD. SPEAKER 580154-1
SELECTOR SWITCH
OUTSIDE VIEW
TRANSFORMER PT. 300707-3
ALTERNATE 120/240 VOLT PRIMARY CIRCUIT
PT. 300707-1
AC OUTLET 500 W MAX.
120 VOLTS AC 50-60 HZ
VOLUME CONTROL LIN.
BASS CONT. LOG.
TREBLE CONTROL LOG.
ULTRA HIGH
ULTRA LO
HUM CONTROL
STAND-BY
POWER SW
THE AMPEG COMPANY
MODEL B 25B
591713-3
591713-2
591713-1
D 591713

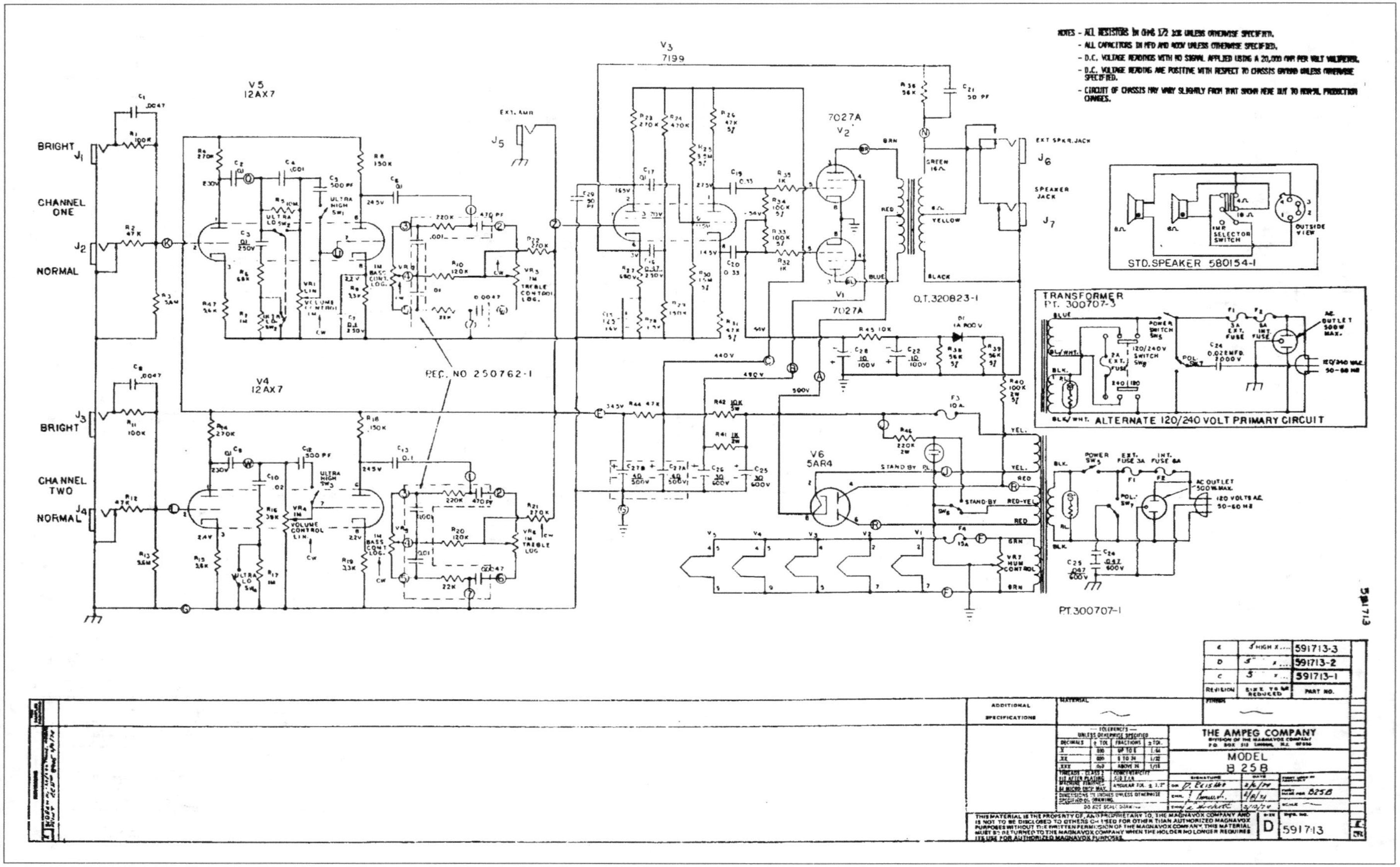
BRIGHT
CHANNEL ONE
NORMAL
BRIGHT
CHANNEL TWO
NORMAL
V5
12AX7
V4
12AX7
V3
7199
7027A
V2
V1
7027A
V6
5AR4
EXT. AMP
J5
EXT SPKR. JACK
J6
SPEAKER JACK
J7
O.T. 320823-1
REC. NO 250762-1
STD. SPEAKER 580154-1
IMP. SELECTOR SWITCH
OUTSIDE VIEW
TRANSFORMER
PT. 300707-3
ALTERNATE 120/240 VOLT PRIMARY CIRCUIT
PT. 300707-1
STAND BY
AC OUTLET 500W MAX.
120 VOLTS AC 50-60 HZ
VOLUME CONTROL LIN
TREBLE CONTROL LOG.
BASS CONT. LOG.
ULTRA HIGH SW1
ULTRA LO SW2
VR7 HUM CONTROL
591713
THE AMPEG COMPANY
MODEL B 25B
591713-3
591713-2
591713-1
PART NO.
ADDITIONAL SPECIFICATIONS
MATERIAL
FINISH
D
591713

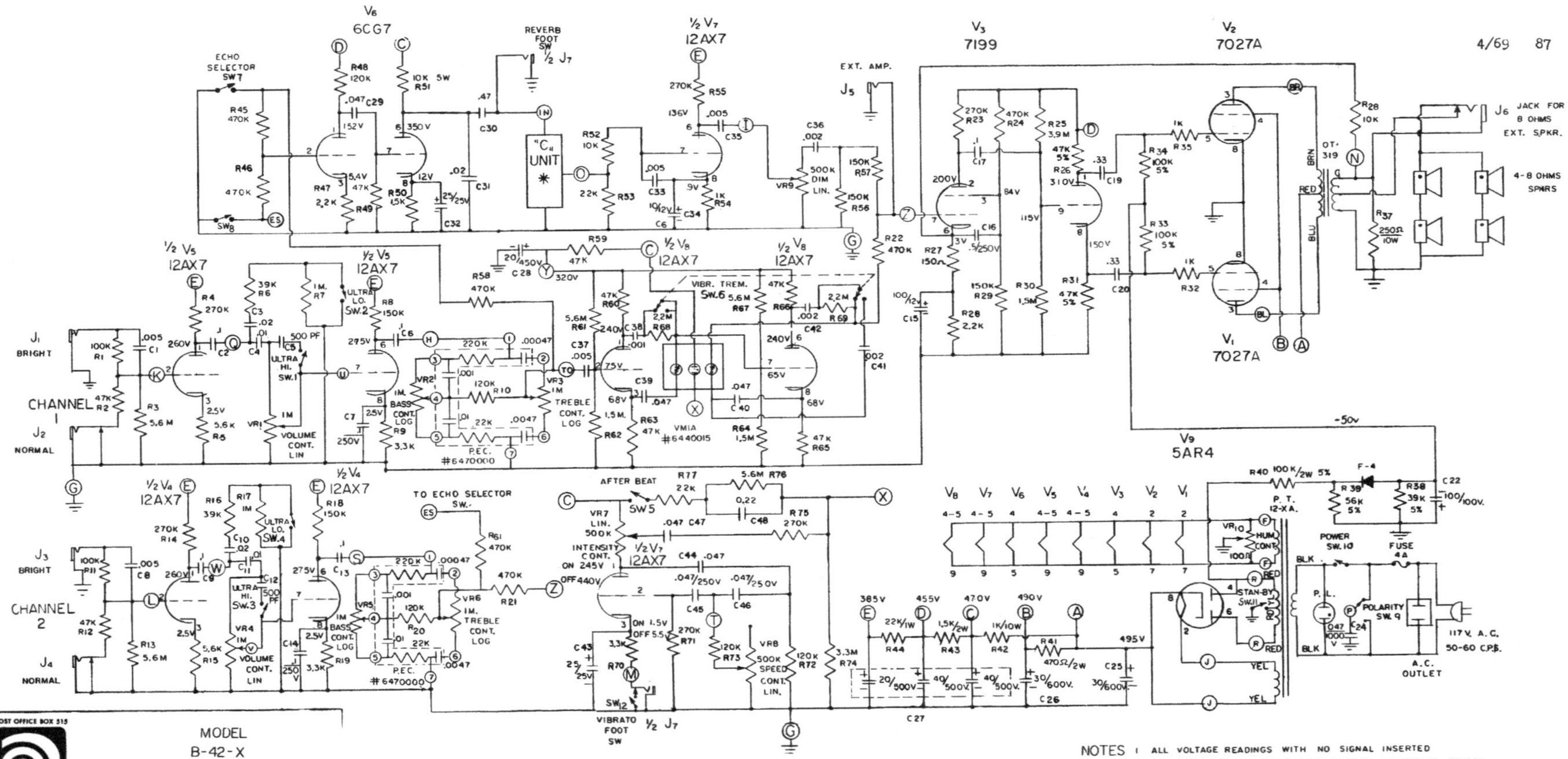

MODEL
B-42-X
U S PATENT 3388257
TUBES
12AX7 12AX7 6CG7 12AX7 12AX7 7199
7027A 7027A 5AR4

DRN CW 768 | DWG NO 04300 | 4010162

POST OFFICE BOX 515
AREA CODE (201) 925-6700
ampeg
LINDEN, NEW JERSEY 07036

NOTES
1. ALL VOLTAGE READINGS WITH NO SIGNAL INSERTED
2. ALL CAPACITORS IN MFD & 400V UNLESS OTHERWISE STATED
3. ALL RESISTORS IN OHMS ½W 10% " " "
* 4. C UNIT WITH INSULATED JACKS OUR NO 6310005

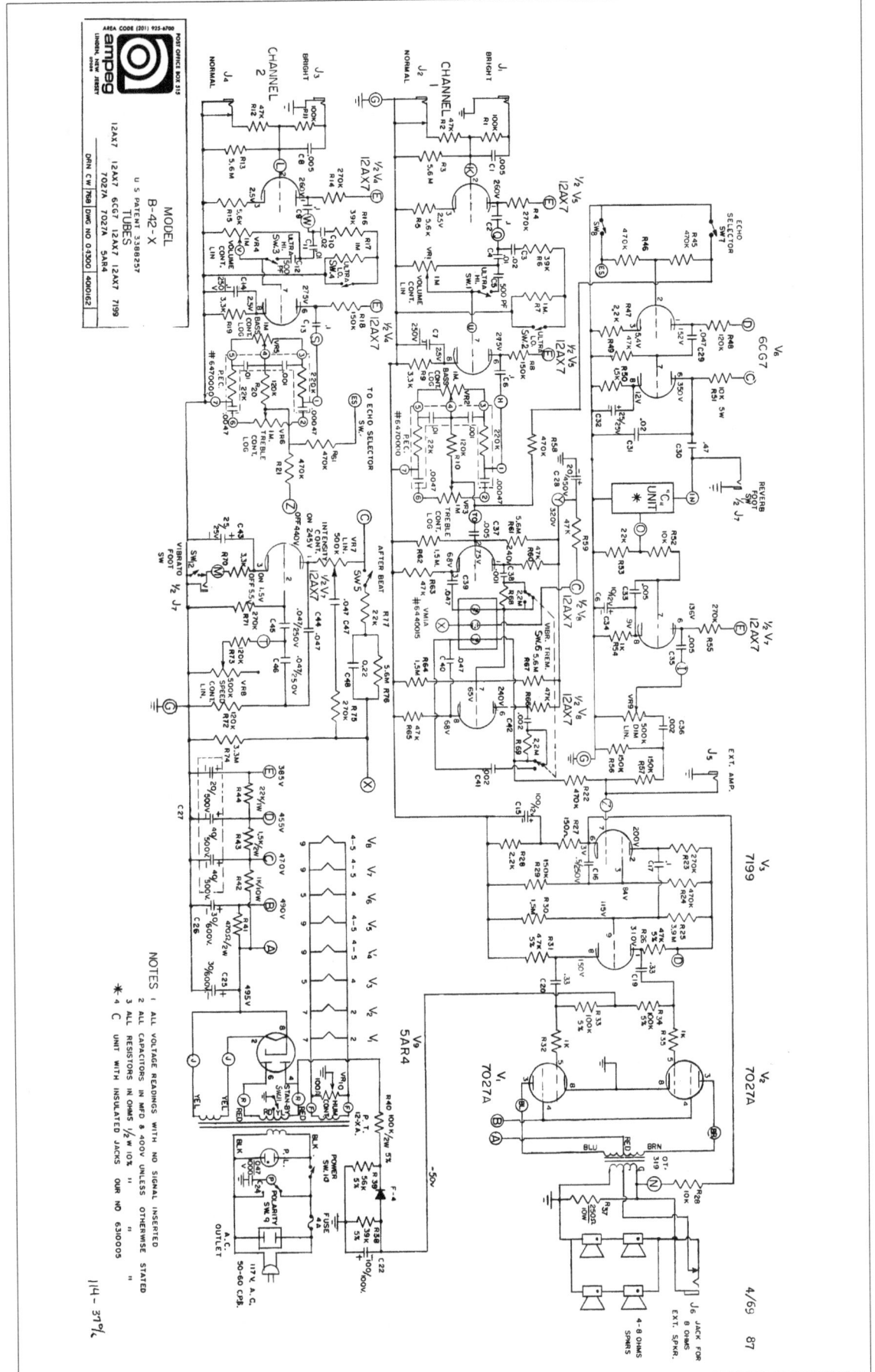

114-37%

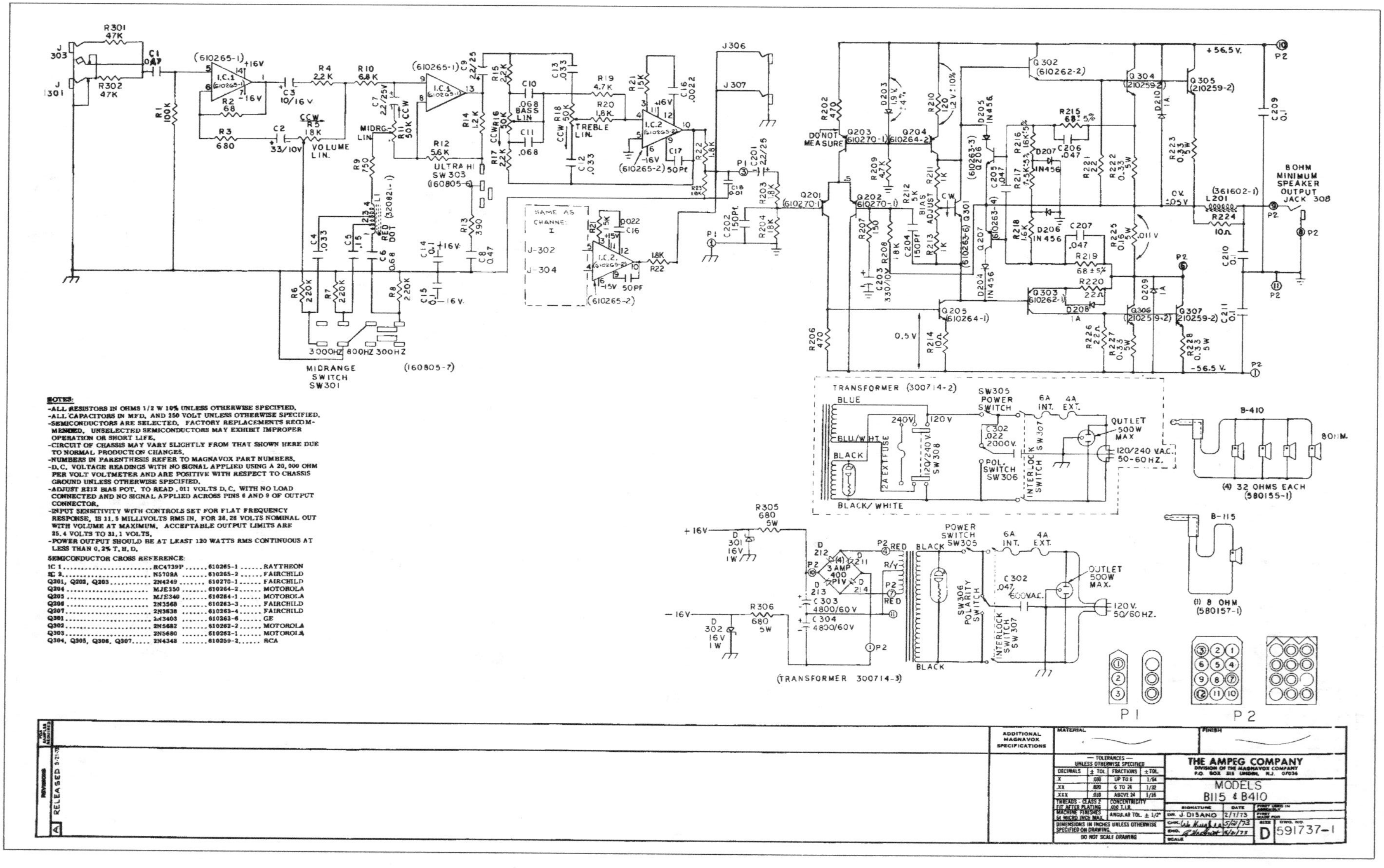

NOTES:

- ALL RESISTORS IN OHMS 1/2 W 10% UNLESS OTHERWISE SPECIFIED.
- ALL CAPACITORS IN MFD. AND 250 VOLT UNLESS OTHERWISE SPECIFIED.
- SEMICONDUCTORS ARE SELECTED. FACTORY REPLACEMENTS RECOMMENDED. UNSELECTED SEMICONDUCTORS MAY EXHIBIT IMPROPER OPERATION OR SHORT LIFE.
- CIRCUIT OF CHASSIS MAY VARY SLIGHTLY FROM THAT SHOWN HERE DUE TO NORMAL PRODUCTION CHANGES.
- NUMBERS IN PARENTHESIS REFER TO MAGNAVOX PART NUMBERS.
- D.C. VOLTAGE READINGS WITH NO SIGNAL APPLIED USING A 20,000 OHM PER VOLT VOLTMETER AND ARE POSITIVE WITH RESPECT TO CHASSIS GROUND UNLESS OTHERWISE SPECIFIED.
- ADJUST R212 BIAS POT. TO READ .011 VOLTS D.C. WITH NO LOAD CONNECTED AND NO SIGNAL APPLIED ACROSS PINS 6 AND 9 OF OUTPUT CONNECTOR.
- INPUT SENSITIVITY WITH CONTROLS SET FOR FLAT FREQUENCY RESPONSE, IS 11.5 MILLIVOLTS RMS IN, FOR 28.28 VOLTS NOMINAL OUT WITH VOLUME AT MAXIMUM. ACCEPTABLE OUTPUT LIMITS ARE 25.4 VOLTS TO 31.1 VOLTS.
- POWER OUTPUT SHOULD BE AT LEAST 120 WATTS RMS CONTINUOUS AT LESS THAN 0.2% T.H.D.

SEMICONDUCTOR CROSS REFERENCE:

IC 1	RC4739P	610265-1	RAYTHEON
IC 2	N5709A	610265-2	FAIRCHILD
Q201, Q202, Q203	2N4249	610270-1	FAIRCHILD
Q204	MJE350	610264-2	MOTOROLA
Q205	MJE340	610264-1	MOTOROLA
Q206	2N3568	610263-3	FAIRCHILD
Q207	2N3638	610263-4	FAIRCHILD
Q301	2N3403	610263-6	GE
Q302	2N5682	610262-2	MOTOROLA
Q303	2N5680	610262-1	MOTOROLA
Q304, Q305, Q306, Q307	2N4348	610259-2	RCA

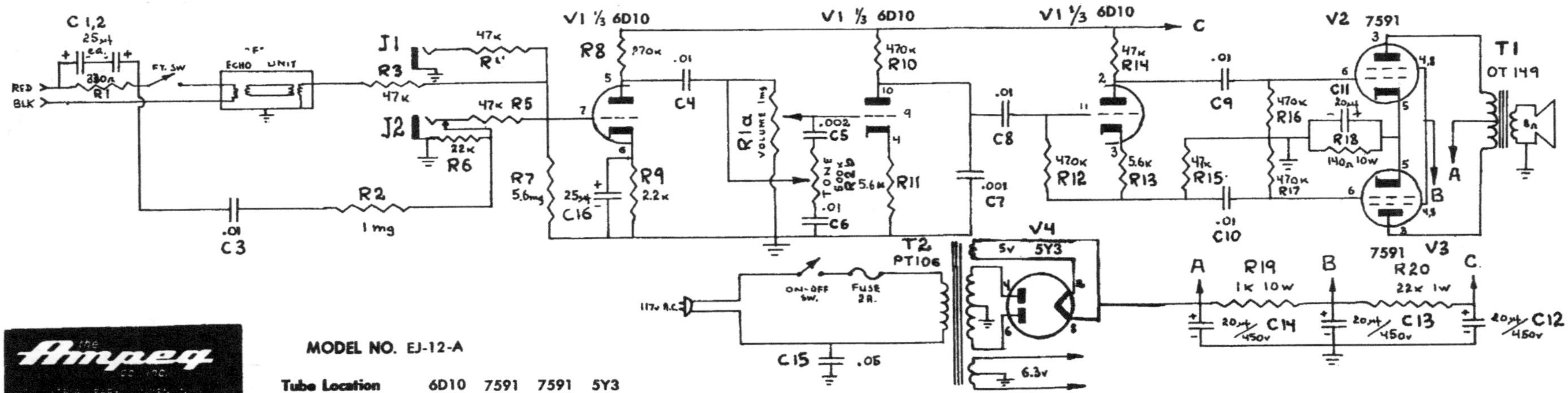

MODEL NO. EJ-12-A

Tube Location 6D10 7591 7591 5Y3

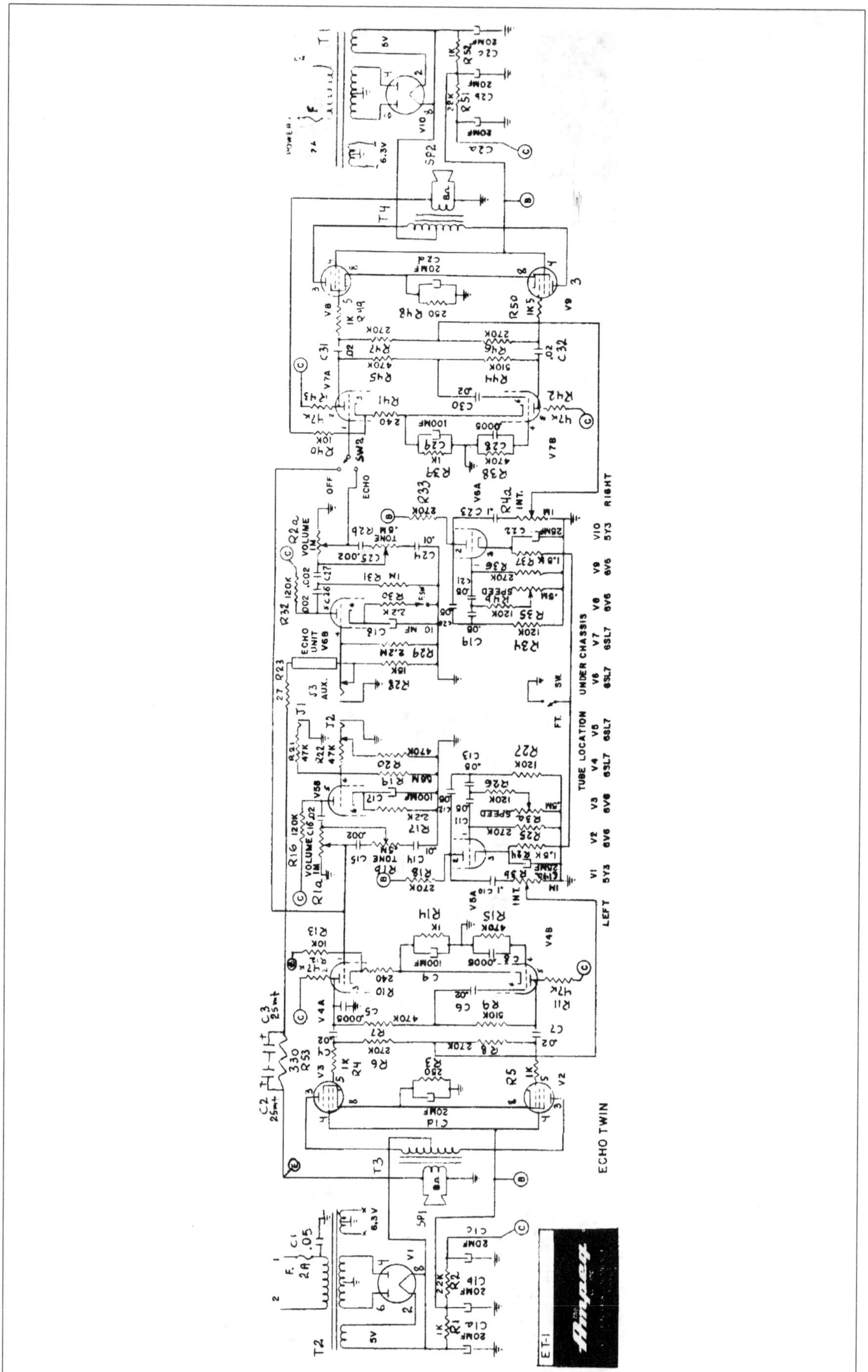
ECHO TWIN
TUBE LOCATION UNDER CHASSIS
V1 5Y3
V2 6V6
V3 6V6
V4 6SL7
V5 6SL7
V6 6SL7
V7 6SL7
V8 6V6
V9 6V6
V10 5Y3
LEFT
RIGHT
ECHO UNIT
ECHO
OFF
VOLUME
TONE
SPEED
INT.
FT. SW.
AUX.
ET-1
Ampeg

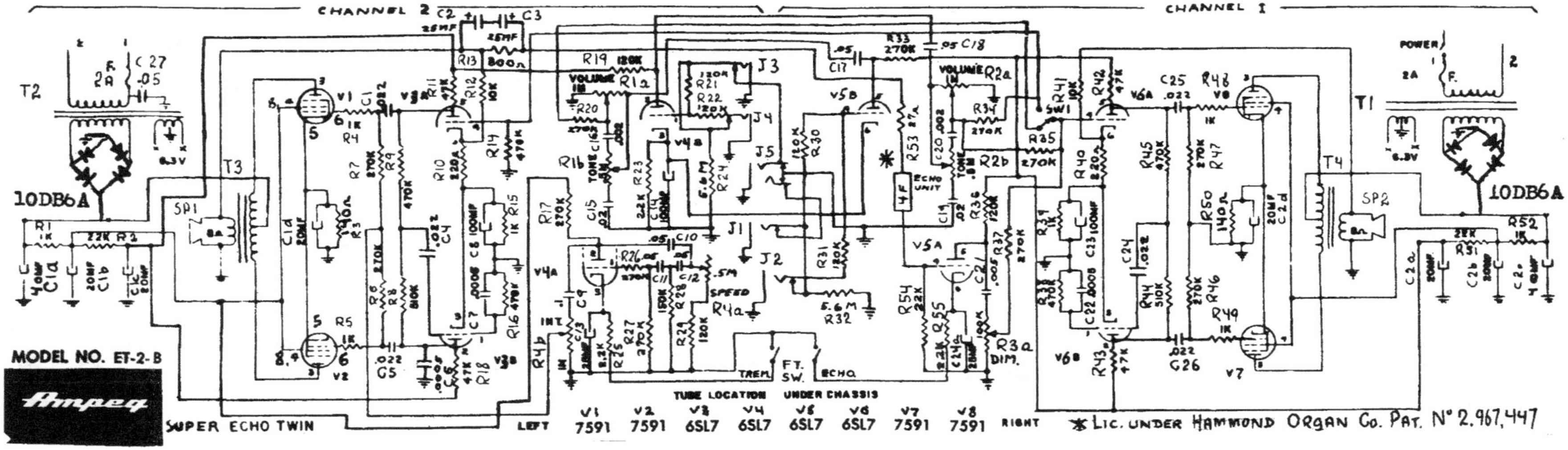

CHANNEL 2
CHANNEL I
MODEL NO. ET-2-B
Ampeg
SUPER ECHO TWIN
10DB6A
LEFT
RIGHT
TUBE LOCATION UNDER CHASSIS
V1 7591
V2 7591
V3 6SL7
V4 6SL7
V5 6SL7
V6 6SL7
V7 7591
V8 7591
TREM
FT. SW.
ECHO
SPEED
INT.
DIM.
POWER
* LIC. UNDER HAMMOND ORGAN Co. PAT. N° 2,967,447

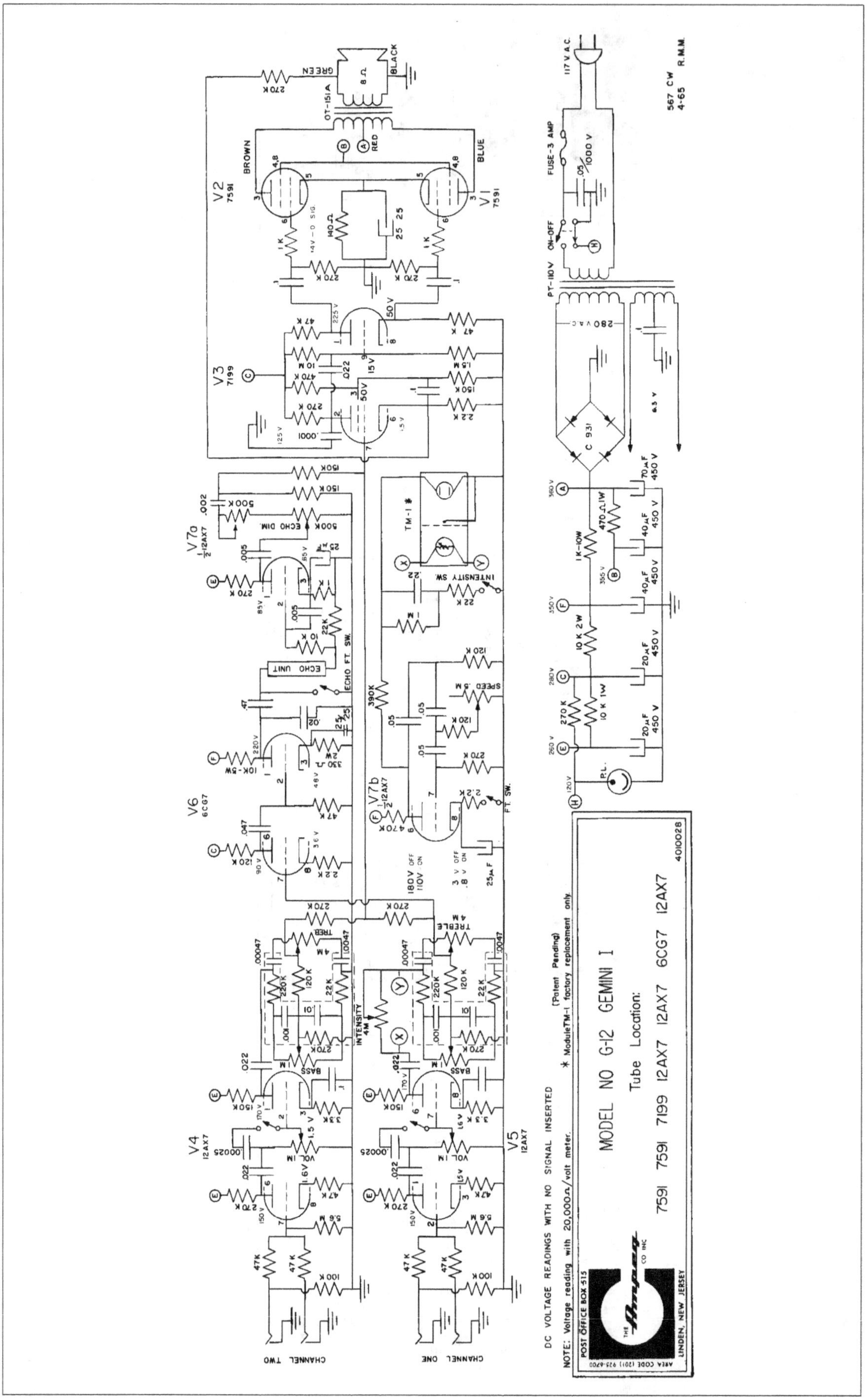
CHANNEL ONE
CHANNEL TWO
DC VOLTAGE READINGS WITH NO SIGNAL INSERTED
NOTE: Voltage reading with 20,000Ω/volt meter.
* ModuleTM-1 factory replacement only.
(Patent Pending)
POST OFFICE BOX 515
LINDEN, NEW JERSEY
AREA CODE (201) 925-6700
MODEL NO G-12 GEMINI I
Tube Location:
7591 7591 7199 12AX7 12AX7 6CG7 12AX7
4010028
567 CW
4-65
R.M.M.

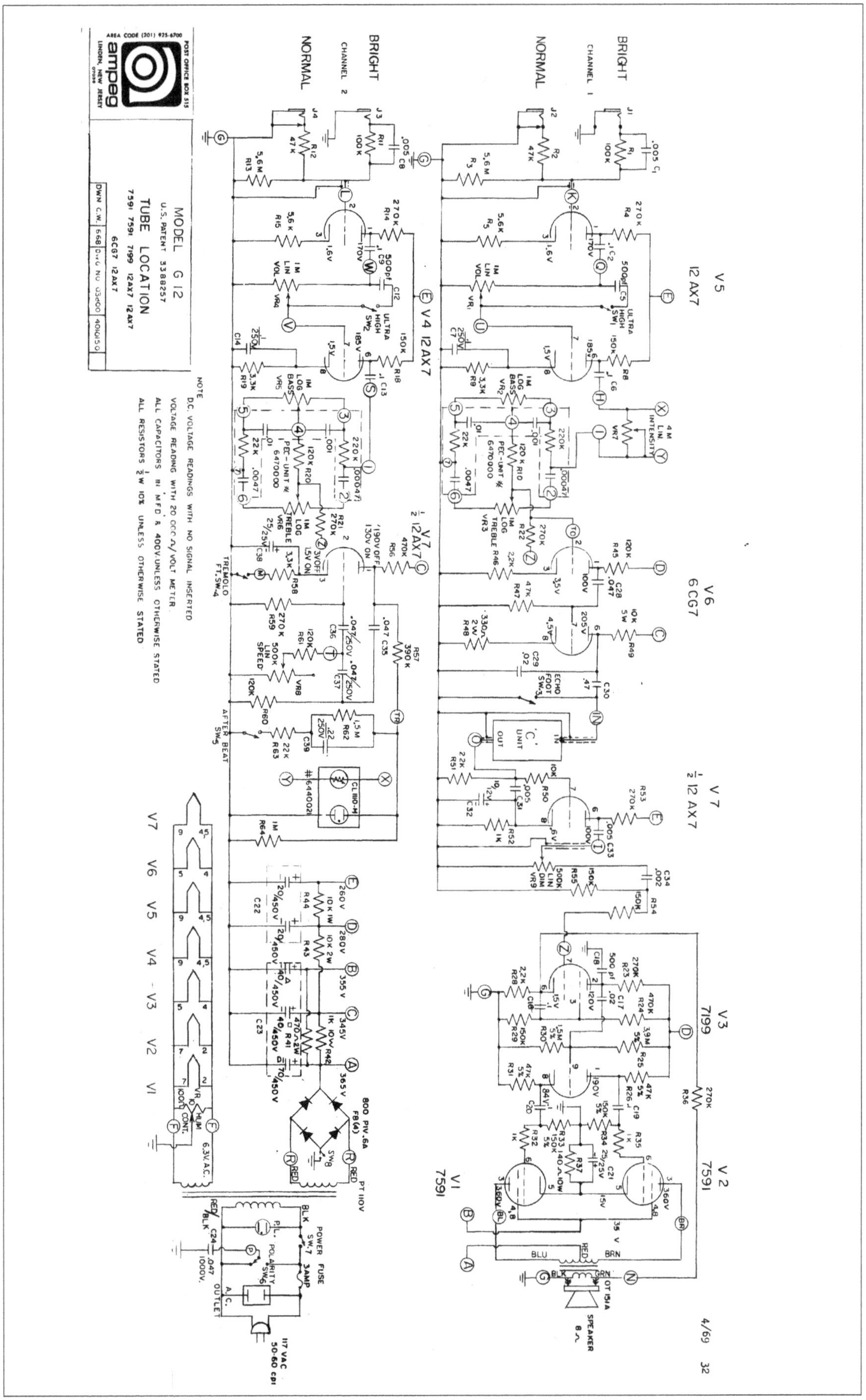

V5
12 AX7
V6
6 CG7
V7
½ 12 AX7
V3
7199
V2
7591
V1
7591
V4 12AX7
½ 12AX7
4/69 32
BRIGHT
CHANNEL 1
NORMAL
BRIGHT
CHANNEL 2
NORMAL
SPEAKER
8 Ω
117 VAC
50-60 CPS
POWER
FUSE
3AMP
POLARITY
A.C. OUTLET
6.3V A.C.
TREMOLO
FT. SW-4
AFTER BEAT
SW5
ECHO
FOOT
SW-3
'C' UNIT
V7 V6 V5 V4 V3 V2 V1
POST OFFICE BOX 515
AREA CODE (201) 925-6700
ampeg
LINDEN, NEW JERSEY
07036
MODEL G 12
U.S. PATENT 3388257
TUBE LOCATION
7591 7591 7199 12AX7 12AX7
6CG7 12AX7
DWN C.W. 668 DWG NO 03600 4010150
NOTE
D.C. VOLTAGE READINGS WITH NO SIGNAL INSERTED
VOLTAGE READING WITH 20 000 Ω/VOLT METER
ALL CAPACITORS IN MFD & 400V UNLESS OTHERWISE STATED
ALL RESISTORS ½W 10% UNLESS OTHERWISE STATED

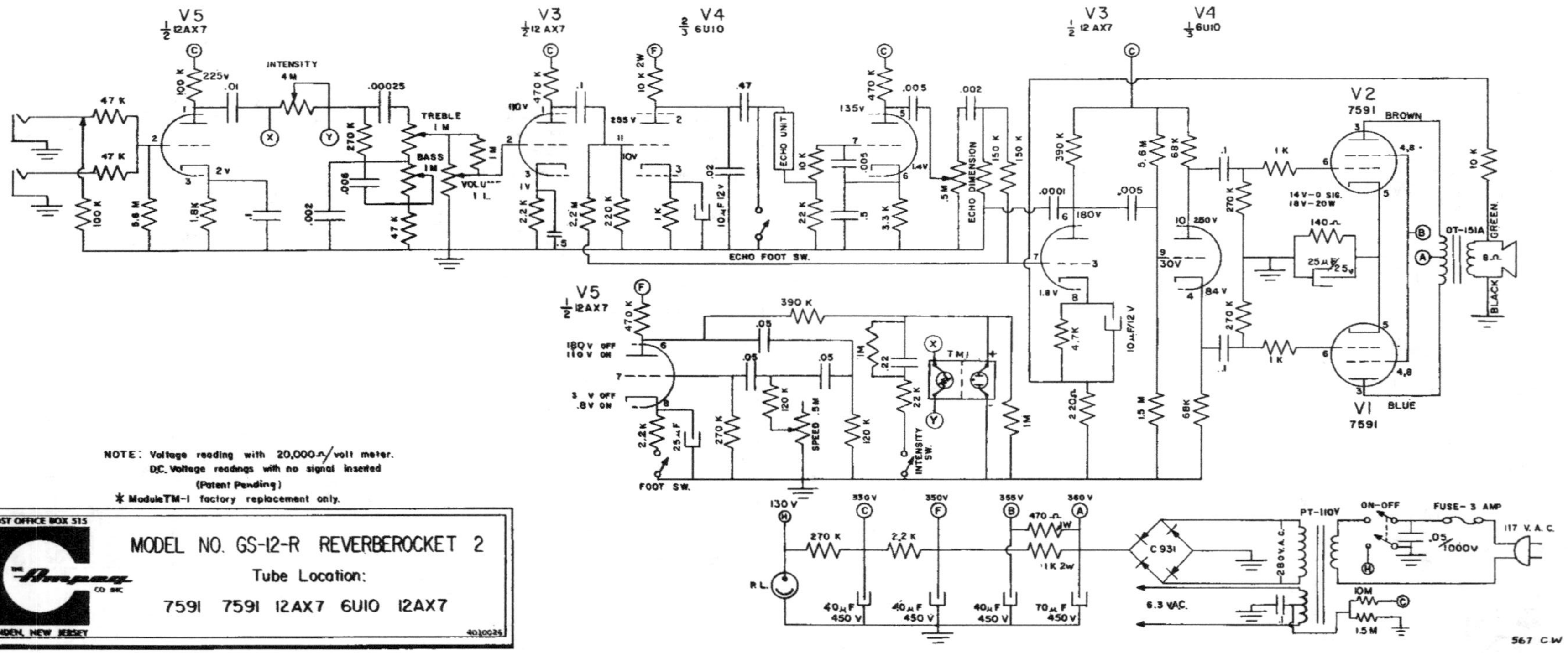

NOTE: Voltage reading with 20,000 Ω/volt meter.
D.C. Voltage readings with no signal inserted
(Patent Pending)
* Module TM-1 factory replacement only.

POST OFFICE BOX 515
THE Ampeg CO. INC.
AREA CODE (201) 925-6700
LINDEN, NEW JERSEY

MODEL NO. GS-12-R REVERBEROCKET 2

Tube Location:

7591 7591 12AX7 6U10 12AX7

4010026

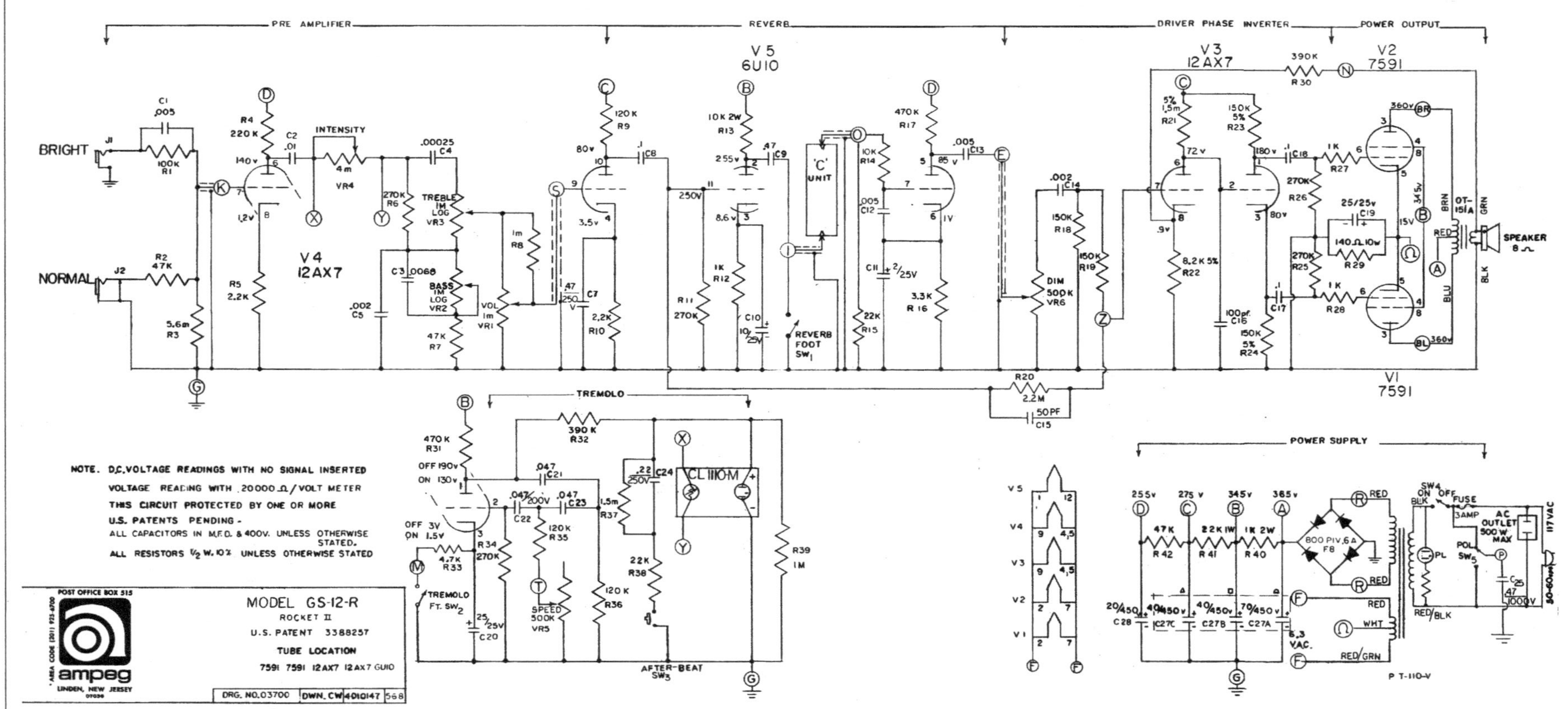
PRE AMPLIFIER
REVERB
DRIVER PHASE INVERTER
POWER OUTPUT
TREMOLO
POWER SUPPLY
BRIGHT
NORMAL
V4
12AX7
V5
6U10
V3
12AX7
V2
7591
V1
7591
INTENSITY
TREBLE
BASS
VOL
'C' UNIT
REVERB FOOT SW1
DIM 500K VR6
SPEAKER 8Ω
OT-151A
SPEED 500K VR5
TREMOLO FT. SW2
AFTER-BEAT SW3
CL 1110-M
800 PIV, 6A F8
AC OUTLET 500 W MAX
3AMP FUSE
117 VAC
P T-110-V
NOTE. D.C. VOLTAGE READINGS WITH NO SIGNAL INSERTED
VOLTAGE READING WITH 20000 Ω / VOLT METER
THIS CIRCUIT PROTECTED BY ONE OR MORE
U.S. PATENTS PENDING -
ALL CAPACITORS IN M.F.D. & 400V. UNLESS OTHERWISE STATED.
ALL RESISTORS 1/2 W. 10% UNLESS OTHERWISE STATED
POST OFFICE BOX 515
AREA CODE (201) 925-6700
ampeg
LINDEN, NEW JERSEY
MODEL GS-12-R
ROCKET II
U.S. PATENT 3388257
TUBE LOCATION
7591 7591 12AX7 12AX7 6U10
DRG. NO. 03700
DWN. CW
4010147
568

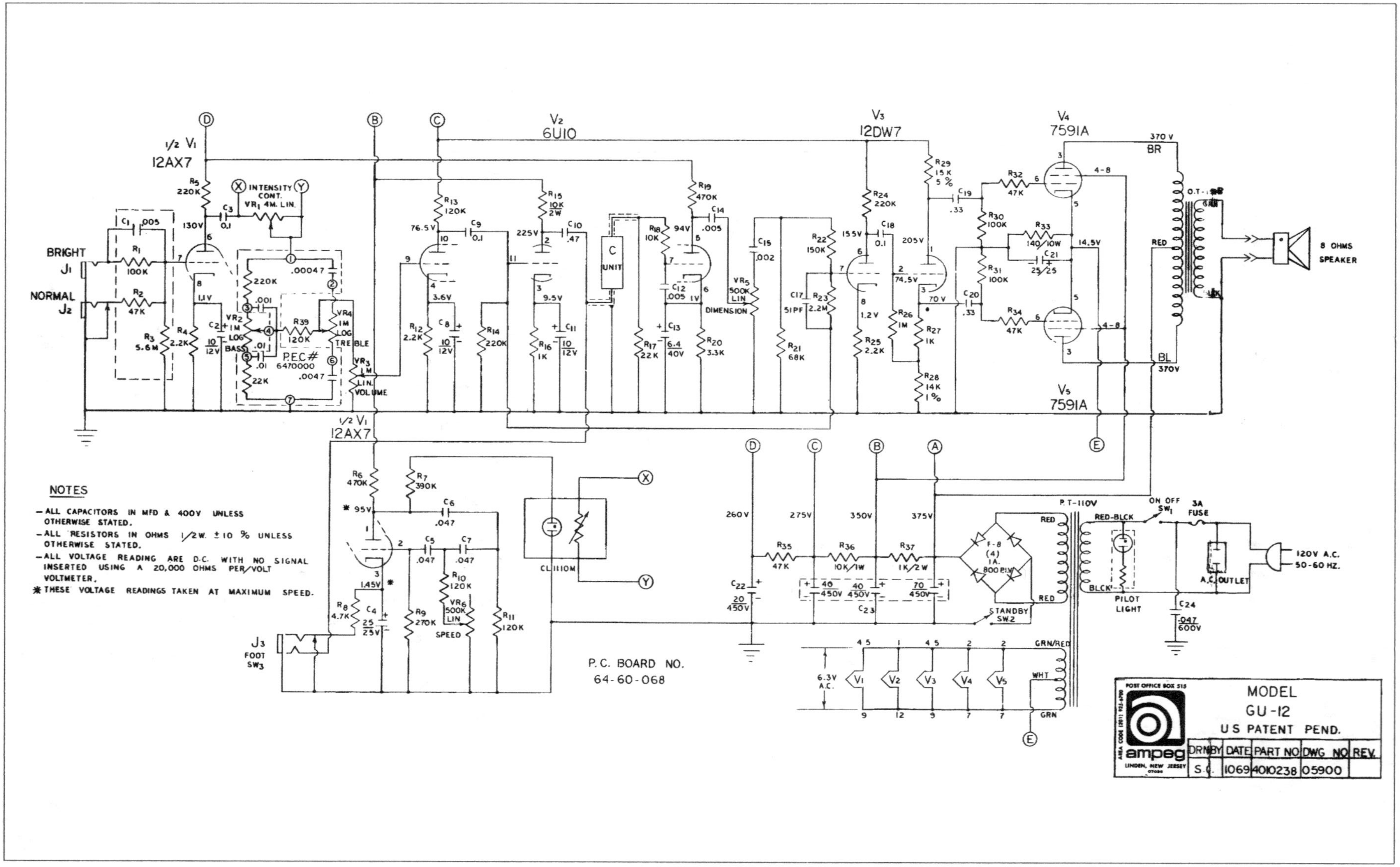
1/2 V1
12AX7
V2
6U10
V3
12DW7
V4
7591A
V5
7591A
BRIGHT
J1
NORMAL
J2
INTENSITY CONT.
VR1 4M. LIN.
P.E.C.#
6470000
VR2 1M LOG BASS
VR4 1M LOG TREBLE
VR3 1M LIN. VOLUME
VR5 500K LIN DIMENSION
C UNIT
VR6 500K LIN SPEED
J3 FOOT SW3
CL1110M
8 OHMS SPEAKER
O.T.
P.T-110V
F-8 (4) 1A. 800 P.I.V.
STANDBY SW.2
ON OFF SW1
3A FUSE
PILOT LIGHT
A.C. OUTLET
120V A.C. 50-60 HZ.
6.3V A.C.
P.C. BOARD NO.
64-60-068
NOTES
– ALL CAPACITORS IN MFD & 400V UNLESS OTHERWISE STATED.
– ALL RESISTORS IN OHMS 1/2W. ±10 % UNLESS OTHERWISE STATED.
– ALL VOLTAGE READING ARE D.C. WITH NO SIGNAL INSERTED USING A 20,000 OHMS PER/VOLT VOLTMETER.
✱ THESE VOLTAGE READINGS TAKEN AT MAXIMUM SPEED.
POST OFFICE BOX 515
ampeg
LINDEN, NEW JERSEY
MODEL
GU-12
US PATENT PEND.
DRN BY | DATE | PART NO | DWG NO | REV.
S.C. | 1069 | 4010238 | 05900

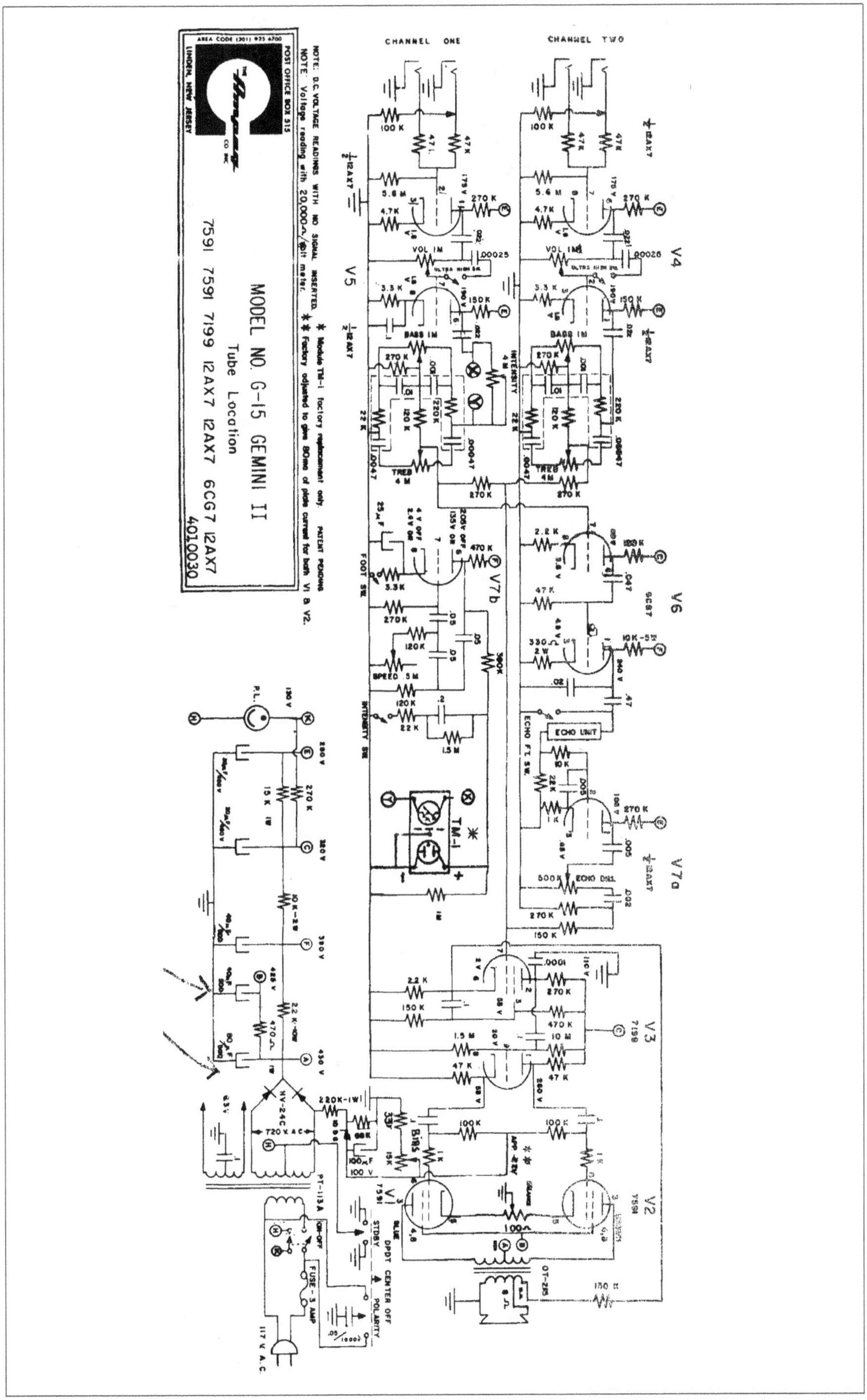

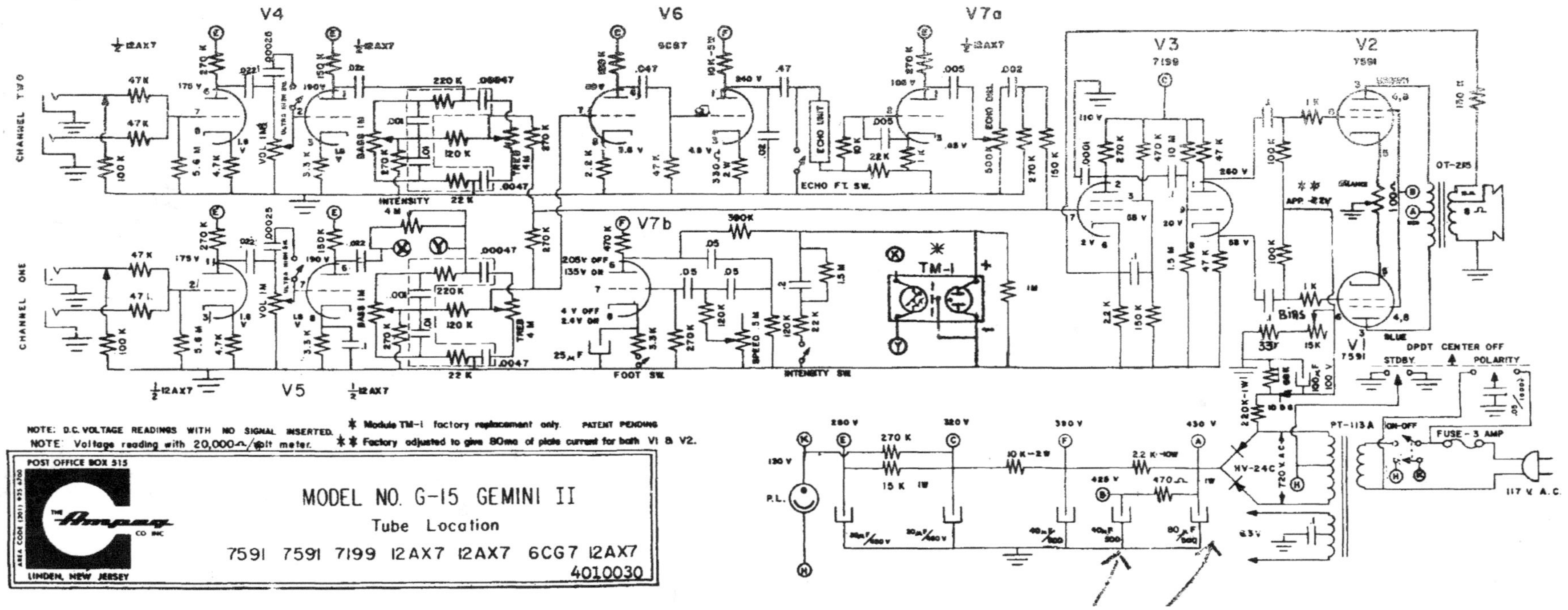
CHANNEL ONE
CHANNEL TWO
V4
V5
V6
V7a
V7b
V3
7199
V2
7591
V1
7591
1/2 12AX7
6CG7
TM-1
ECHO UNIT
ECHO FT. SW.
FOOT SW.
INTENSITY SW.
INTENSITY
SPEED
BASS
TREB
BIAS
STDBY
DPDT CENTER OFF
POLARITY
FUSE - 3 AMP
117 V. A.C.
PT-113A
HV-24C
OT-215
NOTE: D.C. VOLTAGE READINGS WITH NO SIGNAL INSERTED.
NOTE: Voltage reading with 20,000 Ω/volt meter.
* Module TM-1 factory replacement only. PATENT PENDING
** Factory adjusted to give 80ma of plate current for both V1 & V2.
POST OFFICE BOX 315
THE Ampeg CO. INC.
LINDEN, NEW JERSEY
MODEL NO. G-15 GEMINI II
Tube Location
7591 7591 7199 12AX7 12AX7 6CG7 12AX7
4010030

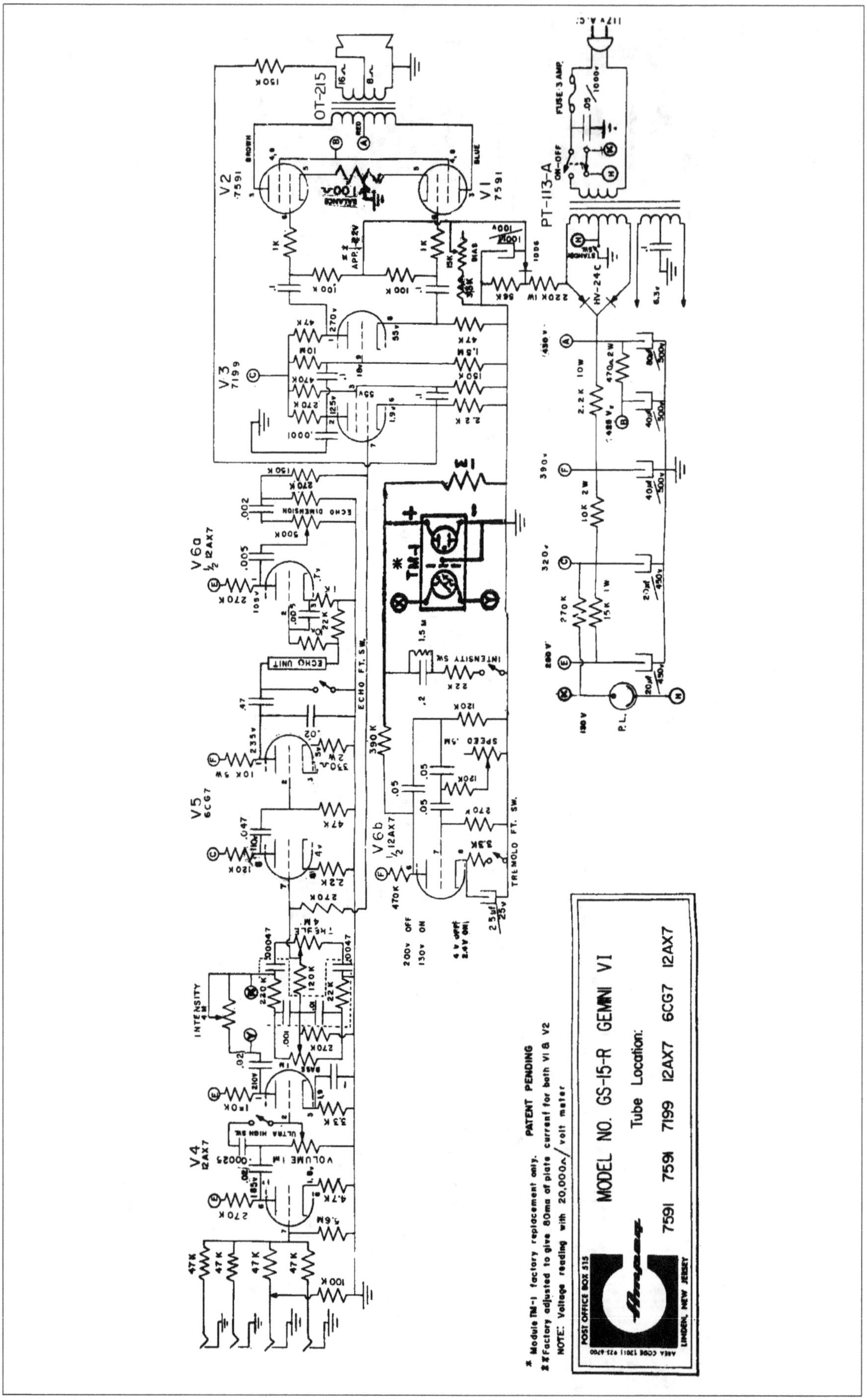
OT-215
PT-113-A
V1 7591
V2 7591
V3 7199
V4 12AX7
V5 6CG7
V6a ½ 12AX7
V6b ½ 12AX7
TM-1
ECHO UNIT
ECHO DIMENSION
ECHO FT. SW.
TREMOLO FT. SW.
INTENSITY SW.
SPEED .5M
INTENSITY 4M
VOLUME 1M
ULTRA HIGH SW.
BASS
TREBLE
BALANCE
BIAS
STANDBY SW.
ON-OFF
FUSE 3 AMP.
117 V.A.C.
HV-24C
P.L.
* Module TM-1 factory replacement only.
PATENT PENDING
** Factory adjusted to give 80ma of plate current for both V1 & V2
NOTE: Voltage reading with 20,000Ω/volt meter
MODEL NO. GS-15-R GEMINI VI
Tube Location:
7591 7591 7199 12AX7 6CG7 12AX7
POST OFFICE BOX 515
LINDEN, NEW JERSEY
AREA CODE (201) 925-6700

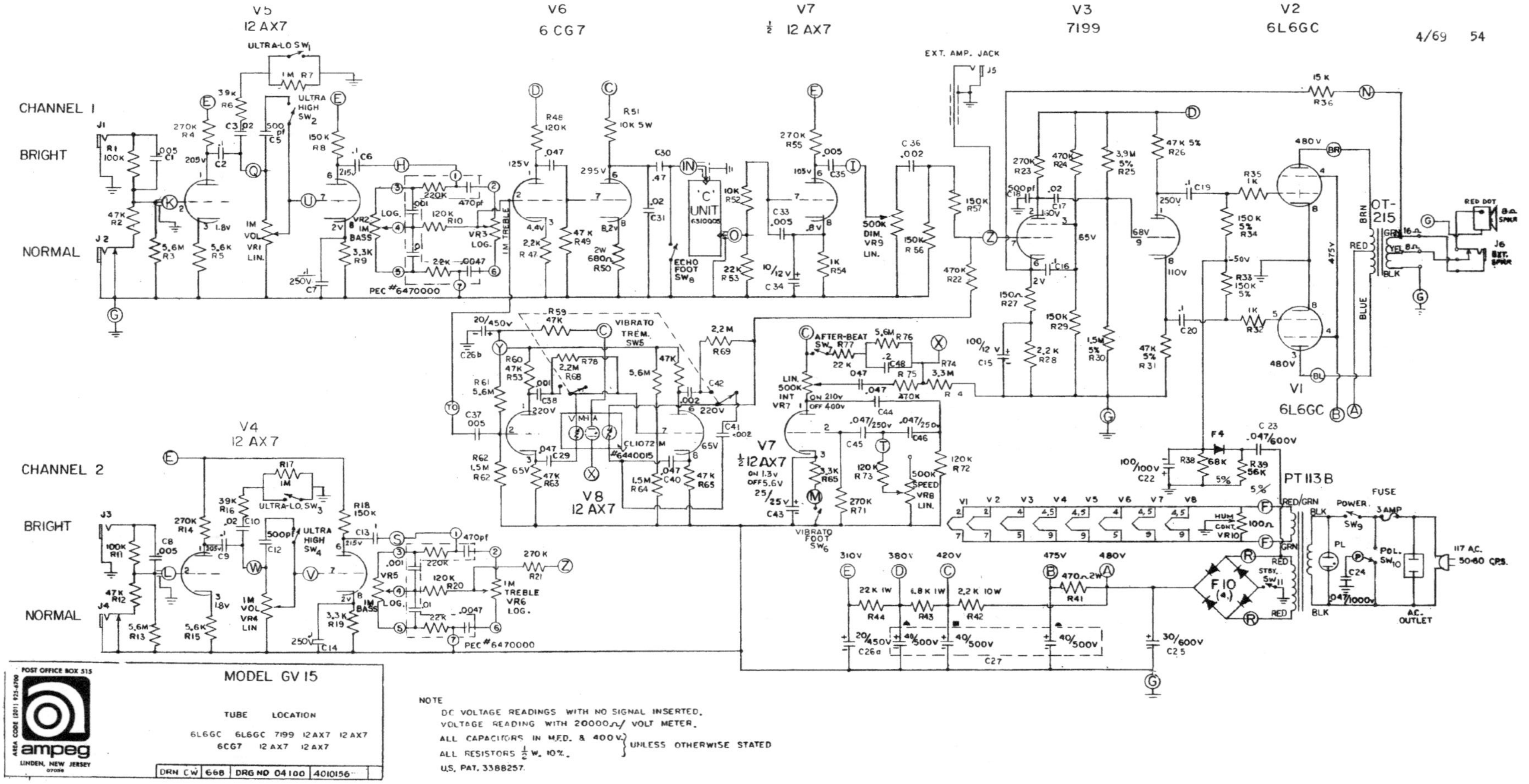
CHANNEL 1
BRIGHT
NORMAL
CHANNEL 2
BRIGHT
NORMAL
V5
12 AX7
V6
6 CG7
V7
½ 12 AX7
V3
7199
V2
6L6GC
4/69 54
V4
12 AX7
V8
12 AX7
V1
6L6GC
ULTRA-LO SW1
ULTRA HIGH SW2
VIBRATO TREM. SW5
ECHO FOOT SW8
'C' UNIT
AFTER-BEAT SW7
VIBRATO FOOT SW6
EXT. AMP. JACK
OT-215
PT 113 B
POWER SW9
FUSE 3 AMP
117 AC. 50-60 CPS.
A.C. OUTLET
PEC #6470000
POST OFFICE BOX 515
AREA CODE (201) 925-6700
ampeg
LINDEN, NEW JERSEY 07036
MODEL GV 15
TUBE LOCATION
6L6GC 6L6GC 7199 12AX7 12AX7
6CG7 12AX7 12AX7
DRN CW 668 DRG NO 04100 4010156
NOTE
DC VOLTAGE READINGS WITH NO SIGNAL INSERTED.
VOLTAGE READING WITH 20000Ω/ VOLT METER.
ALL CAPACITORS IN MFD. & 400V. } UNLESS OTHERWISE STATED
ALL RESISTORS ½ W. 10%.
U.S. PAT. 3388257.

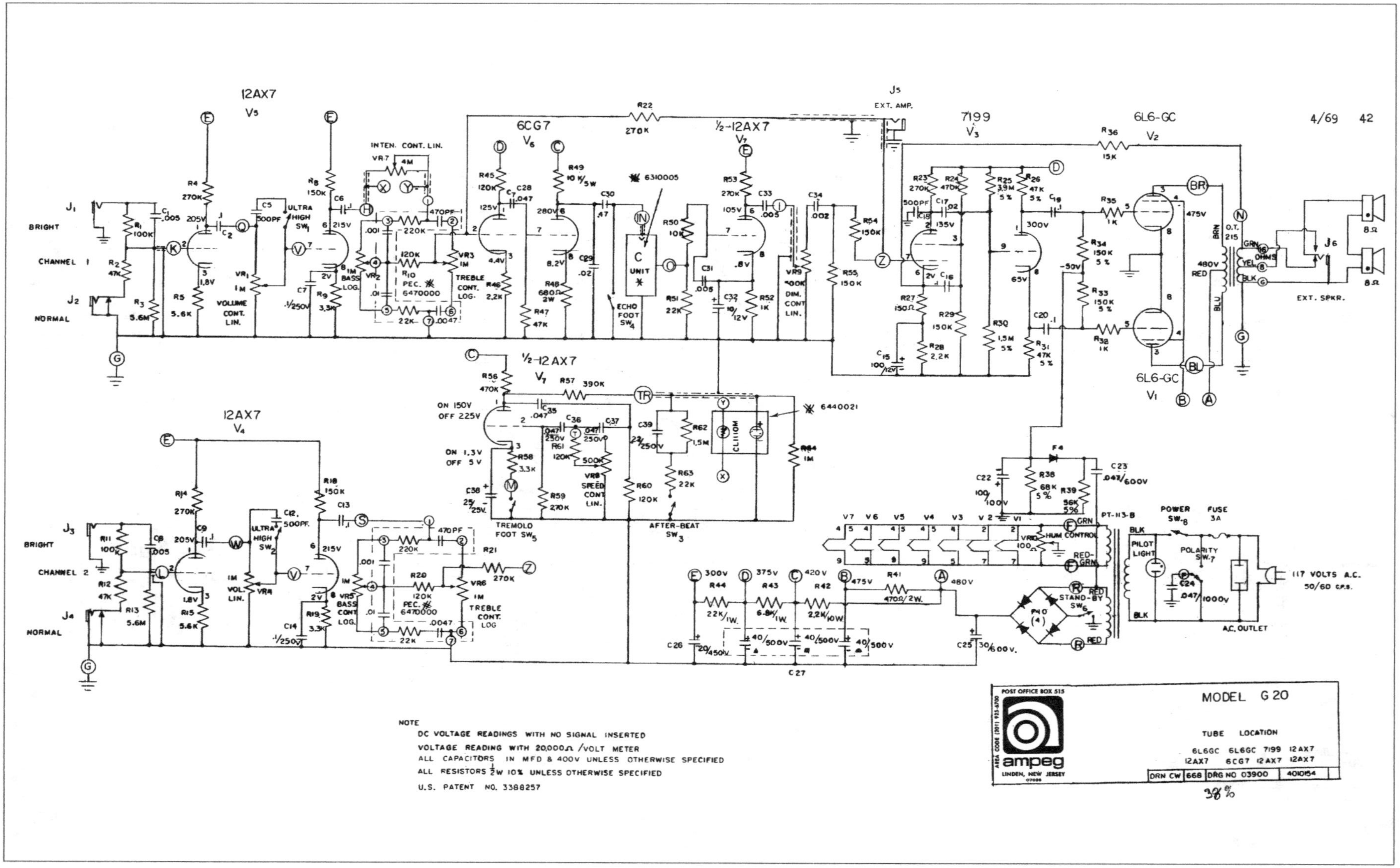
4/69 42
12AX7
V5
12AX7
V4
6CG7
V6
½-12AX7
V7
7199
V3
6L6-GC
V2
6L6-GC
V1
J1
BRIGHT
CHANNEL 1
J2
NORMAL
J3
BRIGHT
CHANNEL 2
J4
NORMAL
J5
EXT. AMP.
J6
EXT. SPKR.
INTEN. CONT. LIN.
VOLUME CONT. LIN.
ULTRA HIGH SW1
TREBLE CONT. LOG.
BASS LOG.
ECHO FOOT SW4
TREMOLO FOOT SW5
AFTER-BEAT SW3
SPEED CONT LIN.
DIM. CONT LIN.
HUM CONTROL
STAND-BY SW6
POWER SW8
FUSE 3A
POLARITY SW7
PILOT LIGHT
A.C. OUTLET
117 VOLTS A.C. 50/60 C.P.S.
PT-113-B
O.T. 215
NOTE
DC VOLTAGE READINGS WITH NO SIGNAL INSERTED
VOLTAGE READING WITH 20,000Ω/VOLT METER
ALL CAPACITORS IN MFD & 400V UNLESS OTHERWISE SPECIFIED
ALL RESISTORS ½W 10% UNLESS OTHERWISE SPECIFIED
U.S. PATENT NO. 3388257
POST OFFICE BOX 515
AREA CODE (201) 925-6700
ampeg
LINDEN, NEW JERSEY
MODEL G 20
TUBE LOCATION
6L6GC 6L6GC 7199 12AX7
12AX7 6CG7 12AX7 12AX7
DRN CW 668 DRG NO 03900 401054

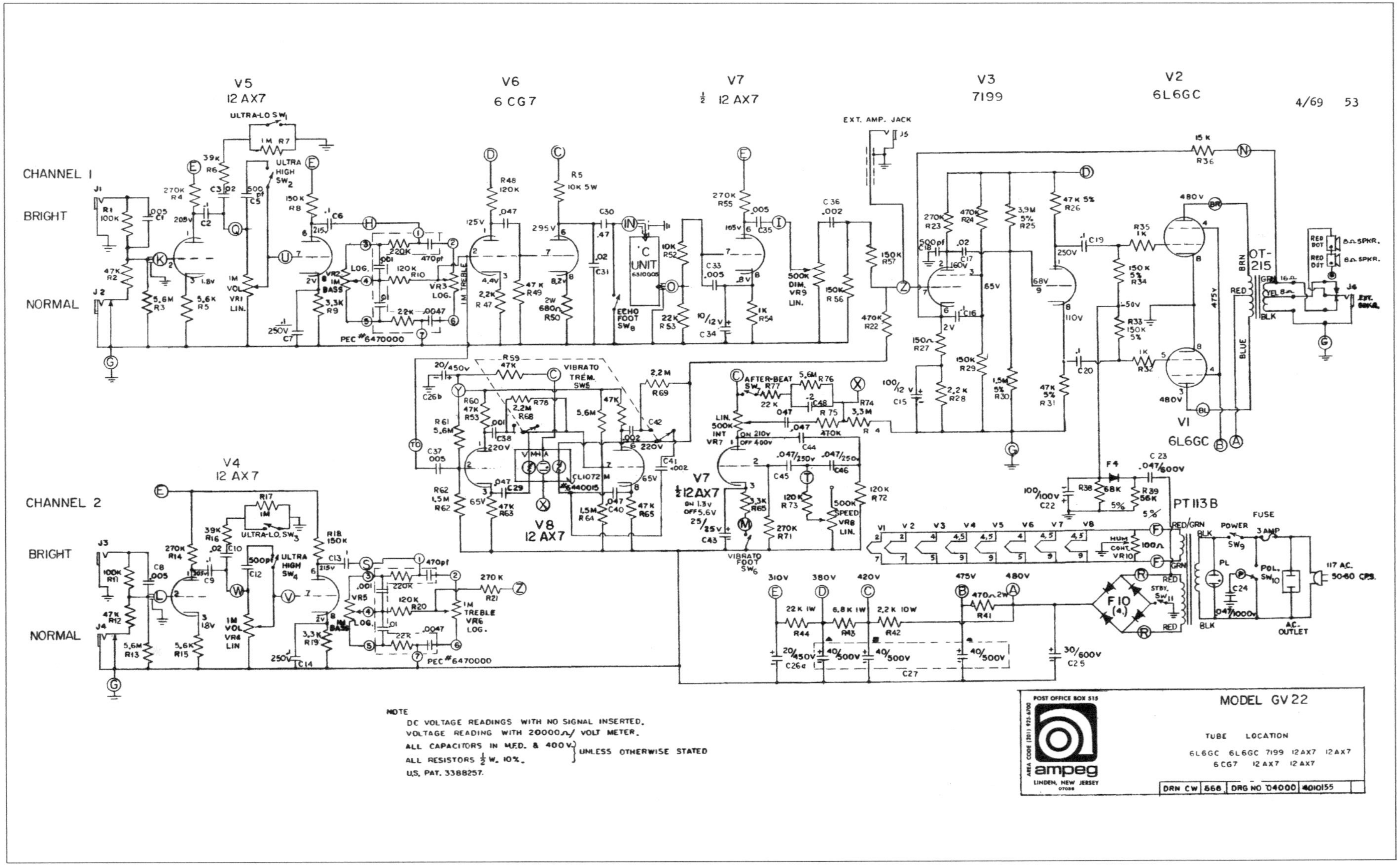

V5
12 AX7
V6
6 CG7
V7
½ 12 AX7
V3
7199
V2
6L6GC
4/69 53
CHANNEL 1
BRIGHT
NORMAL
CHANNEL 2
BRIGHT
NORMAL
V4
12 AX7
V8
12 AX7
V7
½ 12 AX7
V1
6L6GC
EXT. AMP. JACK
ULTRA-LO SW1
ULTRA HIGH SW2
ECHO FOOT SW8
VIBRATO TREM. SW5
AFTER-BEAT SW.
VIBRATO FOOT SW6
PT 113 B
OT-215
POWER SW9
FUSE 3 AMP
117 AC. 50-60 CPS.
AC. OUTLET
PEC #6470000
NOTE
DC VOLTAGE READINGS WITH NO SIGNAL INSERTED.
VOLTAGE READING WITH 20000Ω/ VOLT METER.
ALL CAPACITORS IN MFD. & 400V.
ALL RESISTORS ½ W. 10%.
UNLESS OTHERWISE STATED
U.S. PAT. 3388257.
POST OFFICE BOX 515
ampeg
LINDEN, NEW JERSEY
MODEL GV 22
TUBE LOCATION
6L6GC 6L6GC 7199 12AX7 12AX7
6CG7 12AX7 12AX7
DRN CW 668 DRG NO 04000 4010155

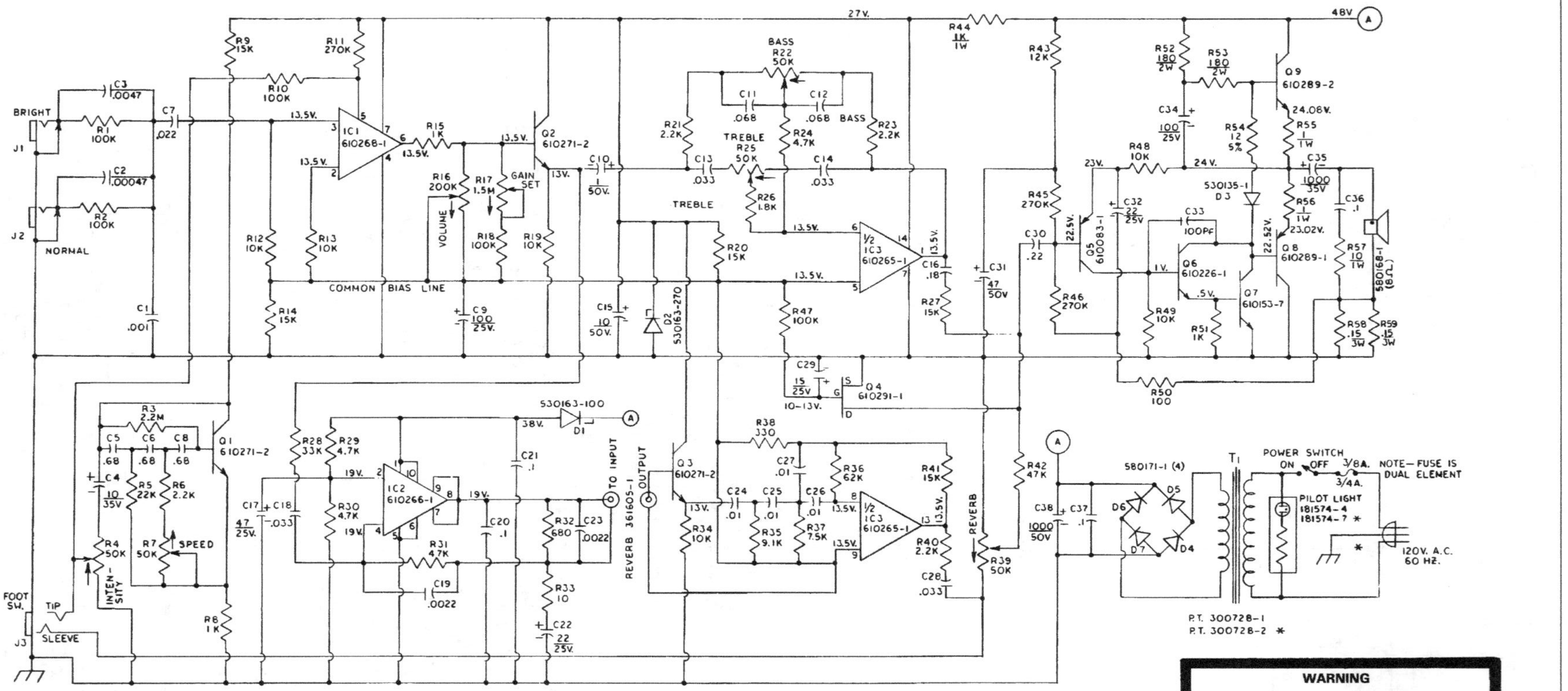

WARNING

For maximum safety, reliability and performance, all parts must be replaced by those having identical specifications. Under no circumstances may the original design be modified or altered without permission from the Ampeg Company.

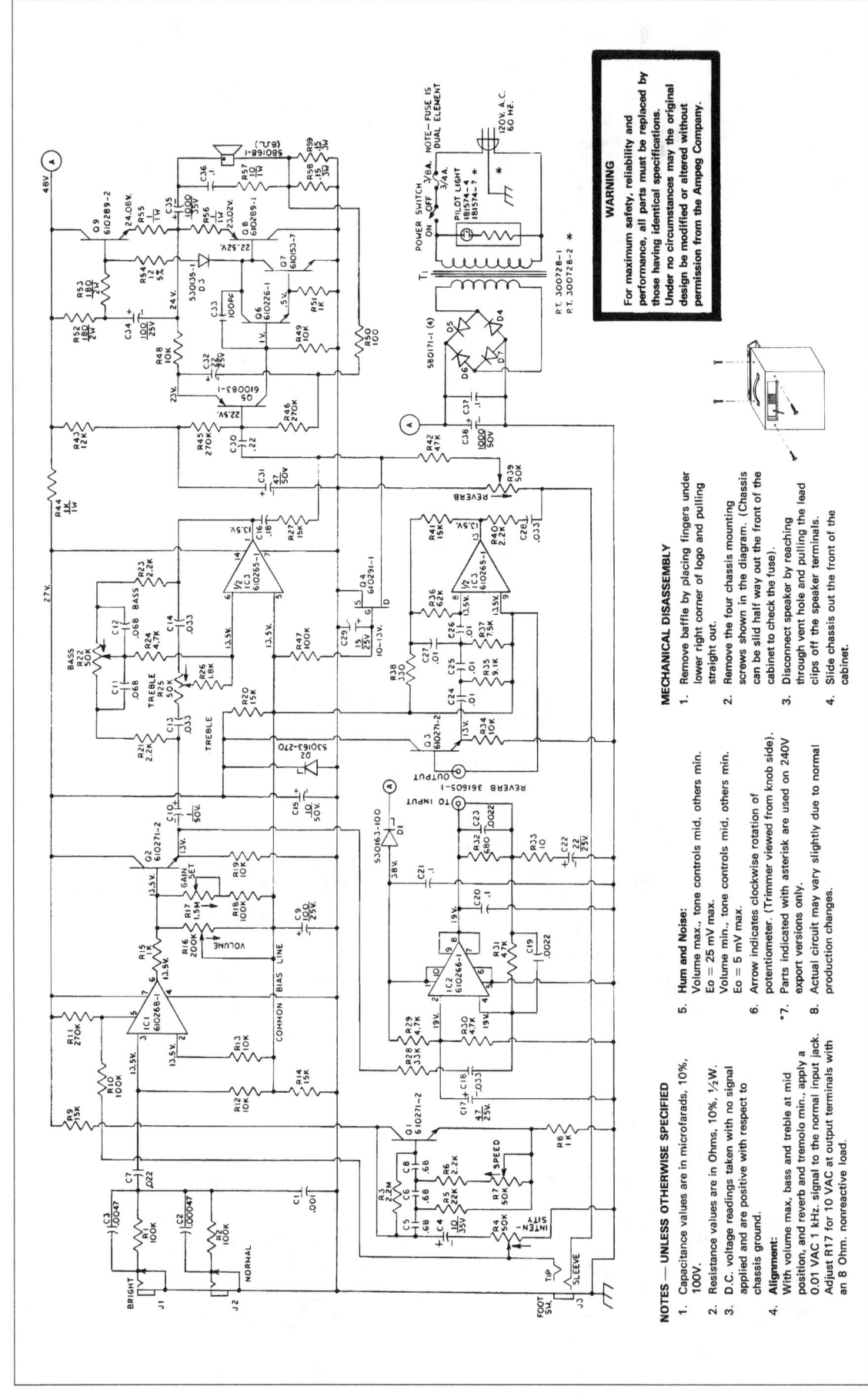

NOTES — UNLESS OTHERWISE SPECIFIED

1. Capacitance values are in microfarads, 10%, 100V.
2. Resistance values are in Ohms, 10%, ½W.
3. D.C. voltage readings taken with no signal applied and are positive with respect to chassis ground.
4. **Alignment:** With volume max, bass and treble at mid position, and reverb and tremolo min., apply a 0.01 VAC 1 kHz. signal to the normal input jack. Adjust R17 for 10 VAC at output terminals with an 8 Ohm. nonreactive load.
5. **Hum and Noise:** Volume max., tone controls mid, others min. Eo = 25 mV max. Volume min., tone controls mid, others min. Eo = 5 mV max.
6. Arrow indicates clockwise rotation of potentiometer. (Trimmer viewed from knob side).
*7. Parts indicated with asterisk are used on 240V export versions only.
8. Actual circuit may vary slightly due to normal production changes.

MECHANICAL DISASSEMBLY

1. Remove baffle by placing fingers under lower right corner of logo and pulling straight out.
2. Remove the four chassis mounting screws shown in the diagram. (Chassis can be slid half way out the front of the cabinet to check the fuse).
3. Disconnect speaker by reaching through vent hole and pulling the lead clips off the speaker terminals.
4. Slide chassis out the front of the cabinet.

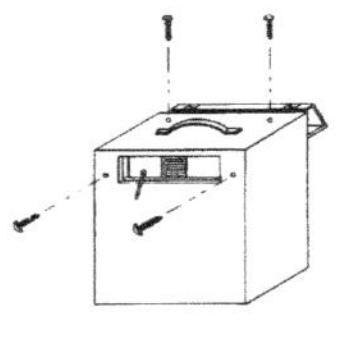

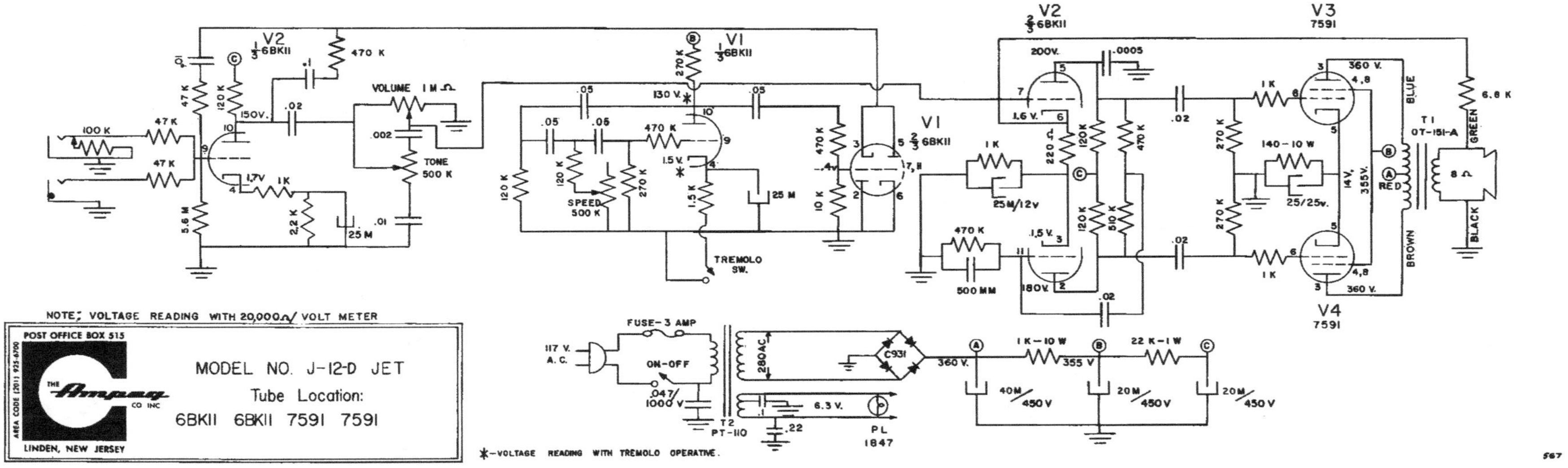
V2
1/3 6BKII
V1
1/3 6BKII
V1
2/3 6BKII
V2
2/3 6BKII
V3
7591
V4
7591
100 K
47 K
47 K
5.6 M
120 K
150V.
1.7V
1 K
2.2 K
25 M
.01
.02
.1
470 K
VOLUME 1 M Ω
.002
TONE
500 K
120 K
.05
.05
120 K
SPEED
500 K
270 K
470 K
270 K
130 V.
1.5 V.
1.5 K
25 M
TREMOLO SW.
.05
470 K
10 K
200V.
.0005
1.6 V.
220 Ω
1 K
25M/12v
470 K
500 MM
1.5 V.
180V.
120 K
120 K
510 K
470 K
.02
.02
.02
270 K
270 K
1 K
1 K
140-10 W
25/25v.
14V.
355V.
360 V.
360 V.
BLUE
RED
BROWN
T1
OT-151-A
GREEN
BLACK
8 Ω
6.8 K
NOTE; VOLTAGE READING WITH 20,000Ω/ VOLT METER
POST OFFICE BOX 515
AREA CODE (201) 925-6700
THE Ampeg CO INC
LINDEN, NEW JERSEY
MODEL NO. J-12-D JET
Tube Location:
6BKII 6BKII 7591 7591
FUSE-3 AMP
117 V. A. C.
ON-OFF
.047/1000 V
T2
PT-110
280AC
C931
6.3 V.
.22
PL
1847
360 V.
1 K-10 W
355 V
22 K-1 W
40M/450 V
20M/450 V
20M/450 V
✱-VOLTAGE READING WITH TREMOLO OPERATIVE.
567

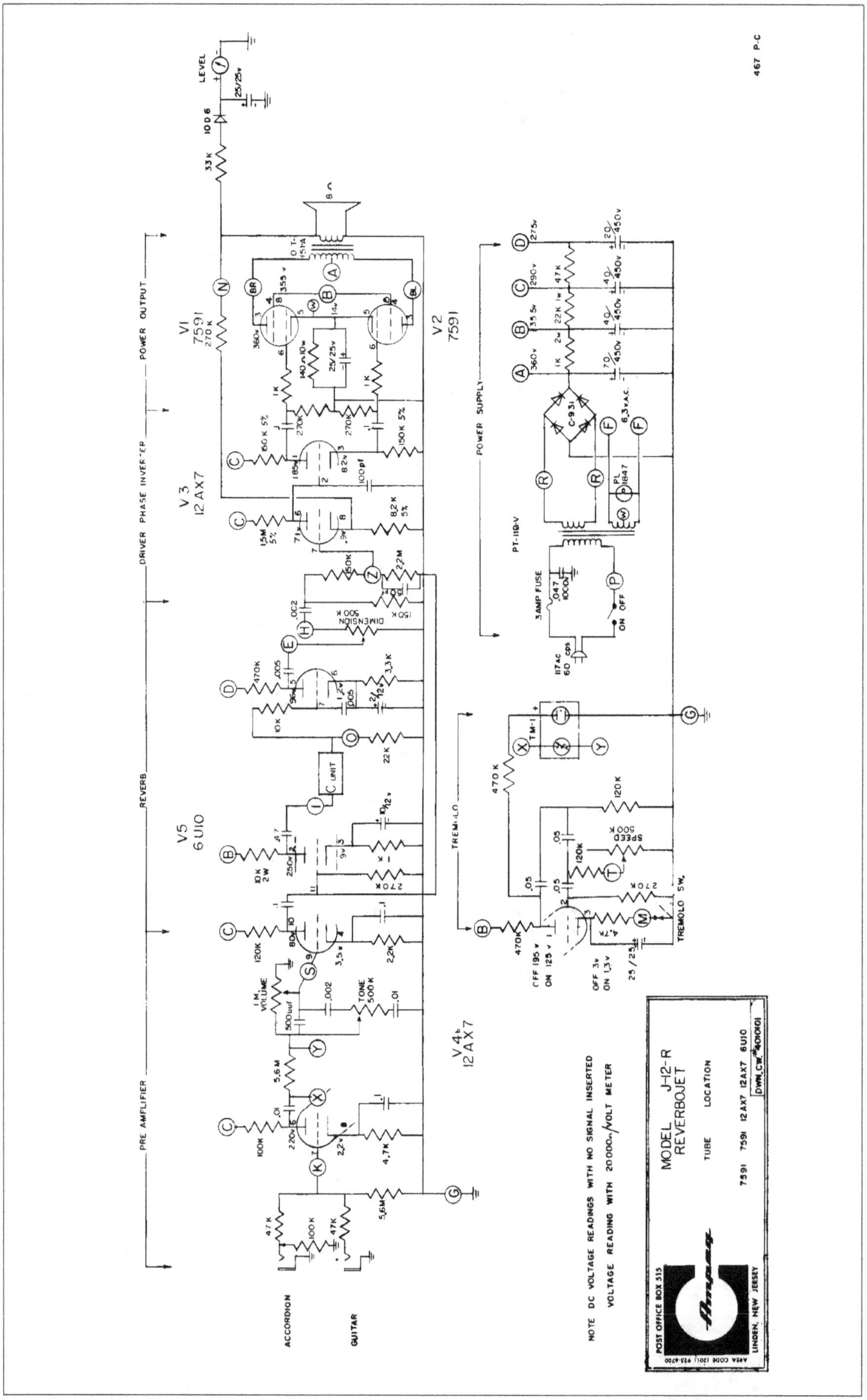

PRE AMPLIFIER
REVERB
DRIVER PHASE INVERTER
POWER OUTPUT
V5
6 U10
V3
12 AX7
V1
7591
V2
7591
V 4b
12 AX7
TREMOLO
POWER SUPPLY
ACCORDION
GUITAR
VOLUME
TONE
DIMENSION
SPEED
LEVEL
C UNIT
TM-1
PT-119-V
3AMP FUSE
117 AC 60 cps
ON OFF
TREMOLO SW.
NOTE DC VOLTAGE READINGS WITH NO SIGNAL INSERTED
VOLTAGE READING WITH 20000Ω/VOLT METER
POST OFFICE BOX 515
AREA CODE (201) 925-6700
LINDEN, NEW JERSEY
MODEL J-12-R
REVERBOJET
TUBE LOCATION
7591 7591 12AX7 12AX7 6U10
DWN. CW. #4010101
467 P-C

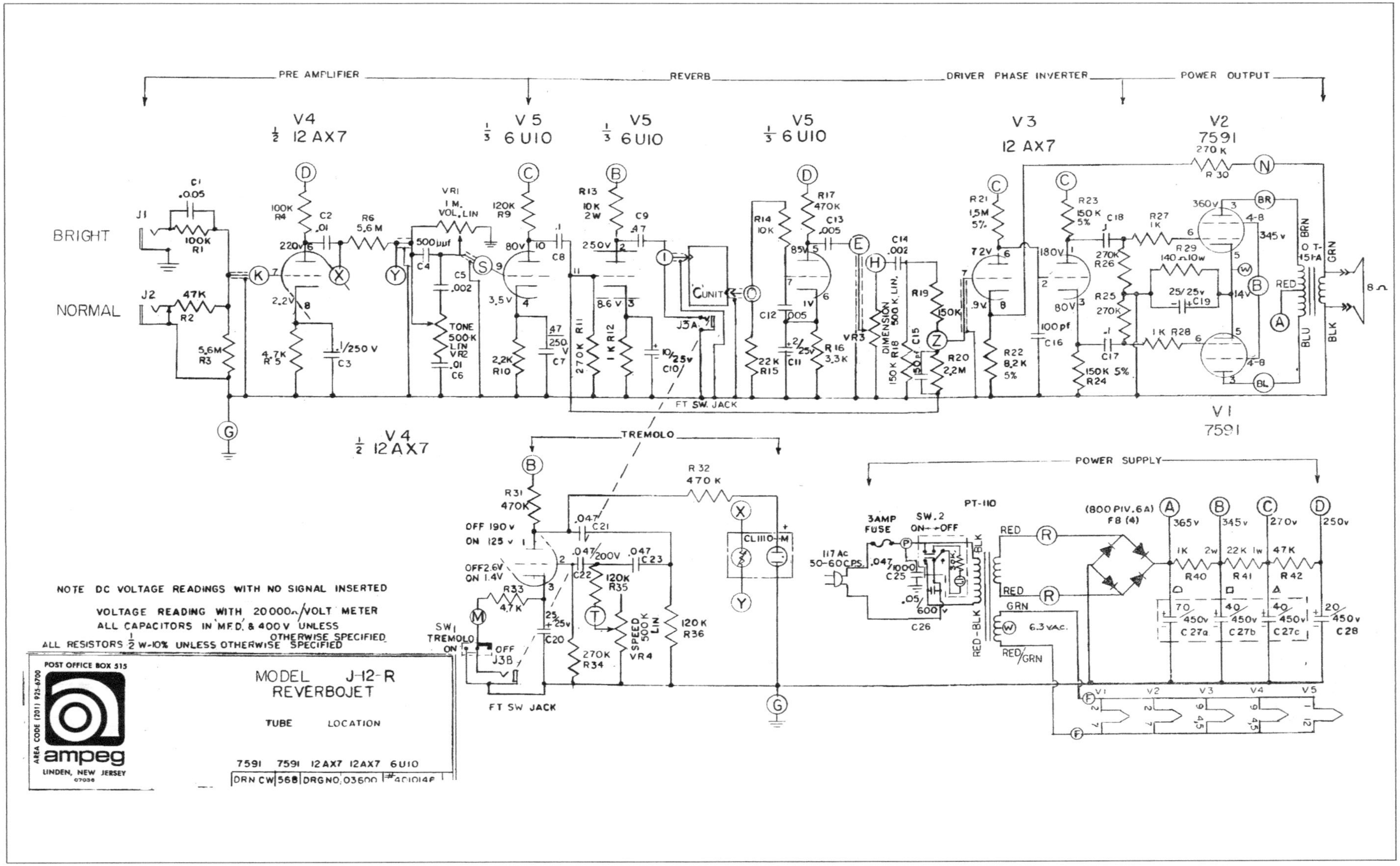
PRE AMPLIFIER
REVERB
DRIVER PHASE INVERTER
POWER OUTPUT
V4 ½ 12AX7
V5 ⅓ 6U10
V3 12AX7
V2 7591
V1 7591
BRIGHT
NORMAL
TREMOLO
POWER SUPPLY
FT SW JACK
PT-110
3AMP FUSE
117 AC 50-60CPS.
NOTE DC VOLTAGE READINGS WITH NO SIGNAL INSERTED
VOLTAGE READING WITH 20000Ω/VOLT METER
ALL CAPACITORS IN MFD. & 400V UNLESS OTHERWISE SPECIFIED
ALL RESISTORS ½ W-10% UNLESS OTHERWISE SPECIFIED
POST OFFICE BOX 315
AREA CODE (201) 925-6700
ampeg
LINDEN, NEW JERSEY 07036
MODEL J-12-R
REVERBOJET
TUBE
LOCATION
7591 7591 12AX7 12AX7 6U10

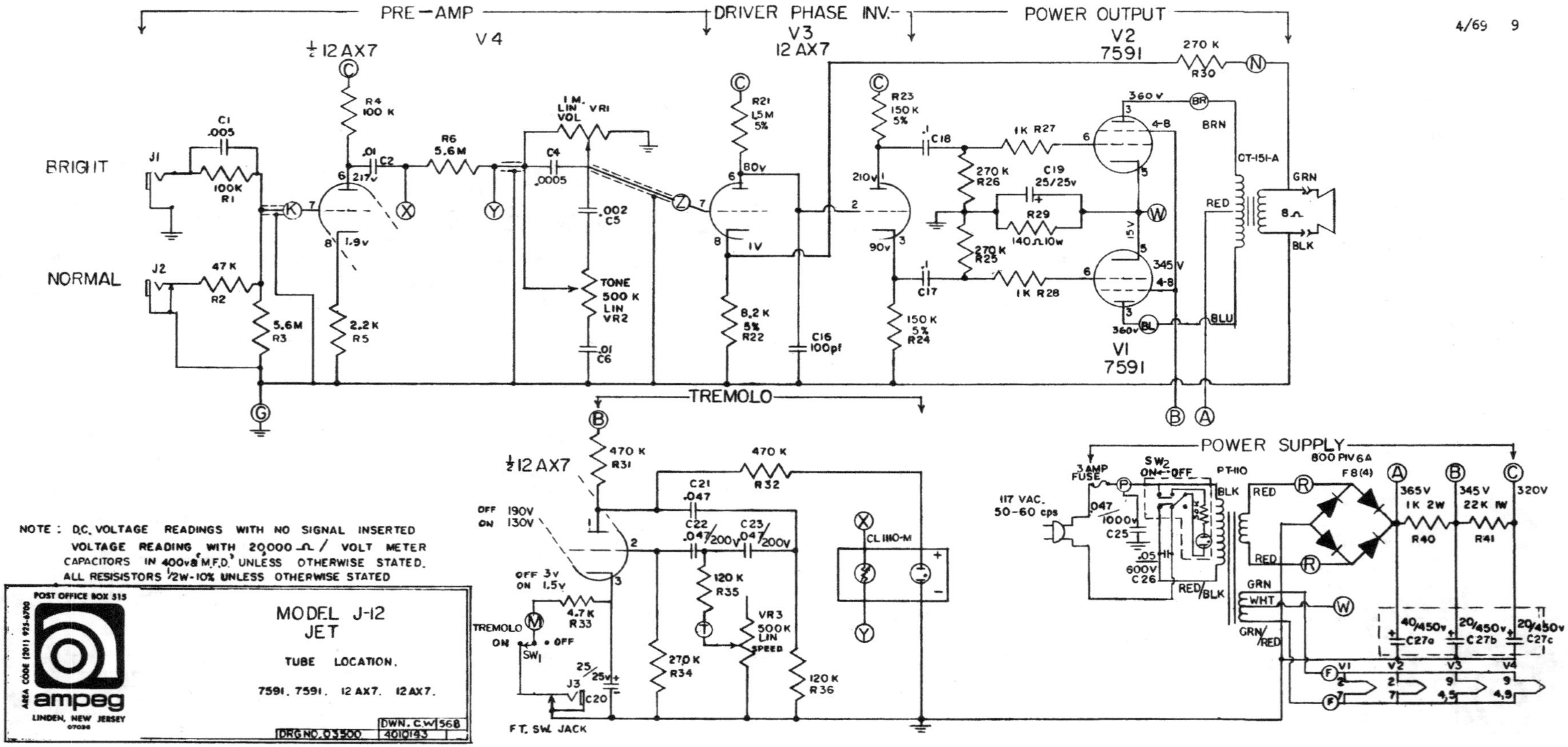
PRE-AMP
V4
½ 12AX7
DRIVER PHASE INV.
V3
12AX7
POWER OUTPUT
V2
7591
V1
7591
4/69 9
BRIGHT
NORMAL
TREMOLO
½ 12AX7
POWER SUPPLY
117 VAC.
50-60 cps
3 AMP FUSE
PT-110
CT-151-A
CL 1110-M
800 PIV 6A
F8(4)
365V
345V
320V
FT. SW. JACK
TREMOLO
NOTE: D.C. VOLTAGE READINGS WITH NO SIGNAL INSERTED
VOLTAGE READING WITH 20,000 Ω / VOLT METER
CAPACITORS IN 400v & M.F.D. UNLESS OTHERWISE STATED.
ALL RESISISTORS ½W-10% UNLESS OTHERWISE STATED
POST OFFICE BOX 315
AREA CODE (201) 925-6700
ampeg
LINDEN, NEW JERSEY 07036
MODEL J-12
JET
TUBE LOCATION.
7591. 7591. 12AX7. 12AX7.
DWN. C.W 568
DRG NO. 03500
4010143

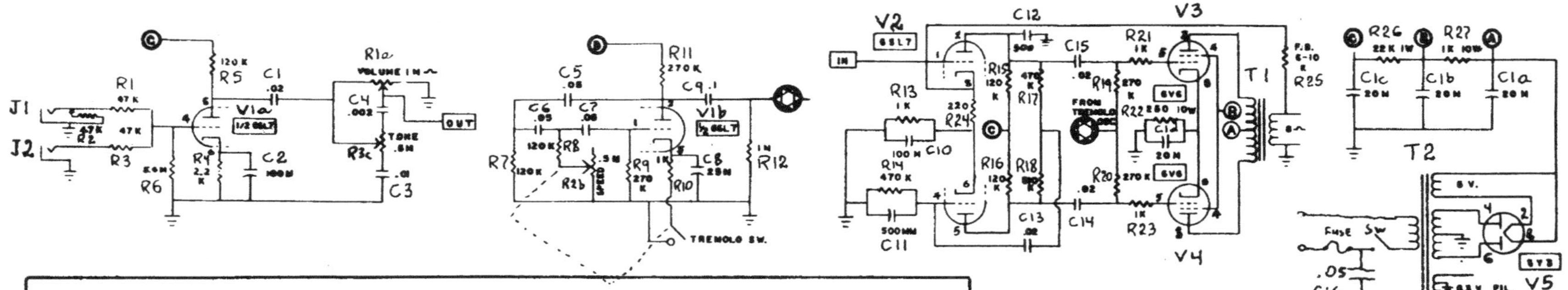

MODEL NO. J-12 B

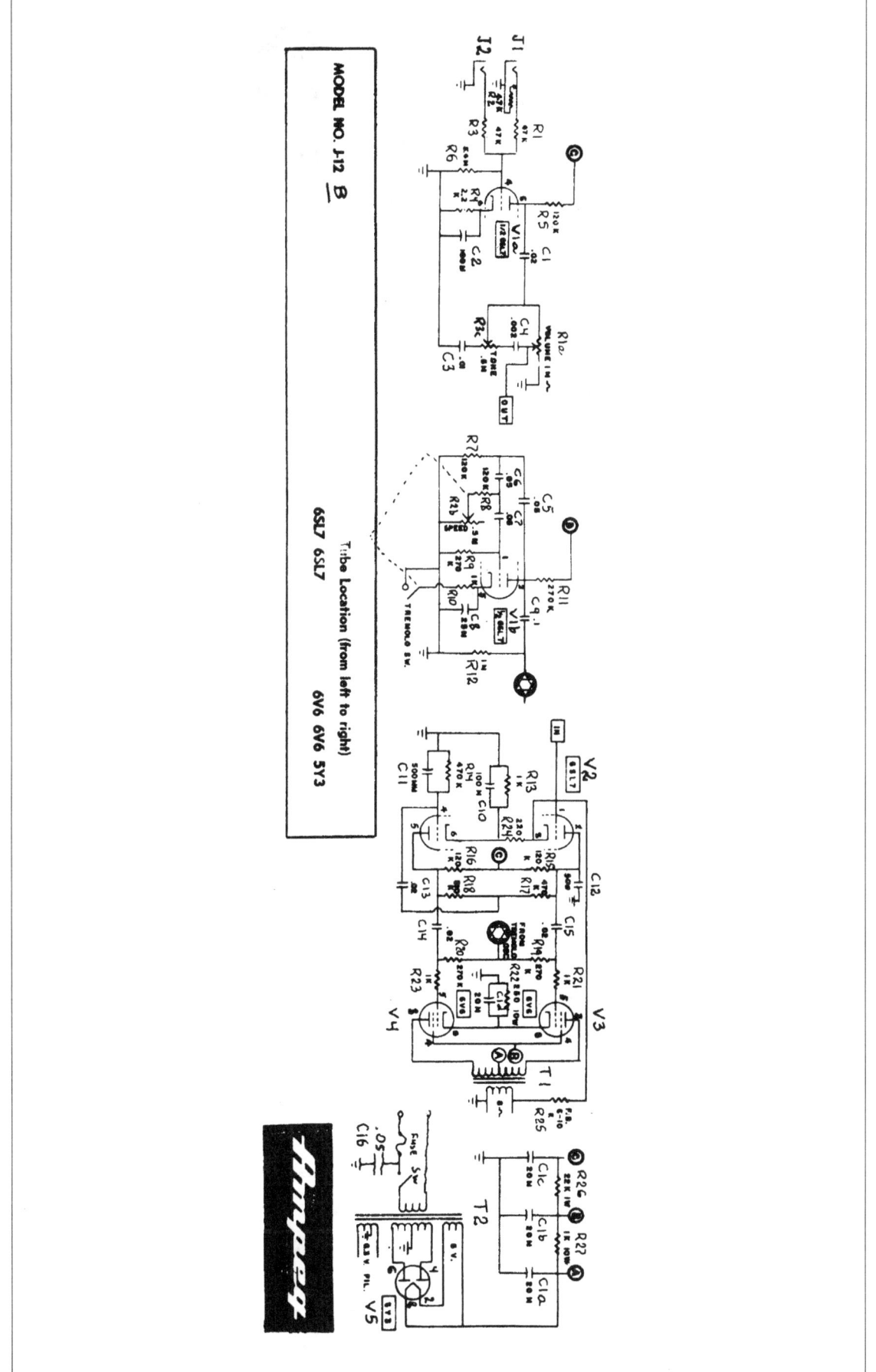

Tube Location (from left to right)

6SL7 6SL7 6V6 6V6 5Y3

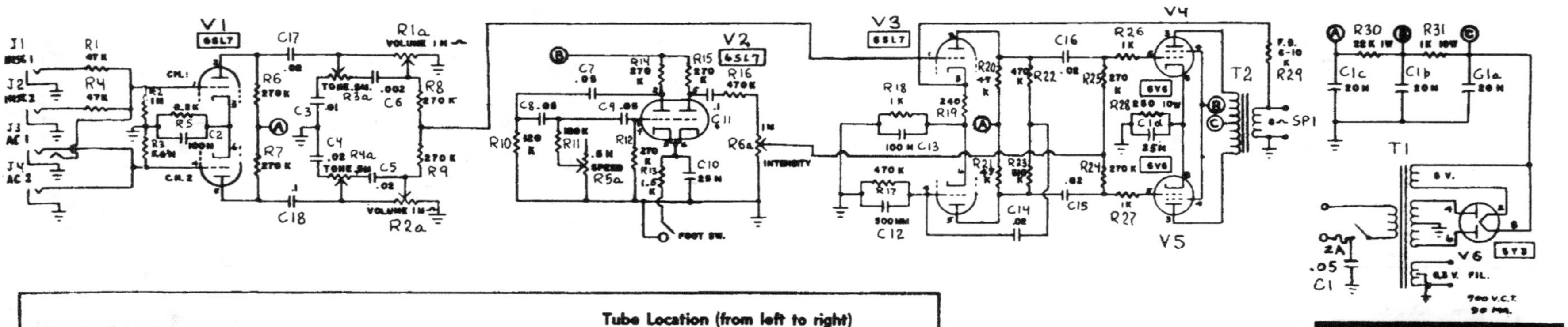

MODEL NO. M-12

Tube Location (from left to right)

6SL7 6SL7 6SL7 6V6 6V6 5Y3

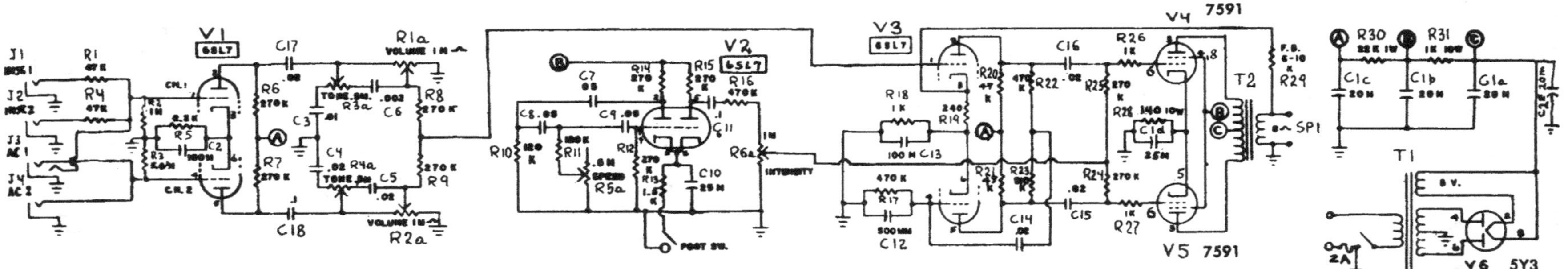

MODEL NO. M-12-A	Tube Location (from left to right)
	6SL7 6SL7 6SL7 7591 7591 5Y3

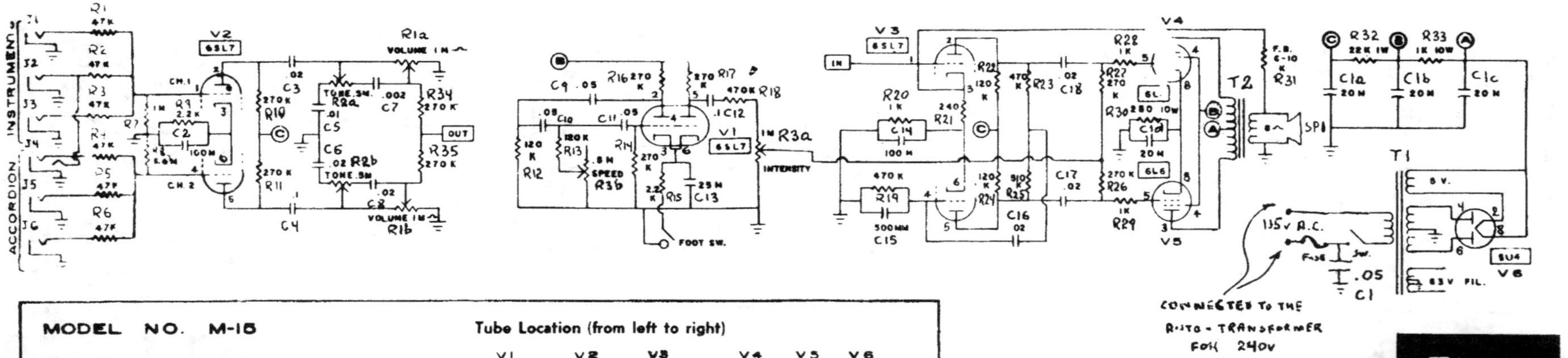

MODEL NO. M-15	**Tube Location (from left to right)**
SERIAL	V1 6SL7, V2 6SL7, V3 6SL7, V4 6L6, V5 6L6, V6 5U4

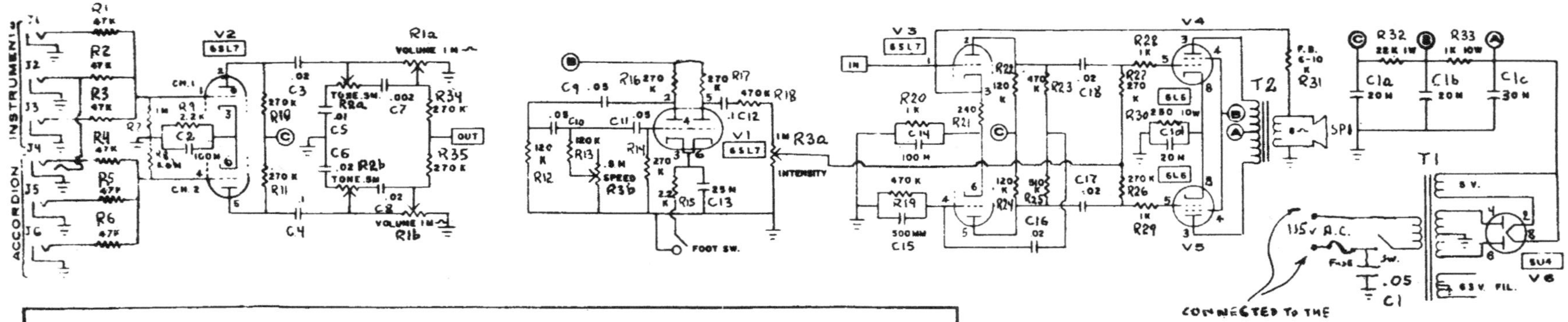

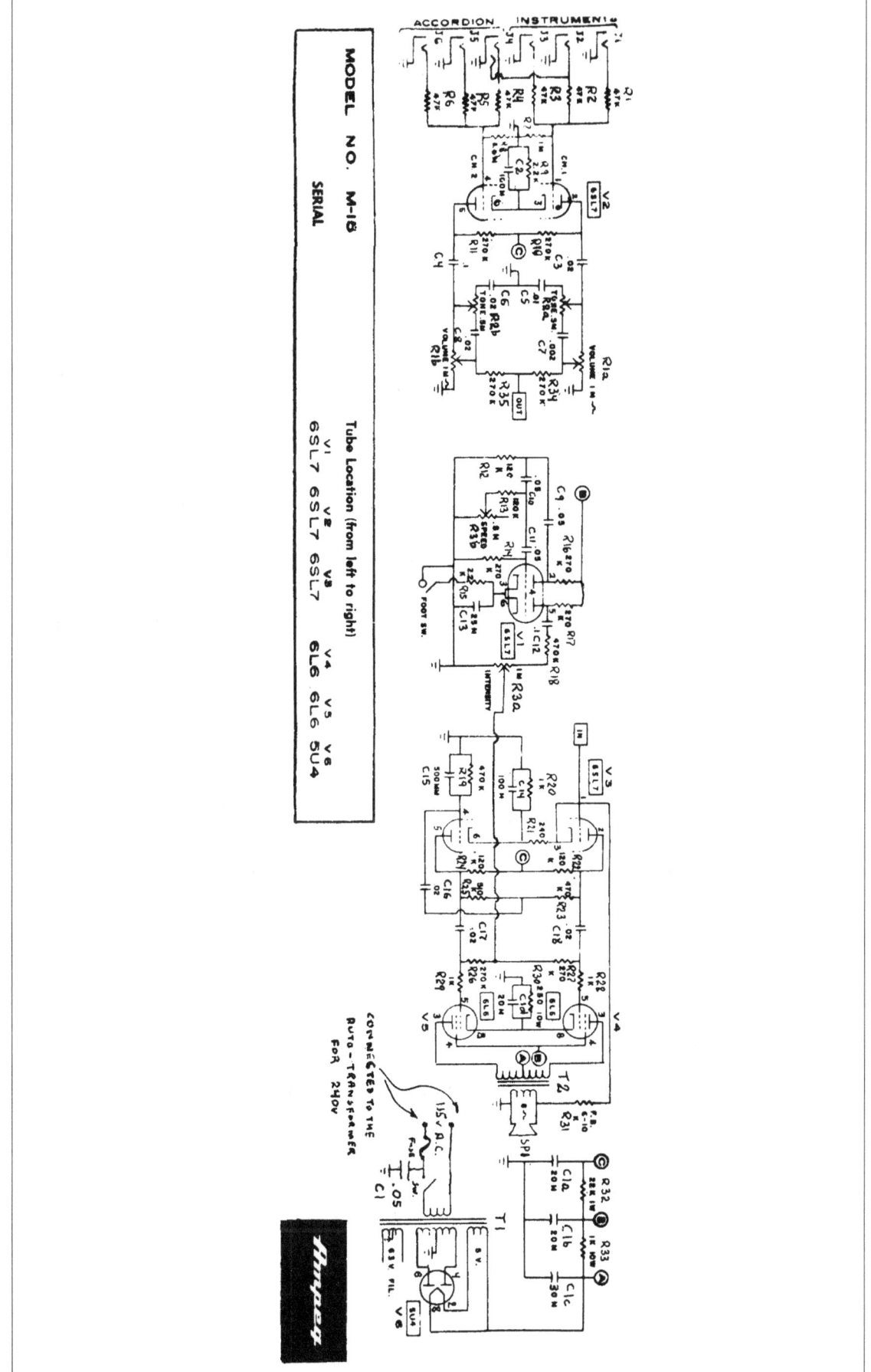

MODEL NO. M-18	Tube Location (from left to right)						
		V1	V2	V3	V4	V5	V6
SERIAL		6SL7	6SL7	6SL7	6L6	6L6	5U4

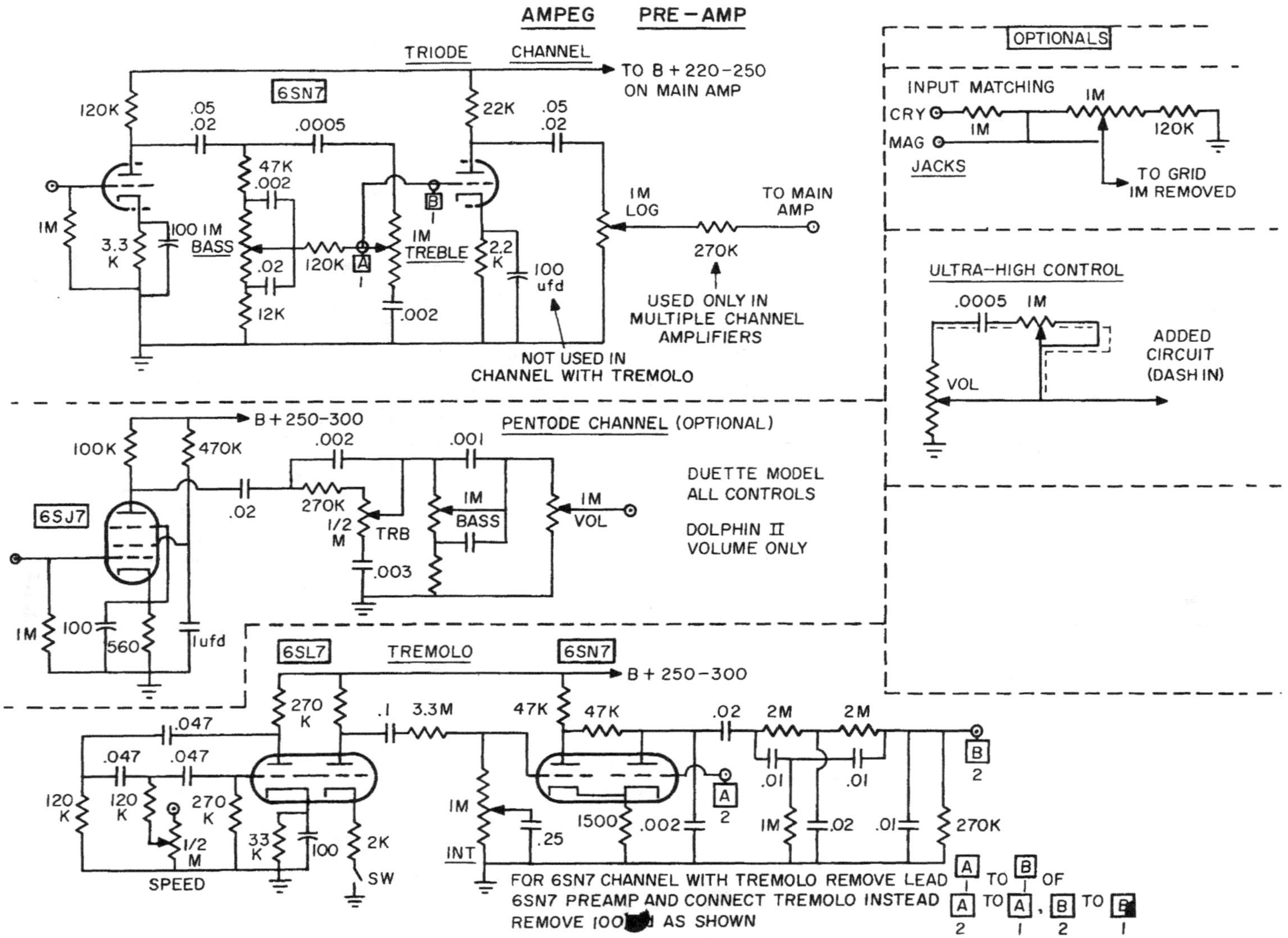
AMPEG PRE-AMP
TRIODE CHANNEL
6SN7
TO B + 220-250 ON MAIN AMP
120K
.05 .02
.0005
22K
47K
.002
IM
3.3 K
100 IM BASS
120K
IM TREBLE
2.2 K
100 ufd
.02
12K
.002
IM LOG
TO MAIN AMP
270K
USED ONLY IN MULTIPLE CHANNEL AMPLIFIERS
NOT USED IN CHANNEL WITH TREMOLO
OPTIONALS
INPUT MATCHING
CRY
MAG
JACKS
120K
TO GRID IM REMOVED
ULTRA-HIGH CONTROL
.0005 IM
VOL
ADDED CIRCUIT (DASH IN)
PENTODE CHANNEL (OPTIONAL)
B + 250-300
100K
470K
.002
.001
6SJ7
270K
1/2 M
TRB
.003
IM BASS
IM VOL
DUETTE MODEL ALL CONTROLS
DOLPHIN II VOLUME ONLY
100
560
1ufd
6SL7
TREMOLO
6SN7
B + 250-300
270 K
.1
3.3M
47K
.047
120 K
1/2 M
SPEED
33 K
2K
SW
INT
.25
1500
.002
2M
.01
.02
270K
FOR 6SN7 CHANNEL WITH TREMOLO REMOVE LEAD A1 TO B1 OF 6SN7 PREAMP AND CONNECT TREMOLO INSTEAD A2 TO A1, B2 TO B1
REMOVE 100 AS SHOWN

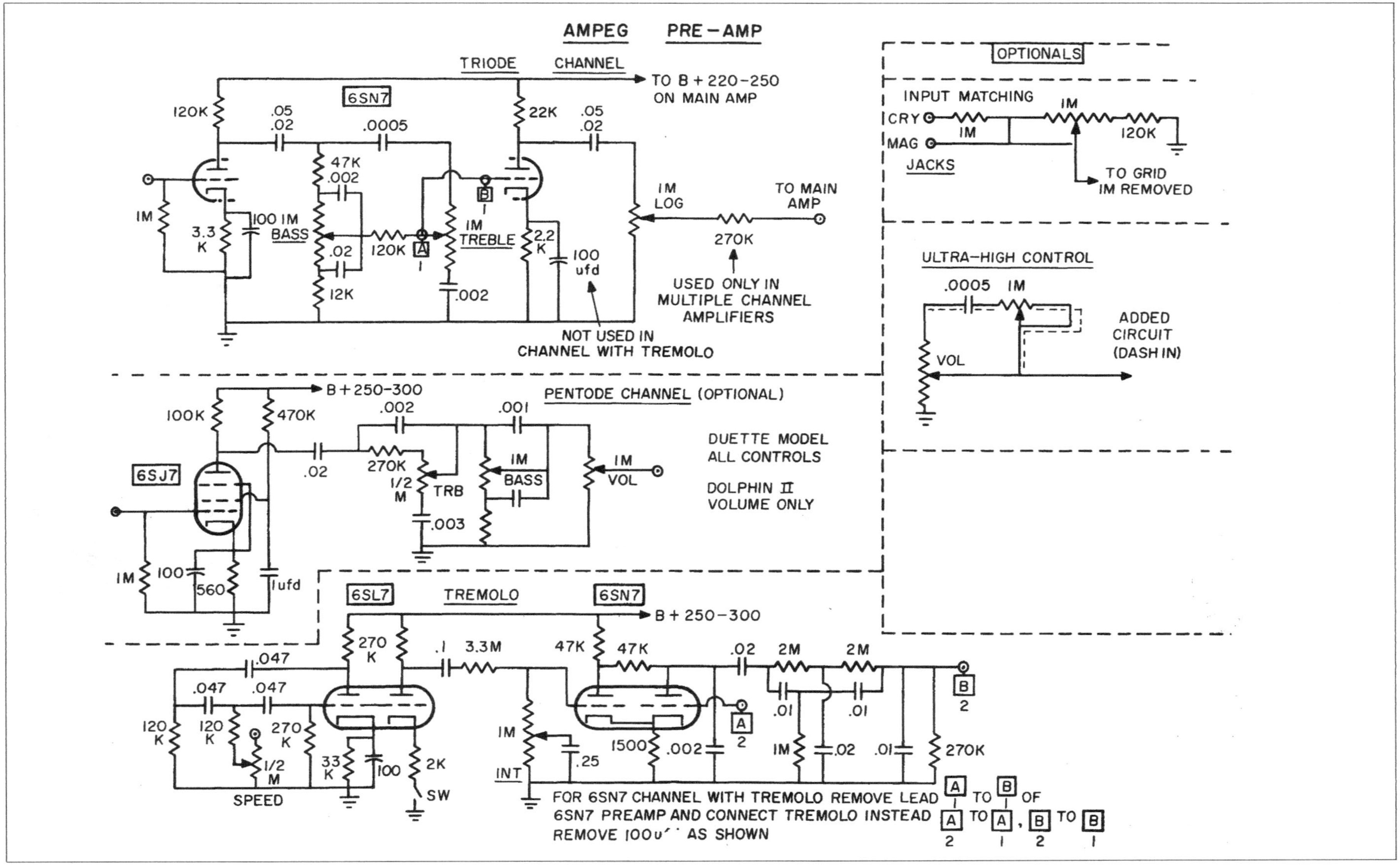
AMPEG PRE-AMP
TRIODE CHANNEL
TO B + 220-250 ON MAIN AMP
6SN7
120K
.05 .02
.0005
22K
47K
.002
IM
3.3 K
100 IM BASS
.02
120K
A 1
B 1
IM TREBLE
2.2 K
100 ufd
12K
.002
IM LOG
TO MAIN AMP
270K
USED ONLY IN MULTIPLE CHANNEL AMPLIFIERS
NOT USED IN CHANNEL WITH TREMOLO
OPTIONALS
INPUT MATCHING
CRY
MAG
JACKS
IM
120K
TO GRID IM REMOVED
ULTRA-HIGH CONTROL
.0005 IM
VOL
ADDED CIRCUIT (DASH IN)
B + 250-300
PENTODE CHANNEL (OPTIONAL)
100K
470K
.002
.001
6SJ7
.02
270K
1/2 M
TRB
IM BASS
IM VOL
.003
DUETTE MODEL ALL CONTROLS
DOLPHIN II VOLUME ONLY
IM
100
560
1ufd
6SL7
TREMOLO
6SN7
B + 250-300
270 K
.1 3.3M
47K
47K
.02
2M
2M
.047
.047
.047
120 K
120 K
270 K
1/2 M
SPEED
33 K
100
2K
SW
IM
INT
.25
1500
.002
A 2
.01
.01
IM
.02
.01
270K
B 2
FOR 6SN7 CHANNEL WITH TREMOLO REMOVE LEAD A 1 TO B 1 OF
6SN7 PREAMP AND CONNECT TREMOLO INSTEAD A 2 TO A 1, B 2 TO B 1
REMOVE 100u'' AS SHOWN

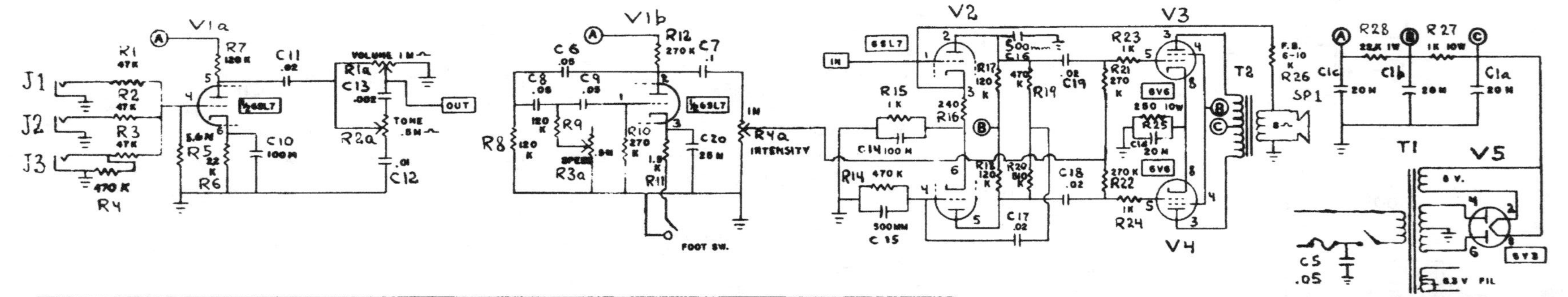

MODEL NO. R-12	**Tube Location (from left to right)**
	6SL7 6SL7 6V6 6V6 5Y3

the **Ampeg** co., inc.
INDUSTRIAL PARK LINDEN N. J.
PHONE

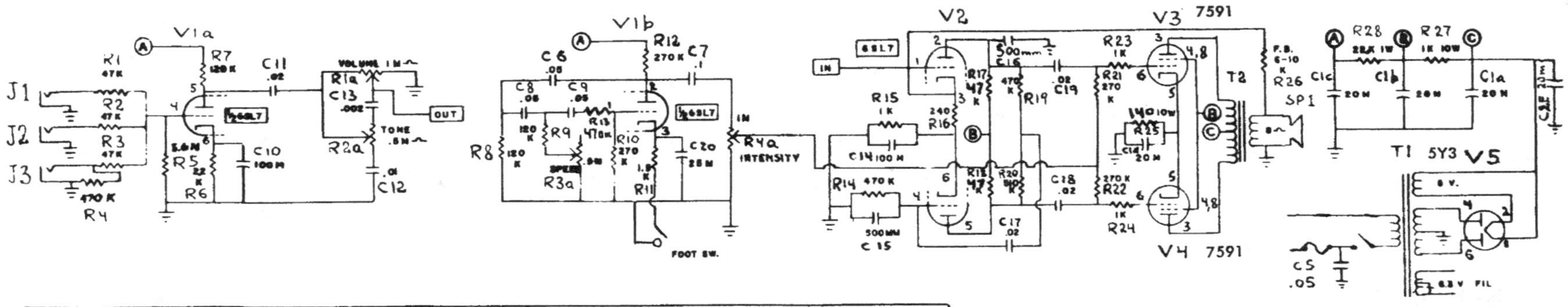

MODEL NO. R-12-A	Tube Location (from left to right)
	6SL7 6SL7 7591 7591 5Y3

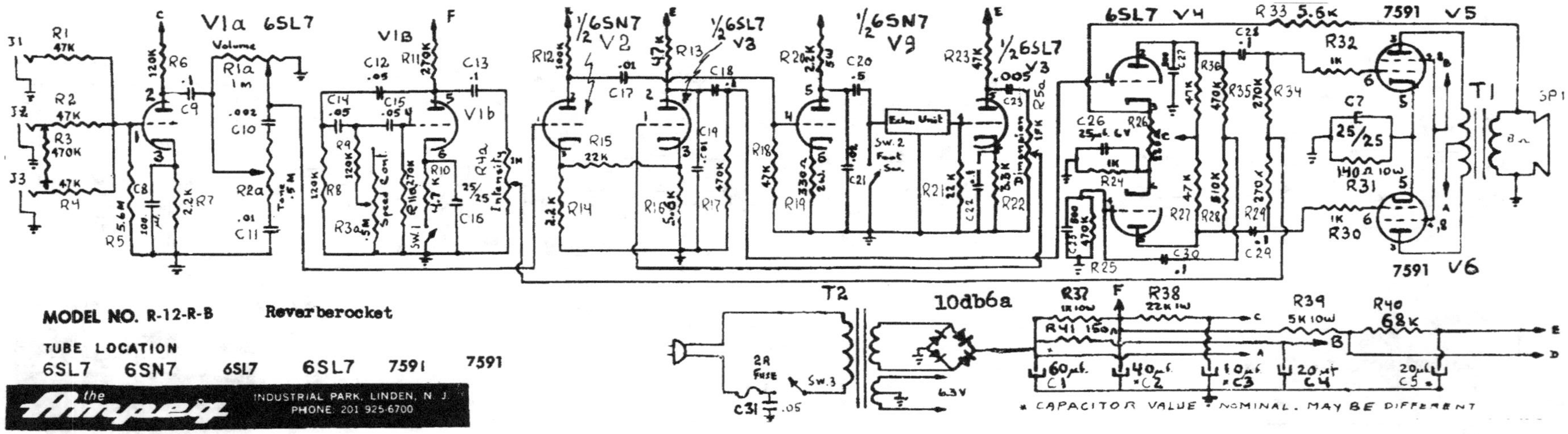

MODEL NO. R-12-R-B   **Reverberocket**

TUBE LOCATION

6SL7 6SN7 6SL7 6SL7 7591 7591

the Ampeg INDUSTRIAL PARK, LINDEN, N. J. PHONE: 201 925-6700

R-12R **REVERB ROCKET**

Ampeg

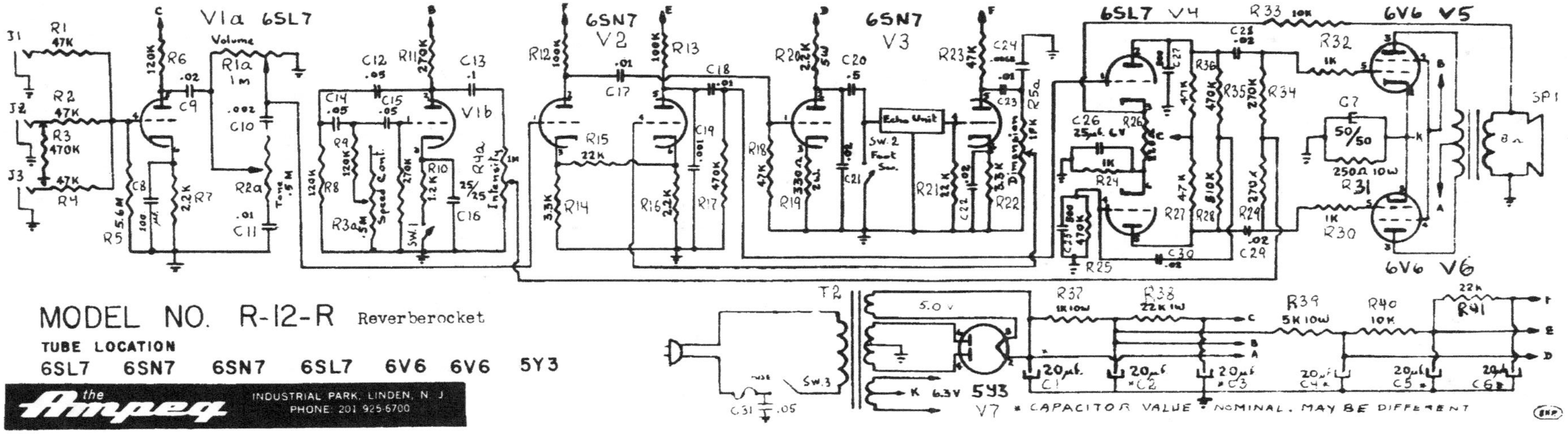

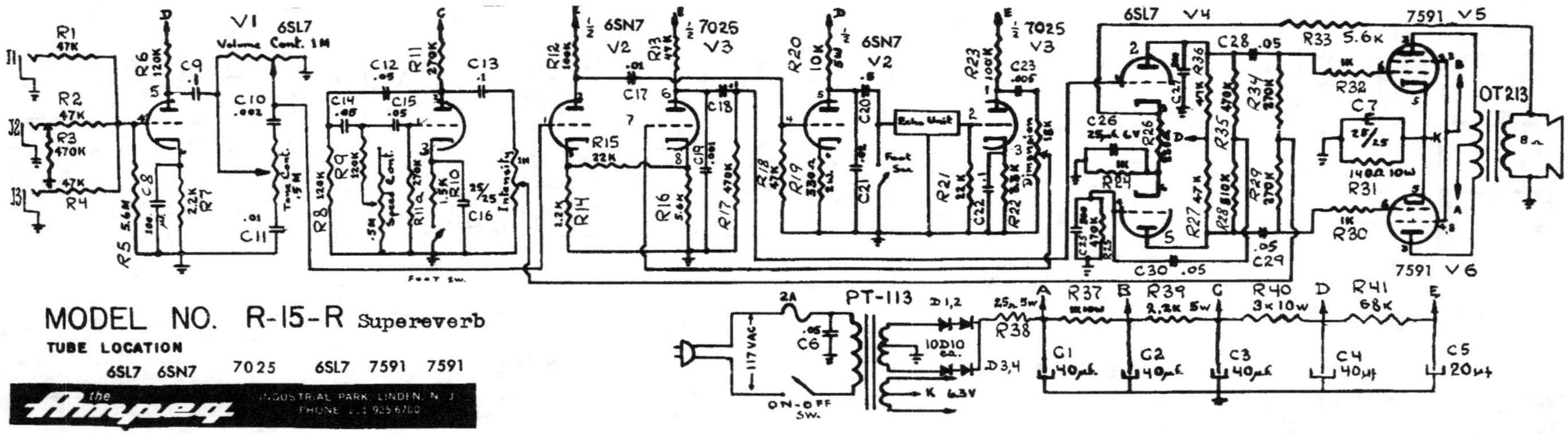
MODEL NO. R-15-R Supereverb
TUBE LOCATION
6SL7 6SN7 7025 6SL7 7591 7591
the Ampeg
INDUSTRIAL PARK LINDEN, N. J.
V1 6SL7
6SN7 V2
½ 7025 V3
6SL7 V4
7591 V5
7591 V6
OT213
PT-113
117 VAC
ON-OFF SW.
Foot sw.
Volume Cont. 1M
Tone Cont. .5M
Speed Cont.
Intensity
Echo Unit
Foot Sw.
Dimension
R1 47K
R2 47K
R3 470K
R4 47K
R5 5.6M
R6 120K
R7 2.2K
R8 120K
R9 120K
R11 270K
R12 100K
R13 47K
R14 1.2K
R15 22K
R16 5.6K
R17 470K
R18 47K
R19 330Ω
R20 10K
R23 100K
R27 47K
R28 810K
R29 270K
R30 1K
R31 140Ω 10w
R32 1K
R33 5.6K
R34 270K
R35 470K
R36 47K
R37
R38 25Ω 5w
R39 2.2K 5w
R40 3K 10w
R41 68K
C1 40μf
C2 40μf
C3 40μf
C4 40μf
C5 20μf
C6 .05
C7 25/25
C9 .1
C10 .002
C11 .01
C12 .05
C13 .1
C14 .05
C15 .05
C16 25/25
C17 .01
C18
C19 .001
C20 .5
C21
C22
C23 .005
C26
C28 .05
C29 .05
C30 .05
D1,2
D3,4
10D10 ea.
K 6.3V
2A
8Ω

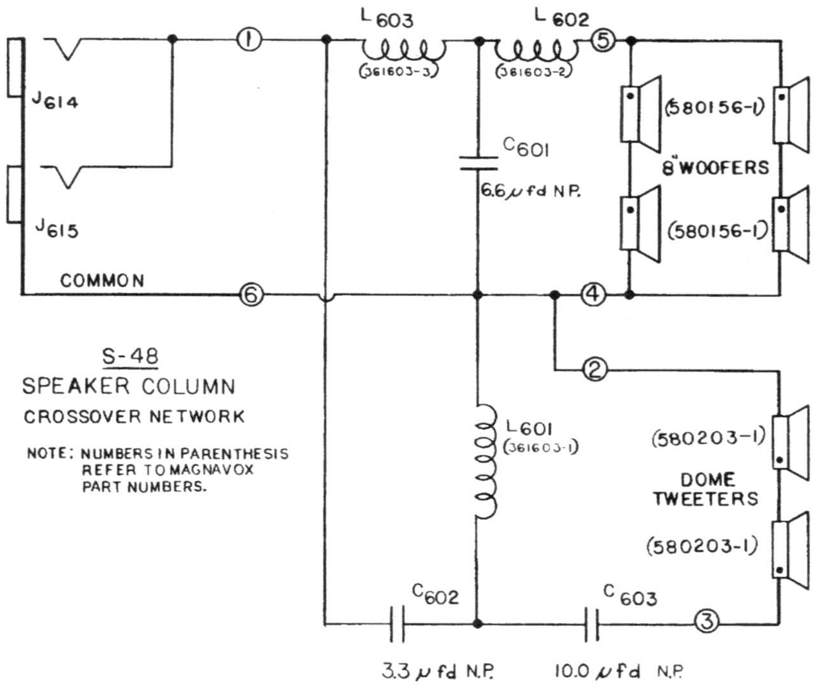

S-48
SPEAKER COLUMN
CROSSOVER NETWORK

NOTE: NUMBERS IN PARENTHESIS REFER TO MAGNAVOX PART NUMBERS.

ampeg

The Ampeg Company
Division of The Magnavox Company
P. O. Box 310, Elkhart, Indiana 46514

Part No. 591744-1

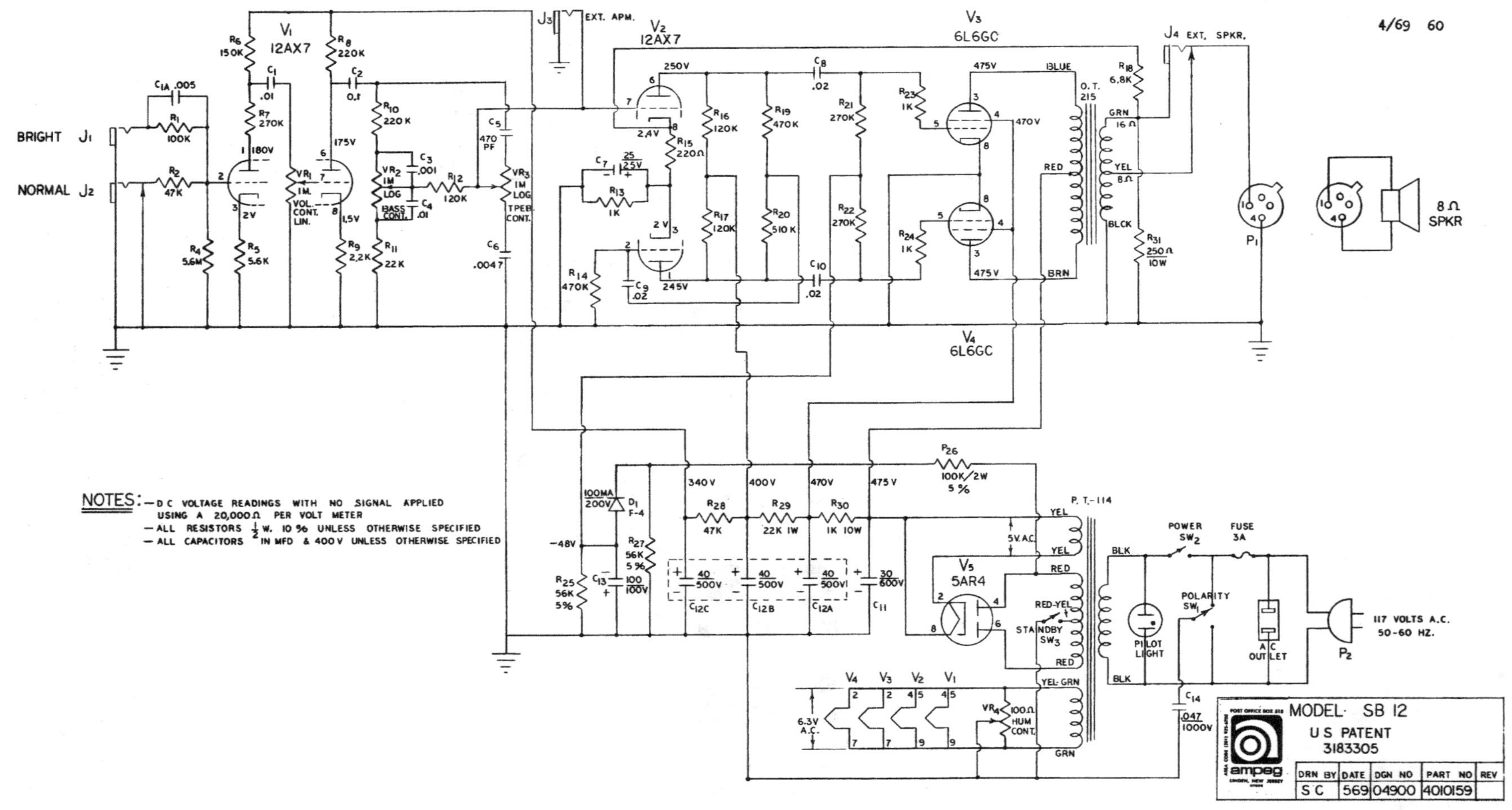
4/69 60
NOTES: — D C VOLTAGE READINGS WITH NO SIGNAL APPLIED USING A 20,000Ω PER VOLT METER
— ALL RESISTORS ½ W. 10 % UNLESS OTHERWISE SPECIFIED
— ALL CAPACITORS IN MFD & 400V UNLESS OTHERWISE SPECIFIED
MODEL · SB 12
U S PATENT
3183305
DRN BY | DATE | DGN NO | PART NO | REV
S C | 569 | 04900 | 4010159 |
117 VOLTS A.C. 50-60 HZ.

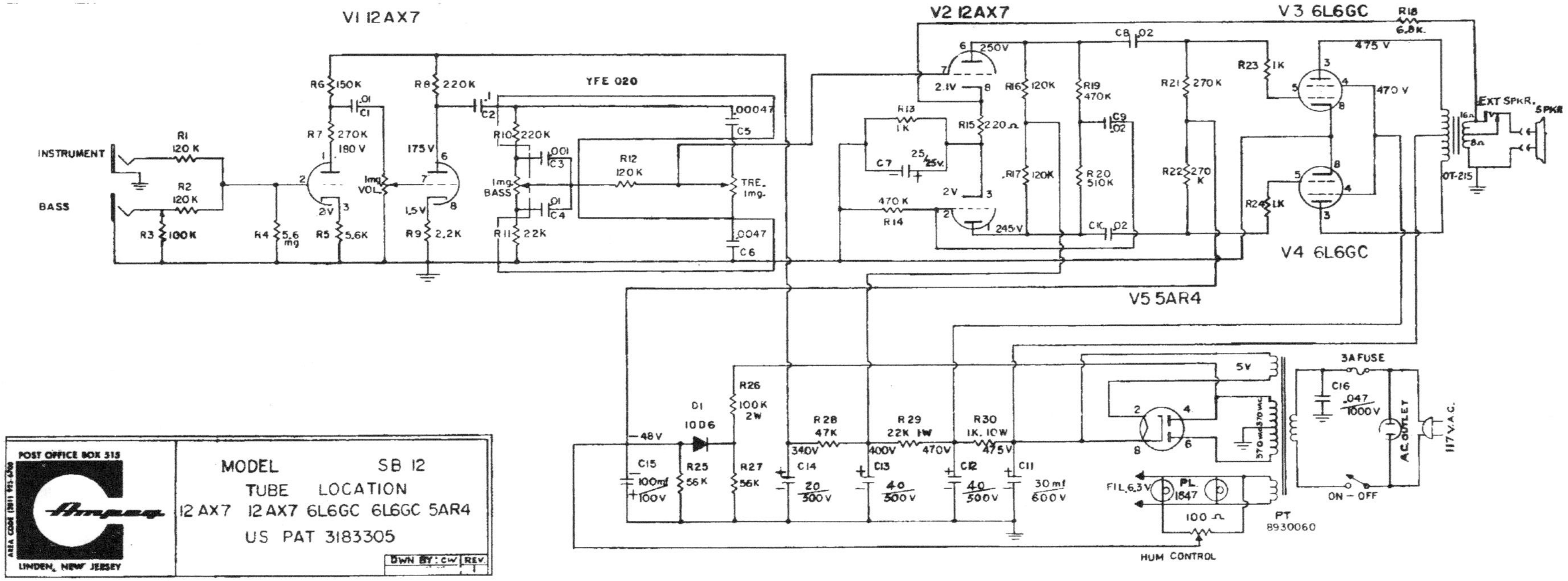
V1 12AX7
V2 12AX7
V3 6L6GC
V4 6L6GC
V5 5AR4
YFE 020
INSTRUMENT
BASS
R1 120 K
R2 120 K
R3 100K
R4 5.6 mg
R5 5.6K
R6 150K
R7 270K
180 V
R8 220K
175 V
R9 2.2K
R10 220K
R11 22K
R12 120 K
R13 1K
R14 470 K
R15 220 Ω
R16 120K
R17 120K
R18 6.8K.
R19 470K
R20 510K
R21 270K
R22 270 K
R23 1K
R24 1K
R25 56K
R26 100K 2W
R27 56K
R28 47K
R29 22K 1W
R30 1K. 10W
C1 .01
C2 .1
C3 .001
C4 .01
C5 .00047
C6 .0047
C7 25/25v.
C8 .02
C9 .02
CK .02
C11 30mf 600V
C12 40/500V
C13 40/500V
C14 20/500V
C15 100mf 100V
C16 .047/1000V
D1 10D6
-48V
340V
400V
470V
475V
250V
245V
2.1V
2V
1.5V
470 V
1mg VOL.
1mg. BASS
TRE. 1mg.
OT-215
EXT SPKR.
SPKR
16Ω
8Ω
5V
370 VAC 370 VAC
3A FUSE
AC OUTLET
117 V.A.C.
ON - OFF
FIL. 6.3V
PL 1847
100 Ω
HUM CONTROL
PT 8930060
POST OFFICE BOX 515
Ampeg
LINDEN, NEW JERSEY
MODEL SB 12
TUBE LOCATION
12 AX7 12 AX7 6L6GC 6L6GC 5AR4
US PAT 3183305
DWN BY: CW
REV.

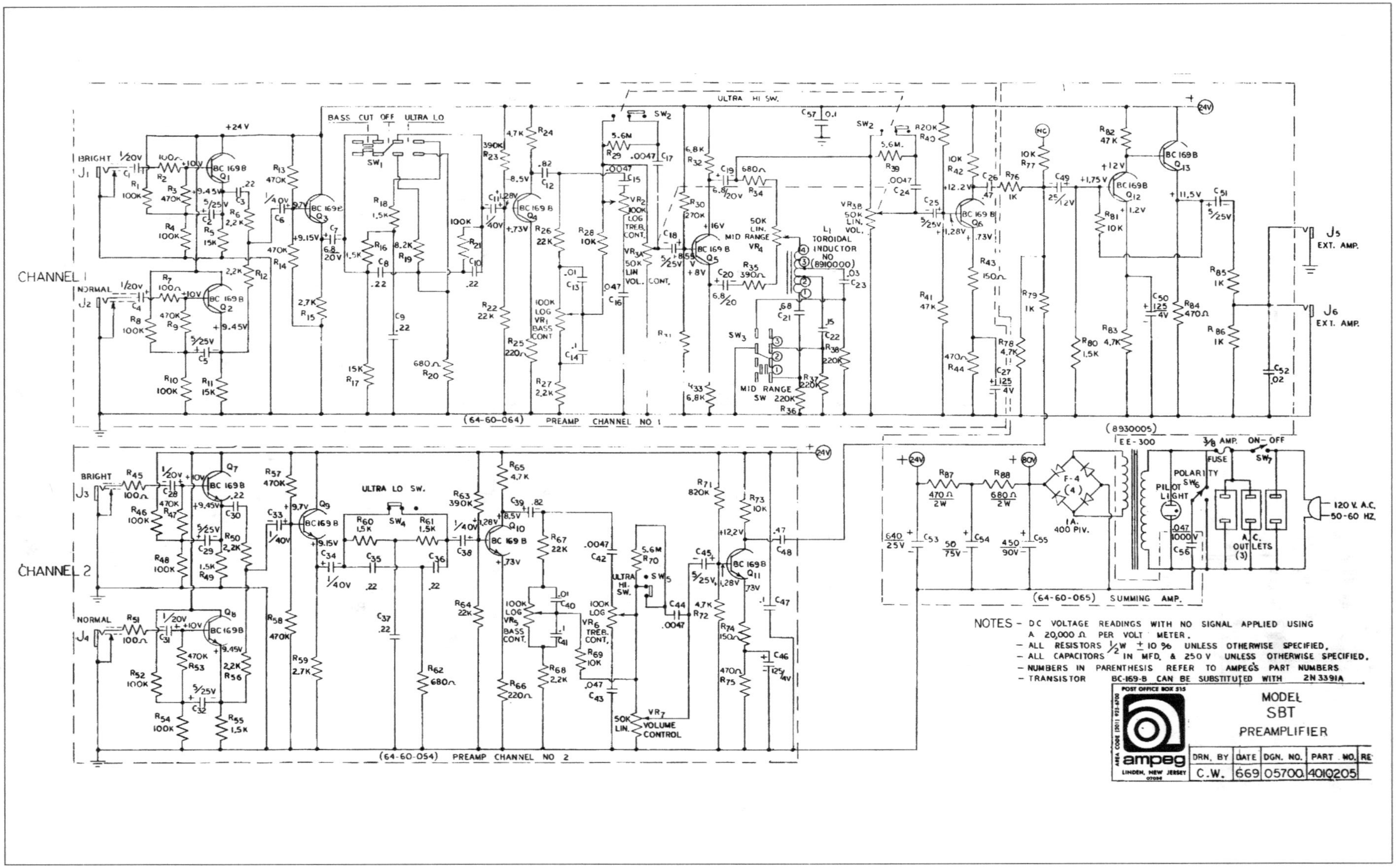
CHANNEL 1
CHANNEL 2
BRIGHT
NORMAL
J1
J2
J3
J4
J5 EXT. AMP.
J6 EXT. AMP.
+24V
BASS CUT OFF ULTRA LO
ULTRA HI SW.
ULTRA LO SW.
MID RANGE SW
L1 TOROIDAL INDUCTOR NO (8910000)
VOLUME CONTROL
(64-60-064) PREAMP CHANNEL NO 1
(64-60-054) PREAMP CHANNEL NO 2
(64-60-065) SUMMING AMP.
(8930005)
EE-300
3/8 AMP. FUSE
ON-OFF SW7
POLARITY SW6
PILOT LIGHT
A.C. OUTLETS (3)
120 V. A.C. 50-60 HZ.
1A. 400 PIV.
NOTES - DC VOLTAGE READINGS WITH NO SIGNAL APPLIED USING A 20,000 Ω PER VOLT METER.
- ALL RESISTORS 1/2 W ± 10 % UNLESS OTHERWISE SPECIFIED.
- ALL CAPACITORS IN MFD. & 250 V UNLESS OTHERWISE SPECIFIED.
- NUMBERS IN PARENTHESIS REFER TO AMPEG'S PART NUMBERS
- TRANSISTOR BC-169-B CAN BE SUBSTITUTED WITH 2N3391A
MODEL SBT PREAMPLIFIER
ampeg
LINDEN, NEW JERSEY 07036
POST OFFICE BOX 315
DRN. BY C.W.
DATE 669
DGN. NO. 05700
PART NO. 4010205

MODEL SR4 SCHEMATIC DIAGRAM

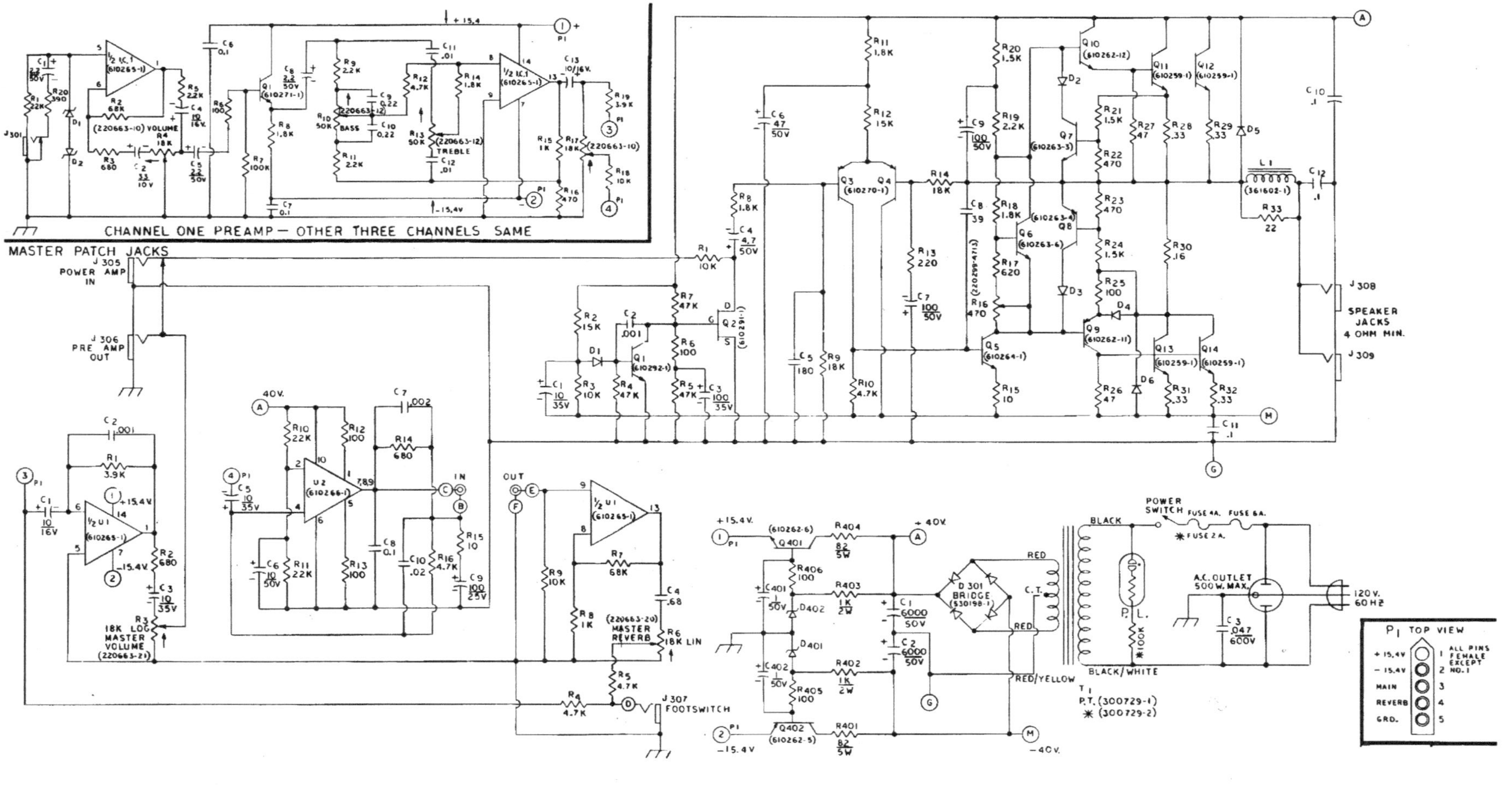

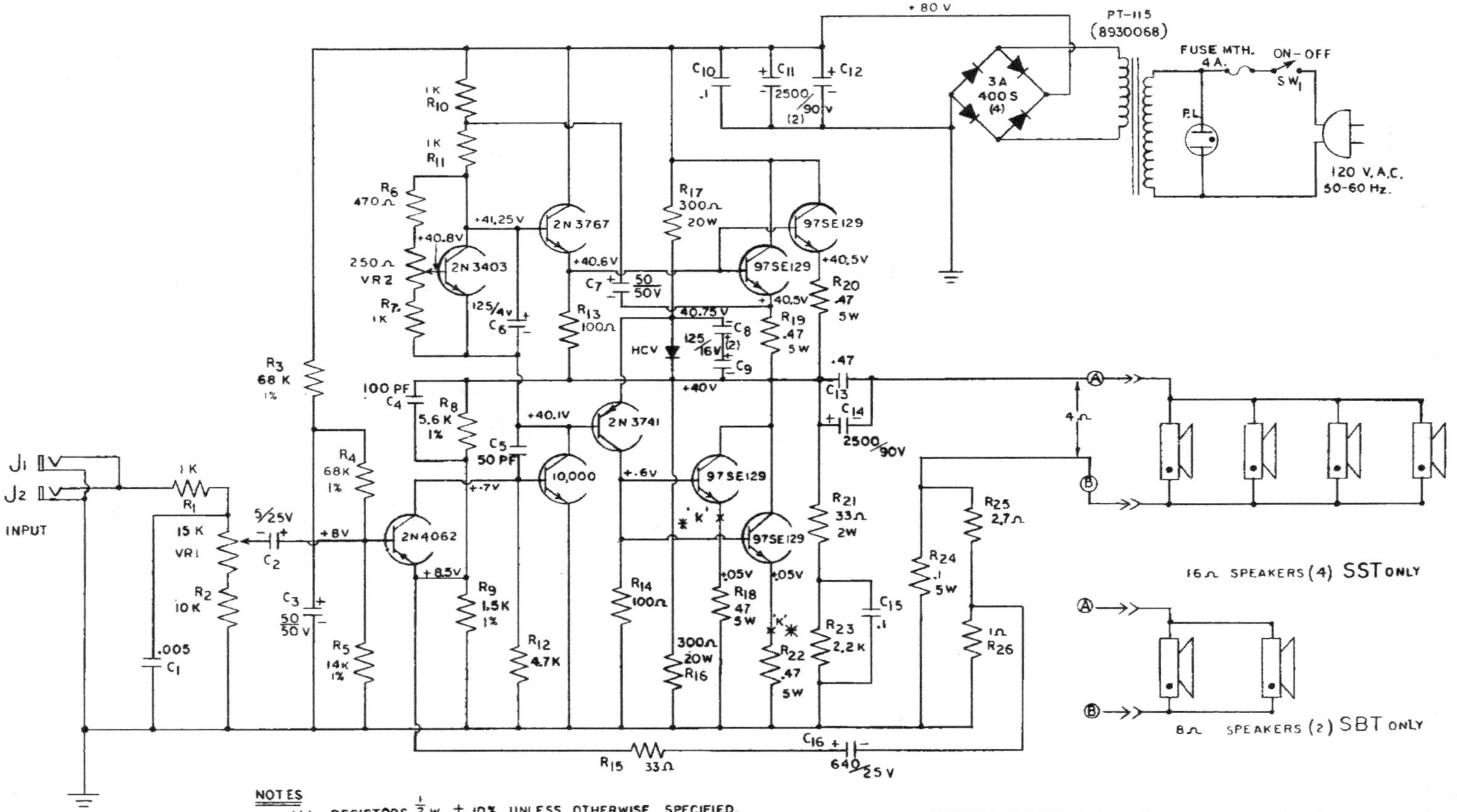

NOTES

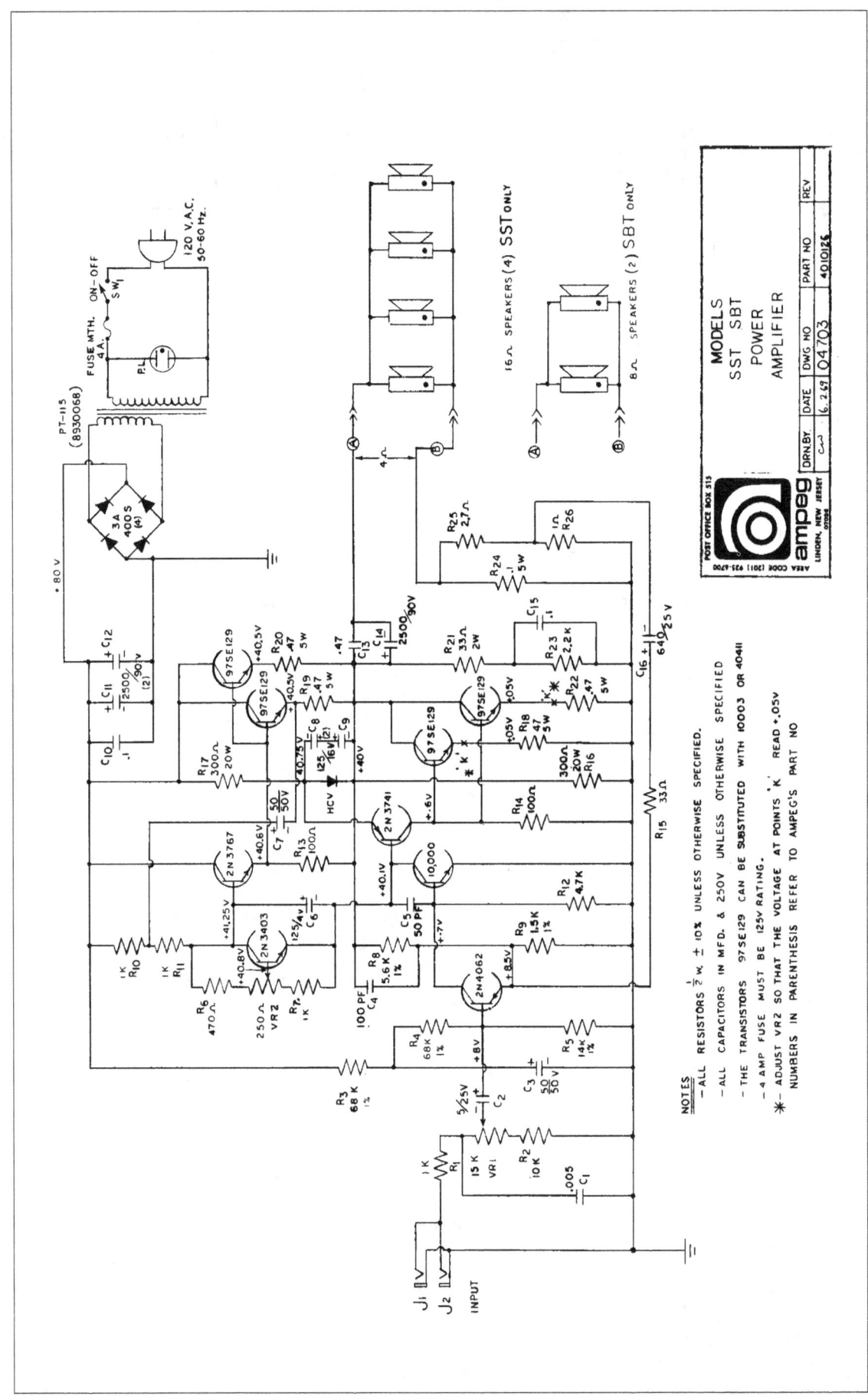

- ALL RESISTORS 1/2 W. ± 10% UNLESS OTHERWISE SPECIFIED.
- ALL CAPACITORS IN MFD. & 250V UNLESS OTHERWISE SPECIFIED
- THE TRANSISTORS 97SE129 CAN BE SUBSTITUTED WITH 10003 OR 40411
- 4 AMP FUSE MUST BE 125V RATING.
- ✱ ADJUST VR2 SO THAT THE VOLTAGE AT POINTS 'K' READ +.05V

NUMBERS IN PARENTHESIS REFER TO AMPEG'S PART NO

POST OFFICE BOX 515
AREA CODE (201) 925-6700
ampeg
LINDEN, NEW JERSEY 07036

MODELS
SST SBT
POWER
AMPLIFIER

DRN.BY.	DATE	DWG NO	PART NO	REV
[illegible]	6.2.69	04703	4010126	

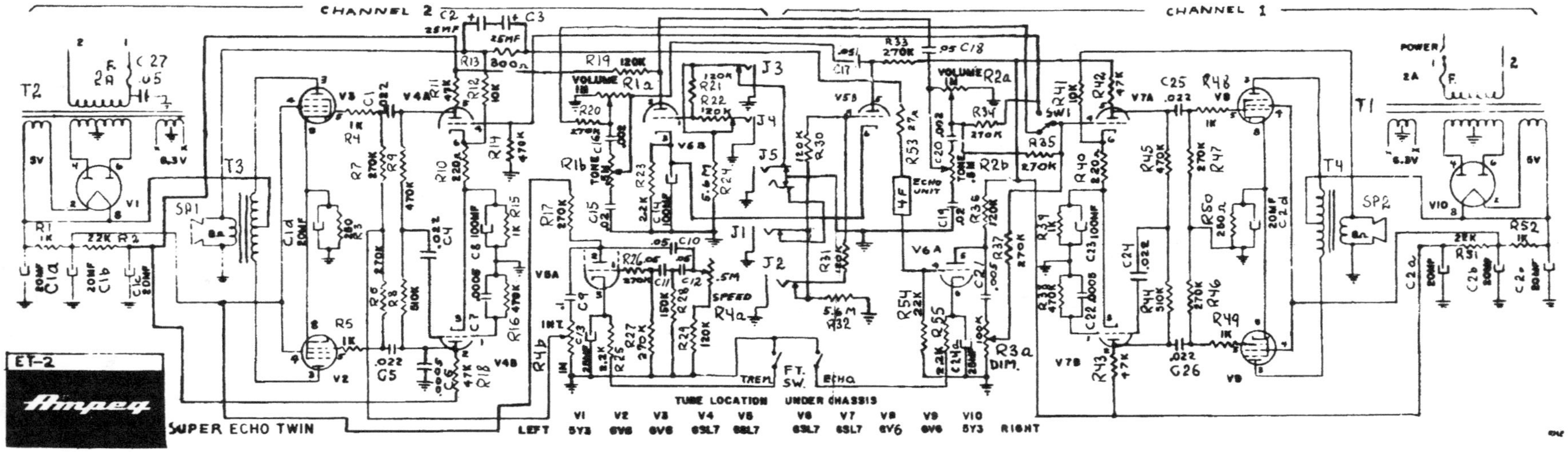
CHANNEL 2
CHANNEL 1
ET-2
SUPER ECHO TWIN
TUBE LOCATION UNDER CHASSIS
LEFT
V1 5Y3
V2 6V6
V3 6V6
V4 6SL7
V5 6SL7
V6 6SL7
V7 6SL7
V8 6V6
V9 6V6
V10 5Y3
RIGHT
TREM
FT. SW.
ECHO
SPEED
DIM.
ECHO UNIT

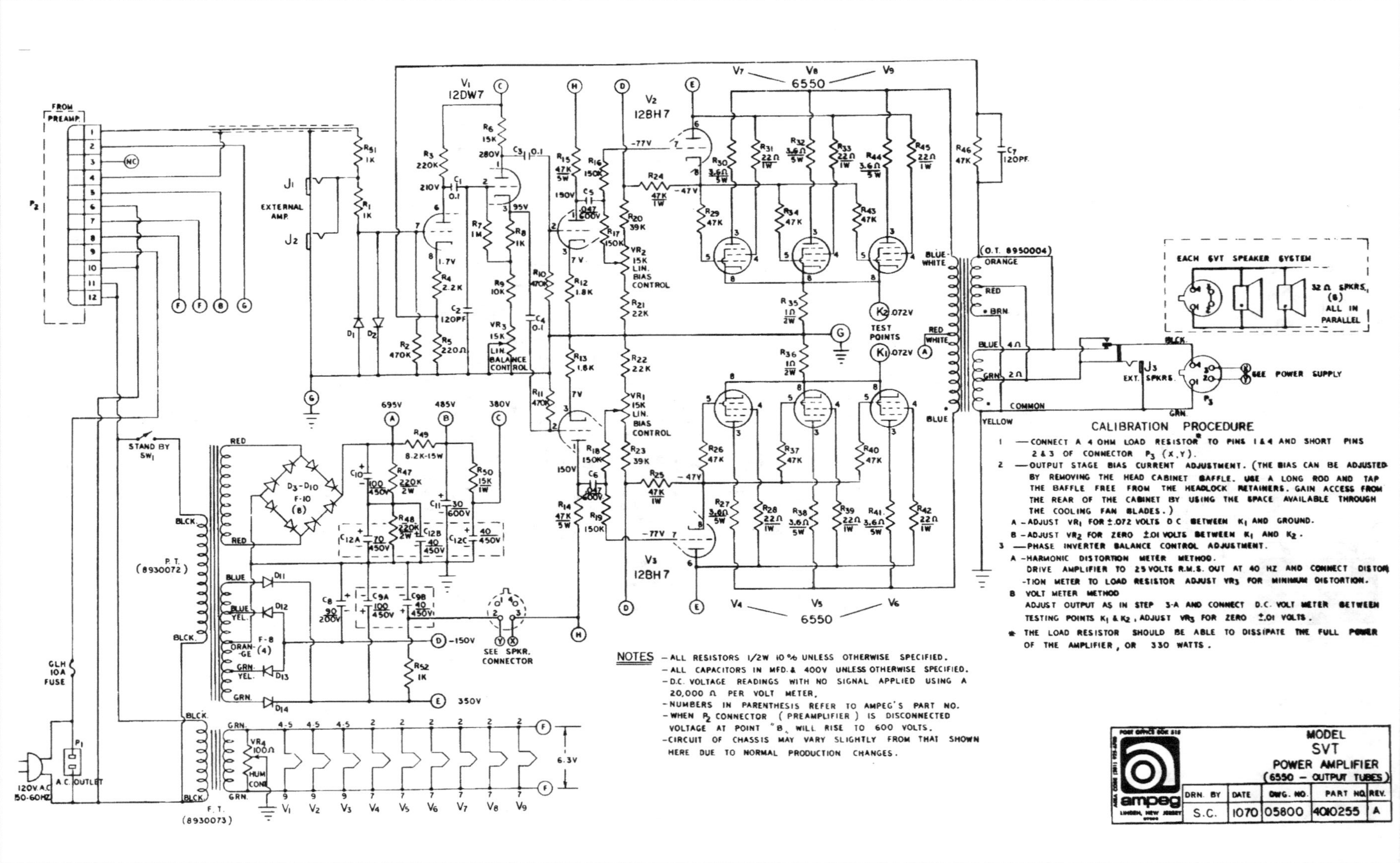

CALIBRATION PROCEDURE
1 —CONNECT A 4 OHM LOAD RESISTOR* TO PINS 1 & 4 AND SHORT PINS 2 & 3 OF CONNECTOR P3 (X, Y).
2 —OUTPUT STAGE BIAS CURRENT ADJUSTMENT. (THE BIAS CAN BE ADJUSTED BY REMOVING THE HEAD CABINET BAFFLE. USE A LONG ROD AND TAP THE BAFFLE FREE FROM THE HEADLOCK RETAINERS. GAIN ACCESS FROM THE REAR OF THE CABINET BY USING THE SPACE AVAILABLE THROUGH THE COOLING FAN BLADES.)
A - ADJUST VR1 FOR ±.072 VOLTS D C BETWEEN K1 AND GROUND.
B - ADJUST VR2 FOR ZERO ±.01 VOLTS BETWEEN K1 AND K2.
3 —PHASE INVERTER BALANCE CONTROL ADJUSTMENT.
A - HARMONIC DISTORTION METER METHOD.
DRIVE AMPLIFIER TO 25 VOLTS R.M.S. OUT AT 40 HZ AND CONNECT DISTORTION METER TO LOAD RESISTOR ADJUST VR3 FOR MINIMUM DISTORTION.
B VOLT METER METHOD
ADJUST OUTPUT AS IN STEP 3-A AND CONNECT D.C. VOLT METER BETWEEN TESTING POINTS K1 & K2, ADJUST VR3 FOR ZERO ±.01 VOLTS.
* THE LOAD RESISTOR SHOULD BE ABLE TO DISSIPATE THE FULL POWER OF THE AMPLIFIER, OR 330 WATTS.
NOTES
- ALL RESISTORS 1/2W 10% UNLESS OTHERWISE SPECIFIED.
- ALL CAPACITORS IN MFD. & 400V UNLESS OTHERWISE SPECIFIED.
- D.C. VOLTAGE READINGS WITH NO SIGNAL APPLIED USING A 20,000 Ω PER VOLT METER.
- NUMBERS IN PARENTHESIS REFER TO AMPEG'S PART NO.
- WHEN P2 CONNECTOR (PREAMPLIFIER) IS DISCONNECTED VOLTAGE AT POINT "B" WILL RISE TO 600 VOLTS.
- CIRCUIT OF CHASSIS MAY VARY SLIGHTLY FROM THAT SHOWN HERE DUE TO NORMAL PRODUCTION CHANGES.
MODEL
SVT
POWER AMPLIFIER
(6550 - OUTPUT TUBES)
DRN. BY S.C.
DATE 1070
DWG. NO. 05800
PART NO. 4010255
REV. A

SVT **POWER AMP (REV D)**

Ampeg

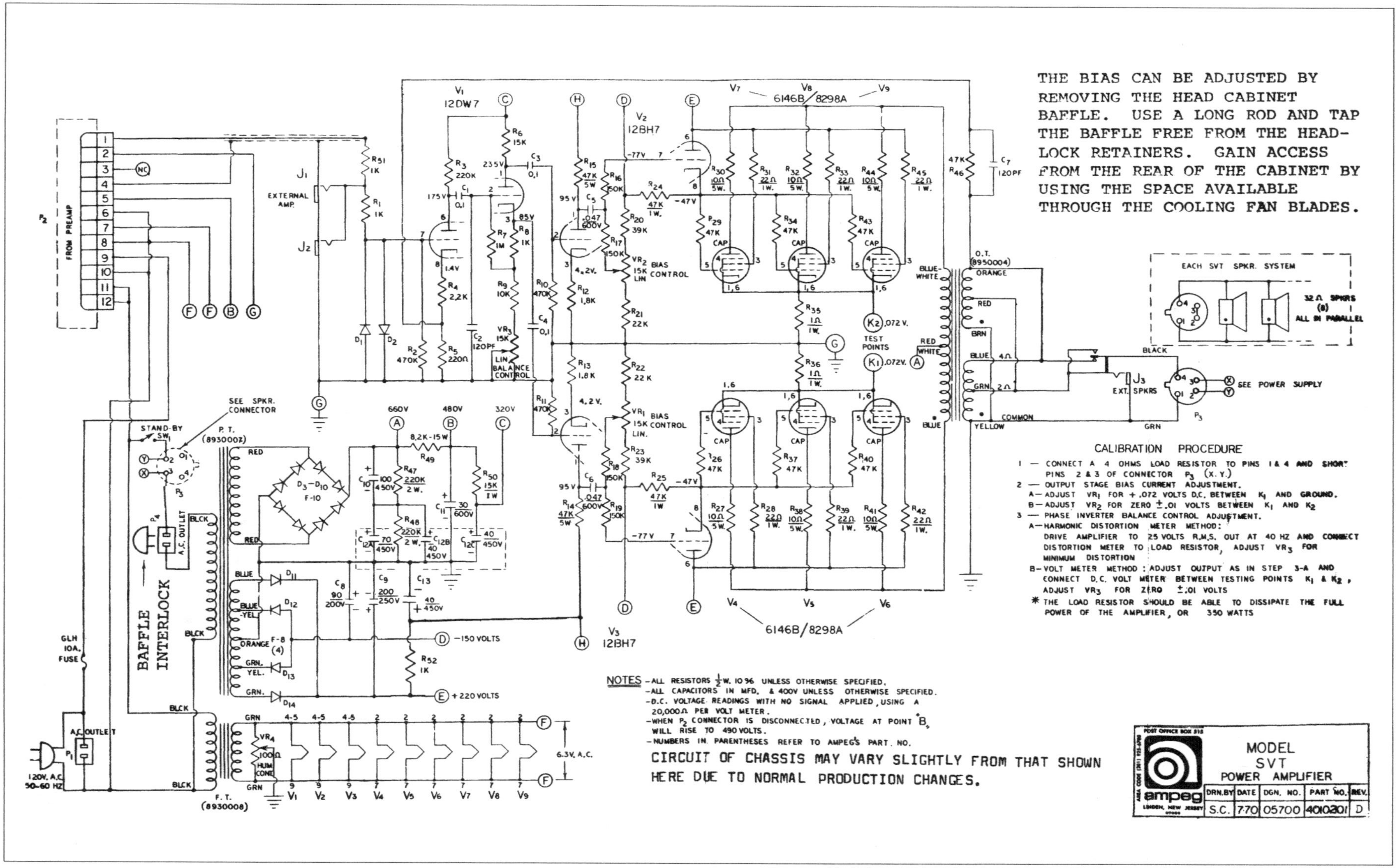

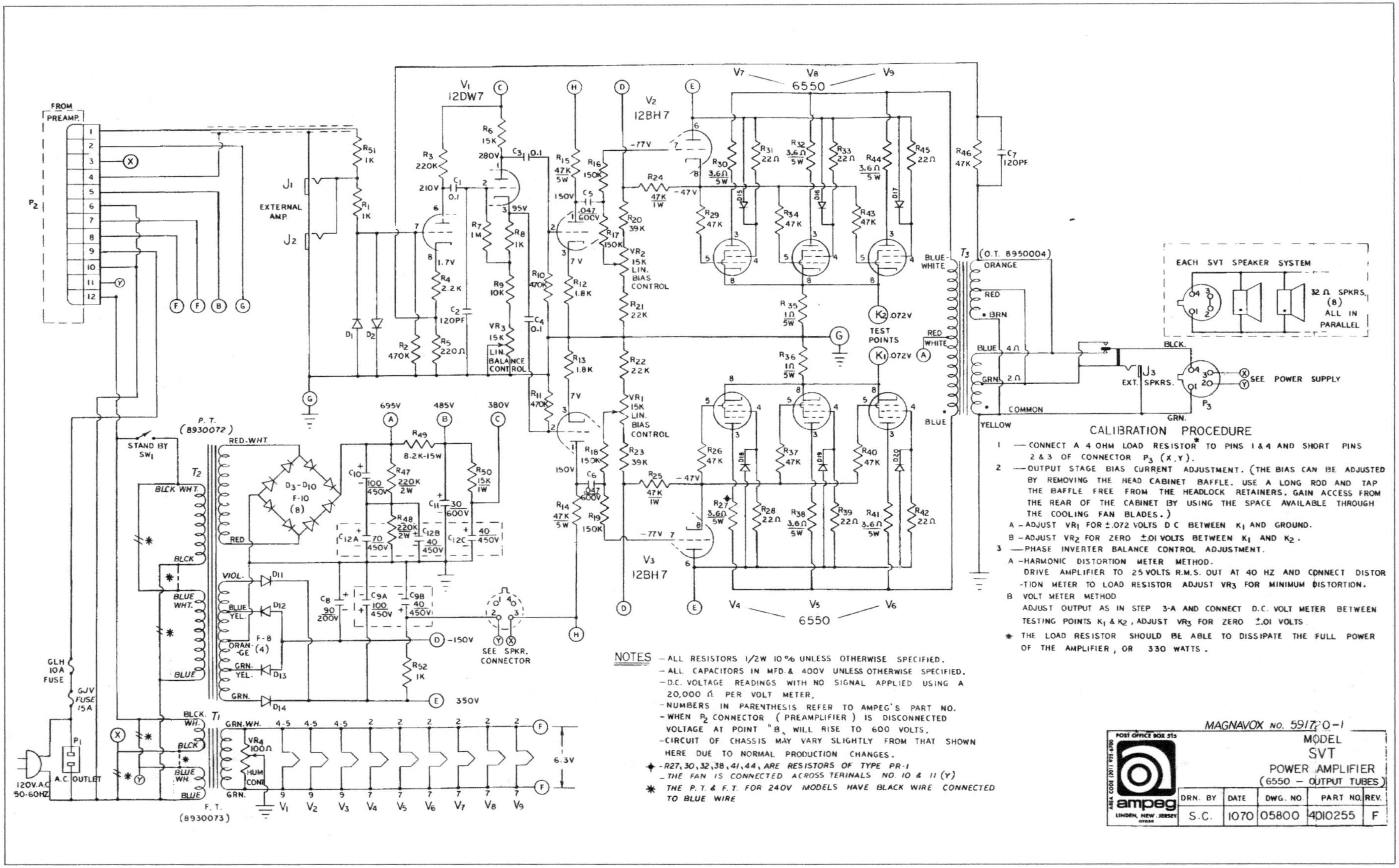
CALIBRATION PROCEDURE
1 —CONNECT A 4 OHM LOAD RESISTOR* TO PINS 1 & 4 AND SHORT PINS 2 & 3 OF CONNECTOR P3 (X, Y).
2 —OUTPUT STAGE BIAS CURRENT ADJUSTMENT. (THE BIAS CAN BE ADJUSTED BY REMOVING THE HEAD CABINET BAFFLE. USE A LONG ROD AND TAP THE BAFFLE FREE FROM THE HEADLOCK RETAINERS. GAIN ACCESS FROM THE REAR OF THE CABINET BY USING THE SPACE AVAILABLE THROUGH THE COOLING FAN BLADES.)
A - ADJUST VR1 FOR ±.072 VOLTS D C BETWEEN K1 AND GROUND.
B - ADJUST VR2 FOR ZERO ±.01 VOLTS BETWEEN K1 AND K2.
3 —PHASE INVERTER BALANCE CONTROL ADJUSTMENT.
A - HARMONIC DISTORTION METER METHOD.
DRIVE AMPLIFIER TO 25 VOLTS R.M.S. OUT AT 40 HZ AND CONNECT DISTORTION METER TO LOAD RESISTOR ADJUST VR3 FOR MINIMUM DISTORTION.
B VOLT METER METHOD
ADJUST OUTPUT AS IN STEP 3-A AND CONNECT D.C. VOLT METER BETWEEN TESTING POINTS K1 & K2, ADJUST VR3 FOR ZERO ±.01 VOLTS
* THE LOAD RESISTOR SHOULD BE ABLE TO DISSIPATE THE FULL POWER OF THE AMPLIFIER, OR 330 WATTS.
NOTES
- ALL RESISTORS 1/2W 10% UNLESS OTHERWISE SPECIFIED.
- ALL CAPACITORS IN MFD & 400V UNLESS OTHERWISE SPECIFIED.
- D.C. VOLTAGE READINGS WITH NO SIGNAL APPLIED USING A 20,000 Ω PER VOLT METER.
- NUMBERS IN PARENTHESIS REFER TO AMPEG'S PART NO.
- WHEN P2 CONNECTOR (PREAMPLIFIER) IS DISCONNECTED VOLTAGE AT POINT "B" WILL RISE TO 600 VOLTS.
- CIRCUIT OF CHASSIS MAY VARY SLIGHTLY FROM THAT SHOWN HERE DUE TO NORMAL PRODUCTION CHANGES.
✦ - R27, 30, 32, 38, 41, 44, ARE RESISTORS OF TYPE PR-1
- THE FAN IS CONNECTED ACROSS TERINALS NO. 10 & 11 (Y)
* THE P.T. & F.T. FOR 240V MODELS HAVE BLACK WIRE CONNECTED TO BLUE WIRE
EACH SVT SPEAKER SYSTEM
32 Ω SPKRS. (8) ALL IN PARALLEL
SEE POWER SUPPLY
EXT. SPKRS.
V1 12DW7
V2 12BH7
V3 12BH7
V7 V8 V9 6550
V4 V5 V6 6550
(O.T. 8950004)
P.T. (8930072)
F.T. (8930073)
FROM PREAMP.
EXTERNAL AMP
STAND BY SW1
BALANCE CONTROL
LIN. BIAS CONTROL
TEST POINTS
SEE SPKR. CONNECTOR
GLH 10A FUSE
GJV FUSE 15A
A.C. OUTLET
120V AC 50-60HZ
MAGNAVOX NO. 5917-0-1
MODEL SVT
POWER AMPLIFIER (6550 — OUTPUT TUBES)
POST OFFICE BOX 315
ampeg
LINDEN, NEW JERSEY
DRN. BY S.C.
DATE 1070
DWG. NO 05800
PART NO. 4010255
REV. F

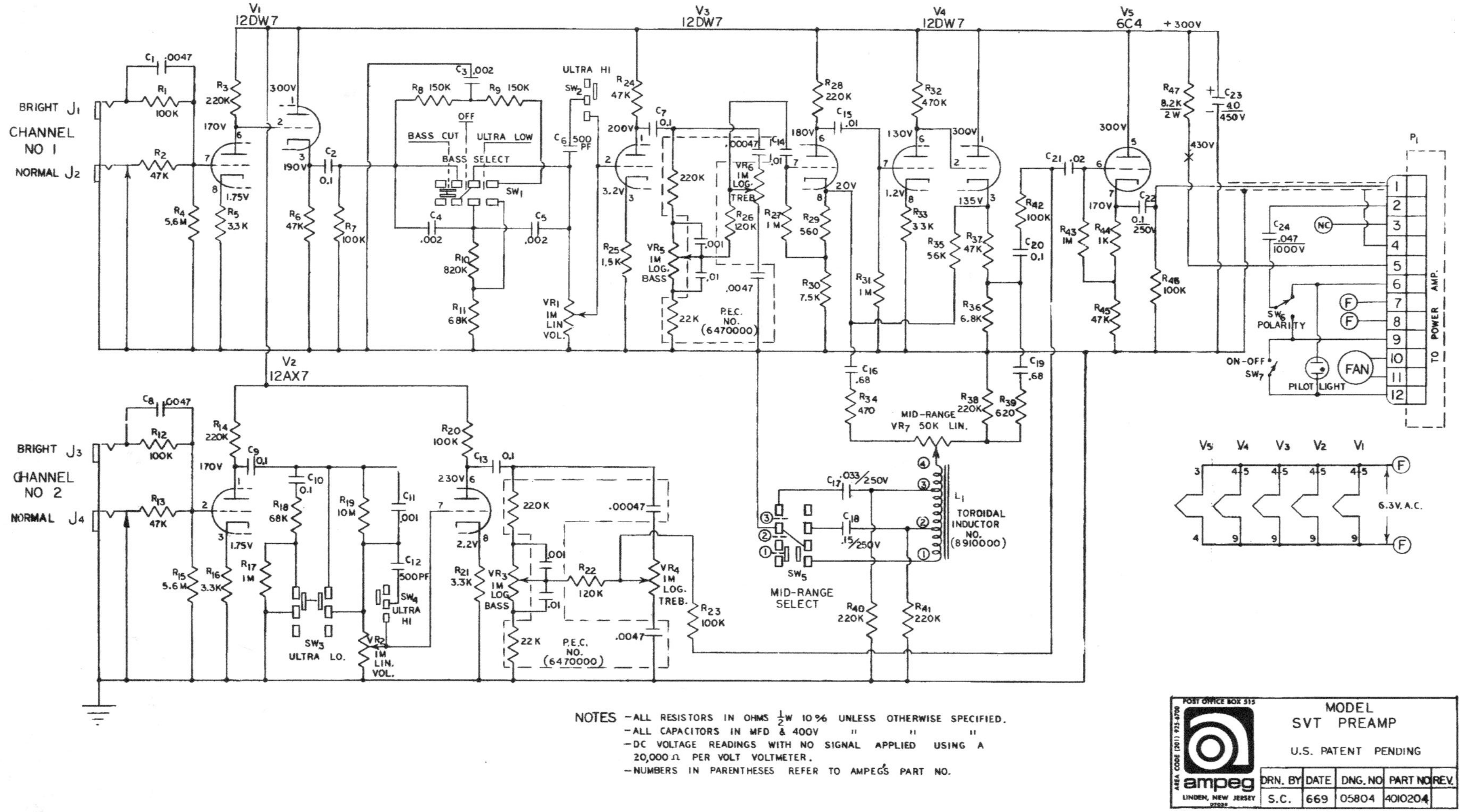

NOTES
- ALL RESISTORS IN OHMS ½W 10% UNLESS OTHERWISE SPECIFIED.
- ALL CAPACITORS IN MFD & 400V " " "
- DC VOLTAGE READINGS WITH NO SIGNAL APPLIED USING A 20,000 Ω PER VOLT VOLTMETER.
- NUMBERS IN PARENTHESES REFER TO AMPEG'S PART NO.

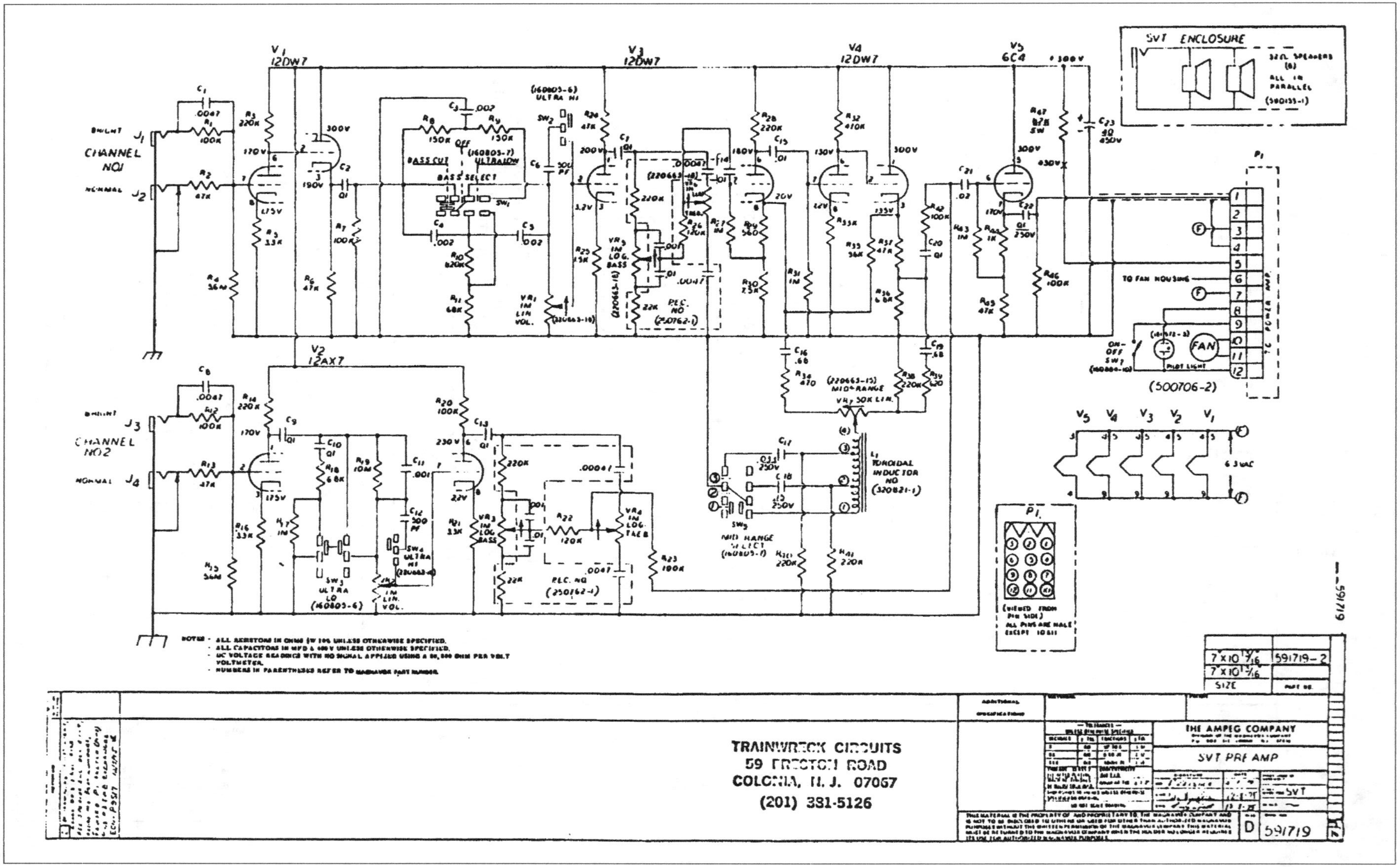
V1 12DW7
V2 12AX7
V3 12DW7
V4 12DW7
V5 6C4
CHANNEL NO1
CHANNEL NO2
J1
J2
J3
J4
SVT ENCLOSURE
(500706-2)
TO FAN HOUSING
PILOT LIGHT
ON-OFF SW1
TOROIDAL INDUCTOR NO (320821-1)
MID RANGE SELECT
BASS SELECT
BASS CUT
ULTRA LOW
6.3 VAC
P1
NOTES: ALL RESISTORS IN OHMS ½W 10% UNLESS OTHERWISE SPECIFIED.
ALL CAPACITORS IN MFD & 600V UNLESS OTHERWISE SPECIFIED.
DC VOLTAGE READINGS WITH NO SIGNAL APPLIED USING A 20,000 OHM PER VOLT VOLTMETER.
NUMBERS IN PARENTHESES REFER TO MAGNAVOX PART NUMBER.
TRAINWRECK CIRCUITS
59 PRESTON ROAD
COLONIA, N. J. 07067
(201) 381-5126
THE AMPEG COMPANY
SVT PRE AMP
591719
591719-2

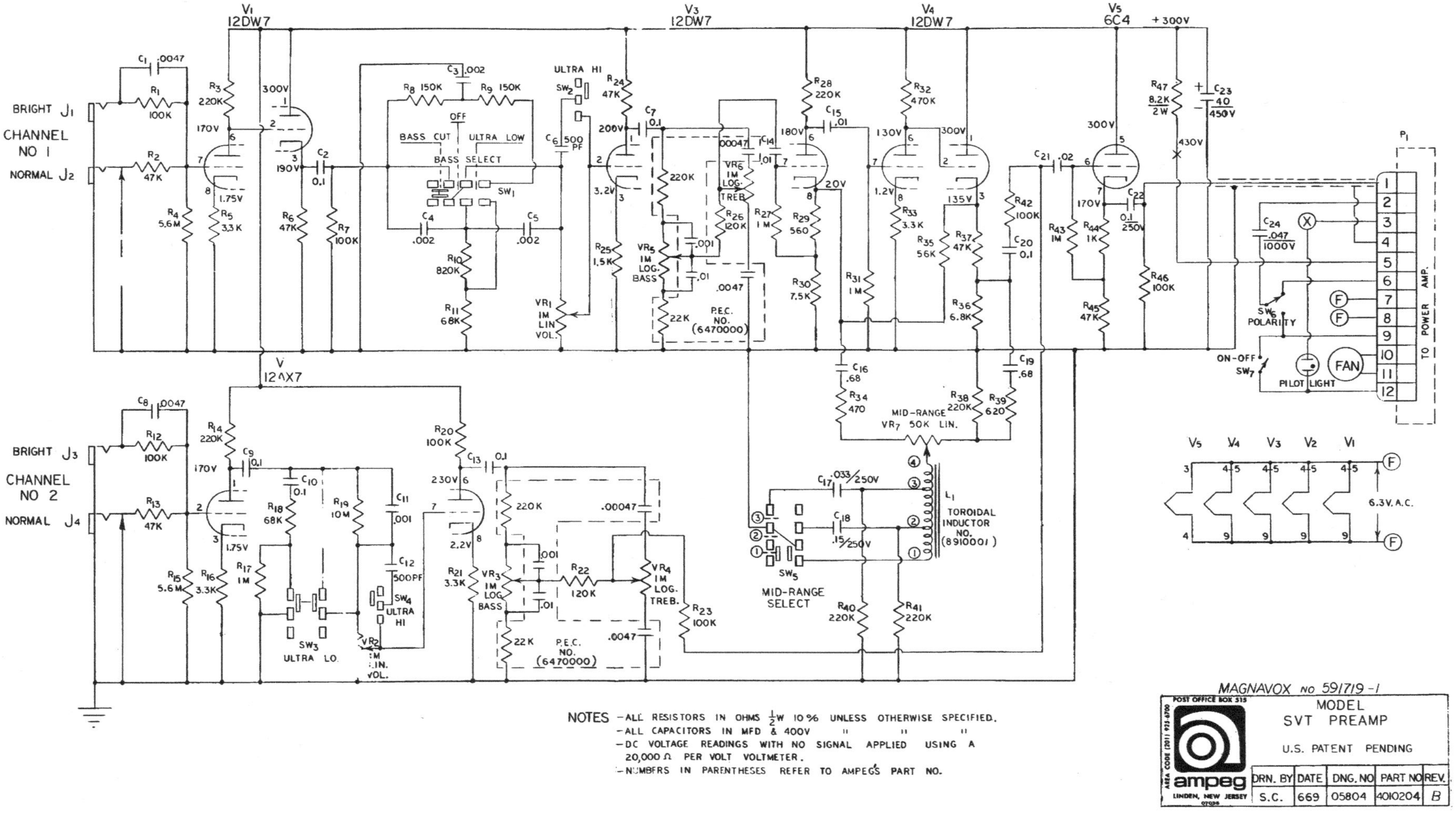

NOTES
- ALL RESISTORS IN OHMS $\frac{1}{2}$W 10% UNLESS OTHERWISE SPECIFIED.
- ALL CAPACITORS IN MFD & 400V " " "
- DC VOLTAGE READINGS WITH NO SIGNAL APPLIED USING A 20,000 Ω PER VOLT VOLTMETER.
- NUMBERS IN PARENTHESES REFER TO AMPEG'S PART NO.

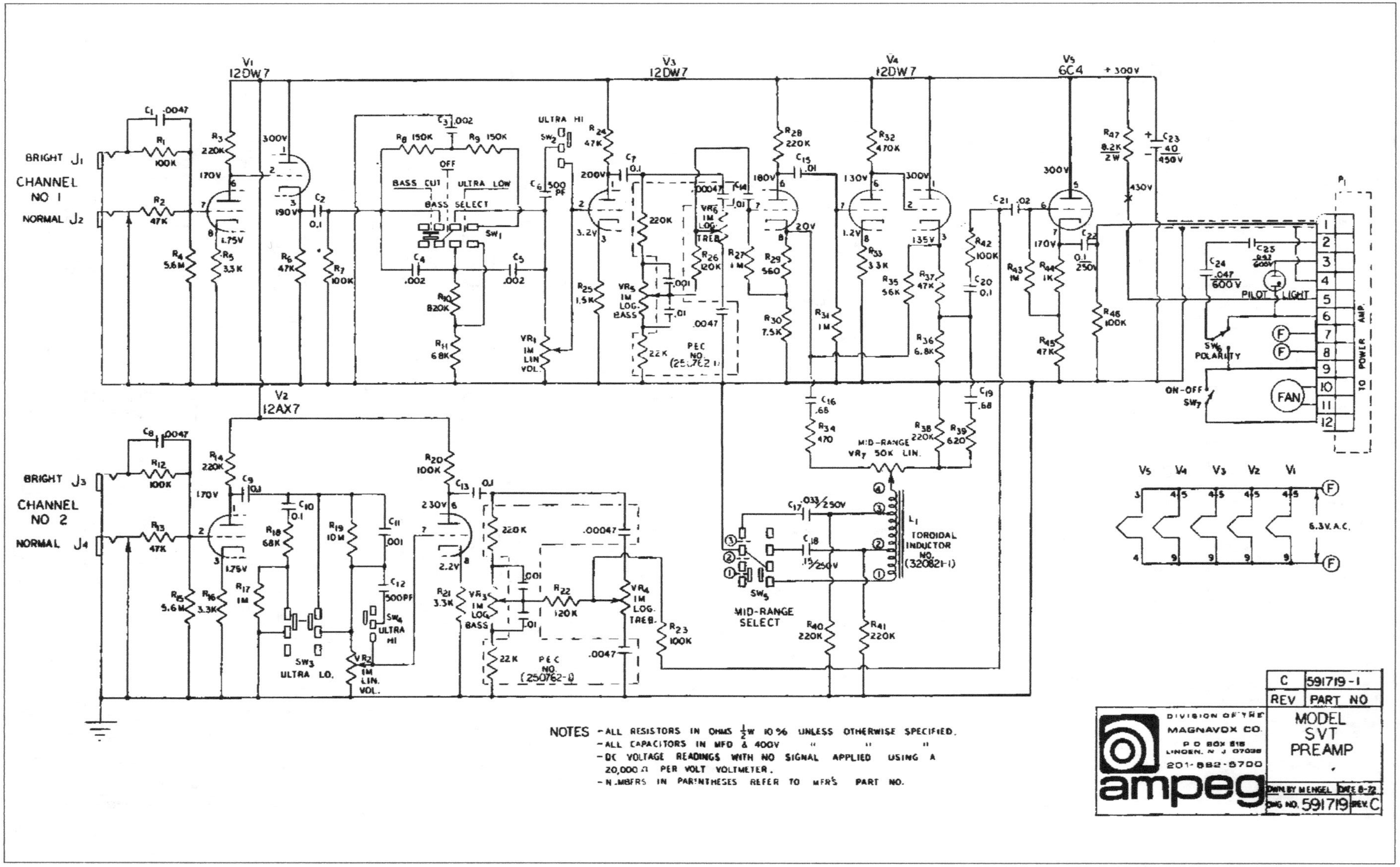
BRIGHT J1
CHANNEL NO 1
NORMAL J2
BRIGHT J3
CHANNEL NO 2
NORMAL J4
V1 12DW7
V2 12AX7
V3 12DW7
V4 12DW7
V5 6C4
+300V
BASS CUT
ULTRA LOW
BASS SELECT
OFF
ULTRA HI
VR1 1M LIN VOL.
VR2 1M LIN. VOL.
VR3 1M LOG. BASS
VR4 1M LOG. TREB.
VR5 1M LOG. BASS
VR6 1M LOG. TREB.
PEC NO. (250762-1)
MID-RANGE VR7 50K LIN.
L1 TOROIDAL INDUCTOR NO. (320821-1)
SW5 MID-RANGE SELECT
SW3 ULTRA LO.
SW4 ULTRA HI
PILOT LIGHT
SW6 POLARITY
ON-OFF SW7
FAN
TO POWER AMP
P1
6.3V.A.C.
NOTES - ALL RESISTORS IN OHMS 1/2W 10% UNLESS OTHERWISE SPECIFIED.
- ALL CAPACITORS IN MFD & 400V " " "
- DC VOLTAGE READINGS WITH NO SIGNAL APPLIED USING A 20,000 Ω PER VOLT VOLTMETER.
- NUMBERS IN PARENTHESES REFER TO MFR'S PART NO.
C 591719-1
REV PART NO
MODEL SVT PREAMP
DIVISION OF THE MAGNAVOX CO.
P O BOX 815 LINDEN, N J 07036
201-882-5700
ampeg
DWN BY MENGEL DATE 8-72
DWG NO. 591719 REV C

SVT **MODEL 6146B**

Ampeg

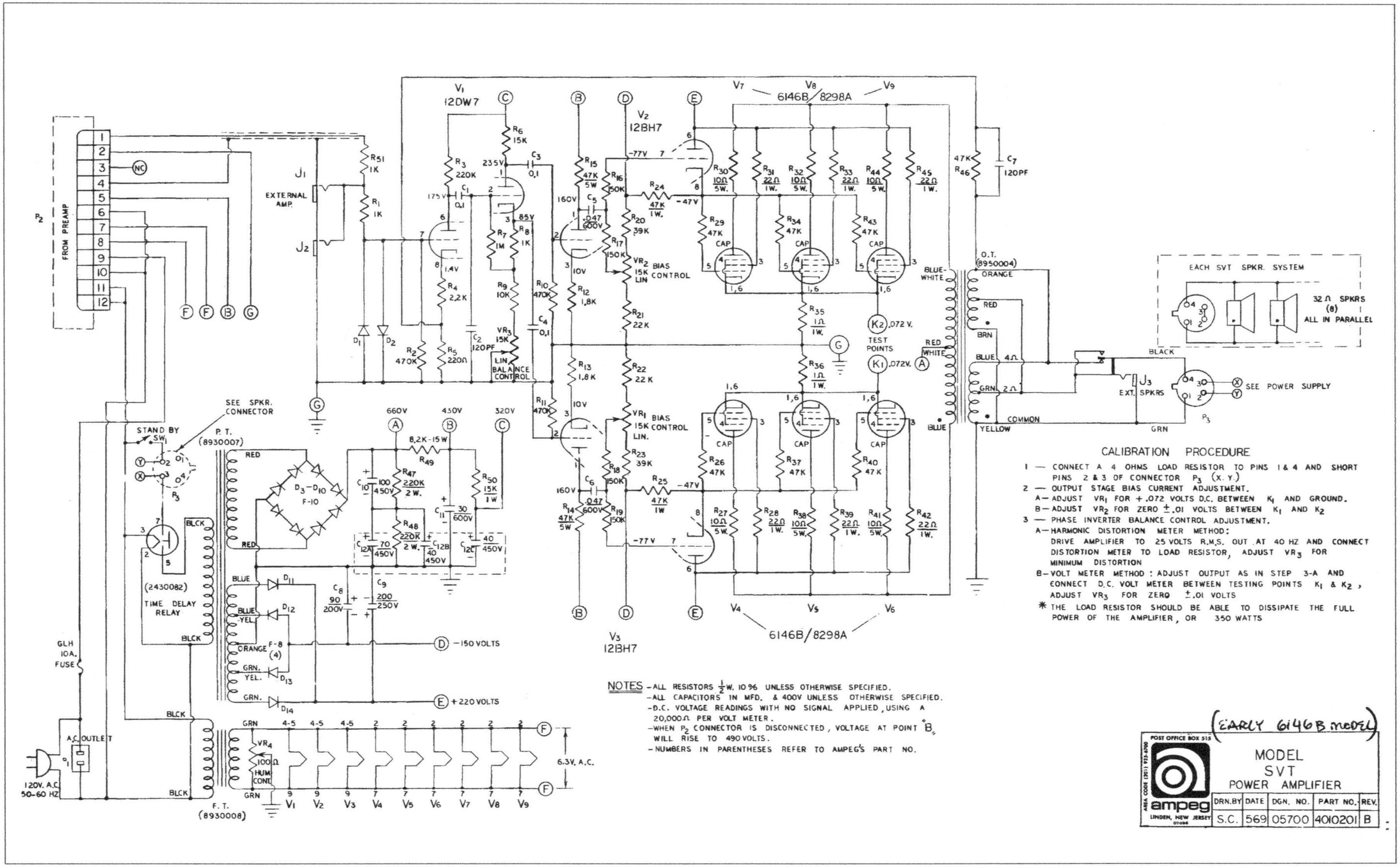

CALIBRATION PROCEDURE

1 — CONNECT A 4 OHMS LOAD RESISTOR TO PINS 1 & 4 AND SHORT PINS 2 & 3 OF CONNECTOR P_3 (X, Y)

2 — OUTPUT STAGE BIAS CURRENT ADJUSTMENT.
A— ADJUST VR_1 FOR +.072 VOLTS D.C. BETWEEN K_1 AND GROUND.
B— ADJUST VR_2 FOR ZERO ±.01 VOLTS BETWEEN K_1 AND K_2

3 — PHASE INVERTER BALANCE CONTROL ADJUSTMENT.
A— HARMONIC DISTORTION METER METHOD:
DRIVE AMPLIFIER TO 25 VOLTS R.M.S. OUT AT 40 HZ AND CONNECT DISTORTION METER TO LOAD RESISTOR, ADJUST VR_3 FOR MINIMUM DISTORTION
B— VOLT METER METHOD: ADJUST OUTPUT AS IN STEP 3-A AND CONNECT D.C. VOLT METER BETWEEN TESTING POINTS K_1 & K_2, ADJUST VR_3 FOR ZERO ±.01 VOLTS

* THE LOAD RESISTOR SHOULD BE ABLE TO DISSIPATE THE FULL POWER OF THE AMPLIFIER, OR 350 WATTS

NOTES
- ALL RESISTORS ½W. 10% UNLESS OTHERWISE SPECIFIED.
- ALL CAPACITORS IN MFD. & 400V UNLESS OTHERWISE SPECIFIED.
- D.C. VOLTAGE READINGS WITH NO SIGNAL APPLIED, USING A 20,000Ω PER VOLT METER.
- WHEN P_2 CONNECTOR IS DISCONNECTED, VOLTAGE AT POINT "B" WILL RISE TO 490 VOLTS.
- NUMBERS IN PARENTHESES REFER TO AMPEG'S PART NO.

(EARLY 6146B MODEL)

POST OFFICE BOX 515
AREA CODE (201) 925-6700
ampeg
LINDEN, NEW JERSEY 07036

MODEL SVT POWER AMPLIFIER

DRN. BY	DATE	DGN. NO.	PART NO.	REV.
S.C.	569	05700	4010201	B

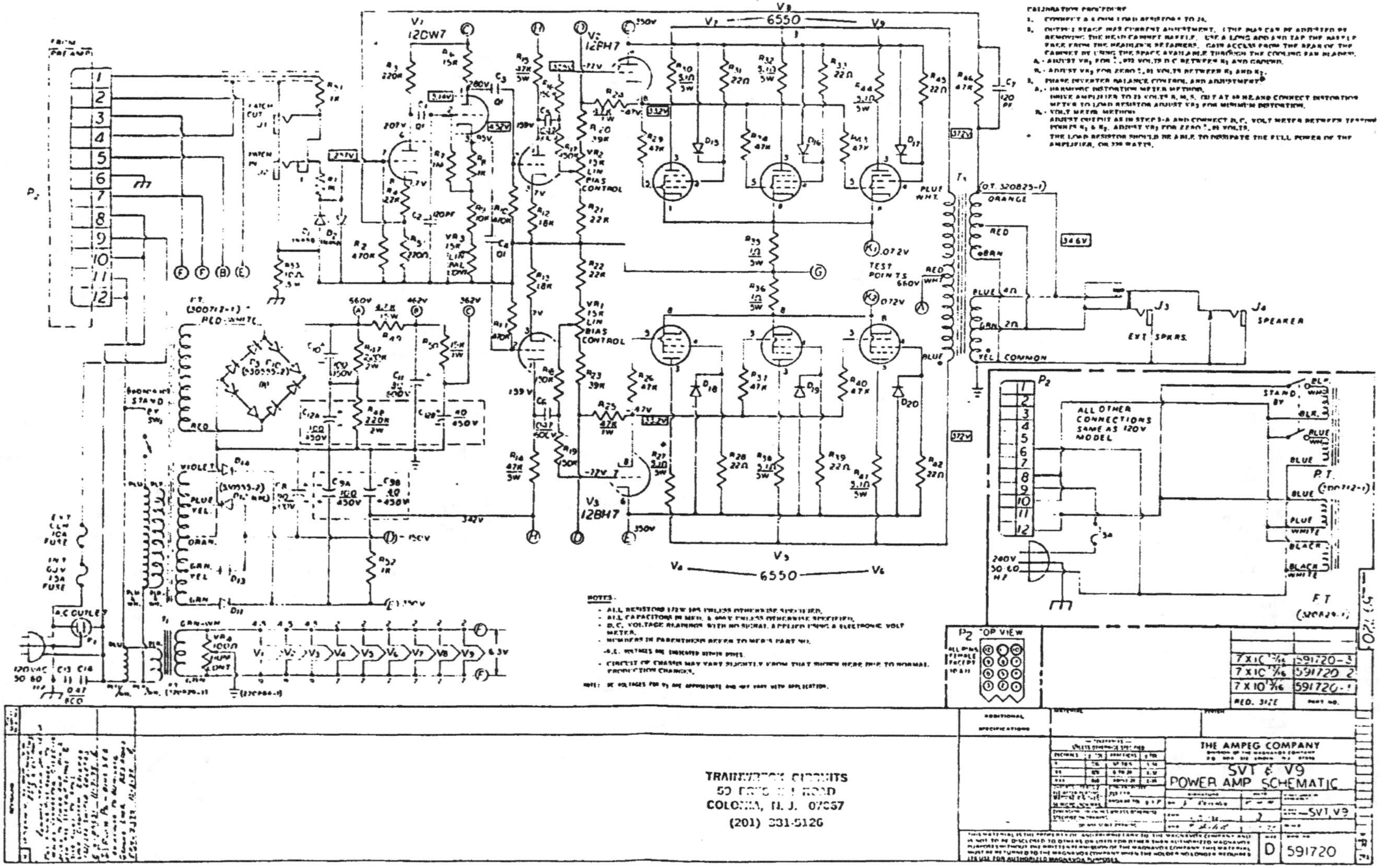
V1
12DW7
V2
12FH7
V3
12BH7
V7 — 6550 — V9
V8
V4 — 6550 — V6
V5
BIAS CONTROL
TEST POINTS
T1
O.T. 320825-1
ORANGE
RED
BRN
BLUE
YEL
COMMON
346V
J3
EXT SPKRS
J4
SPEAKER
P2
ALL OTHER CONNECTIONS SAME AS 120V MODEL
240V 50/60 HZ
P.T.
F.T.
P2 TOP VIEW
7 X 10 5/16
591720-1
591720-2
591720-3
RED. SIZE
THE AMPEG COMPANY
SVT & V9
POWER AMP SCHEMATIC
SVT,V9
D
591720
TRAINWRECK CIRCUITS
COLONIA, N. J. 07067
(201) 331-5126
120VAC 50/60
AC OUTLET
EXT 10A FUSE
INT 15A FUSE

SVT-2 **PWR PRE AMP (REV 7)**

Ampeg

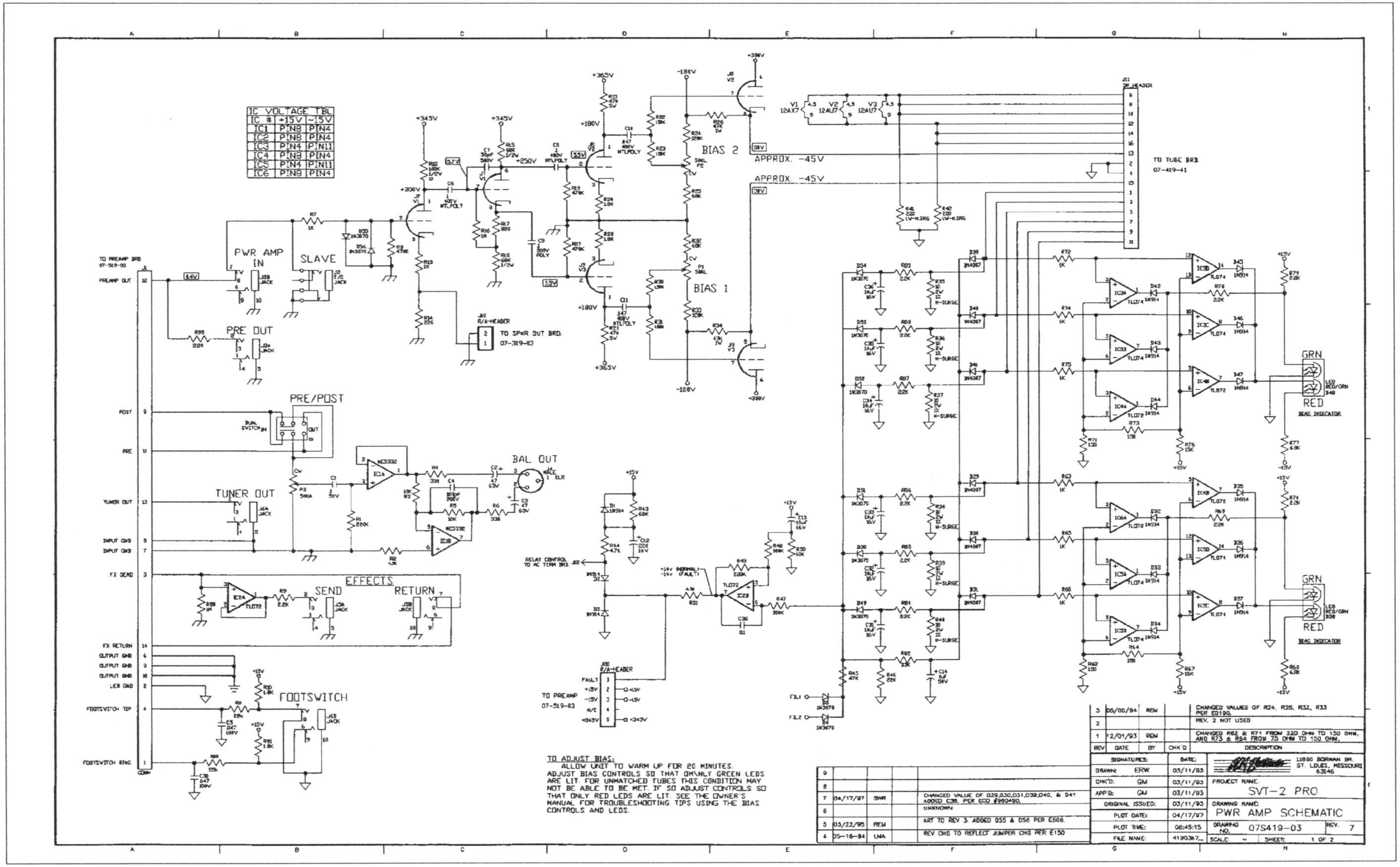

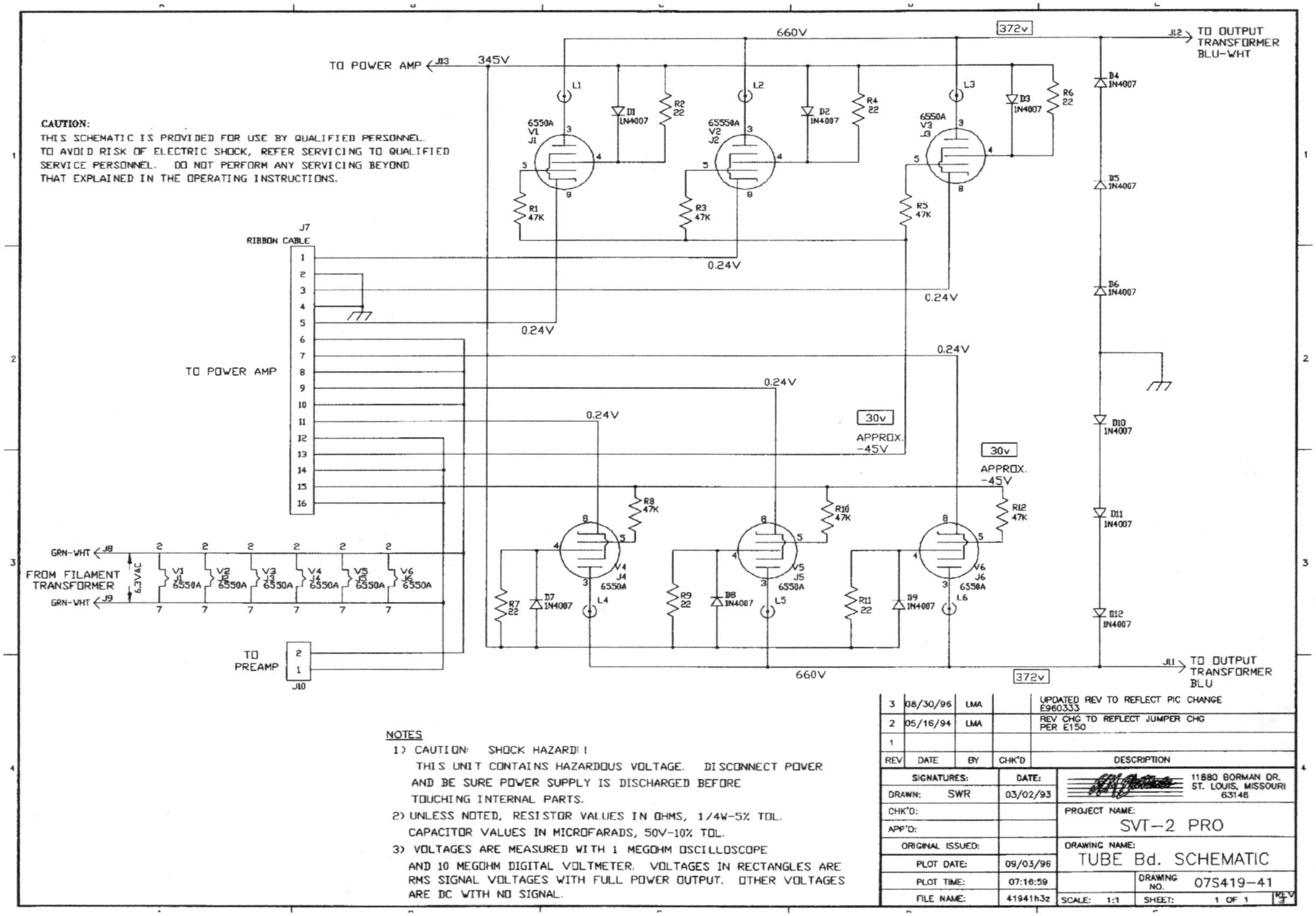

CAUTION:
THIS SCHEMATIC IS PROVIDED FOR USE BY QUALIFIED PERSONNEL. TO AVOID RISK OF ELECTRIC SHOCK, REFER SERVICING TO QUALIFIED SERVICE PERSONNEL. DO NOT PERFORM ANY SERVICING BEYOND THAT EXPLAINED IN THE OPERATING INSTRUCTIONS.

NOTES

1) CAUTION: SHOCK HAZARD! !
THIS UNIT CONTAINS HAZARDOUS VOLTAGE. DISCONNECT POWER AND BE SURE POWER SUPPLY IS DISCHARGED BEFORE TOUCHING INTERNAL PARTS.

2) UNLESS NOTED, RESISTOR VALUES IN OHMS, 1/4W-5% TOL. CAPACITOR VALUES IN MICROFARADS, 50V-10% TOL.

3) VOLTAGES ARE MEASURED WITH 1 MEGOHM OSCILLOSCOPE AND 10 MEGOHM DIGITAL VOLTMETER. VOLTAGES IN RECTANGLES ARE RMS SIGNAL VOLTAGES WITH FULL POWER OUTPUT. OTHER VOLTAGES ARE DC WITH NO SIGNAL.

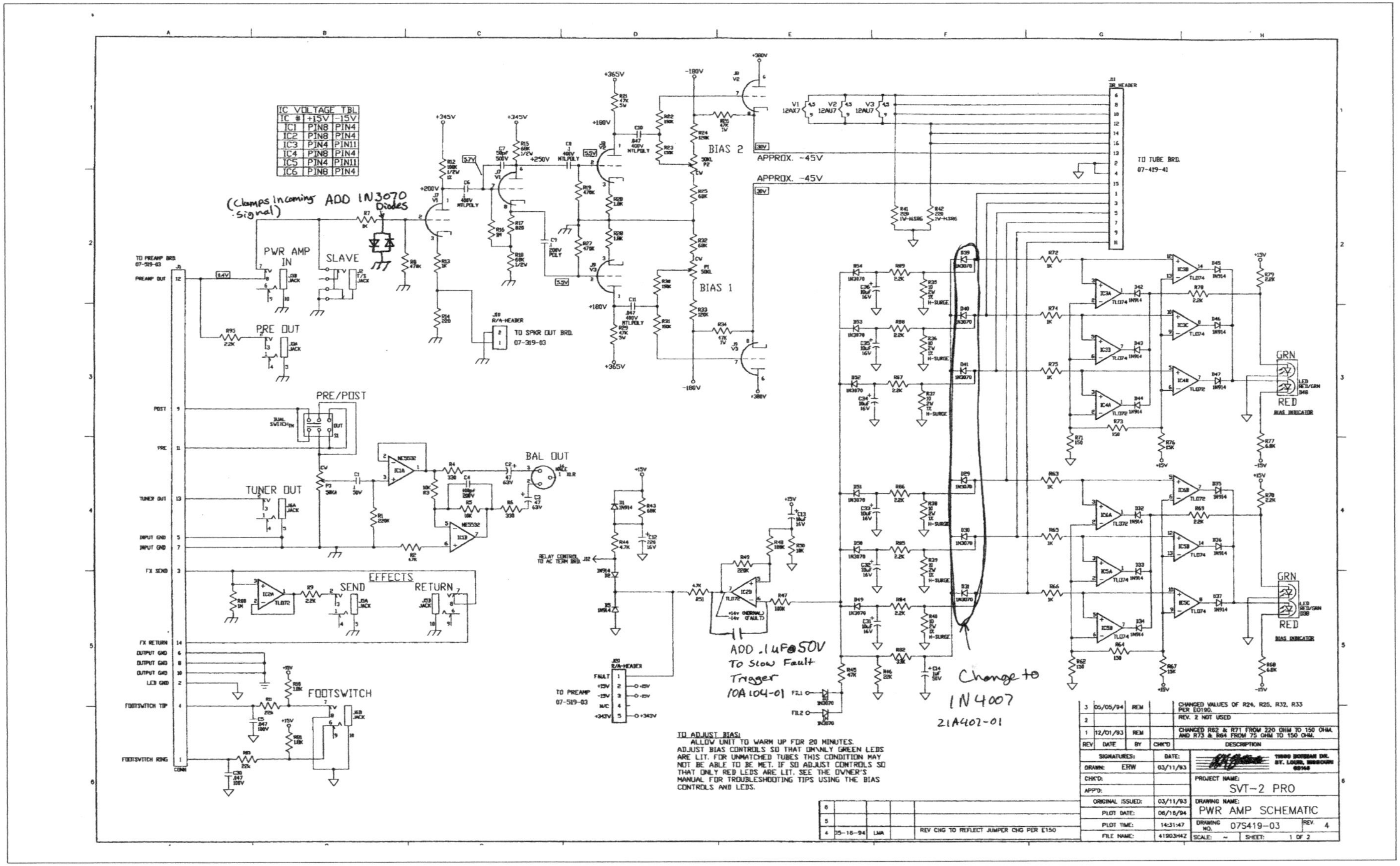
IC VOLTAGE TBL
(Clamps incoming signal) ADD 1N3070 Diodes
PWR AMP IN
SLAVE
PRE OUT
PRE/POST
TUNER OUT
BAL OUT
EFFECTS
SEND
RETURN
FOOTSWITCH
BIAS 2
BIAS 1
APPROX. -45V
APPROX. -45V
TO TUBE BRD. 07-419-41
TO SPKR OUT BRD. 07-319-03
TO PREAMP 07-519-03
ADD .1uF@50V To Slow Fault Trigger 10A104-01
Change to 1N4007 21A407-01
GRN
RED
BIAS INDICATOR
TO ADJUST BIAS:
ALLOW UNIT TO WARM UP FOR 20 MINUTES. ADJUST BIAS CONTROLS SO THAT ONLY GREEN LEDS ARE LIT. FOR UNMATCHED TUBES THIS CONDITION MAY NOT BE ABLE TO BE MET. IF SO ADJUST CONTROLS SO THAT ONLY RED LEDS ARE LIT. SEE THE OWNER'S MANUAL FOR TROUBLESHOOTING TIPS USING THE BIAS CONTROLS AND LEDS.
CHANGED VALUES OF R24, R25, R32, R33 PER E0190.
REV. 2 NOT USED
CHANGED R62 & R71 FROM 220 OHM TO 150 OHM. AND R73 & R64 FROM 75 OHM TO 150 OHM.
REV CHG TO REFLECT JUMPER CHG PER E150
PROJECT NAME: SVT-2 PRO
DRAWING NAME: PWR AMP SCHEMATIC
DRAWING NO. 07S419-03
REV. 4
SHEET: 1 OF 2

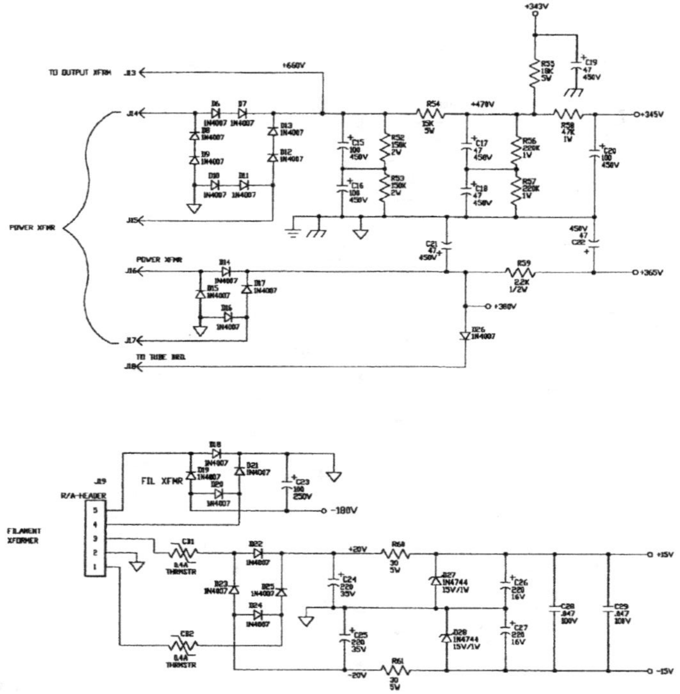

NOTES
1) CAUTION: SHOCK HAZARD!!
THIS UNIT CONTAINS HAZARDOUS VOLTAGE. DISCONNECT POWER AND BE SURE POWER SUPPLY IS DISCHARGED BEFORE TOUCHING INTERNAL PARTS.
2) UNLESS NOTED, RESISTOR VALUES IN OHMS, 1/4W-5% TOL.
CAPACITOR VALUES IN MICROFARADS, 50V-10% TOL.
3) VOLTAGES ARE MEASURED WITH 1 MEGOHM OSCILLOSCOPE AND 10 MEGOHM DIGITAL VOLTMETER.
4) VOLTAGES IN RECTANGLES ARE RMS SIGNAL VOLTAGES WITH 0.4v IN.
OTHER VOLTAGES ARE D.C. IN CONDITIONS STATED.
5) CIRCUIT GROUND ⏚ DIRTY GROUND CHASSIS GROUND ⏚

REV	DATE	BY	CHK'D	DESCRIPTION
4	05/16/94	LMA		REV CHG TO REFLECT JUMPER CHG PER E150
3	05/05/94	REM		CHANGED VALUES OF R24, R25, R32, R33 PER ED190.
2				REV. 2 NOT USED
1	12/01/93	REM		CHANGED R62 & R71 FROM 220 OHM TO 150 OHM. AND R73 & R64 FROM 75 OHM TO 150 OHM.

SIGNATURES:		DATE:		
DRAWN:	ERW	03/11/93		
CHK'D:	GM	03/11/93	PROJECT NAME: SVT-2 PRO	
APP'D:	GM	03/11/93		
ORIGINAL ISSUED:		03/11/93	DRAWING NAME: PWR AMP SCHEMATIC	
PLOT DATE:		06/18/94		
PLOT TIME:		14:31:47	DRAWING NO. 07S419-03	REV. 4
FILE NAME:		D:\419\41903H	SCALE: ~	SHEET: 2 OF 2

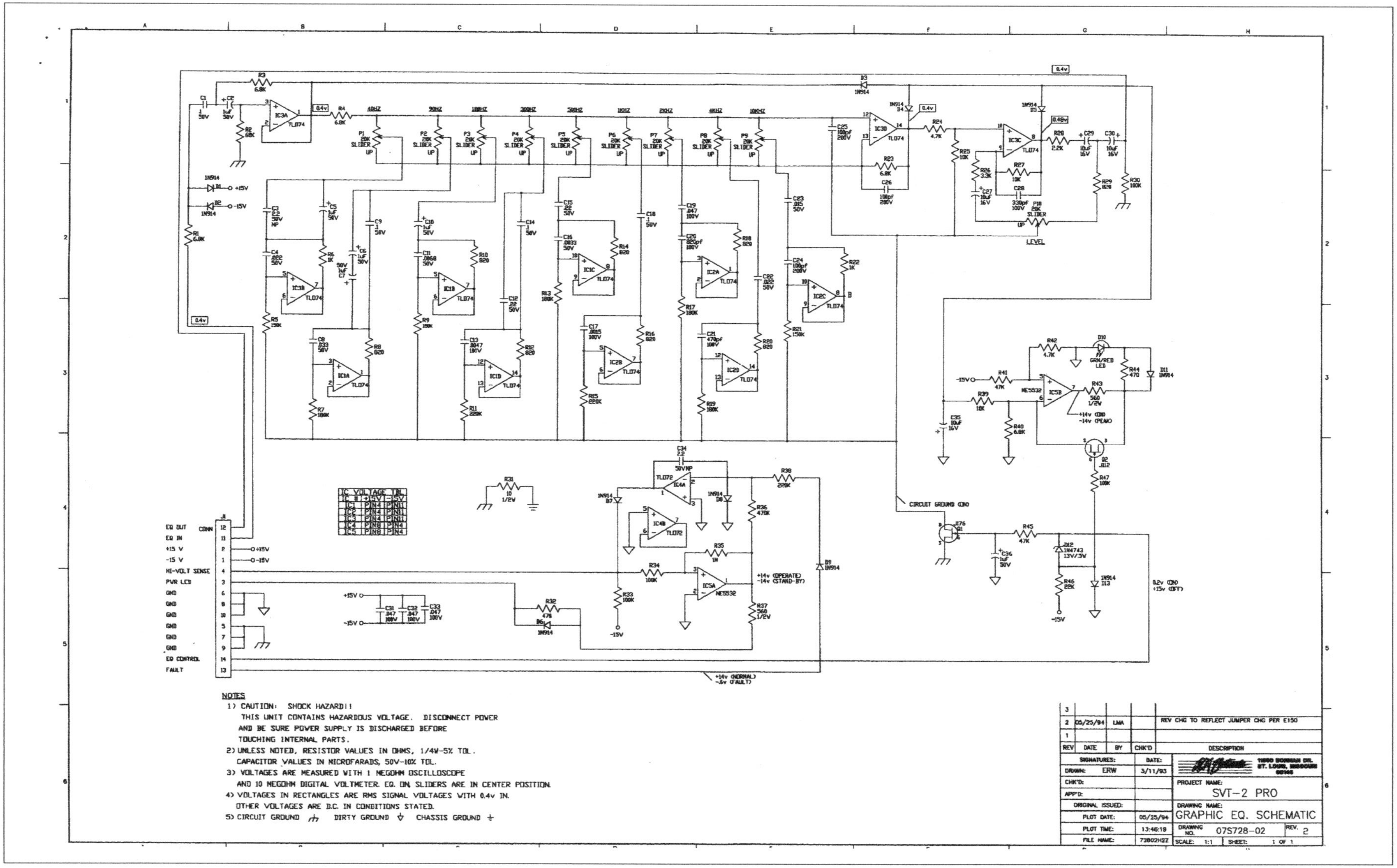
NOTES
1) CAUTION: SHOCK HAZARD!!
THIS UNIT CONTAINS HAZARDOUS VOLTAGE. DISCONNECT POWER
AND BE SURE POWER SUPPLY IS DISCHARGED BEFORE
TOUCHING INTERNAL PARTS.
2) UNLESS NOTED, RESISTOR VALUES IN OHMS, 1/4W-5% TOL.
CAPACITOR VALUES IN MICROFARADS, 50V-10% TOL.
3) VOLTAGES ARE MEASURED WITH 1 MEGOHM OSCILLOSCOPE
AND 10 MEGOHM DIGITAL VOLTMETER. EQ. ON, SLIDERS ARE IN CENTER POSITION.
4) VOLTAGES IN RECTANGLES ARE RMS SIGNAL VOLTAGES WITH 0.4v IN.
OTHER VOLTAGES ARE D.C. IN CONDITIONS STATED.
5) CIRCUIT GROUND DIRTY GROUND CHASSIS GROUND
EQ OUT
EQ IN
+15 V
-15 V
HI-VOLT SENSE
PWR LED
GND
EQ CONTROL
FAULT
CIRCUIT GROUND GND
REV CHG TO REFLECT JUMPER CHG PER E150
DRAWN: ERW 3/11/93
PROJECT NAME: SVT-2 PRO
DRAWING NAME: GRAPHIC EQ. SCHEMATIC
DRAWING NO. 07S728-02 REV. 2
PLOT DATE: 05/25/94
PLOT TIME: 13:46:19
SCALE: 1:1 SHEET: 1 OF 1

SVT-2 PRO PREAMP

Ampeg

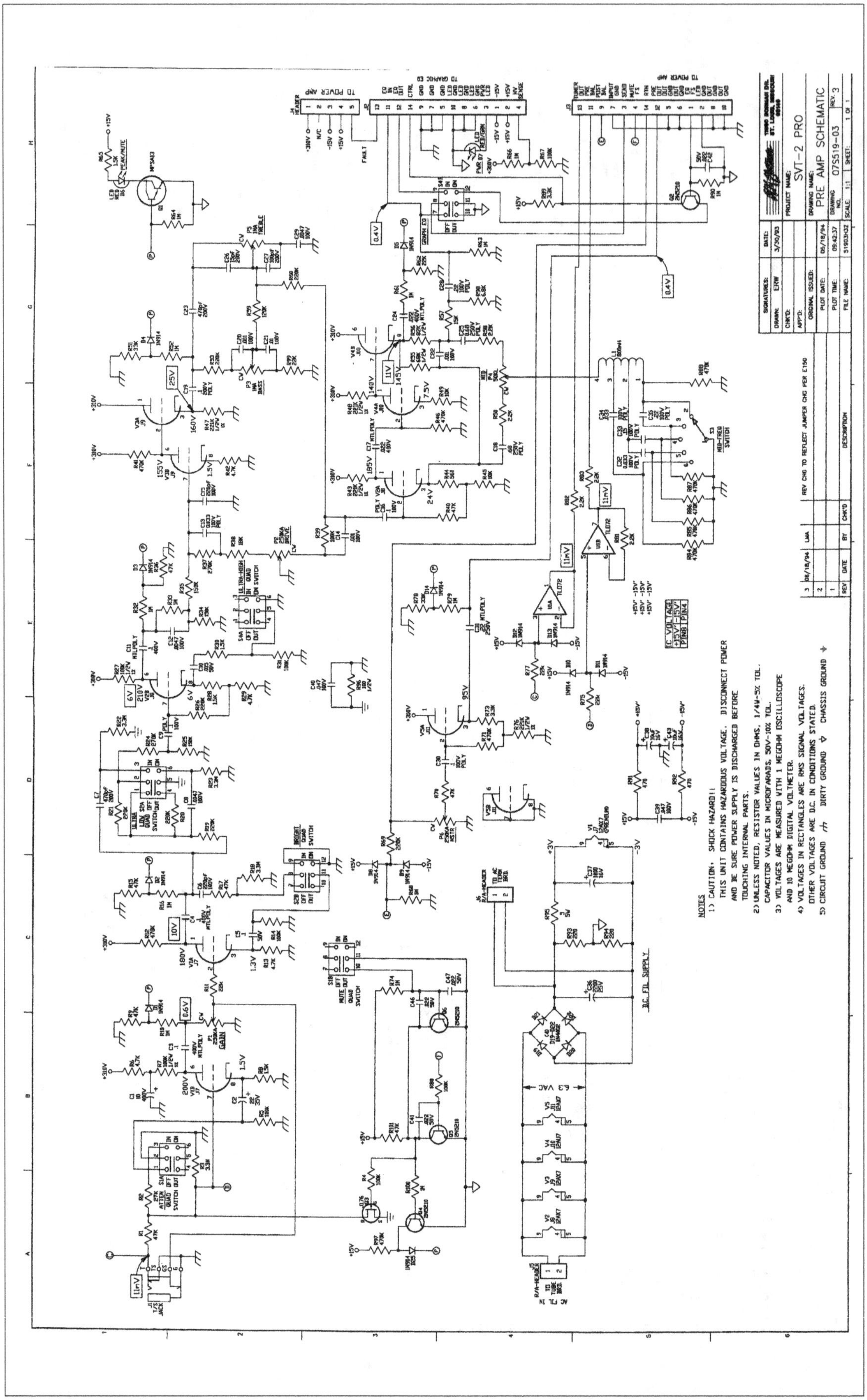

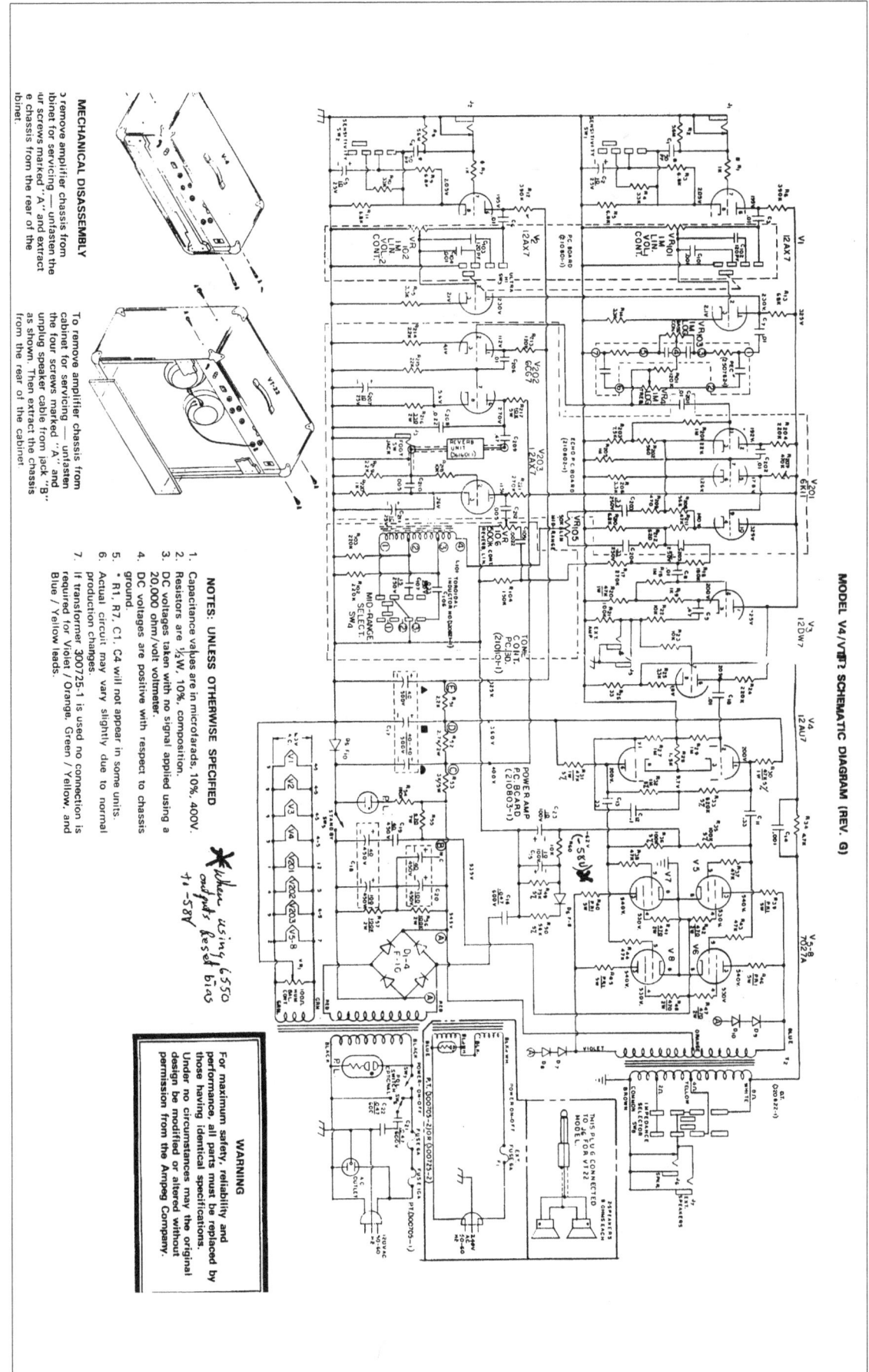

MODEL V4/VT22 SCHEMATIC DIAGRAM (REV. G)

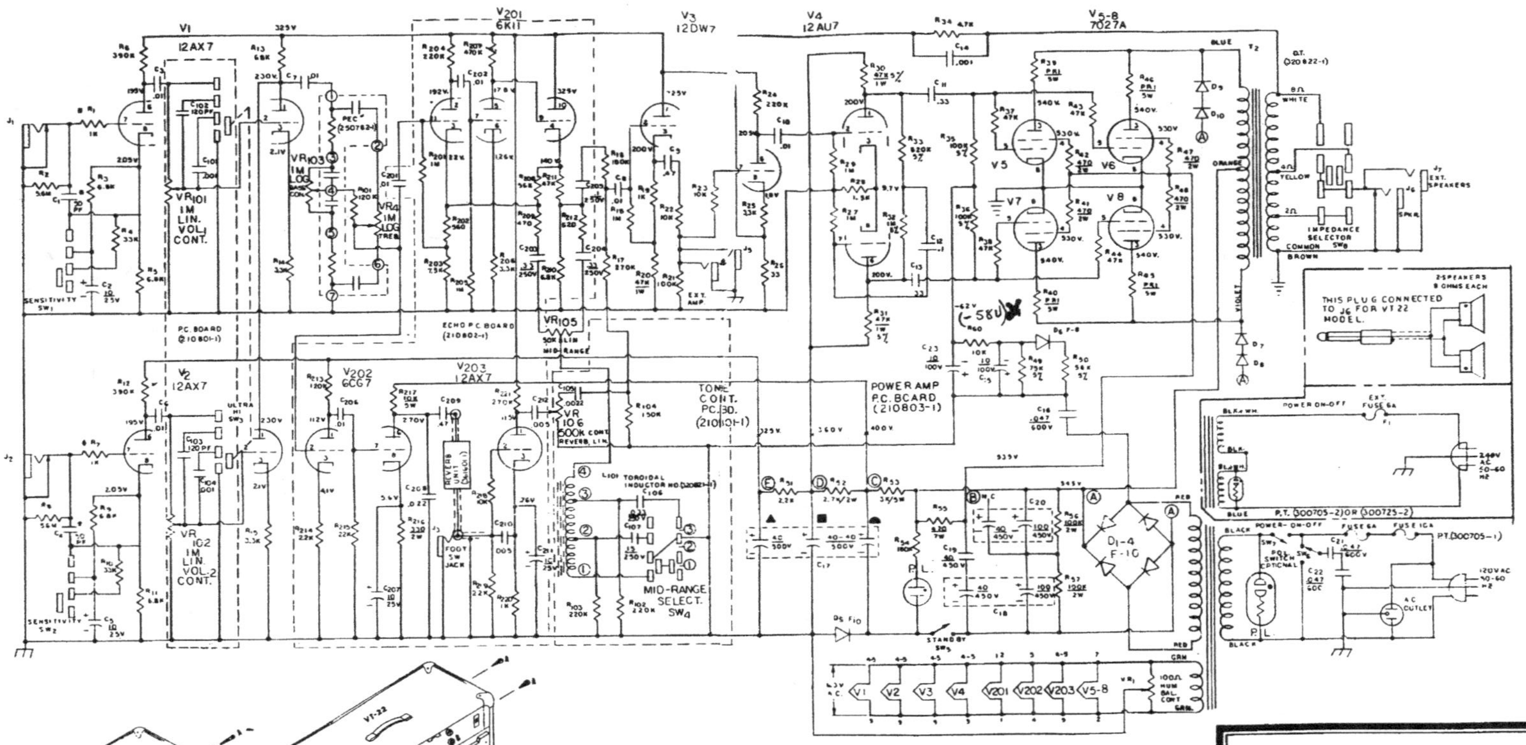

NOTES: UNLESS OTHERWISE SPECIFIED

1. Capacitance values are in microfarads, 10%, 400V.
2. Resistors are ½W, 10%, composition.
3. DC voltages taken with no signal applied using a 20,000 ohm/volt voltmeter.
4. DC voltages are positive with respect to chassis ground.
5. * R1, R7, C1, C4 will not appear in some units.
6. Actual circuit may vary slightly due to normal production changes.
7. If transformer 300725-1 is used no connection is required for Violet / Orange, Green / Yellow, and Blue / Yellow leads.

WARNING

For maximum safety, reliability and performance, all parts must be replaced by those having identical specifications. Under no circumstances may the original design be modified or altered without permission from the Ampeg Company.

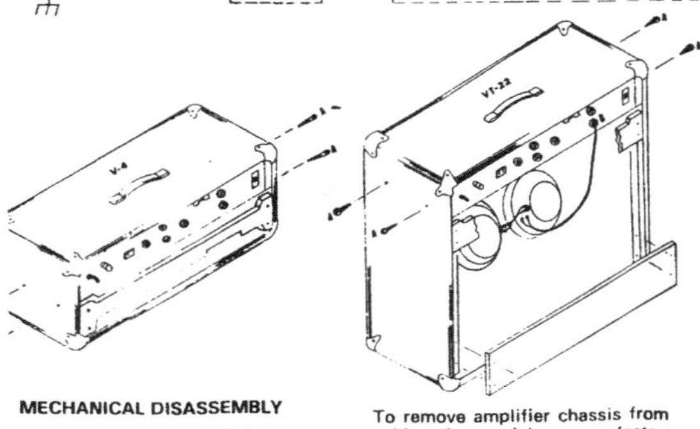

MECHANICAL DISASSEMBLY

ɔ remove amplifier chassis from ıbinet for servicing — unfasten the ıur screws marked "A" and extract e chassis from the rear of the ıbinet.

To remove amplifier chassis from cabinet for servicing — unfasten the four screws marked "A" and unplug speaker cable from jack "B" as shown. Then extract the chassis from the rear of the cabinet.

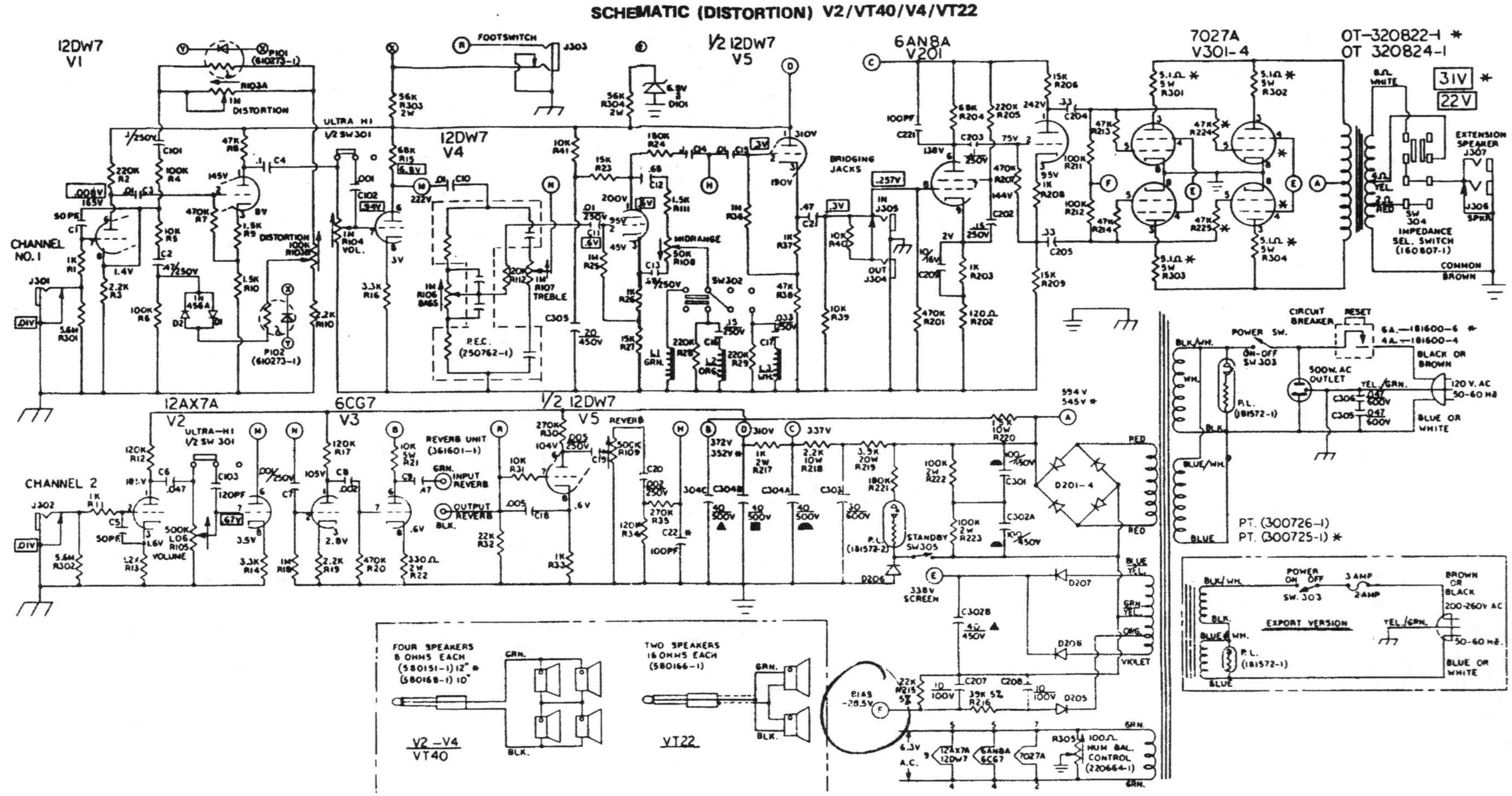
SCHEMATIC (DISTORTION) V2/VT40/V4/VT22
12DW7
V1
12AX7A
V2
6CG7
V3
12DW7
V4
1/2 12DW7
V5
6AN8A
V201
7027A
V301-4
OT-320822-1 *
OT 320824-1
31V *
22V
CHANNEL NO. 1
CHANNEL 2
FOOTSWITCH
DISTORTION
VOL.
VOLUME
BASS
TREBLE
MIDRANGE
BRIDGING JACKS
EXTENSION SPEAKER
IMPEDANCE SEL. SWITCH (160807-1)
COMMON BROWN
CIRCUIT BREAKER
RESET
POWER SW.
ON-OFF SW 303
500W. AC OUTLET
120 V. AC 50-60 Hz
BLUE OR WHITE
BLACK OR BROWN
PT. (300726-1)
PT. (300725-1) *
EXPORT VERSION
STANDBY SW305
338V SCREEN
BIAS
HUM BAL. CONTROL (220664-1)
REVERB UNIT (361601-1)
INPUT REVERB
OUTPUT REVERB
FOUR SPEAKERS 8 OHMS EACH (580151-1) 12" * (580168-1) 10"
TWO SPEAKERS 16 OHMS EACH (580166-1)
V2 - V4
VT40
VT22

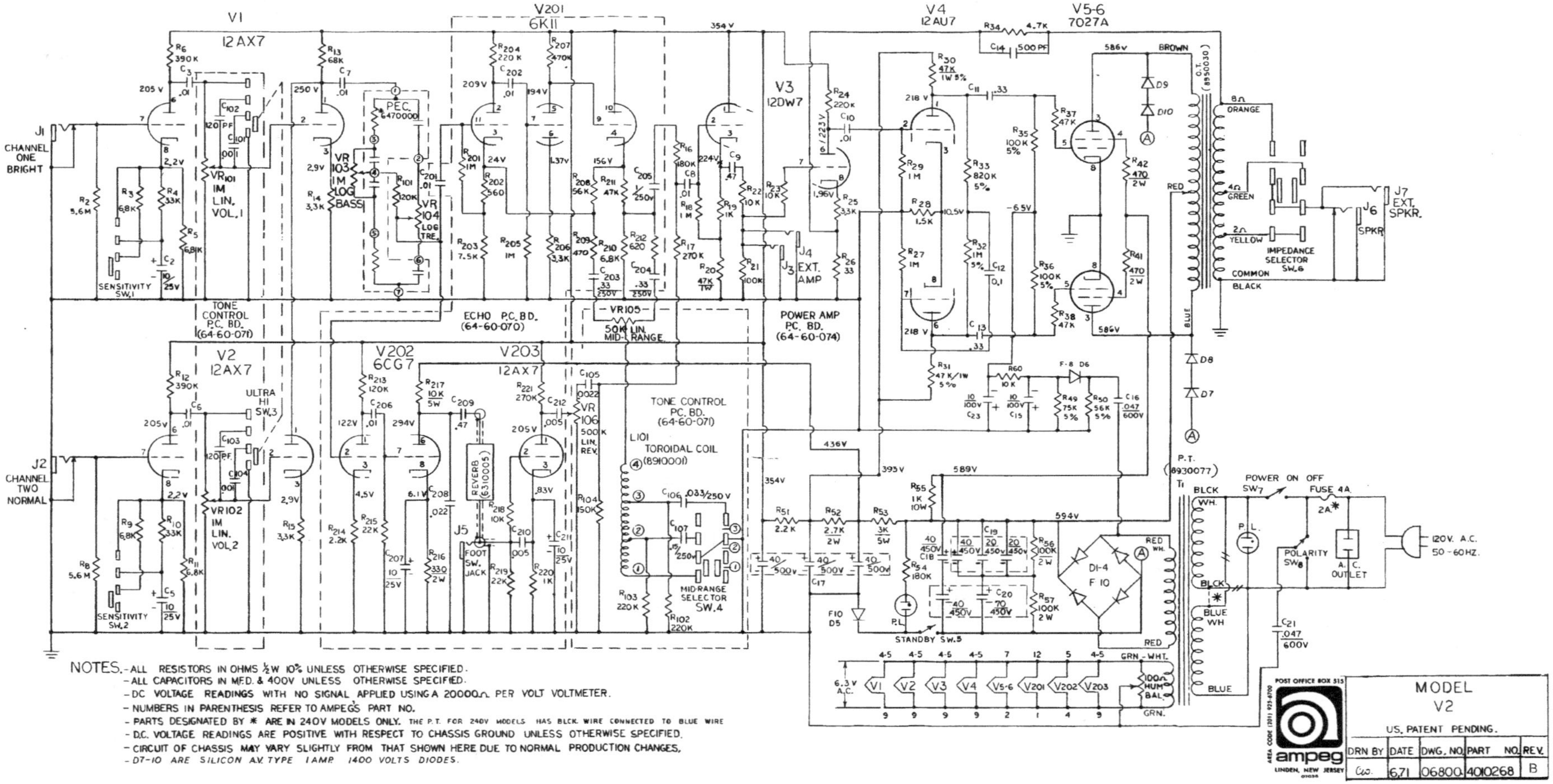

NOTES.
- ALL RESISTORS IN OHMS ½W 10% UNLESS OTHERWISE SPECIFIED.
- ALL CAPACITORS IN MFD. & 400V UNLESS OTHERWISE SPECIFIED.
- DC VOLTAGE READINGS WITH NO SIGNAL APPLIED USING A 20000Ω PER VOLT VOLTMETER.
- NUMBERS IN PARENTHESIS REFER TO AMPEG'S PART NO.
- PARTS DESIGNATED BY * ARE IN 240V MODELS ONLY. THE P.T. FOR 240V MODELS HAS BLCK WIRE CONNECTED TO BLUE WIRE
- D.C. VOLTAGE READINGS ARE POSITIVE WITH RESPECT TO CHASSIS GROUND UNLESS OTHERWISE SPECIFIED.
- CIRCUIT OF CHASSIS MAY VARY SLIGHTLY FROM THAT SHOWN HERE DUE TO NORMAL PRODUCTION CHANGES.
- D7-10 ARE SILICON A.V. TYPE 1 AMP. 1400 VOLTS DIODES.

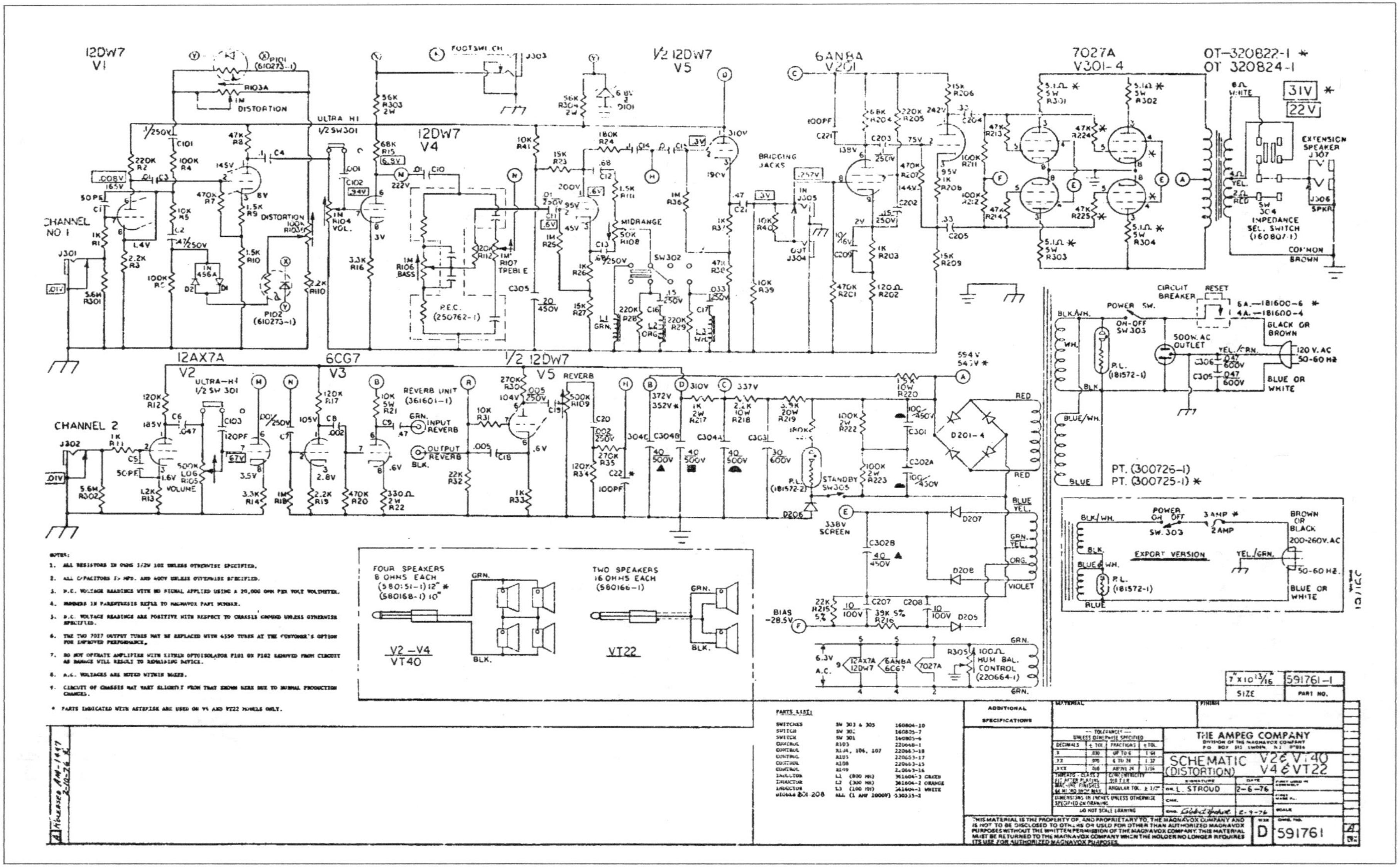
12DW7 V1
CHANNEL NO 1
CHANNEL 2
12AX7A V2
12DW7 V4
6CG7 V3
1/2 12DW7 V5
6ANBA V201
7027A V301-4
OT-320822-1 *
OT 320824-1
EXTENSION SPEAKER J307
IMPEDANCE SEL. SWITCH (160807-1)
COMMON BROWN
BRIDGING JACKS
MIDRANGE
TREBLE
BASS
DISTORTION
VOLUME
REVERB UNIT (361601-1)
INPUT REVERB
OUTPUT REVERB
FOOTSWITCH
CIRCUIT BREAKER
RESET
POWER SW.
ON-OFF SW 303
500W. AC OUTLET
120 V. AC 50-60 Hz
BLUE OR WHITE
PT. (300726-1)
PT. (300725-1) *
EXPORT VERSION
STANDBY SW305
338V SCREEN
BIAS -28.5V
HUM BAL. CONTROL (220664-1)
FOUR SPEAKERS 8 OHMS EACH (580151-1) 12" * (580168-1) 10"
V2 - V4 VT40
TWO SPEAKERS 16 OHMS EACH (580166-1)
VT22
THE AMPEG COMPANY
SCHEMATIC V2 & V140 V4 & VT22 (DISTORTION)
L. STROUD
2-6-76
D 591761
7"x10 13/16
591761-1
SIZE
PART NO.

V2/VT40/V4/VT22

Ampeg

SCHEMATIC (DISTORTION) V2/VT40/V4/VT22

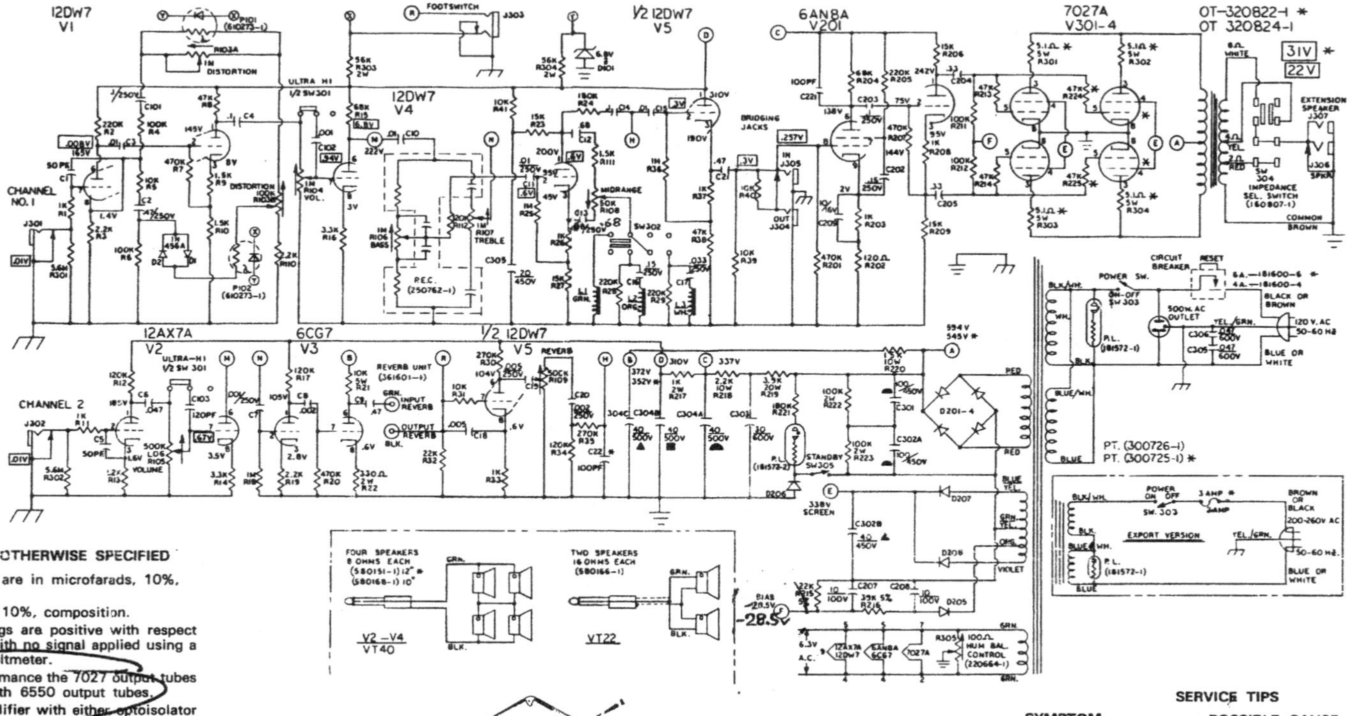

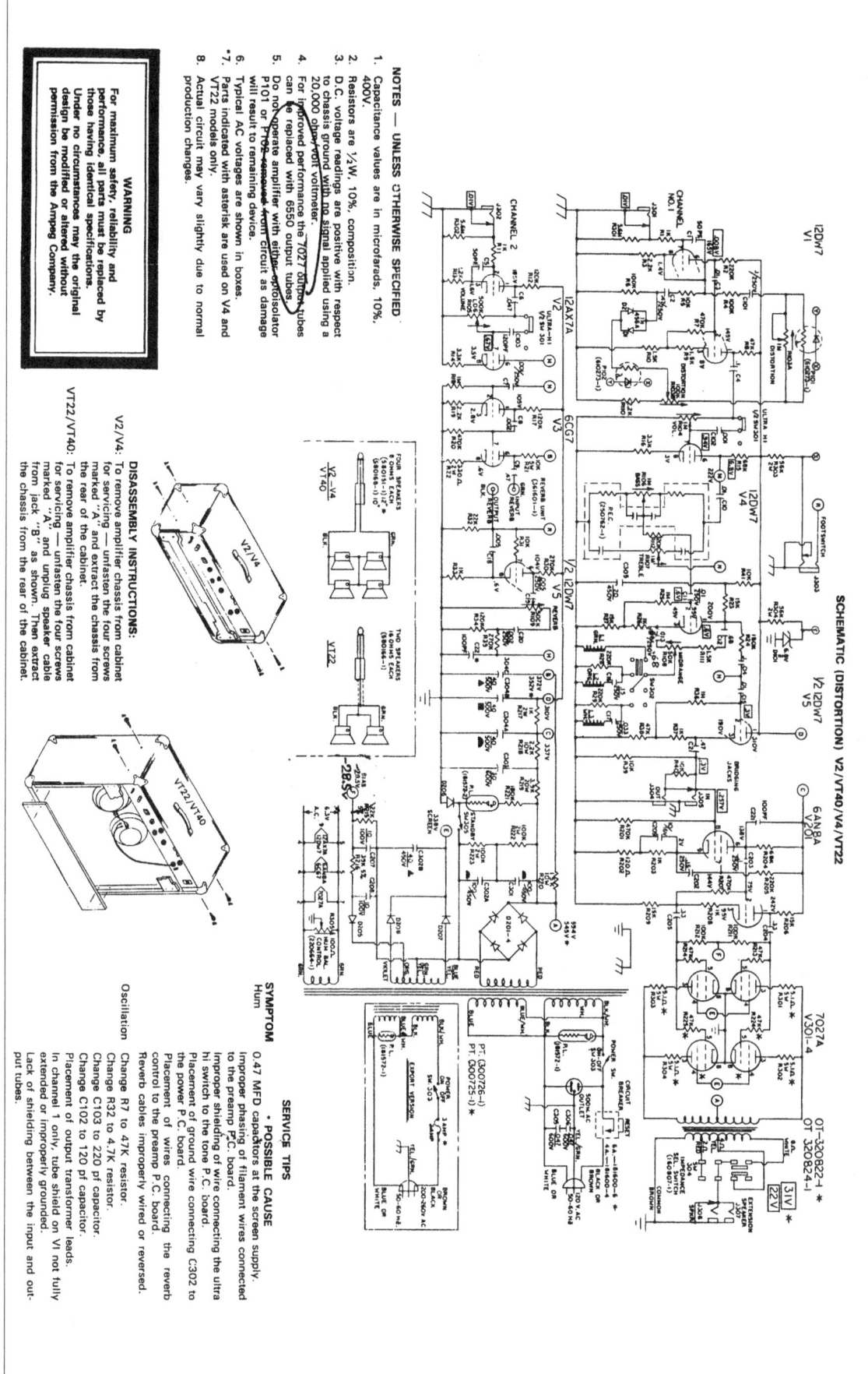

NOTES — UNLESS OTHERWISE SPECIFIED

1. Capacitance values are in microfarads, 10%, 400V.
2. Resistors are ½W, 10%, composition.
3. D.C. voltage readings are positive with respect to chassis ground with no signal applied using a 20,000 ohm/volt voltmeter.
4. For improved performance the 7027 output tubes can be replaced with 6550 output tubes.
5. Do not operate amplifier with either optoisolator P101 or P102 removed from circuit as damage will result to remaining device.
6. Typical AC voltages are shown in boxes.
*7. Parts indicated with asterisk are used on V4 and VT22 models only.
8. Actual circuit may vary slightly due to normal production changes.

WARNING

For maximum safety, reliability and performance, all parts must be replaced by those having identical specifications. Under no circumstances may the original design be modified or altered without permission from the Ampeg Company.

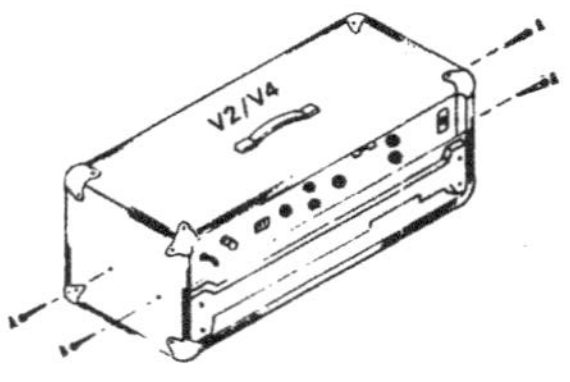

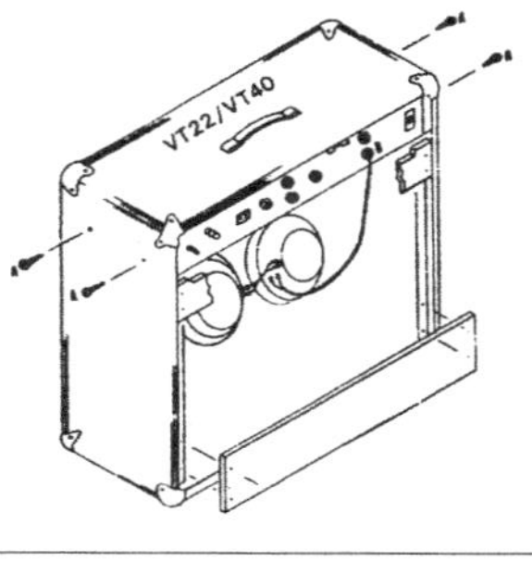

DISASSEMBLY INSTRUCTIONS:

V2/V4: To remove amplifier chassis from cabinet for servicing — unfasten the four screws marked "A" and extract the chassis from the rear of the cabinet.

VT22/VT40: To remove amplifier chassis from cabinet for servicing — unfasten the four screws marked "A" and unplug speaker cable from jack "B" as shown. Then extract the chassis from the rear of the cabinet.

SERVICE TIPS

SYMPTOM	• POSSIBLE CAUSE
Hum	0.47 MFD capacitors at the screen supply.
	Improper phasing of filament wires connected to the preamp P.C. board.
	Improper shielding of wire connecting the ultra hi switch to the tone P.C. board.
	Placement of ground wire connecting C302 to the power P.C. board.
	Placement of wires connecting the reverb control to the preamp P.C. board.
	Reverb cables improperly wired or reversed.
Oscillation	Change R7 to 47K resistor.
	Change R32 to 4.7K resistor.
	Change C103 to 220 pf capacitor.
	Change C102 to 120 pf capacitor.
	Placement of output transformer leads.
	In channel 1 only, tube shield on V1 not fully extended or improperly grounded.
	Lack of shielding between the input and output tubes.

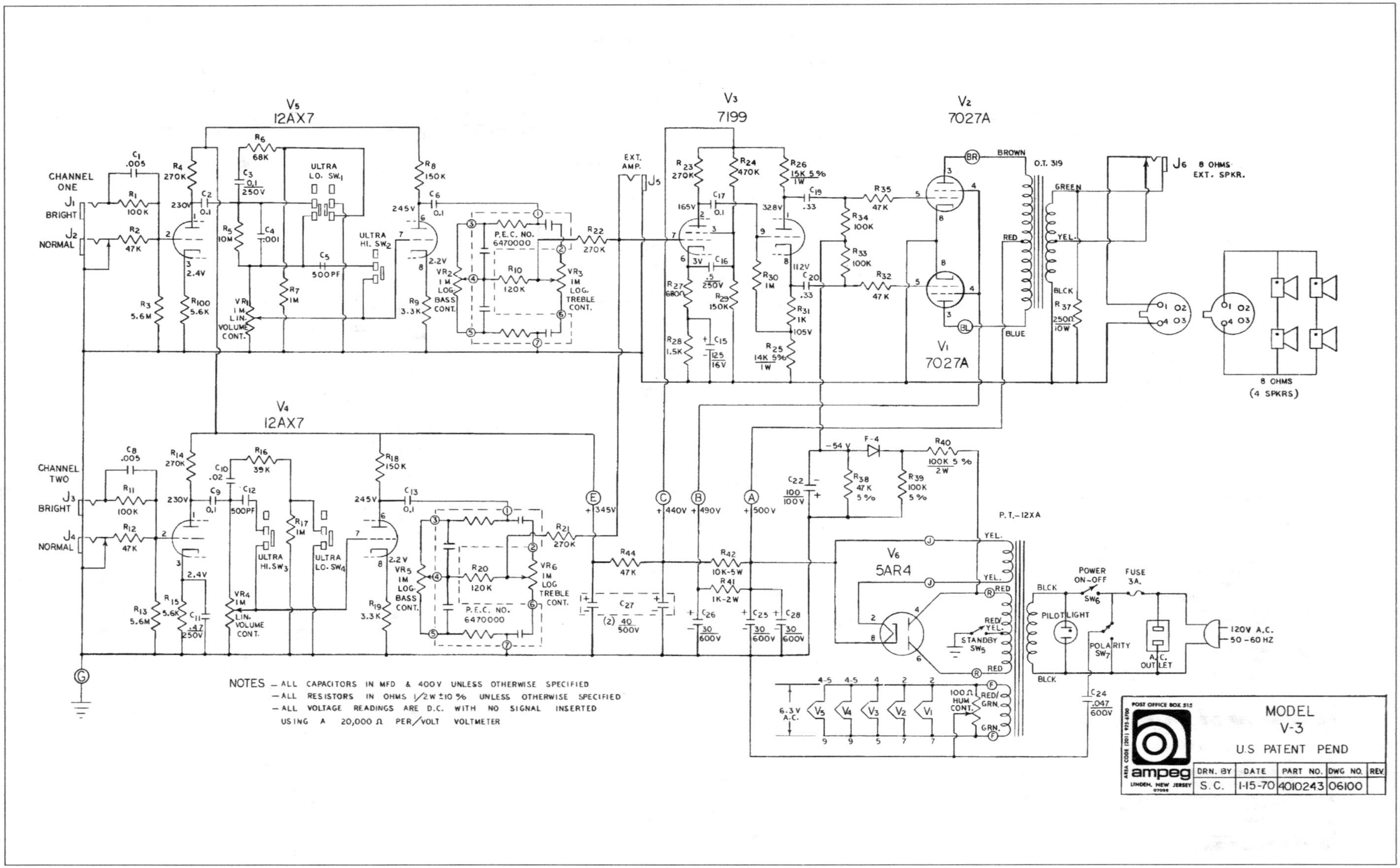
V5 12AX7
V3 7199
V2 7027A
V1 7027A
V4 12AX7
V6 5AR4
CHANNEL ONE
J1 BRIGHT
J2 NORMAL
CHANNEL TWO
J3 BRIGHT
J4 NORMAL
EXT. AMP. J5
J6 8 OHMS EXT. SPKR.
O.T. 319
P.T.-12XA
8 OHMS (4 SPKRS)
POWER ON-OFF SW6
FUSE 3A.
PILOT LIGHT
POLARITY SW7
A.C. OUTLET
120V A.C. 50-60 HZ
STANDBY SW5
100Ω HUM CONT.
6.3V A.C.
ULTRA LO. SW1
ULTRA HI. SW2
ULTRA HI. SW3
ULTRA LO. SW4
VR1 1M LIN. VOLUME CONT.
VR2 1M LOG. BASS CONT.
VR3 1M LOG. TREBLE CONT.
VR4 1M LIN. VOLUME CONT.
VR5 1M LOG. BASS CONT.
VR6 1M LOG TREBLE CONT.
P.E.C. NO. 6470000
NOTES – ALL CAPACITORS IN MFD & 400V UNLESS OTHERWISE SPECIFIED
– ALL RESISTORS IN OHMS 1/2W ±10% UNLESS OTHERWISE SPECIFIED
– ALL VOLTAGE READINGS ARE D.C. WITH NO SIGNAL INSERTED
USING A 20,000 Ω PER/VOLT VOLTMETER
MODEL V-3
U.S PATENT PEND
POST OFFICE BOX 315
ampeg
LINDEN, NEW JERSEY 07036
AREA CODE (201) 925-6700
DRN. BY S.C.
DATE 1-15-70
PART NO. 4010243
DWG NO. 06100
REV.

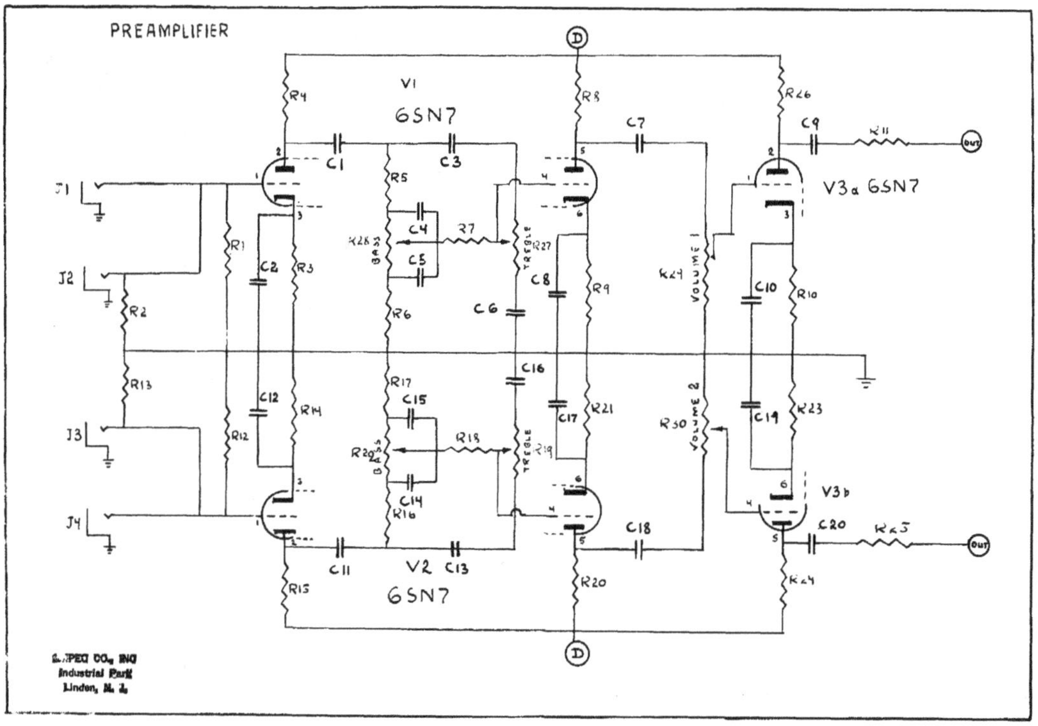
PREAMPLIFIER
V1
6SN7
V2
6SN7
V3a 6SN7
V3b
BASS
TREBLE
VOLUME 1
VOLUME 2
J1
J2
J3
J4
OUT
Industrial Park
Linden, N. J.

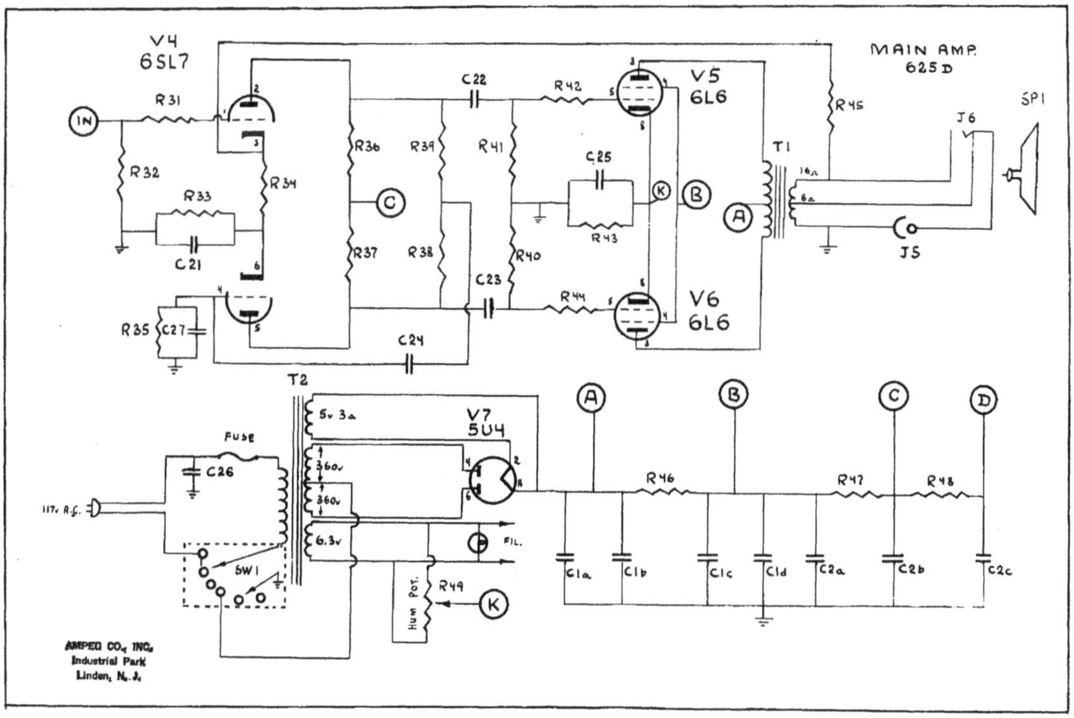
V4
6SL7
V5
6L6
V6
6L6
MAIN AMP.
625D
IN
J6
J5
SP1
T1
T2
V7
5U4
FUSE
117v A.C.
5v 3a
360v
360v
6.3v
SW1
FIL.
HUM POT.
AMPEG CO., INC.
Industrial Park
Linden, N. J.

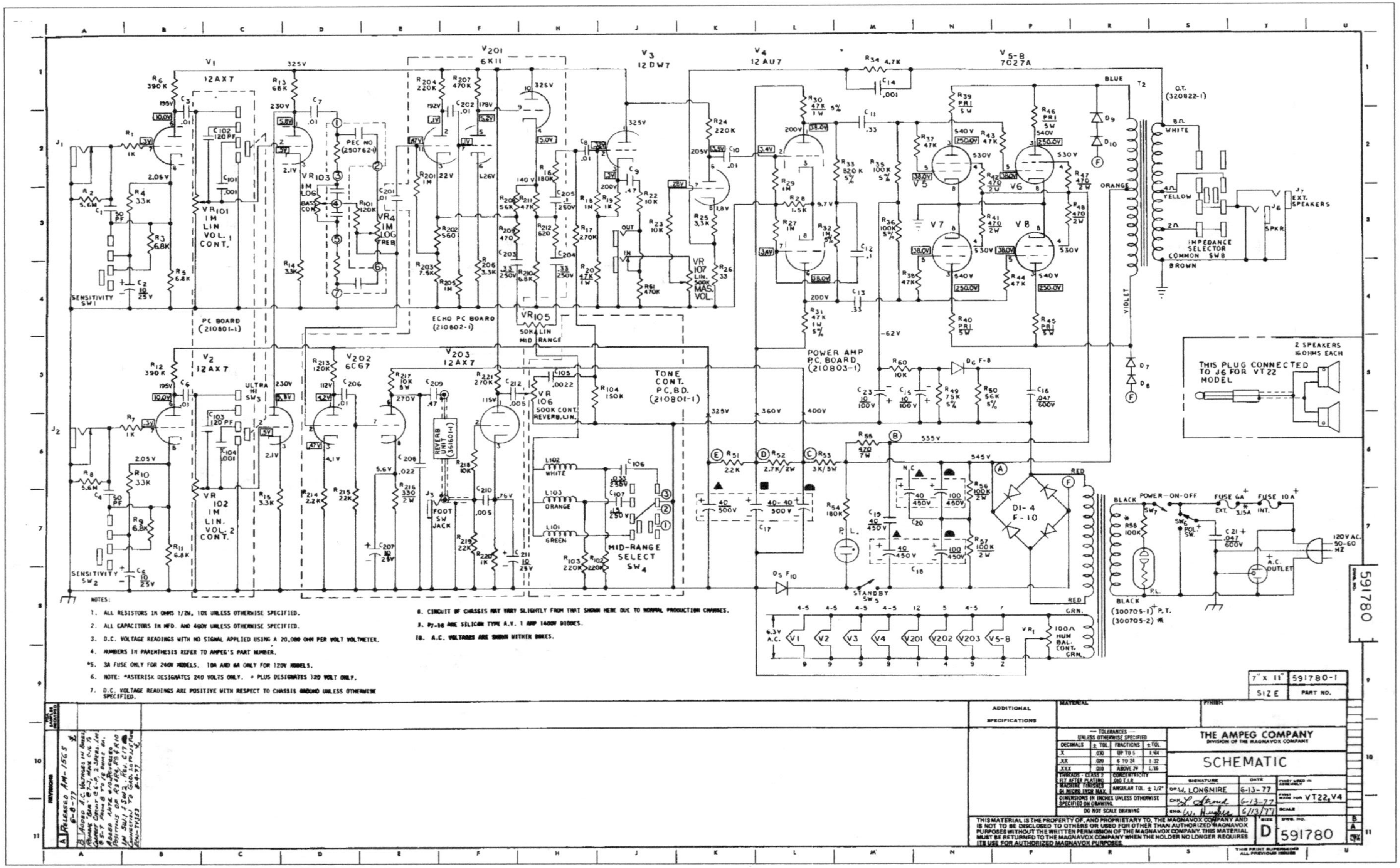

NOTES:
1. ALL RESISTORS IN OHMS 1/2W, 10% UNLESS OTHERWISE SPECIFIED.
2. ALL CAPACITORS IN MFD. AND 400V UNLESS OTHERWISE SPECIFIED.
3. D.C. VOLTAGE READINGS WITH NO SIGNAL APPLIED USING A 20,000 OHM PER VOLT VOLTMETER.
4. NUMBERS IN PARENTHESIS REFER TO AMPEG'S PART NUMBER.
*5. 3A FUSE ONLY FOR 240V MODELS. 10A AND 6A ONLY FOR 120V MODELS.
6. NOTE: *ASTERISK DESIGNATES 240 VOLTS ONLY. + PLUS DESIGNATES 120 VOLT ONLY.
7. D.C. VOLTAGE READINGS ARE POSITIVE WITH RESPECT TO CHASSIS GROUND UNLESS OTHERWISE SPECIFIED.
8. CIRCUIT OF CHASSIS MAY VARY SLIGHTLY FROM THAT SHOWN HERE DUE TO NORMAL PRODUCTION CHANGES.
9. D7-10 ARE SILICON TYPE A.V. 1 AMP 1400V DIODES.
10. A.C. VOLTAGES ARE SHOWN WITHIN BOXES.
POWER AMP P.C. BOARD (210803-1)
TONE CONT. P.C. BD. (210801-1)
ECHO PC BOARD (210802-1)
PC BOARD (210801-1)
THIS PLUG CONNECTED TO J6 FOR VT22 MODEL
2 SPEAKERS 16 OHMS EACH
IMPEDANCE SELECTOR COMMON SW8
MID-RANGE SELECT SW4
THE AMPEG COMPANY
DIVISION OF THE MAGNAVOX COMPANY
SCHEMATIC
DR. W. LONGMIRE 6-13-77
FIRST MADE FOR VT22, V4
SIZE D
DWG. NO. 591780
7" x 11" 591780-1
PART NO.
DO NOT SCALE DRAWING
THIS MATERIAL IS THE PROPERTY OF, AND PROPRIETARY TO, THE MAGNAVOX COMPANY AND IS NOT TO BE DISCLOSED TO OTHERS OR USED FOR OTHER THAN AUTHORIZED MAGNAVOX PURPOSES WITHOUT THE WRITTEN PERMISSION OF THE MAGNAVOX COMPANY. THIS MATERIAL MUST BE RETURNED TO THE MAGNAVOX COMPANY WHEN THE HOLDER NO LONGER REQUIRES ITS USE FOR AUTHORIZED MAGNAVOX PURPOSES.
THIS PRINT SUPERSEDES ALL PREVIOUS ISSUES

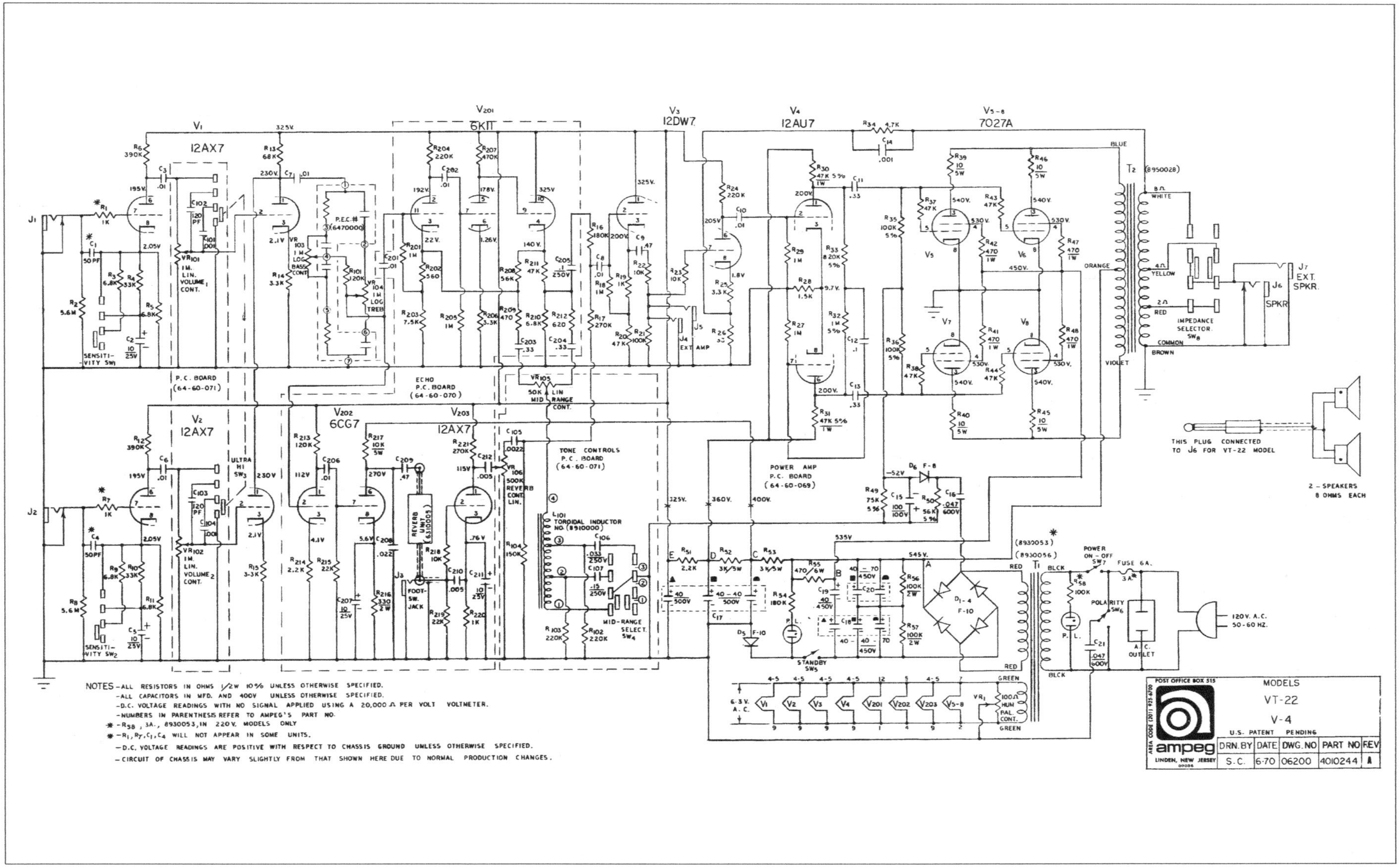
V1 12AX7
V2 12AX7
V201 6K11
V202 6CG7
V203 12AX7
V3 12DW7
V4 12AU7
V5-8 7027A
J1
J2
SENSITIVITY SW1
SENSITIVITY SW2
VR101 1M LIN. VOLUME 1 CONT.
VR102 1M LIN. VOLUME 2 CONT.
ULTRA HI SW3
P.C. BOARD (64-60-071)
ECHO P.C. BOARD (64-60-070)
TONE CONTROLS P.C. BOARD (64-60-071)
POWER AMP P.C. BOARD (64-60-069)
REVERB UNIT (6310005)
FOOT-SW. JACK
L101 TOROIDAL INDUCTOR NO. (8910000)
MID-RANGE SELECT SW4
STANDBY SW5
POLARITY SW6
POWER ON-OFF SW7
FUSE 6A.
IMPEDANCE SELECTOR SW8
T1
T2 (8950028)
J6 SPKR
J7 EXT. SPKR.
THIS PLUG CONNECTED TO J6 FOR VT-22 MODEL
2 - SPEAKERS 8 OHMS EACH
120V. A.C. 50-60 HZ.
A.C. OUTLET
NOTES - ALL RESISTORS IN OHMS 1/2W 10% UNLESS OTHERWISE SPECIFIED.
- ALL CAPACITORS IN MFD. AND 400V UNLESS OTHERWISE SPECIFIED.
- D.C. VOLTAGE READINGS WITH NO SIGNAL APPLIED USING A 20,000 Ω PER VOLT VOLTMETER.
- NUMBERS IN PARENTHESIS REFER TO AMPEG'S PART NO.
✱ - R58, 3A., 8930053, IN 220V. MODELS ONLY
✱ - R1, R7, C1, C4 WILL NOT APPEAR IN SOME UNITS.
- D.C. VOLTAGE READINGS ARE POSITIVE WITH RESPECT TO CHASSIS GROUND UNLESS OTHERWISE SPECIFIED.
- CIRCUIT OF CHASSIS MAY VARY SLIGHTLY FROM THAT SHOWN HERE DUE TO NORMAL PRODUCTION CHANGES.
POST OFFICE BOX 515
AREA CODE (201) 925-6700
ampeg
LINDEN, NEW JERSEY 07036
MODELS VT-22 V-4
U.S. PATENT PENDING
DRN. BY | DATE | DWG. NO | PART NO | REV
S.C. | 6-70 | 06200 | 4010244 | A

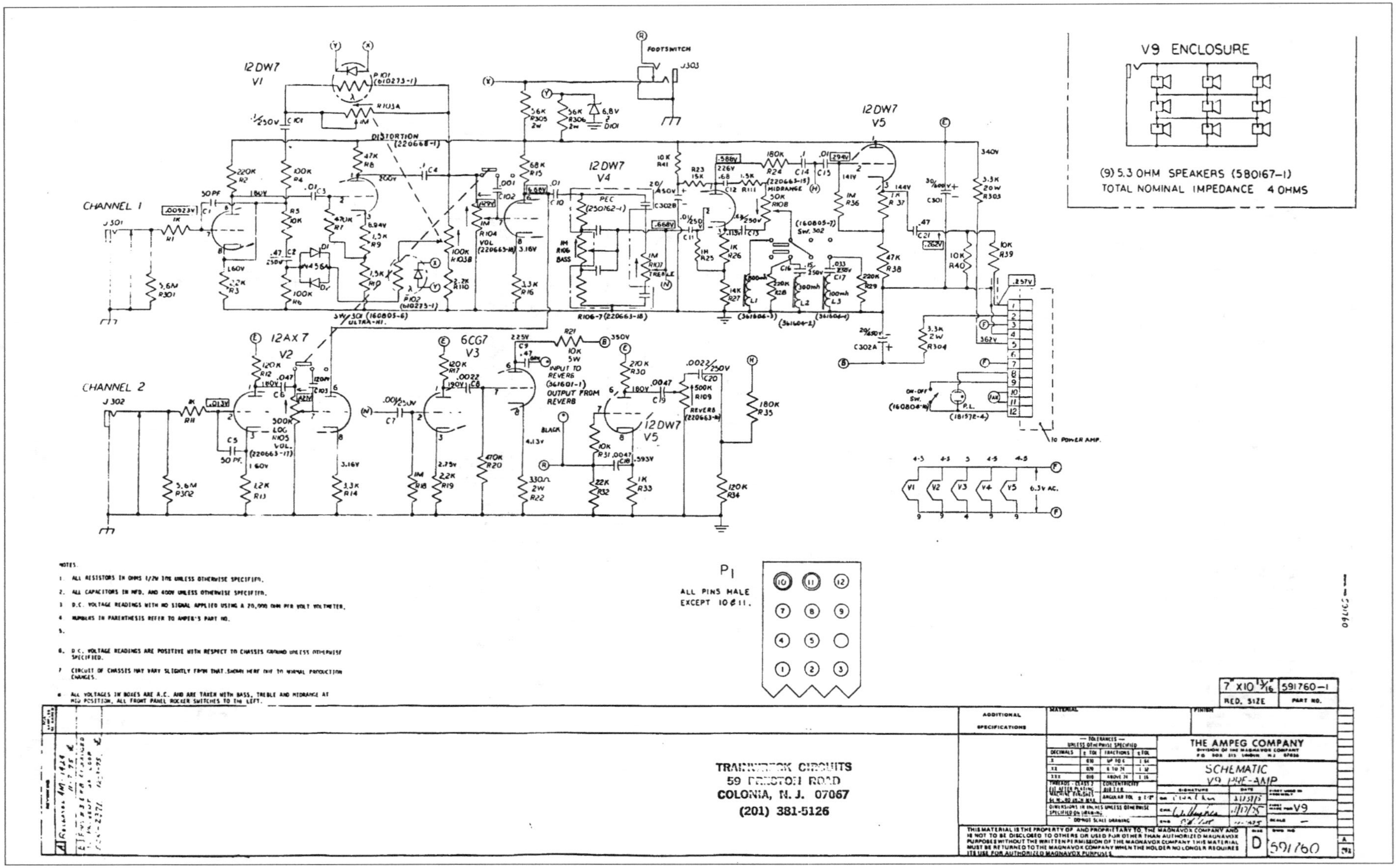

V9 ENCLOSURE
(9) 5.3 OHM SPEAKERS (580167-1)
TOTAL NOMINAL IMPEDANCE 4 OHMS
CHANNEL 1
CHANNEL 2
12DW7 V1
12AX7 V2
6CG7 V3
12DW7 V4
12DW7 V5
FOOTSWITCH
DISTORTION
BASS
TREBLE
MIDRANGE
INPUT TO REVERB
OUTPUT FROM REVERB
TO POWER AMP.
P1
ALL PINS MALE EXCEPT 10 & 11.
NOTES
1. ALL RESISTORS IN OHMS 1/2W 10% UNLESS OTHERWISE SPECIFIED.
2. ALL CAPACITORS IN MFD. AND 400V UNLESS OTHERWISE SPECIFIED.
3. D.C. VOLTAGE READINGS WITH NO SIGNAL APPLIED USING A 20,000 OHM PER VOLT VOLTMETER.
4. NUMBERS IN PARENTHESIS REFER TO AMPEG'S PART NO.
6. D.C. VOLTAGE READINGS ARE POSITIVE WITH RESPECT TO CHASSIS GROUND UNLESS OTHERWISE SPECIFIED.
7. CIRCUIT OF CHASSIS MAY VARY SLIGHTLY FROM THAT SHOWN HERE DUE TO NORMAL PRODUCTION CHANGES.
8. ALL VOLTAGES IN BOXES ARE A.C. AND ARE TAKEN WITH BASS, TREBLE AND MIDRANGE AT MID POSITION, ALL FRONT PANEL ROCKER SWITCHES TO THE LEFT.
TRAINWRECK CIRCUITS
59 PRESTON ROAD
COLONIA, N.J. 07067
(201) 381-5126
THE AMPEG COMPANY
SCHEMATIC
V9 PRE-AMP
591760-1
591760

CHB-20A

Bogen

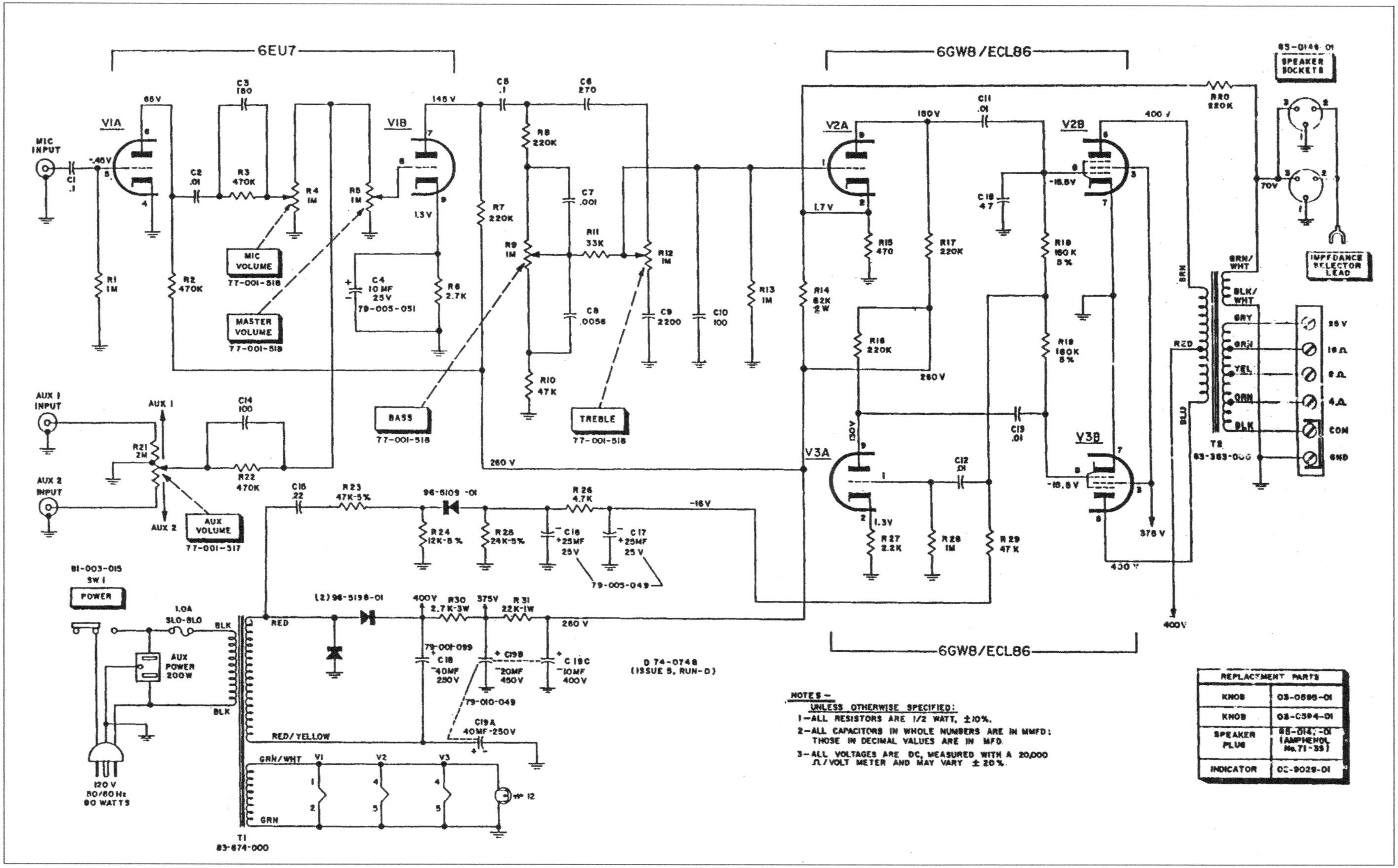

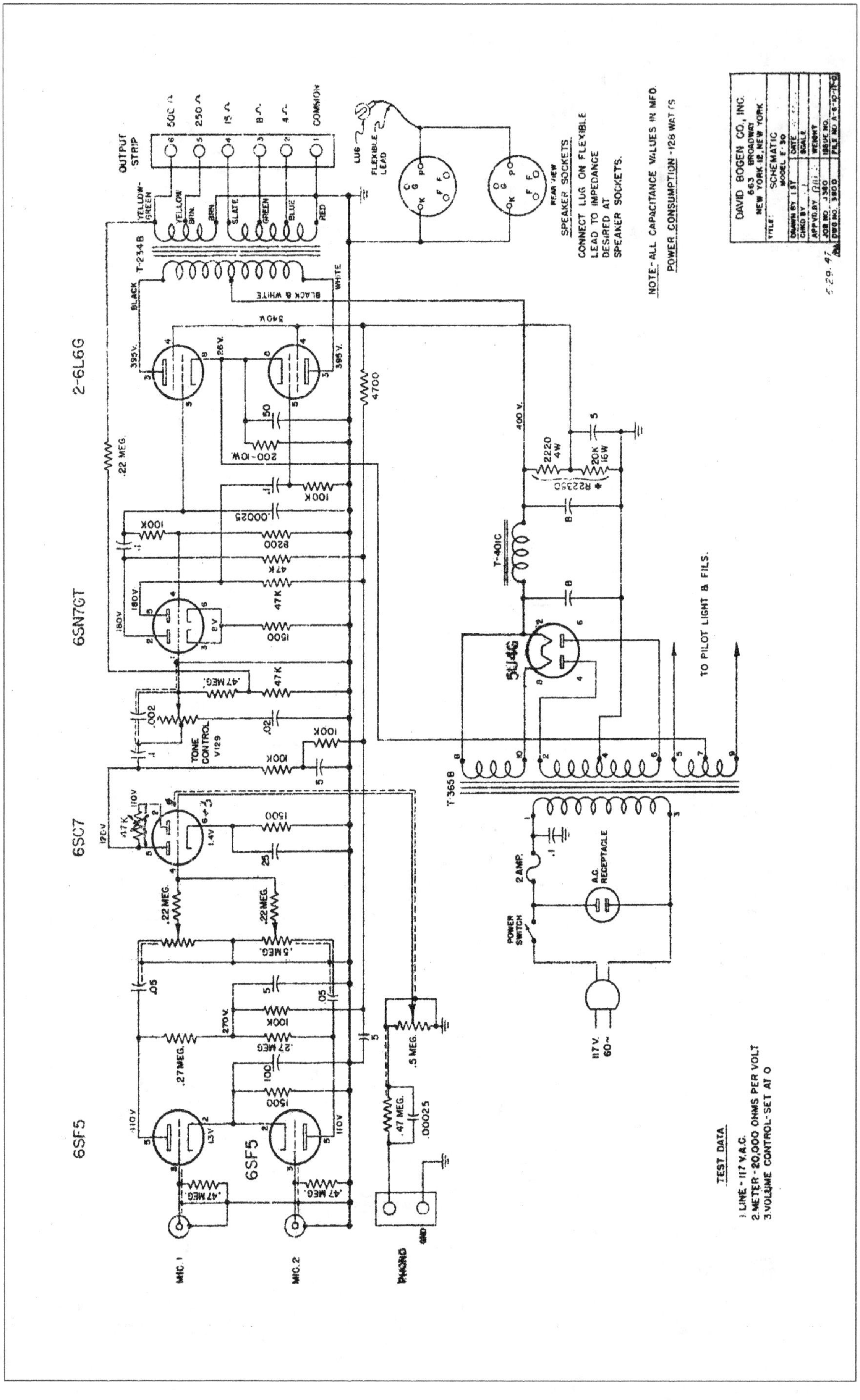
6SF5
6SF5
6SC7
6SN7GT
2-6L6G
5U4G
MIC.1
MIC.2
PHONO
GND
TONE CONTROL V129
OUTPUT STRIP
500 Ω
250 Ω
15 Ω
8 Ω
4 Ω
COMMON
T-234B
T-401C
T 3658
LUG
FLEXIBLE LEAD
REAR VIEW
SPEAKER SOCKETS
CONNECT LUG ON FLEXIBLE LEAD TO IMPEDANCE DESIRED AT SPEAKER SOCKETS.
NOTE - ALL CAPACITANCE VALUES IN MFD.
POWER CONSUMPTION - 128 WATTS
TO PILOT LIGHT & FILS.
POWER SWITCH
2 AMP.
A.C. RECEPTACLE
117 V. 60~
TEST DATA
1. LINE - 117 V.A.C.
2. METER - 20,000 OHMS PER VOLT
3. VOLUME CONTROL - SET AT O
DAVID BOGEN CO., INC.
663 BROADWAY
NEW YORK 12, NEW YORK
SCHEMATIC
MODEL E-30

40 WATTS Hi-Fi AMPLIFIER

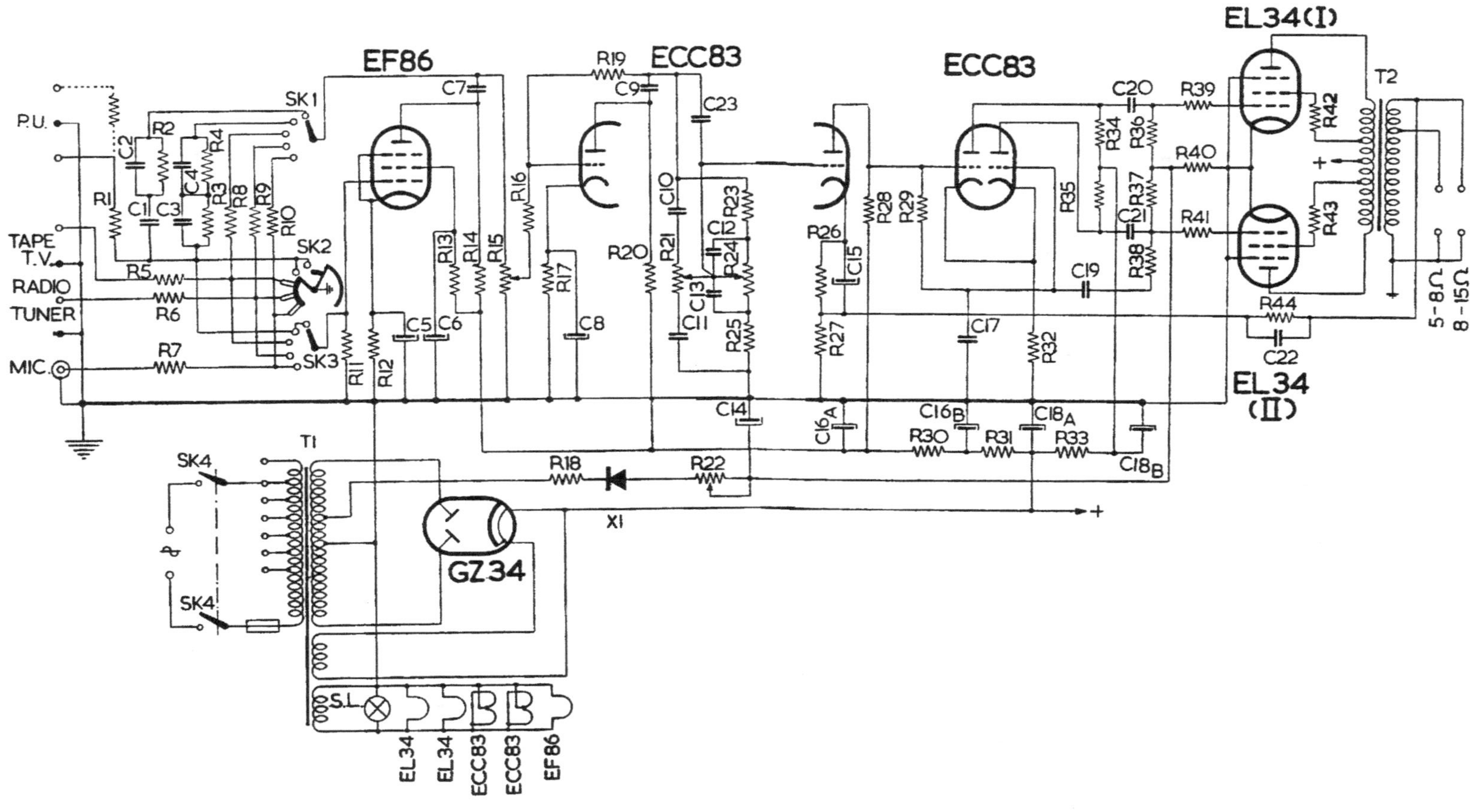

CONTENTS AND CIRCUIT
DIAGRAM WITHOUT ANY
GUARANTEE TO PATENT RIGHTS

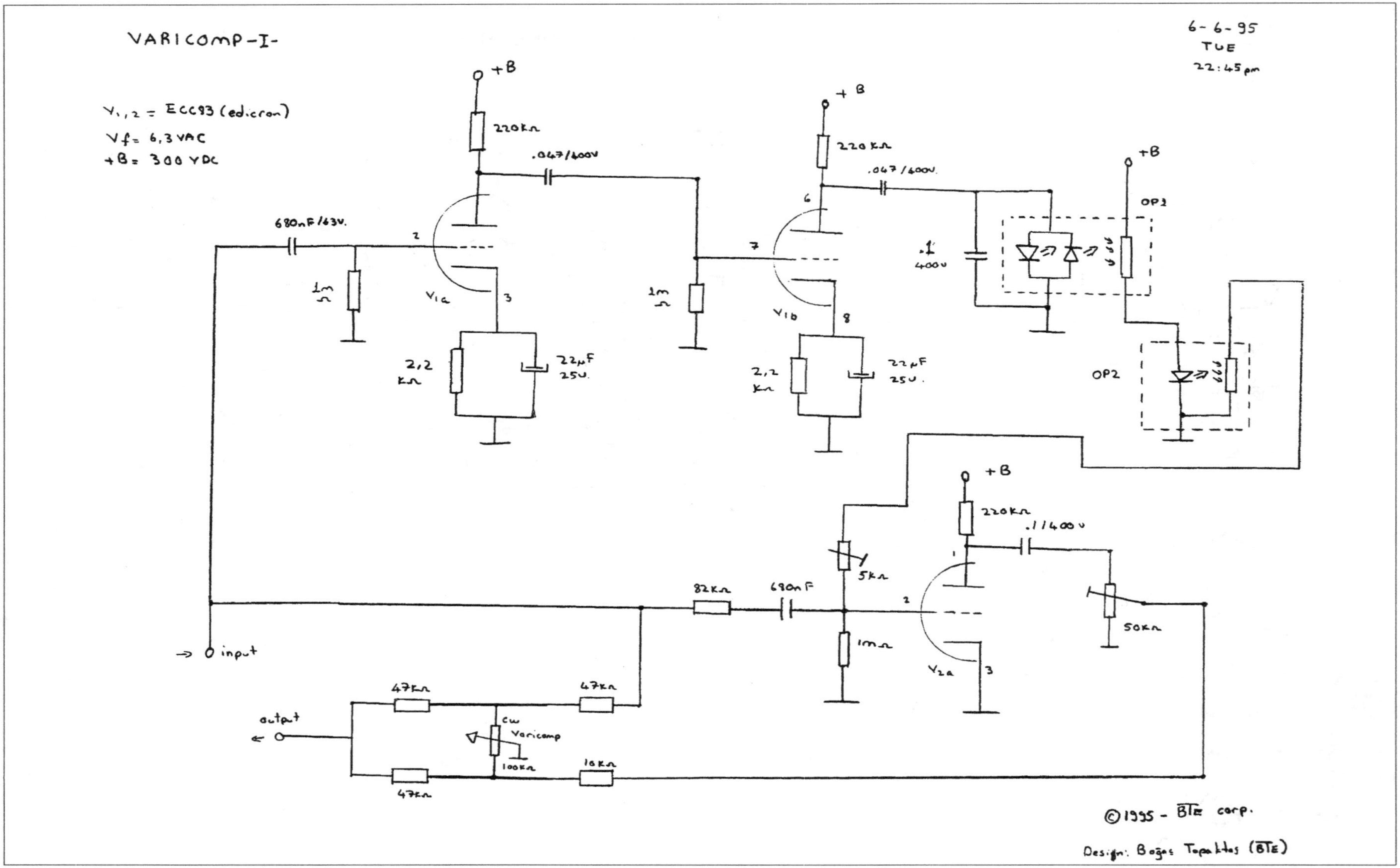
VARICOMP-I-
6-6-95
TUE
22:45pm
V1,2 = ECC83 (edicron)
Vf = 6,3 VAC
+B = 300 VDC
+B
220kΩ
.047/400V
680nF/63V.
1mΩ
V1a
2,2 kΩ
22µF 25V.
220kΩ
.047/400V.
V1b
2,2 kΩ
22µF 25V.
.1 400v
OP1
OP2
220kΩ
.1/400v
5kΩ
1mΩ
V2a
50kΩ
82kΩ
680nF
input
output
47kΩ
47kΩ
47kΩ
10kΩ
cw
Varicomp
100kΩ
©1995 - BTE corp.
Design: Bogos Topaktas (BTE)

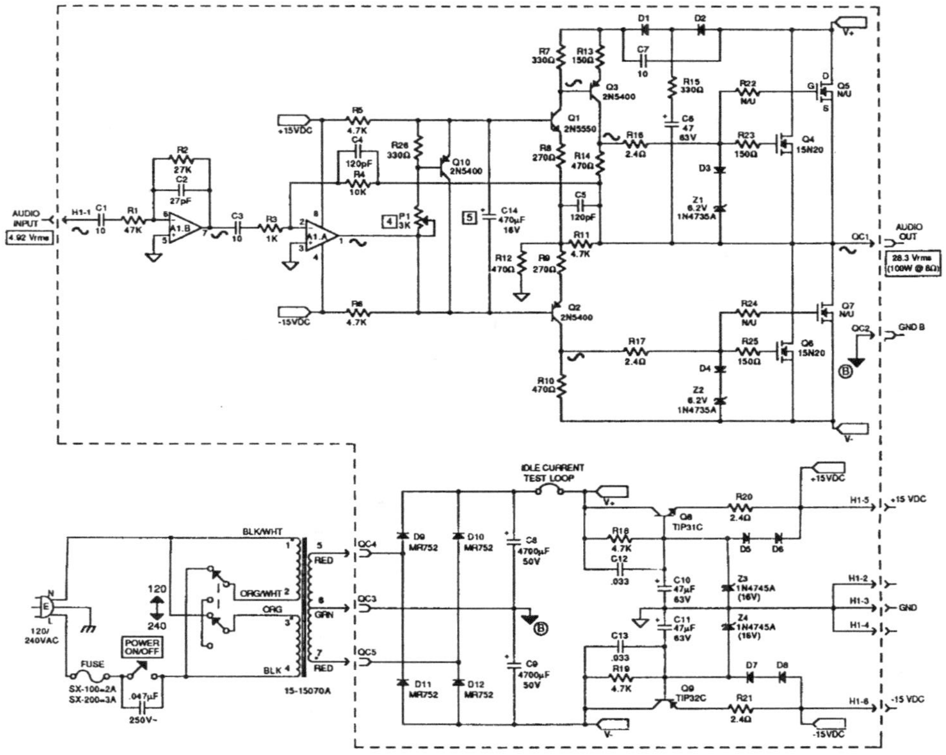

5. ADDED AS SECONDARY.
4. ADJUST P1 FOR 200mA AT IDLE CURRENT TEST LOOP.
3. A1 IS A 4558 TYPE OP AMP.
2. ALL CAPACITORS IN MICROFARADS (EXCEPT AS NOTED).
1. ALL DIODES ARE 1N4003 (EXPECT AS NOTED).
NOTES: UNLESS OTHERWISE SPECIFIED.

LAST USED LIST			
SYM			
A	1		
C	14		
D	12		
H	1		
P	1		
Q	10		
R	26		
Z	4		

CARVIN
1155 INDUSTRIAL AVE.
ESCONDIDO, CA 92029
(619) 747-1710
FAX 747-9065

FET100
POWER MODULE
(USED IN SX-100, SX-200)

DRAWN BY:	C. McEUEN	DATE:	10JAN91
APPROVED:		DATE:	
PCB NO:	30-FET100-8	REV:	A
ASSY. NO:	80-01018	REV:	C
REVISED BY:	C. McEUEN	DATE:	14JAN91

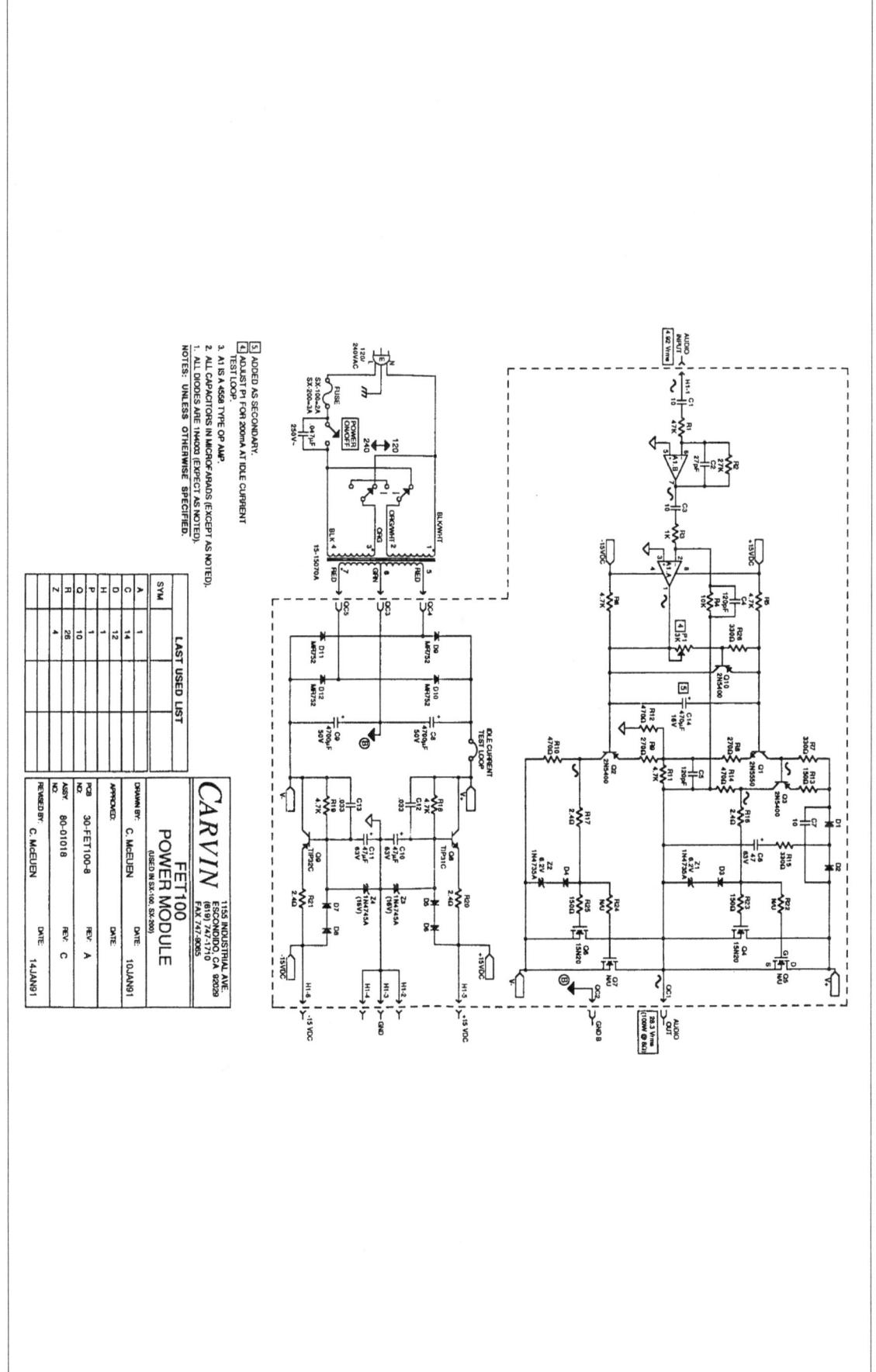

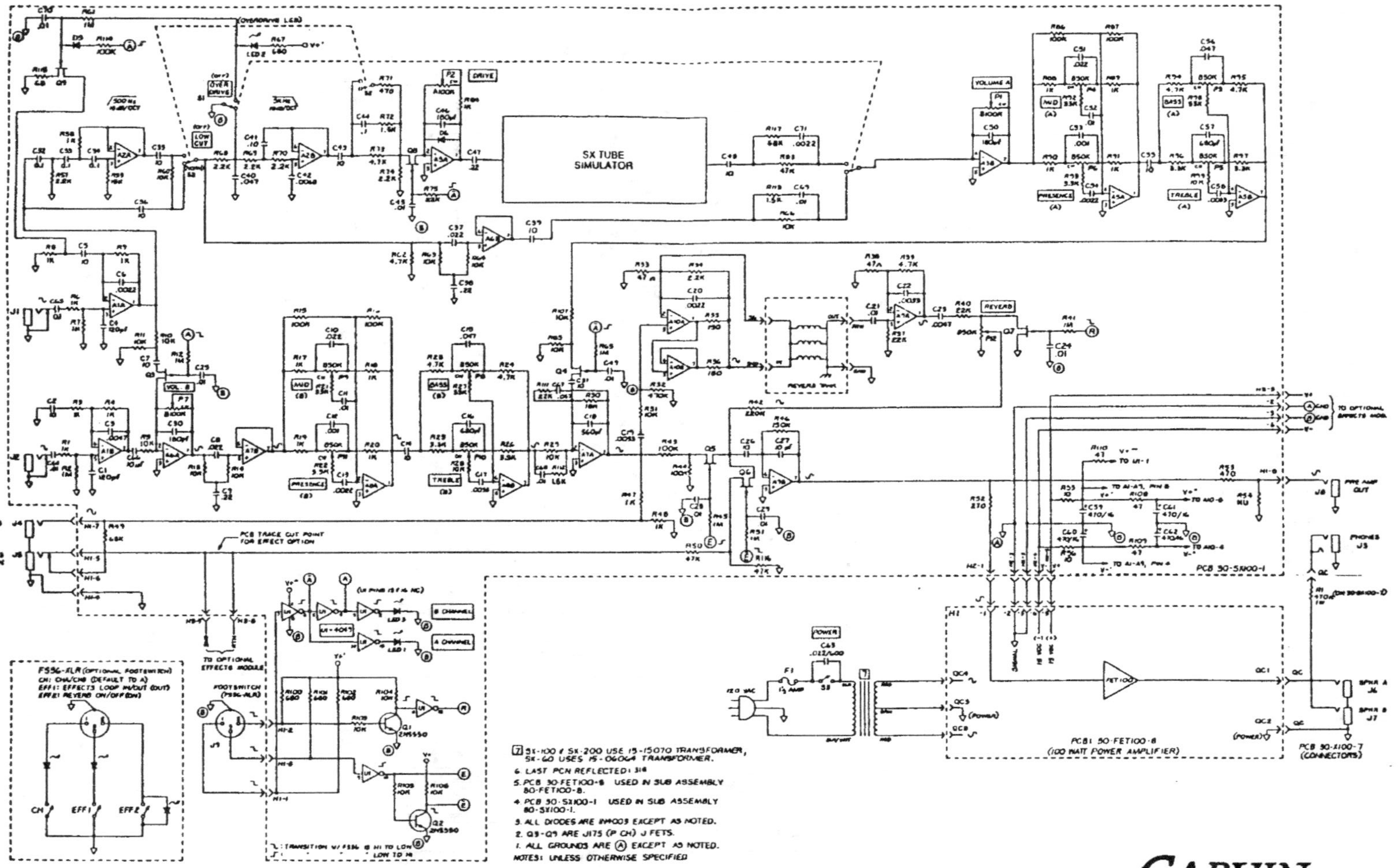

CARVIN
SCHEMATIC
PRE-AMP
Section of SX Series

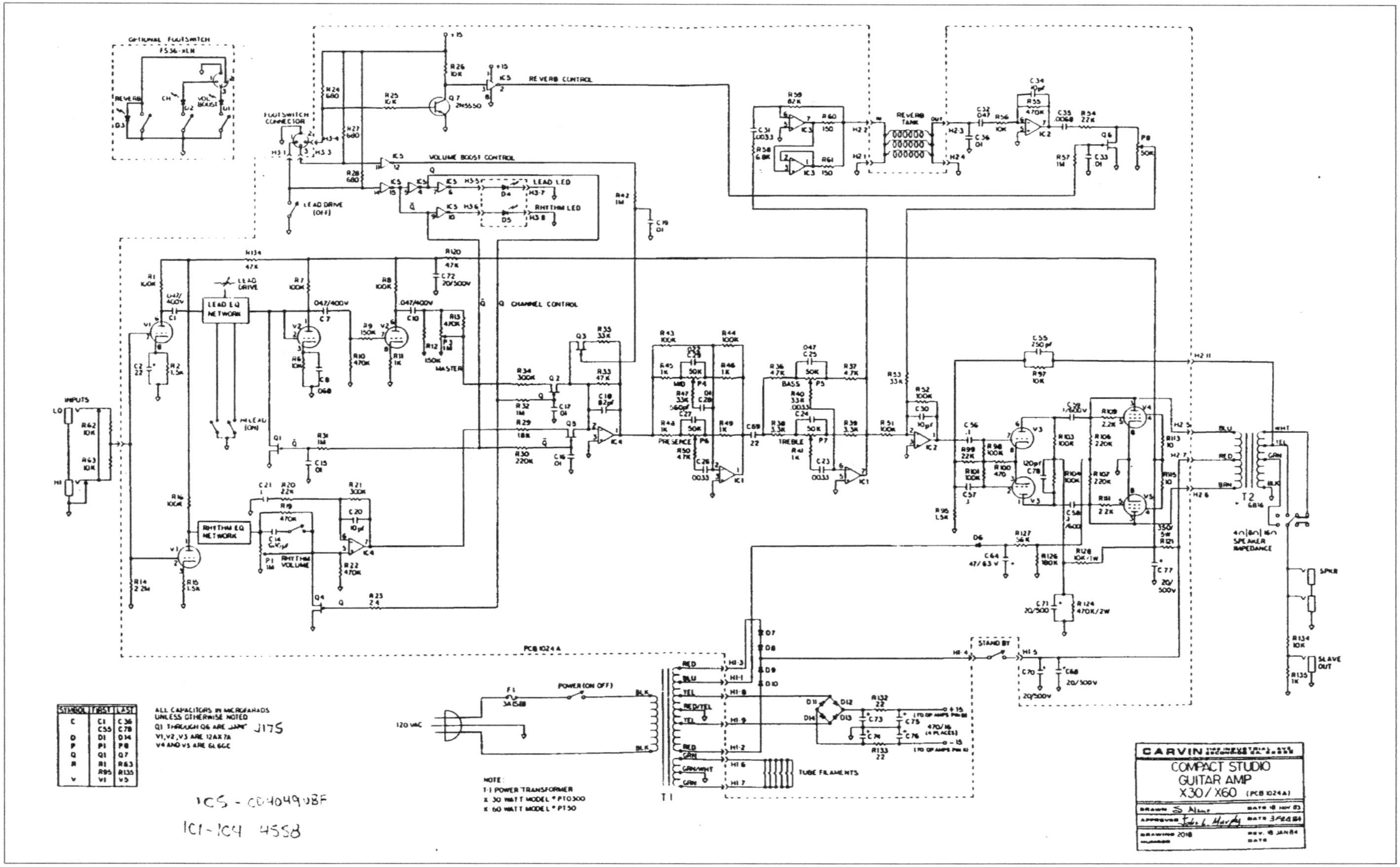
OPTIONAL FOOTSWITCH
FS36-XLN
REVERB
CH
VOL BOOST
FOOTSWITCH CONNECTOR
REVERB CONTROL
VOLUME BOOST CONTROL
LEAD LED
RHYTHM LED
LEAD DRIVE (OFF)
REVERB TANK
LEAD DRIVE
LEAD EQ NETWORK
MASTER
CHANNEL CONTROL
INPUTS
HI LEAD (ON)
RHYTHM EQ NETWORK
RHYTHM VOLUME
MID
PRESENCE
BASS
TREBLE
PCB 1024 A
STAND BY
SPEAKER IMPEDANCE
SPKR
SLAVE OUT
POWER (ON OFF)
120 VAC
TUBE FILAMENTS
SYMBOL | FIRST | LAST
C | C1 C55 | C36 C78
D | D1 | D14
P | P1 | P8
Q | Q1 | Q7
R | R1 R95 | R63 R135
V | V1 | V5
ALL CAPACITORS IN MICROFARADS
UNLESS OTHERWISE NOTED
Q1 THROUGH Q6 ARE J175
V1, V2, V3 ARE 12AX7A
V4 AND V5 ARE 6L6GC
IC5 - CD4049UBF
IC1-IC4 4558
NOTE:
T-1 POWER TRANSFORMER
X-30 WATT MODEL * PT0300
X-60 WATT MODEL * PT50
CARVIN
COMPACT STUDIO
GUITAR AMP
X30 / X60 (PCB 1024A)
DRAWN
DATE 18 NOV 83
APPROVED
DATE 3 FEB 84
DRAWING NUMBER 2018
REV. DATE 18 JAN 84

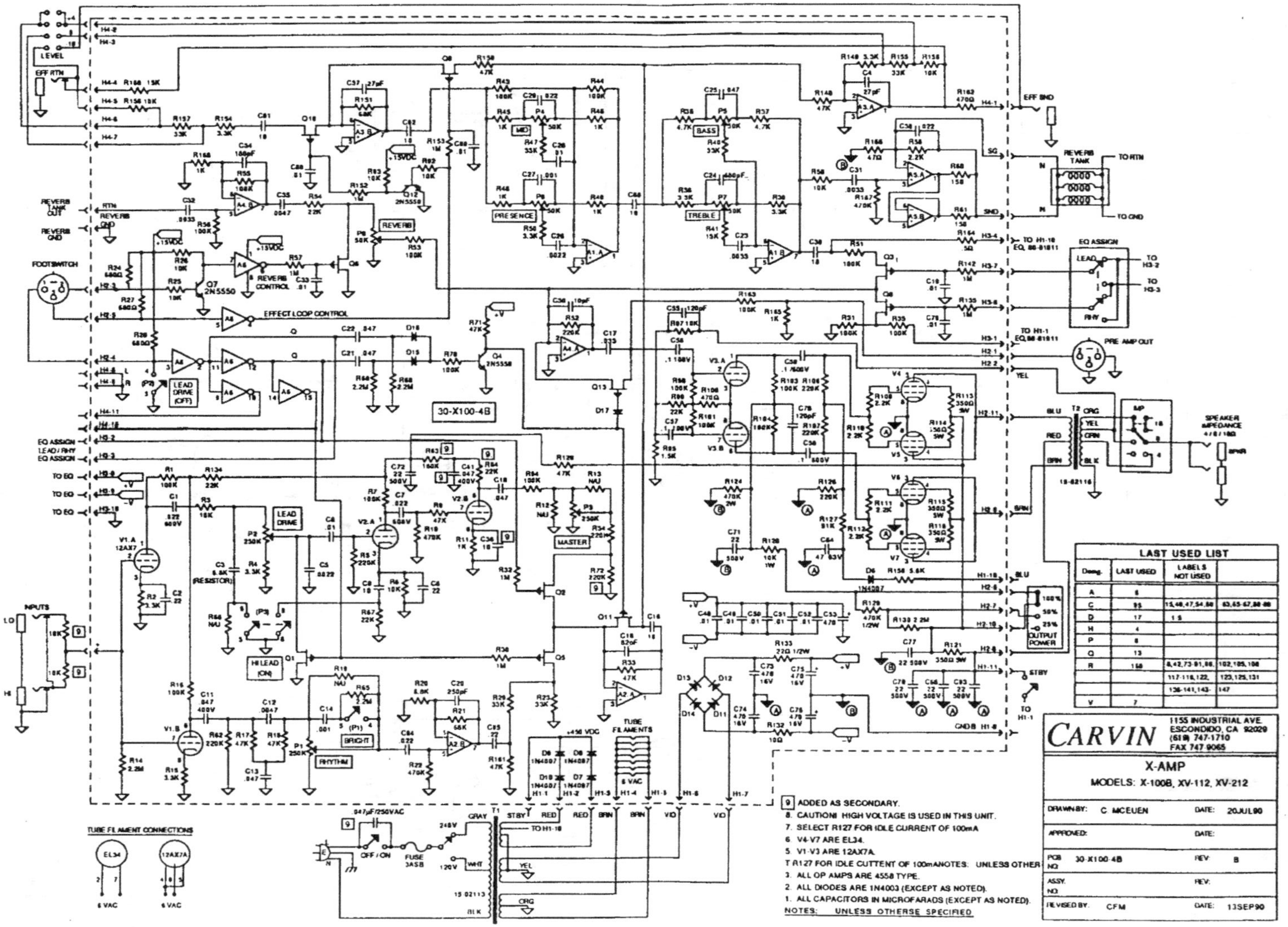
LAST USED LIST
Desig.	LAST USED	LABELS NOT USED	
A	6		
C	95	15,46,47,54,86	63,65-67,86-88
D	17	15	
H	4		
P	8		
Q	13		
R	158	8,42,73-81,88,	102,105,106
		117-118,122,	123,125,131
		136-141,143-	147
V	7		
CARVIN
1155 INDUSTRIAL AVE.
ESCONDIDO, CA 92029
(619) 747-1710
FAX 747 9065
X-AMP
MODELS: X-100B, XV-112, XV-212
DRAWN BY: C MCEUEN
DATE: 20JUL90
APPROVED:
PCB NO. 30-X100-4B
REV: B
ASSY. NO.
REVISED BY: CFM
DATE: 13SEP90
9 ADDED AS SECONDARY.
8. CAUTION! HIGH VOLTAGE IS USED IN THIS UNIT.
7. SELECT R127 FOR IDLE CURRENT OF 100mA
6. V4-V7 ARE EL34.
5. V1-V3 ARE 12AX7A.
T R127 FOR IDLE CUTTENT OF 100mANOTES: UNLESS OTHER
3. ALL OP AMPS ARE 4558 TYPE.
2. ALL DIODES ARE 1N4003 (EXCEPT AS NOTED).
1. ALL CAPACITORS IN MICROFARADS (EXCEPT AS NOTED).
NOTES: UNLESS OTHERSE SPECIFIED
30-X100-4B
TUBE FILAMENT CONNECTIONS
INPUTS
FOOTSWITCH
EFFECT LOOP CONTROL
PRE AMP OUT
SPEAKER IMPEDANCE
EQ ASSIGN
REVERB TANK
TO RTN
TO GND
OUTPUT POWER
TUBE FILAMENTS

REVERB BOX 9100

Danelectro

NOTES:
1. VALUES OF CAPACITORS IN MFD. 2. ALL RESISTORS ARE ½ WATT UNLESS OTHERWISE NOTED. 3. VOLTAGES MEASURED FROM POINTS INDICATED TO CHASSIS WITH 20,000 OHM/VOLT METER.

TUBE LAYOUT

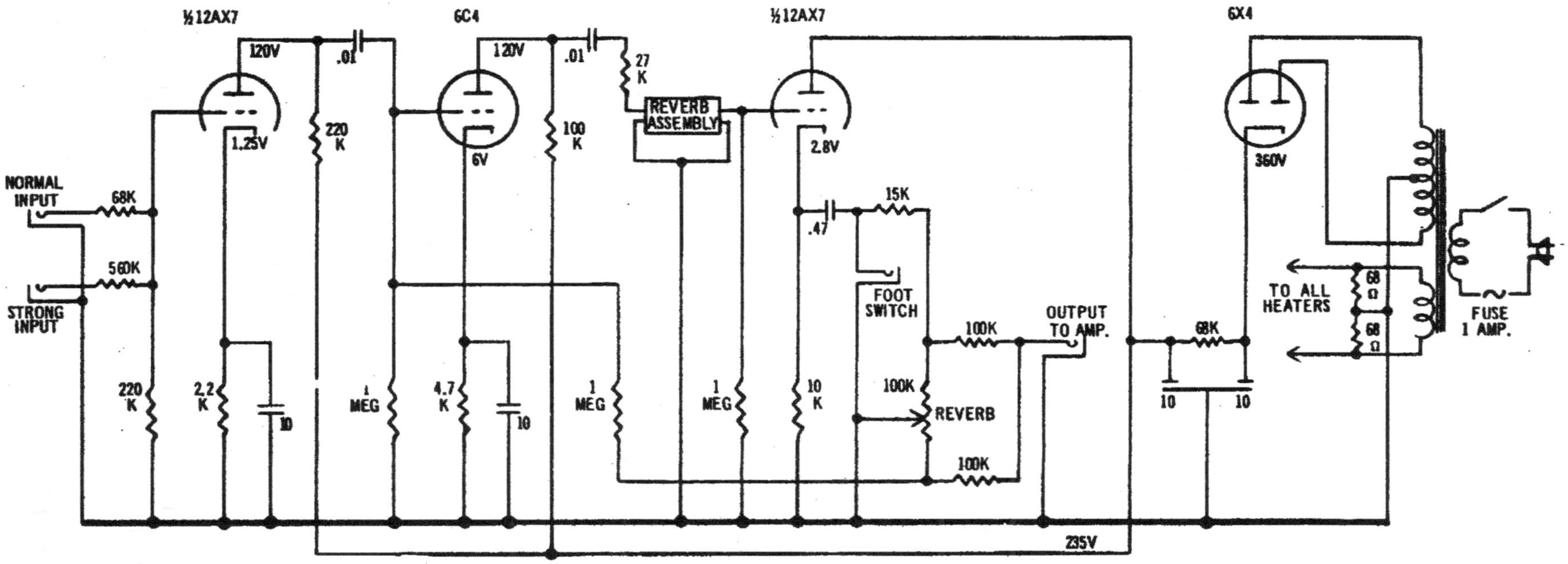

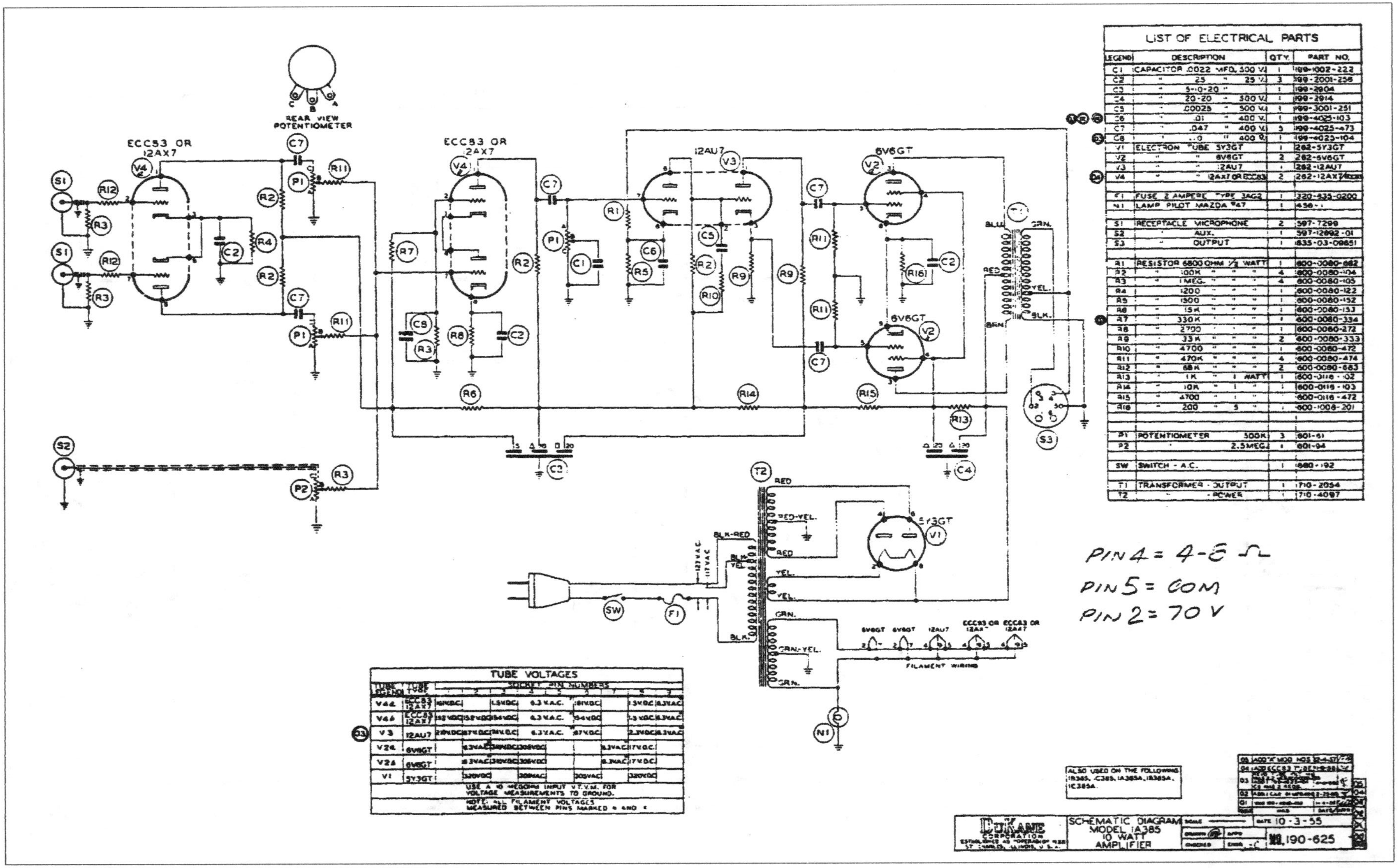

LIST OF ELECTRICAL PARTS

LEGEND	DESCRIPTION	QTY.	PART NO.
C1	CAPACITOR .0022 MFD. 500 V.	1	199-1002-222
C2	" 25 " 25 V.	3	199-2001-256
C3	" 5-10-20 "	1	199-2904
C4	" 20-20 " 500 V.	1	199-2914
C5	" .00025 " 500 V.	1	199-3001-251
C6	" .01 " 400 V.	1	199-4025-103
C7	" .047 " 400 V.	5	199-4025-473
C8	" .10 " 400 V.	1	199-4025-104
V1	ELECTRON TUBE 5Y3GT	1	282-5Y3GT
V2	" " 6V6GT	2	282-6V6GT
V3	" " 12AU7	1	282-12AU7
V4	" " 12AX7 OR ECC83	2	282-12AX7/ECC83
F1	FUSE 2 AMPERE TYPE 3AG2	1	320-835-0200
N1	LAMP PILOT MAZDA #47	1	456-1
S1	RECEPTACLE MICROPHONE	2	597-7299
S2	" AUX.	1	597-12892-01
S3	" OUTPUT	1	835-03-09851
R1	RESISTOR 6800 OHM ½ WATT	1	600-0080-682
R2	" 100K " " "	4	600-0080-104
R3	" 1 MEG. " " "	4	600-0080-105
R4	" 1200 " " "	1	600-0080-122
R5	" 1500 " " "	1	600-0080-152
R6	" 15K " " "	1	600-0080-153
R7	" 330K " " "	1	600-0080-334
R8	" 2700 " " "	1	600-0080-272
R9	" 33K " " "	2	600-0080-333
R10	" 4700 " " "	1	600-0080-472
R11	" 470K " " "	4	600-0080-474
R12	" 68K " " "	2	600-0080-683
R13	" 1K " 1 WATT	1	600-0118-102
R14	" 10K " 1 "	1	600-0118-103
R15	" 4700 " 1 "		600-0118-472
R16	" 200 " 5 "	1	600-1008-201
P1	POTENTIOMETER 500K	3	601-61
P2	" 2.5 MEG.	1	601-94
SW	SWITCH - A.C.	1	680-192
T1	TRANSFORMER - OUTPUT	1	710-2054
T2	" - POWER	1	710-4097

TUBE VOLTAGES

TUBE LEGEND	TUBE TYPE	1	2	3	4	5	6	7	8	9
		SOCKET PIN NUMBERS								
V4a	ECC83 12AX7	161VDC		1.5VDC	6.3 V.A.C.		181VDC		1.5VDC	6.3VAC
V4b	ECC83 12AX7	132VDC	152VDC	154VDC	6.3 V.A.C.		154VDC		1.5VDC	6.3VAC
V3	12AU7	231VDC	87VDC	88VDC	6.3 V.A.C.		87VDC		2.3VDC	6.3VAC
V2a	6V6GT		6.3VAC	310VDC	305VDC			6.3VAC	17VDC	
V2b	6V6GT		6.3VAC	310VDC	305VDC			6.3VAC	17VDC	
V1	5Y3GT		320VDC		300VAC		305VAC		320VDC	

USE A 10 MEGOHM INPUT V.T.V.M. FOR VOLTAGE MEASUREMENTS TO GROUND.

NOTE: ALL FILAMENT VOLTAGES MEASURED BETWEEN PINS MARKED * AND *

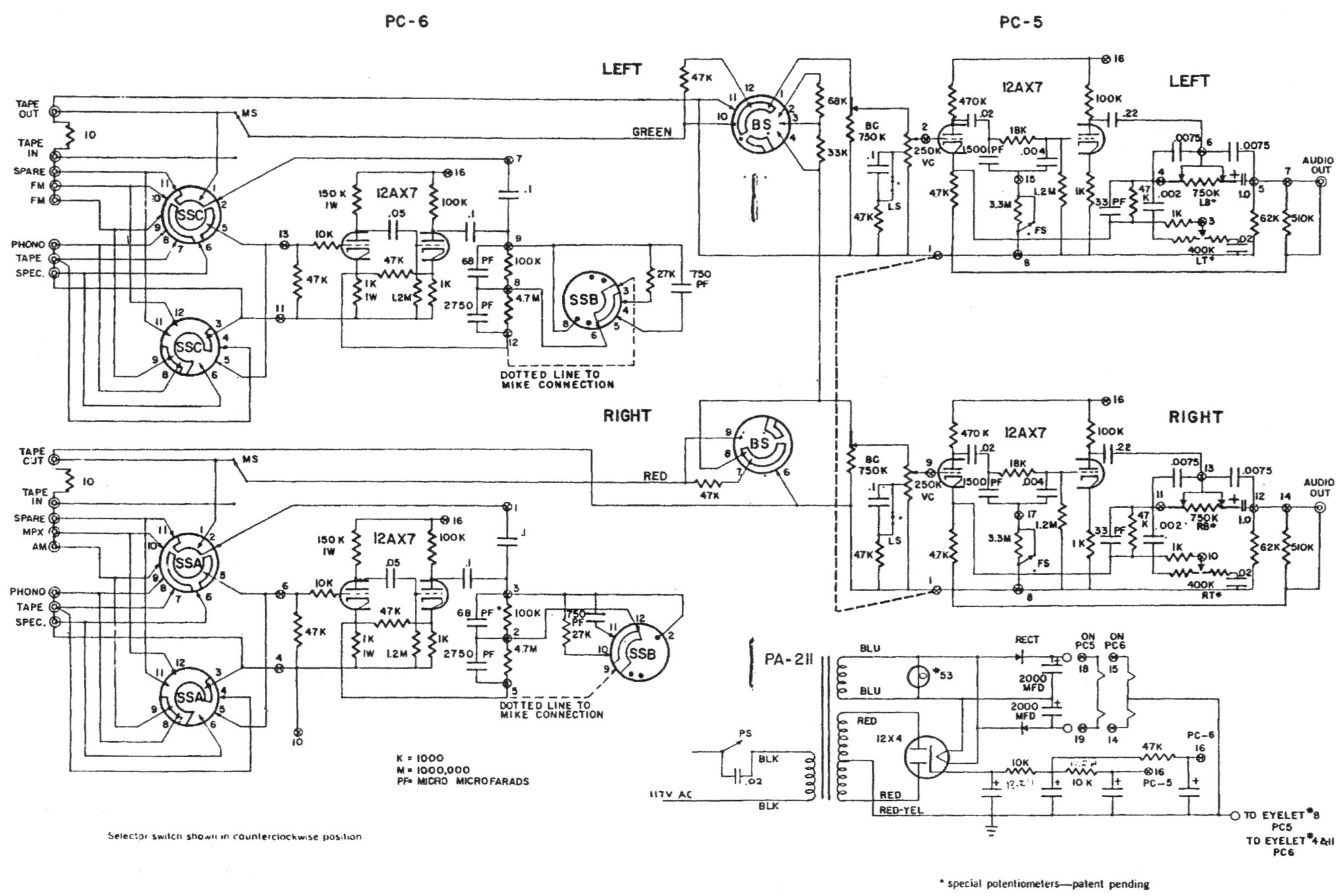

VOLTAGE CHART

Tube pins numbered clockwise viewed from the bottom.

All readings taken from pin to chassis (except #3 and #4 of 12X4) using a vacuum tube voltmeter.

Either tube PC-6

1 115 V DC
2 0
3 .7 V DC
4 0
5 ± 11 V DC
6 135 V DC
7 0
8 .8 V DC
9 ± 5.5 V DC

Either tube PC-5

1 175 V DC
2 0
3 1.45 V DC
4 Less than 1 volt
5 ± 11 V DC
6 200 V DC
7 0
8 1.25 V DC
9 ± 5.5 V DC

12X4

1 335 V AC
2 0
3, 4 10.5 V AC
5 0
6 335 V AC
7 405 V DC

Quad Filter Capacitor

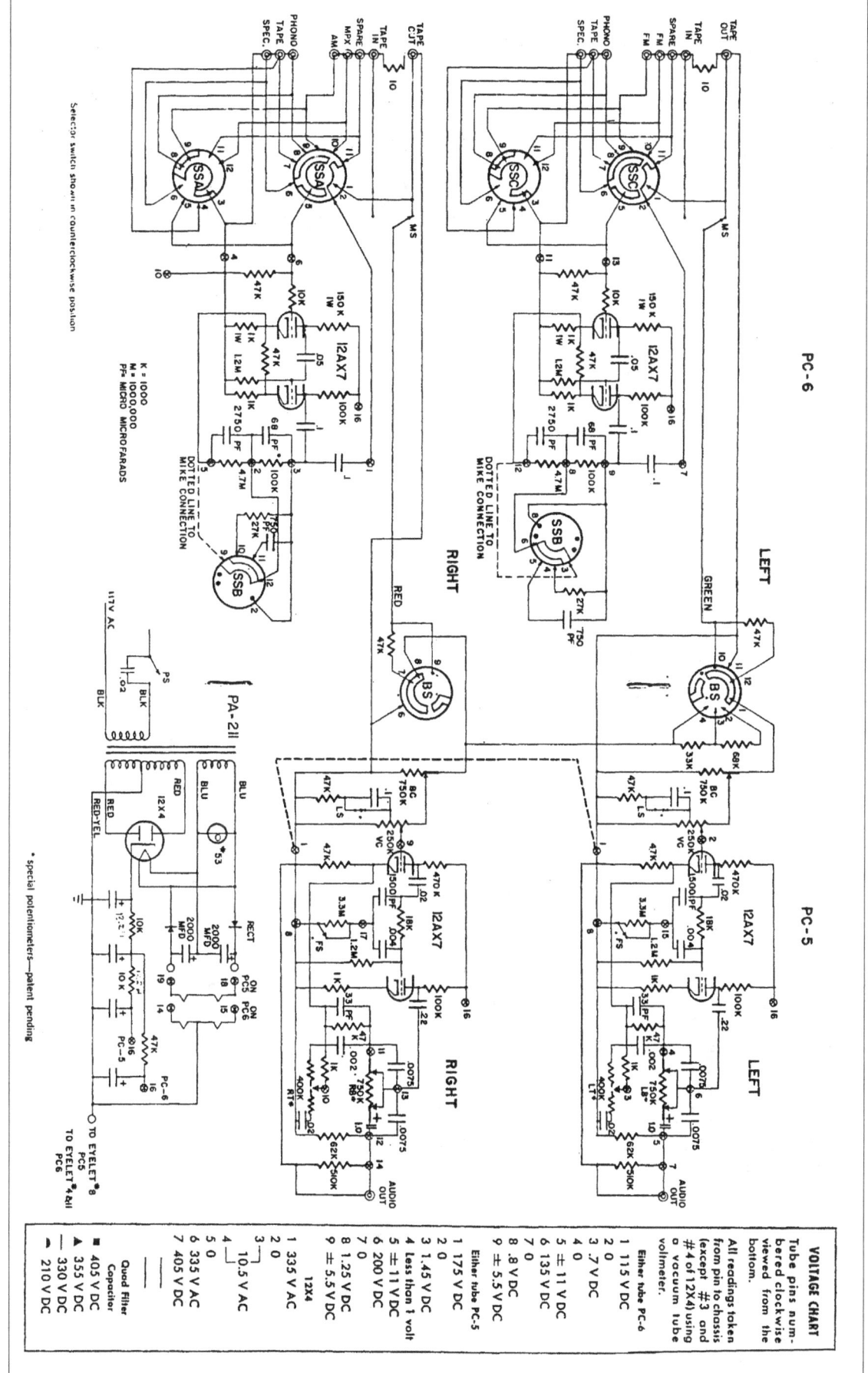

■ 405 V DC
▲ 355 V DC
— 330 V DC
◓ 210 V DC

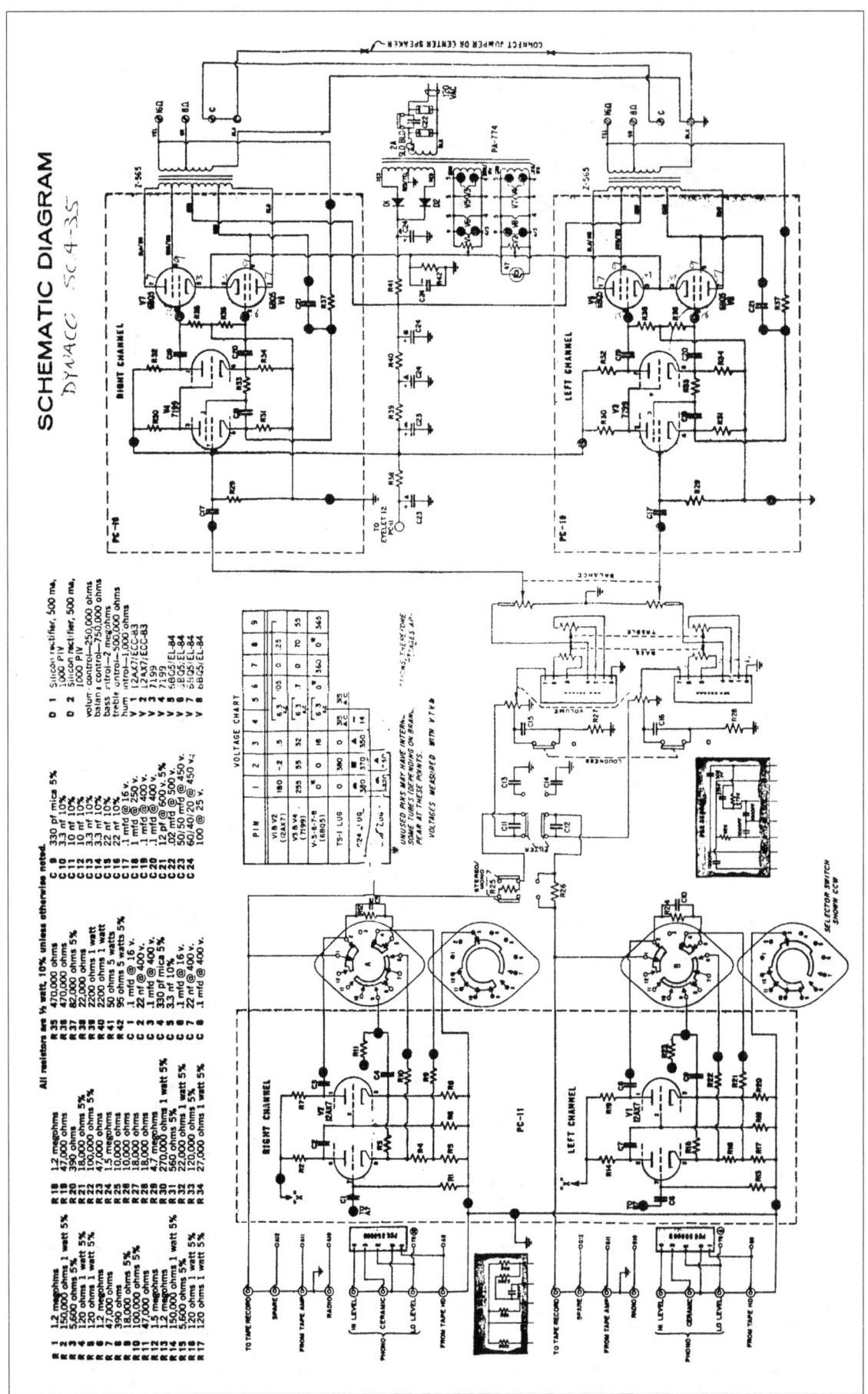

SCHEMATIC DIAGRAM

DYNACO SCA-35

All resistors are ½ watt, 10% unless otherwise noted.

R 1	1.2 megohms
R 2	150,000 ohms 1 watt 5%
R 3	5,600 ohms 5%
R 4	120 ohms 1 watt 5%
R 5	120 ohms 1 watt 5%
R 6	1.2 megohms
R 7	47,000 ohms
R 8	390 ohms
R 9	18,000 ohms 5%
R 10	100,000 ohms 5%
R 11	47,000 ohms
R 12	1.5 megohms
R 13	1.2 megohms
R 14	150,000 ohms 1 watt 5%
R 15	5,600 ohms 5%
R 16	120 ohms 1 watt 5%
R 17	120 ohms 1 watt 5%
R 18	1.2 megohms
R 19	47,000 ohms
R 20	390 ohms
R 21	18,000 ohms 5%
R 22	100,000 ohms 5%
R 23	47,000 ohms
R 24	1.5 megohms
R 25	10,000 ohms
R 26	10,000 ohms
R 27	18,000 ohms
R 28	18,000 ohms
R 29	4.7 megohms
R 30	270,000 ohms 1 watt 5%
R 31	560 ohms 5%
R 32	22,000 ohms 1 watt 5%
R 33	120,000 ohms 5%
R 34	27,000 ohms 1 watt 5%
R 35	470,000 ohms
R 36	470,000 ohms
R 37	82,000 ohms 5%
R 38	22,000 ohms
R 39	2200 ohms 1 watt
R 40	2200 ohms 1 watt
R 41	50 ohms 5 watts
R 42	95 ohms 5 watts 5%
C 1	.1 mfd @ 16 v.
C 2	22 nf @ 400 v.
C 3	.1 mfd @ 400 v.
C 4	330 pf mica 5%
C 5	3.3 nf 10%
C 6	.1 mfd @ 16 v.
C 7	22 nf @ 400 v.
C 8	.1 mfd @ 400 v.
C 9	330 pf mica 5%
C 10	3.3 nf 10%
C 11	10 nf 10%
C 12	10 nf 10%
C 13	3.3 nf 10%
C 14	3.3 nf 10%
C 15	22 nf 10%
C 16	22 nf 10%
C 17	.1 mfd @ 16 v.
C 18	1 mfd @ 250 v.
C 19	.1 mfd @ 400 v.
C 20	.1 mfd @ 400 v.
C 21	12 pf @ 600 v. 5%
C 22	.02 mfd @ 500 v.
C 23	50/50 mfd @ 450 v.
C 24	60/40/20 @ 450 v.; 100 @ 25 v.
D 1	Silicon rectifier, 500 ma, 1000 PIV
D 2	Silicon rectifier, 500 ma, 1000 PIV
volume control	250,000 ohms
balance control	750,000 ohms
bass control	2 megohms
treble control	500,000 ohms
hum control	1,000 ohms
V 1	12AX7/ECC-83
V 2	12AX7/ECC-83
V 3	7199
V 4	7199
V 5	6BQ5/EL-84
V 6	6BQ5/EL-84
V 7	6BQ5/EL-84
V 8	6BQ5/EL-84

VOLTAGE CHART

PIN	1	2	3	4	5	6	7	8	9
V1 & V2 (12AX7)	180	-.2	.5	6.3 AC		105	0	.25	
V3 & V4 (7199)	255	55	32	6.3 AC		.7	0	70	55
V-5-6-7-8 (6BQ5)	0*	0	16	6.3 AC		0*	360	0*	365
TS-1 LUG	0	380	0	315 AC	315 AC				
C24 LUG	● 380	■ 370	▲ 350	– 14					
C23 LUG	[illegible]	▲ [illegible]							

* UNUSED PINS MAY HAVE INTERN[illegible]TIONS, THEREFORE SOME TUBES (DEPENDING ON BRAN[illegible] [illegible]TAGES APPEAR AT THESE POINTS.

VOLTAGES MEASURED WITH VTVM

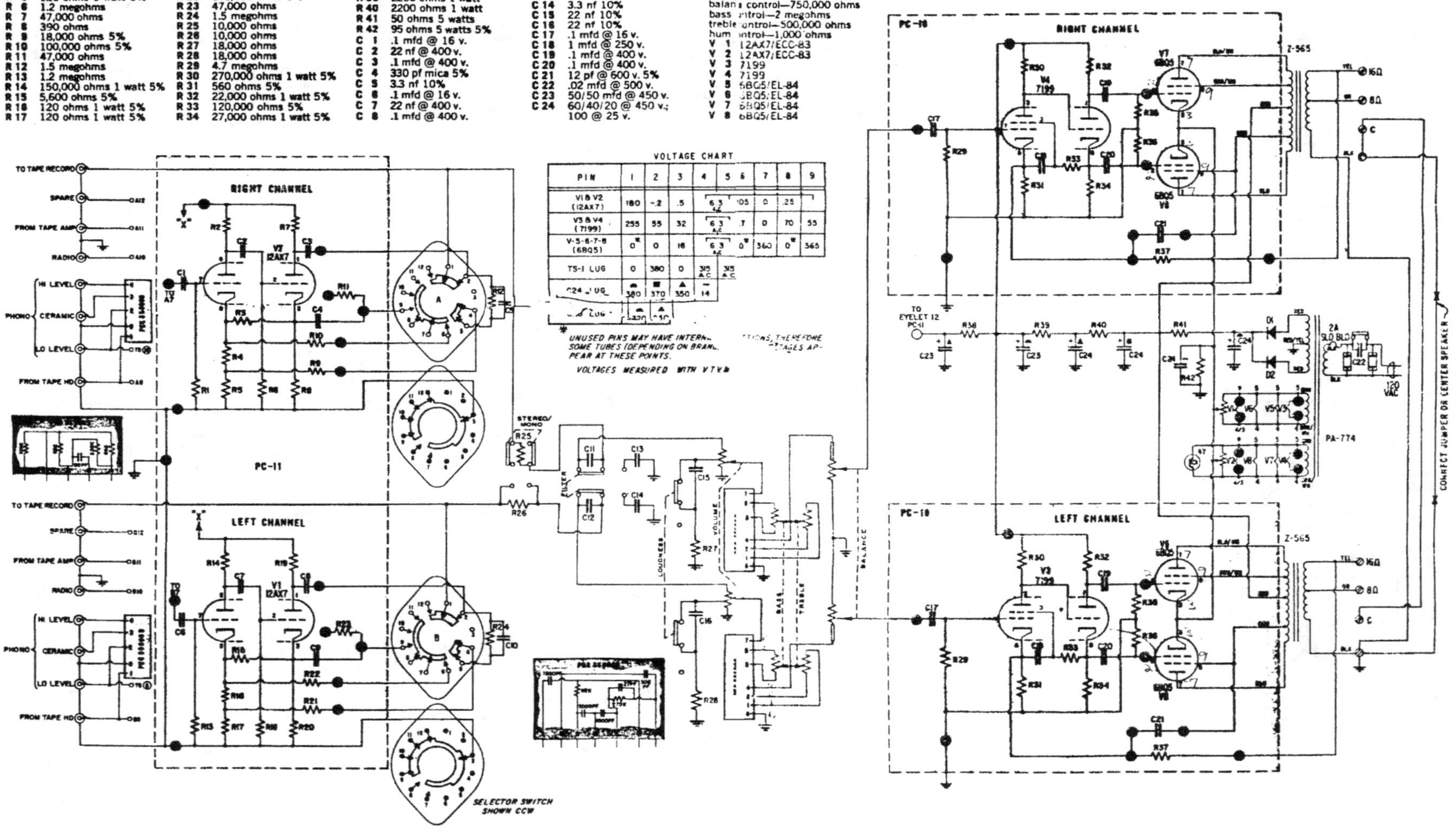

Bass-King T

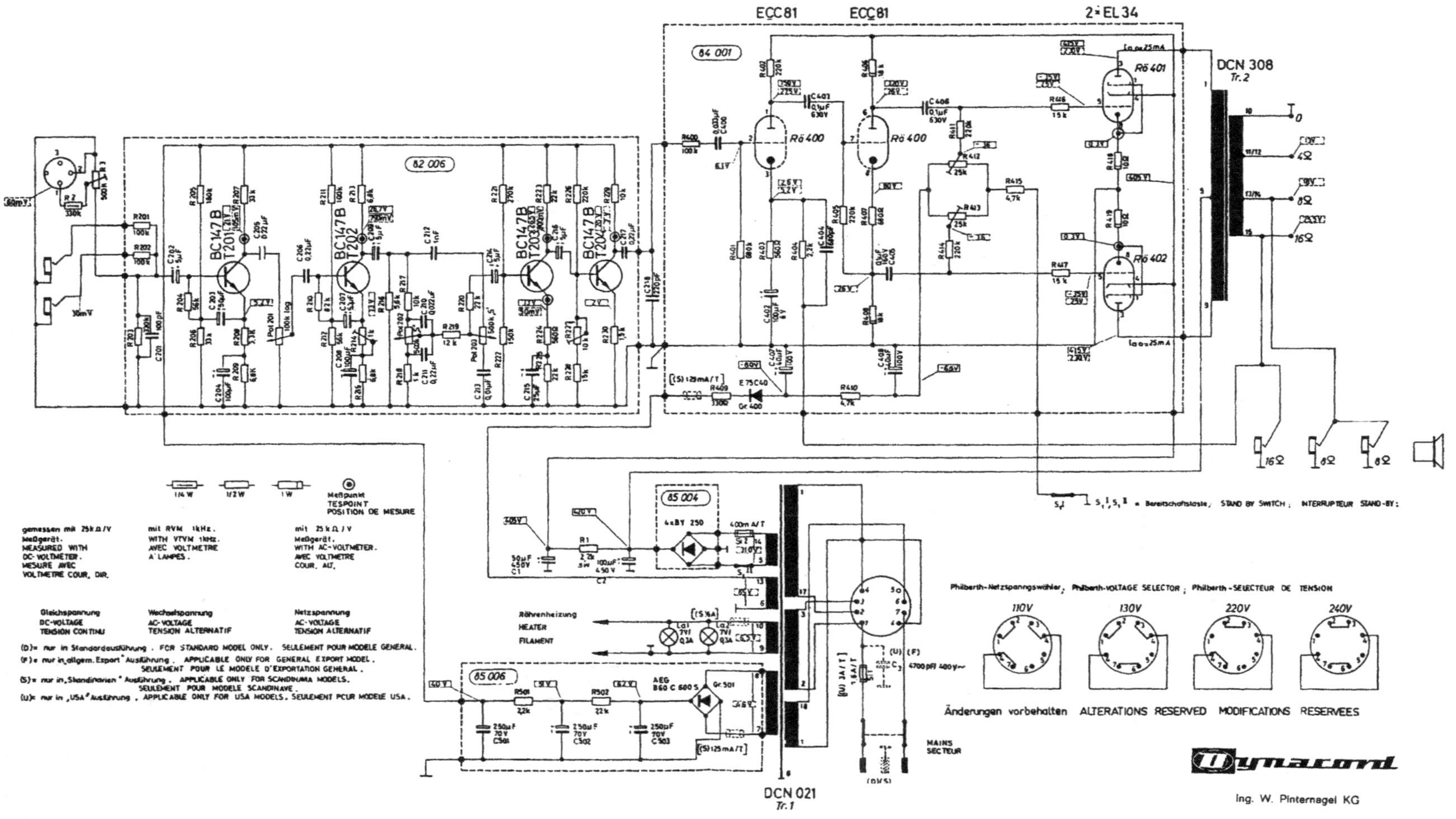
ECC81
ECC81
2× EL34
82 006
84 001
85 004
85 006
DCN 308
Tr. 2
DCN 021
Tr. 1
BC147B T201
BC147B T202
BC147B T203
BC147B T204
Rö 400
Rö 401
Rö 402
1/4 W
1/2 W
1 W
Meßpunkt
TESPOINT
POSITION DE MESURE
gemessen mit 25kΩ/V
Meßgerät.
MEASURED WITH
DC-VOLTMETER.
MESURE AVEC
VOLTMETRE COUR. DIR.
mit RVM 1kHz.
WITH VTVM 1kHz.
AVEC VOLTMETRE
A LAMPES.
mit 25kΩ/V
Meßgerät.
WITH AC-VOLTMETER.
AVEC VOLTMETRE
COUR. ALT.
Gleichspannung
DC-VOLTAGE
TENSION CONTINU
Wechselspannung
AC-VOLTAGE
TENSION ALTERNATIF
Netzspannung
AC-VOLTAGE
TENSION ALTERNATIF
(D)= nur in Standardausführung. FOR STANDARD MODEL ONLY. SEULEMENT POUR MODELE GENERAL.
(F)= nur in allgem. Export Ausführung. APPLICABLE ONLY FOR GENERAL EXPORT MODEL. SEULEMENT POUR LE MODELE D'EXPORTATION GENERAL.
(S)= nur in Skandinavien Ausführung. APPLICABLE ONLY FOR SCANDINAVIA MODELS. SEULEMENT POUR MODELE SCANDINAVE.
(U)= nur in USA Ausführung. APPLICABLE ONLY FOR USA MODELS. SEULEMENT POUR MODELE USA.
Röhrenheizung
HEATER
FILAMENT
MAINS
SECTEUR
S1, S2, S3 = Bereitschaftstaste; STAND BY SWITCH; INTERRUPTEUR STAND-BY;
Philberth-Netzspannungswähler; Philberth-VOLTAGE SELECTOR; Philberth-SELECTEUR DE TENSION
110V
130V
220V
240V
Änderungen vorbehalten ALTERATIONS RESERVED MODIFICATIONS RESERVEES
Dynacord
Ing. W. Pinternagel KG

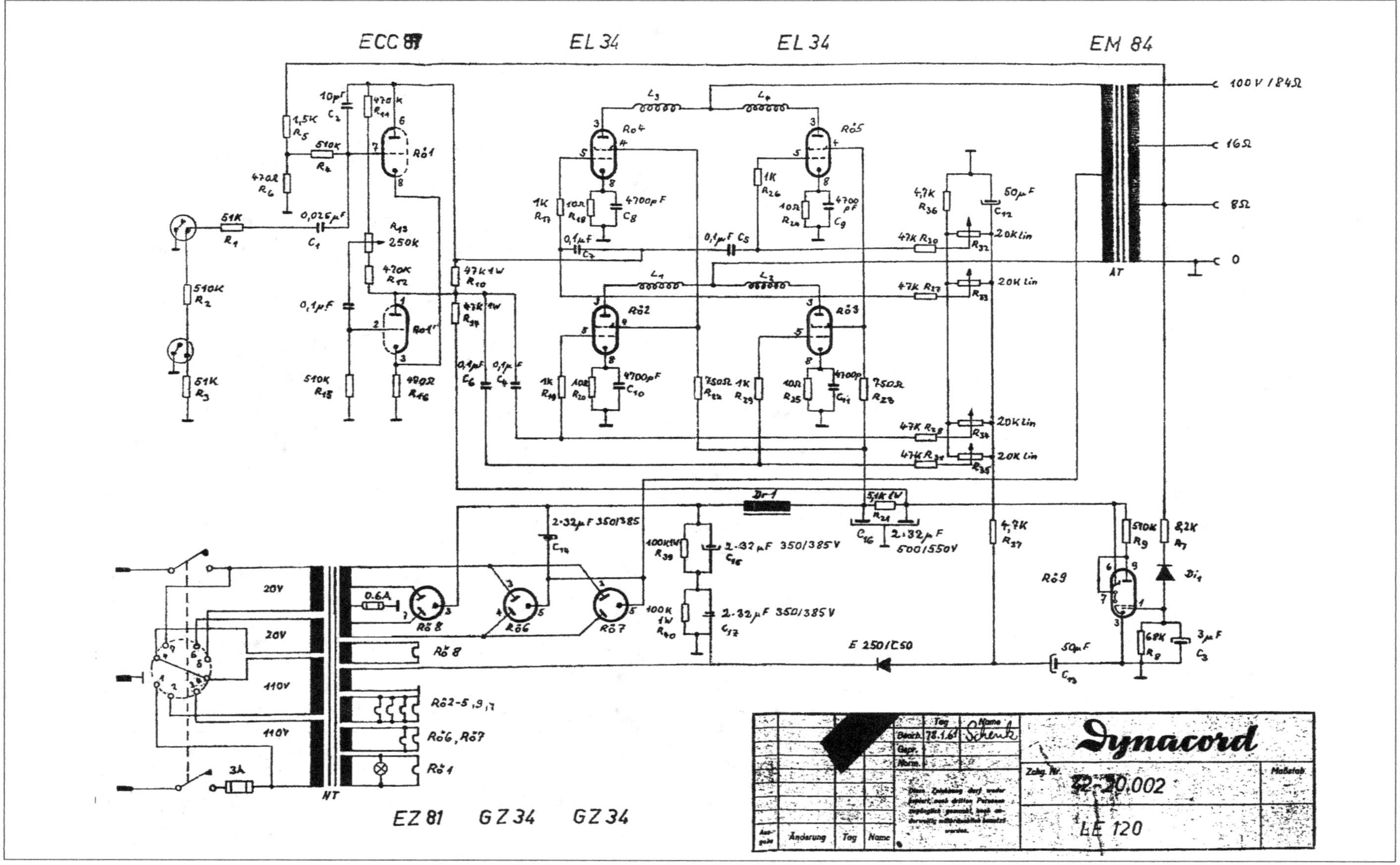
ECC 81
EL 34
EL 34
EM 84
100 V / 84Ω
16Ω
8Ω
0
EZ 81
GZ 34
GZ 34
Dynacord
22-20.002
LE 120

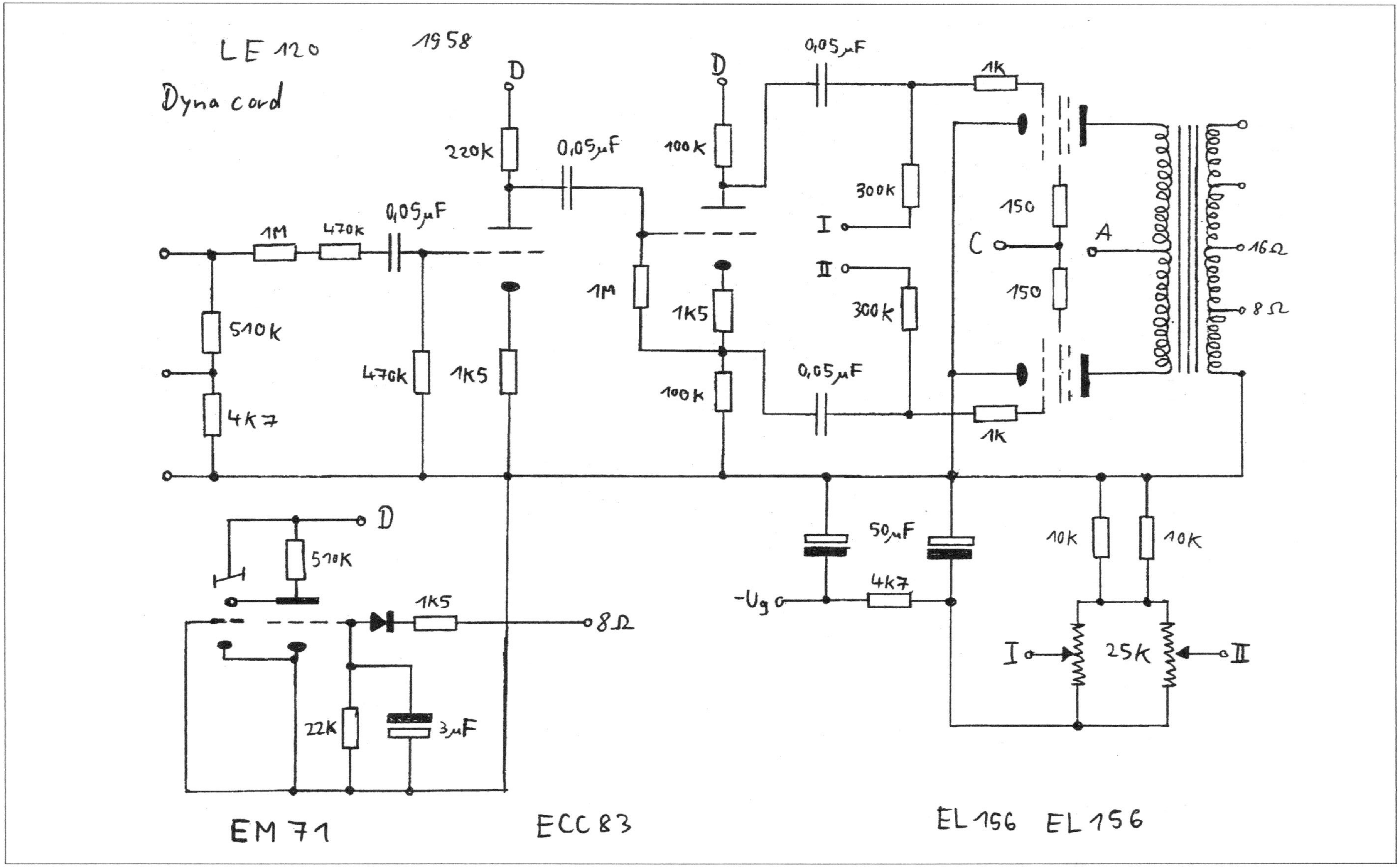
LE 120
Dynacord
1958
D
220K
0,05µF
100K
0,05µF
1K
1M
470K
0,05µF
300K
150
I
C
A
II
16Ω
1M
1K5
150
300K
8Ω
510K
470K
1K5
100K
0,05µF
1K
4K7
D
510K
50µF
10K
10K
1K5
8Ω
-Ug
4K7
I
25K
II
22K
3µF
EM 71
ECC 83
EL 156 EL 156

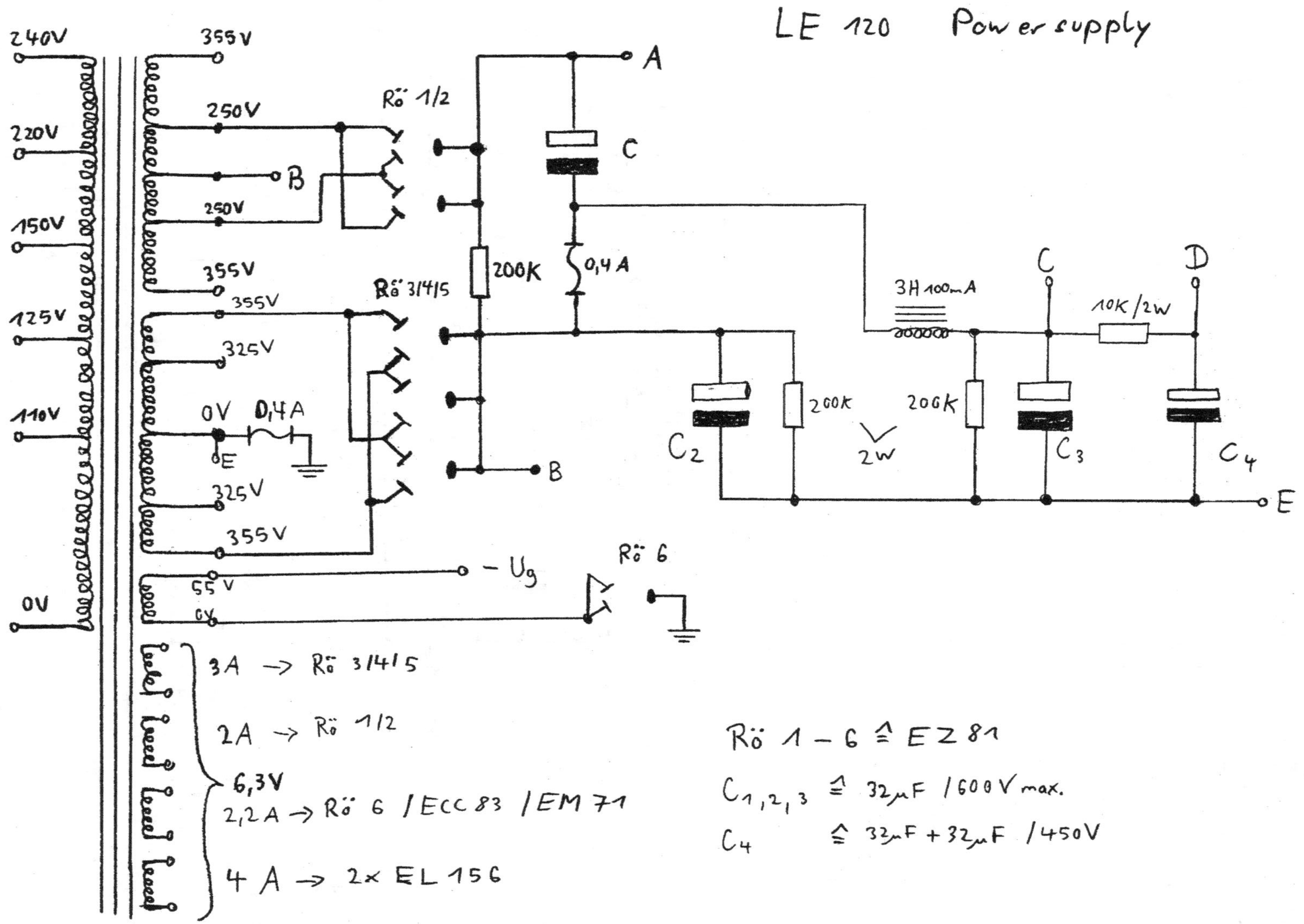
LE 120 Power supply
240V
220V
150V
125V
110V
0V
355V
250V
B
250V
355V
355V
325V
0V
0,4A
E
325V
355V
55 V
0V
Rö 1/2
Rö 3/4/5
200K
A
C
0,4 A
B
3H 100mA
C
D
10K/2W
C2
200K
2W
200K
C3
C4
E
– Ug
Rö 6
3A -> Rö 3/4/5
2A -> Rö 1/2
6,3V
2,2A -> Rö 6 / ECC 83 / EM 71
4 A -> 2x EL 156
Rö 1 – 6 ≙ EZ 81
C1,2,3 ≙ 32µF / 600V max.
C4 ≙ 32µF + 32µF / 450V

G 2000

Dynacord

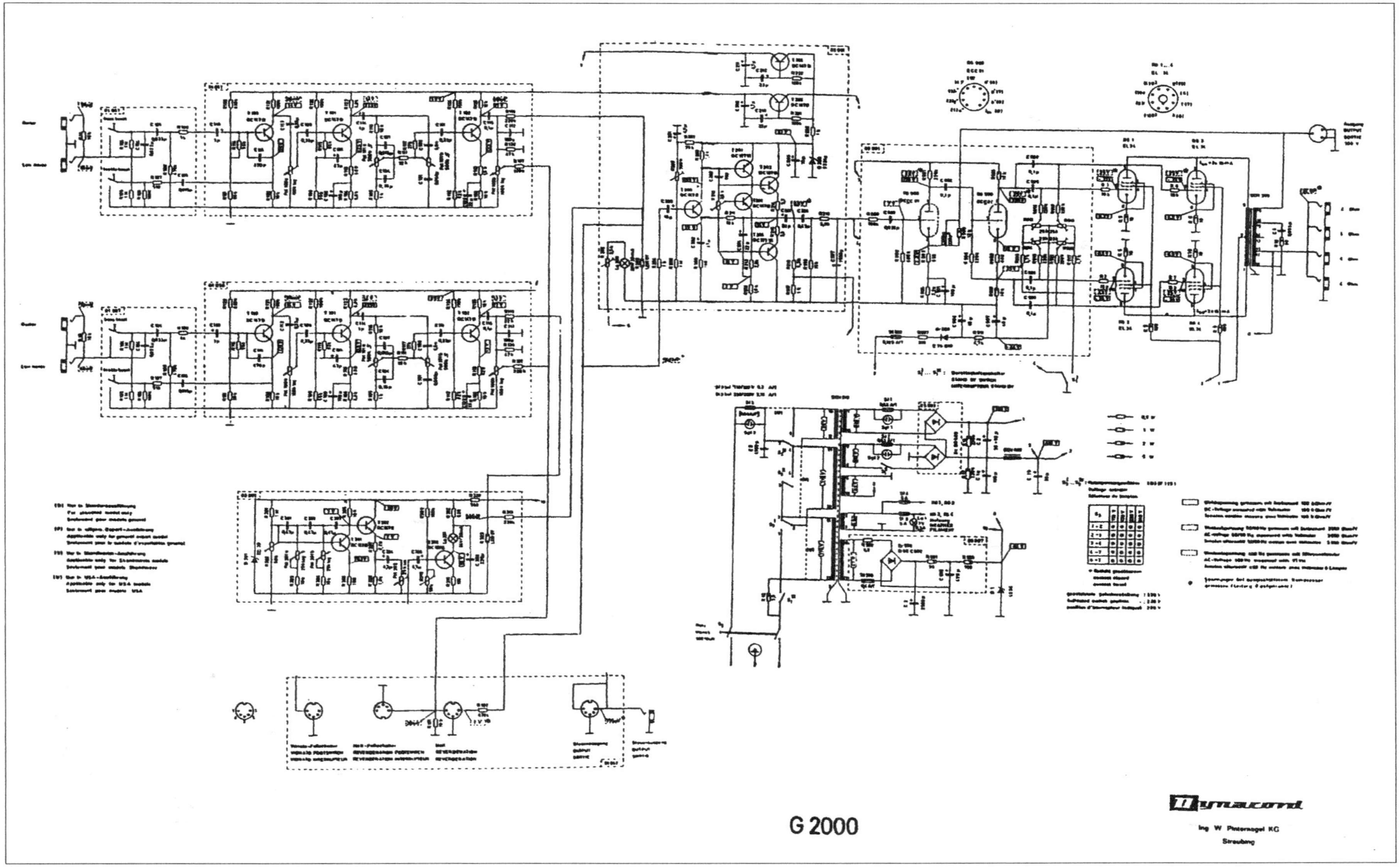

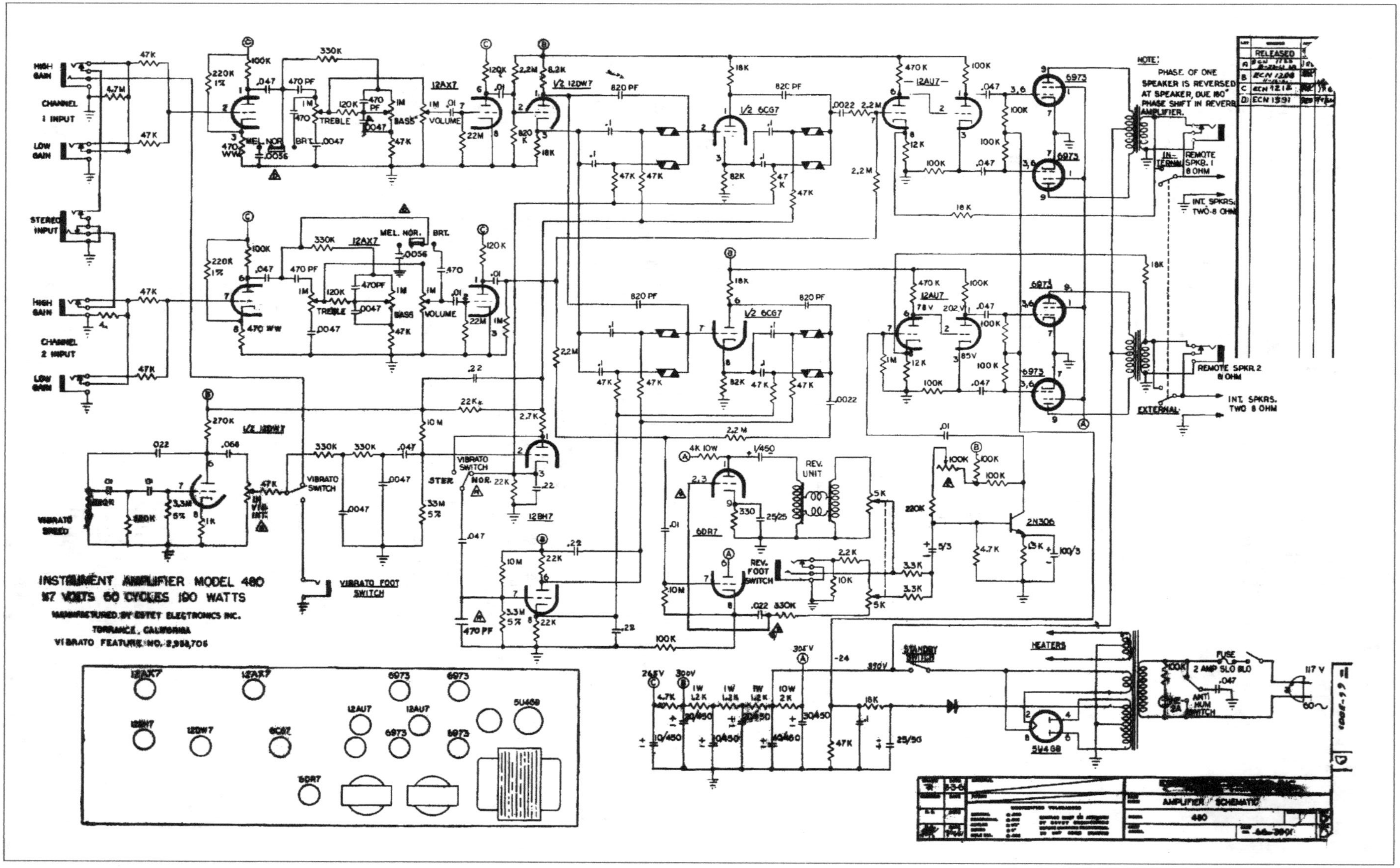
INSTRUMENT AMPLIFIER MODEL 480
117 VOLTS 60 CYCLES 190 WATTS
MANUFACTURED BY ESTEY ELECTRONICS INC.
TORRANCE, CALIFORNIA
NOTE:
PHASE OF ONE SPEAKER IS REVERSED AT SPEAKER, DUE 180° PHASE SHIFT IN REVERB AMPLIFIER.
HIGH GAIN
CHANNEL 1 INPUT
LOW GAIN
STEREO INPUT
CHANNEL 2 INPUT
VIBRATO SPEED
VIBRATO SWITCH
VIBRATO FOOT SWITCH
REV. UNIT
REV. FOOT SWITCH
REMOTE SPKR. 1 8 OHM
INT. SPKRS. TWO-8 OHM
REMOTE SPKR 2 8 OHM
INT. SPKRS. TWO 8 OHM
EXTERNAL
HEATERS
STANDBY SWITCH
FUSE 2 AMP SLO BLO
ANTI HUM SWITCH
117 V
AMPLIFIER SCHEMATIC
480

CONTROL CIRCUIT BOARD

Engl

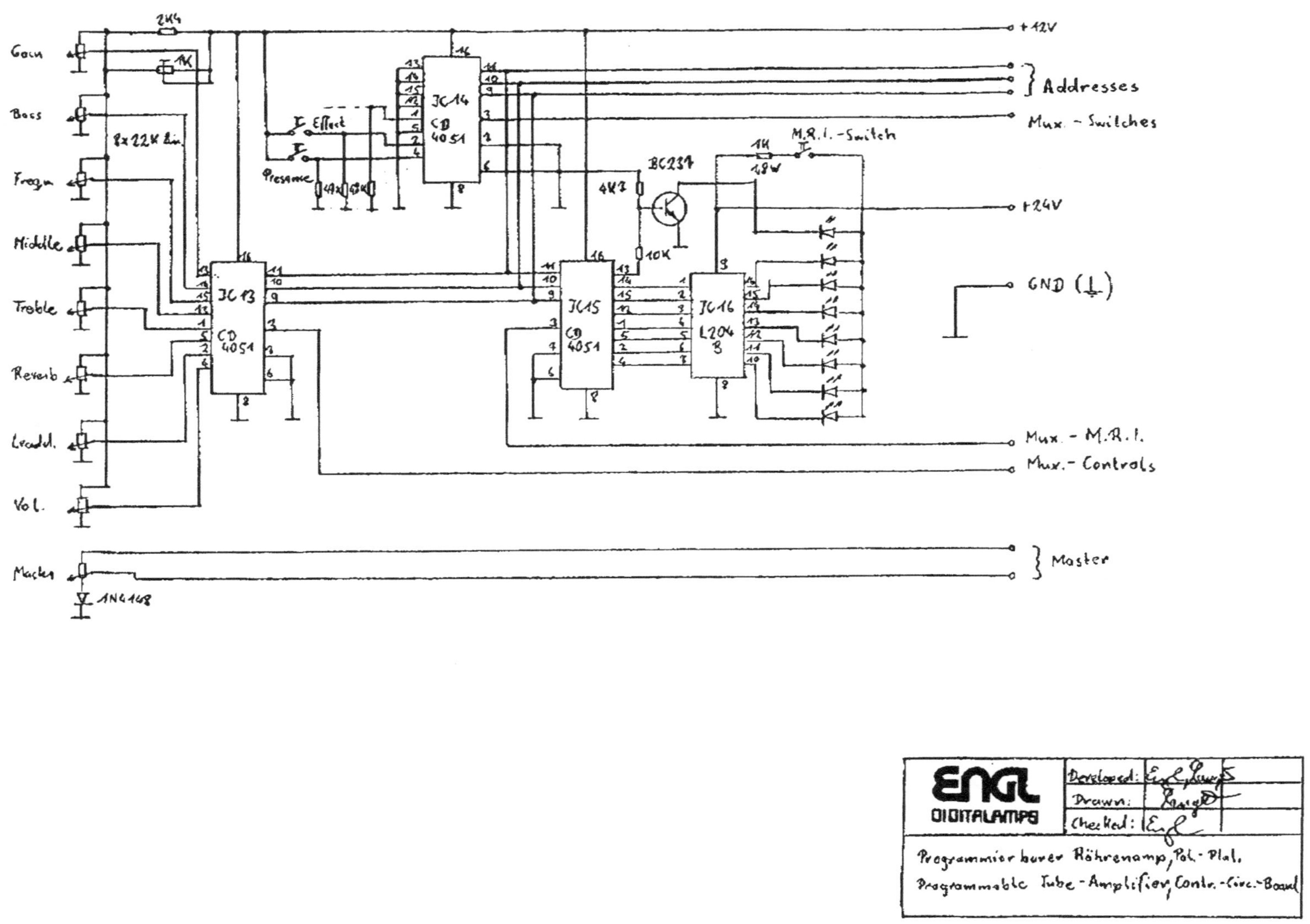

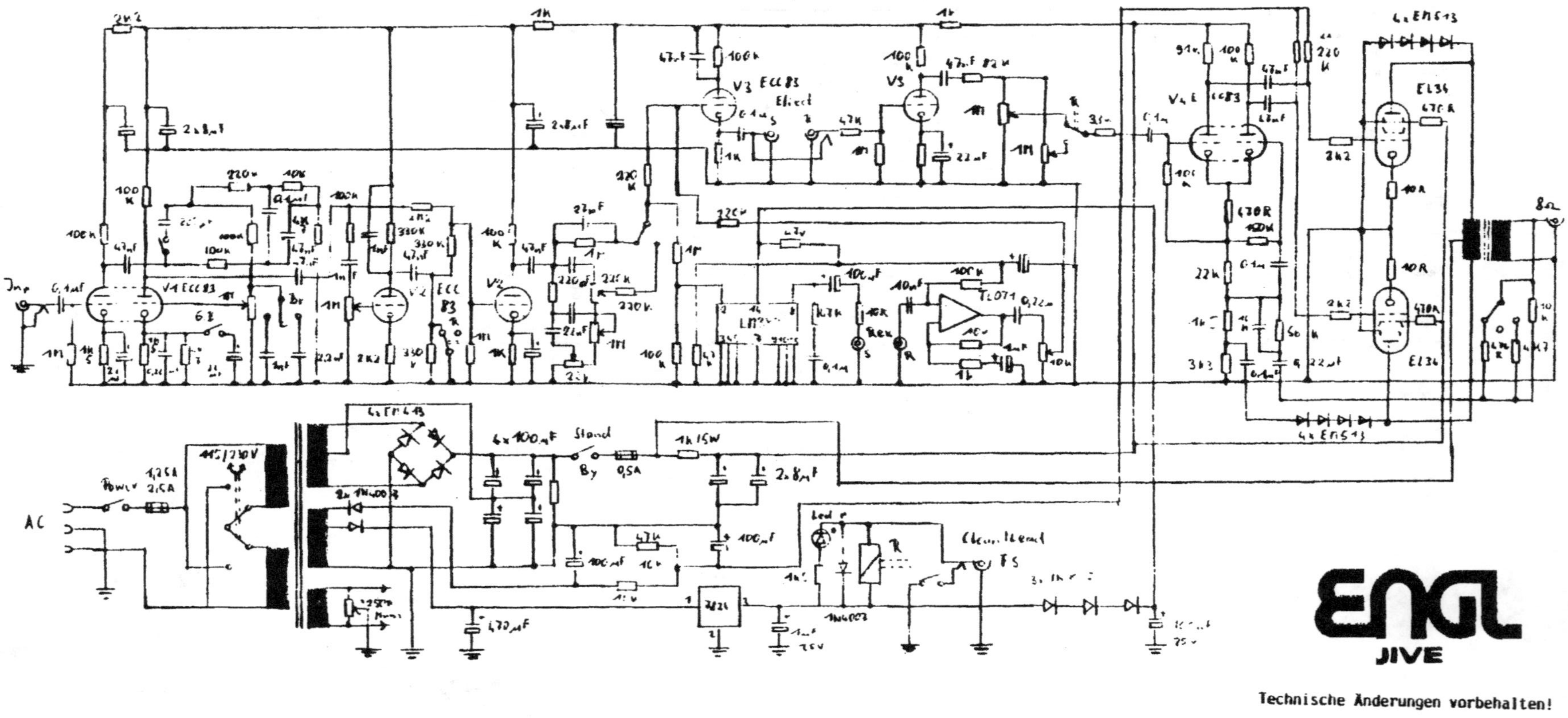
ENGL
JIVE
Technische Änderungen vorbehalten!

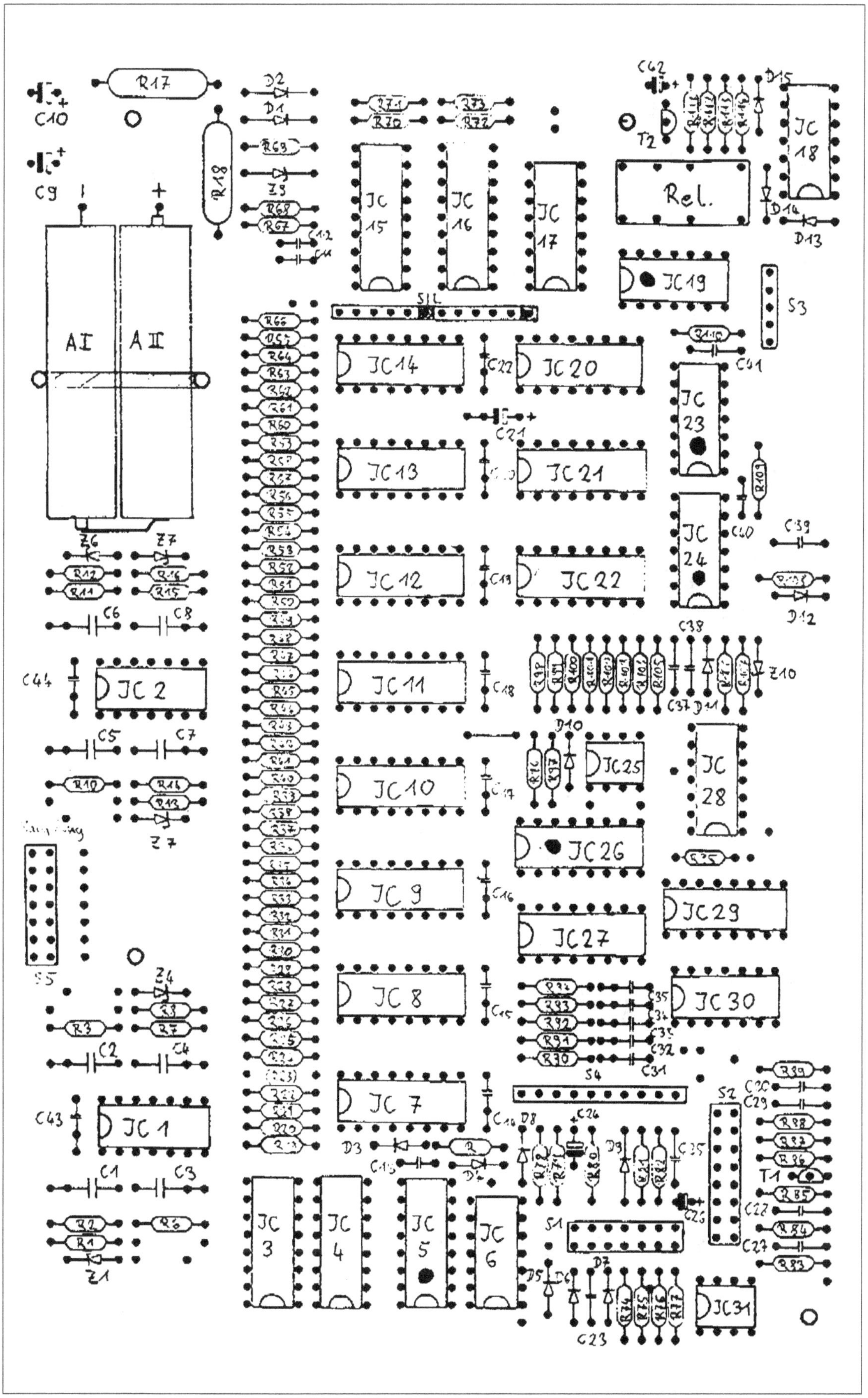
R17
R18
C10
C9
AI
AII
JC 15
JC 16
JC 17
JC 18
Rel.
JC 19
S3
JC 14
JC 20
JC 23
JC 13
JC 21
JC 24
JC 12
JC 22
JC 2
JC 11
JC 25
JC 28
JC 10
JC 26
JC 9
JC 29
JC 27
JC 8
JC 30
S5
S4
JC 7
JC 1
S2
T1
T2
JC 3
JC 4
JC 5
JC 6
S1
JC 31

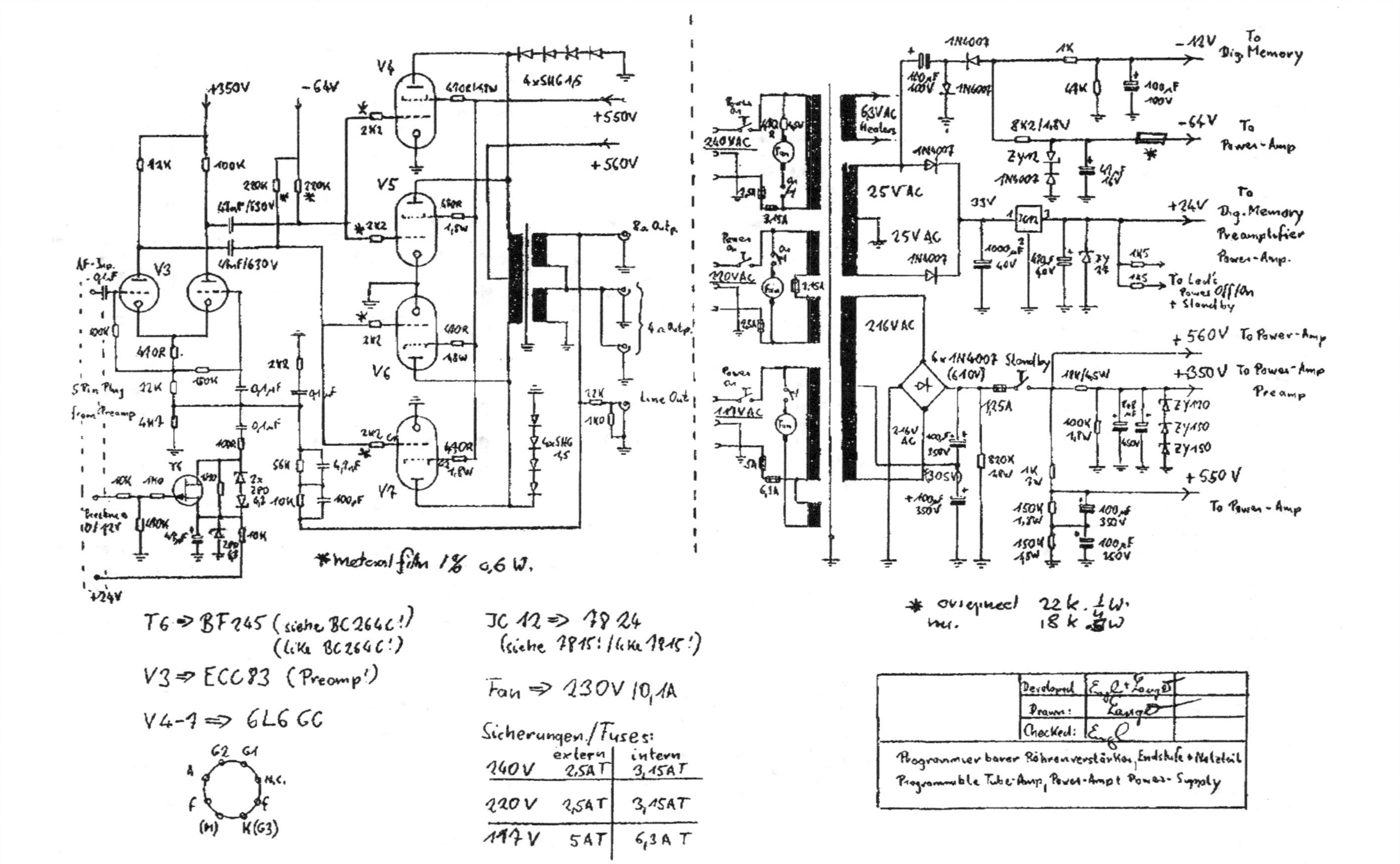
+350V
-64V
+550V
+560V
4xSHG1/5
V3
V4
V5
V6
V7
8Ω Outp.
4Ω Outp.
Line Out
5 Pin Plug from Preamp
*metal film 1% 0,6 W.
-12V To Dig. Memory
-64V To Power-Amp
+24V To Dig. Memory Preamplifier Power-Amp.
To Led's Power Off/On + Standby
+560V To Power-Amp
+350V To Power-Amp Preamp
+550V To Power-Amp
240VAC
220VAC
117VAC
6,3VAC Heaters
25VAC
216VAC
6x1N4007 (610V)
Standby
T6 ⇒ BF245 (siehe BC264C!) (like BC264C!)
V3 ⇒ ECC83 (Preamp!)
V4-7 ⇒ 6L6 GC
G2 G1
A
n.c.
F
f
(M)
K(G3)
IC 12 ⇒ 7824 (siehe 7815! / like 7815!)
Fan ⇒ 230V 10,1A
Sicherungen/Fuses:
extern intern
240V 2,5AT 3,15AT
220V 2,5AT 3,15AT
117V 5AT 6,3AT
Developed Engl + Lange
Drawn: Lange
Checked: Engl
Programmierbarer Röhrenverstärker, Endstufe + Netzteil
Programmable Tube-Amp, Power-Amp + Power-Supply

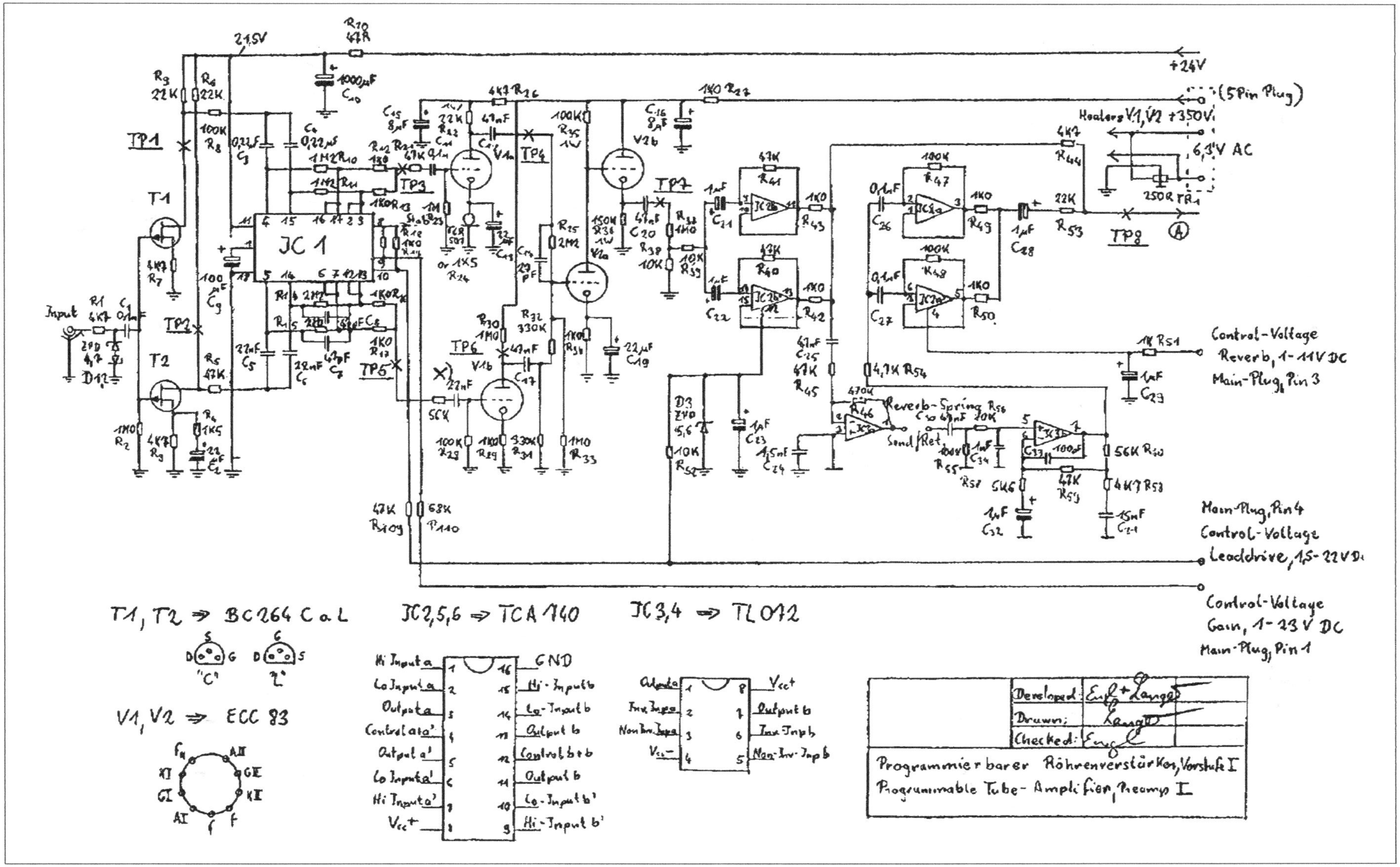

Input
+24V
(5 Pin Plug)
Heaters V1, V2 +350V
6,3V AC
Control-Voltage
Reverb, 1-11V DC
Main-Plug, Pin 3
Main-Plug, Pin 4
Control-Voltage
Leaddrive, 1,5-22V DC
Control-Voltage
Gain, 1-23V DC
Main-Plug, Pin 1
Reverb-Spring
Send
Ret.
T1, T2 ⇒ BC264 C o. L
V1, V2 ⇒ ECC 83
JC2,5,6 ⇒ TCA 740
JC3,4 ⇒ TL072
Developed: Engl + Lange
Drawn: Lange
Checked: Engl
Programmierbarer Röhrenverstärker, Vorstufe I
Programmable Tube-Amplifier, Preamp I

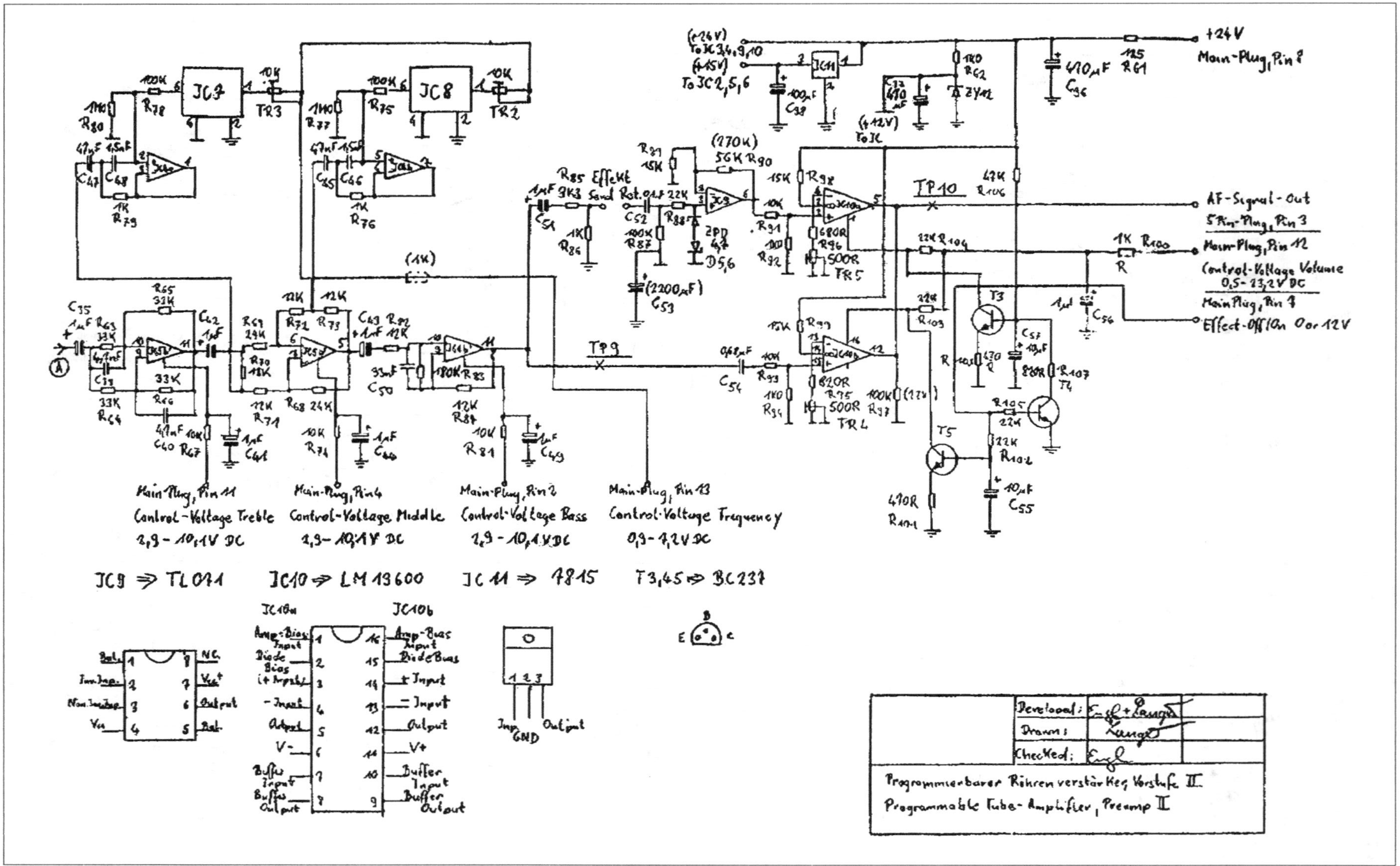
+24V
Main-Plug, Pin 8
AF-Signal-Out
5 Pin-Plug, Pin 3
Main-Plug, Pin 12
Control-Voltage Volume
0,5-13,2V DC
Main Plug, Pin 7
Effect-Off/On 0 or 12V
TP10
TP9
Main-Plug, Pin 11
Control-Voltage Treble
1,9-10,1V DC
Main-Plug, Pin 4
Control-Voltage Middle
1,9-10,1V DC
Main-Plug, Pin 2
Control-Voltage Bass
1,9-10,1V DC
Main-Plug, Pin 13
Control-Voltage Frequency
0,9-1,2V DC
IC9 ⇒ TL071
IC10 ⇒ LM13600
IC11 ⇒ 7815
T3,4,5 ⇒ BC237
Developed: Engl + Lange
Drawn: Lange
Checked: Engl
Programmierbarer Röhrenverstärker, Vorstufe II
Programmable Tube-Amplifier, Preamp II

EA-4T, 4TL, & 6T

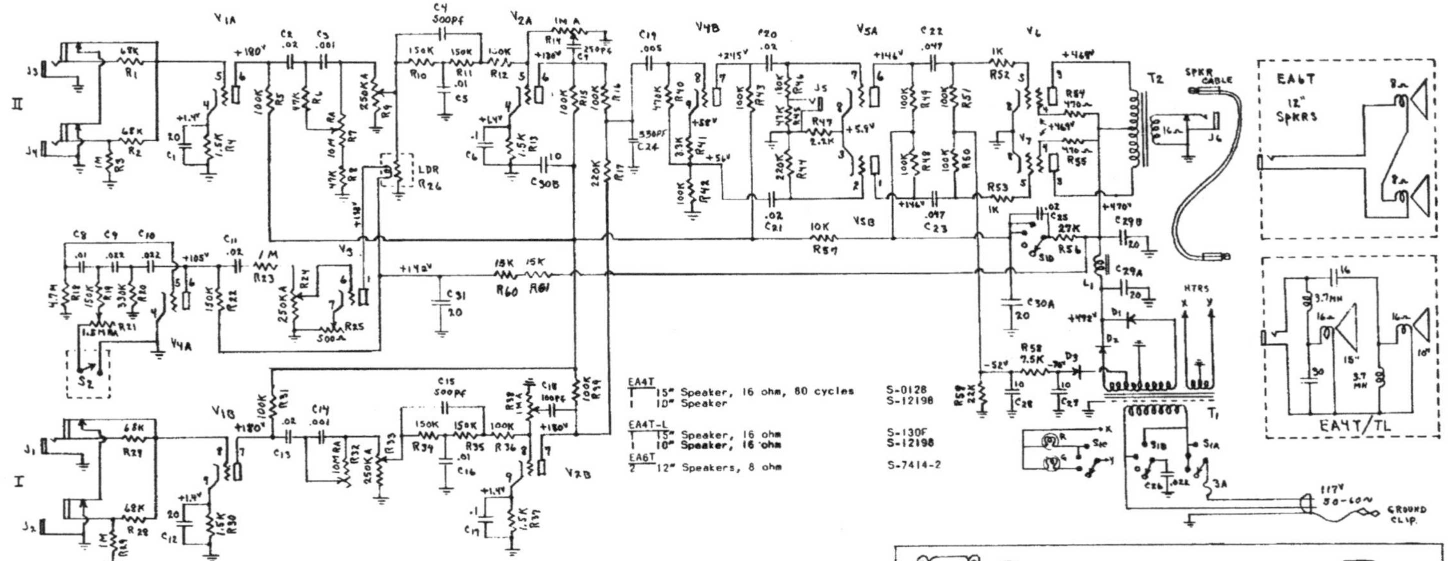

EA4T
1 15" Speaker, 16 ohm, 80 cycles — S-0128
1 10" Speaker — S-12198

EA4T-L
1 15" Speaker, 16 ohm — S-130F
1 10" Speaker, 16 ohm — S-12198

EA6T
2 12" Speakers, 8 ohm — S-7414-2

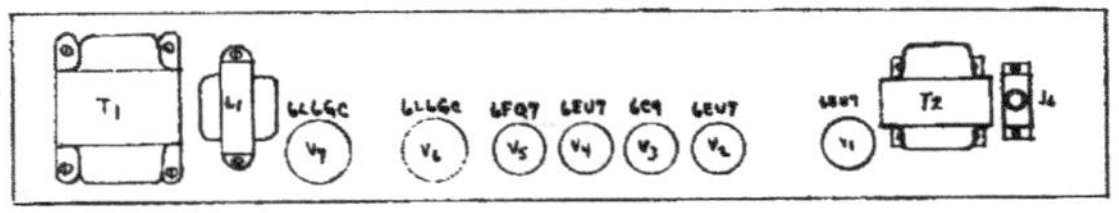

R7	Bass Tone Control, 10 meg. Reversed Audio	C-BA-811-3710
R9	Volume Control, 250K Audio Taper	C-BA-811-3701
R14	Treble Tone Control, 1 meg. Audio Taper	C-BA-811-3709
R21	Tremolo Frequency Control, 15meg. Reversed Audio	C-BA-811-3705
R24	Tremolo Depth Control, 250K Audio Taper	C-BA-811-3701
R25	LDR Gain Control, 500 ohm/200 ohm min.	C-B-23391
R32	Bass Tone Control, 10 meg. Reversed Audio	C-BA-811-3710
R33	Volume Control, 250K Audio Taper	C-BA-811-3701
R38	Treble Tone Control, 1 meg. Audio Taper	C-BA-811-3709
R26	Light Dependent Resistor Lamp Assembly	RE-LDR-430G

D1,D2	Diodes - 1200 PIV, 250 MA	D1-57
D3	Diode - 600 PIV, 250 MA	D1-71B
T1	Power Transformer	TF-77P-S
T2	Output Transformer	TF-300-01
L1	Filter Choke	TF-3021H
S1	Switch, Power, Standby, Polarity	SW-78
S2	Switch, Tremolo	

FUSE

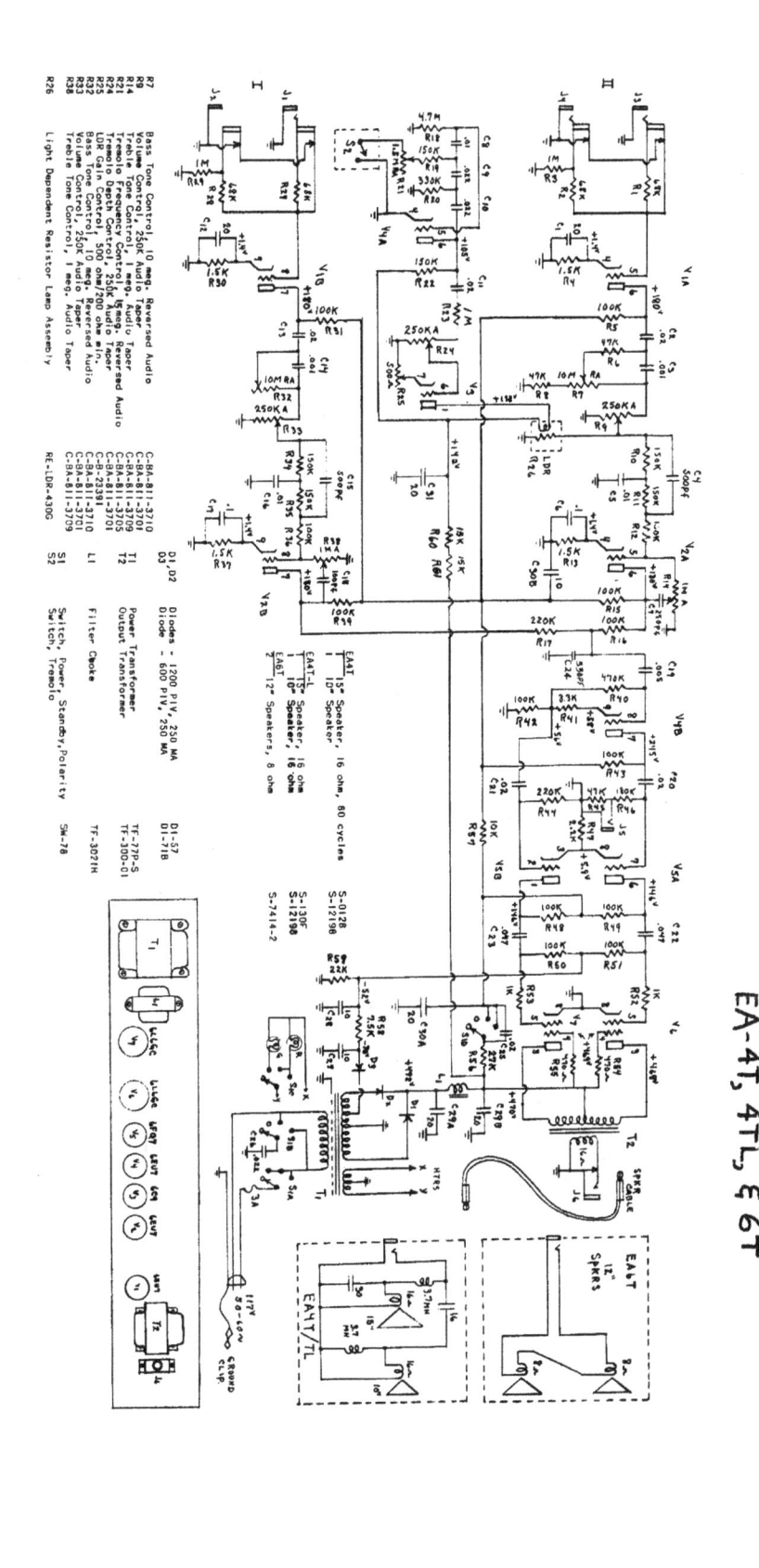

The fuse used in this Amplifier is a type 3AG of three amperes rating.

DO NOT USE FUSES OF HIGHER RATING

SERVICE

If the amplifier is in need of servicing, it should be taken to a reliable radio man. The electrical diagram in this folder should be shown the repairman to assist him in servicing the amplifier.

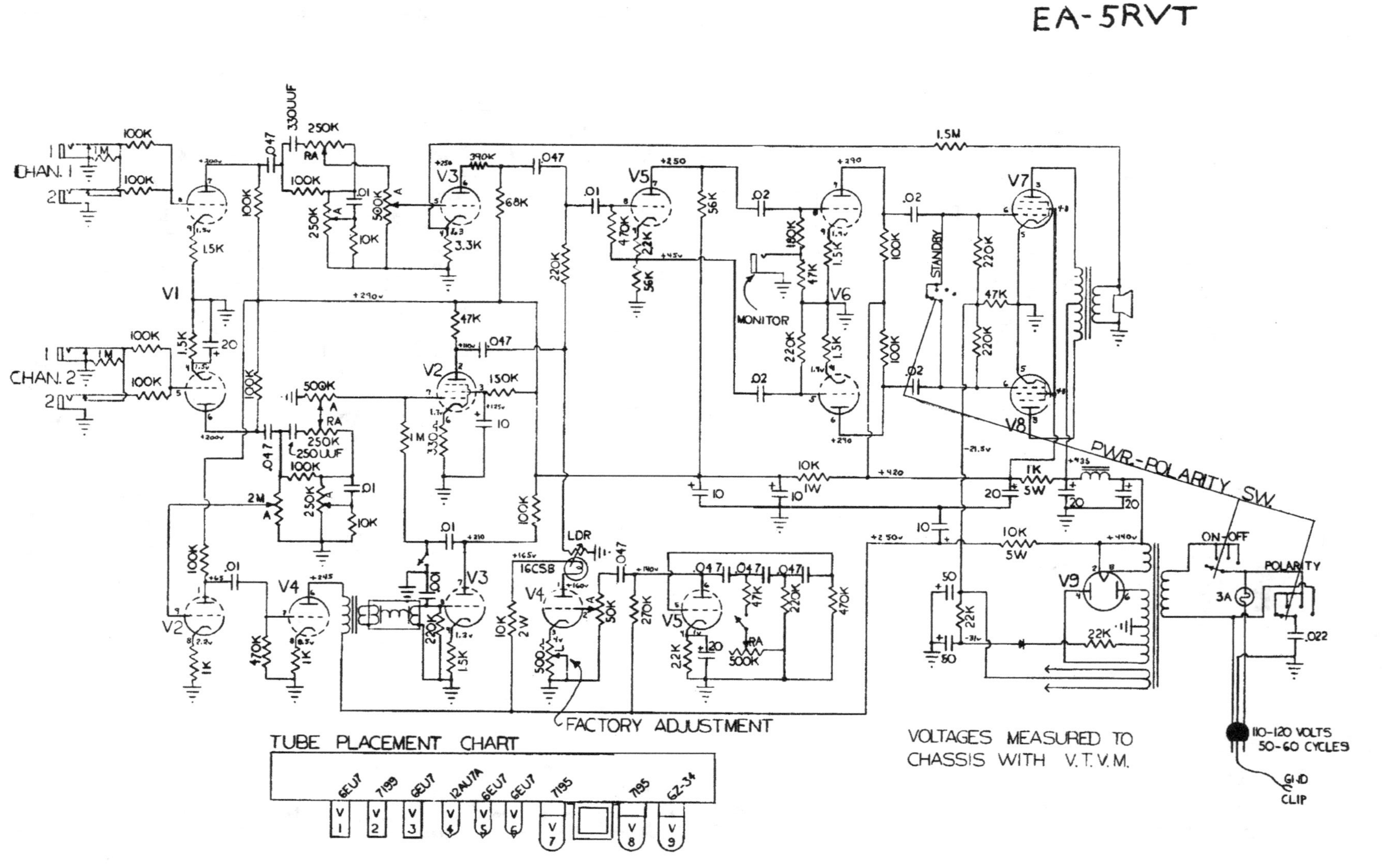
EA-5RVT
CHAN. 1
CHAN. 2
V1
V2
V3
V4
V5
V6
V7
V8
V9
MONITOR
STANDBY
PWR-POLARITY SW.
ON-OFF
POLARITY
3A
110-120 VOLTS
50-60 CYCLES
GND
CLIP
LDR
16CSB
FACTORY ADJUSTMENT
VOLTAGES MEASURED TO
CHASSIS WITH V.T.V.M.
TUBE PLACEMENT CHART
6EU7
7199
6EU7
12AU7A
6EU7
6EU7
7195
7195
6Z-34

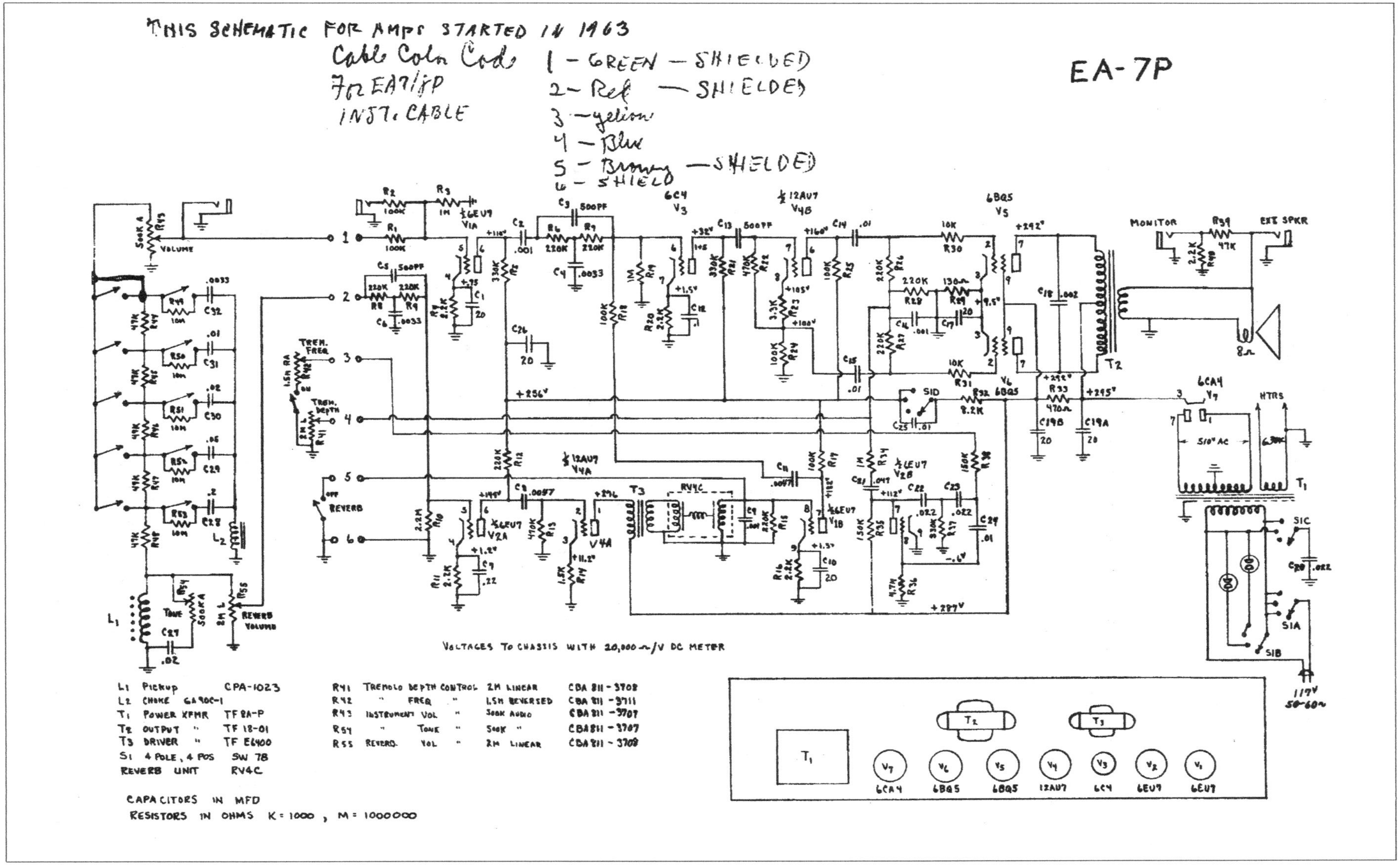
THIS SCHEMATIC FOR AMPS STARTED IN 1963
Cable Color Code For EA7/7P INST. CABLE
1 - GREEN - SHIELDED
2 - Red - SHIELDED
3 - yellow
4 - Blue
5 - Brown - SHIELDED
6 - SHIELD
EA-7P
VOLTAGES TO CHASSIS WITH 20,000 Ω/V DC METER
L1 PICKUP CPA-1023
L2 CHOKE GA90C-1
T1 POWER XFMR TF 8A-P
T2 OUTPUT " TF 18-01
T3 DRIVER " TF E6400
S1 4 POLE, 4 POS SW 78
REVERB UNIT RV4C
R41 TREMOLO DEPTH CONTROL 2M LINEAR CBA 811-3708
R42 " FREQ " 1.5M REVERSED CBA 811-3711
R43 INSTRUMENT VOL " 500K AUDIO CBA 811-3707
R54 " TONE " 500K " CBA 811-3707
R55 REVERB. VOL " 2M LINEAR CBA 811-3708
CAPACITORS IN MFD
RESISTORS IN OHMS K = 1000, M = 1000000
MONITOR
EXT SPKR
HTRS
117V 50-60~
6CA4
6BQ5
6BQ5
12AU7
6C4
6EU7
6EU7

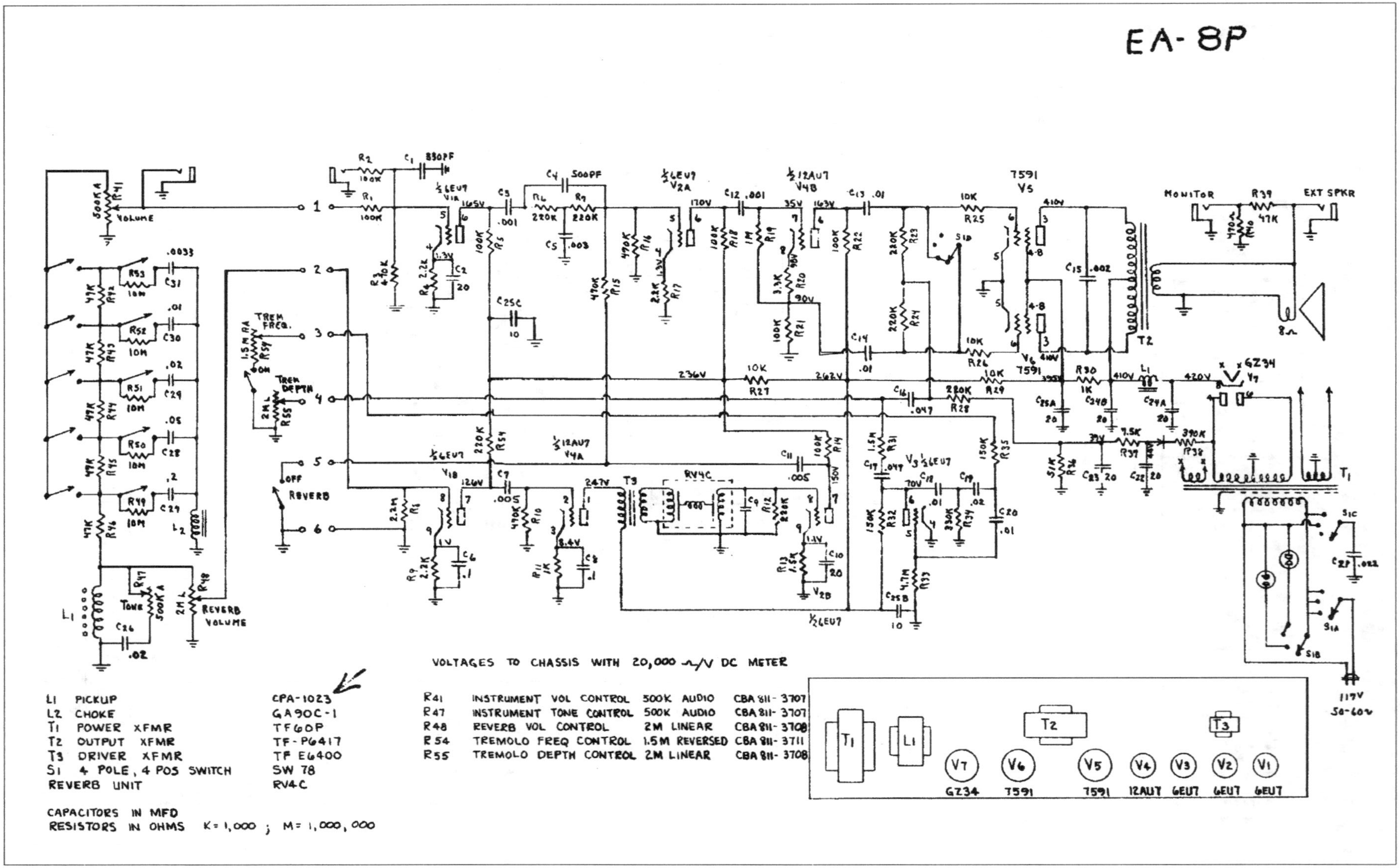
EA-8P
VOLTAGES TO CHASSIS WITH 20,000 Ω/V DC METER
L1 PICKUP CPA-1023
L2 CHOKE GA90C-1
T1 POWER XFMR TF60P
T2 OUTPUT XFMR TF-P6417
T3 DRIVER XFMR TF E6400
S1 4 POLE, 4 POS SWITCH SW 78
REVERB UNIT RV4C
CAPACITORS IN MFD
RESISTORS IN OHMS K = 1,000 ; M = 1,000,000
R41 INSTRUMENT VOL CONTROL 500K AUDIO CBA 811-3707
R47 INSTRUMENT TONE CONTROL 500K AUDIO CBA 811-3707
R48 REVERB VOL CONTROL 2M LINEAR CBA 811-3708
R54 TREMOLO FREQ CONTROL 1.5M REVERSED CBA 811-3711
R55 TREMOLO DEPTH CONTROL 2M LINEAR CBA 811-3708
T1 L1 T2 T3
V7 GZ34
V6 7591
V5 7591
V4 12AU7
V3 6EU7
V2 6EU7
V1 6EU7
MONITOR
EXT SPKR
117V 50-60~

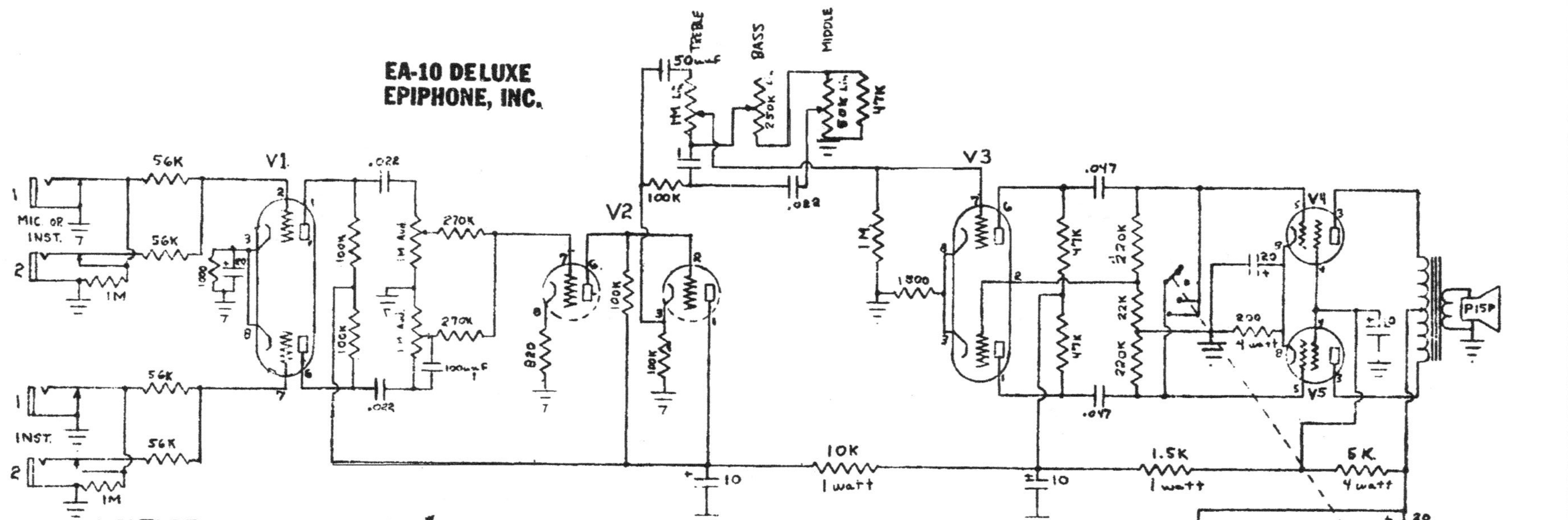

VOLTAGE CHART*

TUBE No.	TUBE TYPE	PIN 1	PIN 2	PIN 3	PIN 4	PIN 6	PIN 7	PIN 8
V1	12AX7	165	0	1.3	—	165	0	1.3
V2	12AX7	235	140	140	—	140	0	.75
V3	12AU7	170	0	8.5	—	170	0	8.5
V4-5	6L6GB	GND.	---	345	286	--	—	21.5
V6	GZ 34	GND.	353	—	300 RMS	300 RMS	—	—

* ALL DC. VOLTAGES MEASURED TO CHASSIS WITH 20,000 Ω/V METER.

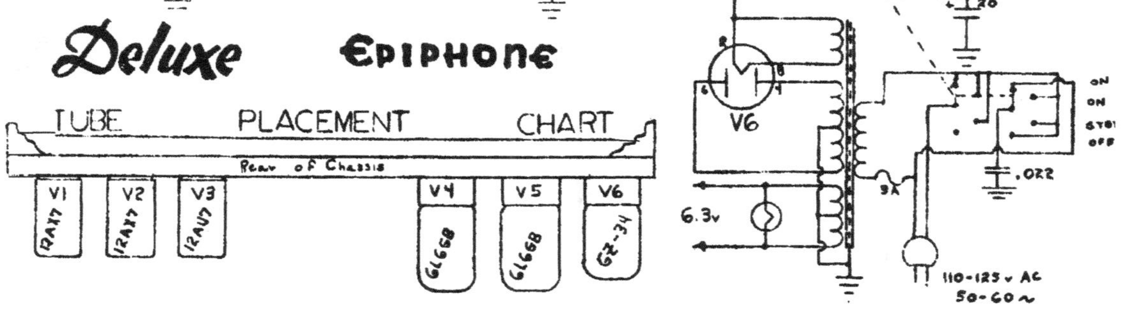

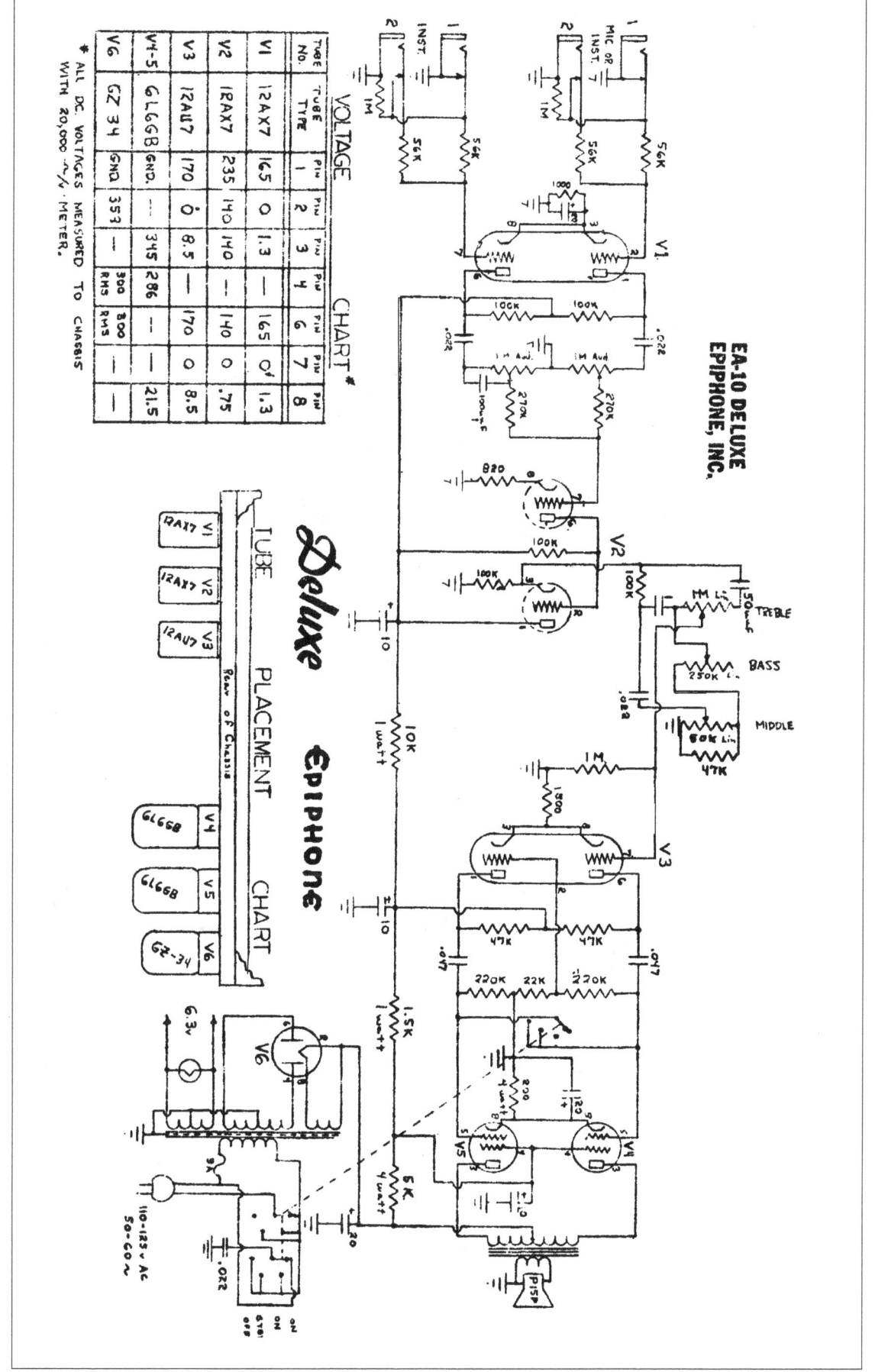

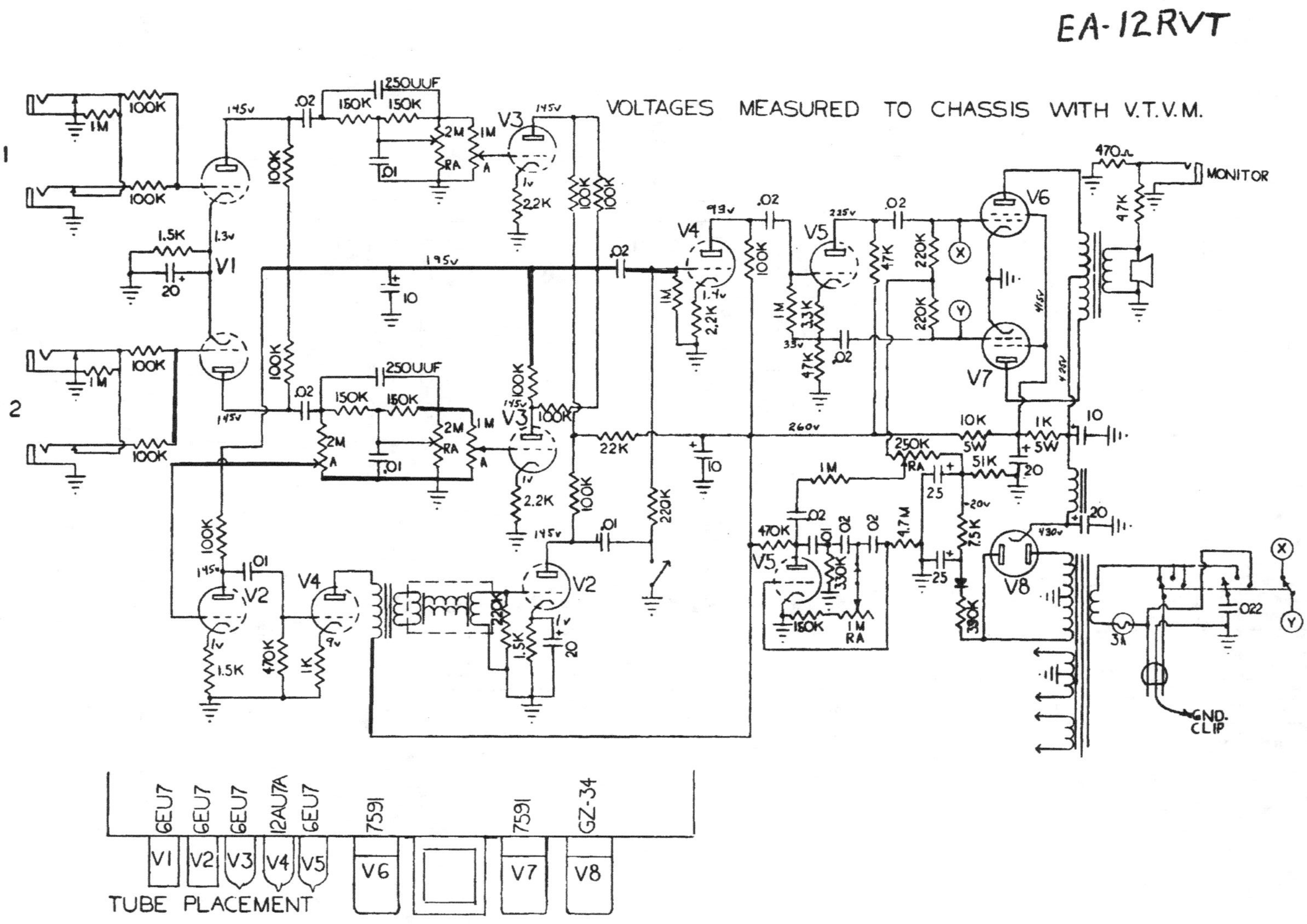
EA-12RVT
VOLTAGES MEASURED TO CHASSIS WITH V.T.V.M.
MONITOR
GND. CLIP
TUBE PLACEMENT
6EU7
6EU7
6EU7
12AU7A
6EU7
7591
7591
GZ-34
V1
V2
V3
V4
V5
V6
V7
V8

EA-15RVT

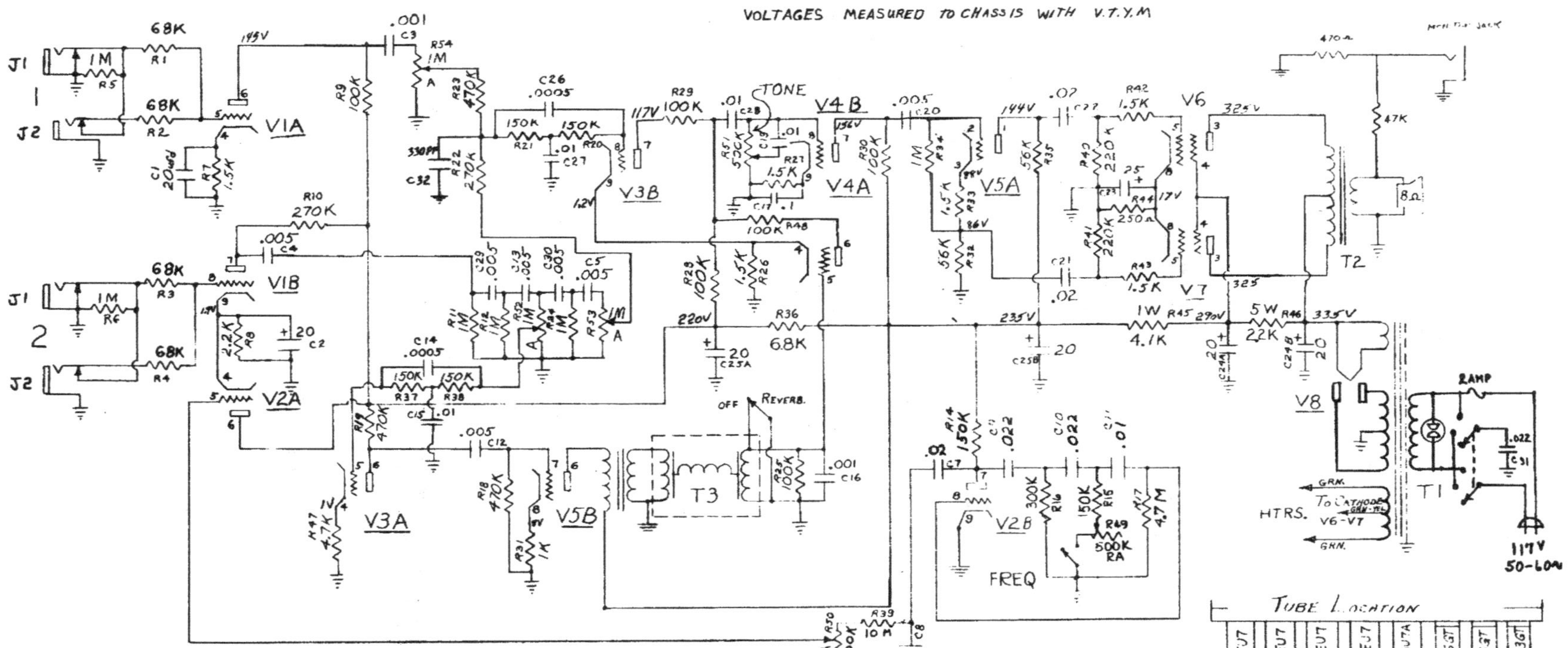

CONTROLS

R52,R53,R54 1 Meg, C2 Audio Taper (Reverb. & Loudness) C-BA-811-3709
R51 (Tone) 500K, C2 Audio Taper C-BA-811-3707
R50 (Depth) 500K, C3 Reversed-Audio C-BA-811-3704
R49 (Frequency) 500K, C3 Reversed Audio C-BA-811-3706

SWITCH

S1 SPST-SPDT 3 prs. Rotary Switch SW-899

SPEAKER

S-7334-2 C15R 15" Speaker with 90 cycle resonance cone. Jensen

TRANSFORMERS

T1	Power Transformer	TF-23-P
T2	Output Transformer	TF-18-01
T3	Reverb. Transformer	TF-E-6400

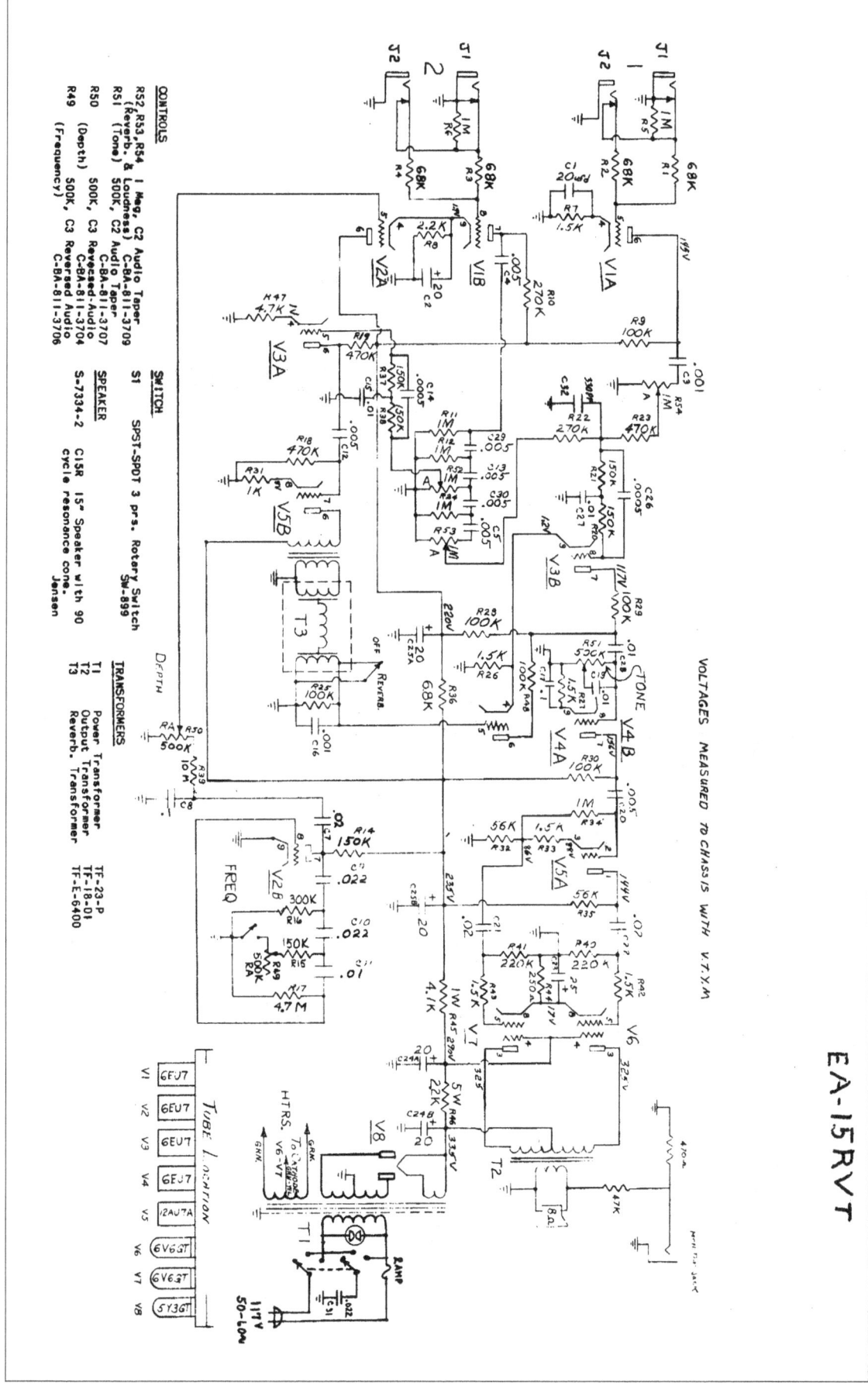

EA 16 RVT

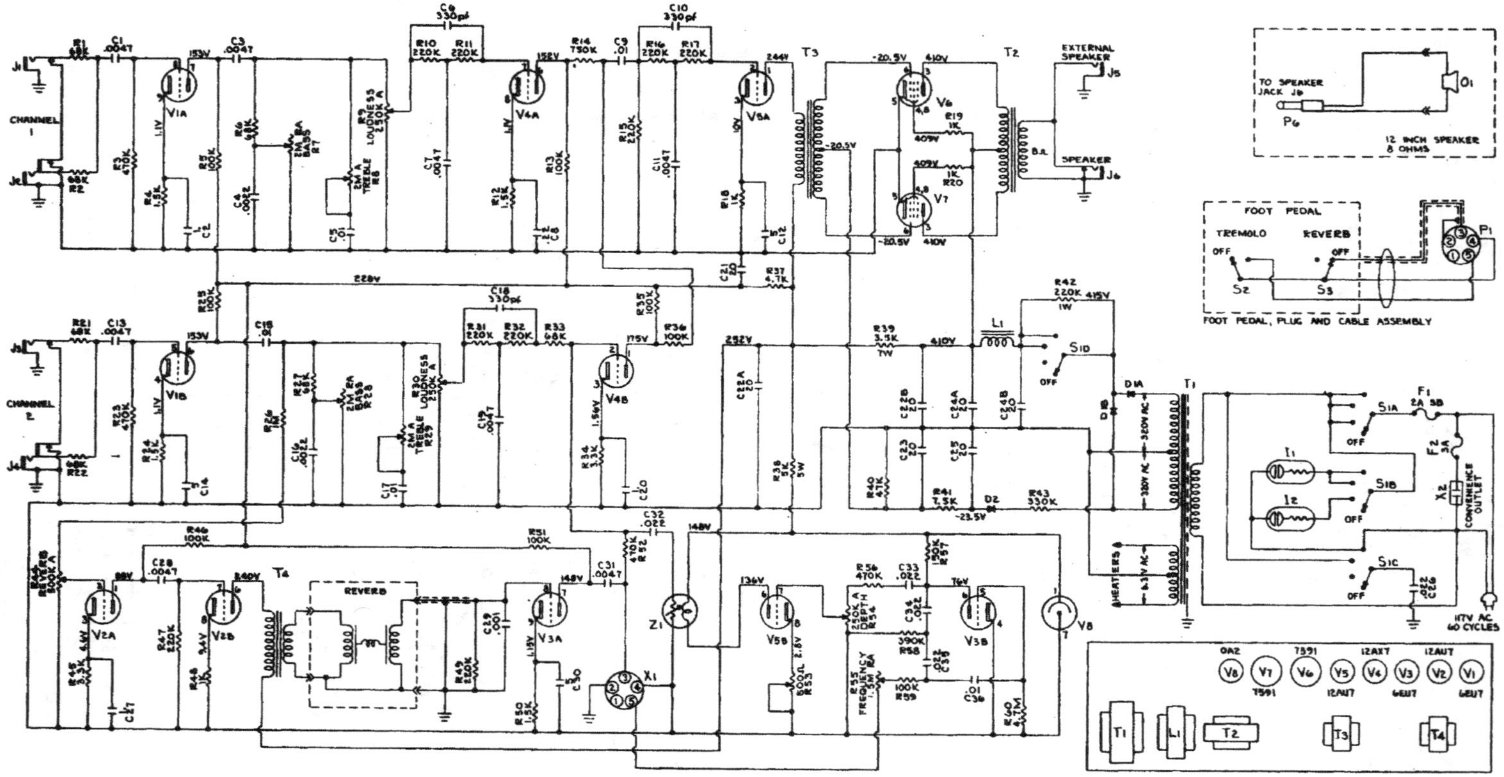

T1	Power Transformer	TF-104P
T2	Output Transformer	TF-P6417
T3	PP Driver Transformer	TF-1001D-1
T4	Reverb Transformer	TF-E6400
L1	Filter Choke	TF-3021H
S1A,B,C,D	Switch - off,standby-on-on	SW-78A-1
S2,S3	Switches - SPST	SW-82403
I1	PL-37R Pilot Light (Red)	
I2	PL-38A Pilot Light (Amber)	
R9,R30,R34	Control - 250K Audio	CBA-4008
R8,R29	Control - 2 meg, audio	CBA-811-3702-1
R44	Control - 500K Audio	CBA-4007
R55	Control - 1.5 meg, RA	CBA-811-3711-1
R7,R28	Control - 1 meg, audio	CBA-811-3709-1

D1A,B	Diode - 1200 PIV	DI-57
D2	Diode - 600 PIV	DI-71B

CN-78PCG-5	Foot Pedal Socket
CN-303	AC Convenience Outlet

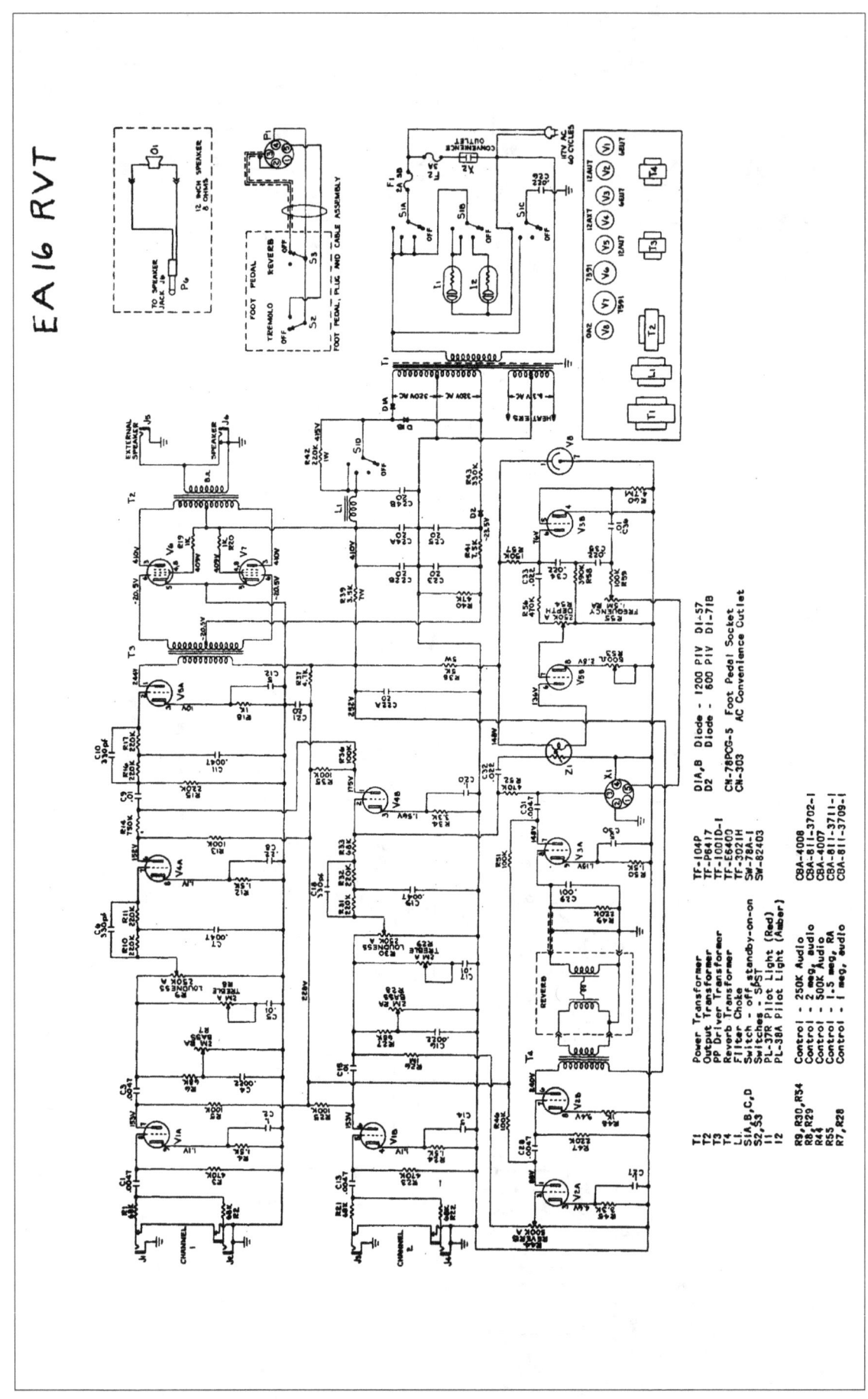

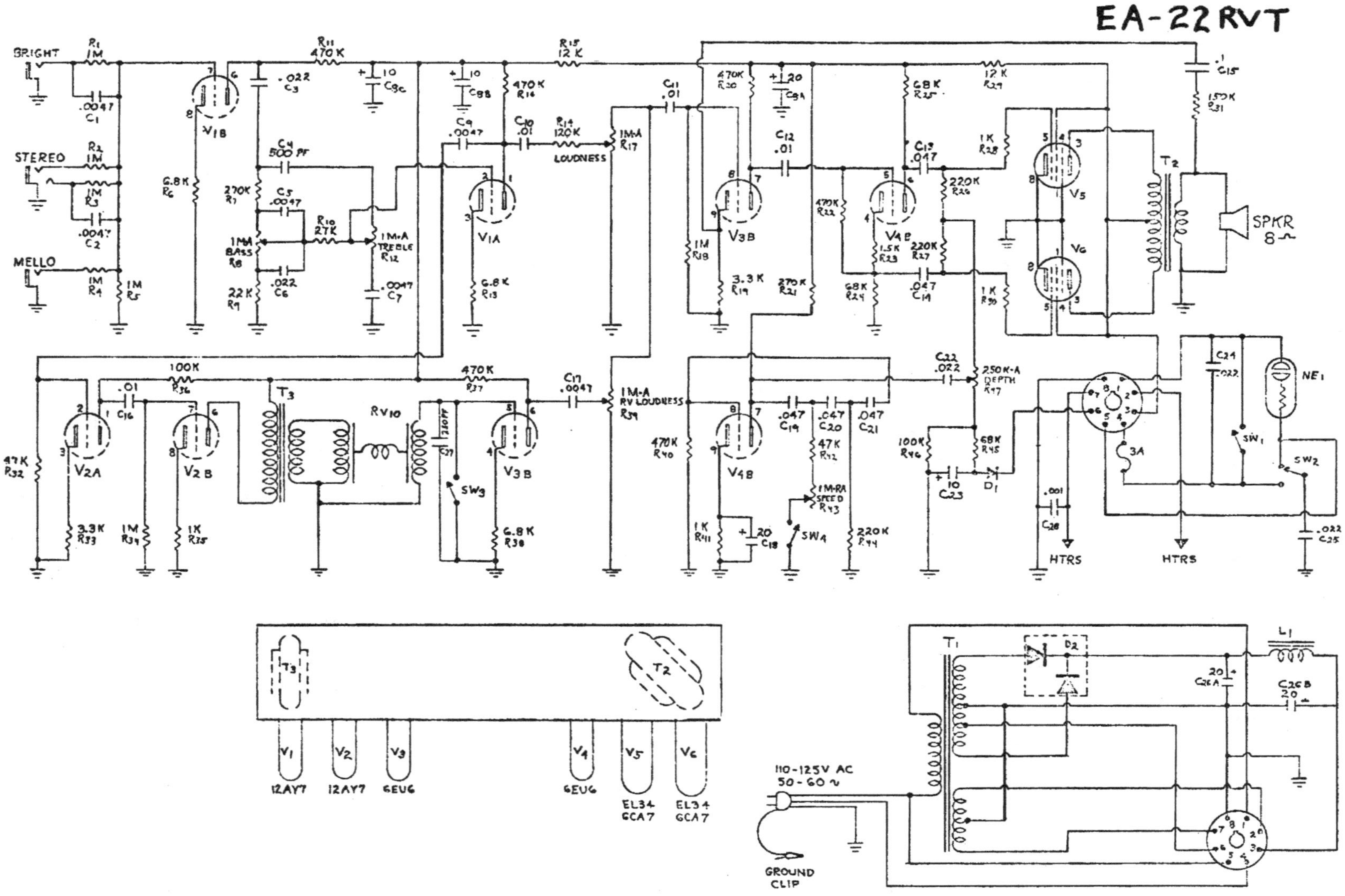
EA-22RVT
BRIGHT
STEREO
MELLO
SPKR 8Ω
V1B
V1A
V2A
V2B
V3B
V4B
V5
V6
T1
T2
T3
RV10
LOUDNESS
1M-A TREBLE R12
1M-A BASS R8
1M-A RV LOUDNESS R39
250K-A DEPTH R47
1M-RA SPEED R43
SW1
SW2
SW3
SW4
NE1
D1
D2
L1
3A
HTRS
12AY7
12AY7
6EU6
6EU6
EL34 6CA7
EL34 6CA7
110-125V AC 50-60 ~
GROUND CLIP

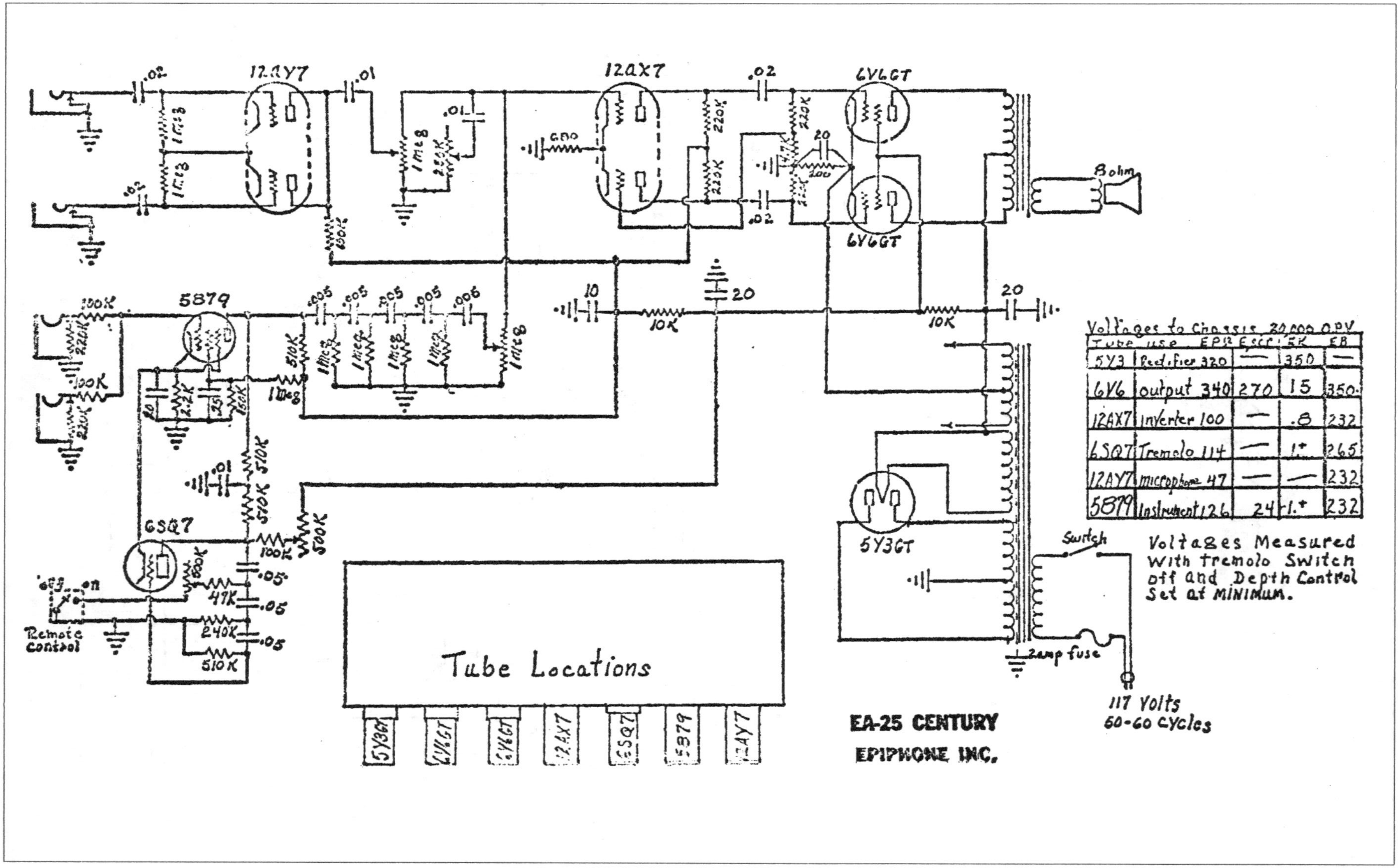

Voltages to Chassis 20,000 O.P.V.

Tube	Use	EP	Esc	EK	EB
5Y3	Rectifier	320	—	350	—
6V6	output	340	270	15	350
12AX7	inverter	100	—	.8	232
6SQ7	Tremolo	114	—	1.+	265
12AY7	microphone	47	—	—	232
5879	Instrument	126	24	1.+	232

Voltages Measured With tremolo Switch off and Depth Control Set at MINIMUM.

EA26-RVT

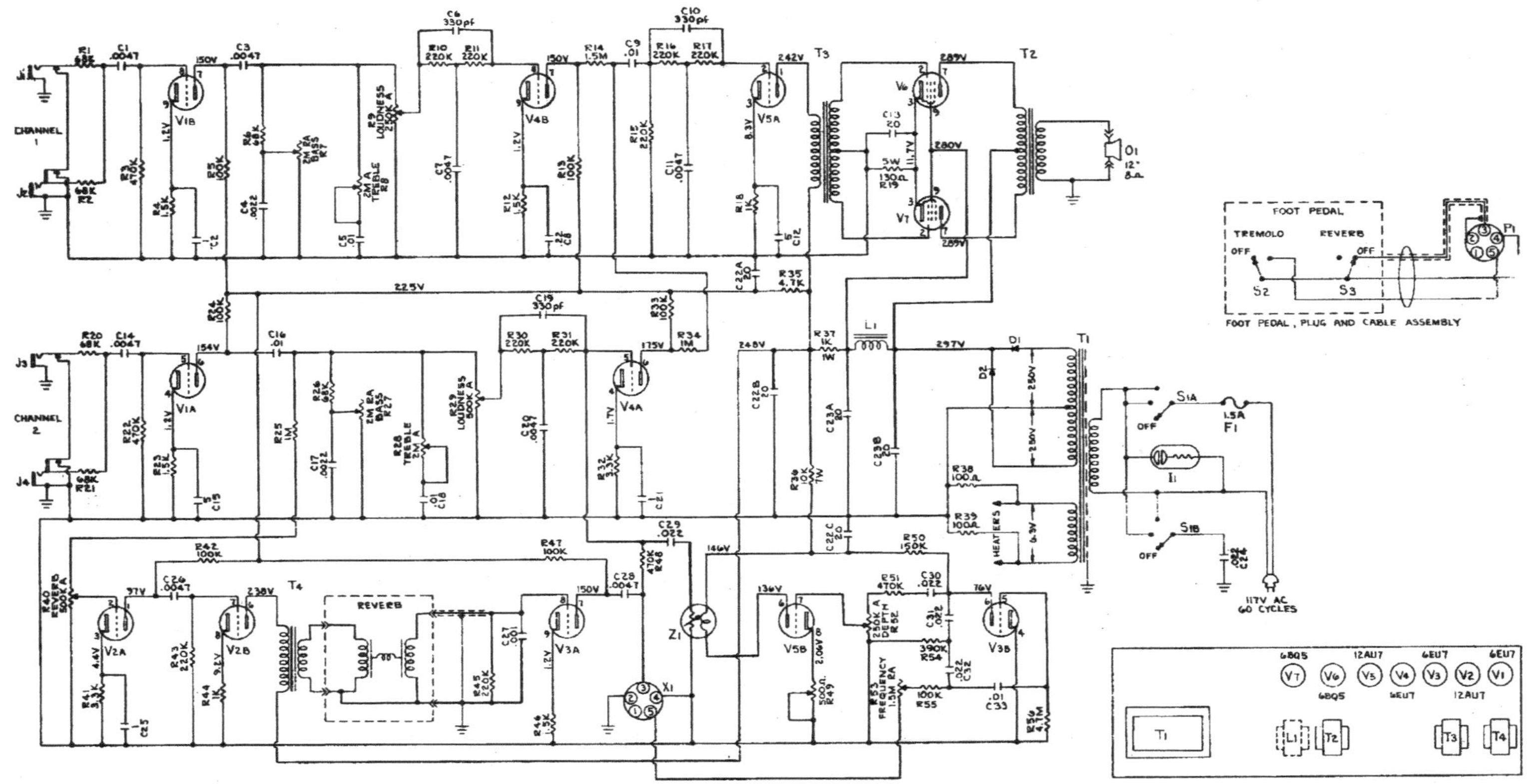

T1	TF-1C5T-1	Power Transformer
T2	TF-504-C	Output Transformer
T3	TF-1001D-1	PP Driver Transformer
T4	TF-E6400	Reverb Transformer
L1	TF-1003-C	Filter Choke
S1A,B	SW-899-1	Switch - SPST Rotary
S2,S3	SW-82403	Switches - SPST
I1	PL-35R	Pilot Light (Neon) w/clip
X1	CN-78PCG-5	Foot Pedal Socket
P1	CN-91MPM-5L	Foot Pedal Plug
D1,D2	DI-404	Diode - 800 PIV

R9,R52	CBA-4008	Control - 250K, Audio (Loudness & Depth)
R8,R28	CBA-311-3702-1	Control - 2 meg, Audio (Treble)
R29,R40	CBA-4007	Control - 500K, Audio (Loudness & Reverb)
R53	CBA-311-3711-1	Control - 1.5 meg, RA (Frequency)
R7,R27	CBA-813-222-1	Control - 2 meg, RA (Bass)
R49	C5-23391	Control - 500 ohm, WW

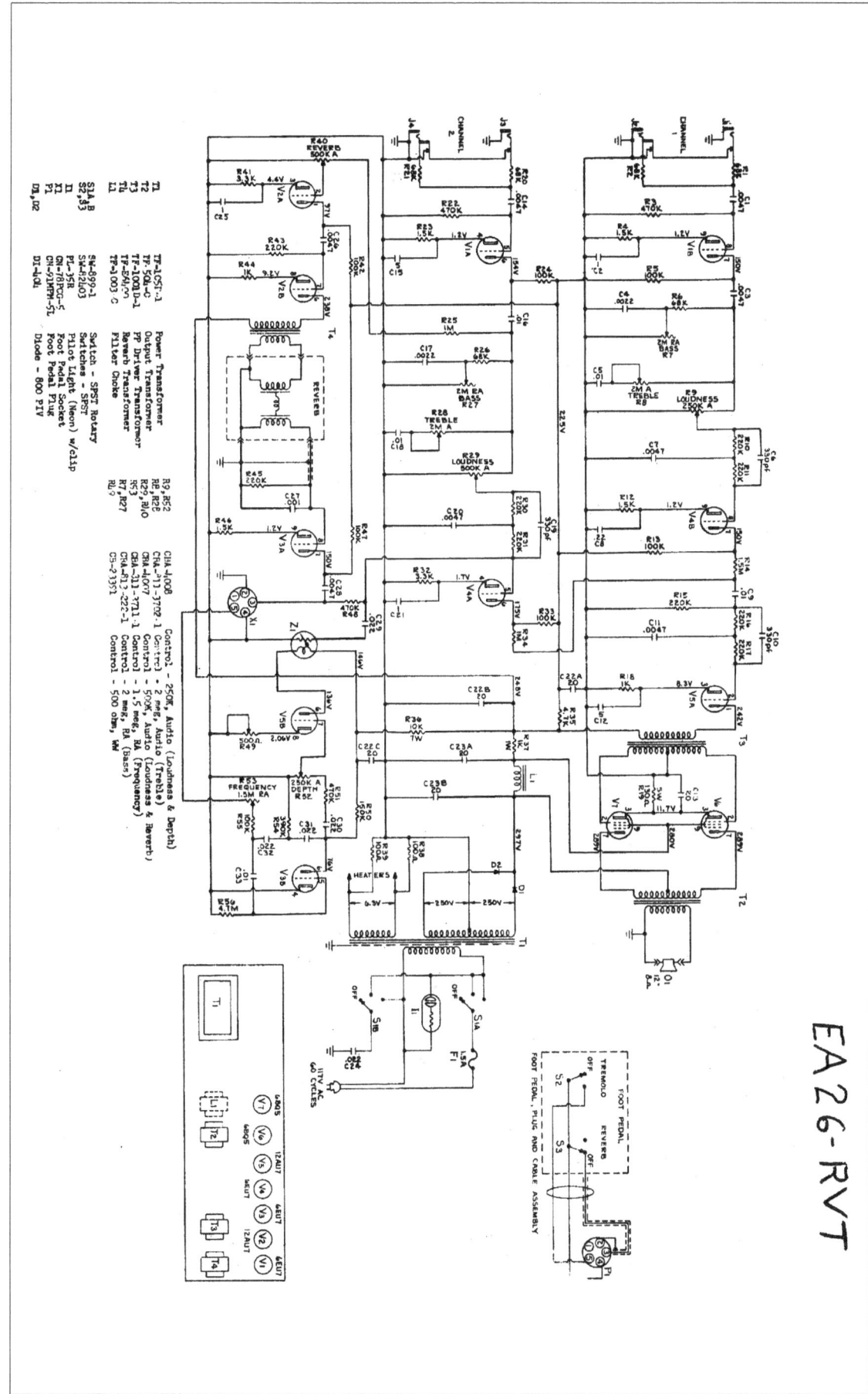

EA28-RVT

R2,R3	Loudness & Reverb. Control, 2 meg audio taper	C-BA-811-3702
R4	Depth Control, 250K reversed audio	C-BA-811-3704
R5	Frequency Control, 500K reversed audio	C-BA-811-3706
R1	Tone Control, 1 meg audio taper	C-BA-811-3709

T1	Output Transformer	TF-18-01
T2	Power Transformer	TF-18P
T3	Reverberation Transformer	TF-E-6400
S1	Toggle Switch (on-off) with 2 nuts	SW-80994
(1)	S-20009 12" Speaker	S-20009

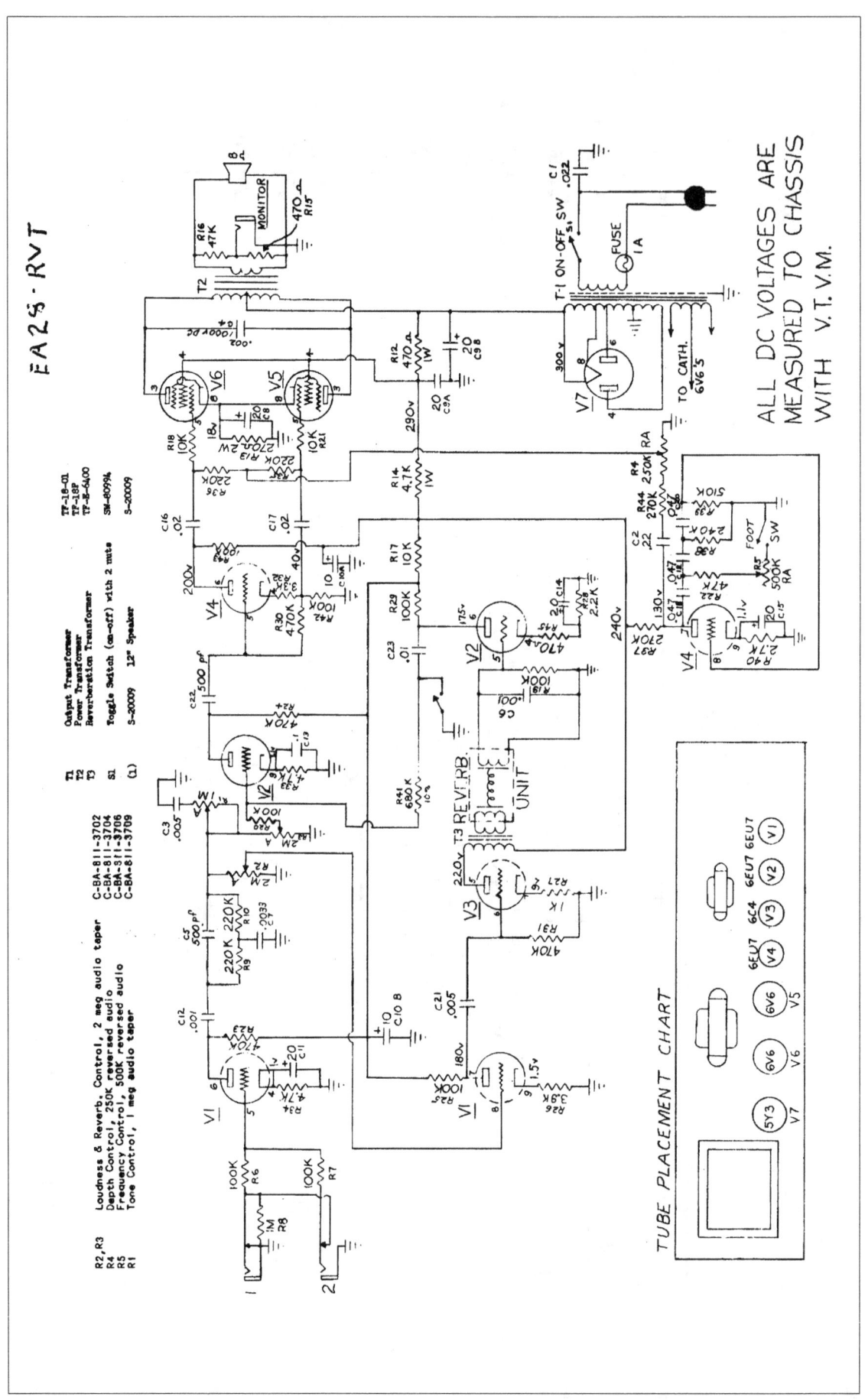

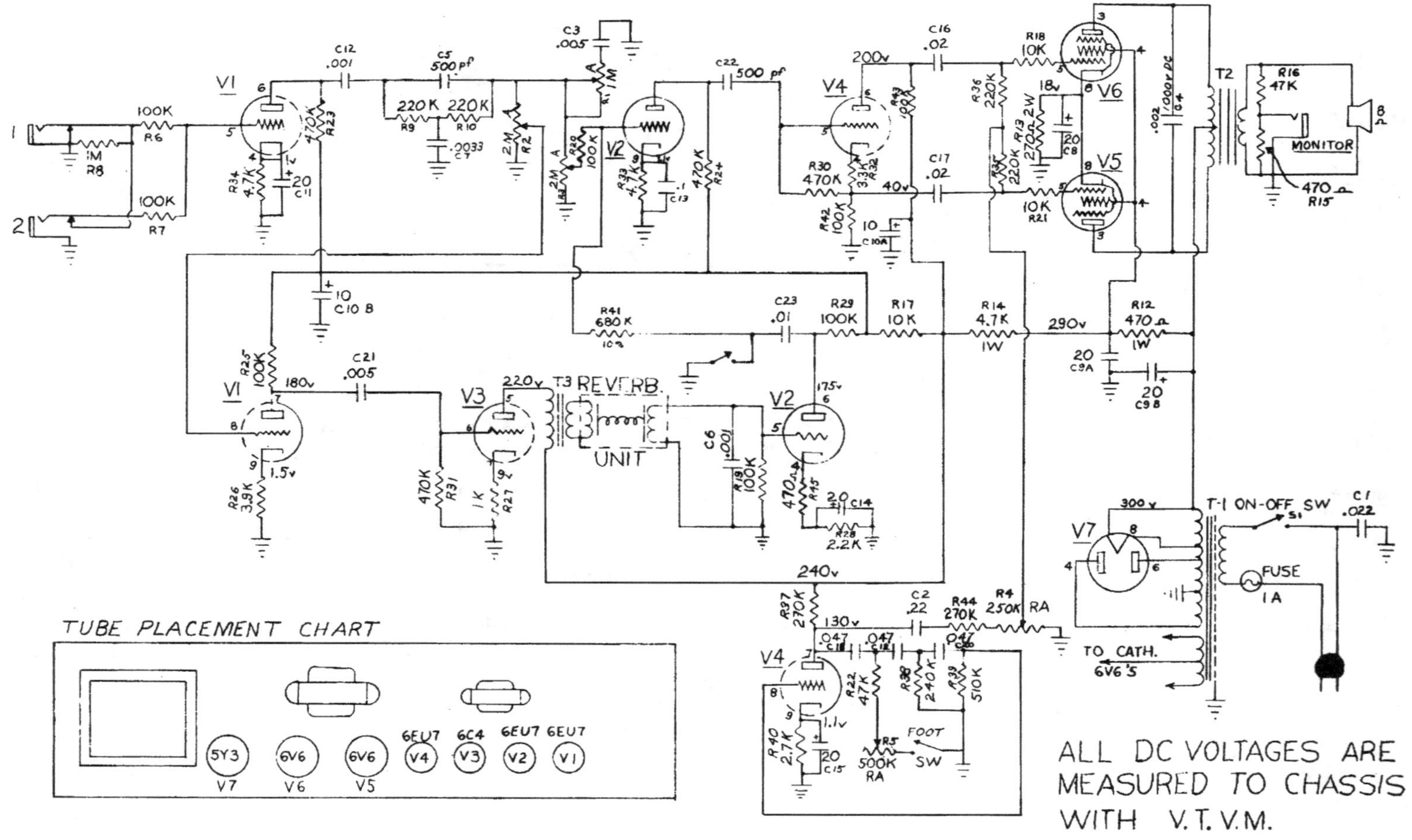

EA-30 **TRIUMPH**

Epiphone

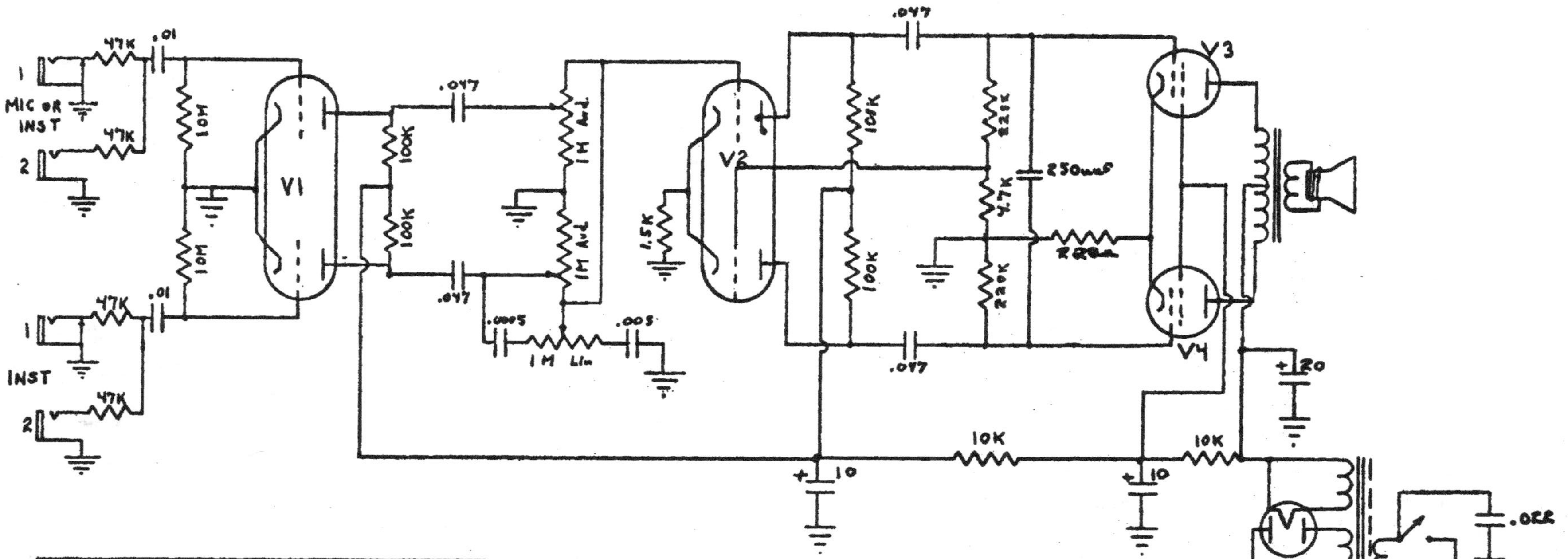

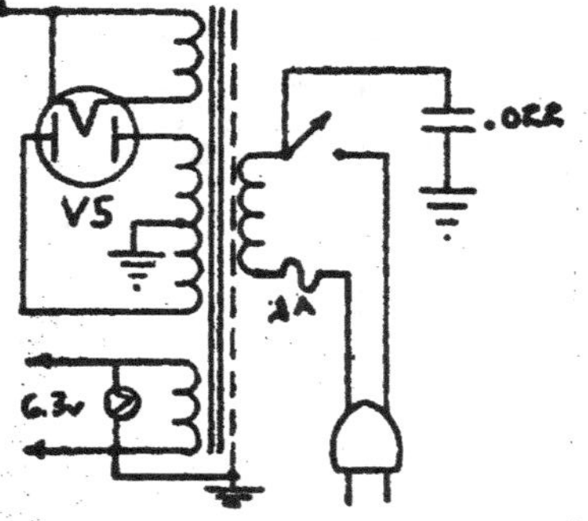

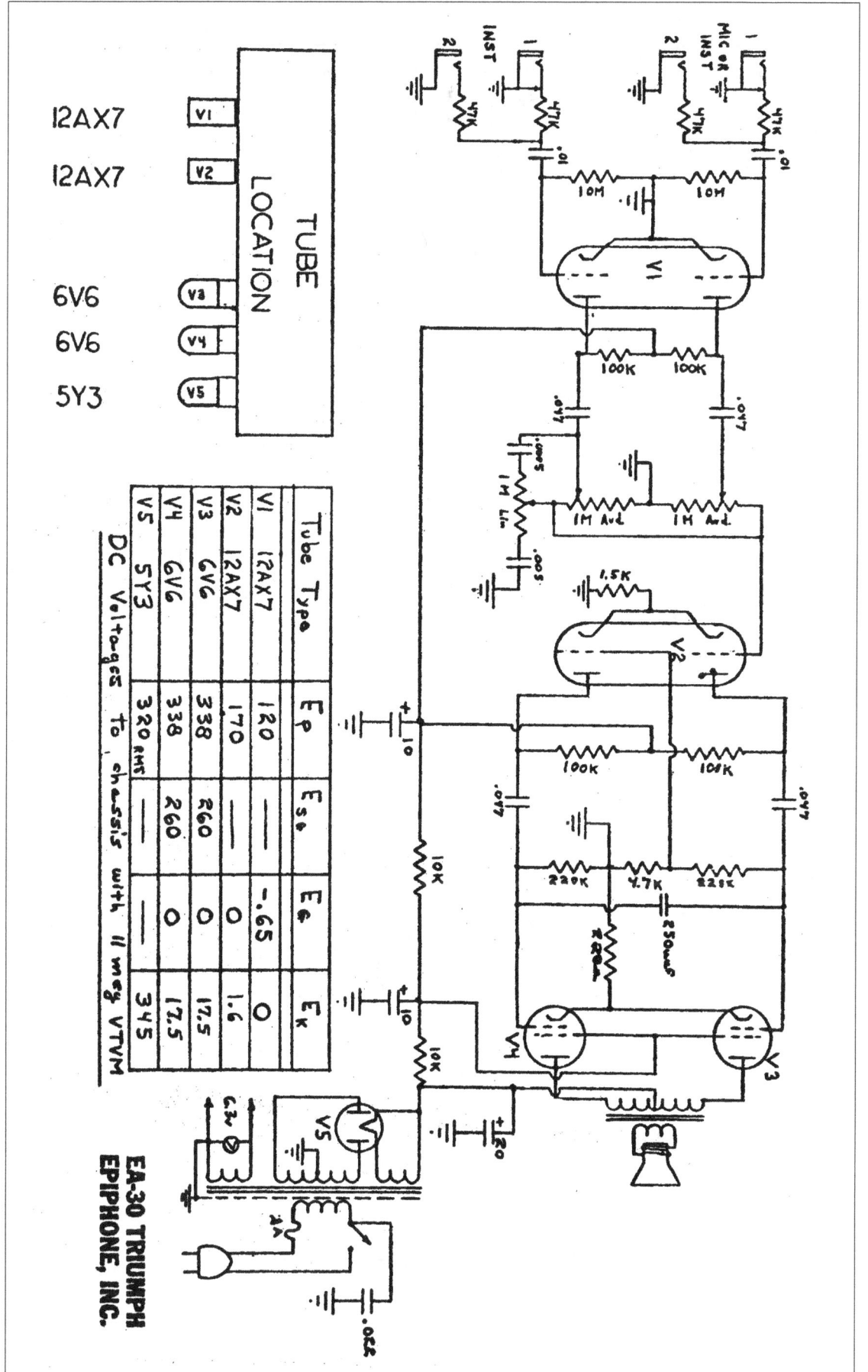

TUBE LOCATION

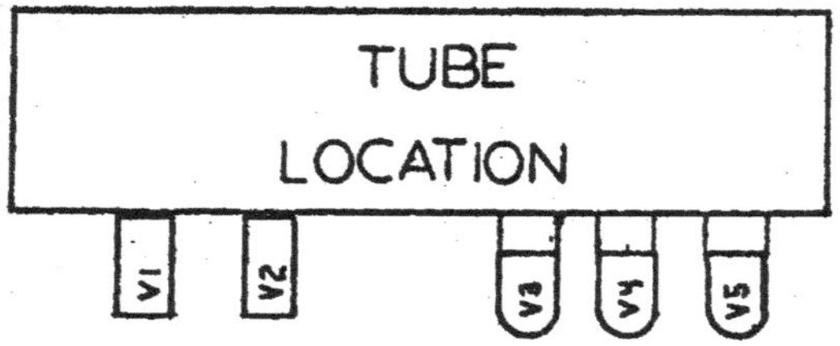

12AX7 12AX7 6V6 6V6 5Y3

Tube Type		E_p	E_{sg}	E_G	E_K
V1	12AX7	120	—	-.65	0
V2	12AX7	170	—	0	1.6
V3	6V6	338	260	0	17.5
V4	6V6	338	260	0	17.5
V5	5Y3	320 RMS	—	—	345

DC Voltages to chassis with 11 meg VTVM

EA-30 TRIUMPH
EPIPHONE, INC.

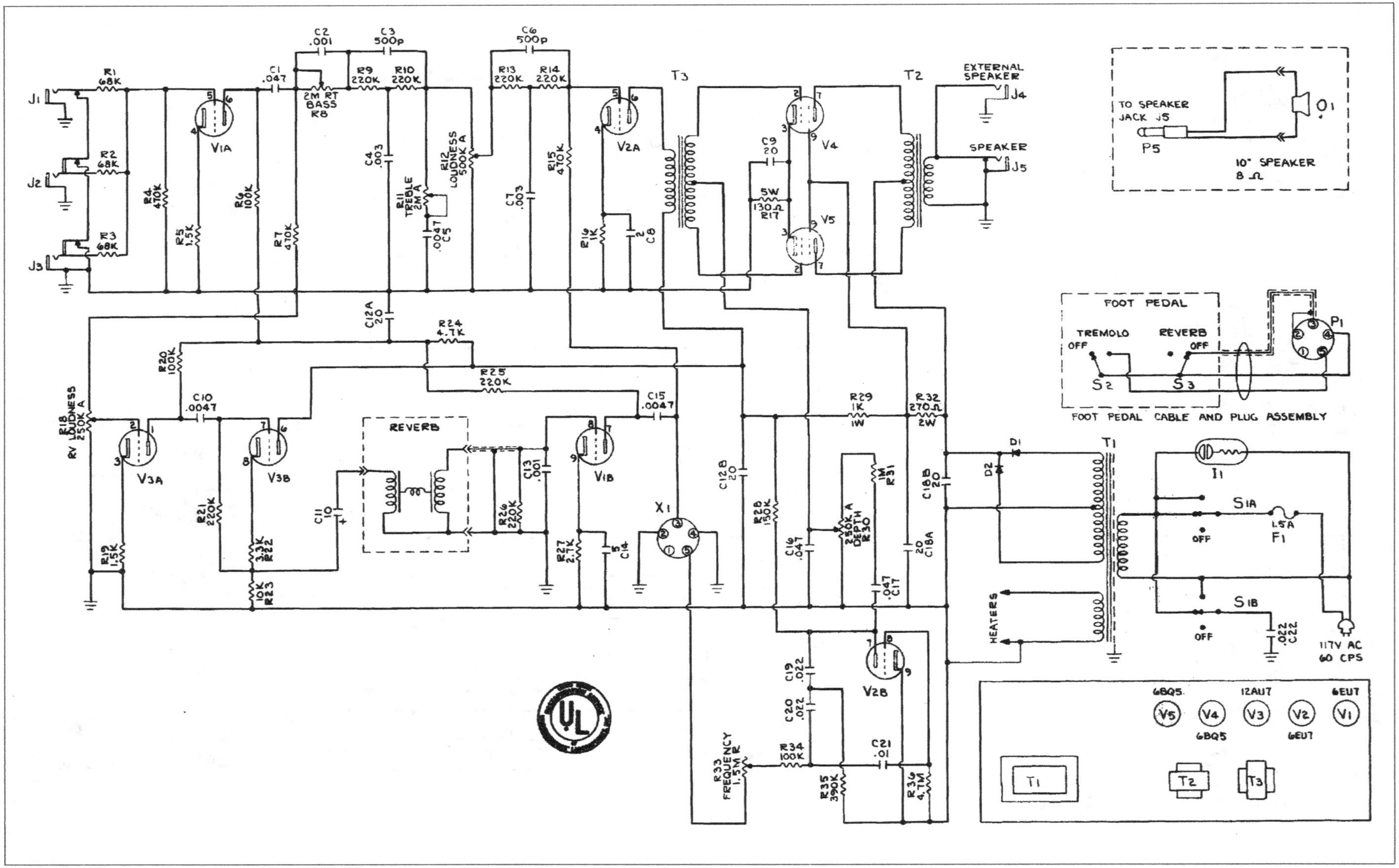

J1
J2
J3
R1 68K
R2 68K
R3 68K
R4 470K
R5 1.5K
R6 100K
R7 470K
V1A
C1 .047
C2 .001
C3 500p
2M RT BASS R8
R9 220K
R10 220K
C4 .003
R11 TREBLE 2M A
.0047 C5
R12 LOUDNESS 500K A
C6 500p
R13 220K
R14 220K
C7 .003
R15 470K
V2A
R16 1K
2 C8
T3
T2
C9 20
5W 130Ω R17
V4
V5
EXTERNAL SPEAKER J4
SPEAKER J5
C12A 20
R24 4.7K
R20 100K
R25 220K
R18 RV LOUDNESS 250K A
C10 .0047
V3A
V3B
R19 1.5K
R21 220K
3.3K R22
10K R23
C11 10
REVERB
R26 220K
C13 .001
V1B
R27 2.7K
5 C14
C15 .0047
X1
C12B 20
R28 150K
C16 .047
R29 1K 1W
R30 250K A DEPTH
1M R31
.047 C17
R32 270Ω 2W
C18B 20
20 C18A
C19 .022
C20 .022
V2B
R33 FREQUENCY 1.5M R
R34 100K
R35 390K
C21 .01
R36 4.7M
D1
D2
T1
HEATERS
I1
S1A
1.5A F1
OFF
S1B
.022 C22
117V AC 60 CPS
TO SPEAKER JACK J5
P5
O1
10" SPEAKER 8 Ω
FOOT PEDAL
TREMOLO
REVERB
OFF
S2
S3
P1
FOOT PEDAL CABLE AND PLUG ASSEMBLY
6BQ5
12AU7
6EU7
V5
V4
V3
V2
V1
T1
T2
T3

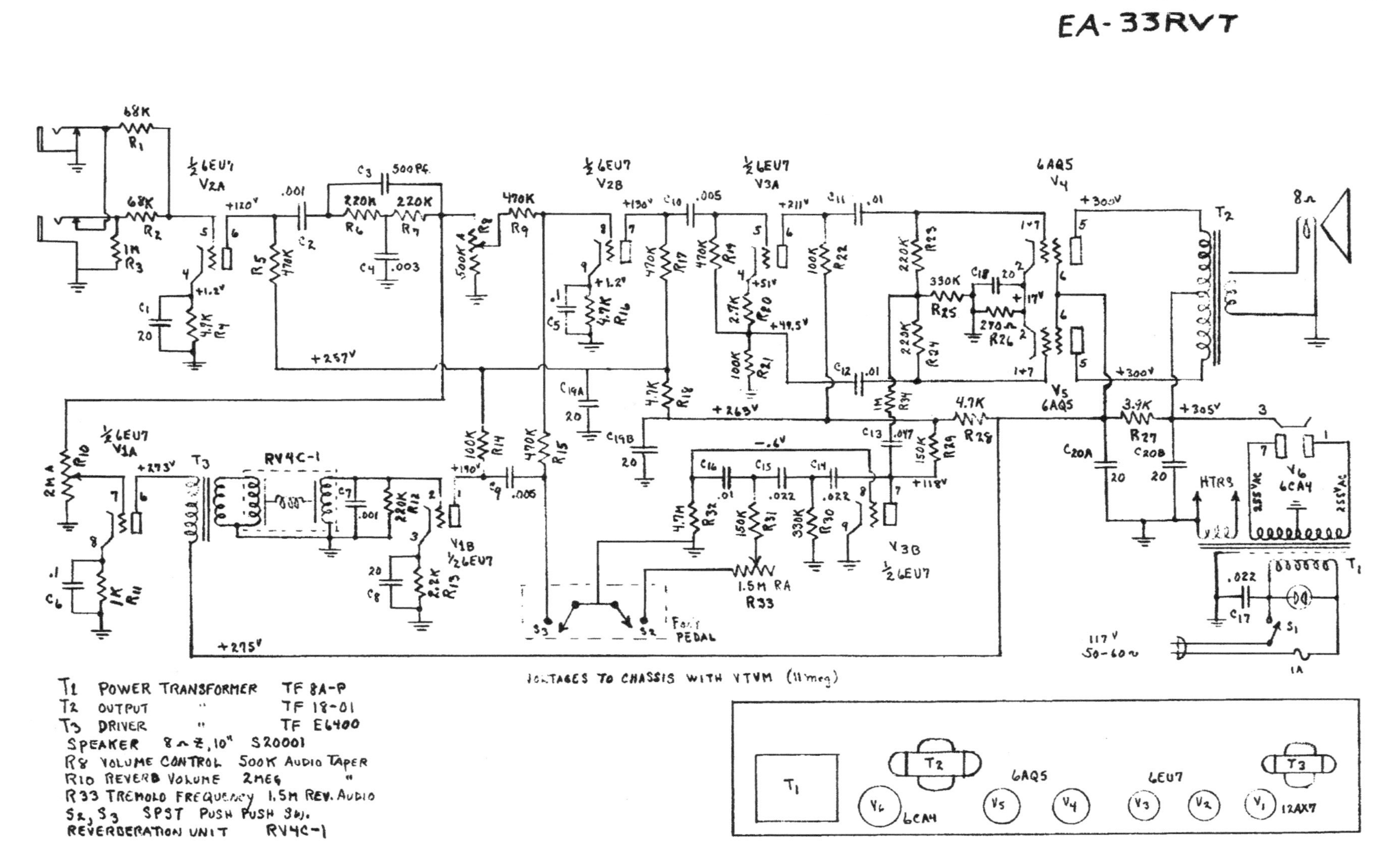
EA-33RVT
T1 POWER TRANSFORMER TF 8A-P
T2 OUTPUT " TF 18-01
T3 DRIVER " TF E6400
SPEAKER 8 Ω Z, 10" S20001
R8 VOLUME CONTROL 500K AUDIO TAPER
R10 REVERB VOLUME 2MEG "
R33 TREMOLO FREQUENCY 1.5M REV. AUDIO
S2, S3 SPST PUSH PUSH SW.
REVERBERATION UNIT RV4C-1
VOLTAGES TO CHASSIS WITH VTVM (11 meg)
FOOT PEDAL
117V 50-60~
6AQS
6EU7
12AX7
6CA4

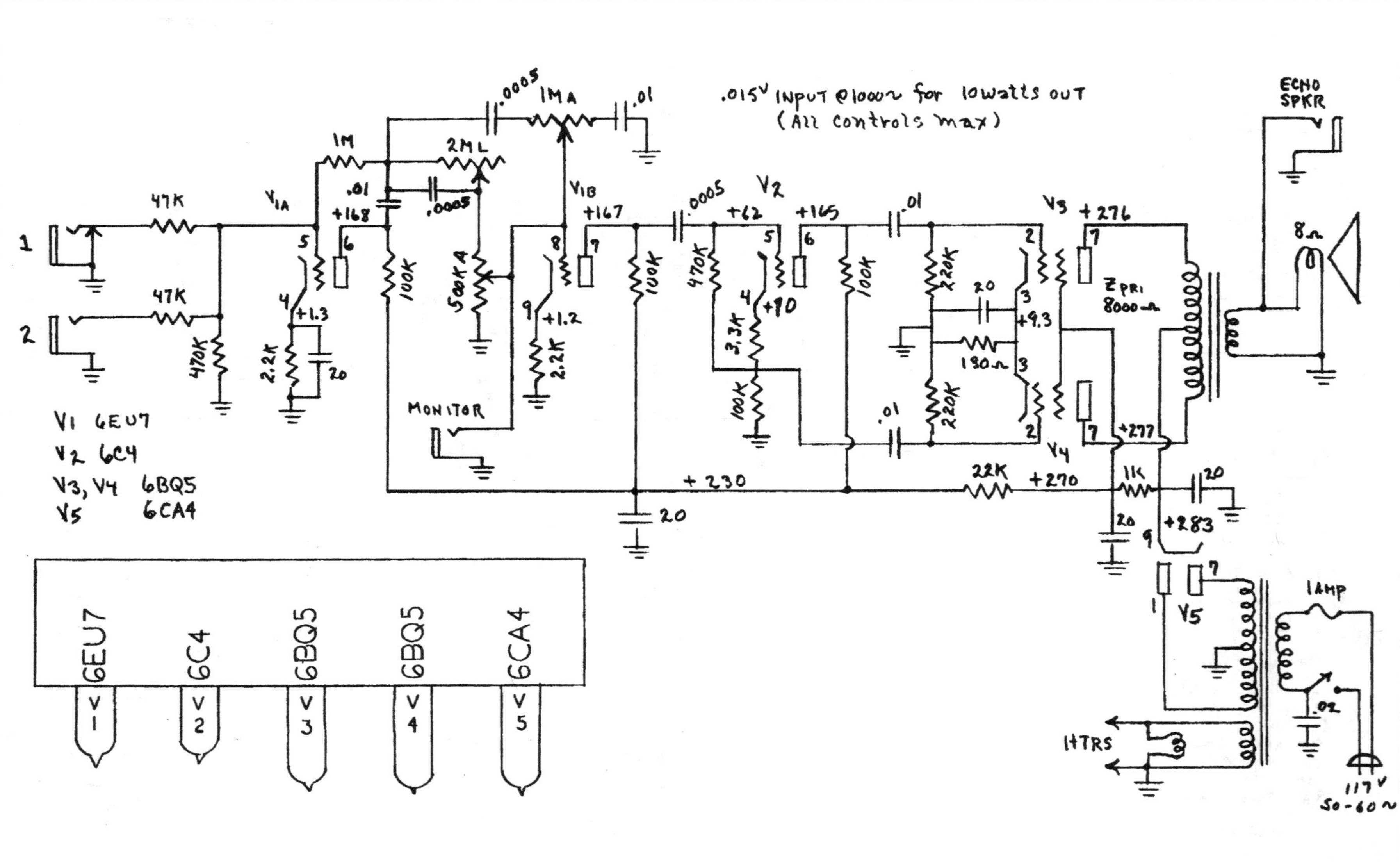
.015V INPUT @1000Ω for 10watts OUT
(All controls max)
ECHO SPKR
8Ω
Z PRI 8000Ω
V1 6EU7
V2 6C4
V3, V4 6BQ5
V5 6CA4
6EU7
6C4
6BQ5
6BQ5
6CA4
MONITOR
HTRS
1AMP
117V 50-60~

EA-35T

Epiphone

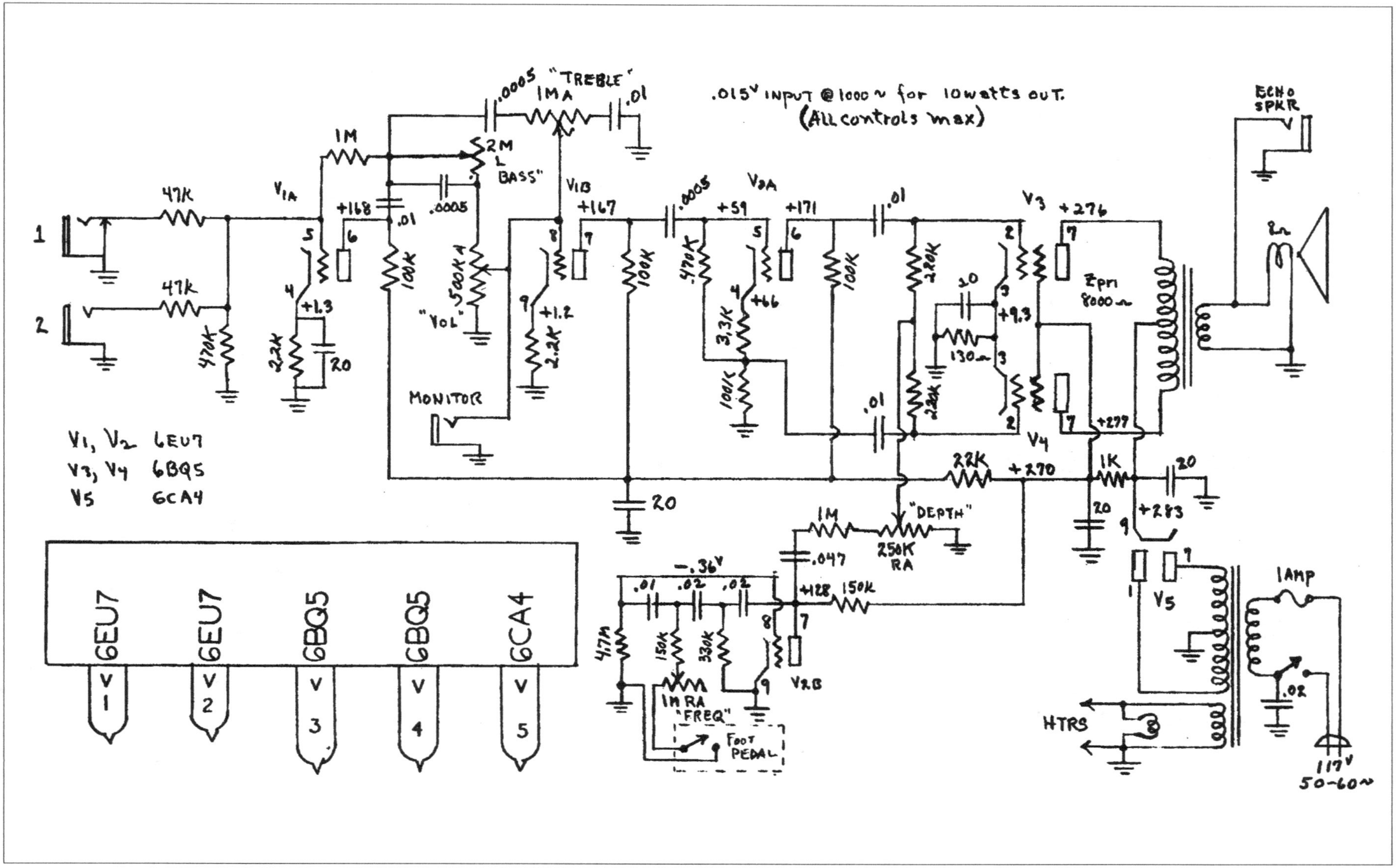

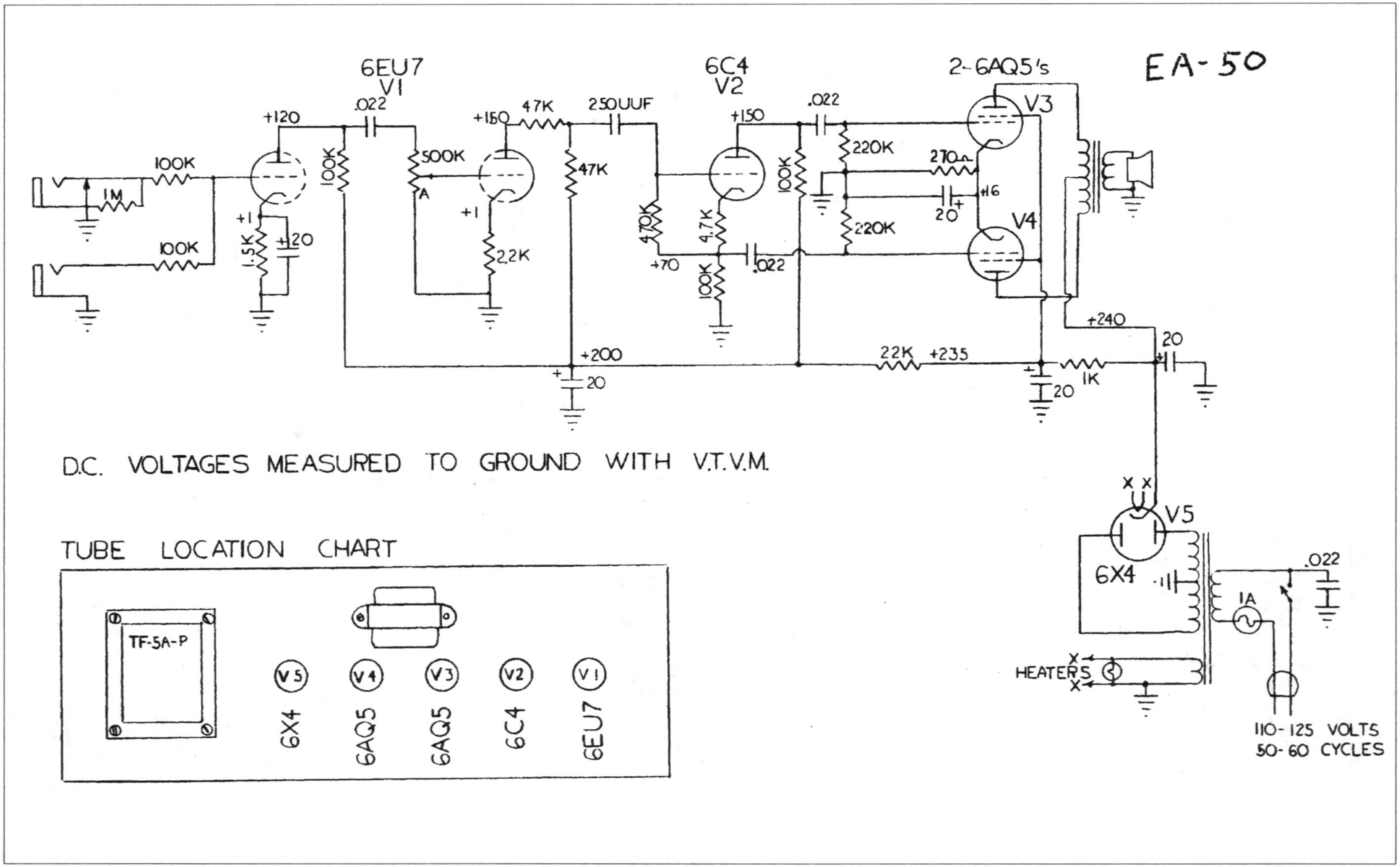
EA-50
6EU7
V1
6C4
V2
2-6AQ5's
V3
V4
V5
6X4
.022
+120
100K
1M
+1
1.5K
+120
500K
A
+150
47K
2.2K
250UUF
+200
20
470K
+70
4.7K
100K
220K
270Ω
+16
+240
22K
+235
1K
HEATERS
X
1A
110-125 VOLTS
50-60 CYCLES
D.C. VOLTAGES MEASURED TO GROUND WITH V.T.V.M.
TUBE LOCATION CHART
TF-5A-P
V5
V4
V3
V2
V1
6X4
6AQ5
6AQ5
6C4
6EU7

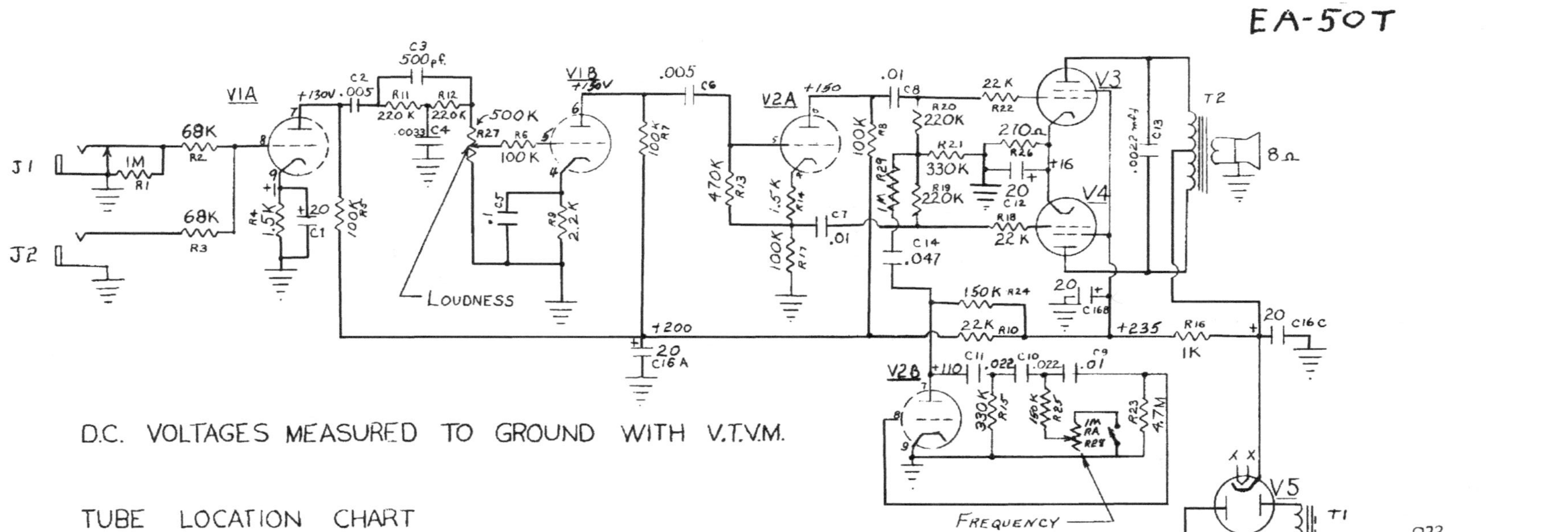

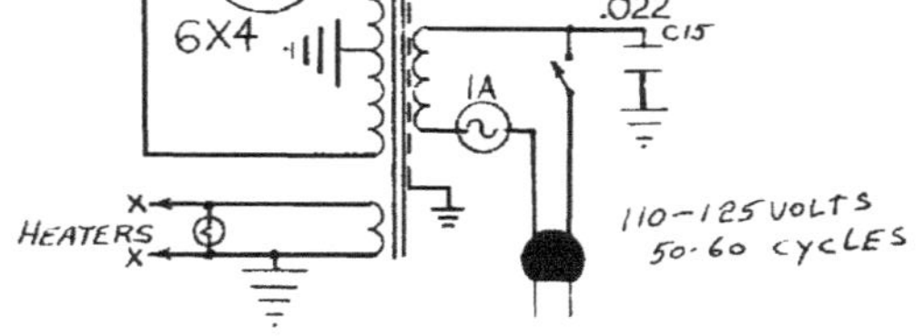

D.C. VOLTAGES MEASURED TO GROUND WITH V.T.V.M.

TUBE LOCATION CHART

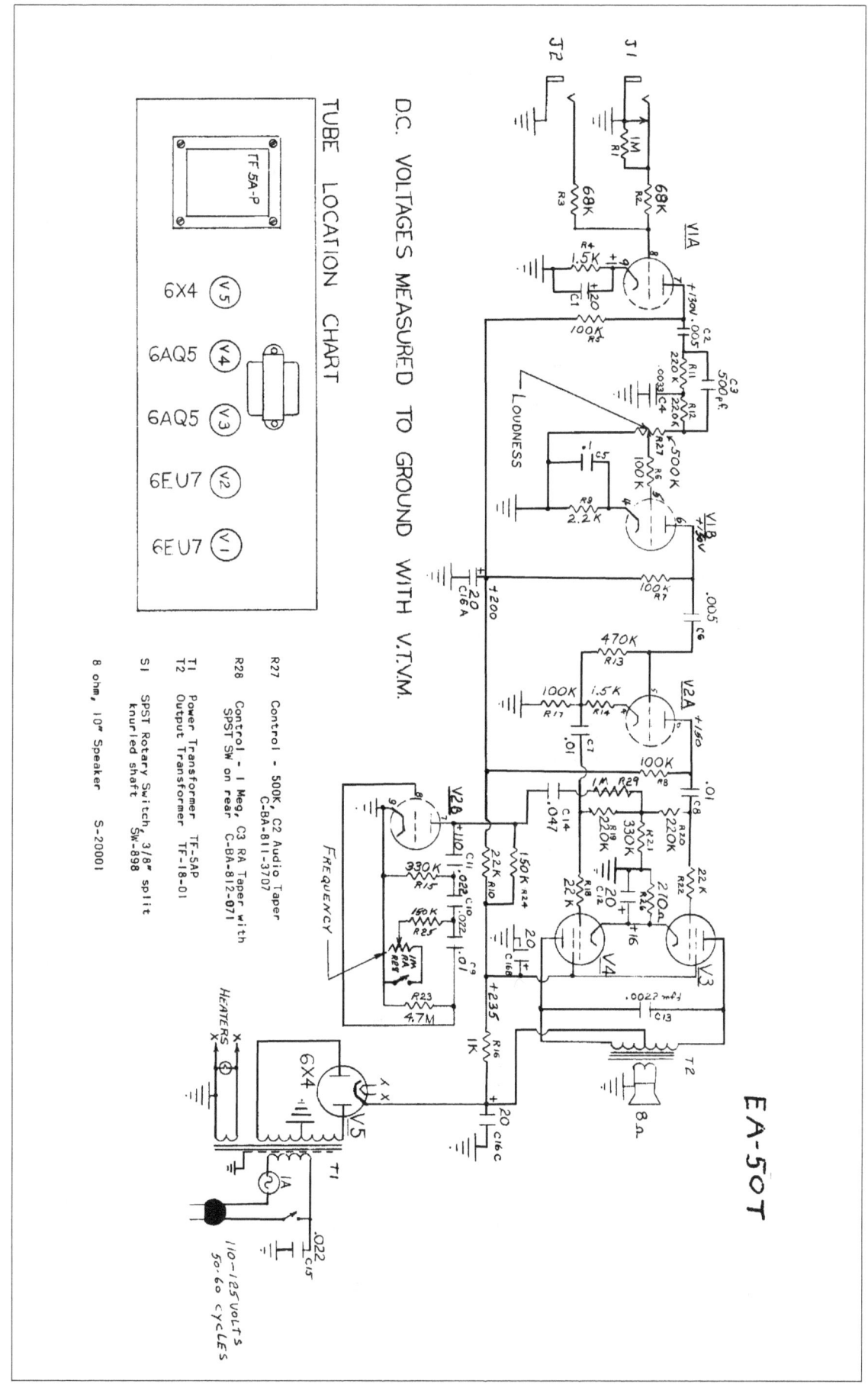

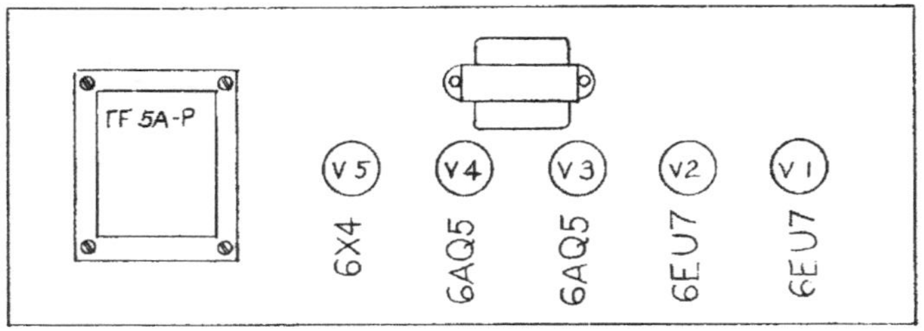

R27 Control - 500K, C2 Audio Taper C-BA-811-3707

R28 Control - 1 Meg, C3 RA Taper with SPST SW on rear C-BA-812-071

T1 Power Transformer TF-5AP
T2 Output Transformer TF-18-01

S1 SPST Rotary Switch, 3/8" split knurled shaft SW-898

8 ohm, 10" Speaker S-20001

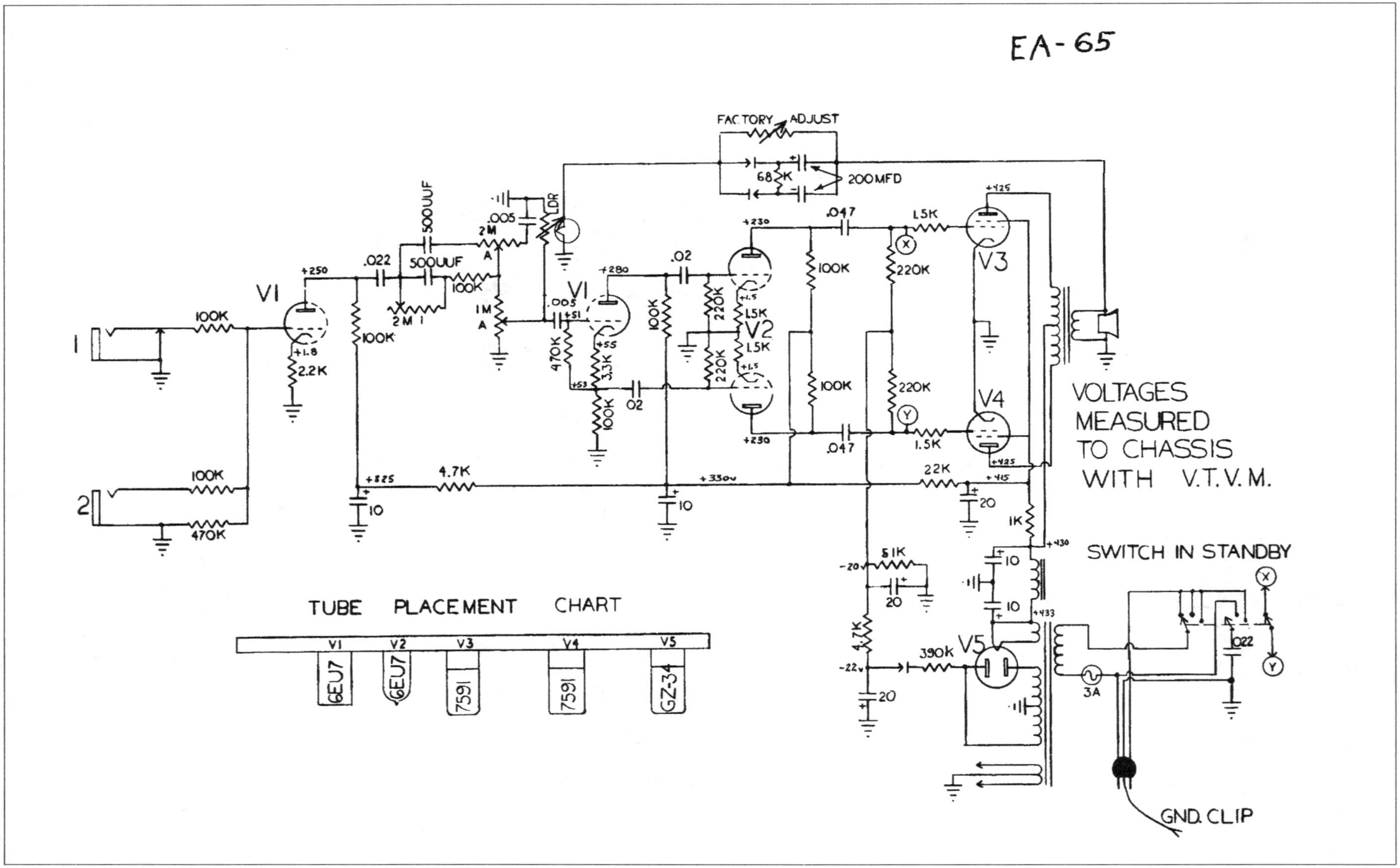
EA-65
VOLTAGES MEASURED TO CHASSIS WITH V.T.V.M.
SWITCH IN STANDBY
GND. CLIP
FACTORY ADJUST
200MFD
LDR
V1
V2
V3
V4
V5
TUBE PLACEMENT CHART
V1 6EU7
V2 6EU7
V3 7591
V4 7591
V5 GZ-34

EA-70

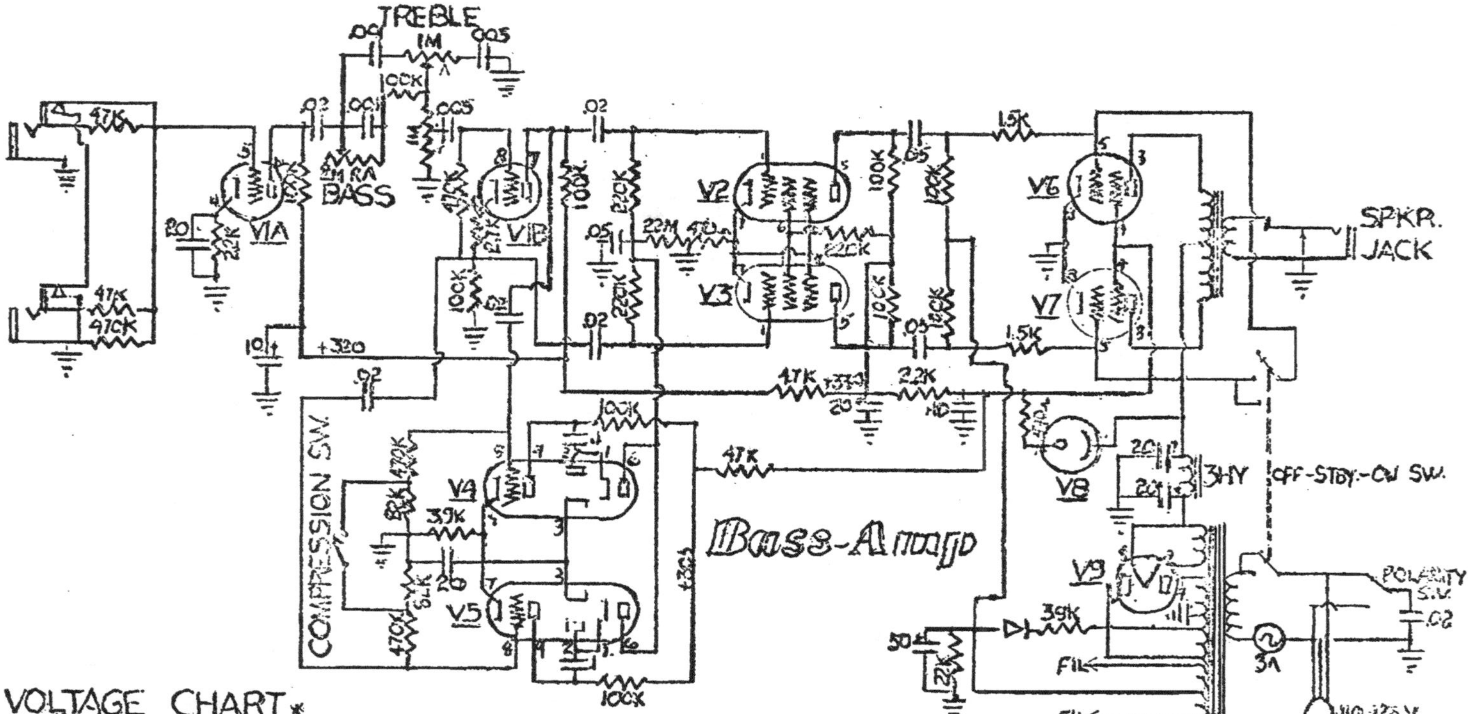

VOLTAGE CHART*

[illegible]	TYPE	PIN1	PIN2	PIN3	PIN4	PIN5	PIN6	PIN7	PIN8	PIN9
V1	6EU7	FIL	FIL	—	1.9	0	235	235	39	1/2
V2	6BD6	0	2.7	FIL	FIL	120	45	2.7	—	—
V3	6BD6	0	2.7	FIL	FIL	120	45	2.7	—	—
V4	6RM8	1.6	1.6	3.8	FIL	FIL	0	7.5	0	260
V5	6EM8	1.6	1.6	3.6	FIL	FIL	0	3.6	0	260
V6	6L6GC	GND	FIL	423	345	-33	—	FIL	GND	—
V7	6L6GC	GND	FIL	423	345	-33	—	FIL	GND	—
V8	OC2	425	—	—	—	—	—	350	—	—
V9	GZ 3P	GND	FIL	—	350	—	350	—	430	—

*ALL DC VOLTAGES MEASURED TO CHASSIS WITH V.T.V.M.

TUBE PLACEMENT CHART

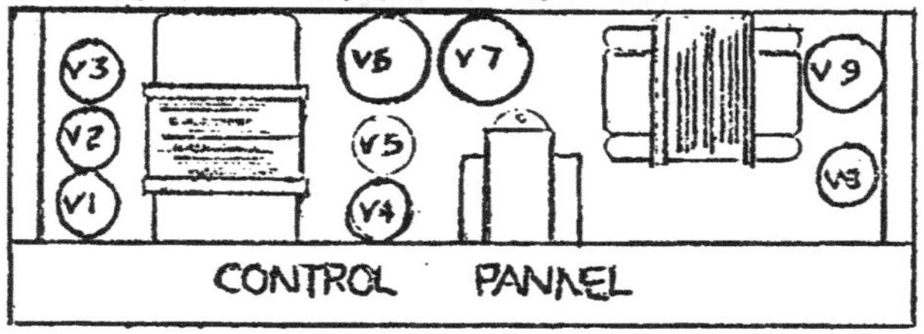

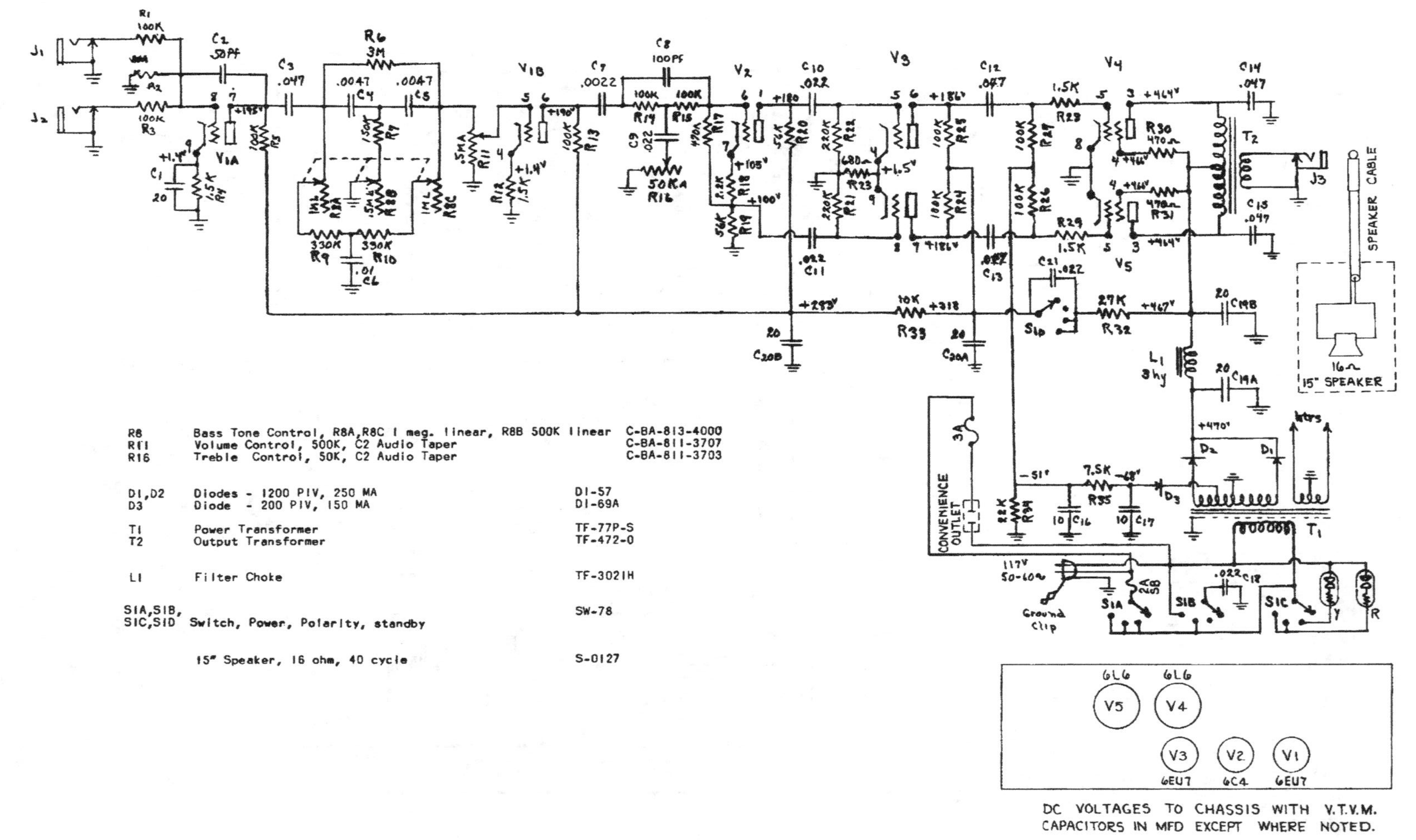

R8 R11 R16	Bass Tone Control, R8A,R8C 1 meg. linear, R8B 500K linear Volume Control, 500K, C2 Audio Taper Treble Control, 50K, C2 Audio Taper	C-BA-813-4000 C-BA-811-3707 C-BA-811-3703
D1,D2 D3	Diodes - 1200 PIV, 250 MA Diode - 200 PIV, 150 MA	D1-57 D1-69A
T1 T2	Power Transformer Output Transformer	TF-77P-S TF-472-0
L1	Filter Choke	TF-3021H
S1A,S1B, S1C,S1D	Switch, Power, Polarity, standby	SW-78
	15" Speaker, 16 ohm, 40 cycle	S-0127

EA-72

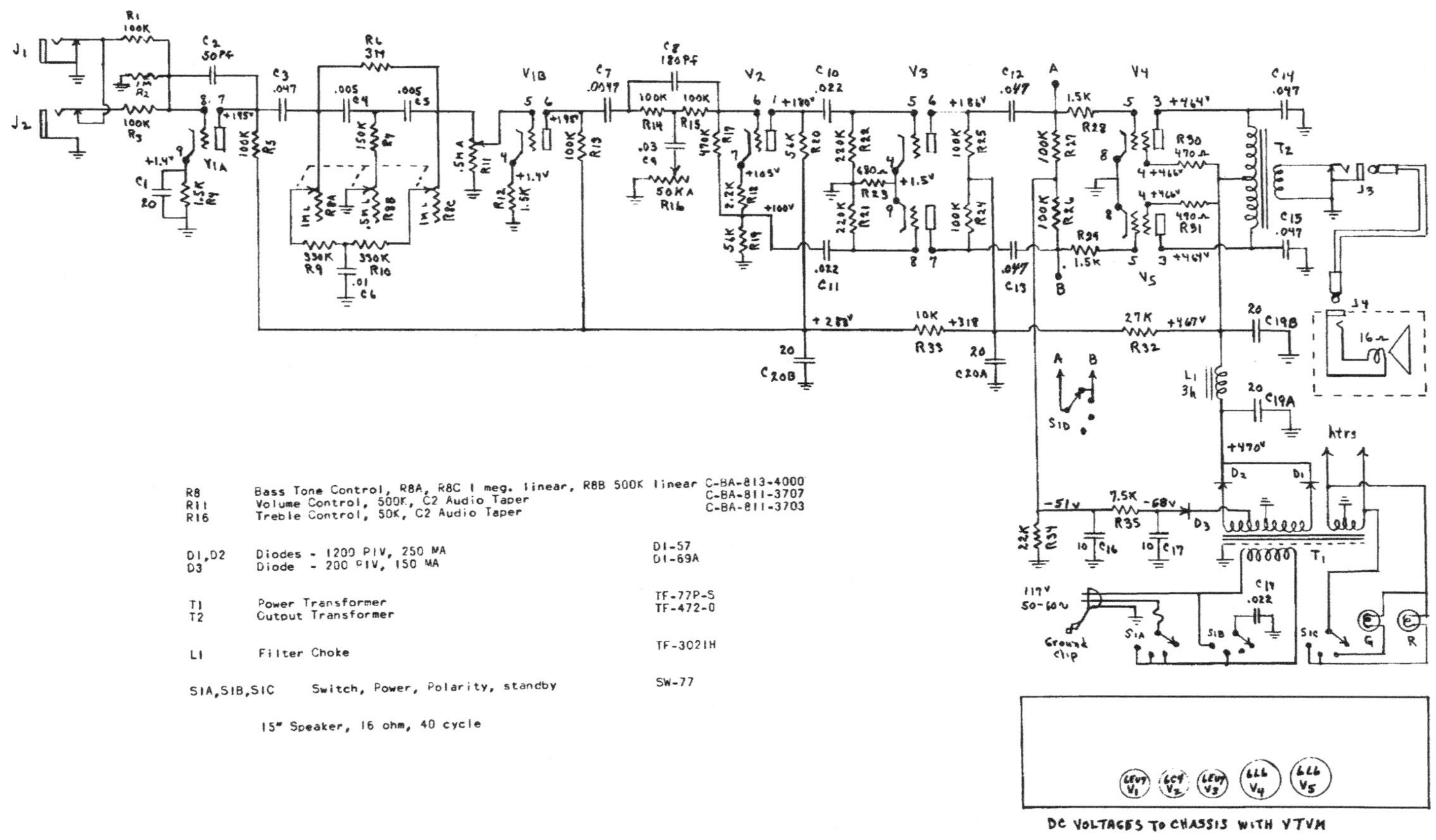

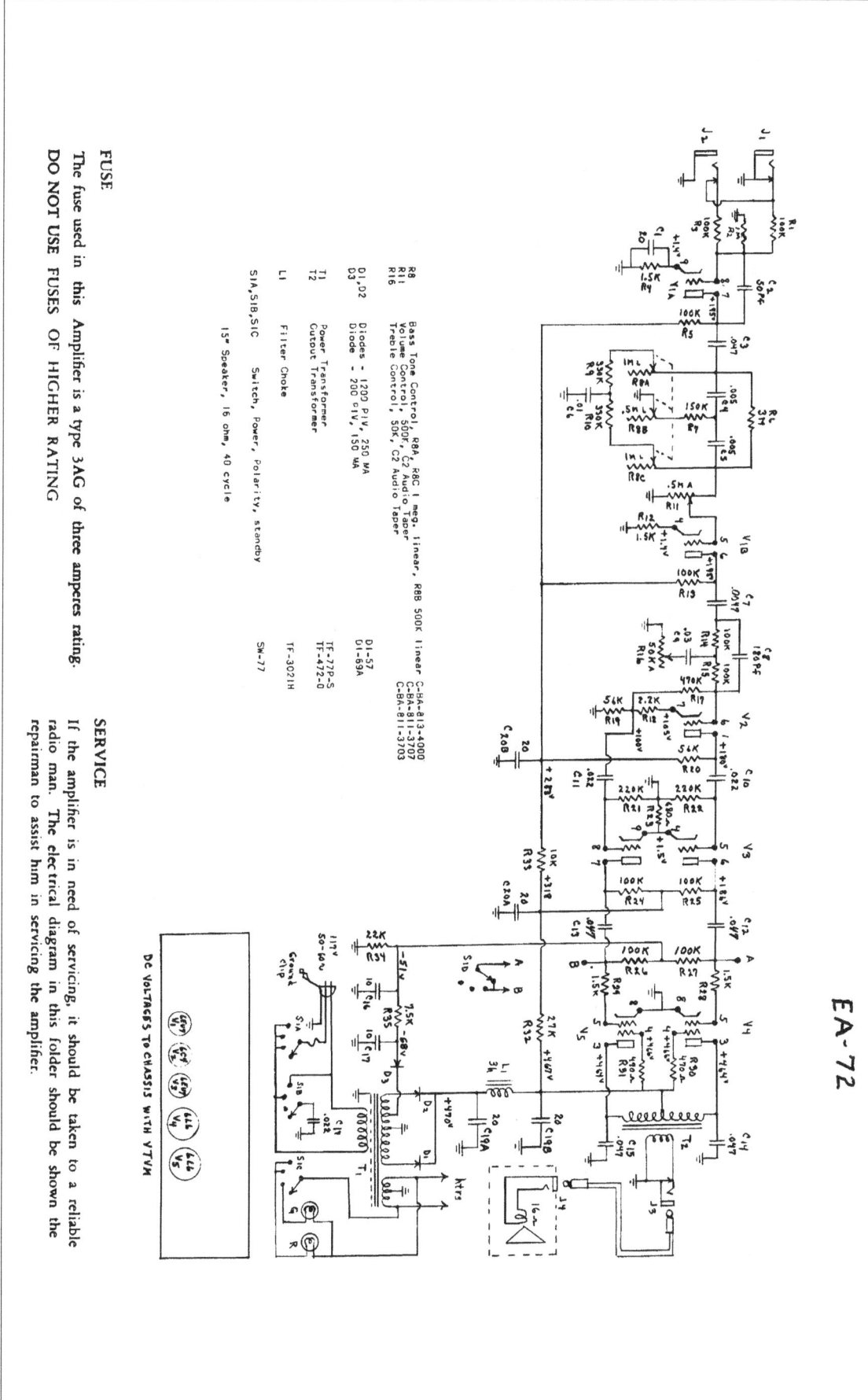

DC VOLTAGES TO CHASSIS WITH VTVM

R8	Bass Tone Control, R8A, R8C 1 meg. linear, R8B 500K linear	C-BA-813-4000
R11	Volume Control, 500K, C2 Audio Taper	C-BA-811-3707
R16	Treble Control, 50K, C2 Audio Taper	C-BA-811-3703
D1,D2	Diodes - 1200 PIV, 250 MA	D1-57
D3	Diode - 200 PIV, 150 MA	D1-69A
T1	Power Transformer	TF-77P-S
T2	Output Transformer	TF-472-0
L1	Filter Choke	TF-3021H
S1A,S1B,S1C	Switch, Power, Polarity, standby	SW-77
	15" Speaker, 16 ohm, 40 cycle	

FUSE

The fuse used in this Amplifier is a type 3AG of three amperes rating. DO NOT USE FUSES OF HIGHER RATING

SERVICE

If the amplifier is in need of servicing, it should be taken to a reliable radio man. The electrical diagram in this folder should be shown the repairman to assist him in servicing the amplifier.

EA-300RVT

Epiphone

EA-300 RVT

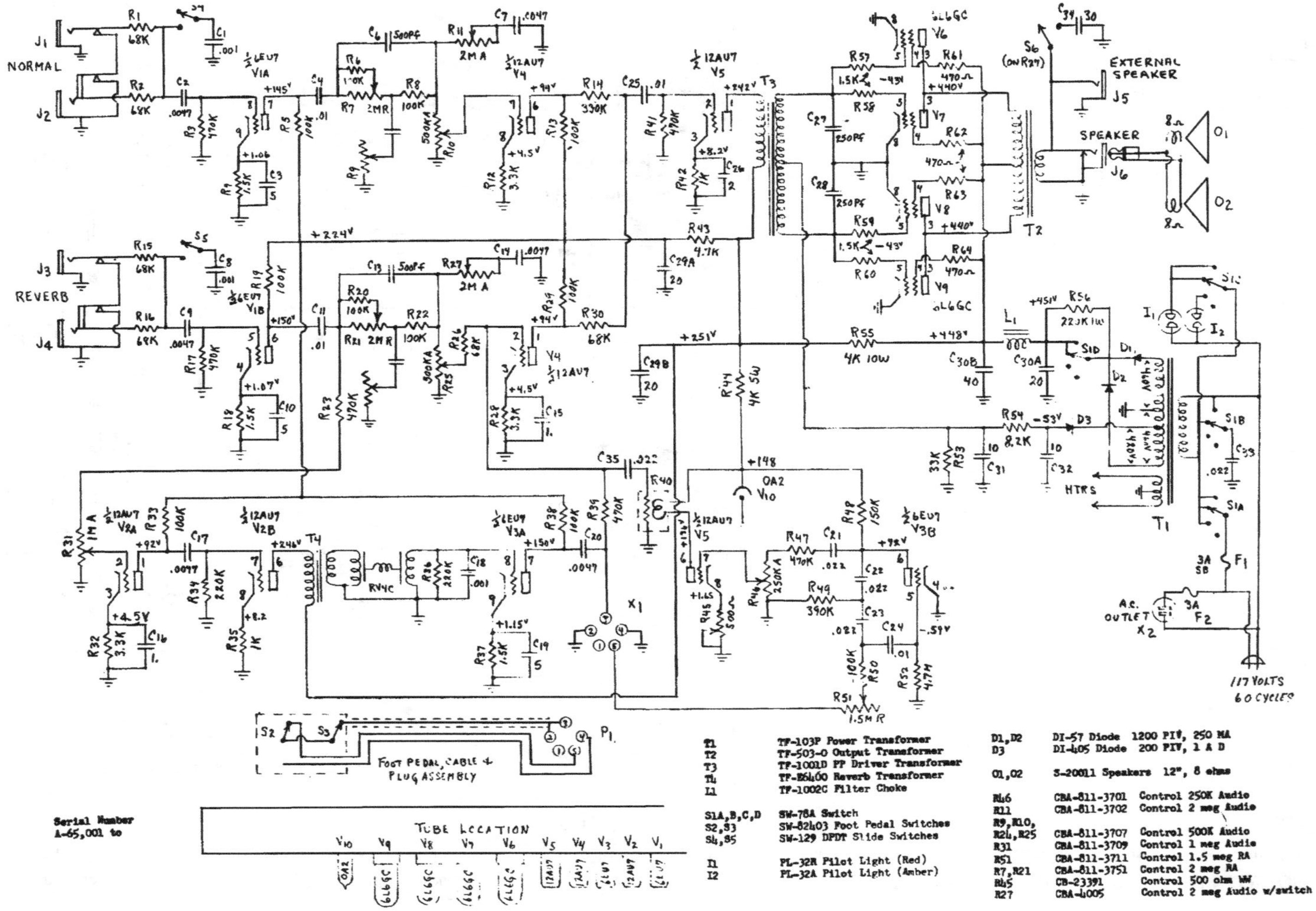

Serial Number A-65,001 to

T1	TF-103P Power Transformer
T2	TF-503-0 Output Transformer
T3	TF-1001D PP Driver Transformer
T4	TF-E6400 Reverb Transformer
L1	TF-1002C Filter Choke
S1A,B,C,D	SW-78A Switch
S2,S3	SW-82403 Foot Pedal Switches
S4,S5	SW-129 DPDT Slide Switches
I1	PL-32R Pilot Light (Red)
I2	PL-32A Pilot Light (Amber)

D1,D2	DI-57 Diode	1200 PIV, 250 MA
D3	DI-405 Diode	200 PIV, 1 A D
O1,O2	S-20011 Speakers	12", 8 ohms
R46	CBA-811-3701	Control 250K Audio
R11	CBA-811-3702	Control 2 meg Audio
R9,R10, R24,R25	CBA-811-3707	Control 500K Audio
R31	CBA-811-3709	Control 1 meg Audio
R51	CBA-811-3711	Control 1.5 meg RA
R7,R21	CBA-811-3751	Control 2 meg RA
R45	CB-23391	Control 500 ohm WW
R27	CBA-4005	Control 2 meg Audio w/switch

T1	Power Transformer	TF-135P
T2	Output Transformer	TF-531-0
L1	Filter Choke	TF-3021H
L1, L2	Inductances 3.7 mh	
S1	Switch - off, standby, on-on	SW-78
LDR-1	Light Dependent Resistor Lamp Assembly	RE-LDR-430G

(2)	Speakers, 10" 8 ohm	S-12197
(1)	Speaker, 15", 16 ohm	S-0128
R61	10 meg reversed audio (bass) control	C-BA-811-3710
R58, R66	250K audio taper (volume and depth) control	C-BA-811-3701
R57	2 meg reversed audio (BASS) control	C-BA-811-3751
R62, R60	500K audio taper (volume and treble) control	C-BA-811-[illegible]07
R64	1 meg audio taper (treble) control	C-BA-8[illegible]9
R67	1.5 meg reversed audio (frequency) control	C-BA-8[illegible]1
R65	500 ohm wire wound control	C-B-233[illegible]

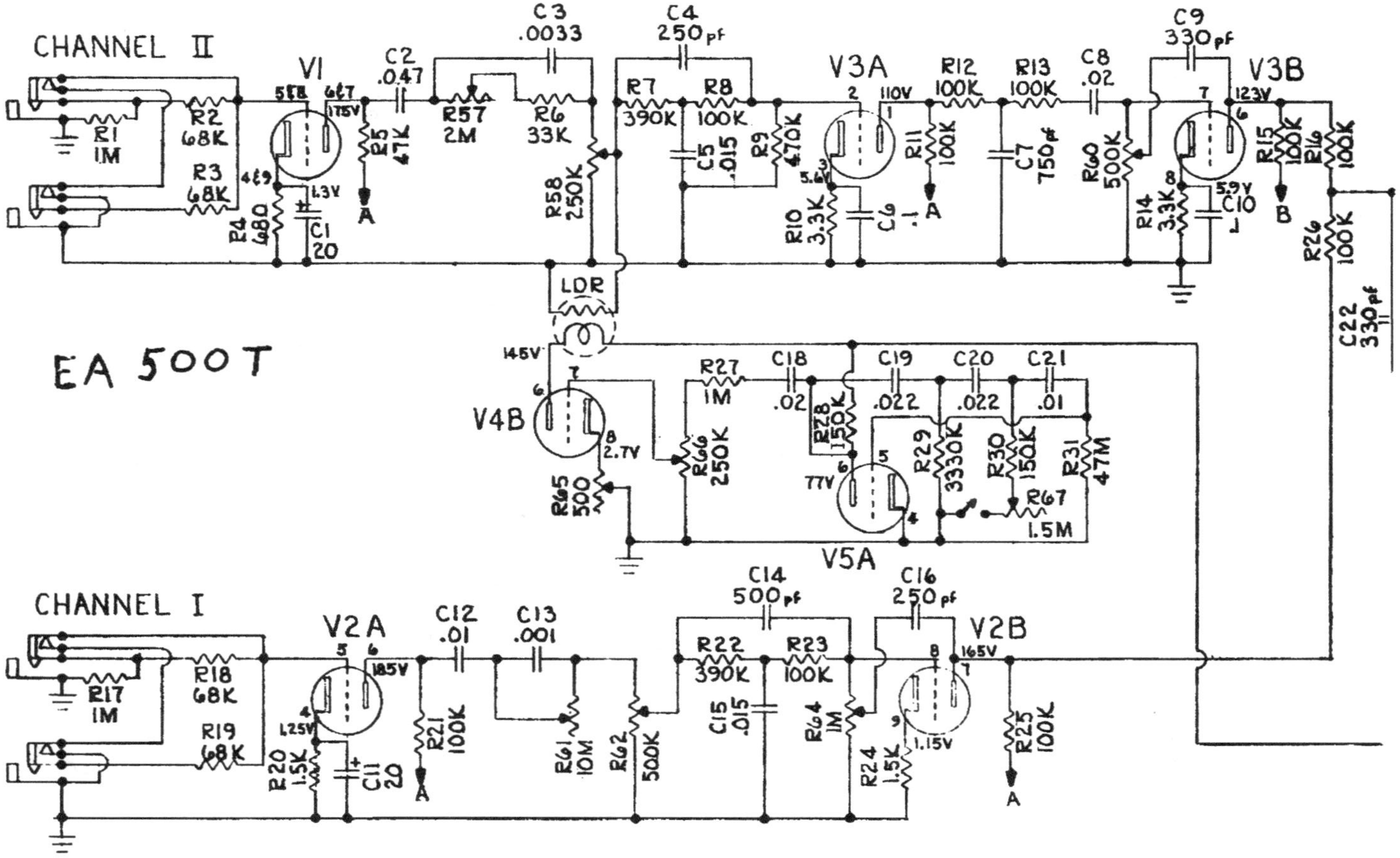

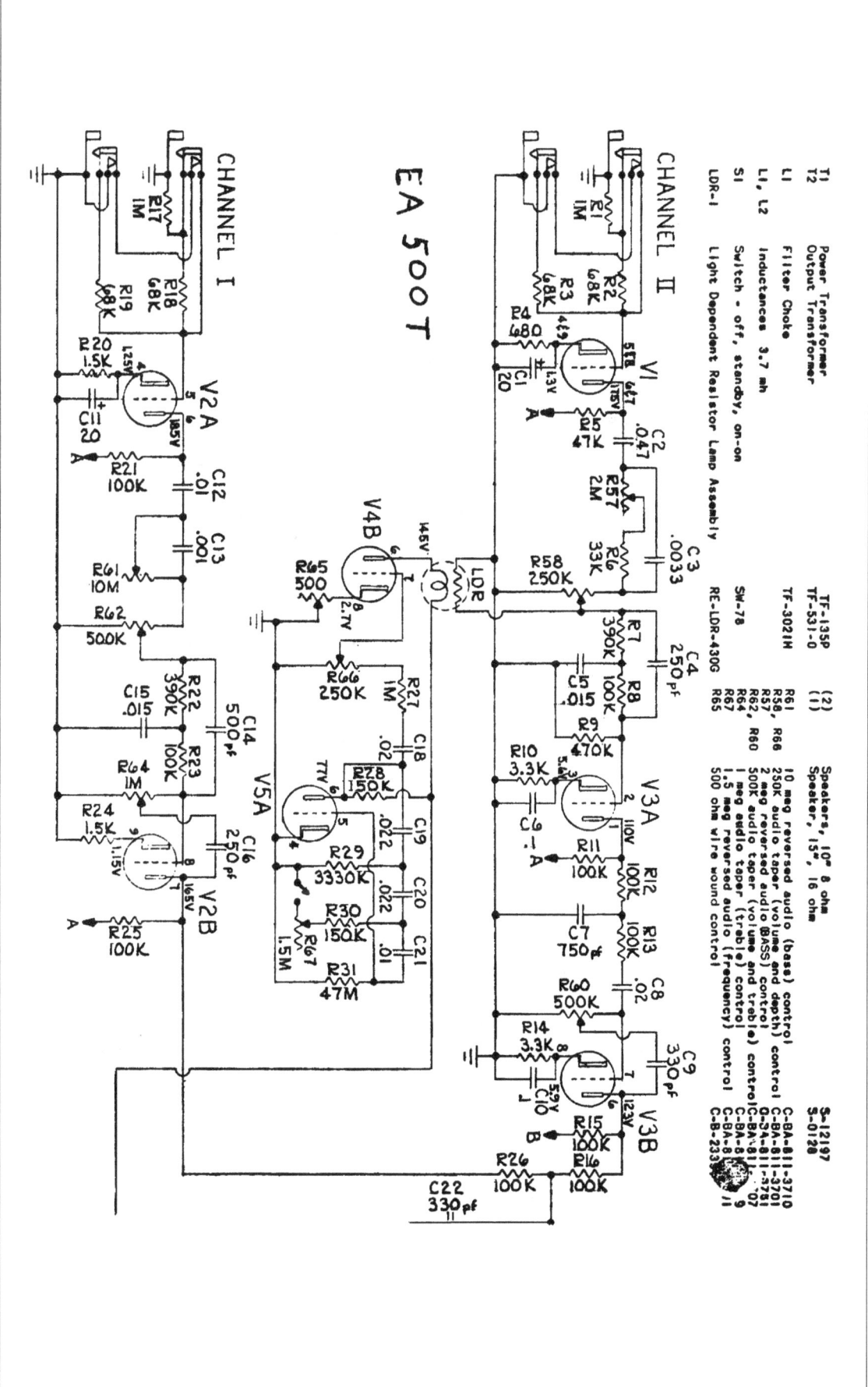

EA-500T

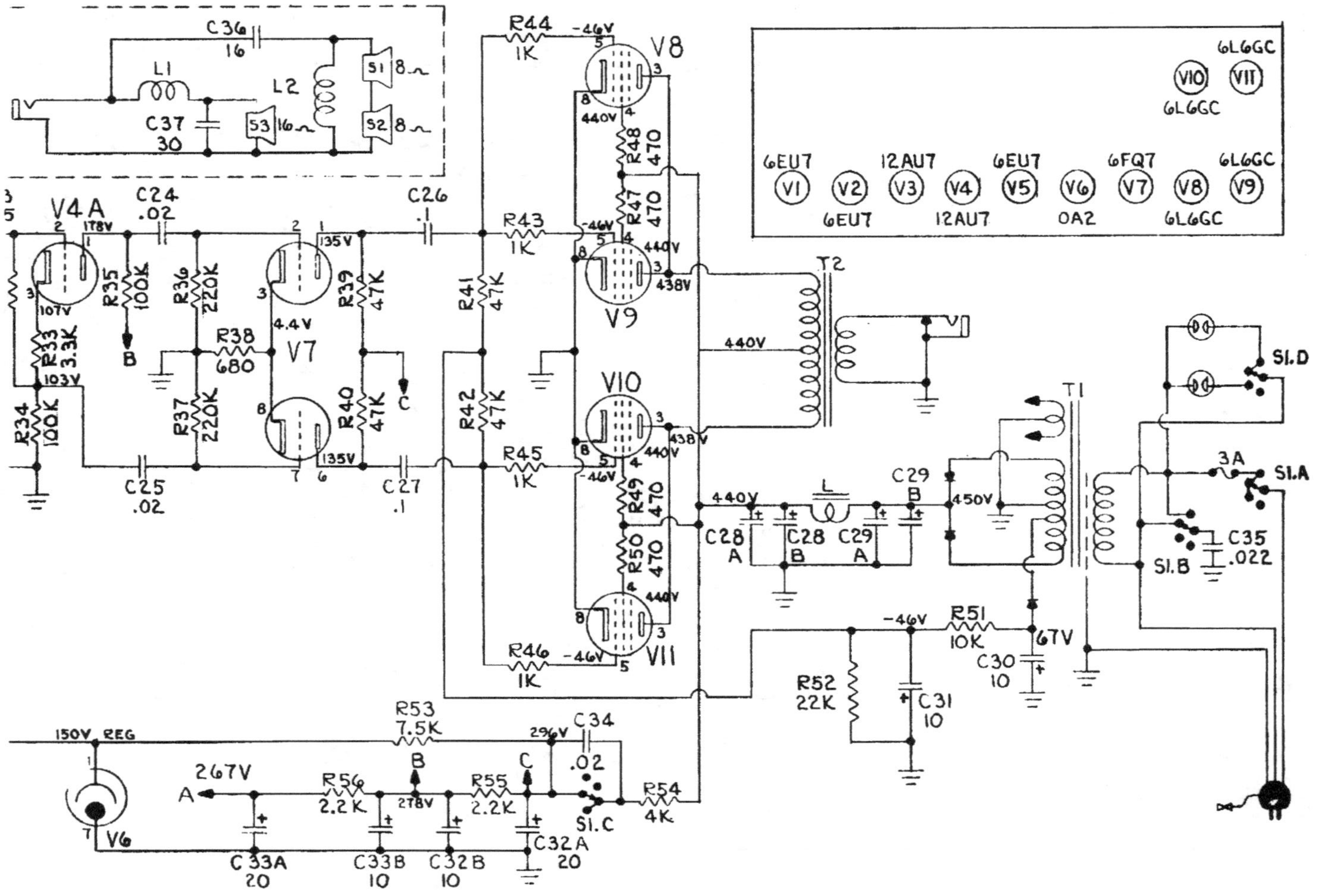
6EU7 V1
6EU7 V2
12AU7 V3
12AU7 V4
6EU7 V5
OA2 V6
6FQ7 V7
6L6GC V8
6L6GC V9
6L6GC V10
6L6GC V11
V4A
V7
V8
V9
V10
V11
V6
T1
T2
L1
L2
L
C36 16
C37 30
S1 8~
S2 8~
S3 16~
C24 .02
C25 .02
C26 .1
C27 .1
R33 3.3K
R34 100K
R35 100K
R36 220K
R37 220K
R38 680
R39 47K
R40 47K
R41 47K
R42 47K
R43 1K
R44 1K
R45 1K
R46 1K
R47 470
R48 470
R49 470
R50 470
R51 10K
R52 22K
R53 7.5K
R54 4K
R55 2.2K
R56 2.2K
C28 A
C28 B
C29 A
C29 B
C30 10
C31 10
C32A 20
C32B 10
C33A 20
C33B 10
C34 .02
C35 .022
S1.A
S1.B
S1.C
S1.D
3A
150V REG
267V
278V
296V
440V
450V
438V
-46V
67V
135V
4.4V
178V
107V
103V

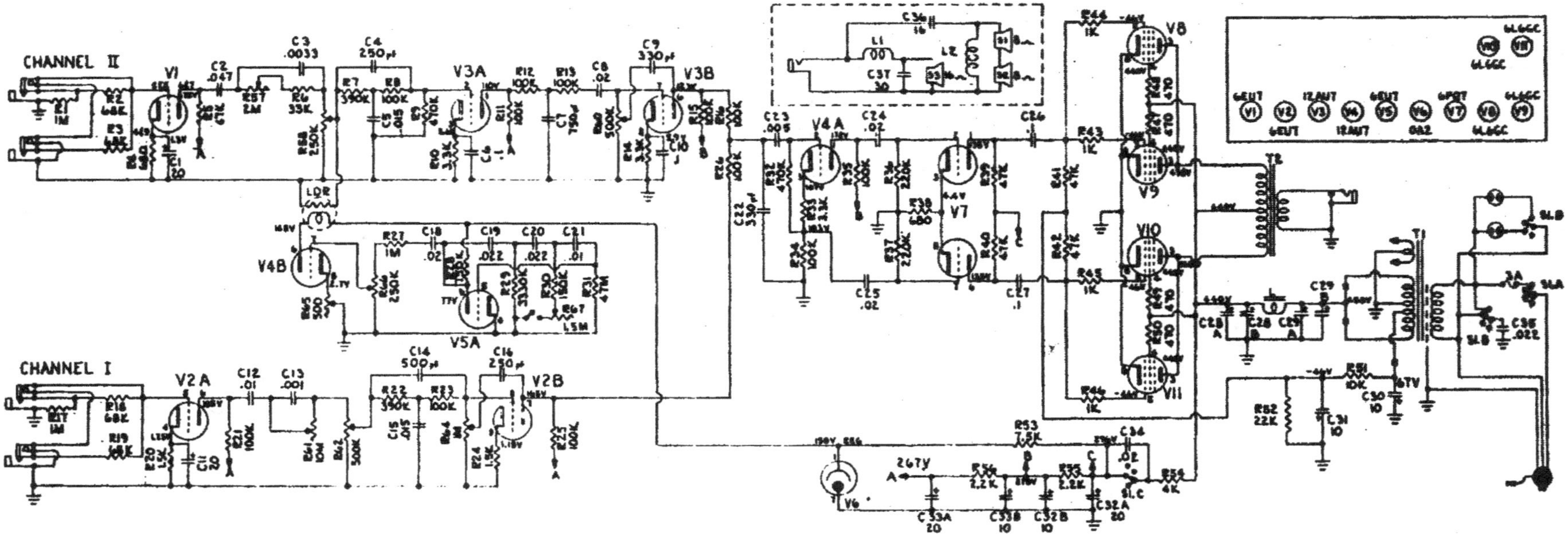

Ediphone Model EA-500T

FENDER "BANDMASTER" LAYOUT
MODEL 5E7

NOTICE
VOLTAGES READ TO GROUND
WITH ELECTRONIC VOLTMETER
VALUES SHOWN + OR − 20 %

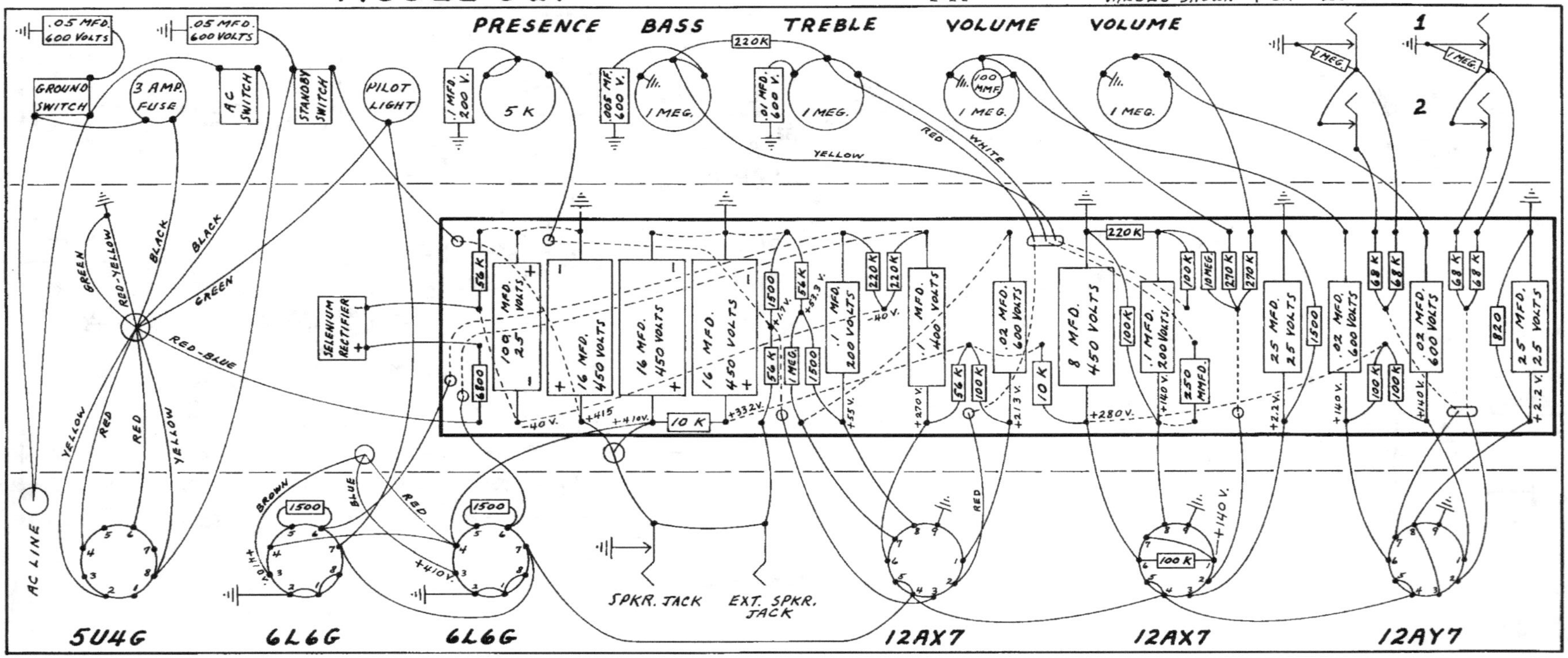

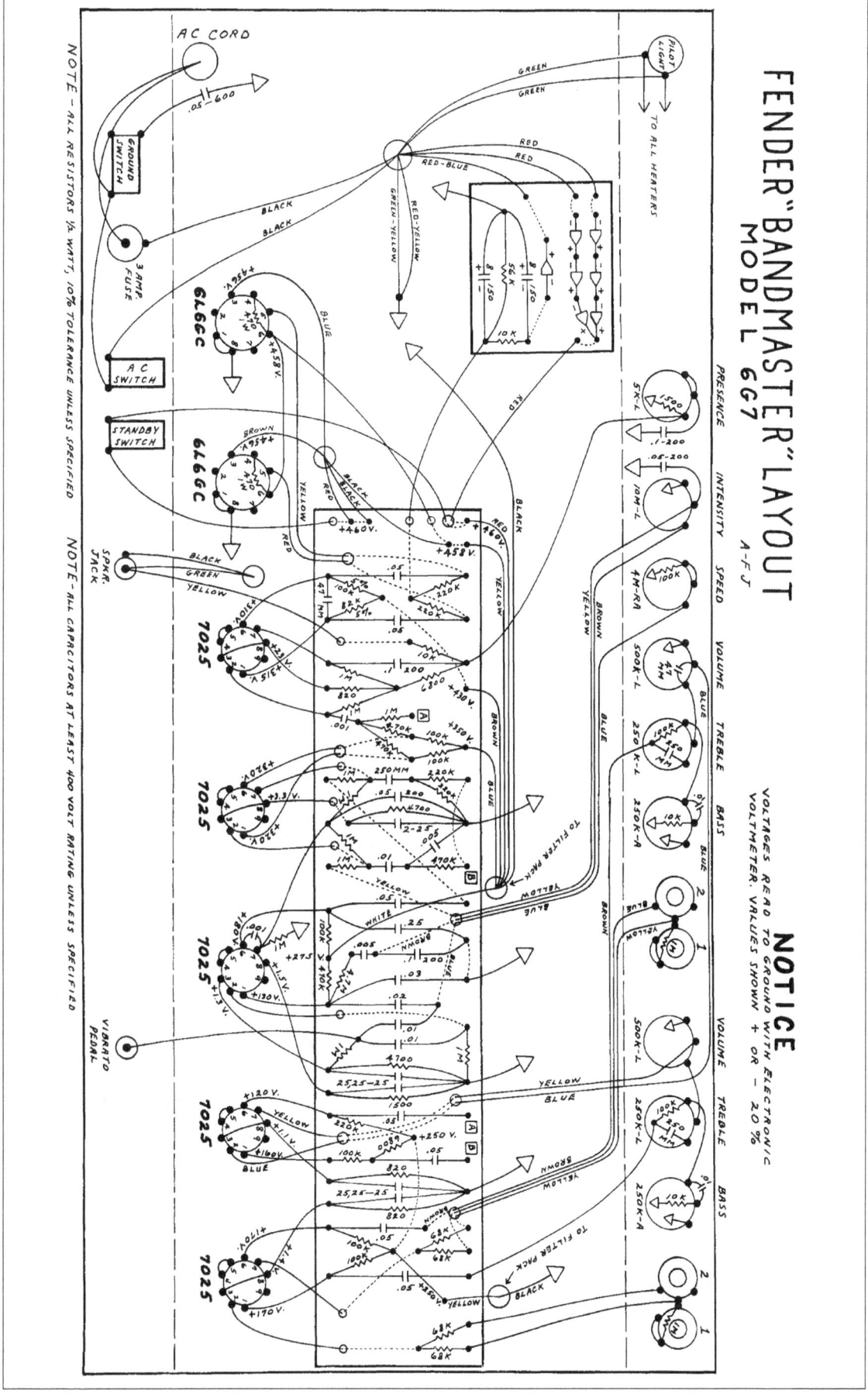

FENDER "BANDMASTER" LAYOUT
MODEL 6G7

A-F-J

NOTICE
VOLTAGES READ TO GROUND WITH ELECTRONIC VOLTMETER. VALUES SHOWN + OR − 20%

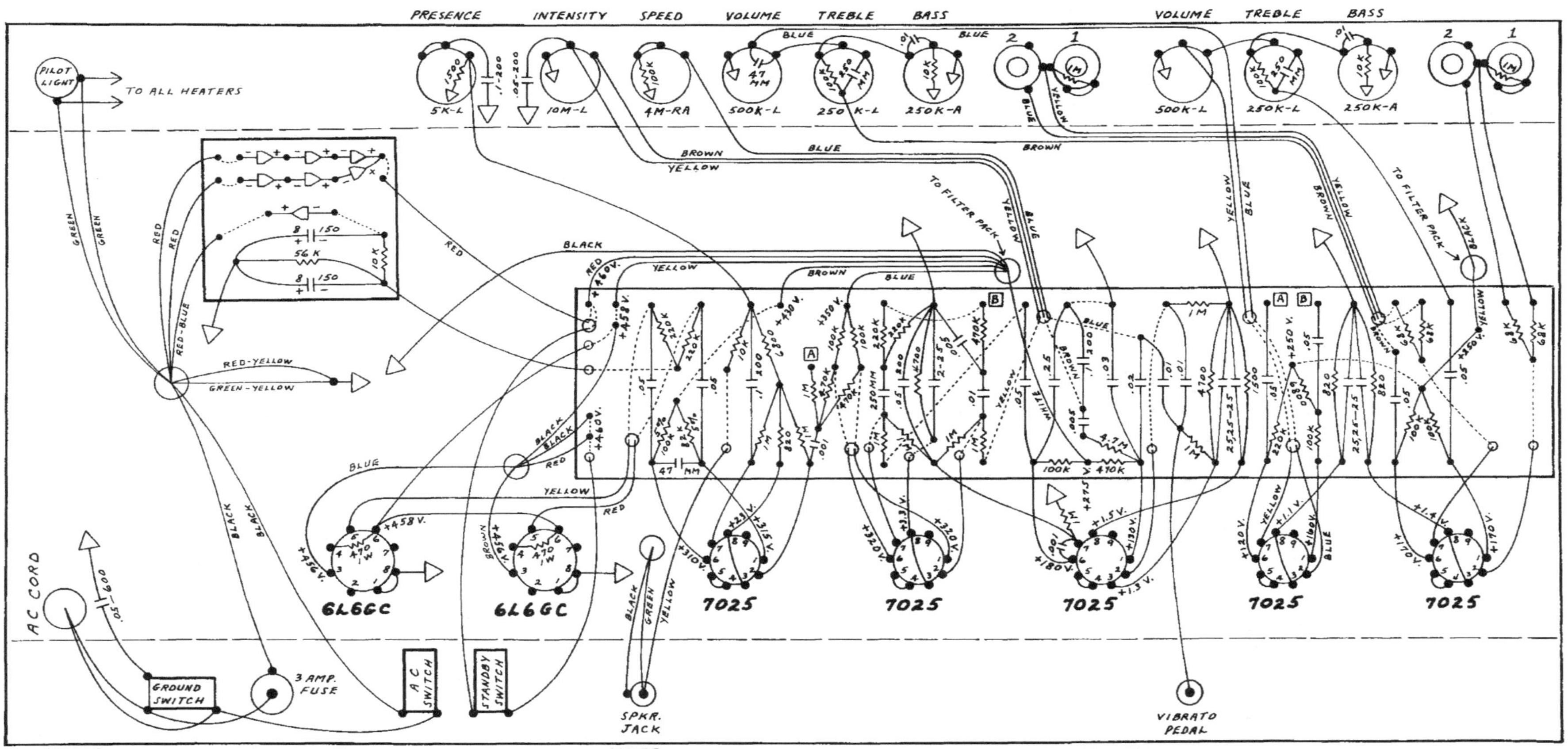

NOTE – ALL RESISTORS ½ WATT, 10% TOLERANCE UNLESS SPECIFIED

NOTE – ALL CAPACITORS AT LEAST 400 VOLT RATING UNLESS SPECIFIED

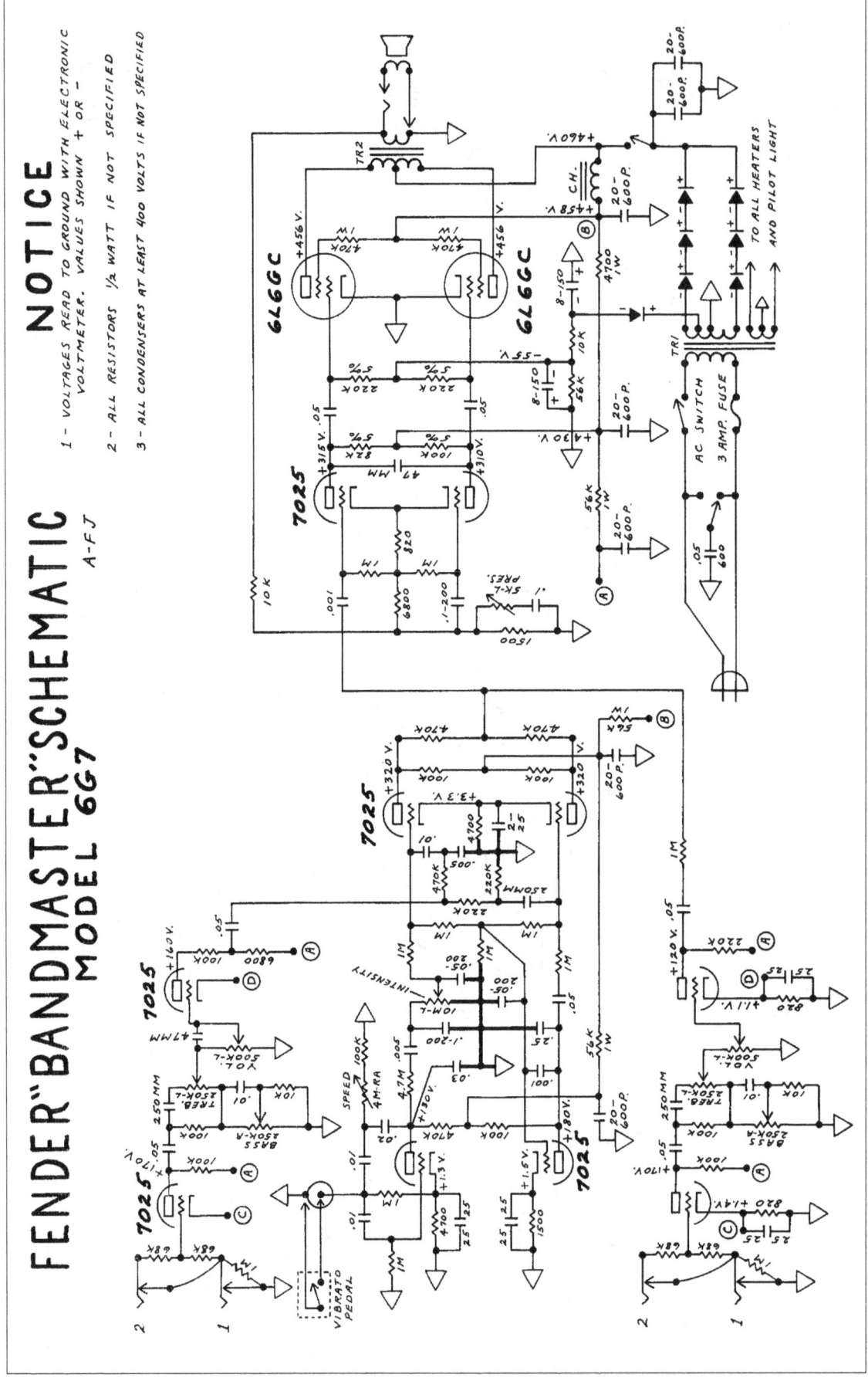

FENDER "BANDMASTER" SCHEMATIC
MODEL 6G7

A-F J

NOTICE

1 - VOLTAGES READ TO GROUND WITH ELECTRONIC VOLTMETER. VALUES SHOWN + OR -

2 - ALL RESISTORS ½ WATT IF NOT SPECIFIED

3 - ALL CONDENSERS AT LEAST 400 VOLTS IF NOT SPECIFIED

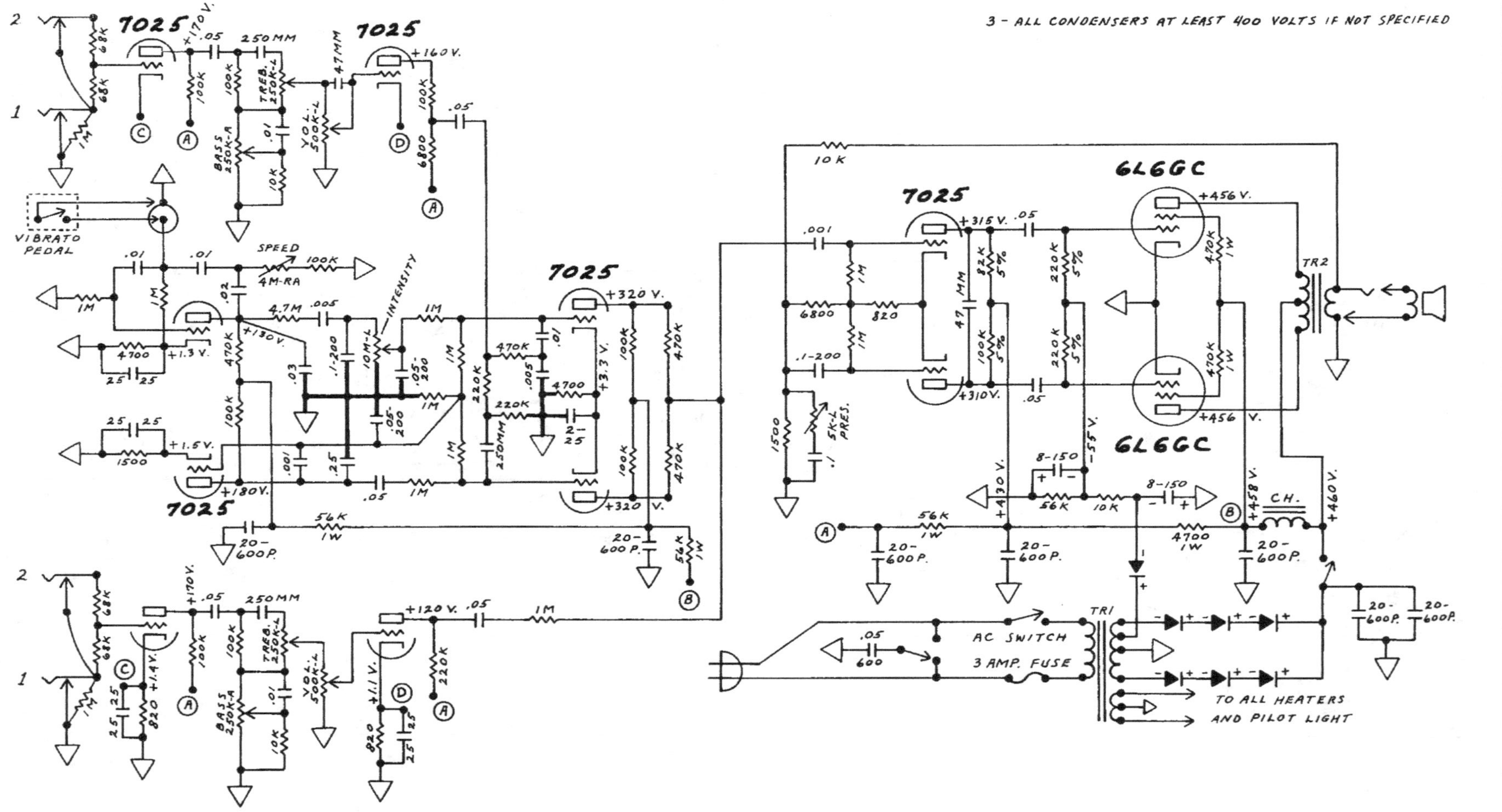

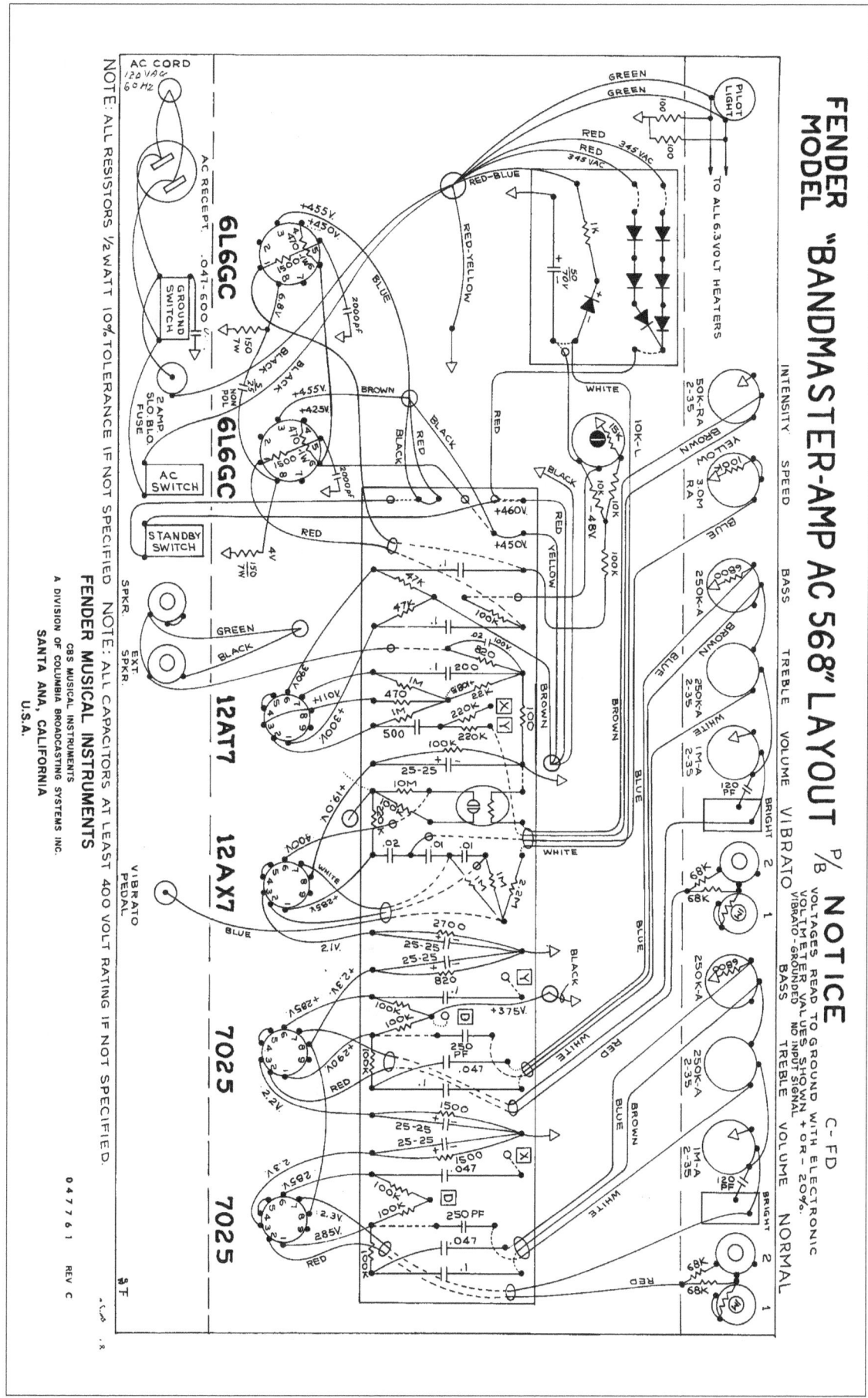

FENDER MODEL "BANDMASTER-AMP AC 568" LAYOUT P/B

NOTICE C-FD
VOLTAGES READ TO GROUND WITH ELECTRONIC VOLTMETER VALUES SHOWN + OR − 20%.
VIBRATO - GROUNDED NO INPUT SIGNAL

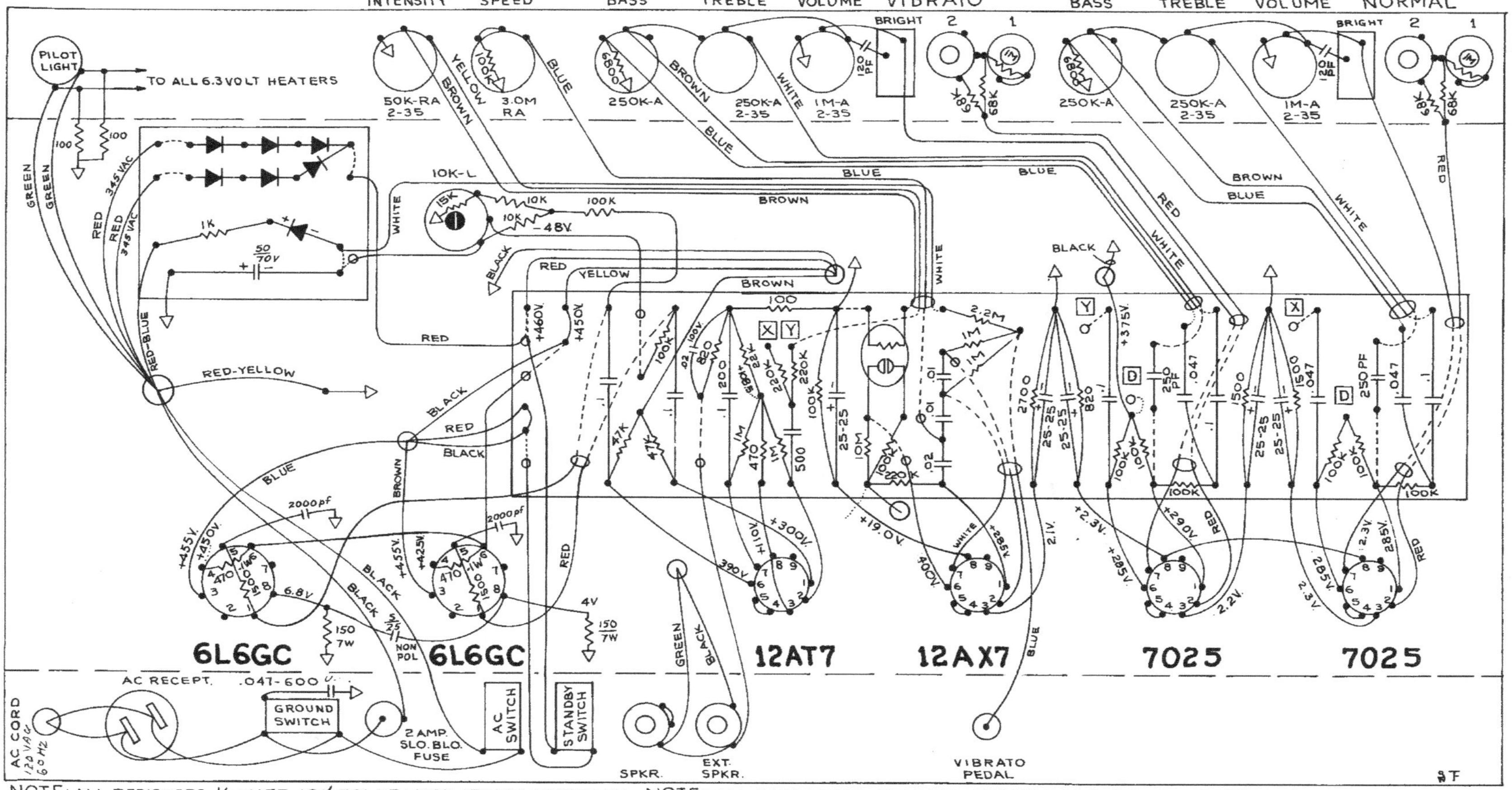

NOTE: ALL RESISTORS ½ WATT 10% TOLERANCE IF NOT SPECIFIED NOTE: ALL CAPACITORS AT LEAST 400 VOLT RATING IF NOT SPECIFIED.

FENDER MUSICAL INSTRUMENTS
CBS MUSICAL INSTRUMENTS
A DIVISION OF COLUMBIA BROADCASTING SYSTEMS INC.
SANTA ANA, CALIFORNIA
U.S.A.

0 4 7 7 6 1 REV C

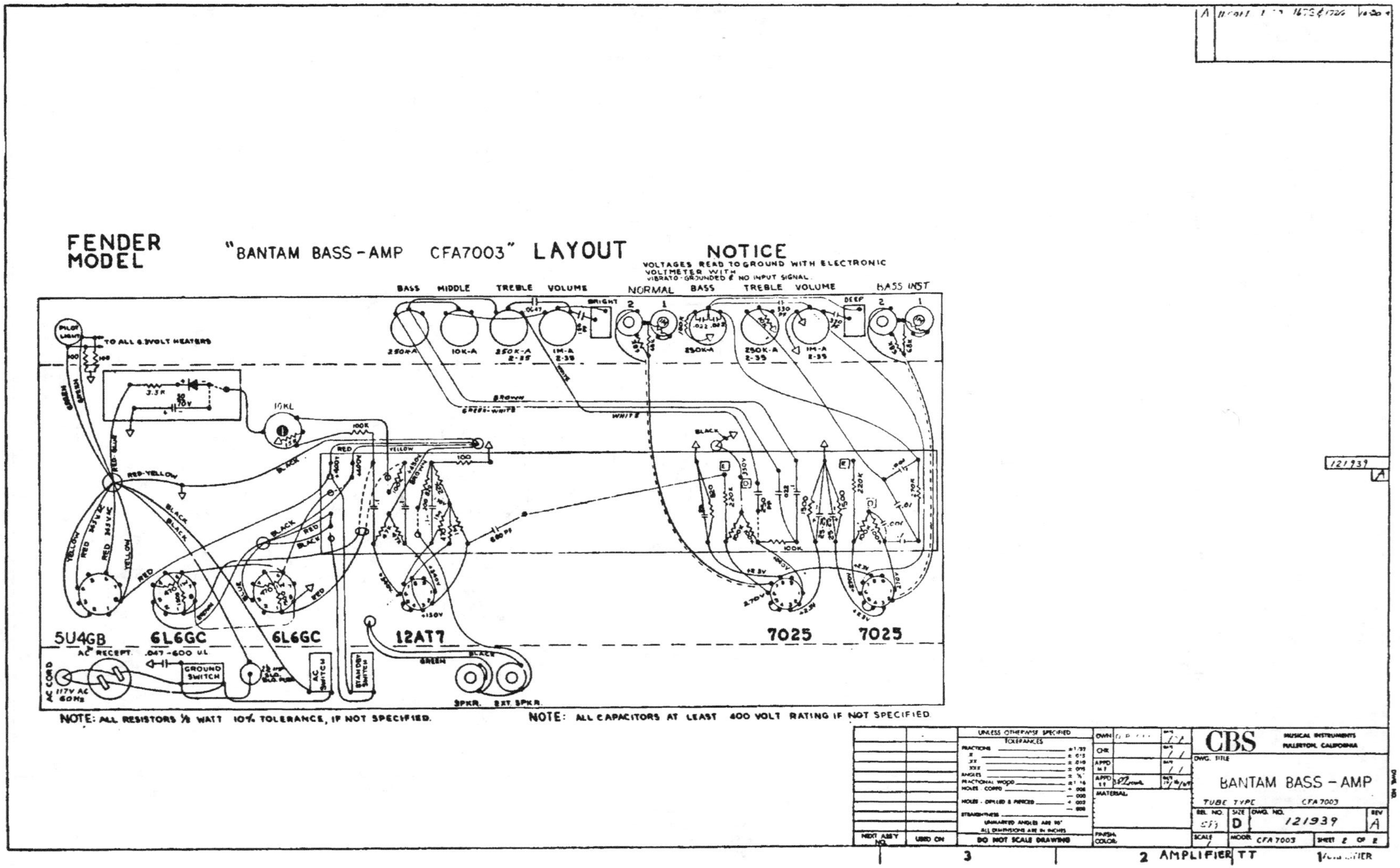
FENDER MODEL
"BANTAM BASS-AMP CFA7003" LAYOUT
NOTICE
VOLTAGES READ TO GROUND WITH ELECTRONIC VOLTMETER WITH VIBRATO GROUNDED & NO INPUT SIGNAL.
BASS MIDDLE TREBLE VOLUME
NORMAL BASS TREBLE VOLUME
BASS INST
BRIGHT
DEEP
5U4GB
6L6GC
6L6GC
12AT7
7025
7025
AC RECEPT.
.047-600 UL
GROUND SWITCH
AC SWITCH
STANDBY SWITCH
SPKR.
EXT. SPKR.
AC CORD 117V AC 60Hz
PILOT LIGHT
TO ALL 6.3VOLT HEATERS
NOTE: ALL RESISTORS ½ WATT 10% TOLERANCE, IF NOT SPECIFIED.
NOTE: ALL CAPACITORS AT LEAST 400 VOLT RATING IF NOT SPECIFIED
CBS MUSICAL INSTRUMENTS FULLERTON, CALIFORNIA
BANTAM BASS - AMP
TUBE TYPE CFA7003
D
121939
A
DO NOT SCALE DRAWING
2 AMPLIFIER TT

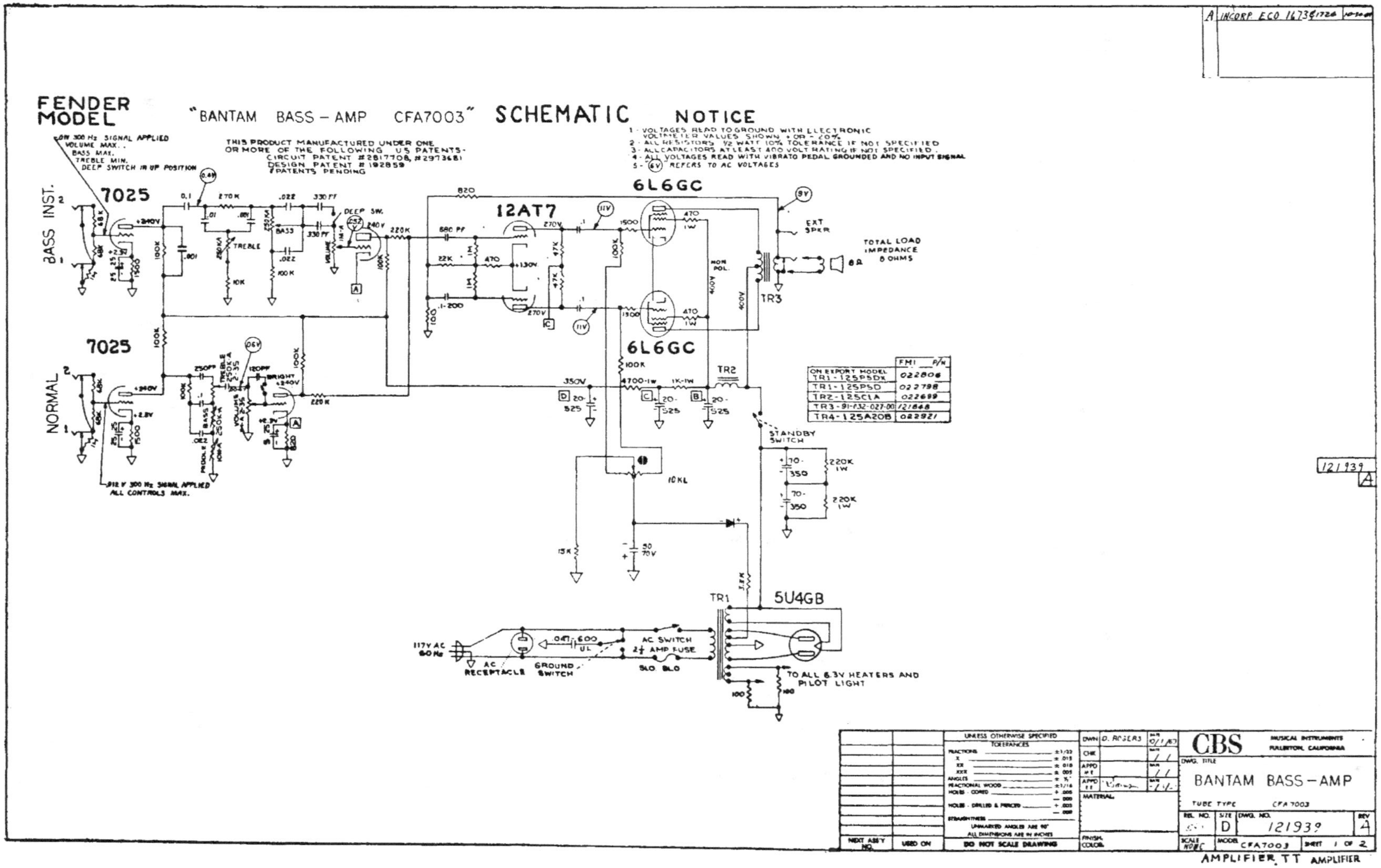
FENDER MODEL
"BANTAM BASS – AMP CFA7003" SCHEMATIC
THIS PRODUCT MANUFACTURED UNDER ONE OR MORE OF THE FOLLOWING U.S. PATENTS - CIRCUIT PATENT #2817708, #2973681 DESIGN PATENT #192859 PATENTS PENDING
NOTICE
1 - VOLTAGES READ TO GROUND WITH ELECTRONIC VOLTMETER. VALUES SHOWN + OR - 20%
2 - ALL RESISTORS 1/2 WATT 10% TOLERANCE IF NOT SPECIFIED
3 - ALL CAPACITORS AT LEAST 400 VOLT RATING IF NOT SPECIFIED.
4 - ALL VOLTAGES READ WITH VIBRATO PEDAL GROUNDED AND NO INPUT SIGNAL
5 - (6V) REFERS TO AC VOLTAGES
BASS INST.
NORMAL
7025
7025
12AT7
6L6GC
6L6GC
5U4GB
TR1
TR2
TR3
DEEP SW.
TREBLE
BASS
VOLUME
MIDDLE
BRIGHT
EXT SPKR
TOTAL LOAD IMPEDANCE 8 OHMS
STANDBY SWITCH
117V AC 60 Hz
AC RECEPTACLE
GROUND SWITCH
AC SWITCH
2½ AMP FUSE SLO BLO
TO ALL 6.3V HEATERS AND PILOT LIGHT
ON EXPORT MODEL TR1-125P5DX 022806
TR1-125P5D 022798
TR2-125C1A 022699
TR3-91-032-027-00 121848
TR4-125A20B 022921
A INCORP ECO 1673 & 1726
CBS MUSICAL INSTRUMENTS FULLERTON, CALIFORNIA
BANTAM BASS – AMP
TUBE TYPE CFA 7003
DWG. NO. 121939
SIZE D
REV A
MODEL CFA7003
SHEET 1 OF 2
AMPLIFIER, TT AMPLIFIER
DO NOT SCALE DRAWING

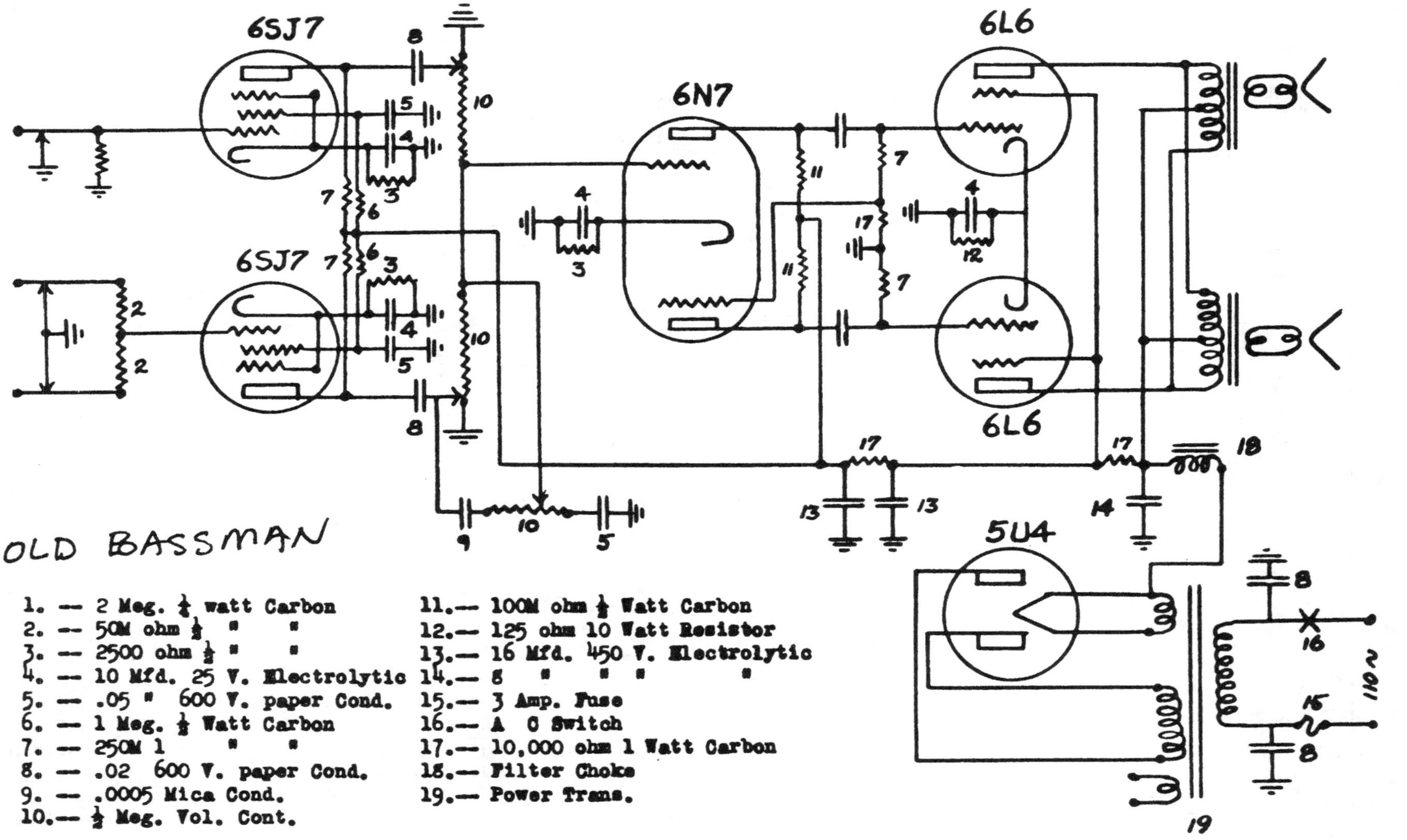

1. — 2 Meg. ½ watt Carbon
2. — 50M ohm ½ " "
3. — 2500 ohm ½ " "
4. — 10 Mfd. 25 V. Electrolytic
5. — .05 " 600 V. paper Cond.
6. — 1 Meg. ½ Watt Carbon
7. — 250M 1 " "
8. — .02 600 V. paper Cond.
9. — .0005 Mica Cond.
10. — ½ Meg. Vol. Cont.
11. — 100M ohm ½ Watt Carbon
12. — 125 ohm 10 Watt Resistor
13. — 16 Mfd. 450 V. Electrolytic
14. — 8 " " "
15. — 3 Amp. Fuse
16. — A C Switch
17. — 10,000 ohm 1 Watt Carbon
18. — Filter Choke
19. — Power Trans.

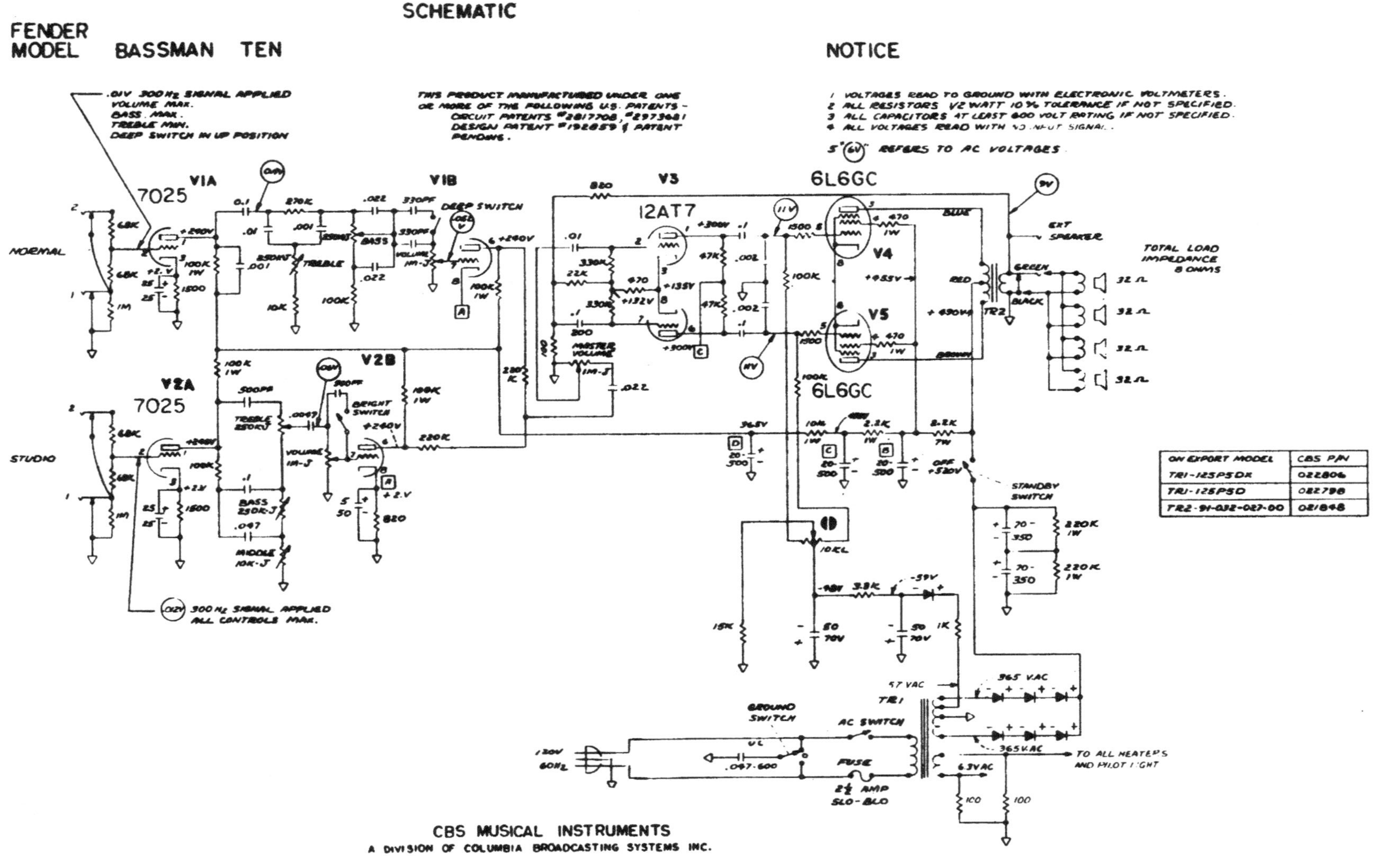

ON EXPORT MODEL	CBS P/N
TR1-125PSDX	022806
TR1-125PSD	022798
TR2-91-032-027-00	021848

CBS MUSICAL INSTRUMENTS
A DIVISION OF COLUMBIA BROADCASTING SYSTEMS INC.
FULLERTON, CALIFORNIA
U.S.A.

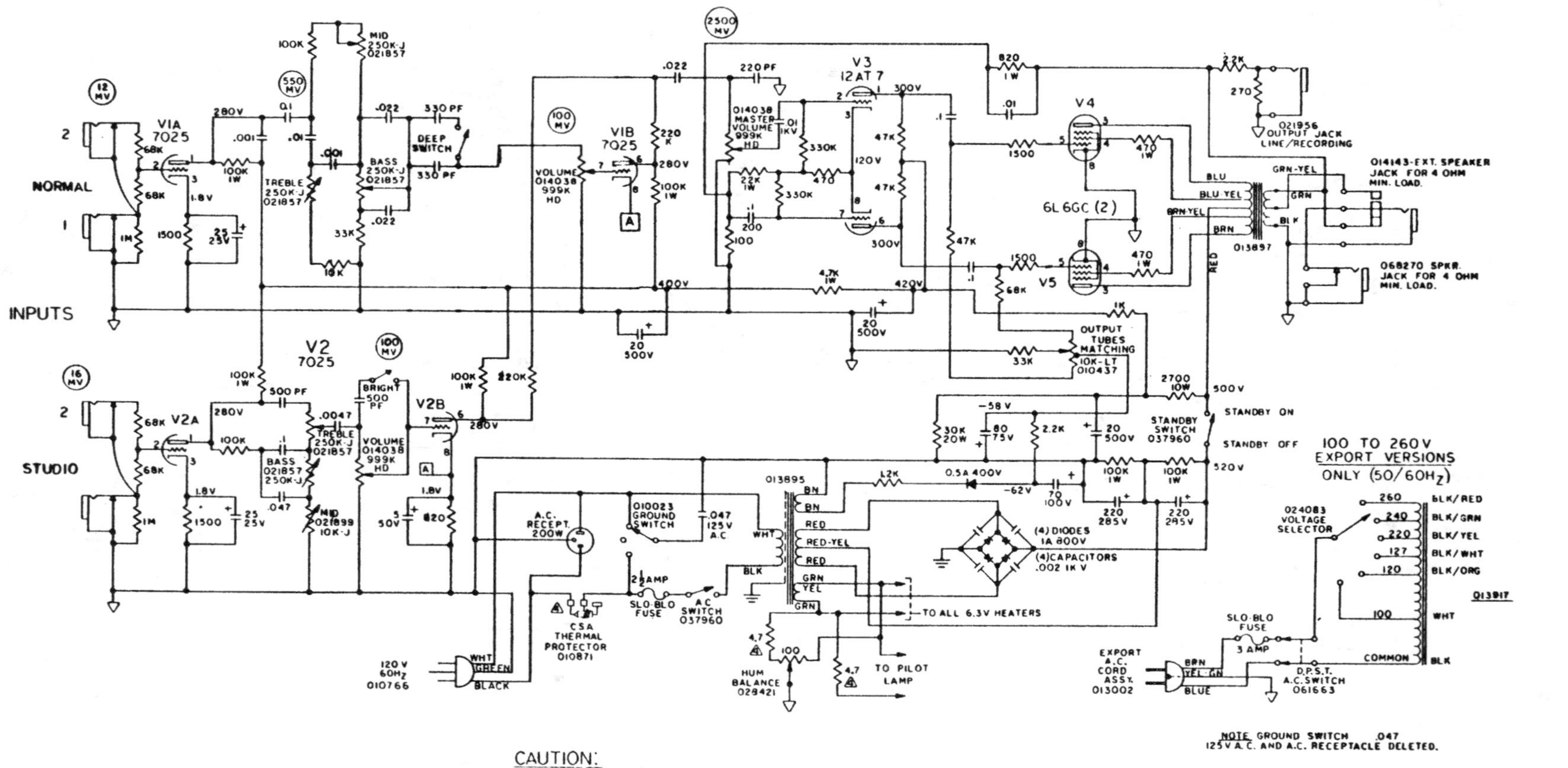

NOTE: GROUND SWITCH .047 125V A.C. AND A.C. RECEPTACLE DELETED.

CAUTION:
500 VOLTS AVAILABLE. DISCONNECT POWER SUPPLY AND ALLOW AT LEAST 20 SECONDS FOR VOLTAGES TO BLEED OFF. BEFORE SERVICING AMPLIFIER.

▲ USED IN SOME VERSIONS.

○ DENOTES TYPICAL 300Hz SIGNAL FOR 10W OUTPUT WITH VOLUME MAX. BASS MAX. TREBLE MIN AND DEEP SWITCH IN UP POSITION.

3. WHEN INSTALLING NEW 6L6'S RESET O.T. MATCHING FOR MIN. HUM.

2. ALL CAPACITORS AT LEAST 600 VOLT RATING IF NOT SPECIFIED.

1. ALL RESISTORS 1/2 WATT, 10% TOLERANCE IF NOT SPECIFIED.

NOTES: UNLESS OTHERWISE SPECIFIED

Fender/Rogers/Rhodes

DWN RON RICO 5/9/77

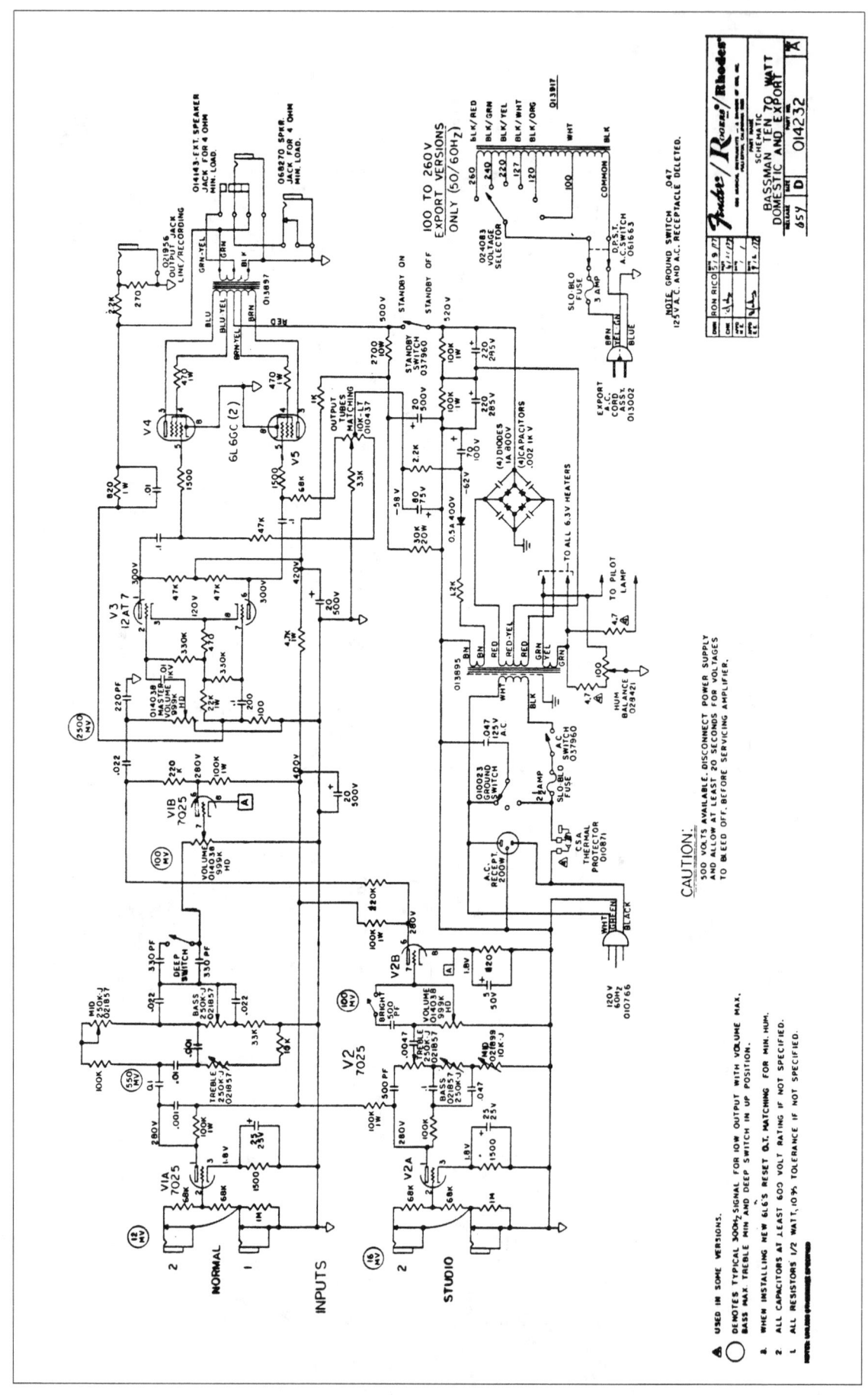

SCHEMATIC
BASSMAN TEN 70 WATT
DOMESTIC AND EXPORT

RELEASE	SIZE	PART NO.	REV
654	D	014232	A

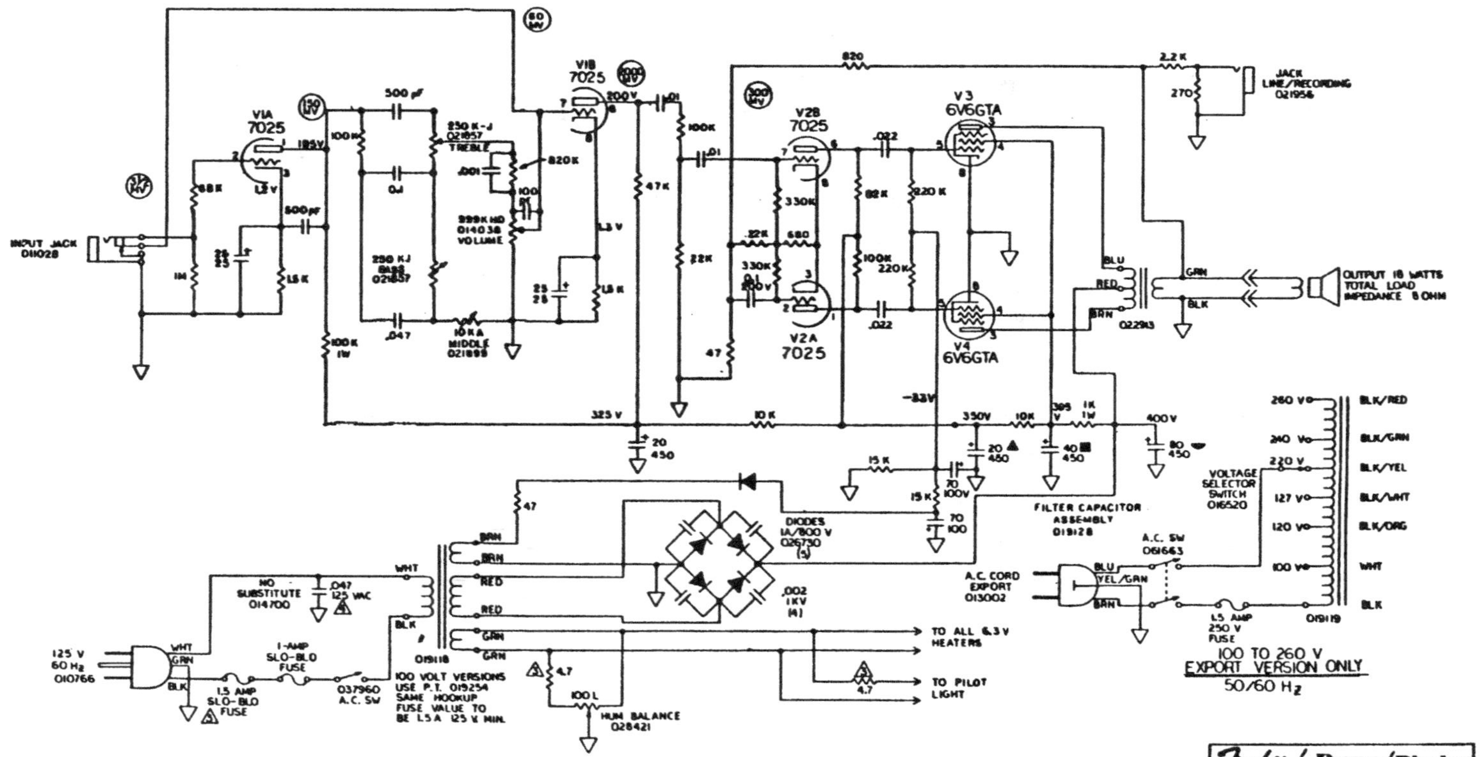

4. SOME VERSION USE .01/125 VAC. P/N 018764 (NO SUBSTITUTE)
3. USED IN SOME VERSIONS.
2. ALL CAPACITORS 600 V MIN.
1. ALL RESISTORS ½ WATT ±10%.
(MV) DENOTES TYPICAL 1KHz SIGNAL REQUIRED FOR 1 WATT R.M.S. OUTPUT ALL CONTROLS MAX.

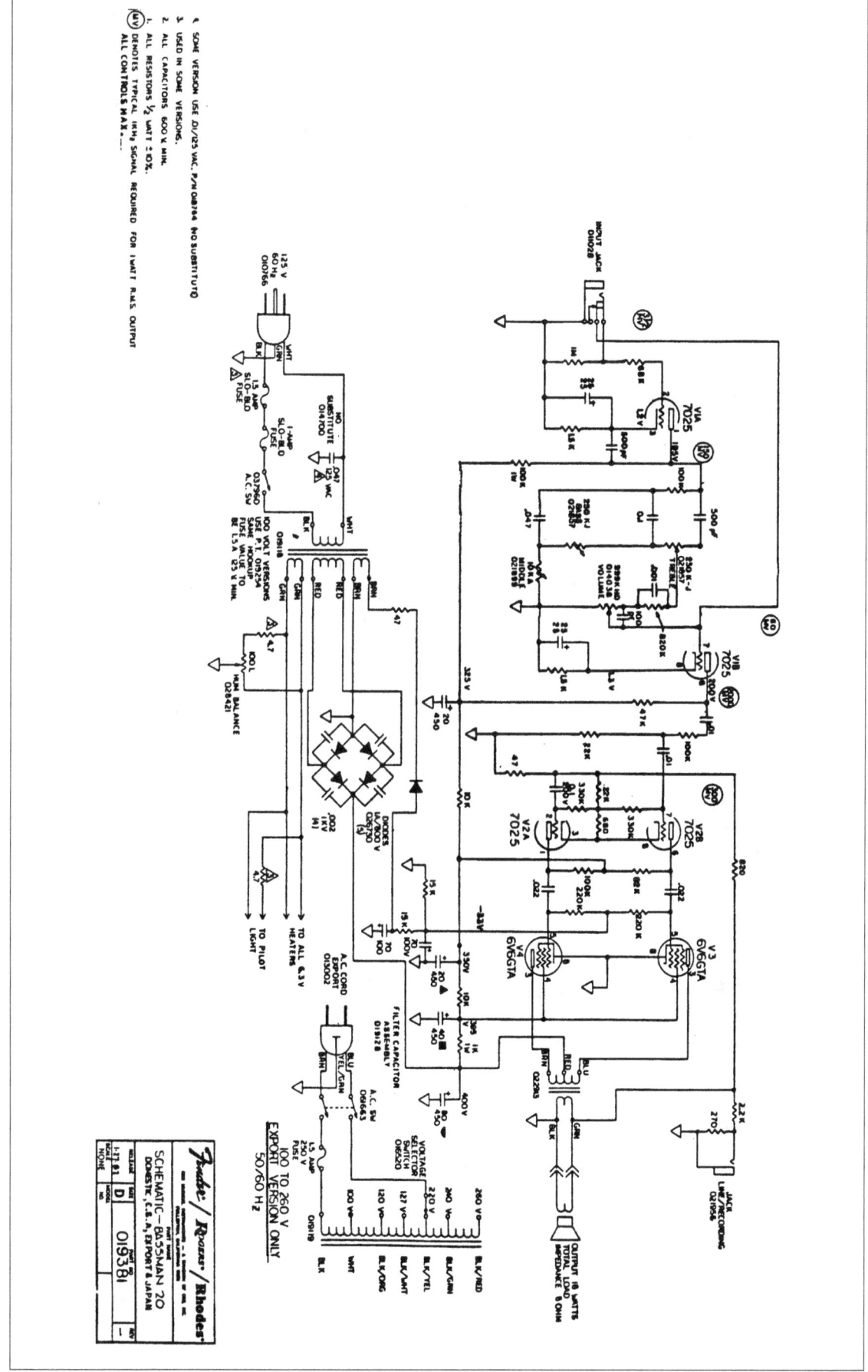

Fender / Rogers / Rhodes
SCHEMATIC—BASSMAN 20
DOMESTIC, C.S.A., EXPORT & JAPAN

RELEASE	SIZE	PART NO	REV
1-27-81	D	019381	–
SCALE NONE	MODEL NO.		

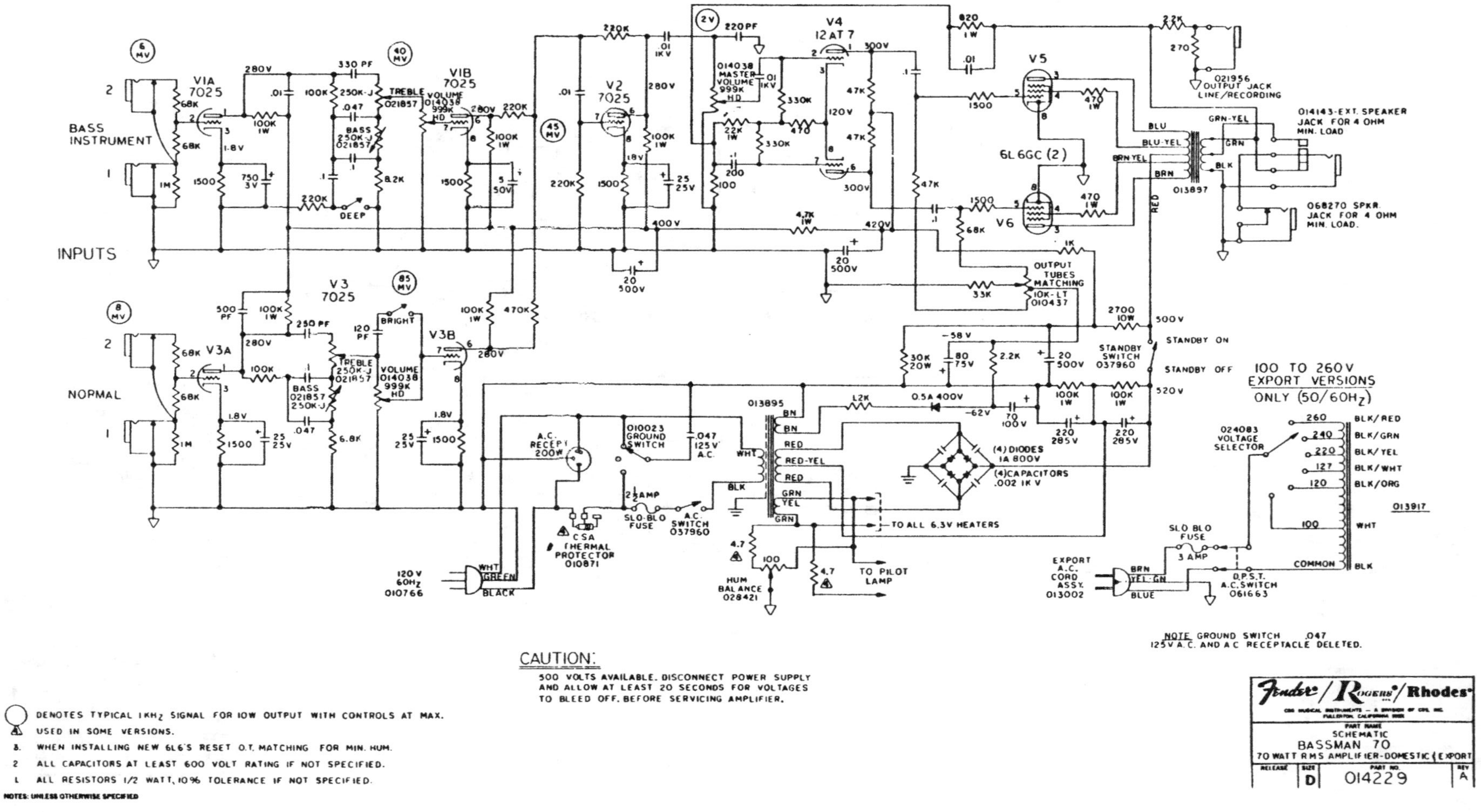
BASS INSTRUMENT
NORMAL
INPUTS
V1A 7025
V1B 7025
V2 7025
V3 7025
V3A
V3B
V4 12AT7
V5
V6
6L6GC (2)
TREBLE
BASS
VOLUME
MASTER VOLUME
BRIGHT
DEEP
OUTPUT TUBES MATCHING 10K-LT 010437
021956 OUTPUT JACK LINE/RECORDING
014143-EXT. SPEAKER JACK FOR 4 OHM MIN. LOAD
068270 SPKR. JACK FOR 4 OHM MIN. LOAD.
STANDBY ON
STANDBY OFF
STANDBY SWITCH 037960
100 TO 260V EXPORT VERSIONS ONLY (50/60Hz)
024083 VOLTAGE SELECTOR
(4) DIODES 1A 800V
(4) CAPACITORS .002 1KV
TO ALL 6.3V HEATERS
TO PILOT LAMP
HUM BALANCE 028421
A.C. RECEPT 200W
010023 GROUND SWITCH
SLO-BLO FUSE
A.C. SWITCH 037960
CSA THERMAL PROTECTOR 010871
120V 60Hz 010766
EXPORT A.C. CORD ASSY. 013002
D.P.S.T. A.C. SWITCH 061663
NOTE GROUND SWITCH .047 125V A.C. AND A.C. RECEPTACLE DELETED.
CAUTION:
500 VOLTS AVAILABLE. DISCONNECT POWER SUPPLY AND ALLOW AT LEAST 20 SECONDS FOR VOLTAGES TO BLEED OFF. BEFORE SERVICING AMPLIFIER.
DENOTES TYPICAL 1KHz SIGNAL FOR 10W OUTPUT WITH CONTROLS AT MAX.
USED IN SOME VERSIONS.
3. WHEN INSTALLING NEW 6L6'S RESET O.T. MATCHING FOR MIN. HUM.
2 ALL CAPACITORS AT LEAST 600 VOLT RATING IF NOT SPECIFIED.
1. ALL RESISTORS 1/2 WATT, 10% TOLERANCE IF NOT SPECIFIED.
NOTES: UNLESS OTHERWISE SPECIFIED
Fender/Rogers/Rhodes
SCHEMATIC
BASSMAN 70
70 WATT RMS AMPLIFIER-DOMESTIC (EXPORT
D
014229
A

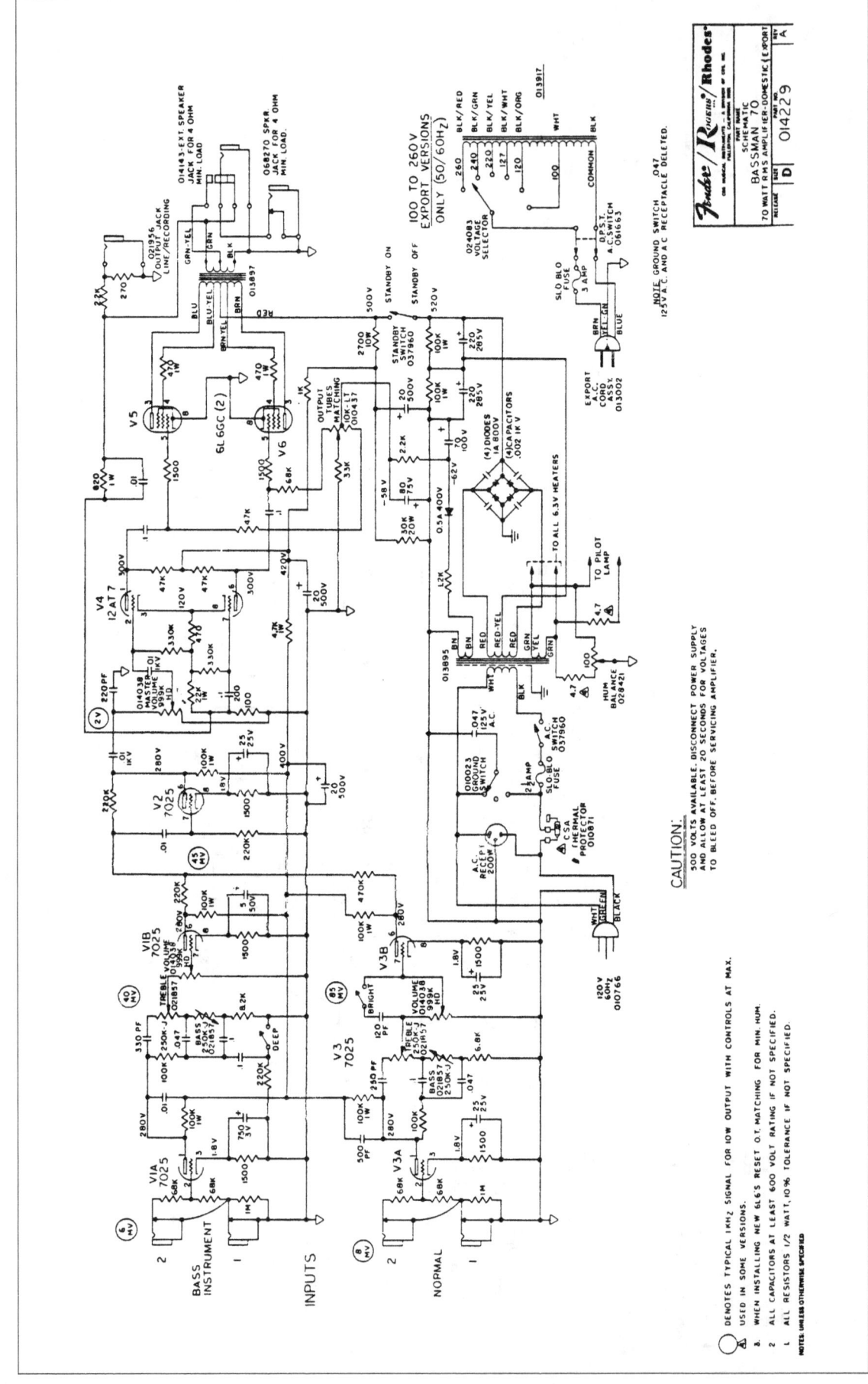

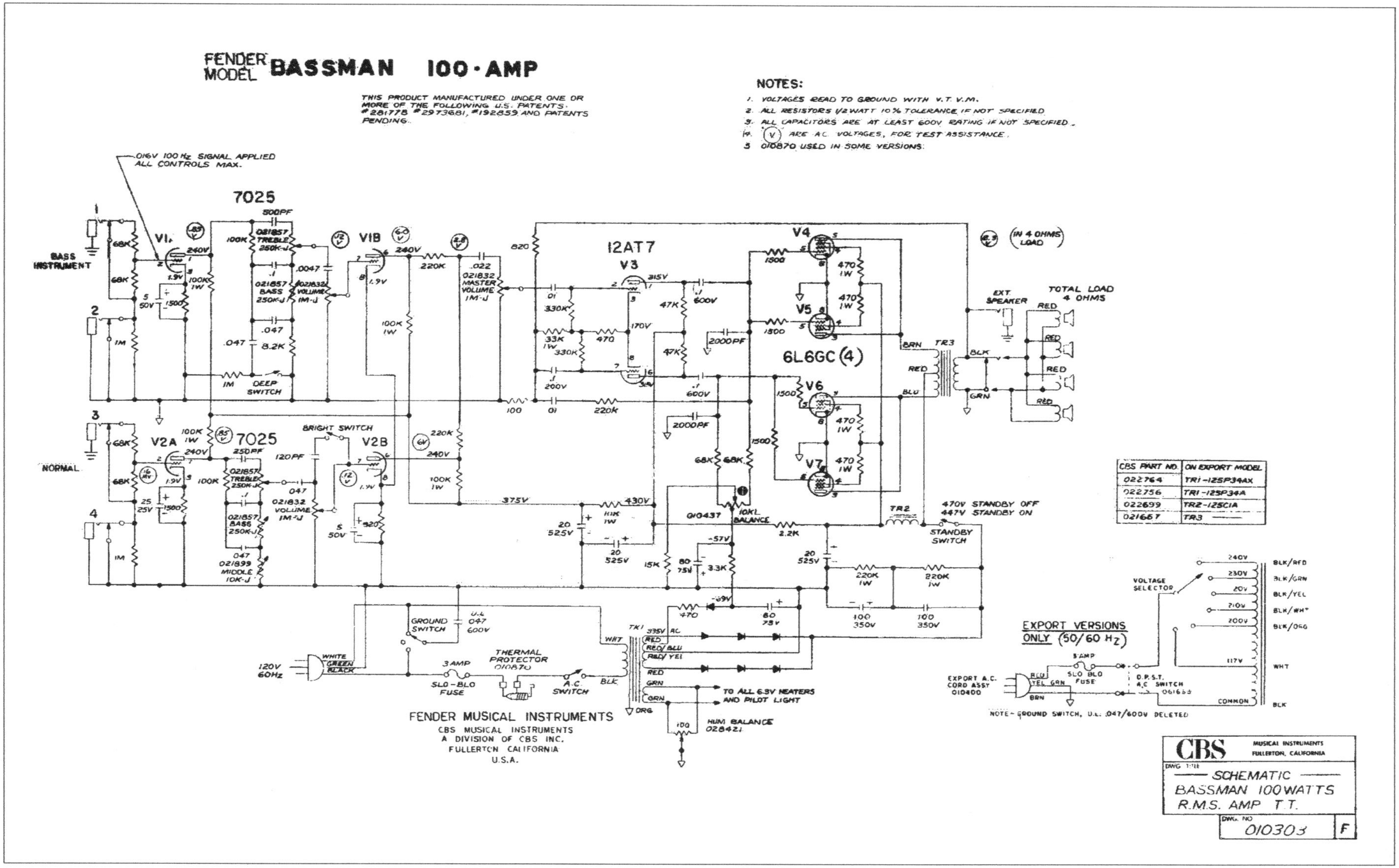
FENDER MODEL BASSMAN 100·AMP
THIS PRODUCT MANUFACTURED UNDER ONE OR MORE OF THE FOLLOWING U.S. PATENTS: #281778 #2973681, #192859 AND PATENTS PENDING.
NOTES:
1. VOLTAGES READ TO GROUND WITH V.T.V.M.
2. ALL RESISTORS 1/2 WATT 10% TOLERANCE IF NOT SPECIFIED
3. ALL CAPACITORS ARE AT LEAST 600V RATING IF NOT SPECIFIED.
4. (V) ARE A.C. VOLTAGES, FOR TEST ASSISTANCE.
5. 010870 USED IN SOME VERSIONS.
.016V 100 Hz SIGNAL APPLIED ALL CONTROLS MAX.
BASS INSTRUMENT
NORMAL
7025
V1A
V1B
V2A
V2B
12AT7
V3
V4
V5
V6
V7
6L6GC (4)
DEEP SWITCH
BRIGHT SWITCH
021857 TREBLE 250K-J
021857 BASS 250K-J
021832 VOLUME 1M-J
021899 MIDDLE 10K-J
021832 MASTER VOLUME 1M-J
010437
10KL BALANCE
375V
430V
-57V
-69V
EXT SPEAKER
TOTAL LOAD 4 OHMS
(IN 4 OHMS LOAD)
TR1
TR2
TR3
470V STANDBY OFF
447V STANDBY ON
STANDBY SWITCH
TO ALL 6.3V HEATERS AND PILOT LIGHT
HUM BALANCE 028421
120V 60Hz
WHITE
GREEN
BLACK
GROUND SWITCH
U.L. .047 600V
3 AMP SLO-BLO FUSE
THERMAL PROTECTOR 010870
A.C. SWITCH
FENDER MUSICAL INSTRUMENTS
CBS MUSICAL INSTRUMENTS
A DIVISION OF CBS INC.
FULLERTON CALIFORNIA
U.S.A.
CBS PART NO. | ON EXPORT MODEL
022764 | TR1-125P34AX
022756 | TR1-125P34A
022699 | TR2-125C1A
021667 | TR3
EXPORT VERSIONS ONLY (50/60 Hz)
VOLTAGE SELECTOR
EXPORT A.C. CORD ASSY 010400
D.P.S.T. A.C. SWITCH
NOTE - GROUND SWITCH, U.L. .047/600V DELETED
CBS MUSICAL INSTRUMENTS FULLERTON, CALIFORNIA
DWG TITLE
SCHEMATIC
BASSMAN 100 WATTS
R.M.S. AMP T.T.
DWG NO
010303
F

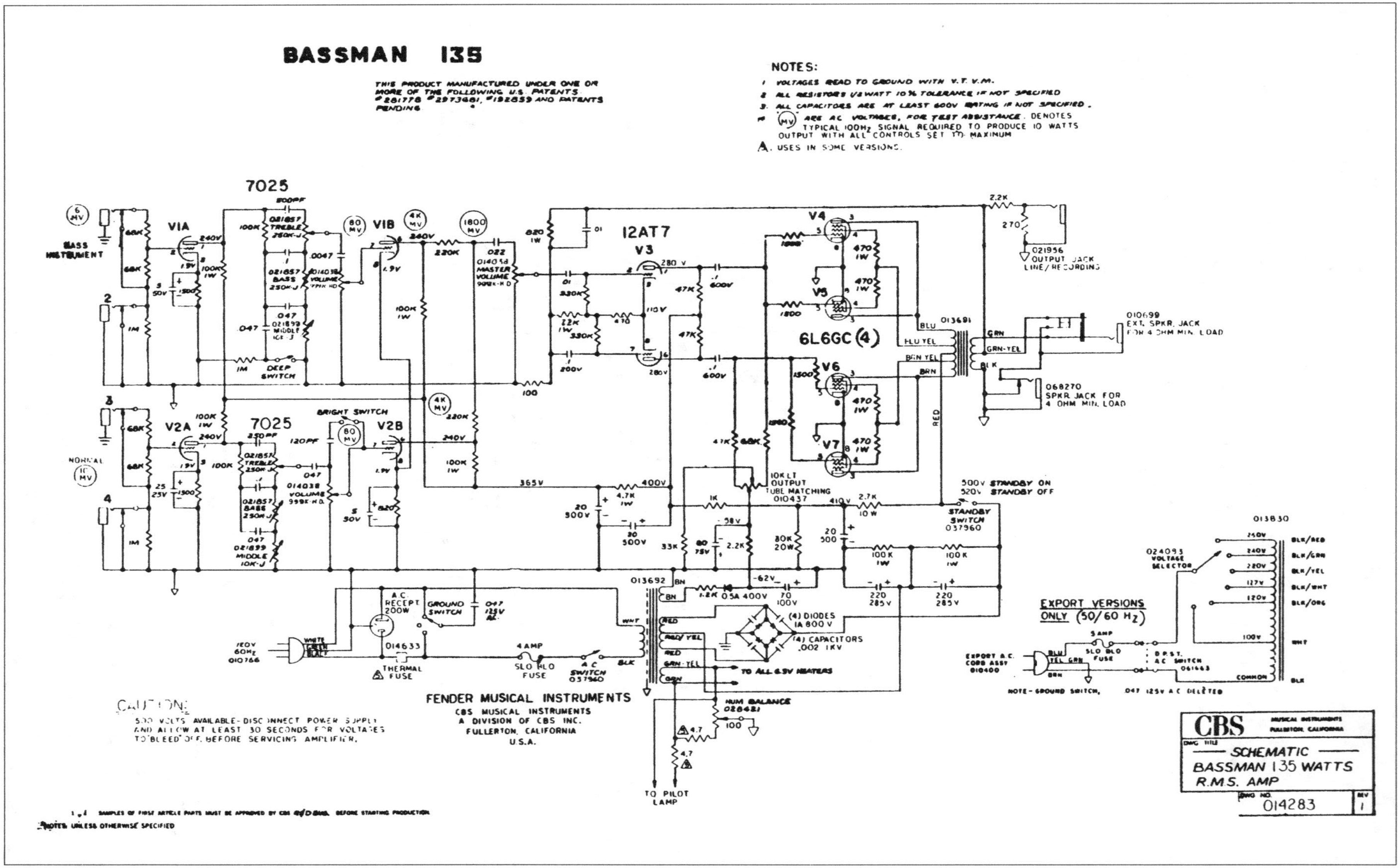
BASSMAN 135
THIS PRODUCT MANUFACTURED UNDER ONE OR MORE OF THE FOLLOWING U.S. PATENTS. #281778 #2973681, #192859 AND PATENTS PENDING
NOTES:
1 VOLTAGES READ TO GROUND WITH V.T.V.M.
2 ALL RESISTORS 1/2 WATT 10% TOLERANCE IF NOT SPECIFIED
3. ALL CAPACITORS ARE AT LEAST 600V RATING IF NOT SPECIFIED.
4 (MV) ARE AC VOLTAGES, FOR TEST ASSISTANCE. DENOTES TYPICAL 100Hz SIGNAL REQUIRED TO PRODUCE 10 WATTS OUTPUT WITH ALL CONTROLS SET TO MAXIMUM
A. USES IN SOME VERSIONS.
7025
V1A
V1B
BASS INSTRUMENT
NORMAL
V2A
V2B
BRIGHT SWITCH
DEEP SWITCH
12AT7
V3
V4
V5
V6
V7
6L6GC (4)
MASTER VOLUME
10K LT OUTPUT TUBE MATCHING 010437
021956 OUTPUT JACK LINE/RECORDING
010699 EXT. SPKR. JACK FOR 4 OHM MIN. LOAD
068270 SPKR JACK FOR 4 OHM MIN. LOAD
500V STANDBY ON
520V STANDBY OFF
STANDBY SWITCH 037960
(4) DIODES 1A 800 V
(4) CAPACITORS .002 1KV
TO ALL 6.3V HEATERS
HUM BALANCE 028481
TO PILOT LAMP
A.C. RECEPT 200W
GROUND SWITCH
014633 THERMAL FUSE
4 AMP SLO BLO FUSE
A C SWITCH 037960
120V 60Hz 010766
CAUTION:
500 VOLTS AVAILABLE-DISCONNECT POWER SUPPLY AND ALLOW AT LEAST 30 SECONDS FOR VOLTAGES TO 'BLEED' OFF BEFORE SERVICING AMPLIFIER.
FENDER MUSICAL INSTRUMENTS
CBS MUSICAL INSTRUMENTS
A DIVISION OF CBS INC.
FULLERTON, CALIFORNIA
U.S.A.
EXPORT VERSIONS ONLY (50/60 Hz)
024093 VOLTAGE SELECTOR
013830
EXPORT A.C. CORD ASSY 010400
NOTE-GROUND SWITCH, .047 125V A.C. DELETED
CBS MUSICAL INSTRUMENTS FULLERTON, CALIFORNIA
SCHEMATIC
BASSMAN 135 WATTS
R.M.S. AMP
DWG NO. 014283
REV 1
NOTES UNLESS OTHERWISE SPECIFIED

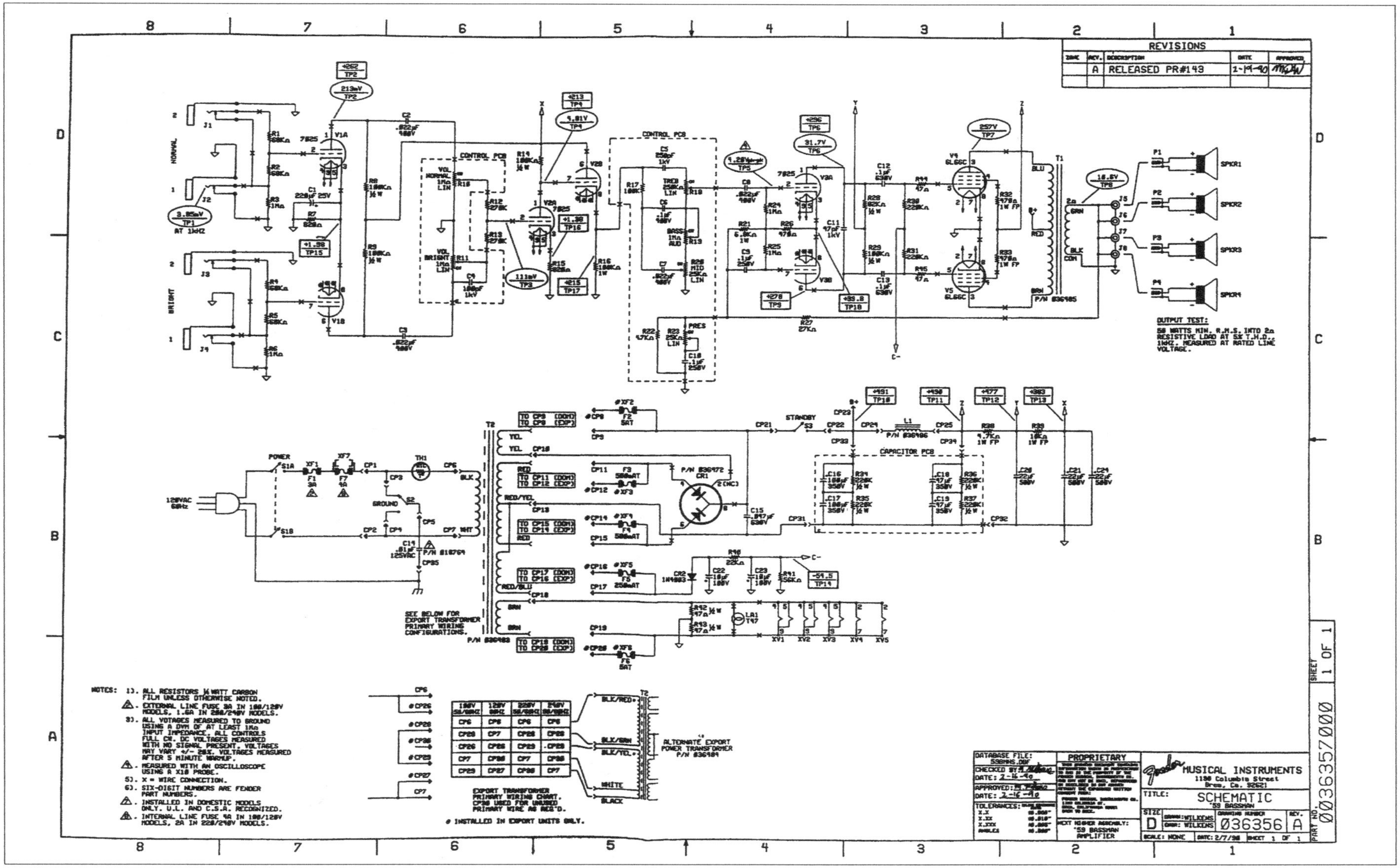
REVISIONS
A RELEASED PR#143
OUTPUT TEST:
PROPRIETARY
MUSICAL INSTRUMENTS
SCHEMATIC
'59 BASSMAN
036356
PART NO. 0036357000
SHEET 1 OF 1

BASSMAN 59 (REV E)

Fender

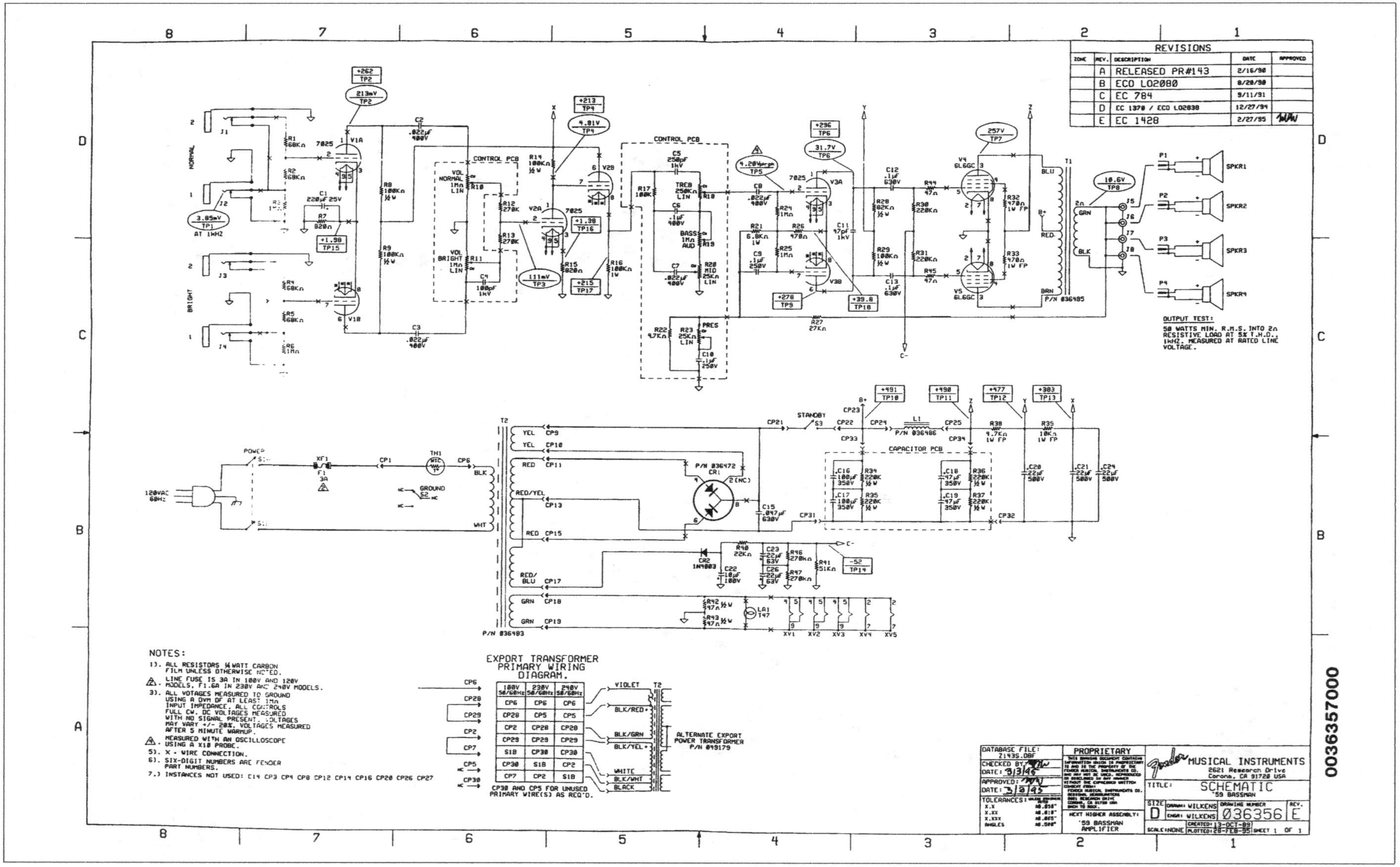

FENDER "BASSMAN" AMPLIFIER PARTS LAYOUT

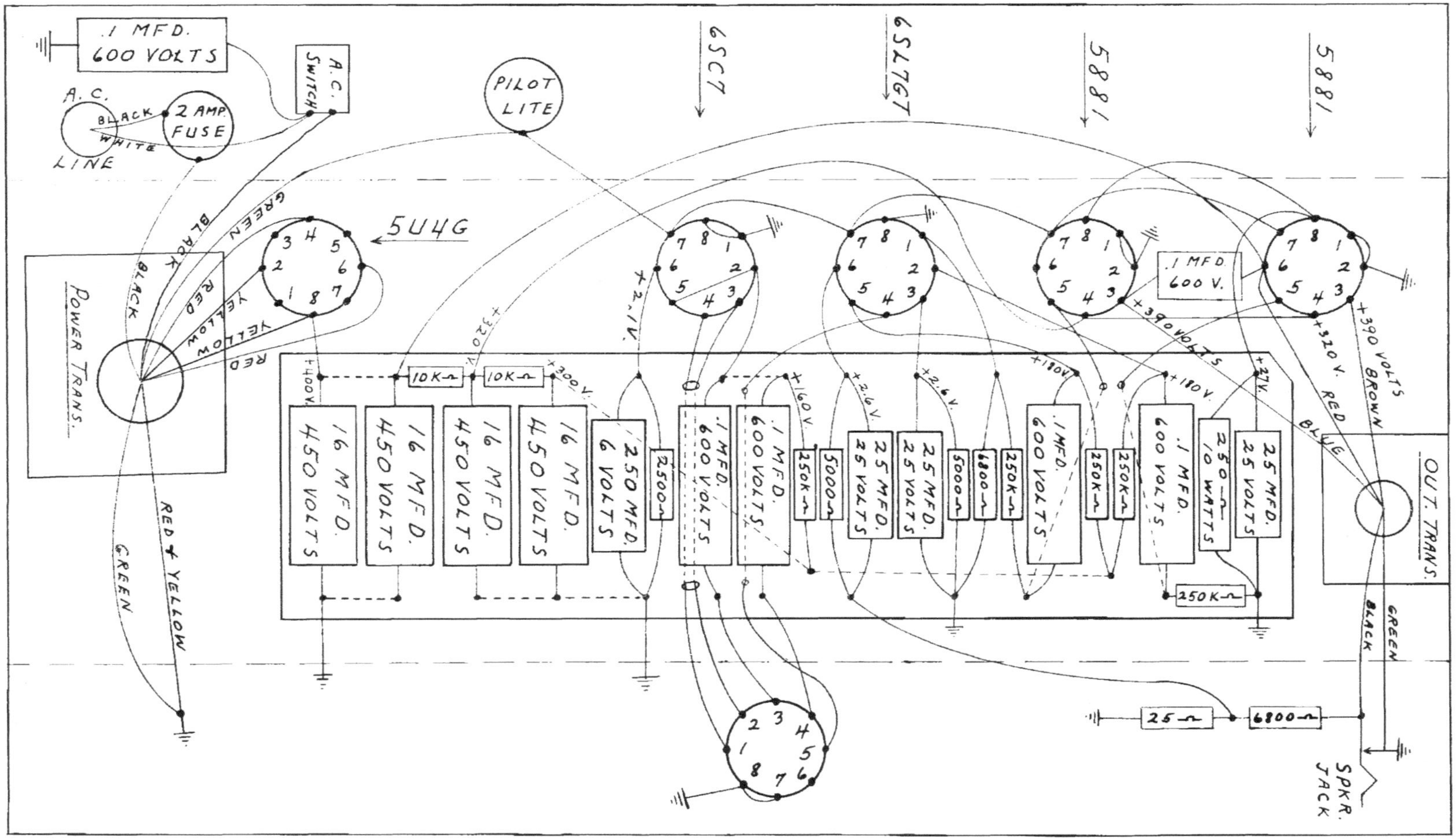

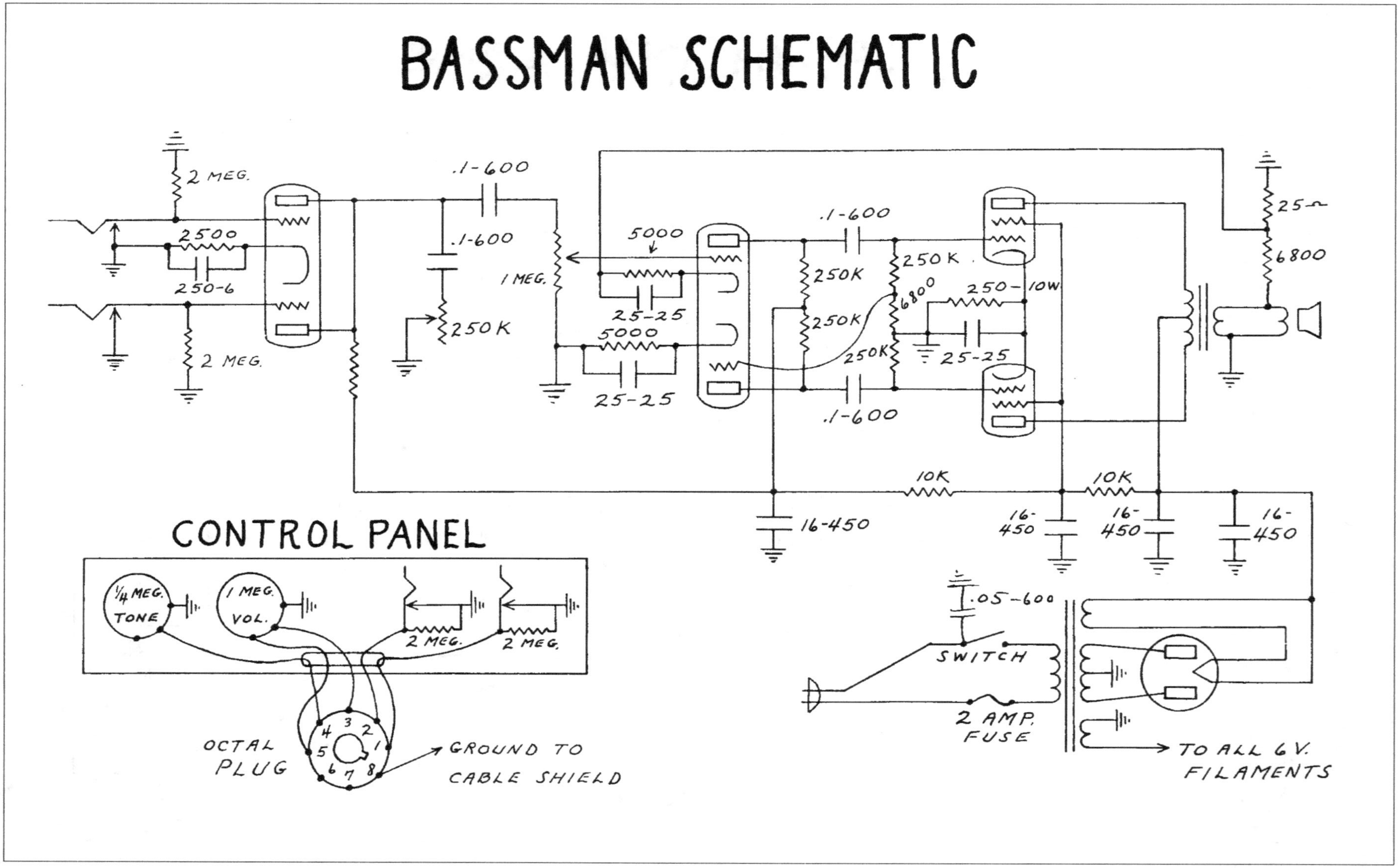
BASSMAN SCHEMATIC
2 MEG.
2500
250-6
2 MEG.
.1-600
.1-600
1 MEG.
250K
5000
25-25
5000
25-25
.1-600
250K
250K
250K
6800
250K
250-
10W
25-25
.1-600
25
6800
10K
10K
16-450
16-450
16-450
16-450
.05-600
SWITCH
2 AMP. FUSE
TO ALL 6V. FILAMENTS
CONTROL PANEL
1/4 MEG. TONE
1 MEG. VOL.
2 MEG.
2 MEG.
OCTAL PLUG
GROUND TO CABLE SHIELD

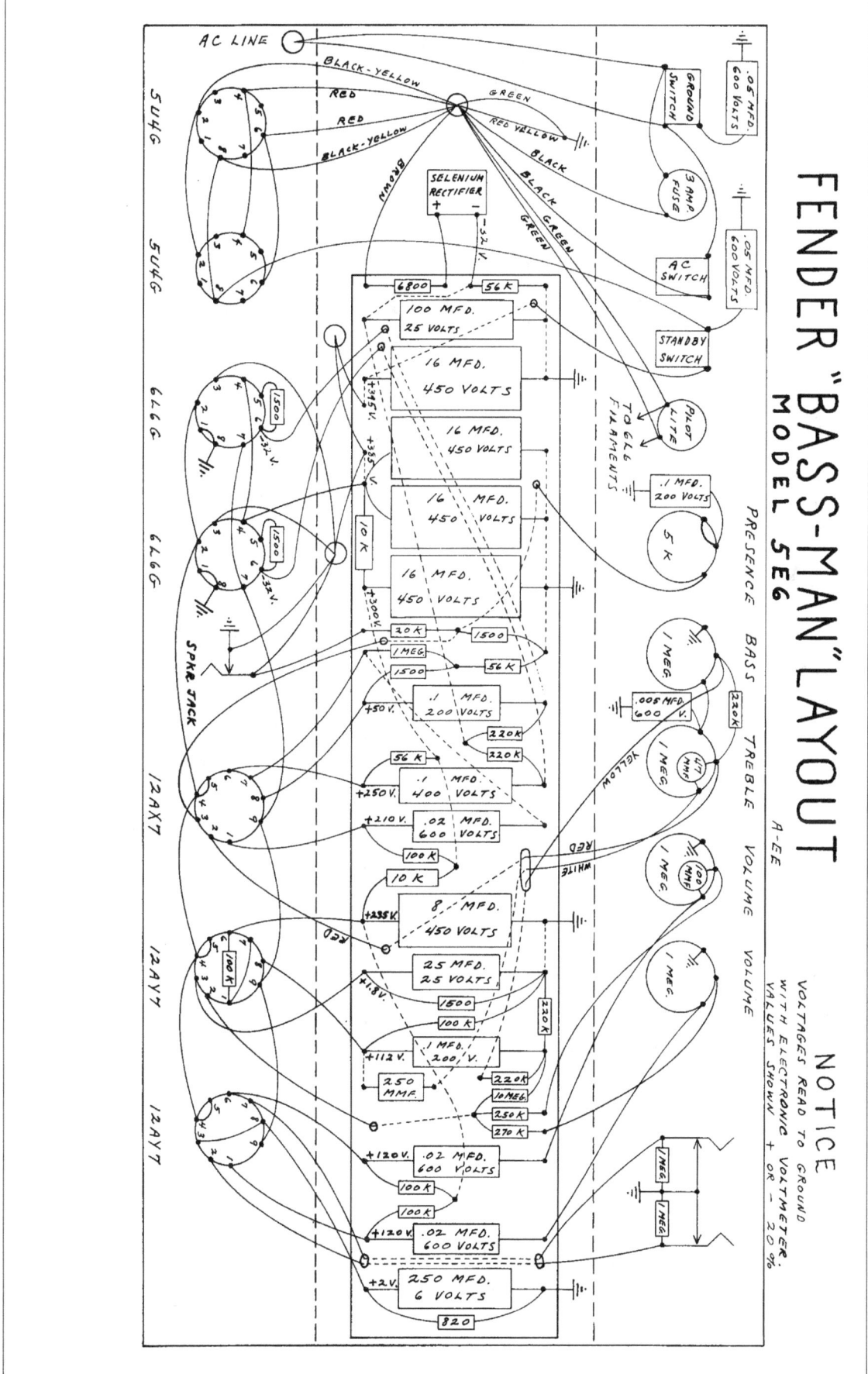

FENDER "BASS-MAN" LAYOUT

MODEL 5E6

A-EE

NOTICE

VOLTAGES READ TO GROUND WITH ELECTRONIC VOLTMETER. VALUES SHOWN + OR − 20%

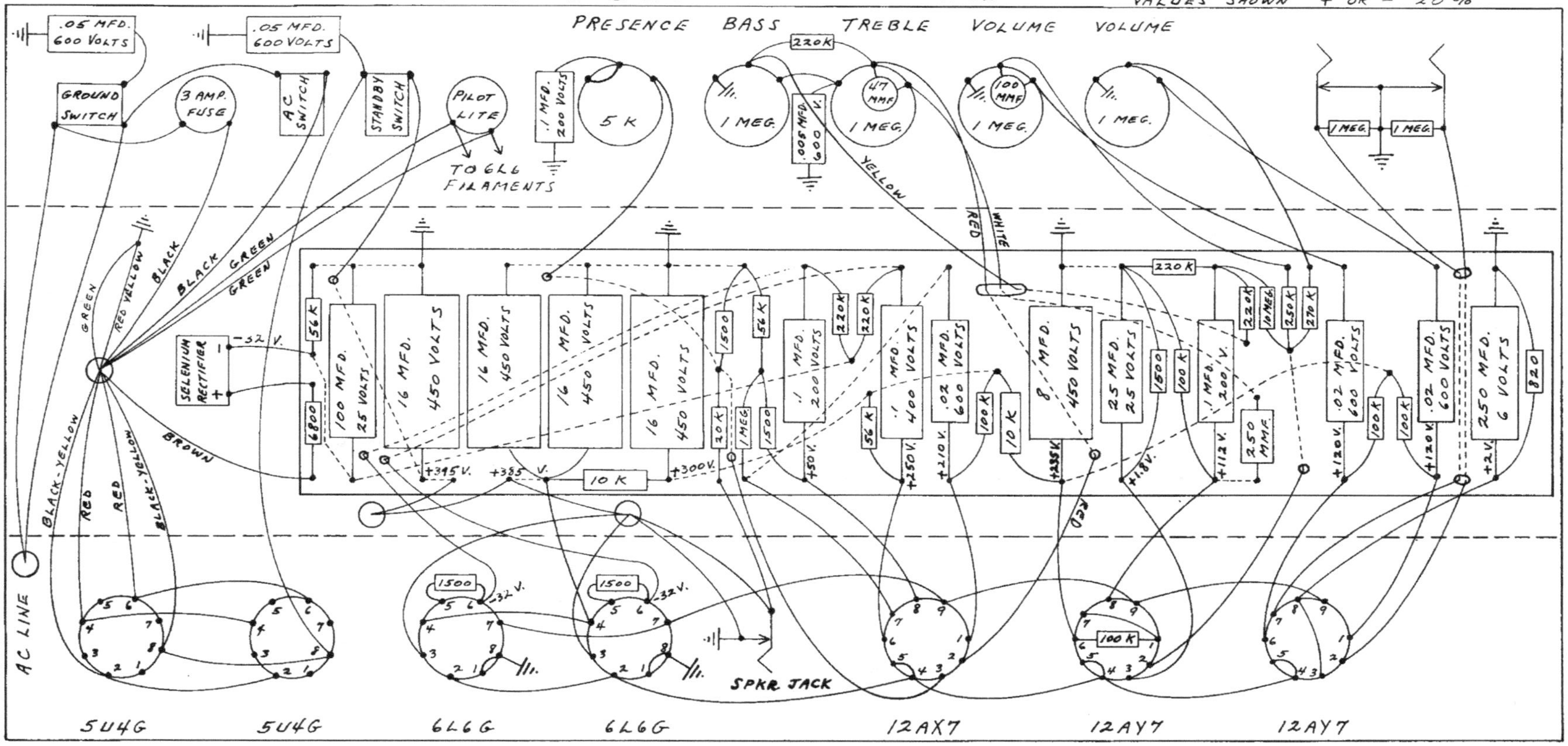

FENDER "BASSMAN" SCHEMATIC
MODEL 5E6

A-EE

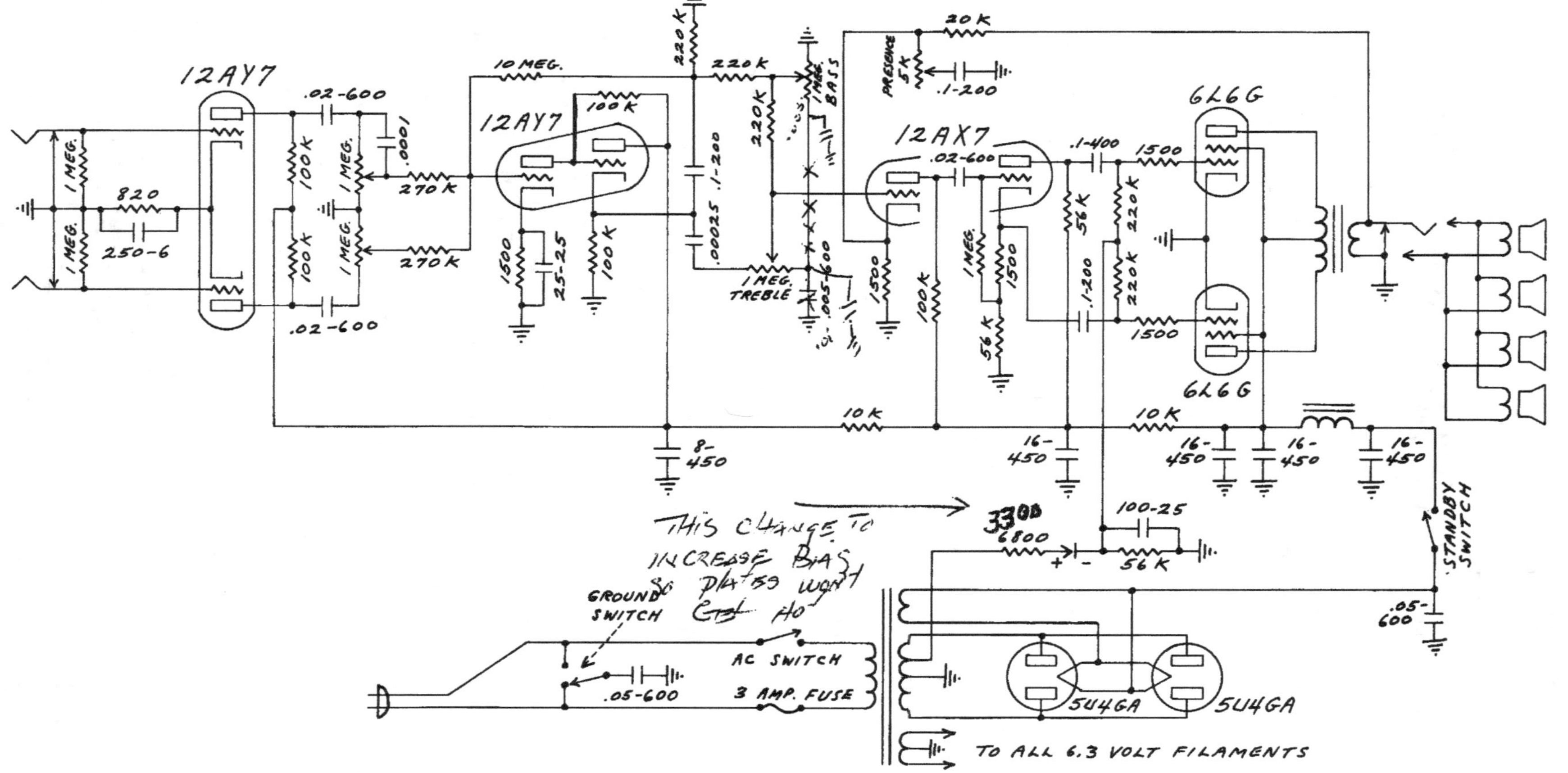

FENDER "BASS-MAN" LAYOUT

MODEL 5E6-A

A-EE

NOTICE
VOLTAGES READ TO GROUND
WITH ELECTRONIC VOLTMETER.
VALUES SHOWN + OR − 20%

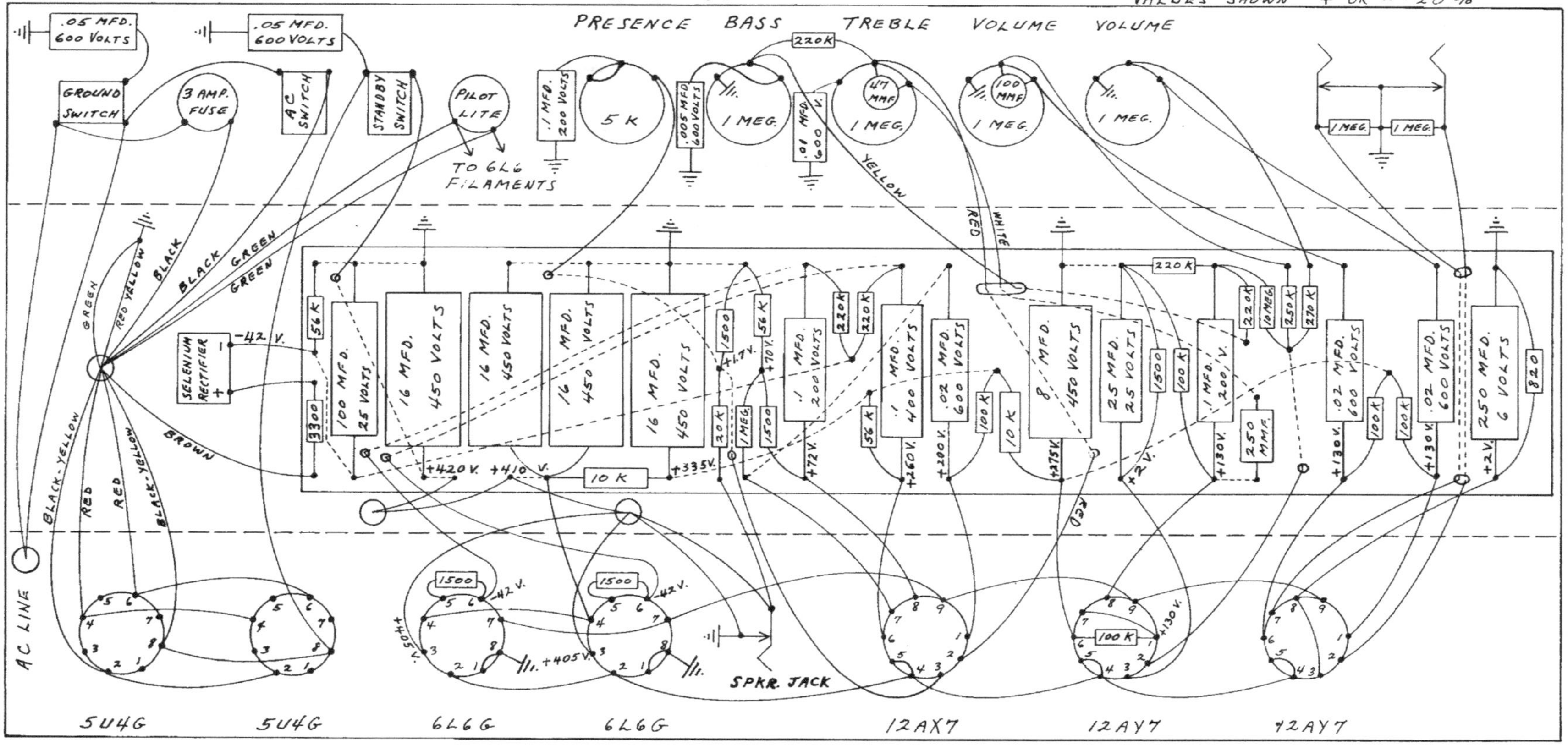

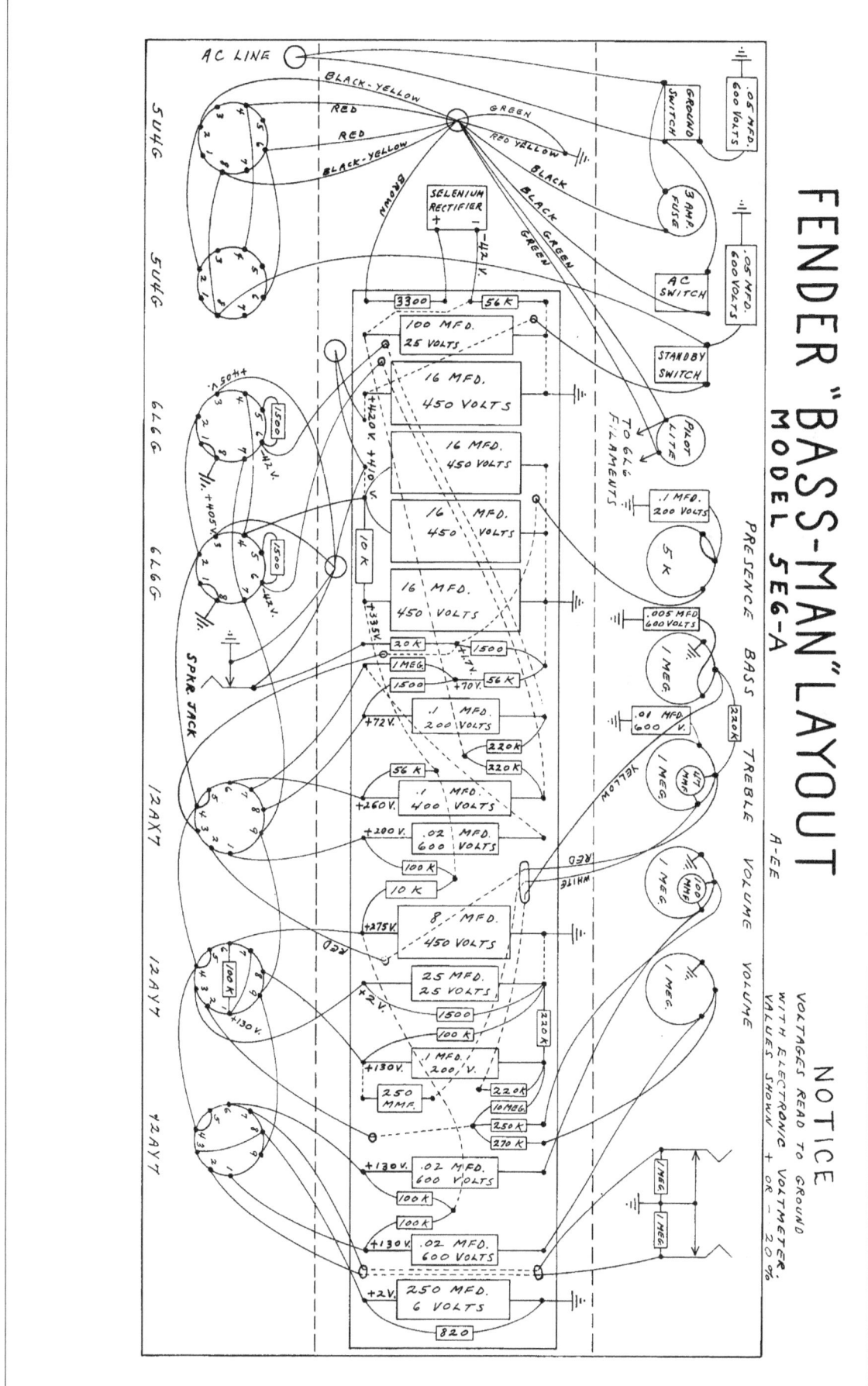

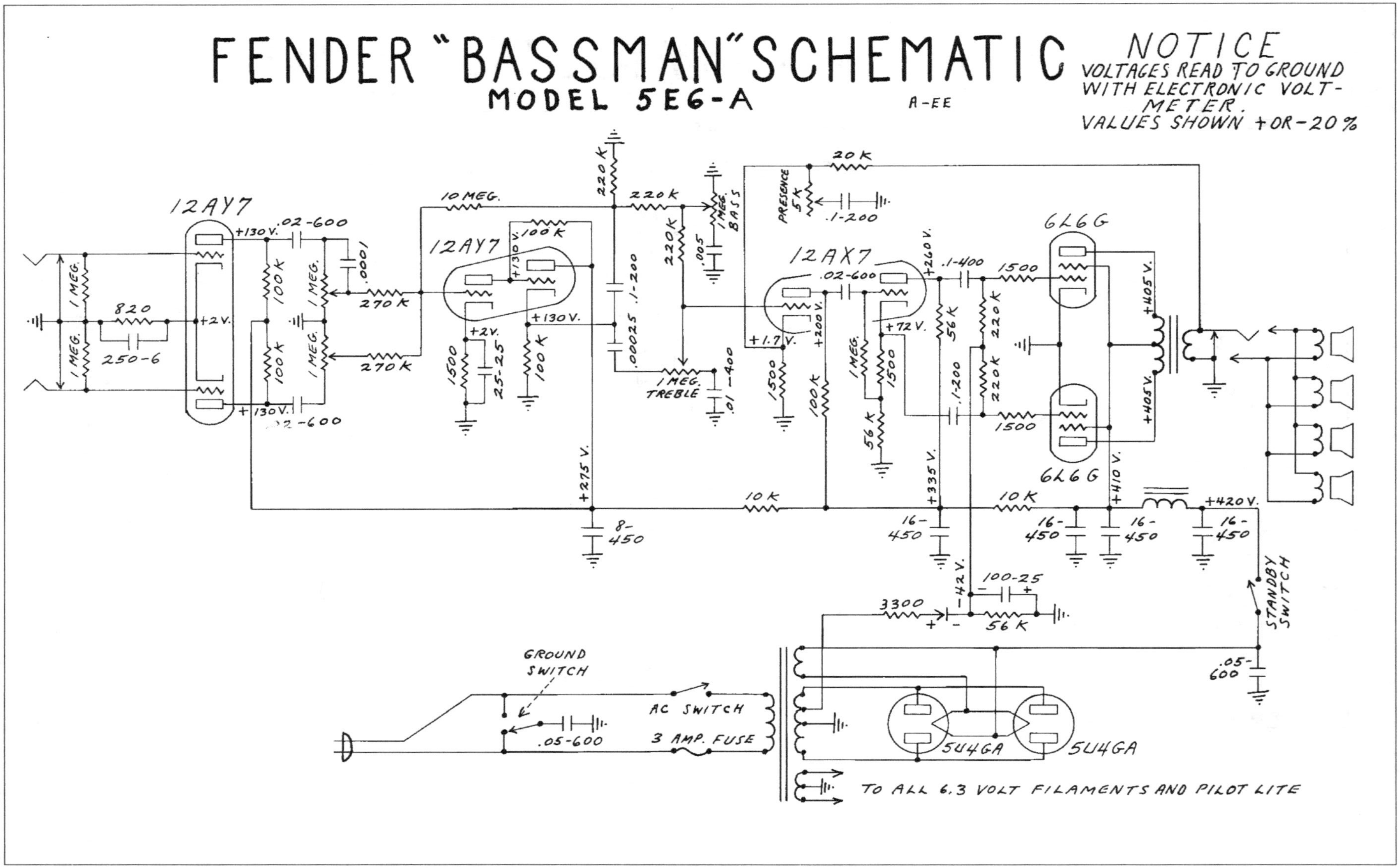
FENDER "BASSMAN" SCHEMATIC
MODEL 5E6-A
A-EE
NOTICE
VOLTAGES READ TO GROUND WITH ELECTRONIC VOLT-METER.
VALUES SHOWN + OR -20%
12AY7
12AY7
12AX7
6L6G
6L6G
5U4GA
5U4GA
1 MEG.
820
250-6
+2V.
+130V.
.02-600
100 K
.0001
270 K
10 MEG.
220 K
220 K
220 K
1 MEG. BASS
.005
.1-200
.00025
1 MEG. TREBLE
.01-400
1500
25-25
+275 V.
8-450
10 K
PRESENCE 5 K
20 K
+1.7 V.
+200 V.
+72 V.
+260 V.
.1-400
56 K
+335 V.
+405 V.
+410 V.
+420 V.
16-450
-42 V.
100-25
3300
STANDBY SWITCH
.05-600
GROUND SWITCH
AC SWITCH
3 AMP. FUSE
TO ALL 6.3 VOLT FILAMENTS AND PILOT LITE

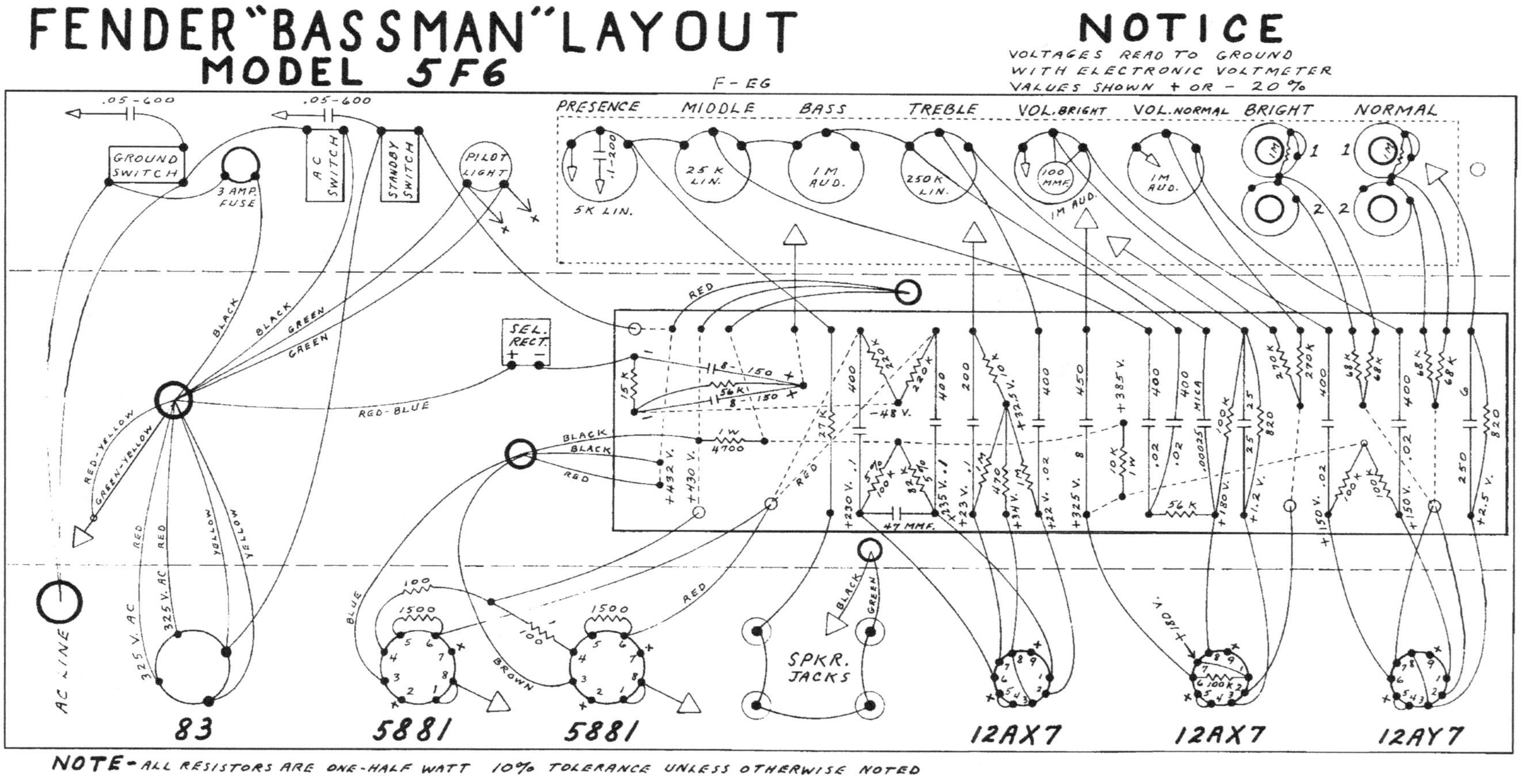
FENDER "BASSMAN" LAYOUT
MODEL 5F6
NOTICE
VOLTAGES READ TO GROUND
WITH ELECTRONIC VOLTMETER
VALUES SHOWN + OR - 20%
F-EG
PRESENCE
MIDDLE
BASS
TREBLE
VOL. BRIGHT
VOL. NORMAL
BRIGHT
NORMAL
5K LIN.
25 K LIN.
1M AUD.
250K LIN.
100 MMF.
1M AUD.
GROUND SWITCH
3 AMP. FUSE
AC SWITCH
STANDBY SWITCH
PILOT LIGHT
.05-600
SEL. RECT.
RED-BLUE
RED-YELLOW
GREEN-YELLOW
BLACK
GREEN
RED
YELLOW
BLUE
BROWN
325 V. AC
AC LINE
+432 V.
+430 V.
+230 V.
+235 V.
+23 V.
+34 V.
+32.5 V.
+22 V.
+325 V.
+385 V.
+180 V.
+1.2 V.
+150 V.
+2.5 V.
-48 V.
47 MMF.
MICA
SPKR. JACKS
83
5881
5881
12AX7
12AX7
12AY7
NOTE- ALL RESISTORS ARE ONE-HALF WATT 10% TOLERANCE UNLESS OTHERWISE NOTED

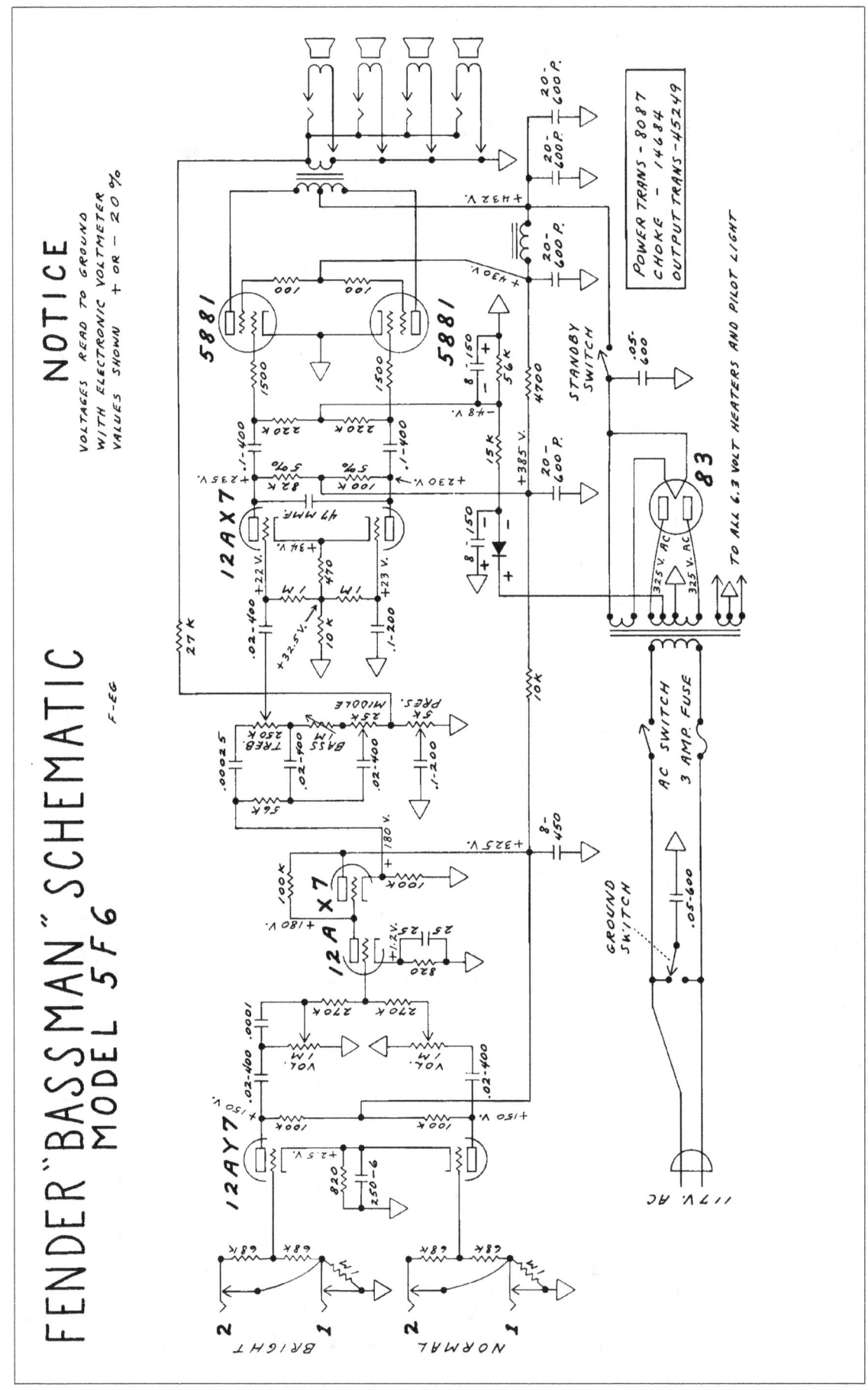

FENDER "BASSMAN" SCHEMATIC
MODEL 5F6

F-EG

NOTICE

VOLTAGES READ TO GROUND
WITH ELECTRONIC VOLTMETER
VALUES SHOWN + OR − 20 %

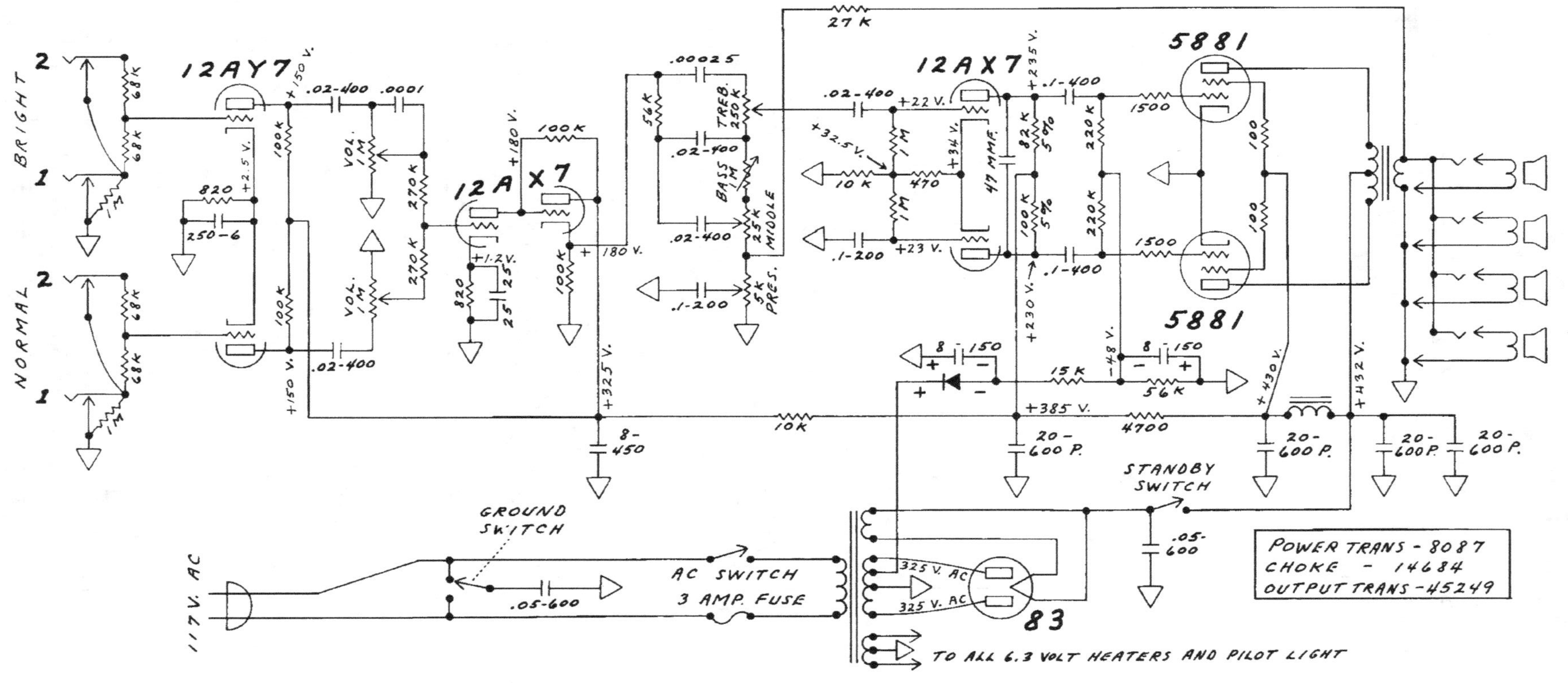

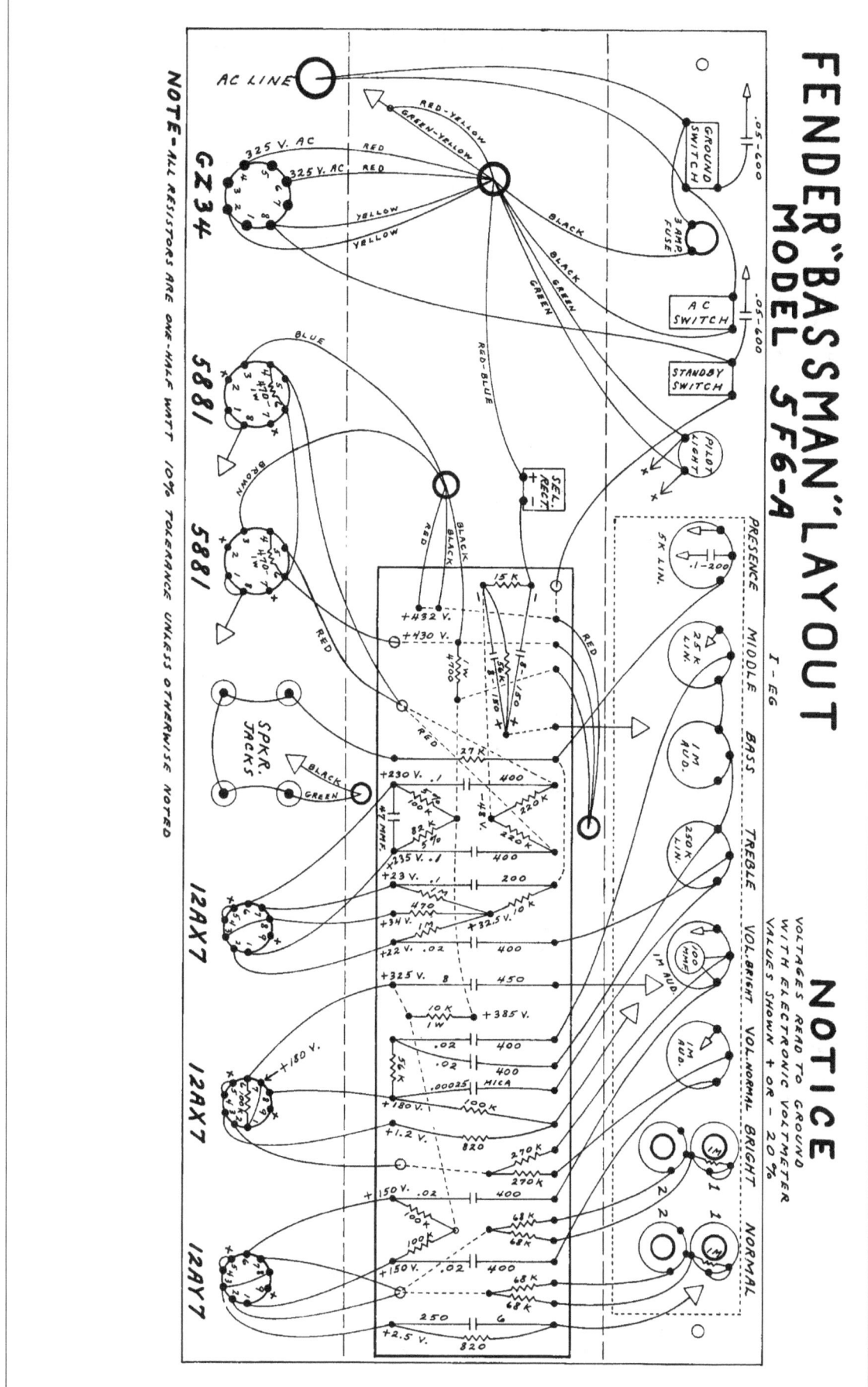

FENDER "BASSMAN" LAYOUT
MODEL 5F6-A

NOTICE

VOLTAGES READ TO GROUND
WITH ELECTRONIC VOLTMETER
VALUES SHOWN + OR − 20%

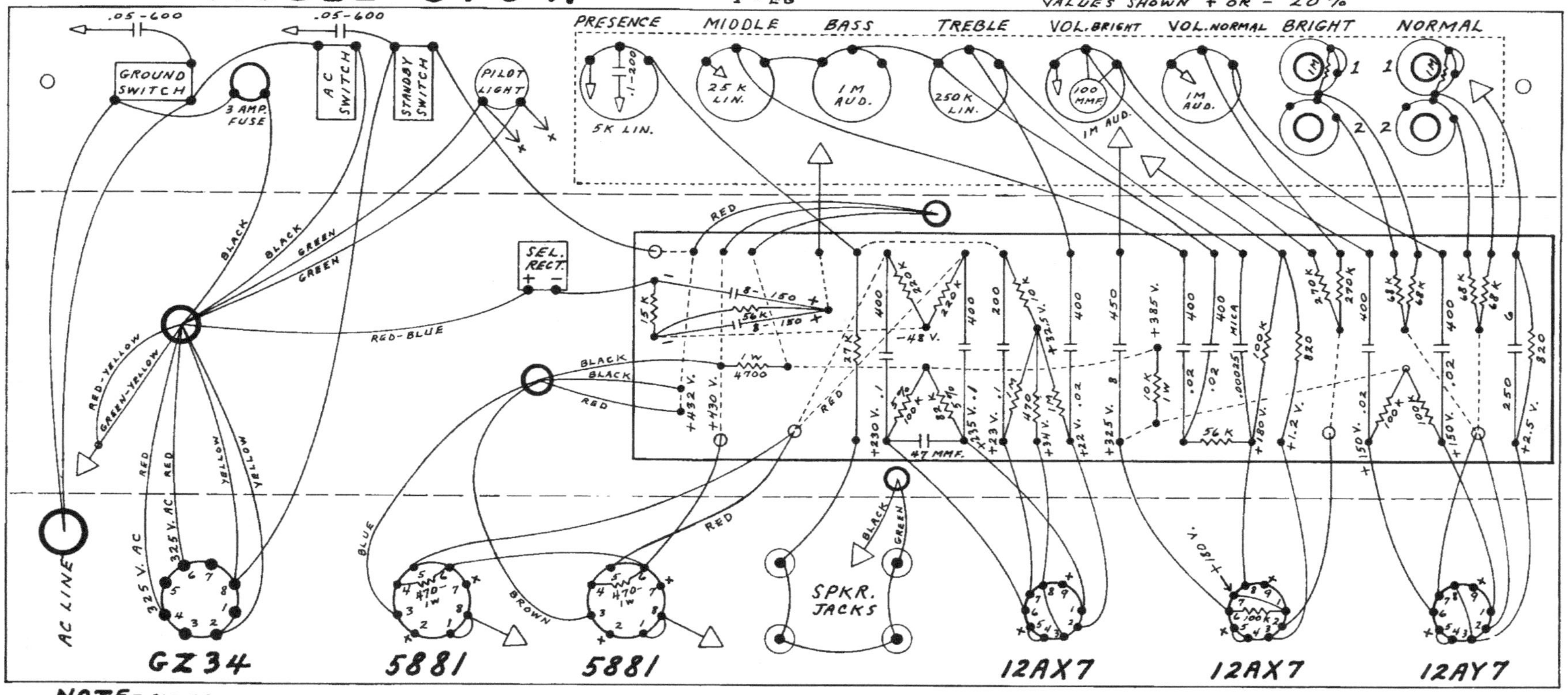

NOTE - ALL RESISTORS ARE ONE-HALF WATT 10% TOLERANCE UNLESS OTHERWISE NOTED

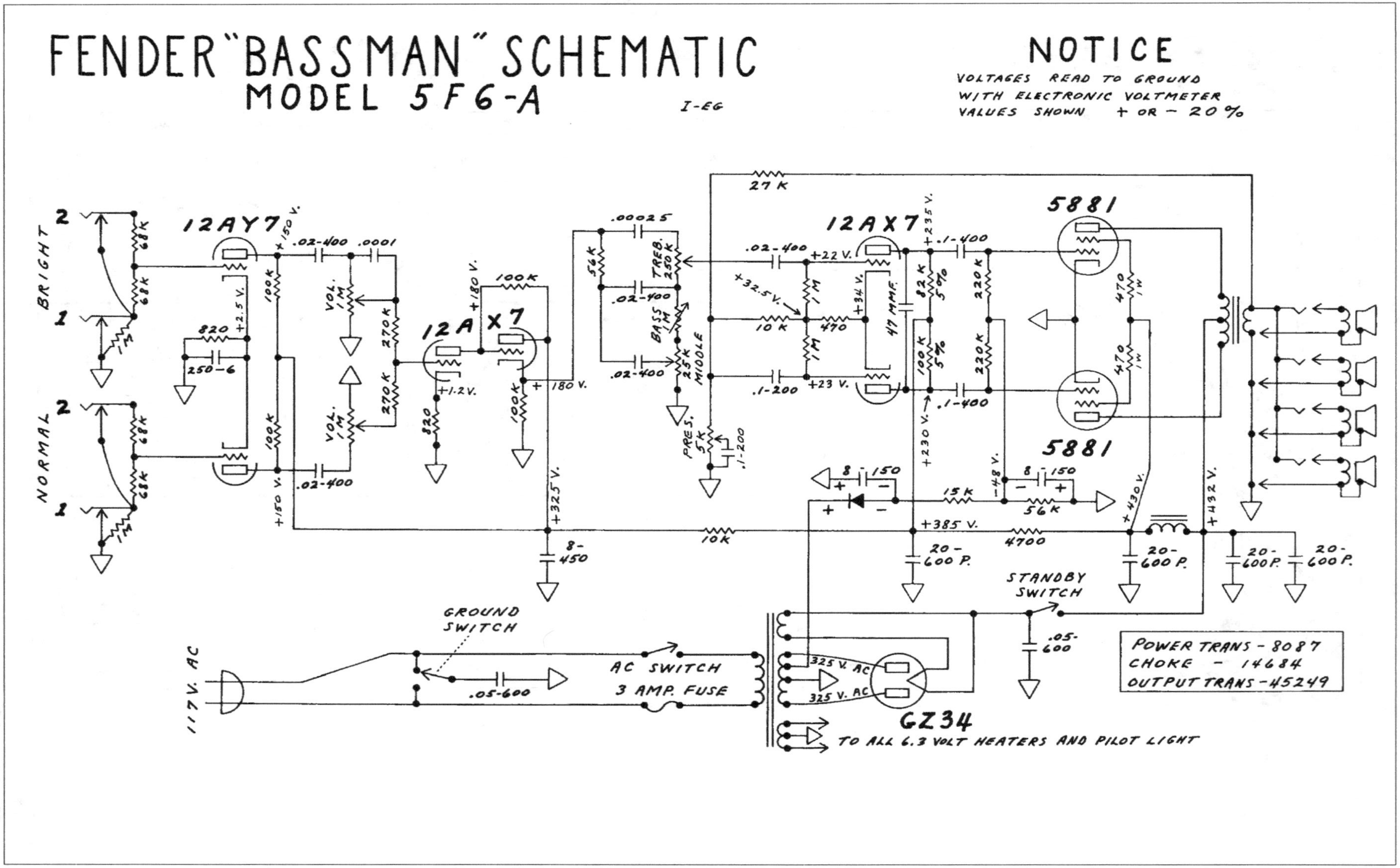
FENDER "BASSMAN" SCHEMATIC
MODEL 5F6-A
I-EG
NOTICE
VOLTAGES READ TO GROUND
WITH ELECTRONIC VOLTMETER
VALUES SHOWN + OR − 20 %
BRIGHT
NORMAL
12AY7
12AX7
12AX7
5881
5881
GZ34
POWER TRANS - 8087
CHOKE - 14684
OUTPUT TRANS - 45249
STANDBY SWITCH
GROUND SWITCH
AC SWITCH
3 AMP. FUSE
117 V. AC
TO ALL 6.3 VOLT HEATERS AND PILOT LIGHT
TREB. 250 K
BASS 1M
MIDDLE 25 K
PRES. 5K
VOL. 1M
+432 V.
+430 V.
+385 V.
+325 V.
+235 V.
+230 V.
+180 V.
+150 V.
-48 V.
325 V. AC

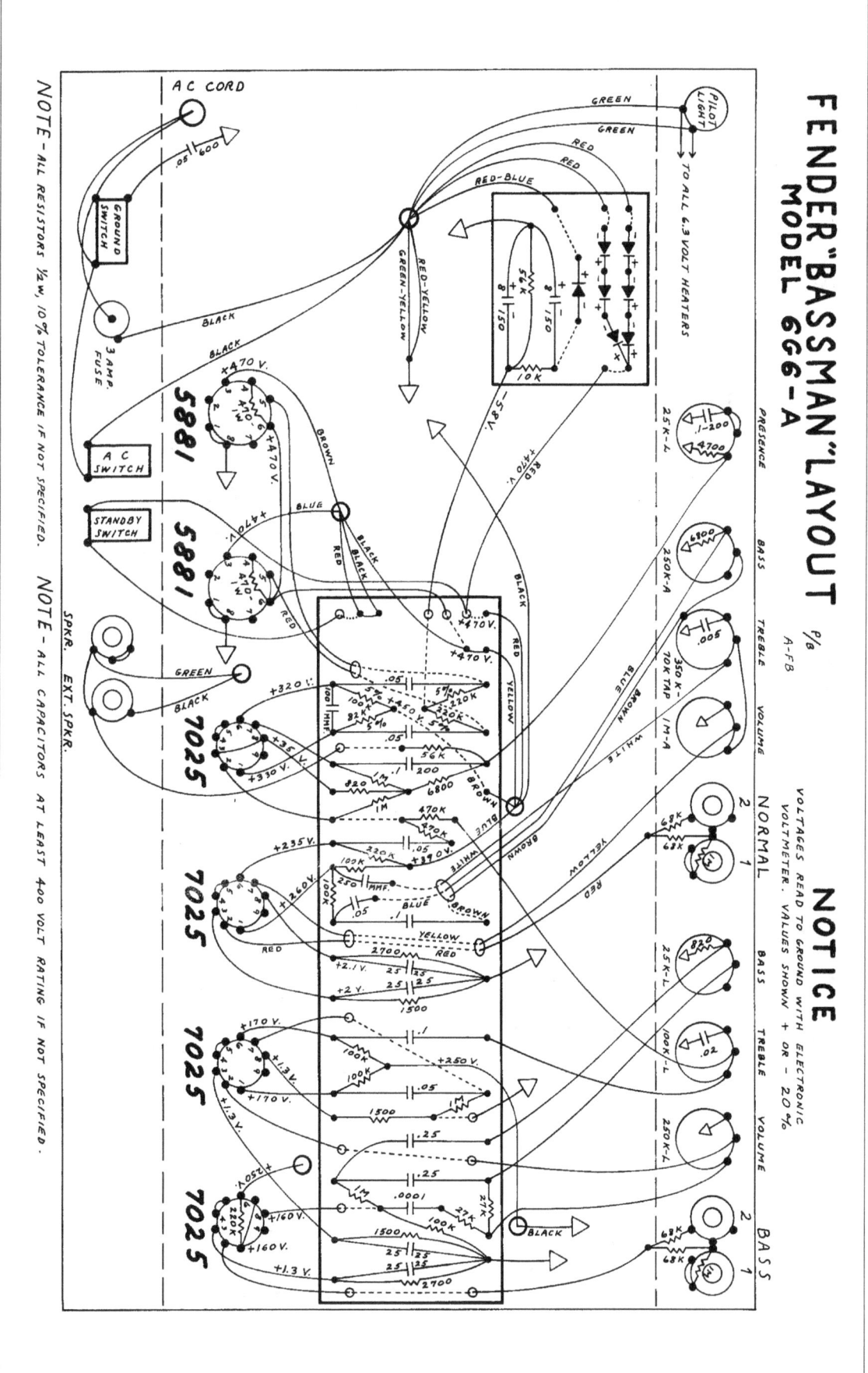

FENDER "BASSMAN" LAYOUT

P/B

MODEL 6G6-A

A-FB

NOTICE

VOLTAGES READ TO GROUND WITH ELECTRONIC VOLTMETER. VALUES SHOWN + OR - 20%

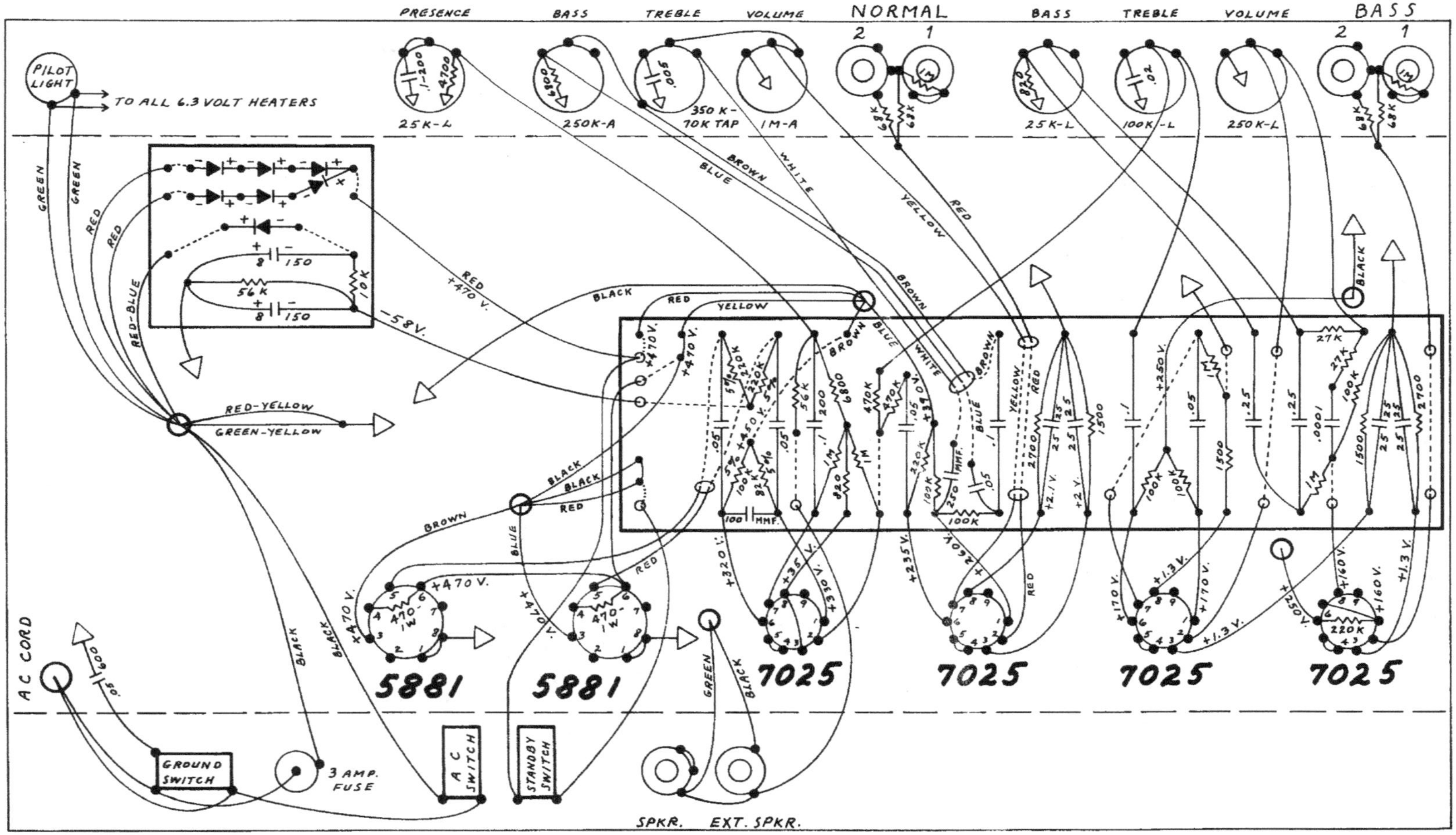

NOTE - ALL RESISTORS ½W, 10% TOLERANCE IF NOT SPECIFIED. NOTE - ALL CAPACITORS AT LEAST 400 VOLT RATING IF NOT SPECIFIED.

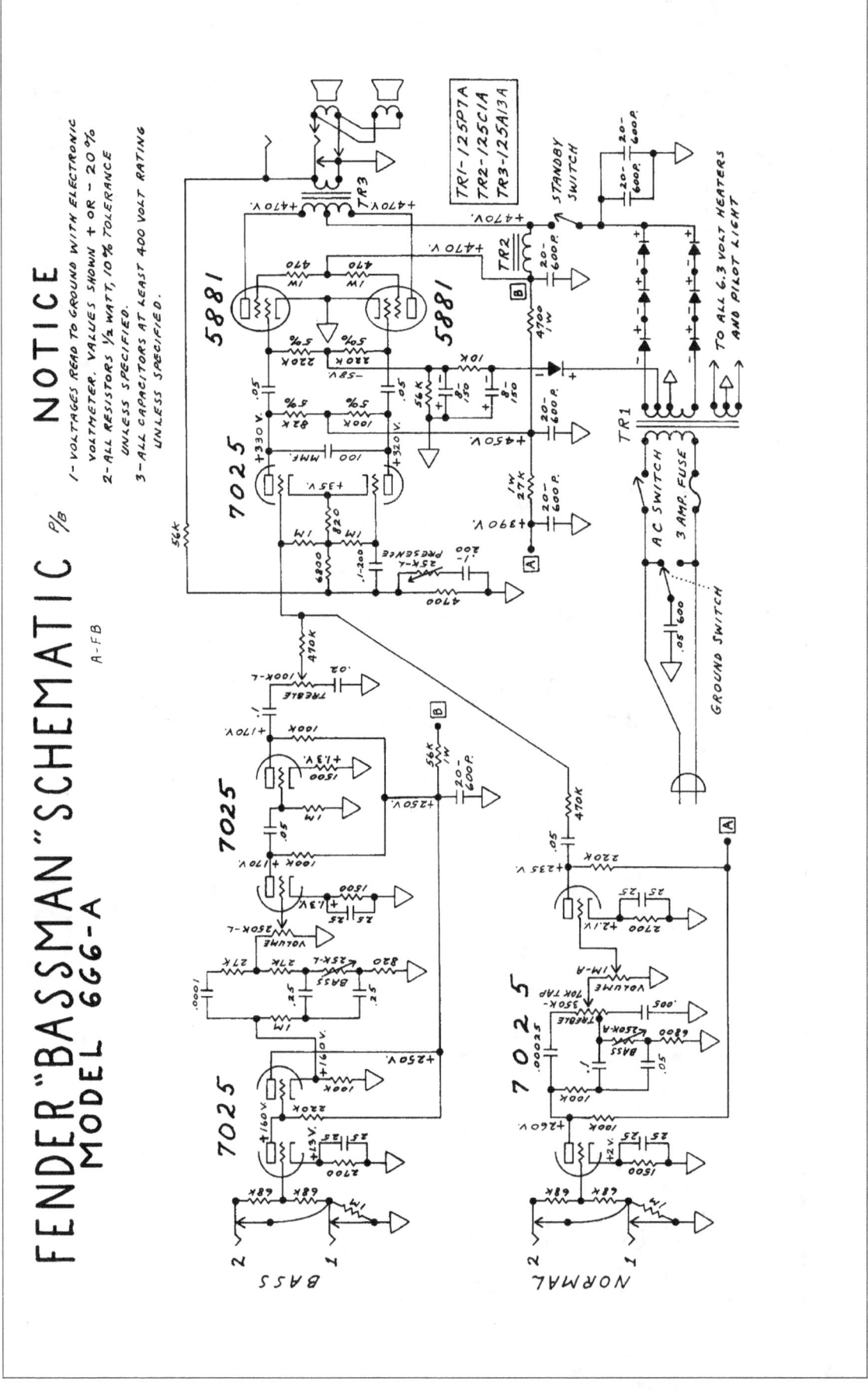

FENDER "BASSMAN" SCHEMATIC P/B
MODEL 6G6-A

A-FB

NOTICE

1- VOLTAGES READ TO GROUND WITH ELECTRONIC VOLTMETER. VALUES SHOWN + OR - 20%
2- ALL RESISTORS ½ WATT, 10% TOLERANCE UNLESS SPECIFIED.
3- ALL CAPACITORS AT LEAST 400 VOLT RATING UNLESS SPECIFIED.

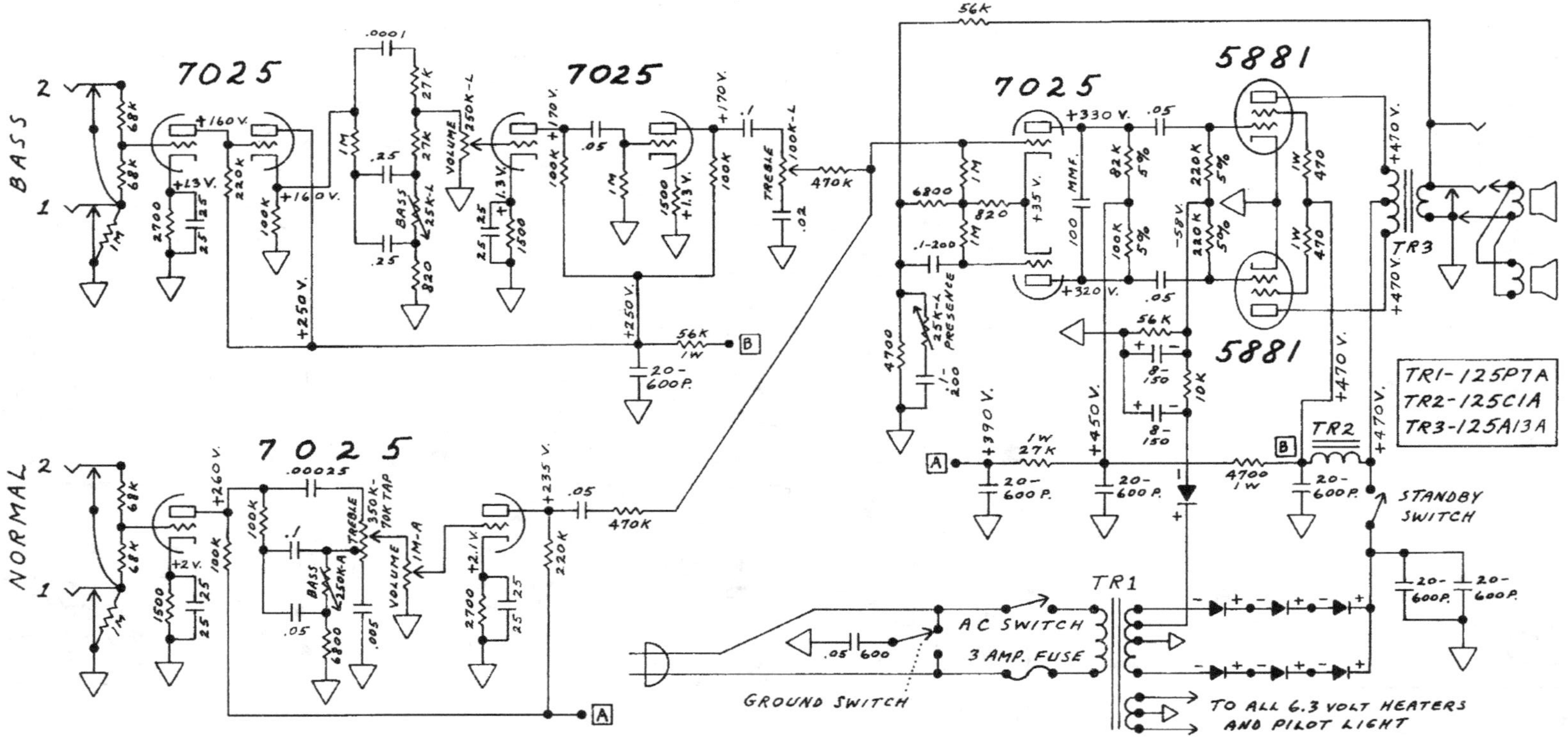

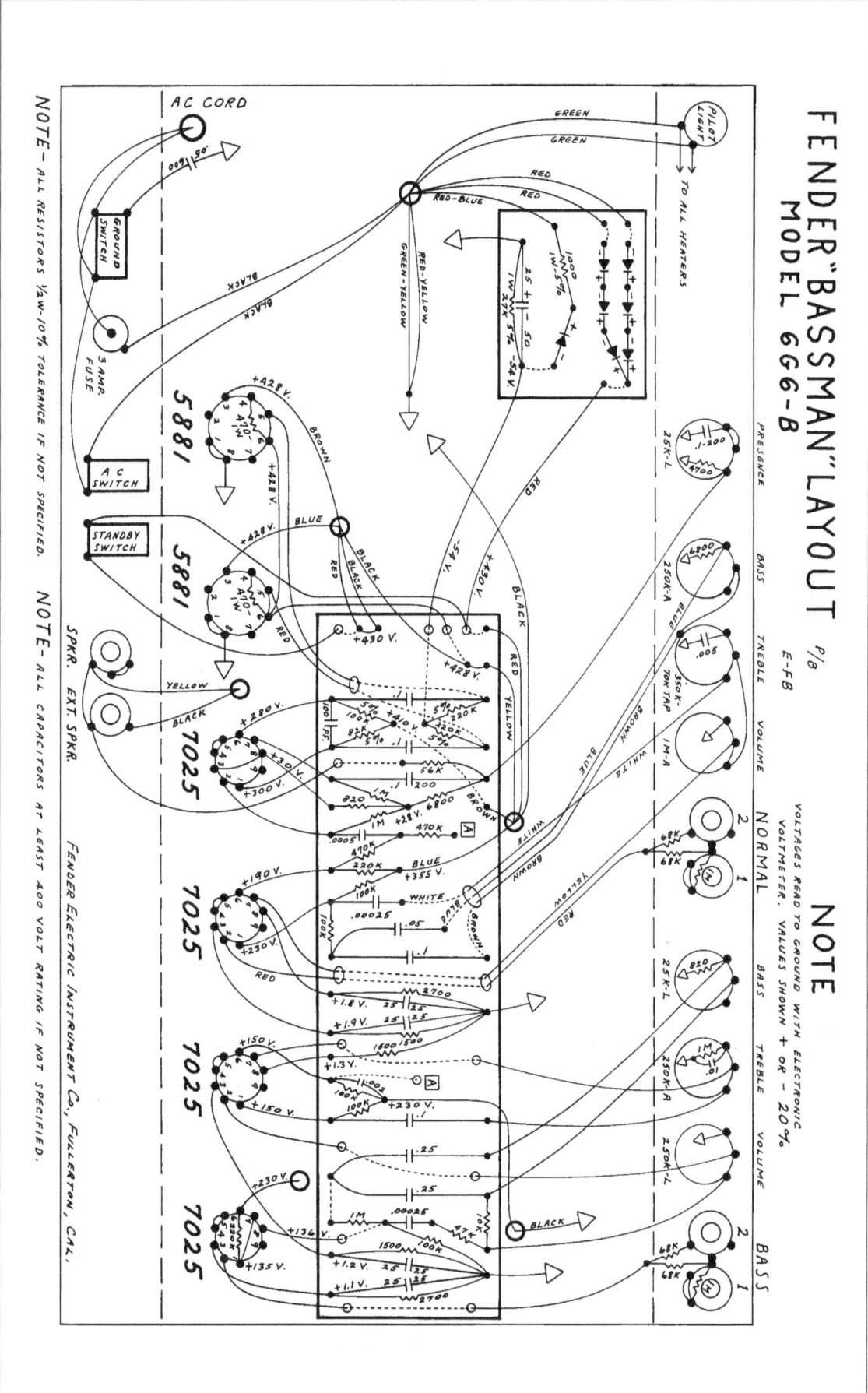

FENDER "BASSMAN" LAYOUT P/B
MODEL 6G6-B

E-FB

NOTE
VOLTAGES READ TO GROUND WITH ELECTRONIC VOLTMETER. VALUES SHOWN + OR - 20%

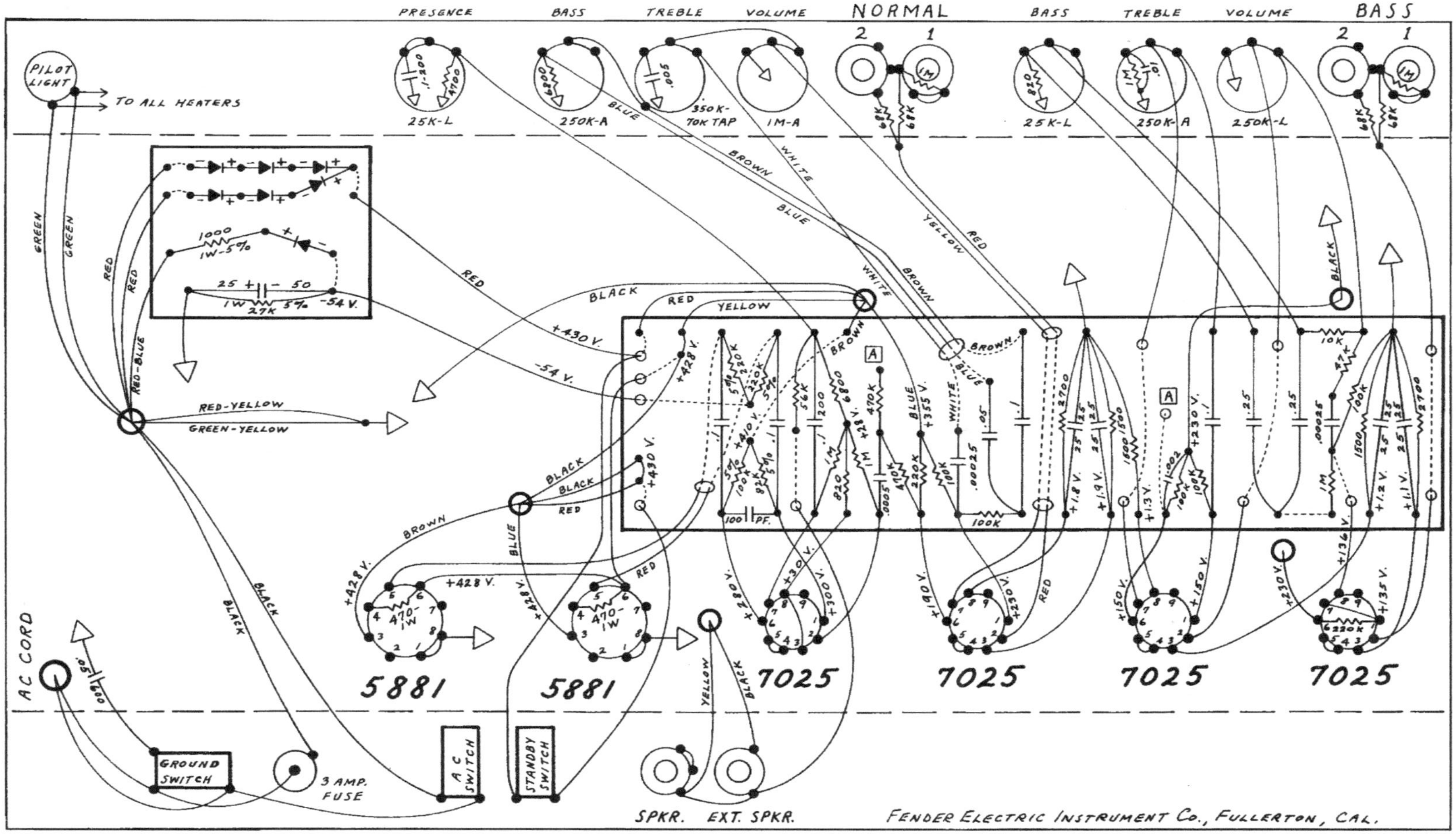

NOTE– ALL RESISTORS 1/2W-10% TOLERANCE IF NOT SPECIFIED. NOTE– ALL CAPACITORS AT LEAST 400 VOLT RATING IF NOT SPECIFIED.

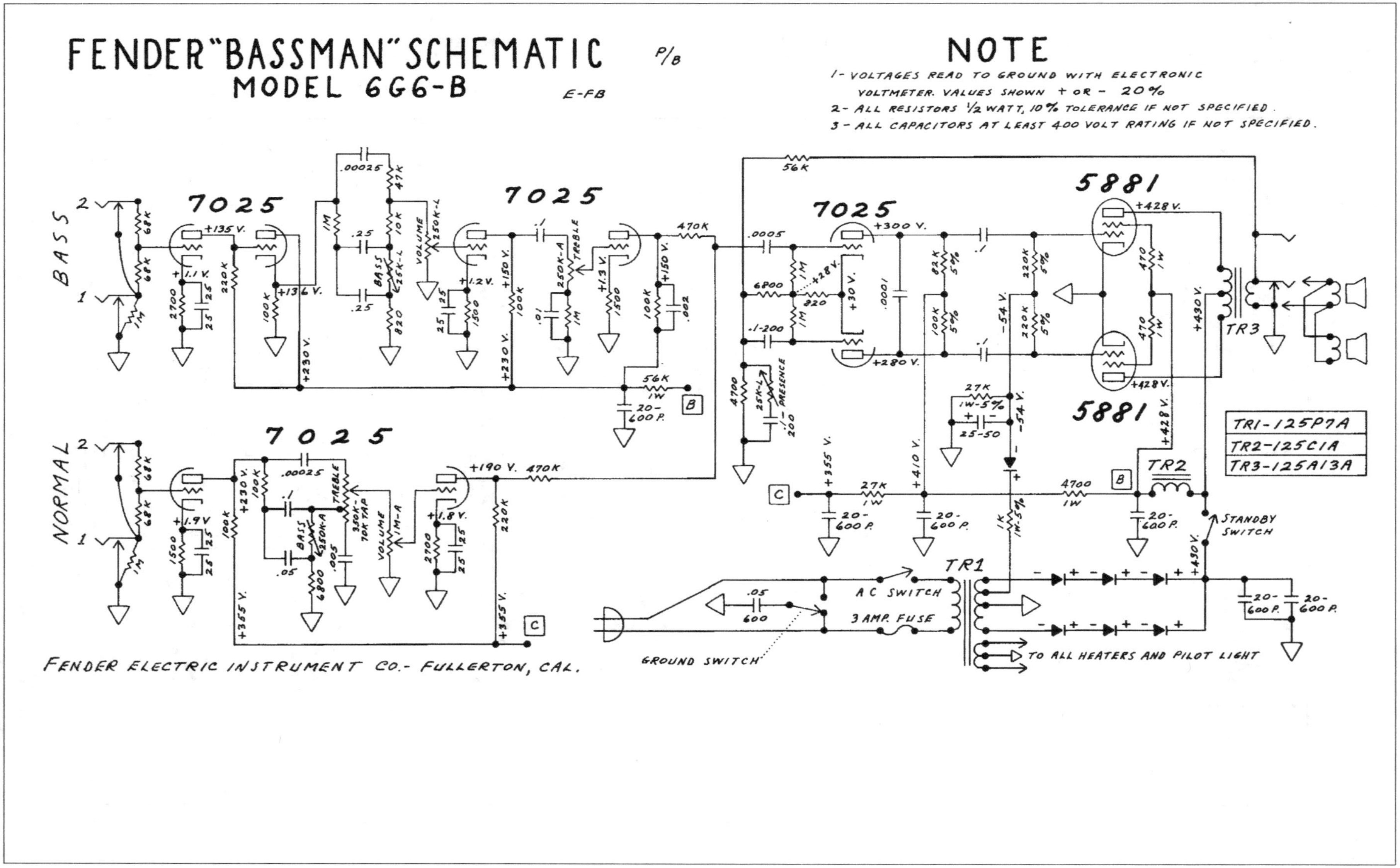
FENDER "BASSMAN" SCHEMATIC
MODEL 6G6-B
P/B
E-FB
NOTE
1- VOLTAGES READ TO GROUND WITH ELECTRONIC VOLTMETER. VALUES SHOWN + OR - 20%
2- ALL RESISTORS ½ WATT, 10% TOLERANCE IF NOT SPECIFIED.
3- ALL CAPACITORS AT LEAST 400 VOLT RATING IF NOT SPECIFIED.
BASS
NORMAL
7025
7025
7025
7025
5881
5881
TR1-125P7A
TR2-125C1A
TR3-125A13A
TR1
TR2
TR3
STANDBY SWITCH
AC SWITCH
3 AMP. FUSE
GROUND SWITCH
TO ALL HEATERS AND PILOT LIGHT
FENDER ELECTRIC INSTRUMENT CO.- FULLERTON, CAL.

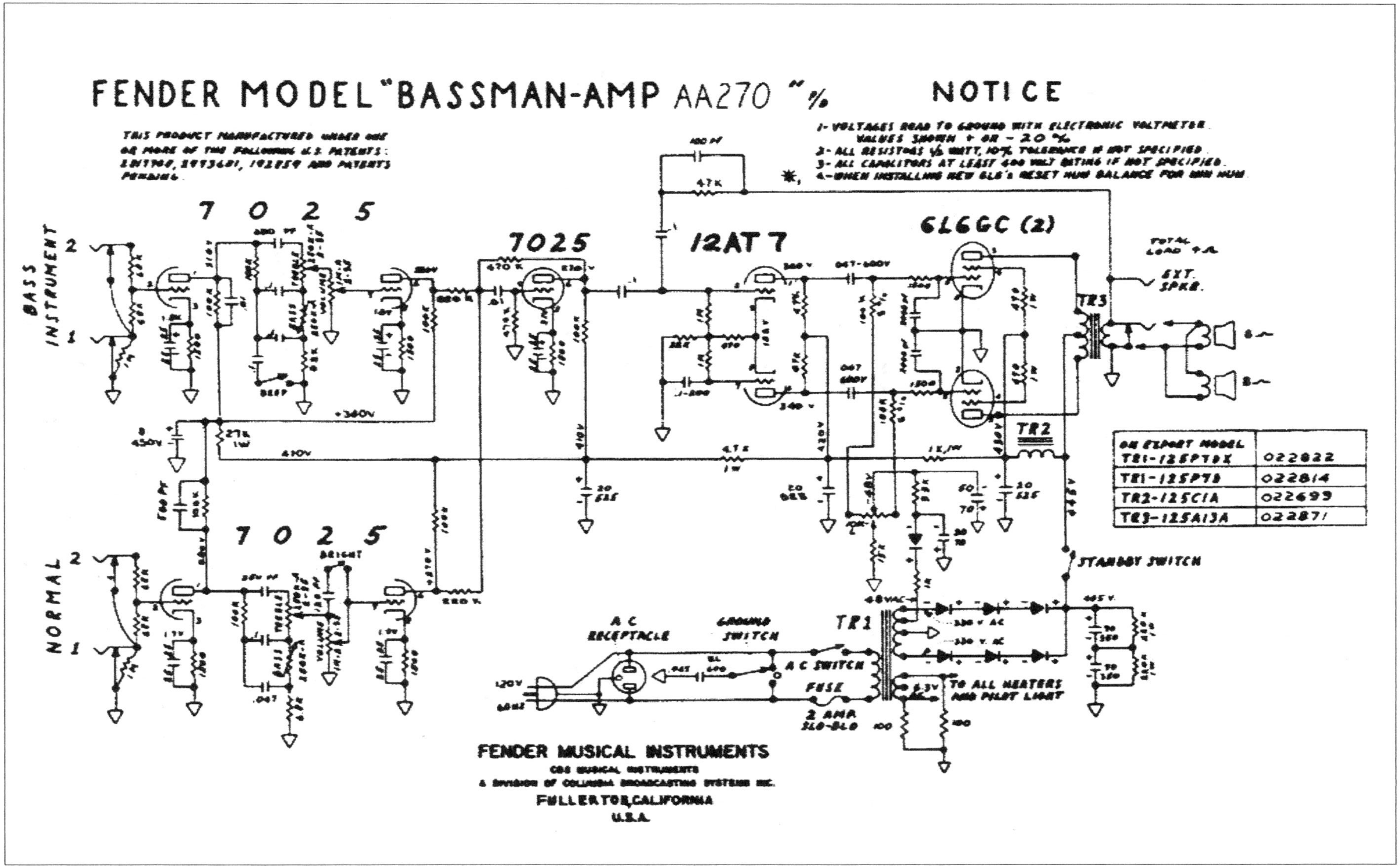
FENDER MODEL "BASSMAN-AMP AA270"
NOTICE
1- VOLTAGES READ TO GROUND WITH ELECTRONIC VOLTMETER. VALUES SHOWN + OR - 20 %
2- ALL RESISTORS ½ WATT, 10% TOLERANCE IF NOT SPECIFIED
3- ALL CAPACITORS AT LEAST 400 VOLT RATING IF NOT SPECIFIED.
4- WHEN INSTALLING NEW 6L6'S RESET HUM BALANCE FOR MIN HUM
BASS INSTRUMENT
NORMAL
7025
7025
7025
12AT7
6L6GC (2)
TR1
TR2
TR3
EXT. SPKR.
TOTAL LOAD 4 Ω
STANDBY SWITCH
A C RECEPTACLE
GROUND SWITCH
A C SWITCH
FUSE
2 AMP SLO-BLO
TO ALL HEATERS AND PILOT LIGHT
BRIGHT
DEEP
ON EXPORT MODEL TR1-125P7DX 022822
TR1-125P7B 022814
TR2-125C1A 022699
TR3-125A13A 022871
FENDER MUSICAL INSTRUMENTS
CBS MUSICAL INSTRUMENTS
A DIVISION OF COLUMBIA BROADCASTING SYSTEMS INC.
FULLERTON, CALIFORNIA
U.S.A.

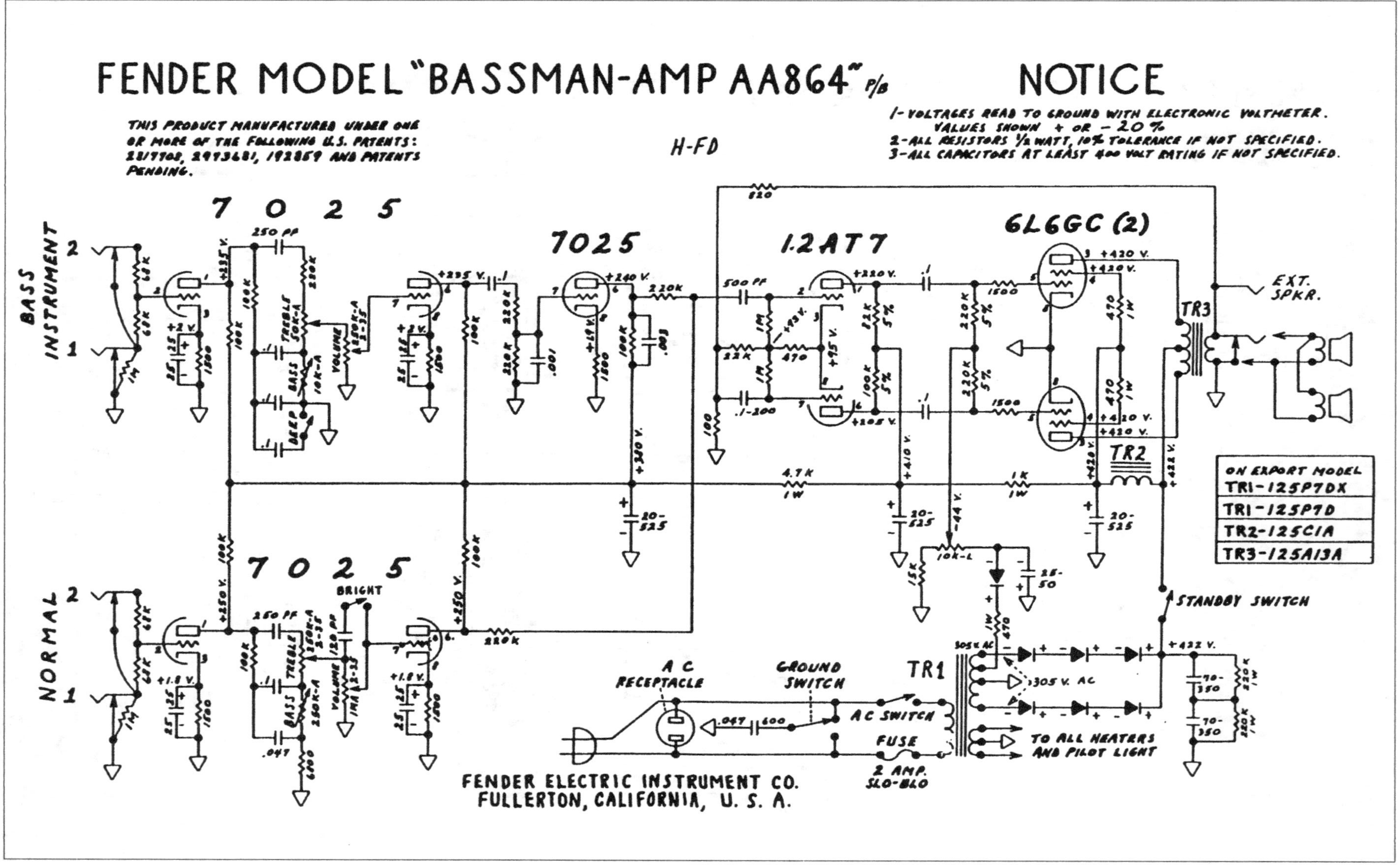
FENDER MODEL "BASSMAN-AMP AA864" P/B
NOTICE
1- VOLTAGES READ TO GROUND WITH ELECTRONIC VOLTMETER. VALUES SHOWN + OR - 20%
2- ALL RESISTORS ½ WATT, 10% TOLERANCE IF NOT SPECIFIED.
3- ALL CAPACITORS AT LEAST 400 VOLT RATING IF NOT SPECIFIED.
THIS PRODUCT MANUFACTURED UNDER ONE OR MORE OF THE FOLLOWING U.S. PATENTS: 2817708, 2973681, 192859 AND PATENTS PENDING.
H-FD
7025
7025
1.2AT7
6L6GC (2)
7025
BASS INSTRUMENT
NORMAL
BRIGHT
TREBLE
BASS
DEEP
VOLUME
EXT. SPKR.
TR3
TR2
TR1
ON EXPORT MODEL
TR1-125P7DX
TR1-125P7D
TR2-125C1A
TR3-125A13A
STANDBY SWITCH
305 V. AC
TO ALL HEATERS AND PILOT LIGHT
A C RECEPTACLE
GROUND SWITCH
AC SWITCH
FUSE
2 AMP. SLO-BLO
FENDER ELECTRIC INSTRUMENT CO.
FULLERTON, CALIFORNIA, U. S. A.

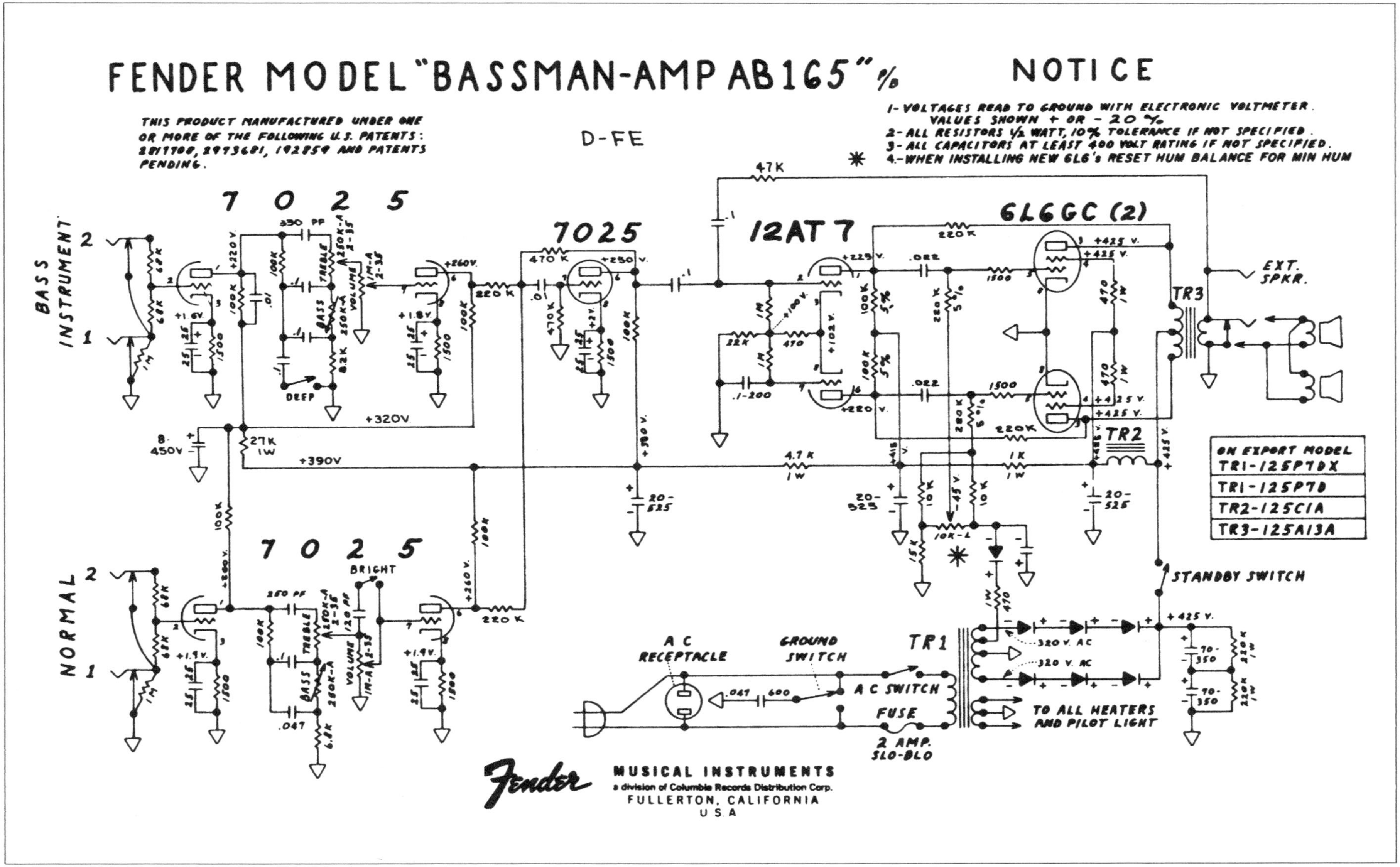
FENDER MODEL "BASSMAN-AMP AB165" %
NOTICE
1- VOLTAGES READ TO GROUND WITH ELECTRONIC VOLTMETER. VALUES SHOWN + OR - 20%
2- ALL RESISTORS ½ WATT, 10% TOLERANCE IF NOT SPECIFIED.
3- ALL CAPACITORS AT LEAST 400 VOLT RATING IF NOT SPECIFIED.
4.-WHEN INSTALLING NEW 6L6's RESET HUM BALANCE FOR MIN HUM
THIS PRODUCT MANUFACTURED UNDER ONE OR MORE OF THE FOLLOWING U.S. PATENTS: 2817708, 2973681, 192859 AND PATENTS PENDING.
D-FE
BASS INSTRUMENT
NORMAL
7025
12AT7
6L6GC (2)
DEEP
BRIGHT
TR1
TR2
TR3
EXT. SPKR.
STANDBY SWITCH
A C RECEPTACLE
GROUND SWITCH
A C SWITCH
FUSE
2 AMP. SLO-BLO
TO ALL HEATERS AND PILOT LIGHT
ON EXPORT MODEL TR1-125P7DX
TR1-125P7D
TR2-125C1A
TR3-125A13A
MUSICAL INSTRUMENTS
a division of Columbia Records Distribution Corp.
FULLERTON, CALIFORNIA
U S A

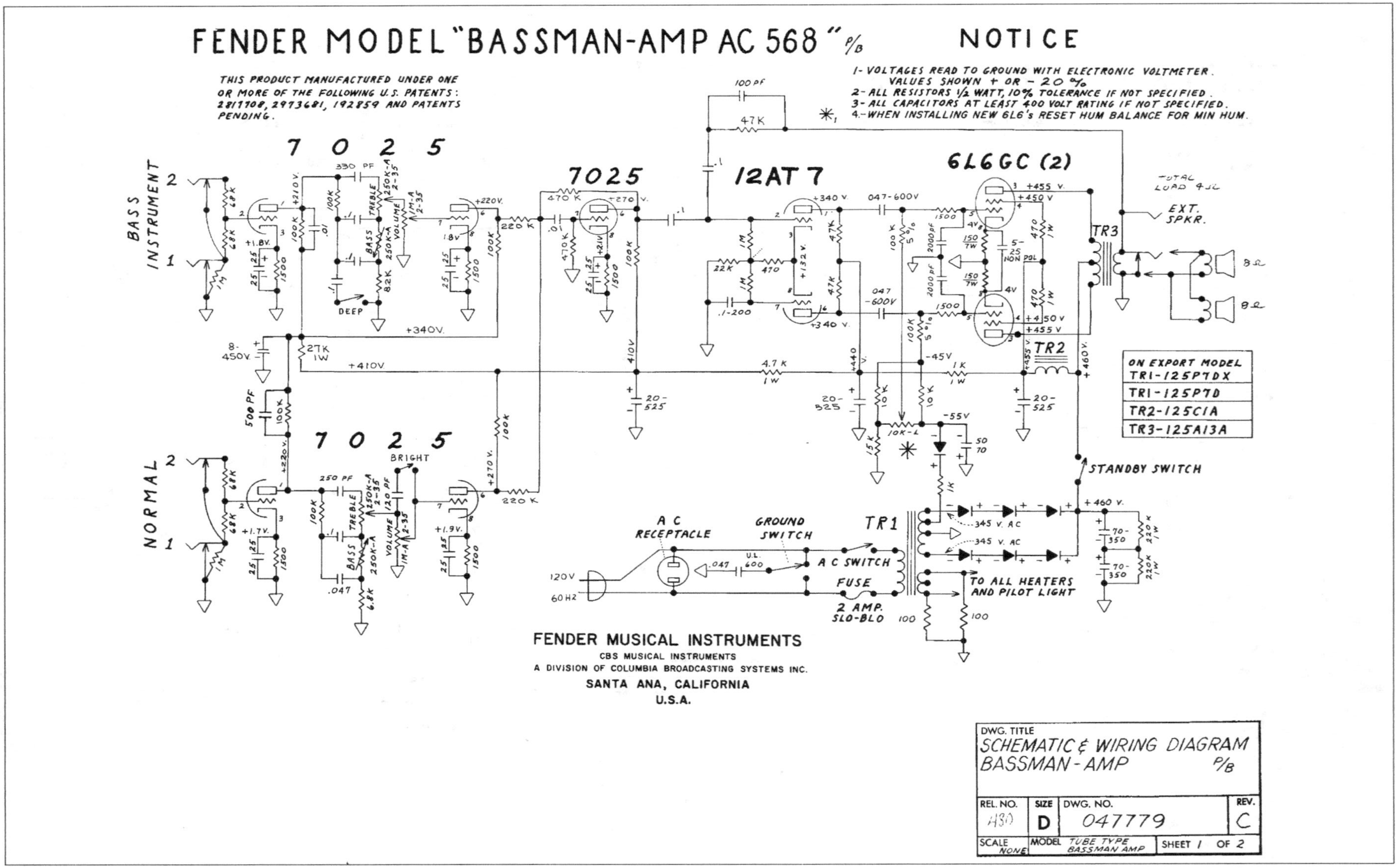
FENDER MODEL "BASSMAN-AMP AC 568" P/B
NOTICE
1- VOLTAGES READ TO GROUND WITH ELECTRONIC VOLTMETER. VALUES SHOWN + OR - 20 %
2- ALL RESISTORS 1/2 WATT, 10% TOLERANCE IF NOT SPECIFIED.
3- ALL CAPACITORS AT LEAST 400 VOLT RATING IF NOT SPECIFIED.
4.-WHEN INSTALLING NEW 6L6's RESET HUM BALANCE FOR MIN HUM.
THIS PRODUCT MANUFACTURED UNDER ONE OR MORE OF THE FOLLOWING U.S. PATENTS: 2817708, 2973681, 192859 AND PATENTS PENDING.
7025
7025
12AT7
6L6GC (2)
7025
BASS INSTRUMENT
NORMAL
BRIGHT
DEEP
TREBLE
BASS
VOLUME
EXT. SPKR.
TR1
TR2
TR3
ON EXPORT MODEL
TR1-125P7DX
TR1-125P7D
TR2-125C1A
TR3-125A13A
STANDBY SWITCH
A C RECEPTACLE
GROUND SWITCH
A C SWITCH
FUSE
2 AMP. SLO-BLO
TO ALL HEATERS AND PILOT LIGHT
120V
60 HZ
FENDER MUSICAL INSTRUMENTS
CBS MUSICAL INSTRUMENTS
A DIVISION OF COLUMBIA BROADCASTING SYSTEMS INC.
SANTA ANA, CALIFORNIA
U.S.A.
DWG. TITLE
SCHEMATIC & WIRING DIAGRAM
BASSMAN-AMP P/B
REL. NO.
SIZE
D
DWG. NO.
047779
REV.
C
SCALE
NONE
MODEL
TUBE TYPE BASSMAN AMP
SHEET 1 OF 2

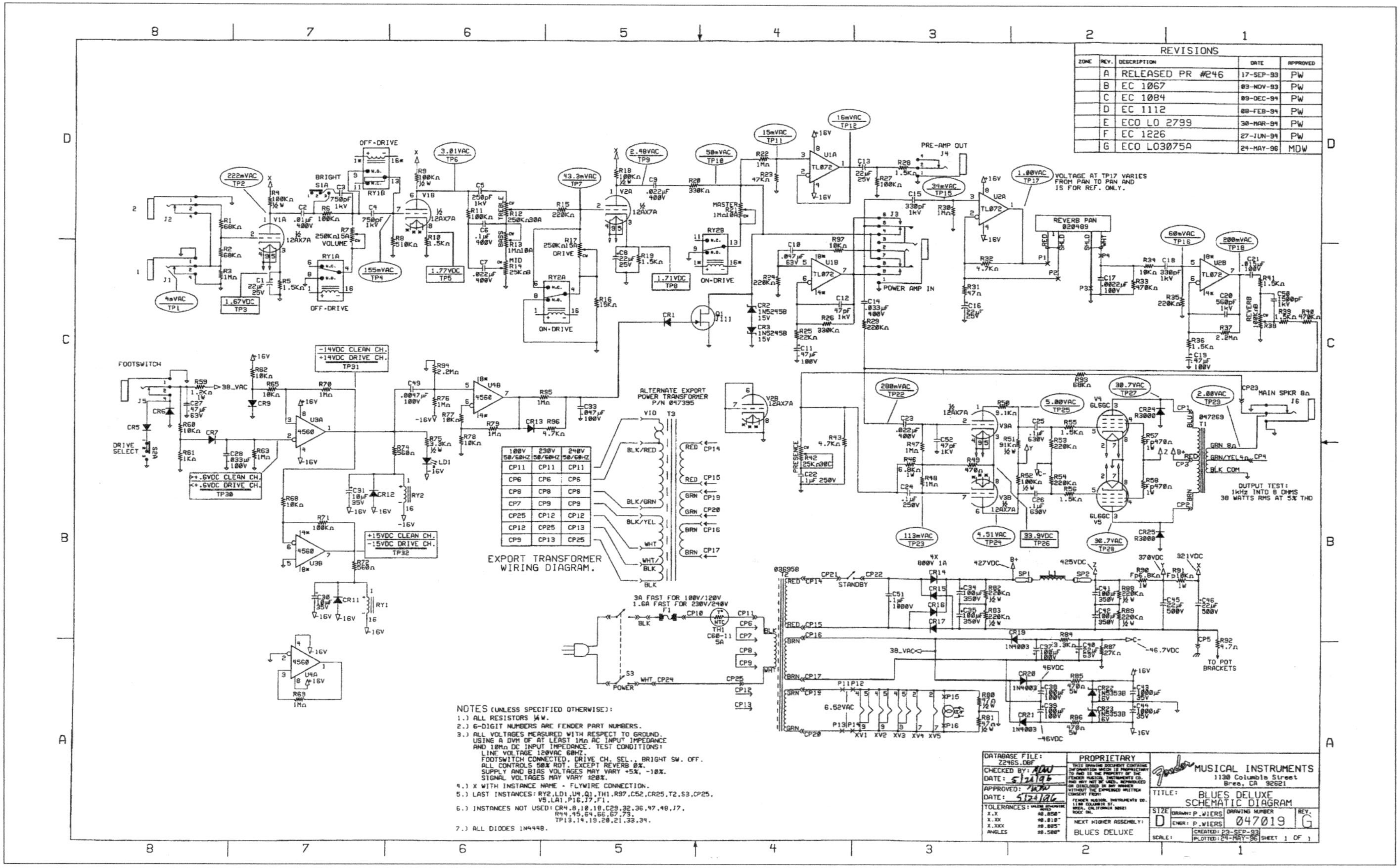

REVISIONS
A RELEASED PR #246 17-SEP-93 PW
B EC 1067 03-NOV-93 PW
C EC 1084 09-DEC-93 PW
D EC 1112 08-FEB-94 PW
E ECO LO 2799 30-MAR-94 PW
F EC 1226 27-JUN-94 PW
G ECO LO3075A 24-MAY-96 MDW
VOLTAGE AT TP17 VARIES FROM PAN TO PAN AND IS FOR REF. ONLY.
FOOTSWITCH
DRIVE SELECT
PRE-AMP OUT
POWER AMP IN
REVERB PAN 020409
MAIN SPKR 8Ω
OUTPUT TEST: 1KHZ INTO 8 OHMS 30 WATTS RMS AT 5% THD
ALTERNATE EXPORT POWER TRANSFORMER P/N 047395
EXPORT TRANSFORMER WIRING DIAGRAM.
3A FAST FOR 100V/120V 1.6A FAST FOR 230V/240V
STANDBY
POWER
TO POT BRACKETS
NOTES (UNLESS SPECIFIED OTHERWISE):
1.) ALL RESISTORS ¼W.
2.) 6-DIGIT NUMBERS ARE FENDER PART NUMBERS.
3.) ALL VOLTAGES MEASURED WITH RESPECT TO GROUND. USING A DVM OF AT LEAST 1MΩ AC INPUT IMPEDANCE AND 10MΩ DC INPUT IMPEDANCE. TEST CONDITIONS:
LINE VOLTAGE 120VAC 60HZ.
FOOTSWITCH CONNECTED. DRIVE CH. SEL., BRIGHT SW. OFF.
ALL CONTROLS 50% ROT. EXCEPT REVERB 0%.
SUPPLY AND BIAS VOLTAGES MAY VARY +5%, -10%.
SIGNAL VOLTAGES MAY VARY ±20%.
4.) X WITH INSTANCE NAME - FLYWIRE CONNECTION.
5.) LAST INSTANCES: RY2,LD1,U4,Q1,TH1,R97,C52,CR25,T2,S3,CP25, V5,LA1,P16,J7,F1.
6.) INSTANCES NOT USED: CR4,8,10,18,C29,32,36,47,48,J7, R44,45,64,66,67,73, TP13,14,19,20,21,33,34.
7.) ALL DIODES 1N4448.
PROPRIETARY
MUSICAL INSTRUMENTS
1130 Columbia Street
Brea, CA 92621
TITLE: BLUES DELUXE SCHEMATIC DIAGRAM
SIZE D
DRAWN: P.WIERS
ENGR: P.WIERS
DRAWING NUMBER 047019
REV. G
NEXT HIGHER ASSEMBLY: BLUES DELUXE
CREATED: 23-SEP-93
PLOTTED: 24-MAY-96
SHEET 1 OF 1

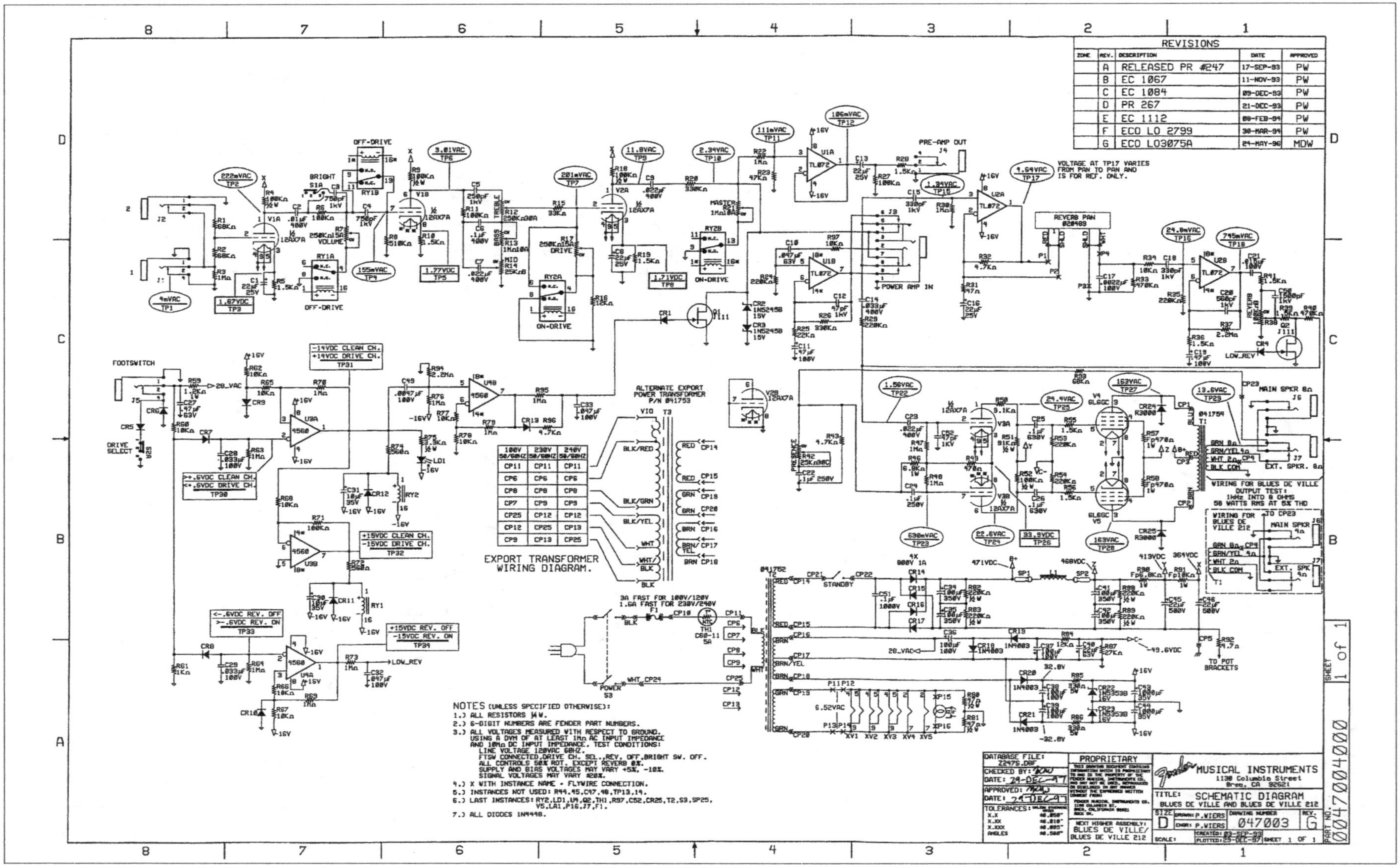
REVISIONS
ZONE | REV. | DESCRIPTION | DATE | APPROVED
A | RELEASED PR #247 | 17-SEP-93 | PW
B | EC 1067 | 11-NOV-93 | PW
C | EC 1084 | 09-DEC-93 | PW
D | PR 267 | 21-DEC-93 | PW
E | EC 1112 | 08-FEB-94 | PW
F | ECO LO 2799 | 30-MAR-94 | PW
G | ECO LO3075A | 24-MAY-96 | MDW
VOLTAGE AT TP17 VARIES FROM PAN TO PAN AND IS FOR REF. ONLY.
PRE-AMP OUT
POWER AMP IN
REVERB PAN
FOOTSWITCH
DRIVE SELECT
OFF-DRIVE
ON-DRIVE
BRIGHT
VOLUME
TREBLE
BASS
MID
DRIVE
MASTER
PRESENCE
STANDBY
POWER
LOW_REV
ALTERNATE EXPORT POWER TRANSFORMER P/N 041753
EXPORT TRANSFORMER WIRING DIAGRAM.
100V 50/60HZ | 230V 50/60HZ | 240V 50/60HZ
3A FAST FOR 100V/120V
1.6A FAST FOR 230V/240V
MAIN SPKR 8Ω
EXT. SPKR. 8Ω
WIRING FOR BLUES DE VILLE
OUTPUT TEST: 1kHz INTO 8 OHMS 50 WATTS RMS AT 5% THD
WIRING FOR BLUES DE VILLE 212
TO CP23
TO POT BRACKETS
NOTES (UNLESS SPECIFIED OTHERWISE):
1.) ALL RESISTORS ¼W.
2.) 6-DIGIT NUMBERS ARE FENDER PART NUMBERS.
3.) ALL VOLTAGES MEASURED WITH RESPECT TO GROUND. USING A DVM OF AT LEAST 1MΩ AC INPUT IMPEDANCE AND 10MΩ DC INPUT IMPEDANCE. TEST CONDITIONS: LINE VOLTAGE 120VAC 60HZ. FTSW CONNECTED,DRIVE CH. SEL.,REV. OFF,BRIGHT SW. OFF. ALL CONTROLS 50% ROT. EXCEPT REVERB 0%. SUPPLY AND BIAS VOLTAGES MAY VARY +5%, -10%. SIGNAL VOLTAGES MAY VARY ±20%.
4.) X WITH INSTANCE NAME = FLYWIRE CONNECTION.
5.) INSTANCES NOT USED: R44,45,C47,48,TP13,14.
6.) LAST INSTANCES: RY2,LD1,U4,Q2,TH1,R97,C52,CR25,T2,S3,SP25, V5,LA1,P16,J7,F1.
7.) ALL DIODES 1N4448.
DATABASE FILE: 22475.DBF
PROPRIETARY
MUSICAL INSTRUMENTS
1130 Columbia Street
Brea, CA 92621
TITLE: SCHEMATIC DIAGRAM
BLUES DE VILLE AND BLUES DE VILLE 212
DRAWN: P.WIERS
CHKR: P.WIERS
DRAWING NUMBER 047003
REV. G
NEXT HIGHER ASSEMBLY: BLUES DE VILLE/ BLUES DE VILLE 212
SHEET 1 OF 1
PART NO. 0047004000

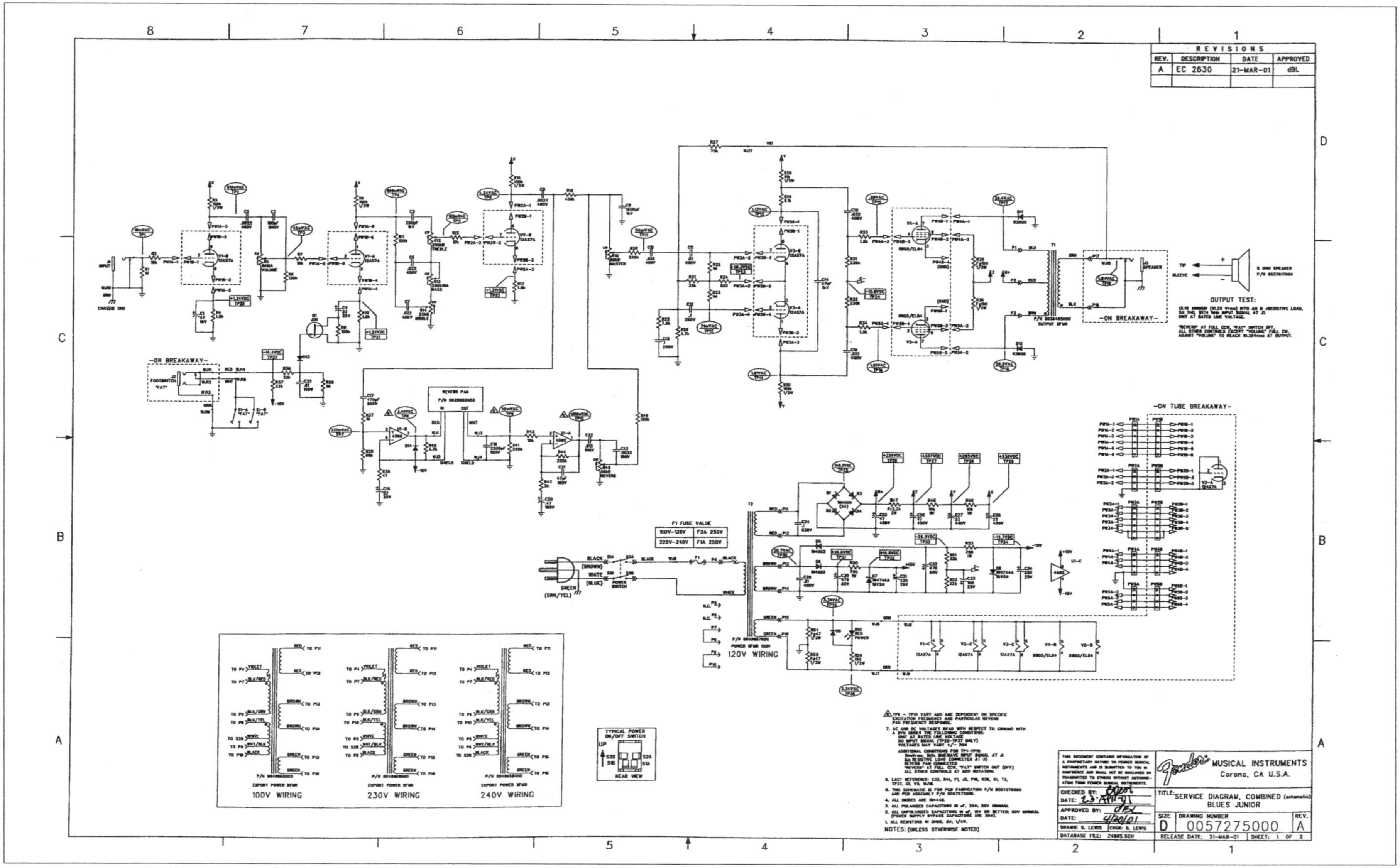
REVISIONS
REV. DESCRIPTION DATE APPROVED
A EC 2630 21-MAR-01 dBL
–ON BREAKAWAY–
–ON TUBE BREAKAWAY–
OUTPUT TEST:
F1 FUSE VALUE
100V–120V F2A 250V
220V–240V F1A 250V
120V WIRING
100V WIRING
230V WIRING
240V WIRING
TYPICAL POWER ON/OFF SWITCH
REAR VIEW
NOTES: (UNLESS OTHERWISE NOTED)
MUSICAL INSTRUMENTS
Corona, CA U.S.A.
TITLE: SERVICE DIAGRAM, COMBINED
BLUES JUNIOR
D 0057275000 A

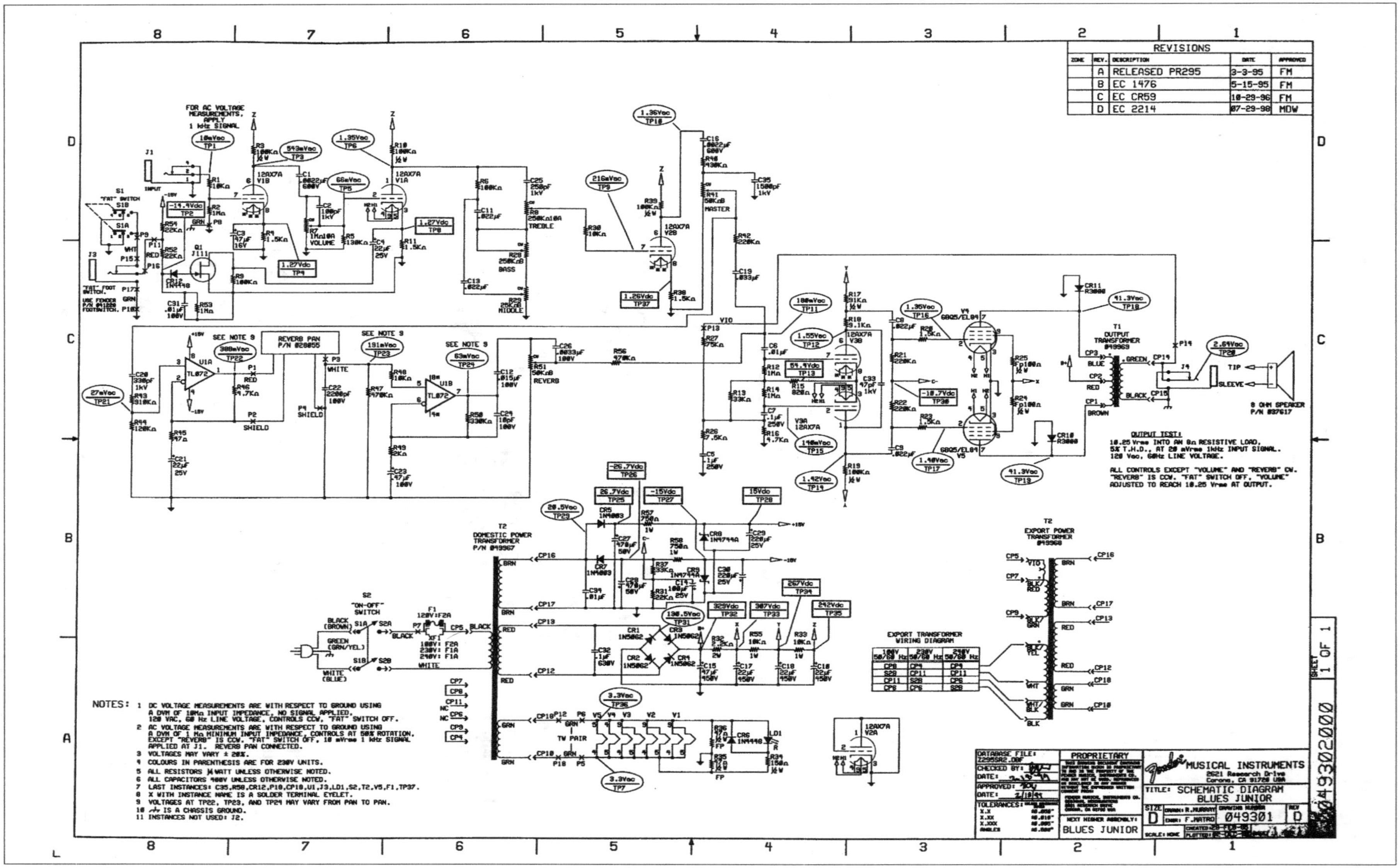
REVISIONS
A RELEASED PR295 3-3-95 FM
B EC 1476 5-15-95 FM
C EC CR59 10-29-96 FM
D EC 2214 07-29-98 MDW
REVERB PAN P/N 028055
OUTPUT TEST:
8 OHM SPEAKER
DOMESTIC POWER TRANSFORMER
EXPORT POWER TRANSFORMER
EXPORT TRANSFORMER WIRING DIAGRAM
NOTES:
MUSICAL INSTRUMENTS
TITLE: SCHEMATIC DIAGRAM BLUES JUNIOR
049301
BLUES JUNIOR
SHEET 1 OF 1

FENDER CHAMP-AMP LAYOUT

MODEL 5C1

F-DH

NOTICE

VOLTAGES READ TO GROUND
WITH ELECTRONIC VOLT METER
VALUES SHOWN + OR − 20 %

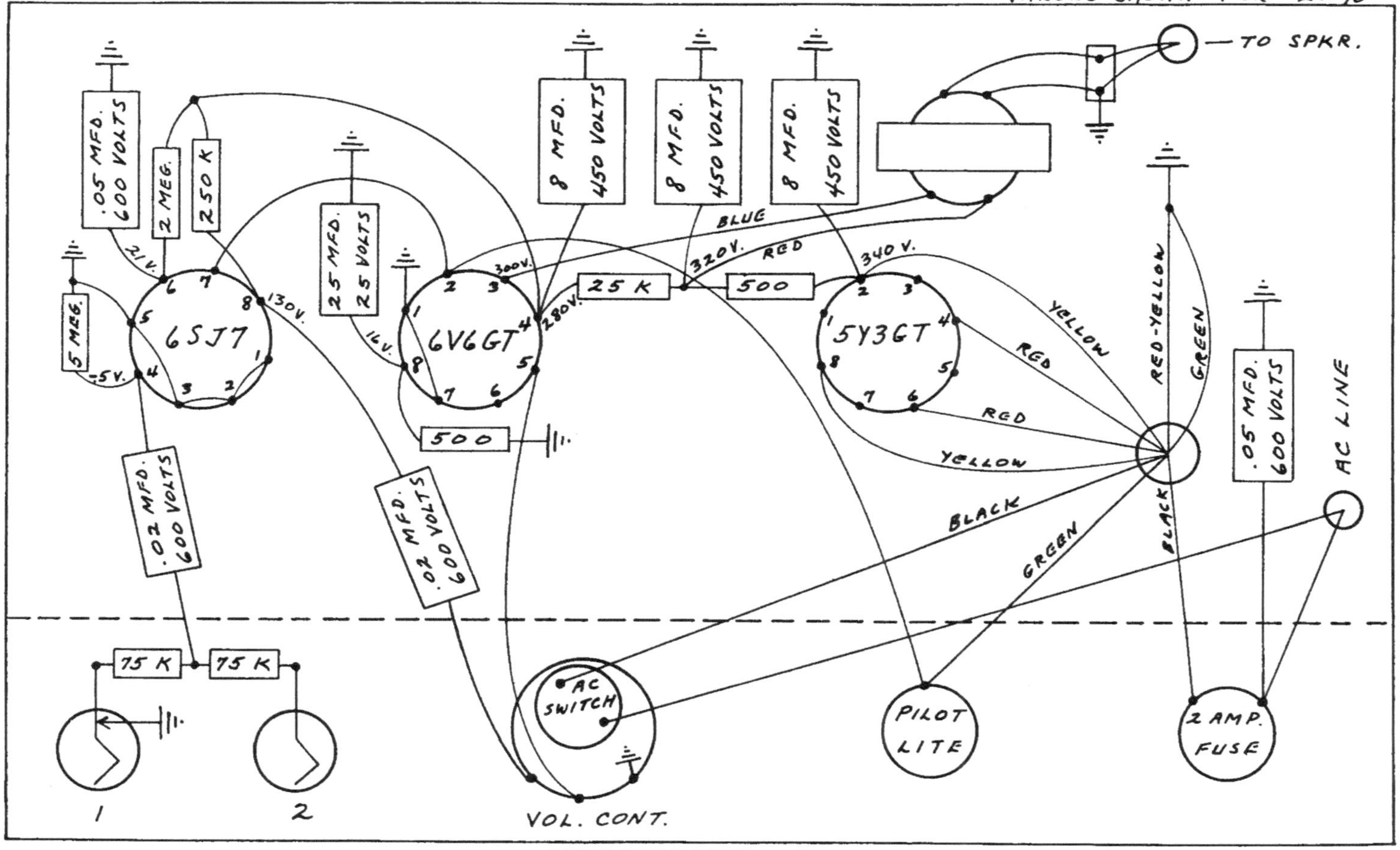

FENDER MUSICAL INSTRUMENTS
A DIVISION OF COLUMBIA RECORDS DISTRIBUTION CORP.
SANTA ANA, CALIFORNIA

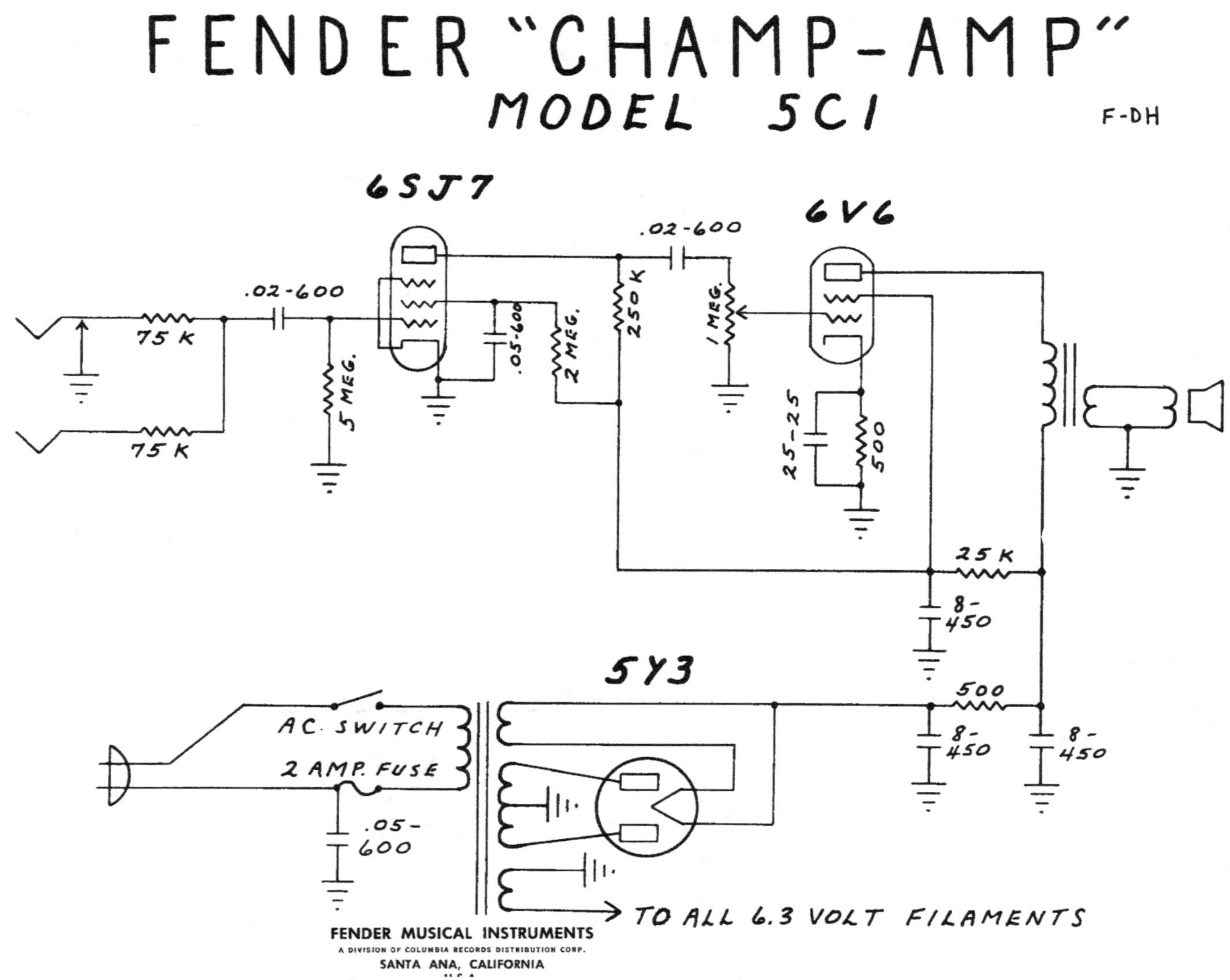
FENDER "CHAMP-AMP"
MODEL 5C1
F-DH
6SJ7
6V6
5Y3
.02-600
.02-600
75 K
75 K
5 MEG.
.05-600
2 MEG.
250 K
1 MEG.
25-25
500
25 K
8-450
8-450
8-450
500
AC. SWITCH
2 AMP. FUSE
.05-600
TO ALL 6.3 VOLT FILAMENTS
FENDER MUSICAL INSTRUMENTS
A DIVISION OF COLUMBIA RECORDS DISTRIBUTION CORP.
SANTA ANA, CALIFORNIA

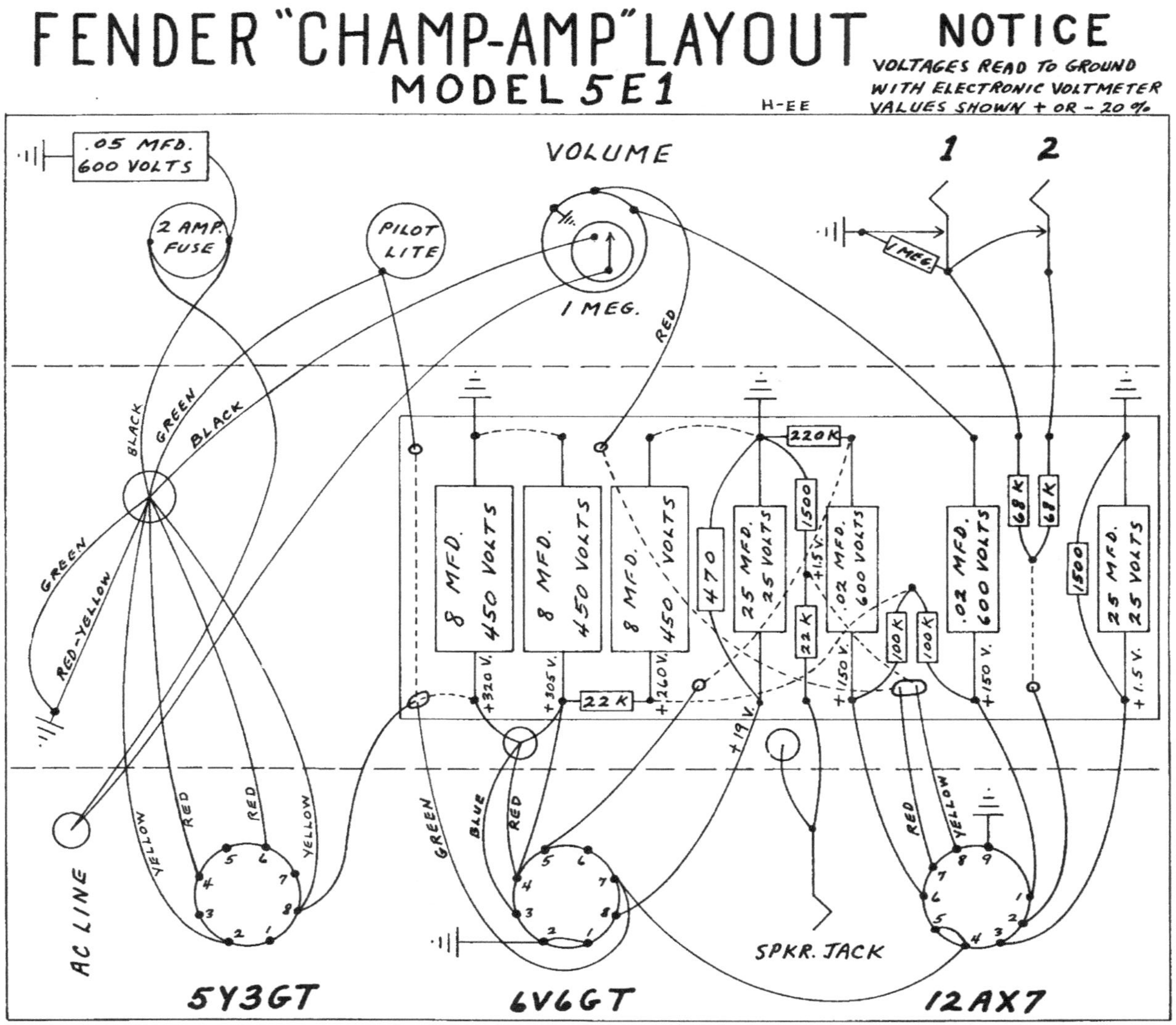
FENDER "CHAMP-AMP" LAYOUT
MODEL 5E1
H-EE
NOTICE
VOLTAGES READ TO GROUND
WITH ELECTRONIC VOLTMETER
VALUES SHOWN + OR - 20 %
.05 MFD.
600 VOLTS
2 AMP. FUSE
PILOT LITE
VOLUME
1 MEG.
1
2
1 MEG.
RED
BLACK
GREEN
BLACK
GREEN
RED-YELLOW
8 MFD.
450 VOLTS
+320 V.
+305 V.
+260 V.
22 K
470
25 MFD.
25 VOLTS
+19 V.
220 K
1500
+1.5 V.
22 K
.02 MFD.
600 VOLTS
+150 V.
100 K
68 K
+1.5 V.
AC LINE
YELLOW
RED
5Y3GT
GREEN
BLUE
6V6GT
SPKR. JACK
YELLOW
12AX7

FENDER "CHAMP-AMP" SCHEMATIC
MODEL 5E1

H-EE

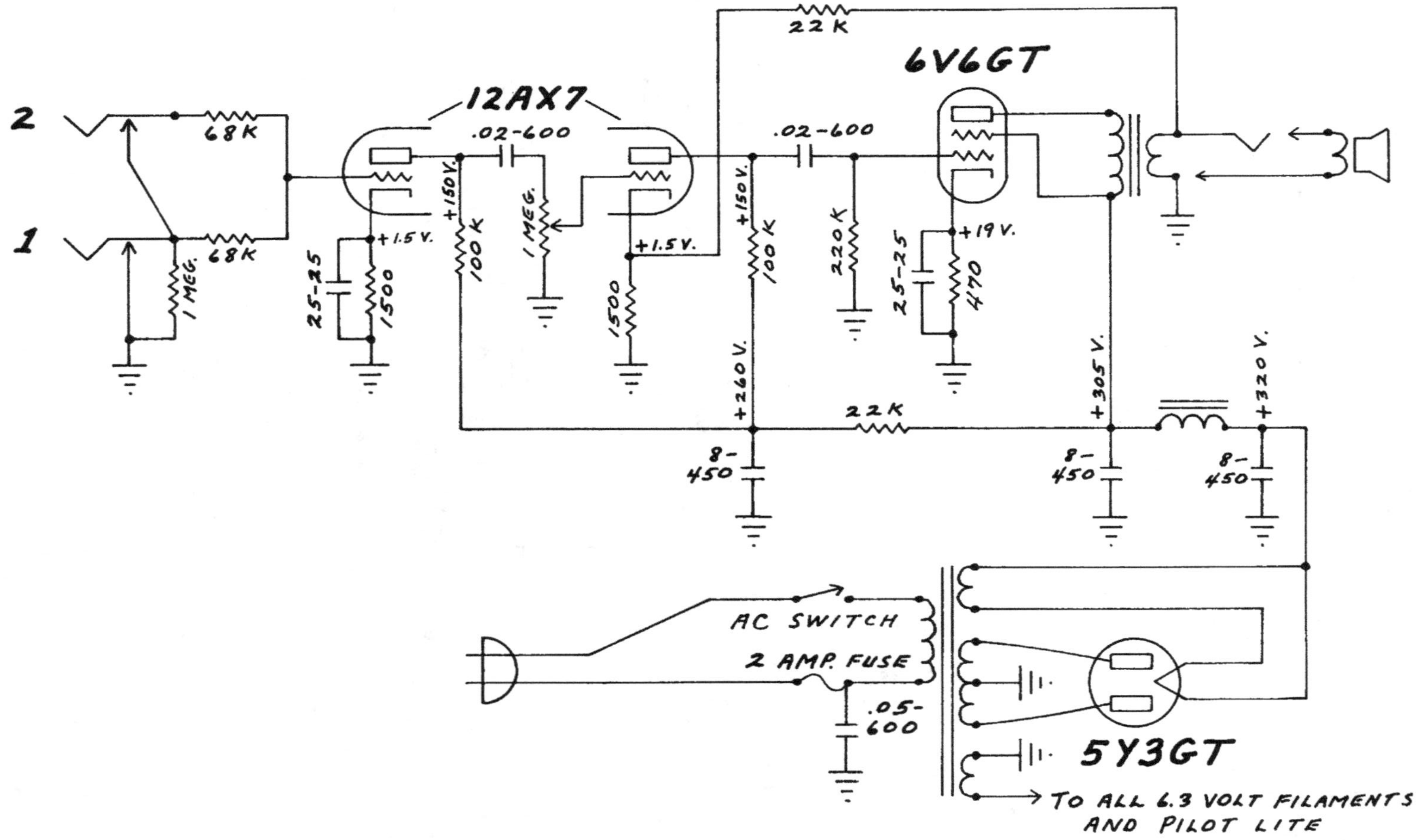

FENDER MODEL "VIBRO-CHAMP-AMP AA764" LAYOUT

NOTICE

1- VOLTAGES READ TO GROUND WITH ELECTRONIC VOLTMETER. VALUES SHOWN + OR − 20%
2- ALL RESISTORS ½ WATT 10% TOLERANCE IF NOT SPECIFIED.
3- ALL CAPACITORS AT LEAST 400 VOLT RATING IF NOT SPECIFIED.

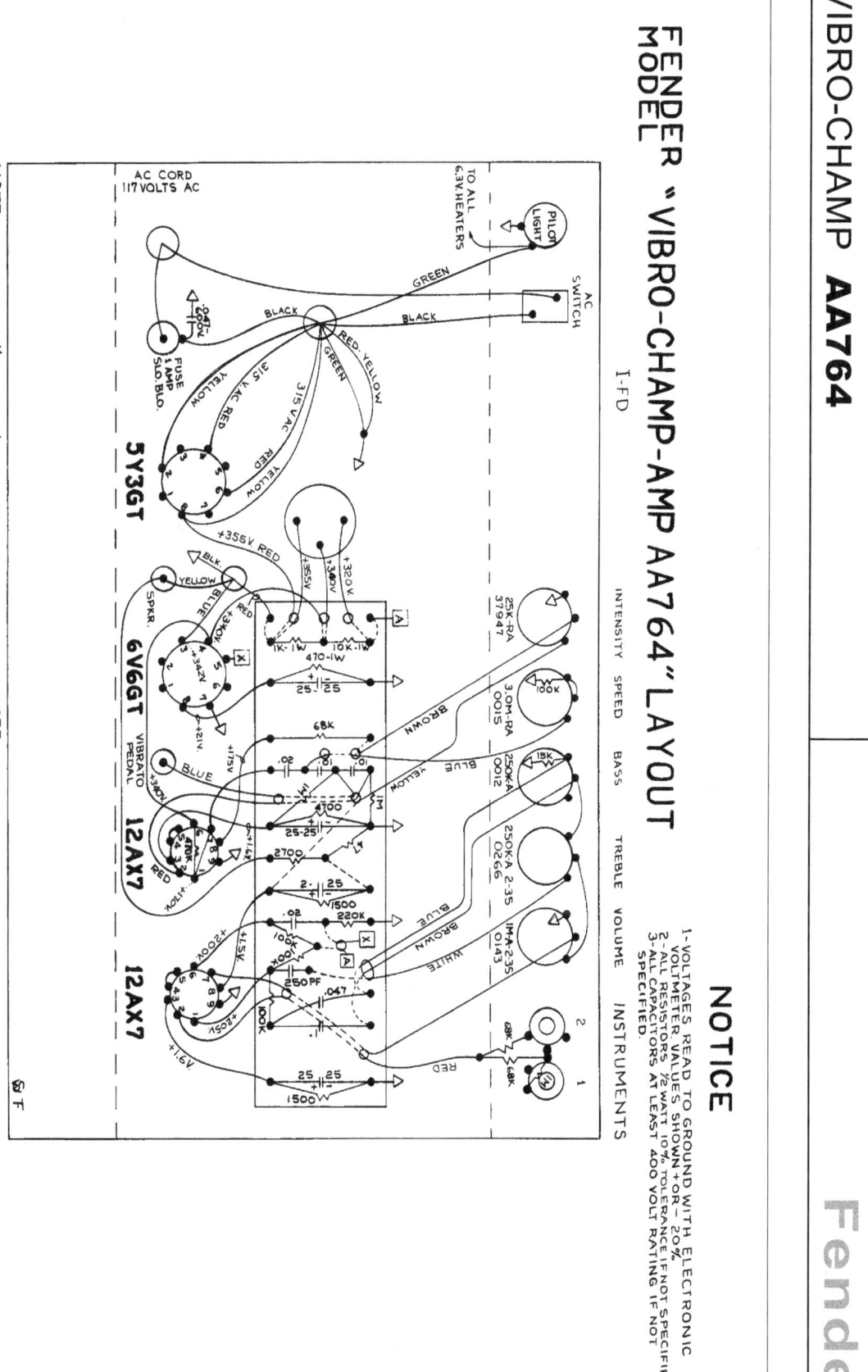

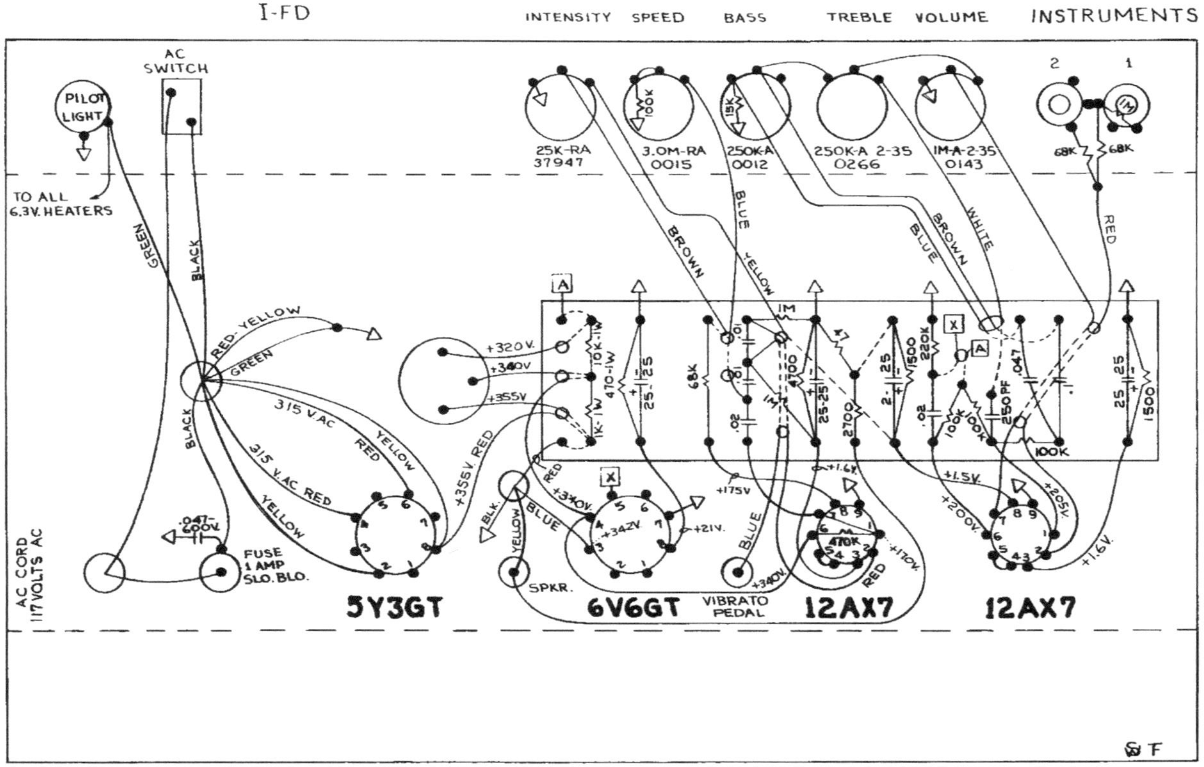

NOTE- ALL RESISTORS ½ WATT 10% TOLERANCE, IF NOT SPECIFIED. NOTE- ALL CAPACITORS AT LEAST 400 VOLT RATING IF NOT SPECIFIED.

FENDER ELECTRIC INSTRUMENT COMPANY
FULLERTON, CALIFORNIA
U.S.A.

FENDER MODEL

"VIBRO-CHAMP AMP AA764" SCHEMATIC

THIS PRODUCT MANUFACTURED UNDER ONE OR MORE OF THE FOLLOWING U.S. PATENTS - #2817708, # 2973681, 192859 & PATENTS PENDING.

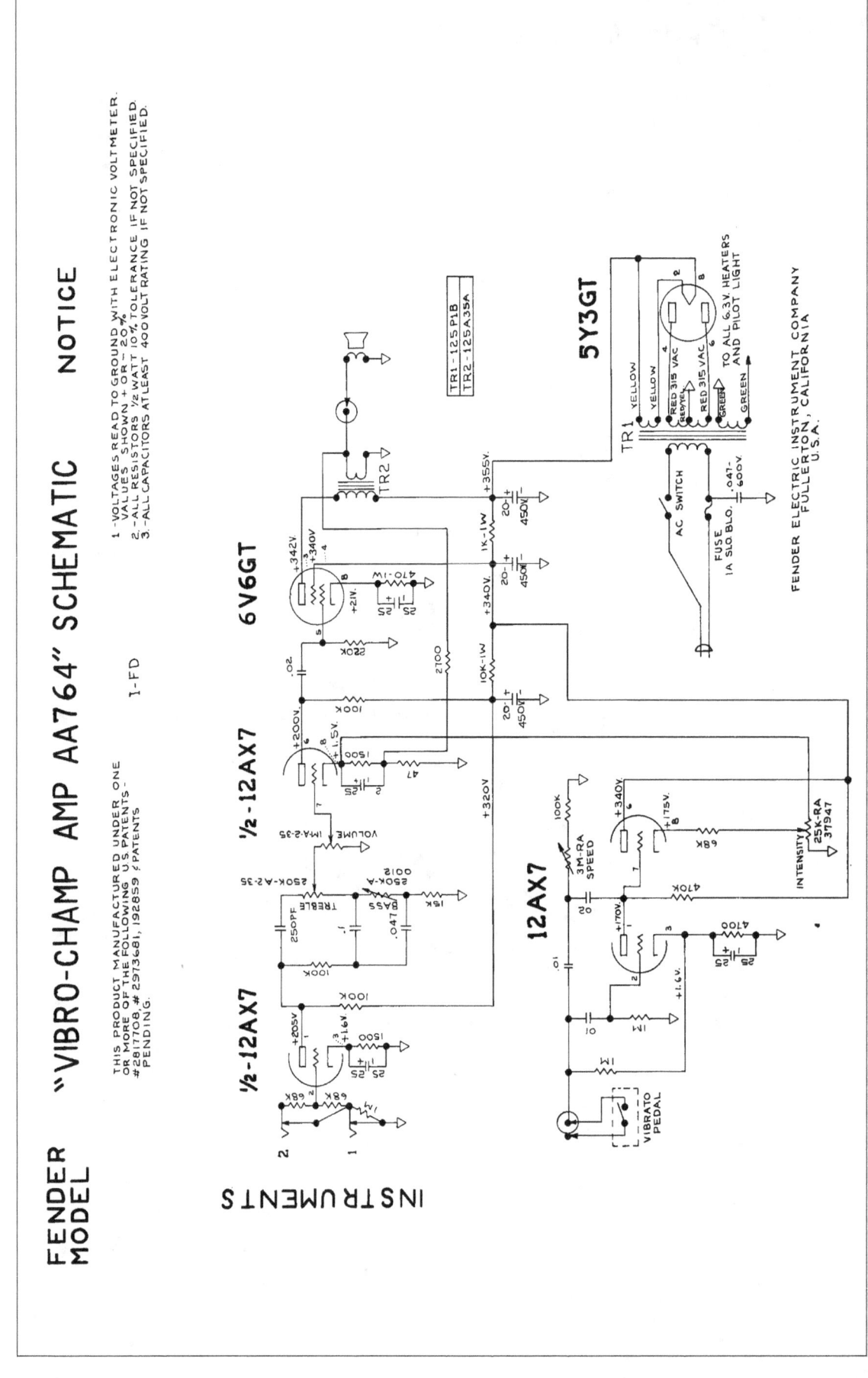

I-FD

NOTICE

1.-VOLTAGES READ TO GROUND WITH ELECTRONIC VOLTMETER. VALUES SHOWN + OR − 20%
2.-ALL RESISTORS ½ WATT 10% TOLERANCE IF NOT SPECIFIED.
3.-ALL CAPACITORS AT LEAST 400 VOLT RATING IF NOT SPECIFIED.

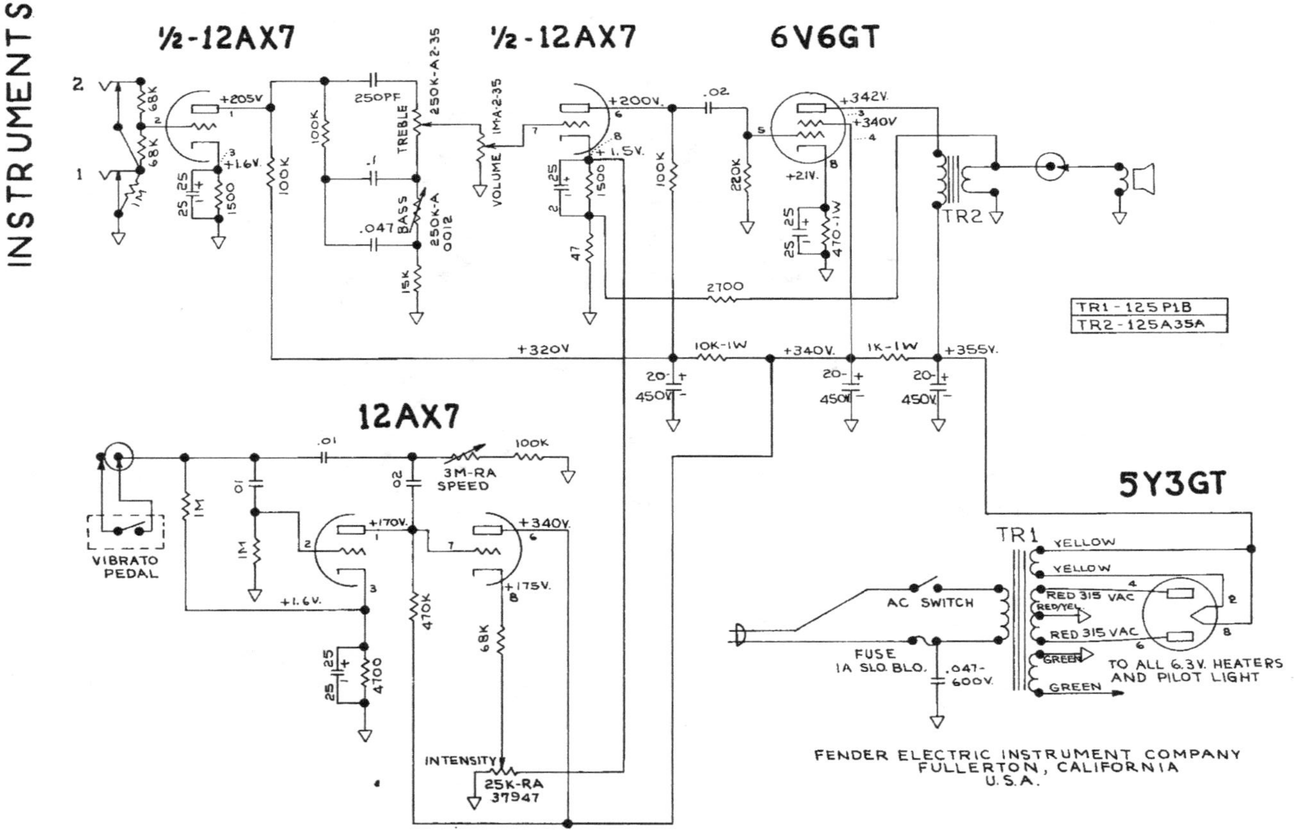

FENDER ELECTRIC INSTRUMENT COMPANY
FULLERTON, CALIFORNIA
U.S.A.

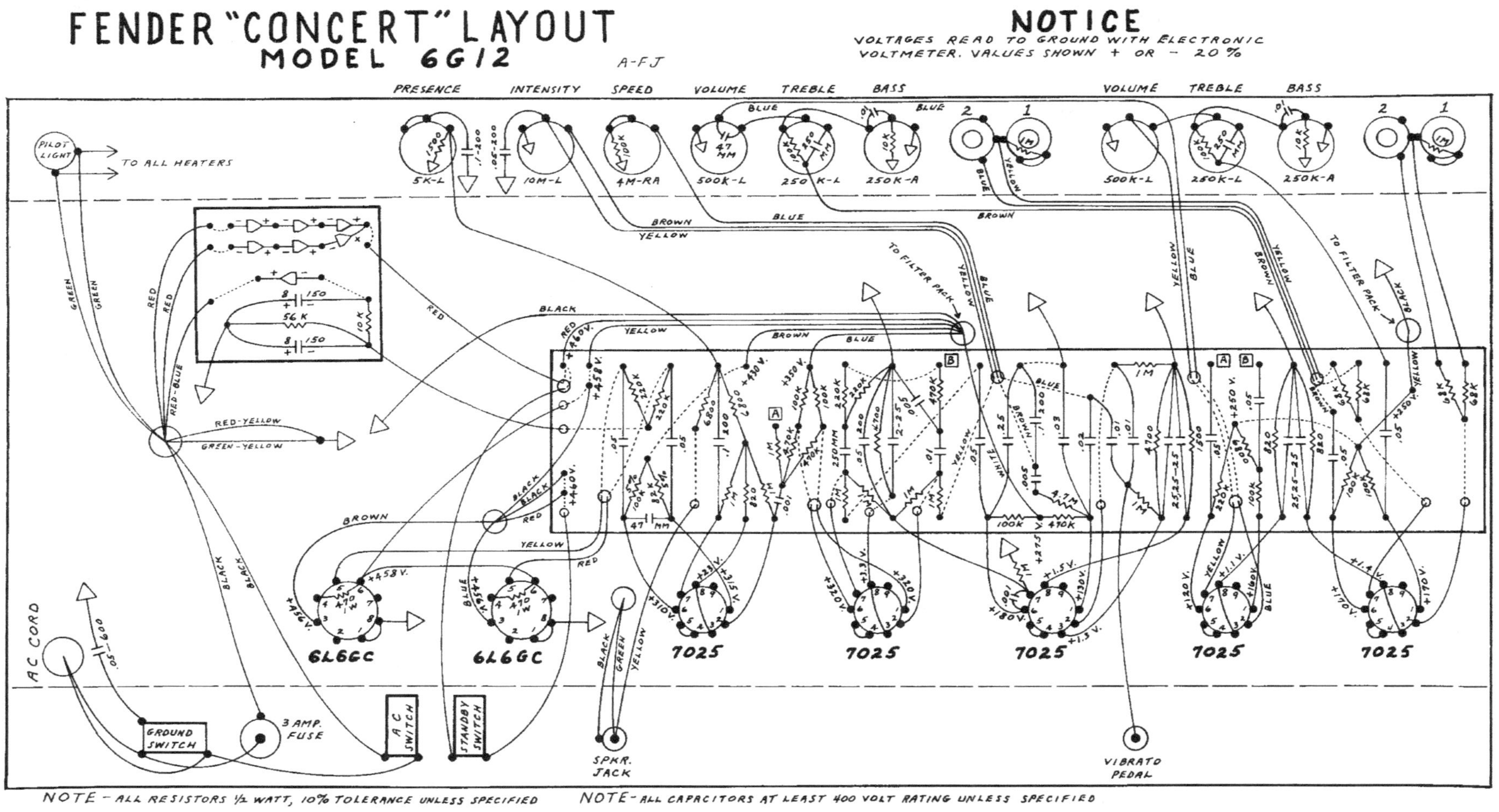
FENDER "CONCERT" LAYOUT
MODEL 6G12
A-F-J
NOTICE
VOLTAGES READ TO GROUND WITH ELECTRONIC VOLTMETER. VALUES SHOWN + OR − 20%
PRESENCE
INTENSITY
SPEED
VOLUME
TREBLE
BASS
VOLUME
TREBLE
BASS
5K-L
10M-L
4M-RA
500K-L
250 K-L
250K-A
500K-L
250K-L
250K-A
PILOT LIGHT
TO ALL HEATERS
TO FILTER PACK
TO FILTER PACK
6L6GC
6L6GC
7025
7025
7025
7025
7025
AC CORD
GROUND SWITCH
3 AMP. FUSE
AC SWITCH
STANDBY SWITCH
SPKR. JACK
VIBRATO PEDAL
NOTE - ALL RESISTORS ½ WATT, 10% TOLERANCE UNLESS SPECIFIED
NOTE - ALL CAPACITORS AT LEAST 400 VOLT RATING UNLESS SPECIFIED

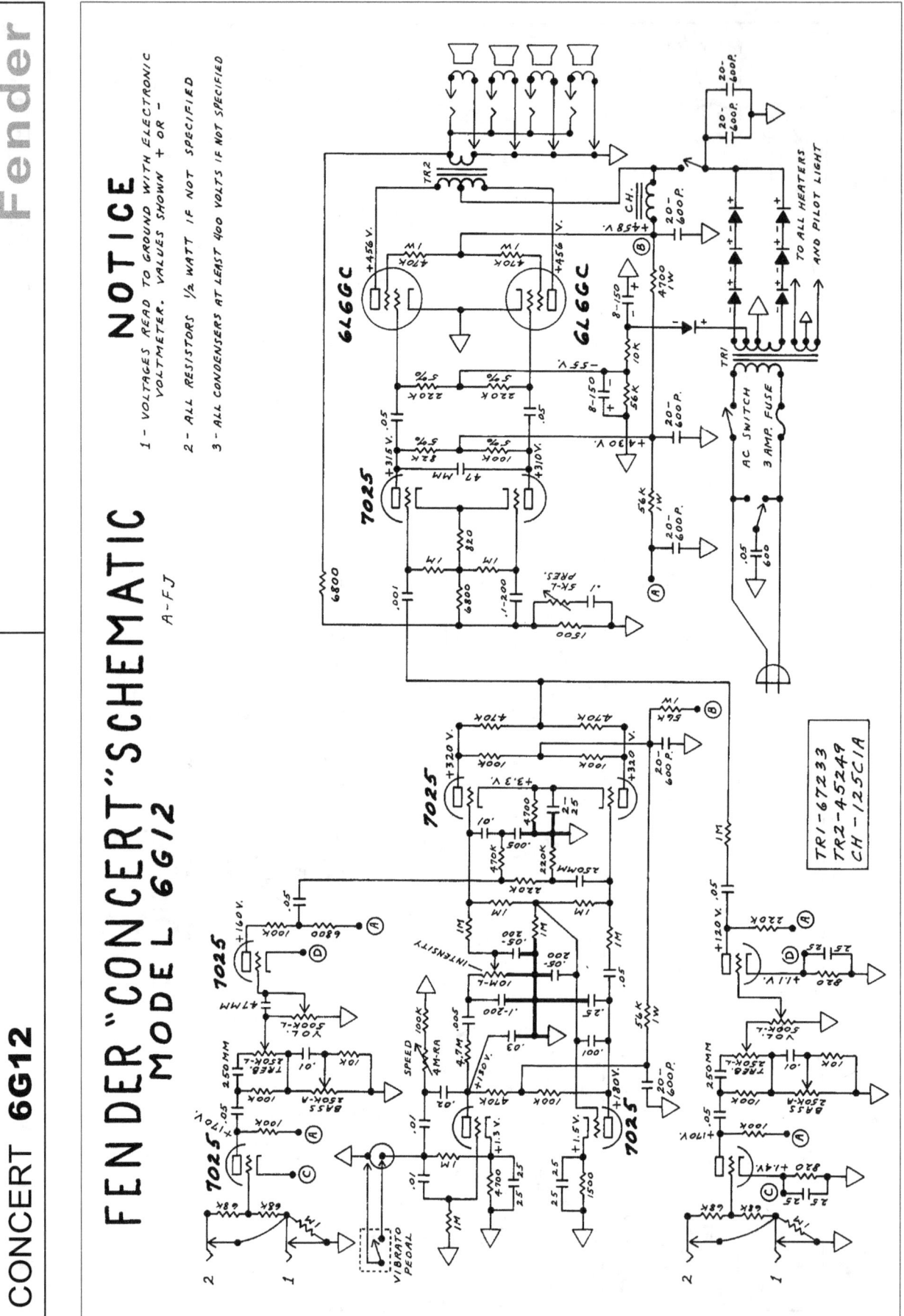

FENDER "CONCERT" SCHEMATIC
MODEL 6G12

A-FJ

NOTICE

1 - VOLTAGES READ TO GROUND WITH ELECTRONIC VOLTMETER. VALUES SHOWN + OR -

2 - ALL RESISTORS ½ WATT IF NOT SPECIFIED

3 - ALL CONDENSERS AT LEAST 400 VOLTS IF NOT SPECIFIED

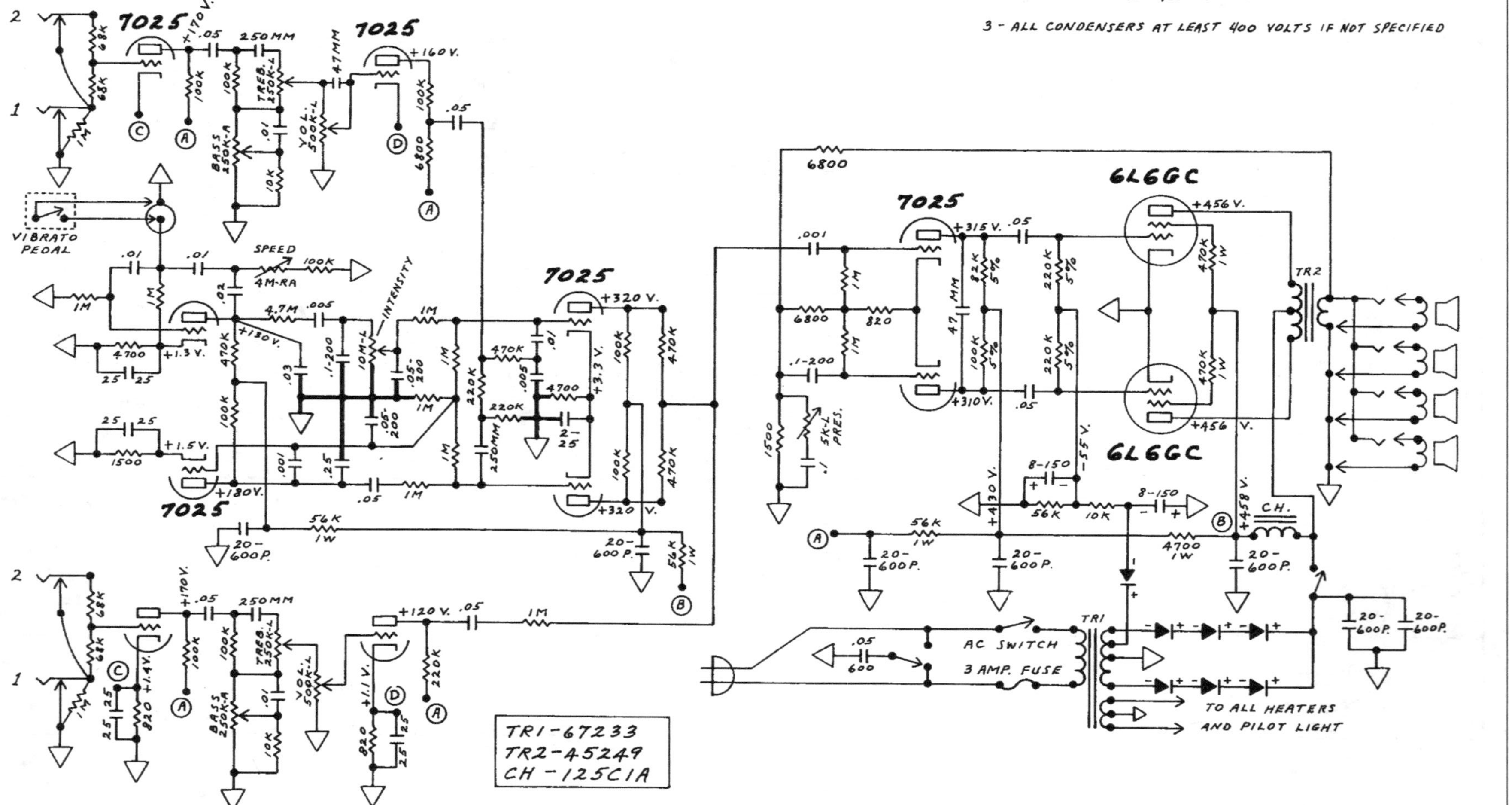

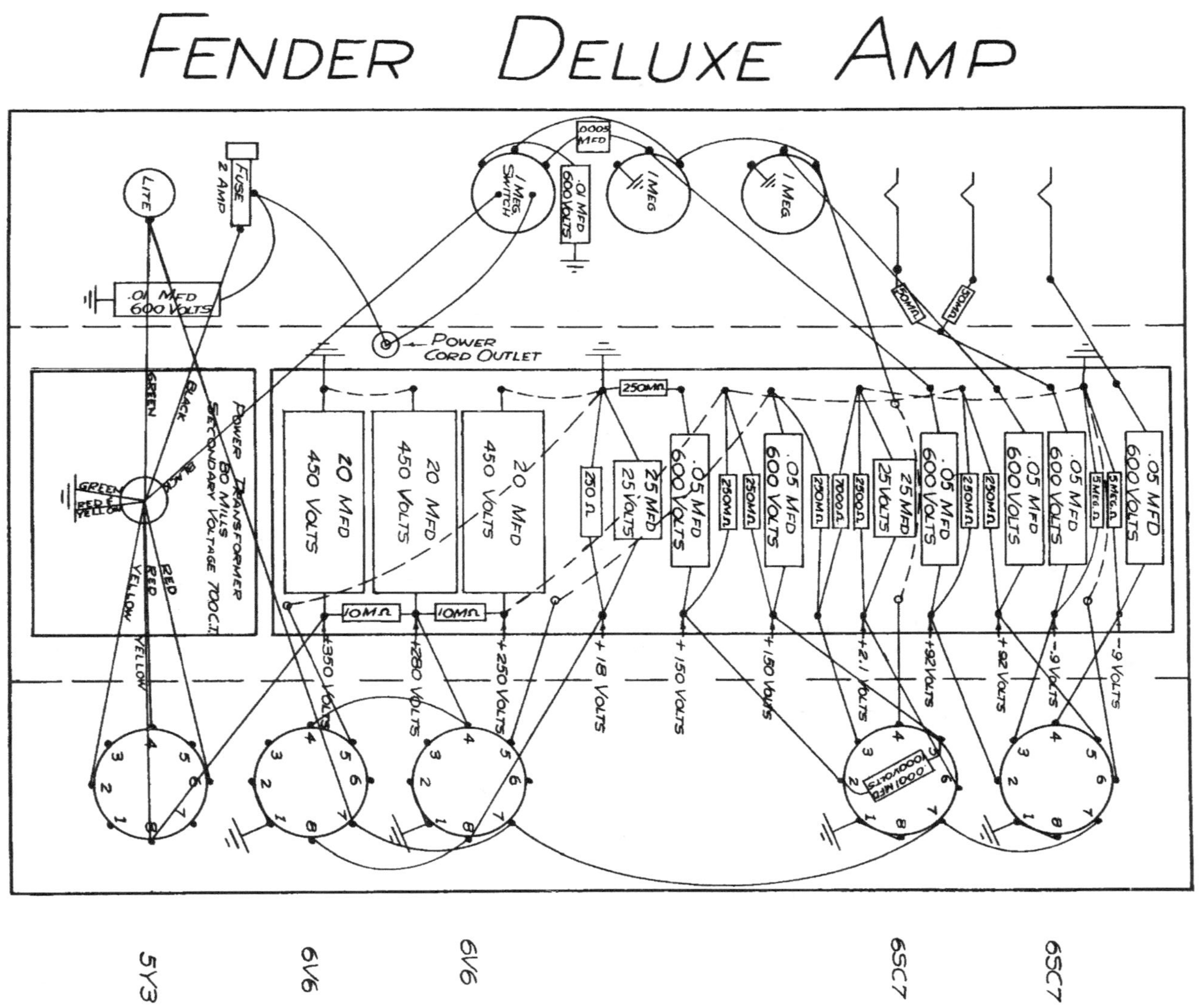
FENDER DELUXE AMP
LITE
FUSE 2 AMP
.01 MFD 600 VOLTS
1 MEG. SWITCH
.01 MFD 600 VOLTS
.0005 MFD
1 MEG
1 MEG
50M.Ω
50M.Ω
POWER CORD OUTLET
POWER TRANSFORMER 80 MILLS SECONDARY VOLTAGE 700C.T.
GREEN
BLACK
RED
RED
YELLOW
YELLOW
20 MFD 450 VOLTS
20 MFD 450 VOLTS
20 MFD 450 VOLTS
10MΩ
10MΩ
250MΩ
250 Ω
25 MFD 25 VOLTS
.05 MFD 600 VOLTS
250MΩ
250MΩ
.05 MFD 600 VOLTS
250MΩ
700Ω
2500Ω
25 MFD 25 VOLTS
.05 MFD 600 VOLTS
250MΩ
250MΩ
.05 MFD 600 VOLTS
.05 MFD 600 VOLTS
5 MEG.Ω
5 MEG.Ω
.05 MFD 600 VOLTS
+350 VOLTS
+280 VOLTS
+250 VOLTS
+18 VOLTS
+150 VOLTS
+150 VOLTS
+2.1 VOLTS
+92 VOLTS
+92 VOLTS
-.9 VOLTS
-.9 VOLTS
.0001 MFD 1000 VOLTS
5Y3
6V6
6V6
6SC7
6SC7

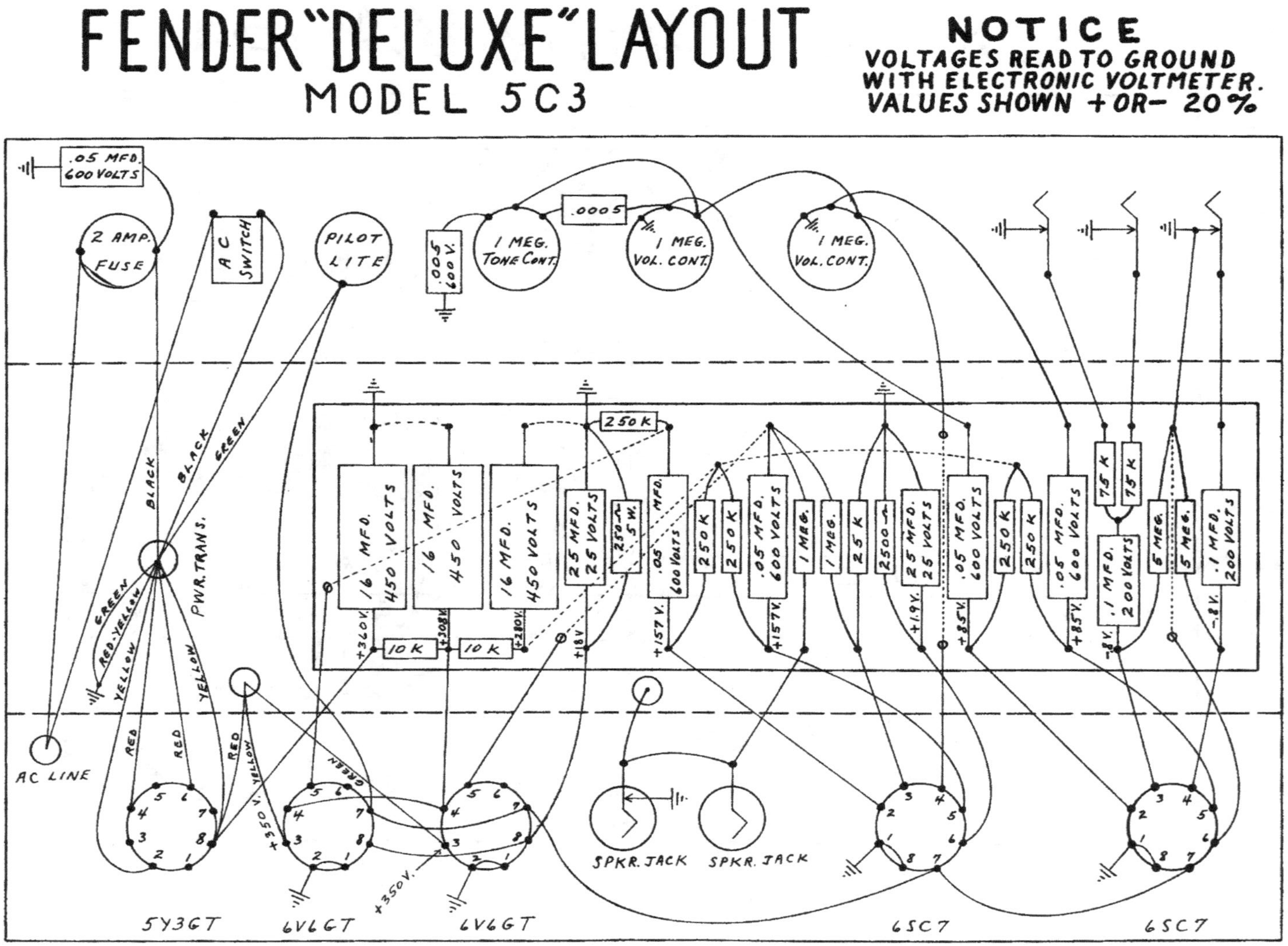
FENDER "DELUXE" LAYOUT
MODEL 5C3
NOTICE
VOLTAGES READ TO GROUND
WITH ELECTRONIC VOLTMETER.
VALUES SHOWN + OR – 20%
.05 MFD. 600 VOLTS
2 AMP. FUSE
AC SWITCH
PILOT LITE
.005 600 V.
1 MEG. TONE CONT.
.0005
1 MEG. VOL. CONT.
1 MEG. VOL. CONT.
250 K
16 MFD. 450 VOLTS
16 MFD. 450 VOLTS
16 MFD. 450 VOLTS
25 MFD. 25 VOLTS
250Ω 5 W.
.05 MFD. 600 VOLTS
250 K
250 K
.05 MFD. 600 VOLTS
1 MEG.
1 MEG.
25 K
2500Ω
25 MFD. 25 VOLTS
.05 MFD. 600 VOLTS
250 K
250 K
.05 MFD. 600 VOLTS
75 K
75 K
.1 MFD. 200 VOLTS
5 MEG.
5 MEG.
.1 MFD. 200 VOLTS
+360 V.
10 K
+308 V.
10 K
+280 V.
+18 V.
+157 V.
+157 V.
+1.9 V.
+85 V.
+85 V.
-8 V.
-8 V.
BLACK
BLACK
GREEN
GREEN
RED-YELLOW
YELLOW
YELLOW
PWR. TRANS.
RED
RED
RED
YELLOW
+350 V.
GREEN
+350 V.
AC LINE
SPKR. JACK
SPKR. JACK
5Y3GT
6V6GT
6V6GT
6SC7
6SC7

FENDER "DELUXE" SCHEMATIC

MODEL 5C3

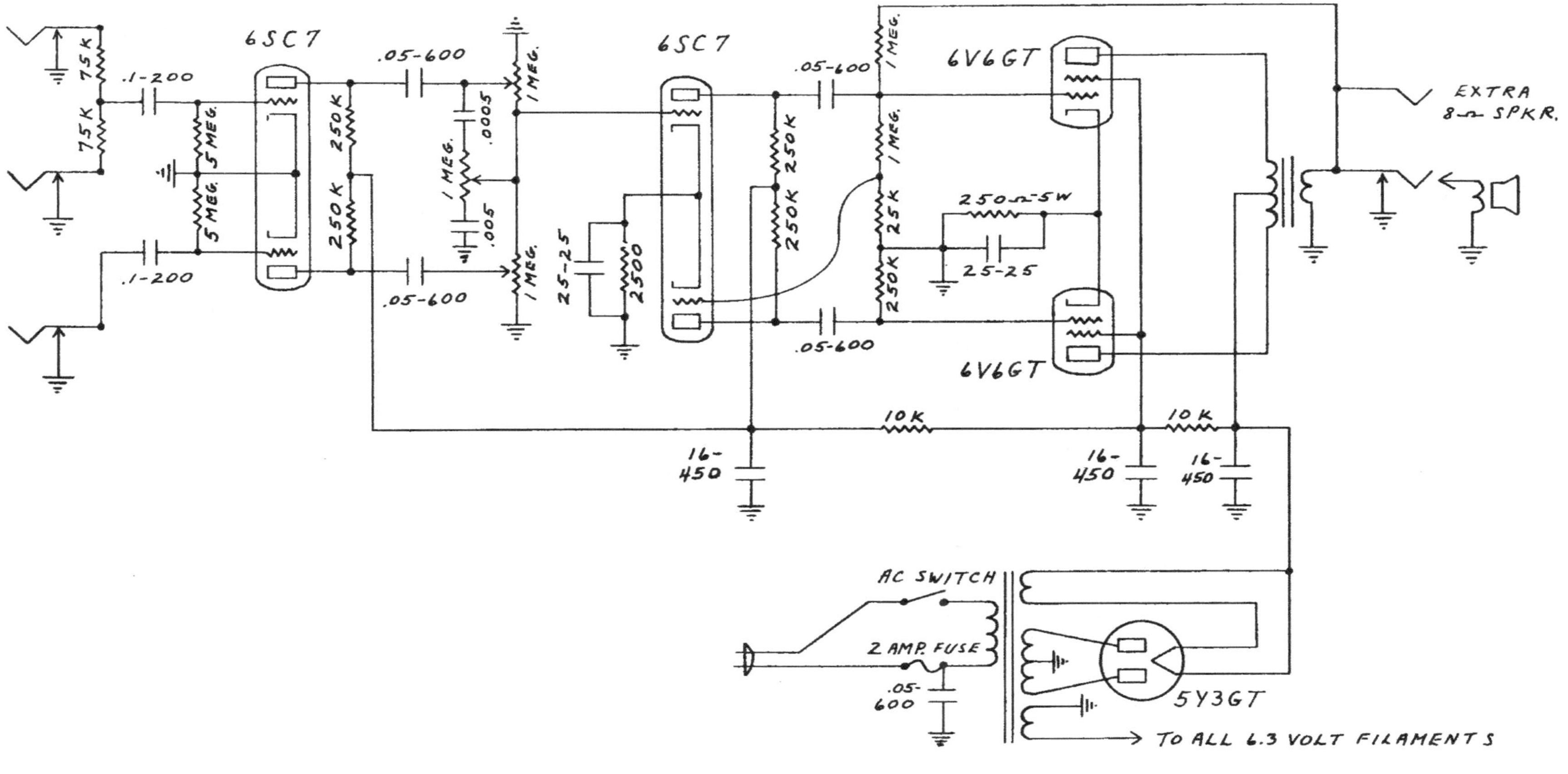

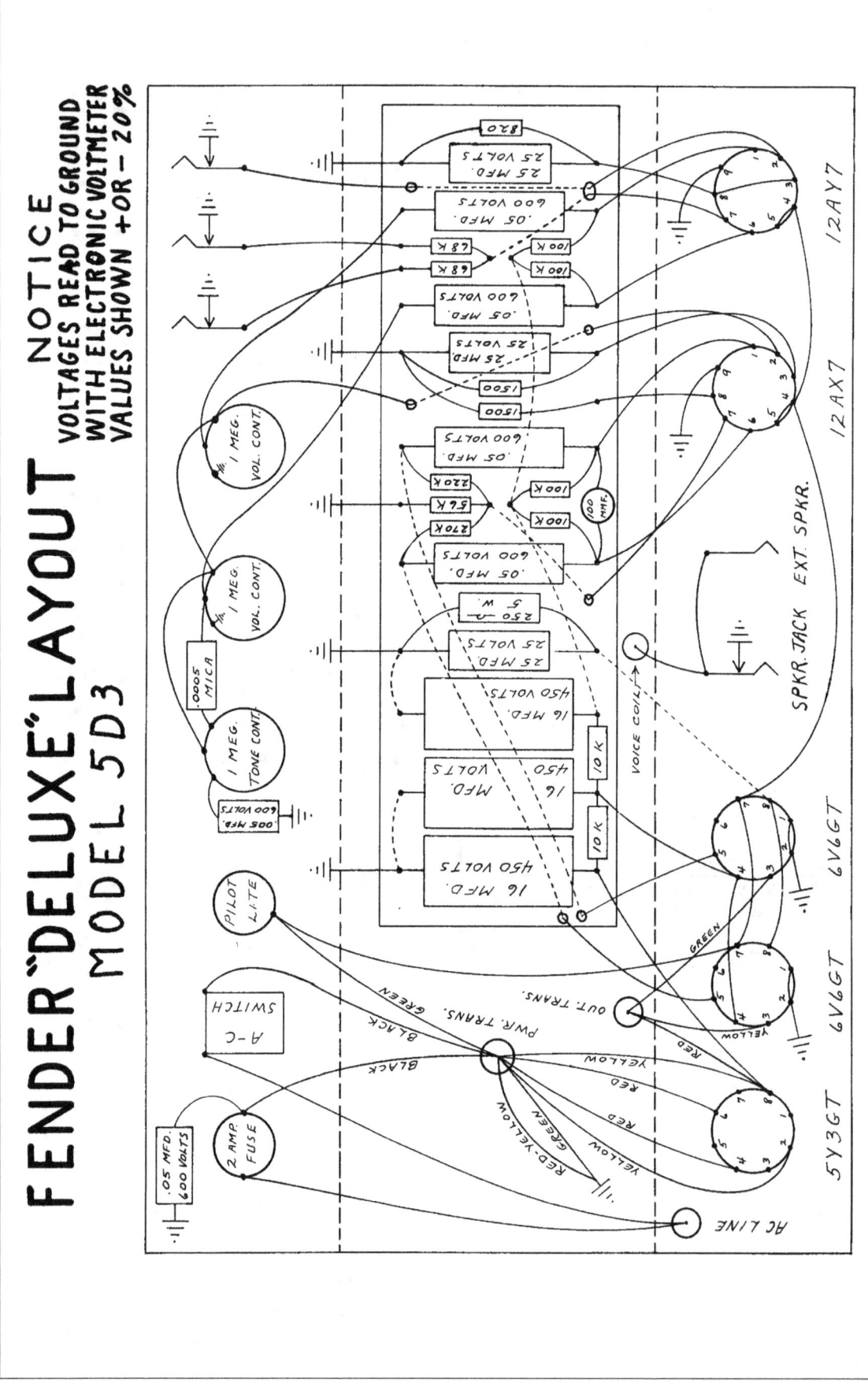

FENDER "DELUXE" LAYOUT
MODEL 5D3

NOTICE
VOLTAGES READ TO GROUND WITH ELECTRONIC VOLTMETER VALUES SHOWN +OR – 20%

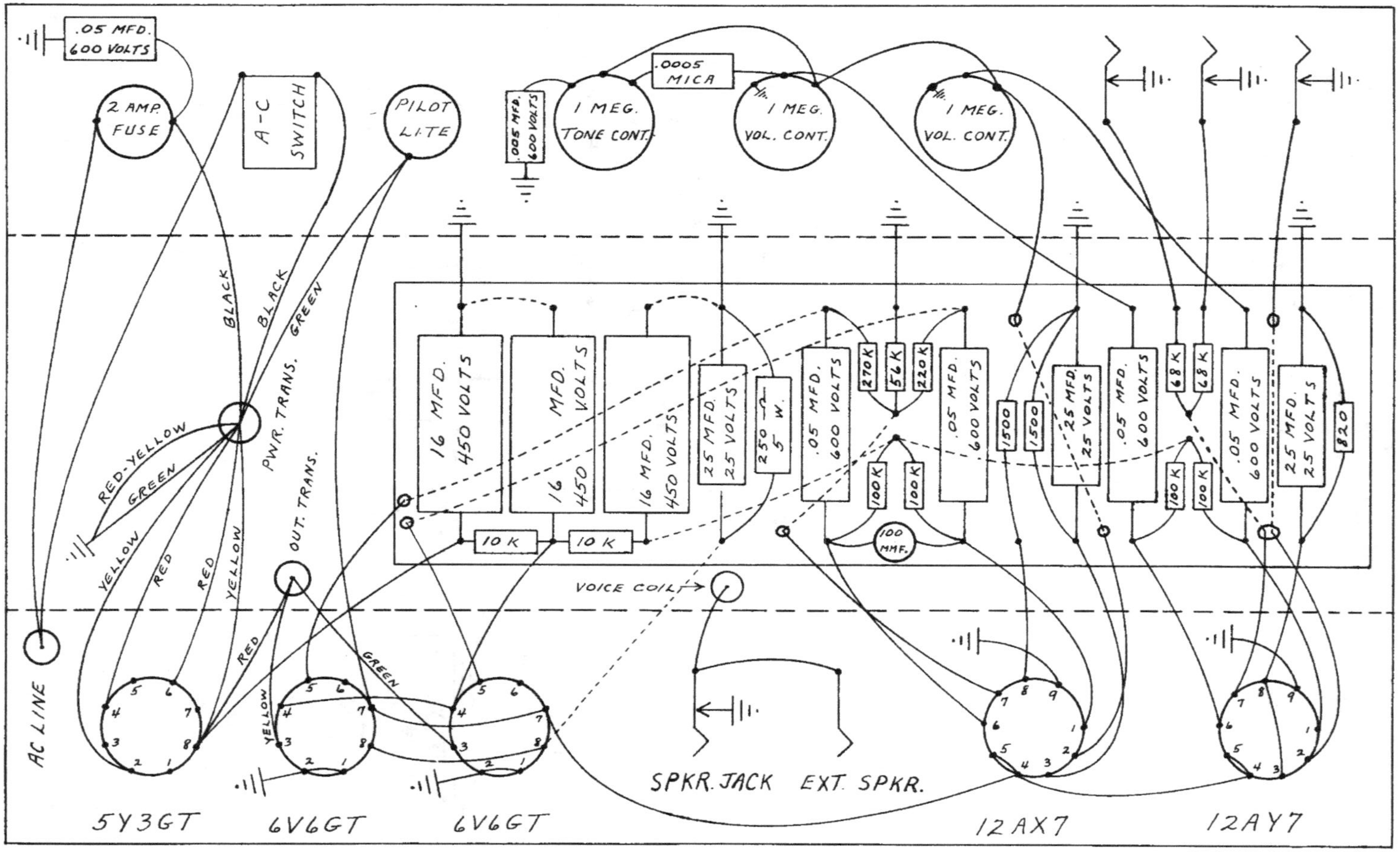

FENDER DELUXE SCHEMATIC
MODEL 5D3

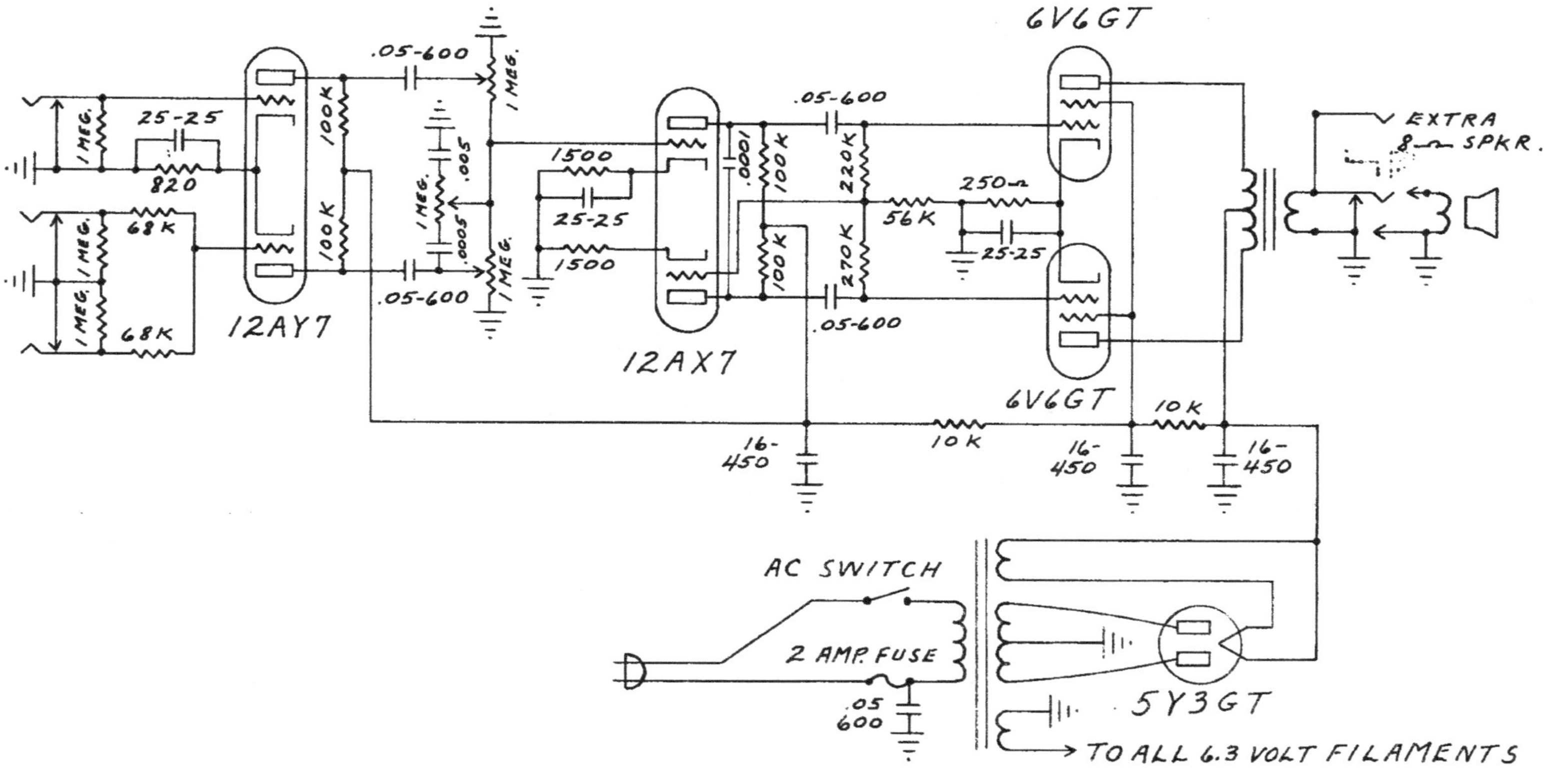

FENDER "DELUXE" LAYOUT
MODEL 5D4

NOTICE
VOLTAGES READ TO GROUND
WITH ELECTRONIC VOLTMETER
VALUES SHOWN +OR – 20%

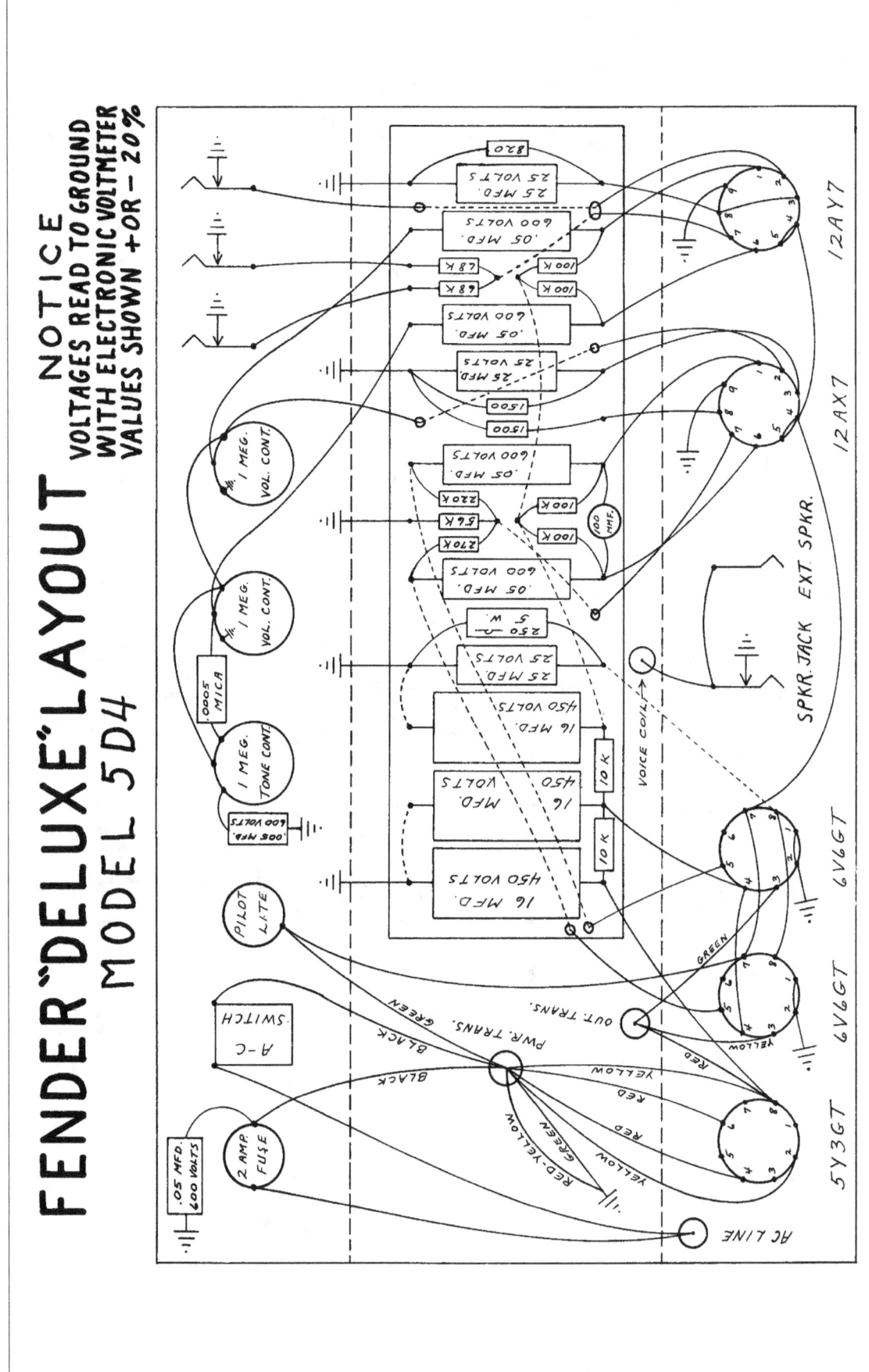

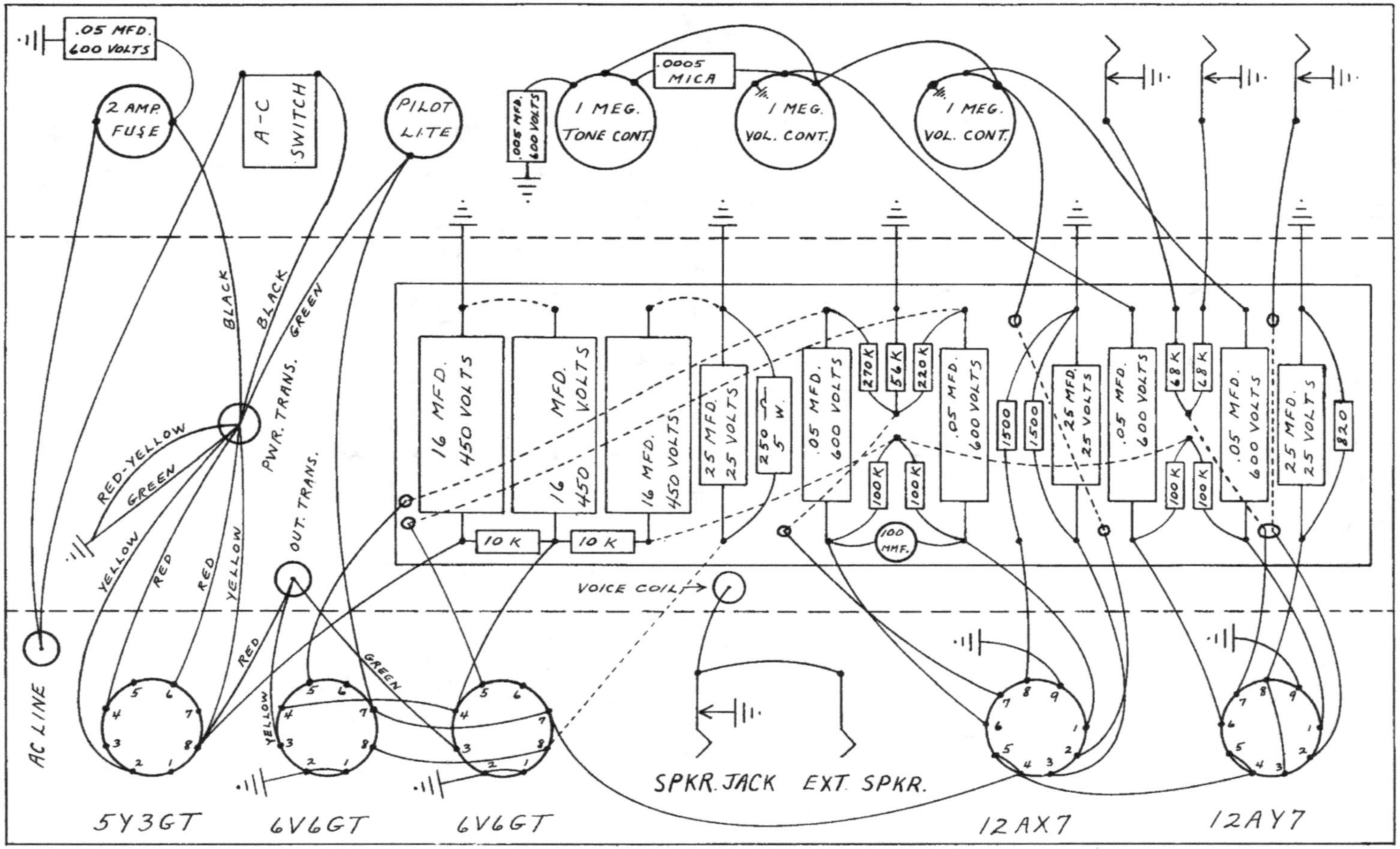

FENDER DELUXE SCHEMATIC

MODEL 5D4

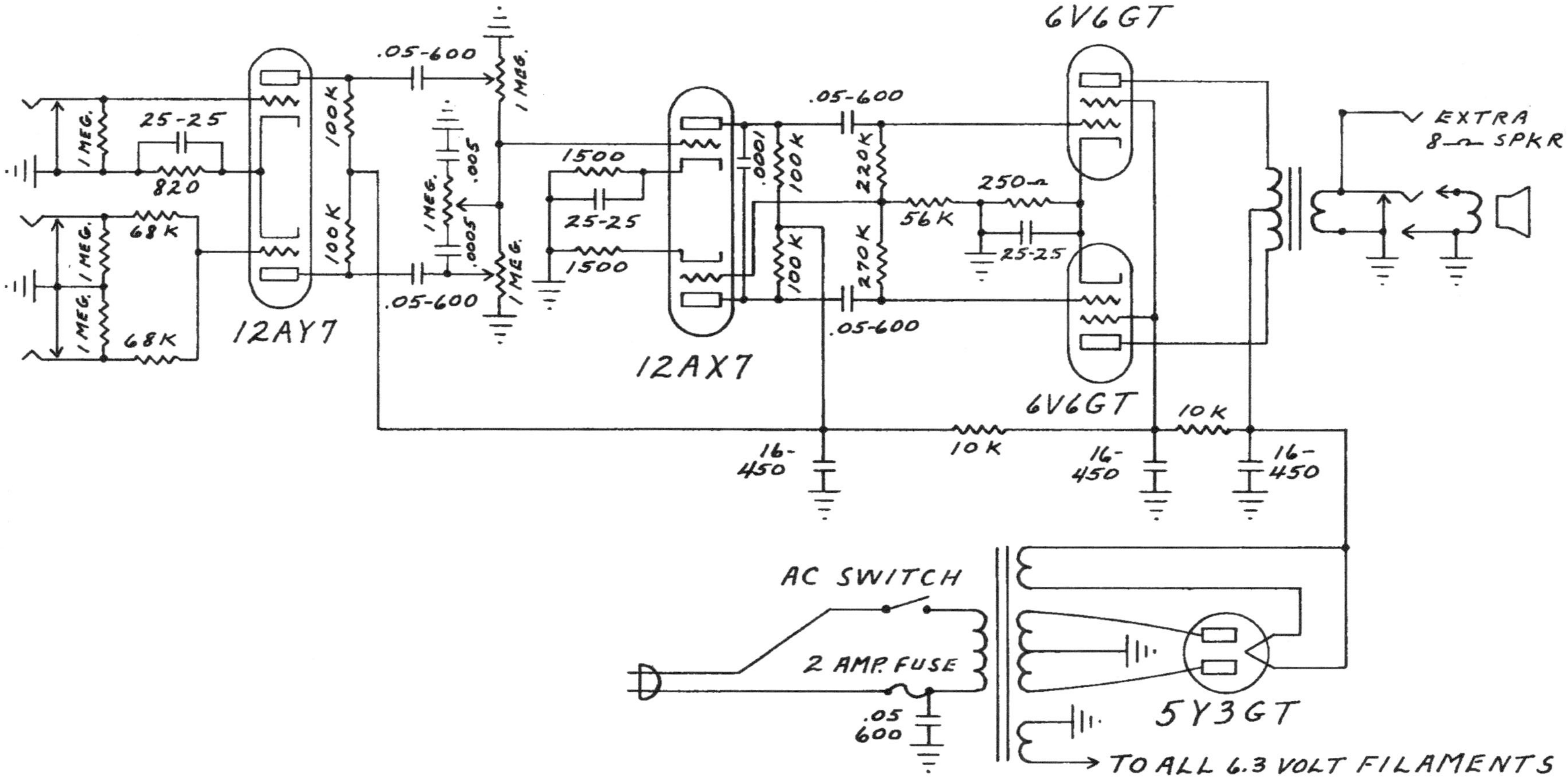

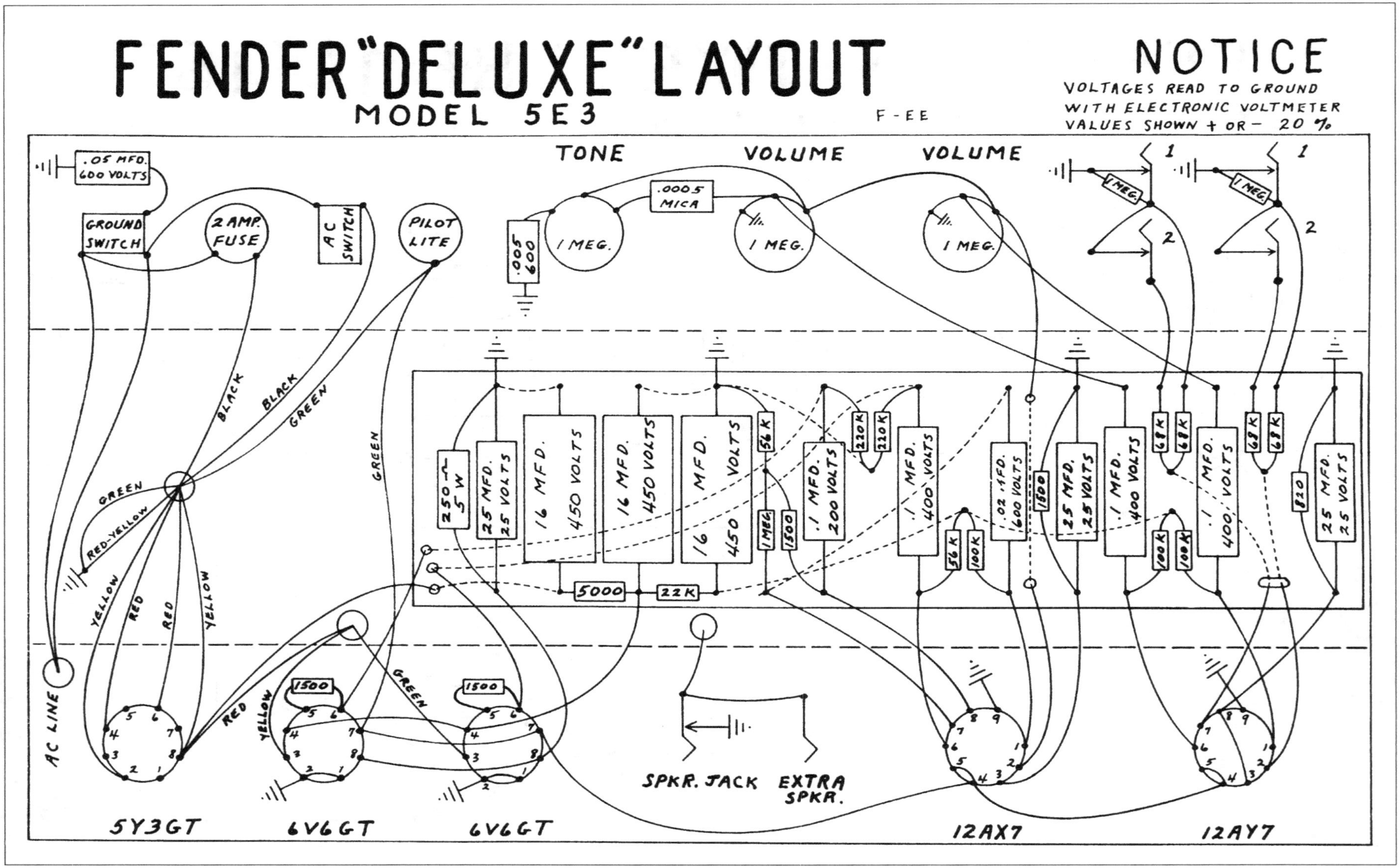
FENDER "DELUXE" LAYOUT
MODEL 5E3
F-EE
NOTICE
VOLTAGES READ TO GROUND
WITH ELECTRONIC VOLTMETER
VALUES SHOWN + OR − 20 %
TONE
VOLUME
VOLUME
.05 MFD. 600 VOLTS
GROUND SWITCH
2 AMP. FUSE
AC SWITCH
PILOT LITE
.005 600
1 MEG.
.0005 MICA
1 MEG.
1 MEG.
1 MEG
1 MEG
250 Ω 5 W
25 MFD. 25 VOLTS
16 MFD. 450 VOLTS
16 MFD. 450 VOLTS
16 MFD. 450 VOLTS
56 K
1 MEG
1500
.1 MFD. 200 VOLTS
220 K
220 K
.1 MFD. 400 VOLTS
56 K
100 K
.02 .1 FD. 600 VOLTS
1800
25 MFD. 25 VOLTS
.1 MFD. 400 VOLTS
100 K
100 K
68 K
68 K
.1 MFD. 400 VOLTS
68 K
68 K
820
25 MFD. 25 VOLTS
5000
22 K
GREEN
BLACK
BLACK
GREEN
GREEN
RED-YELLOW
YELLOW
RED
RED
YELLOW
RED
YELLOW
GREEN
AC LINE
1500
1500
SPKR. JACK
EXTRA SPKR.
5Y3GT
6V6GT
6V6GT
12AX7
12AY7

FENDER "DELUXE" SCHEMATIC

MODEL 5E3

F-EE

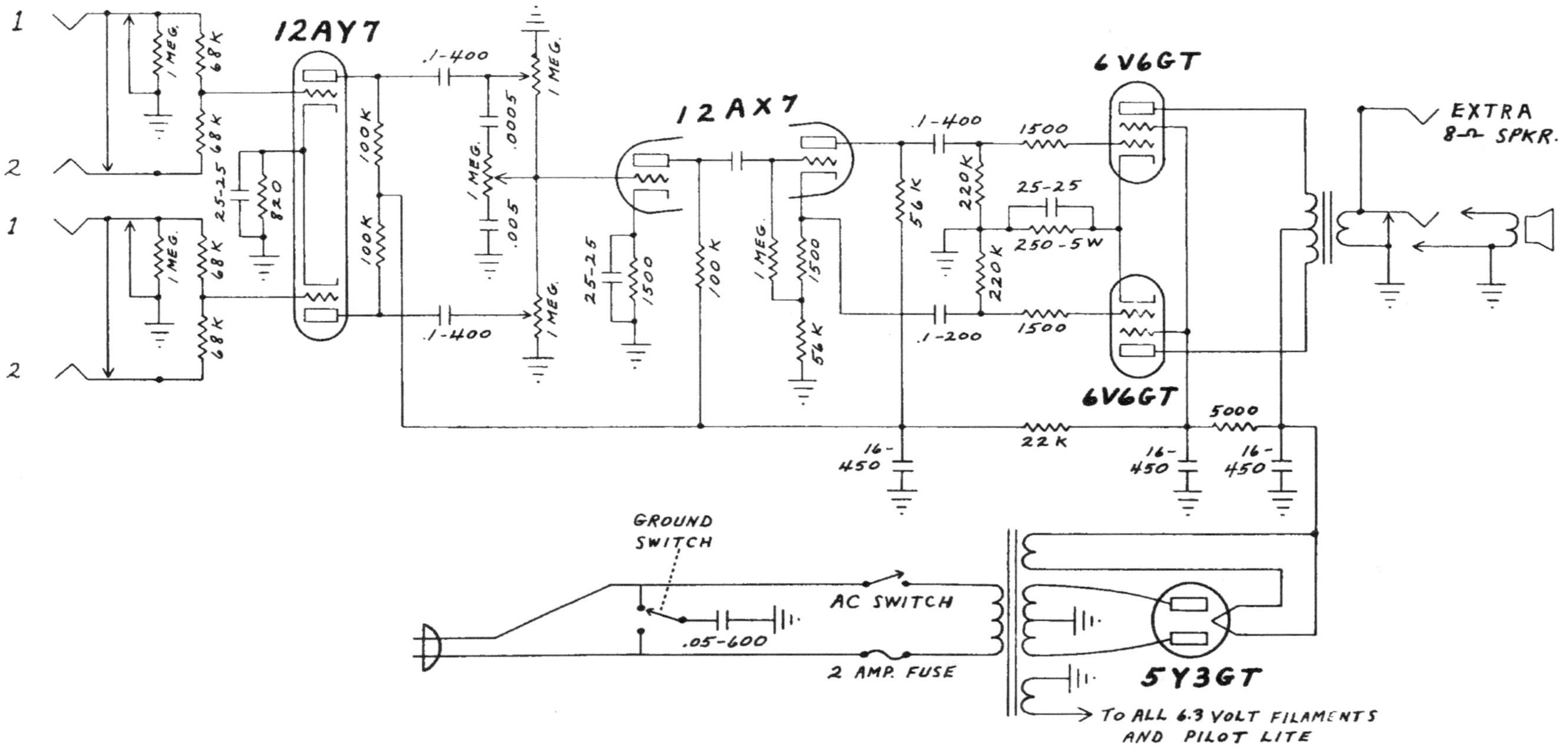

FENDER "DELUXE" LAYOUT
MODEL 6G3

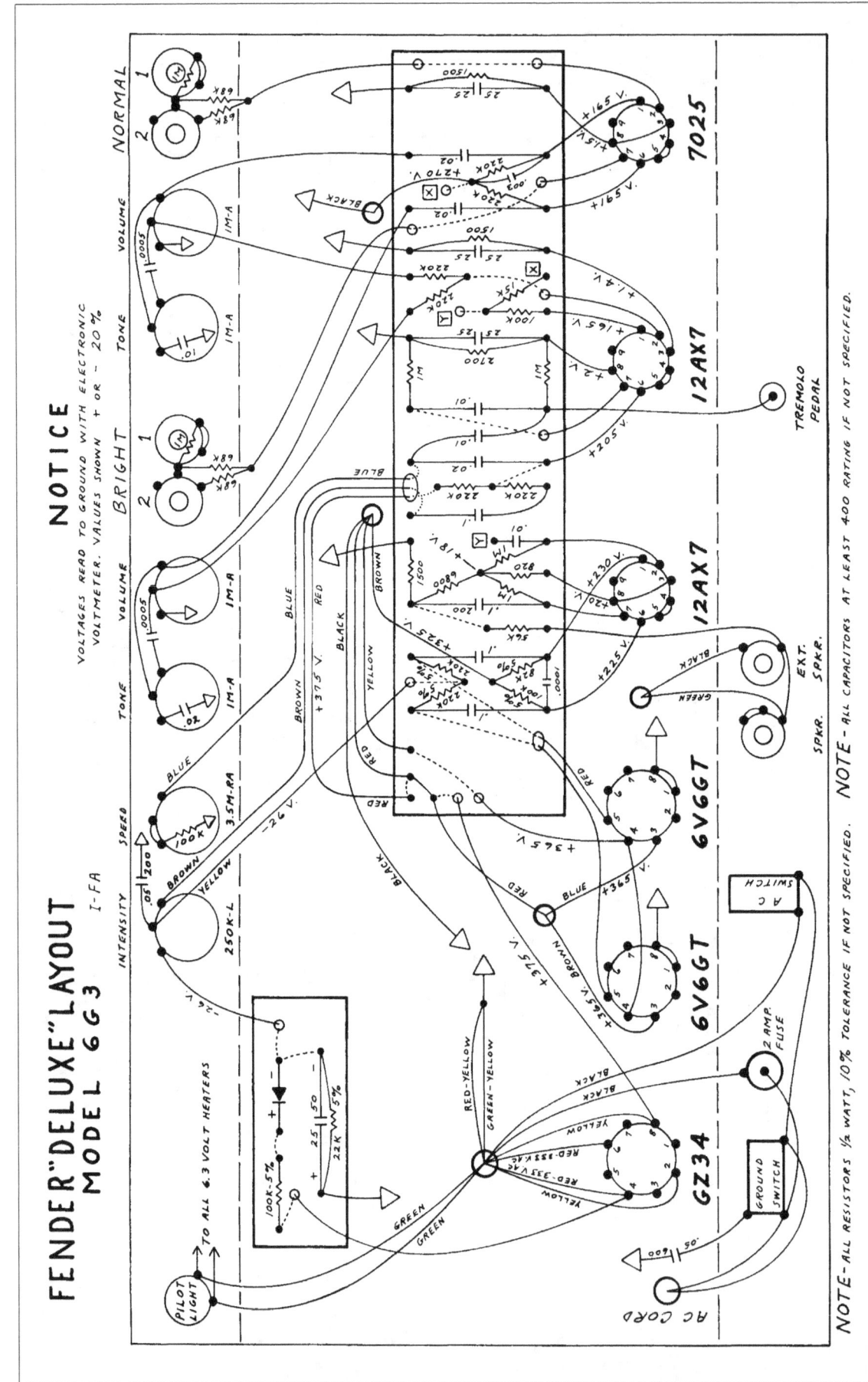

I-FA

NOTICE

VOLTAGES READ TO GROUND WITH ELECTRONIC VOLTMETER. VALUES SHOWN + OR − 20%

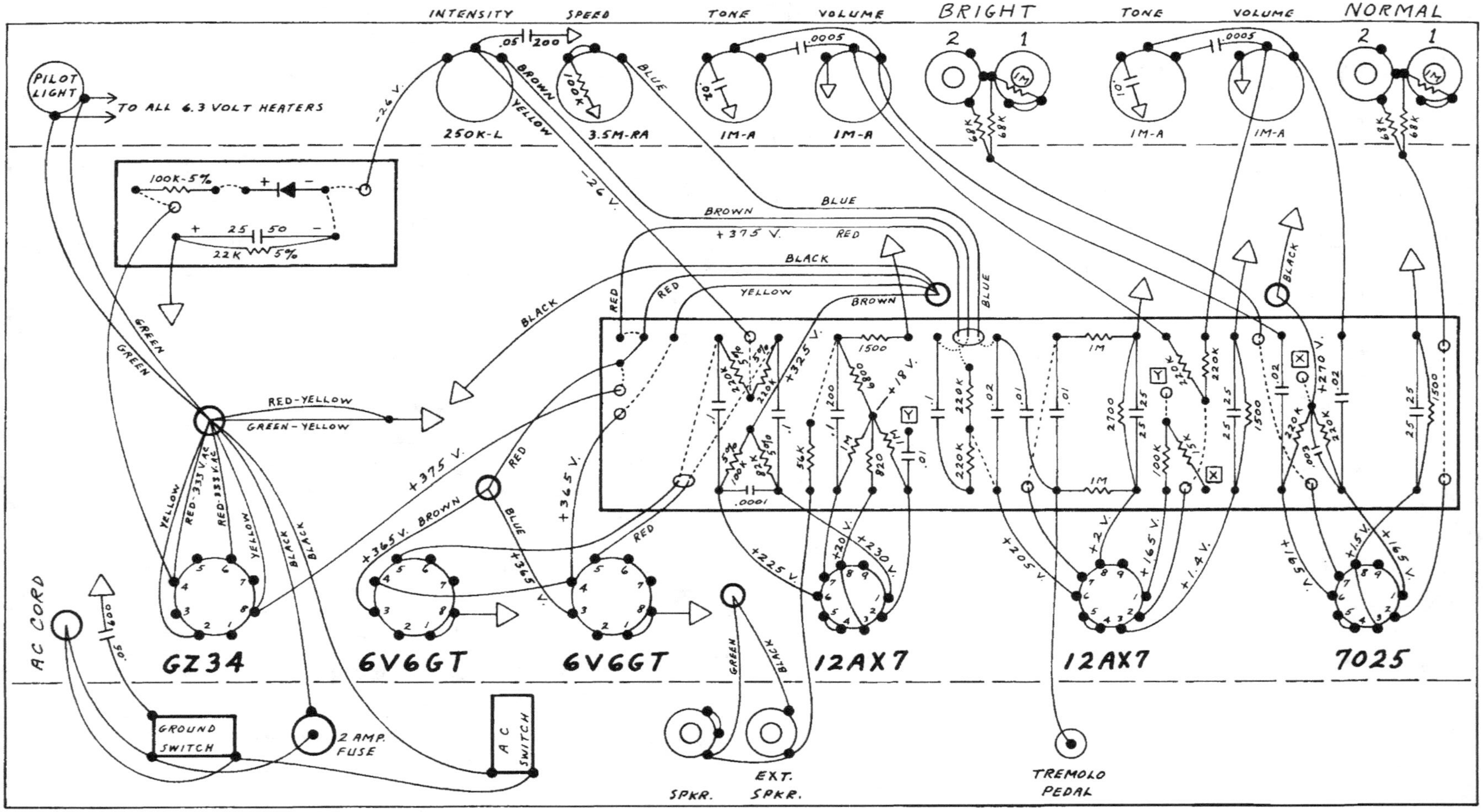

NOTE – ALL RESISTORS ½ WATT, 10% TOLERANCE IF NOT SPECIFIED. NOTE – ALL CAPACITORS AT LEAST 400 RATING IF NOT SPECIFIED.

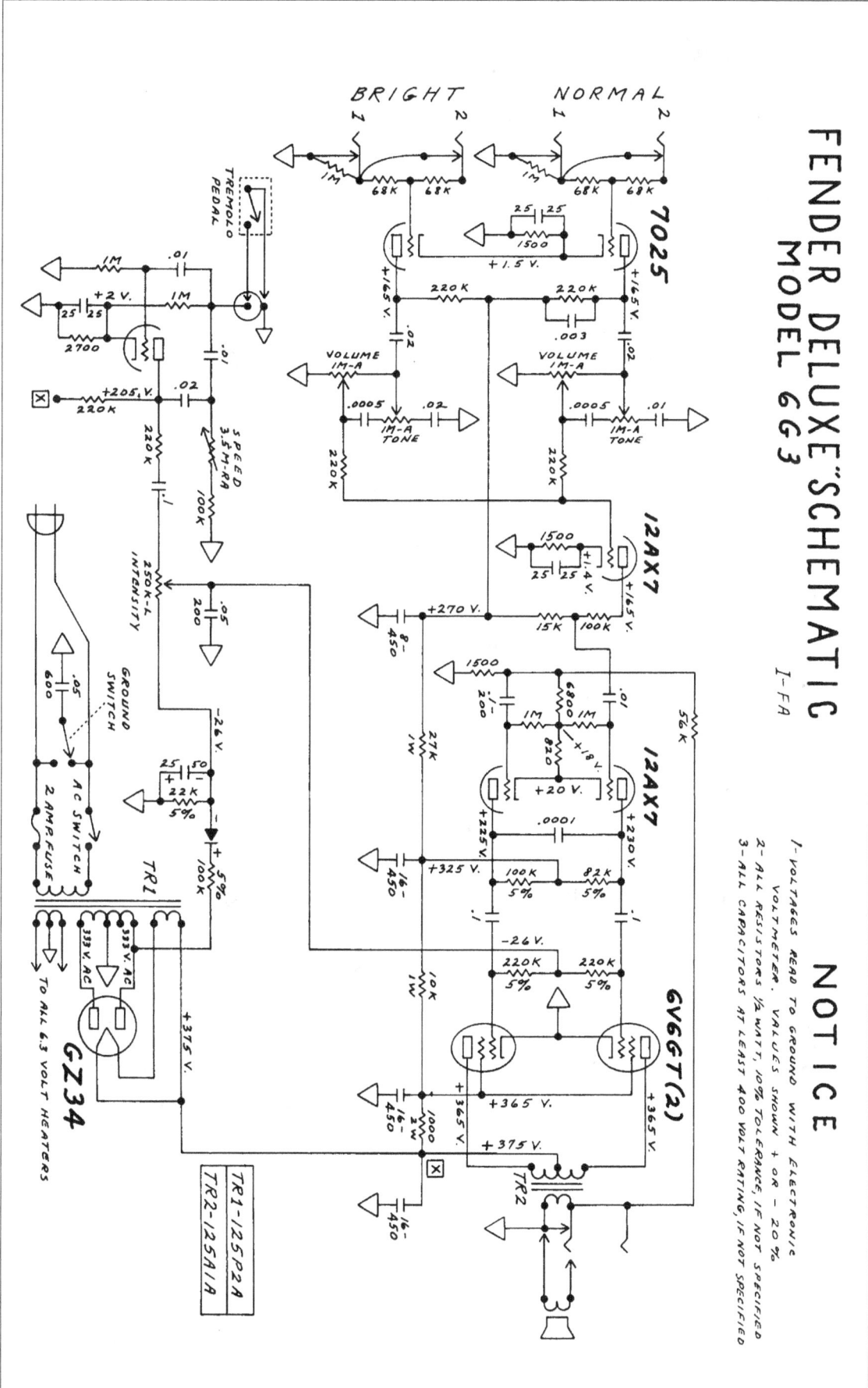

FENDER DELUXE" SCHEMATIC
MODEL 6G3

I-FA

NOTICE

1- VOLTAGES READ TO GROUND WITH ELECTRONIC VOLTMETER. VALUES SHOWN + OR - 20%
2- ALL RESISTORS ½ WATT, 10% TOLERANCE, IF NOT SPECIFIED
3- ALL CAPACITORS AT LEAST 400 VOLT RATING, IF NOT SPECIFIED

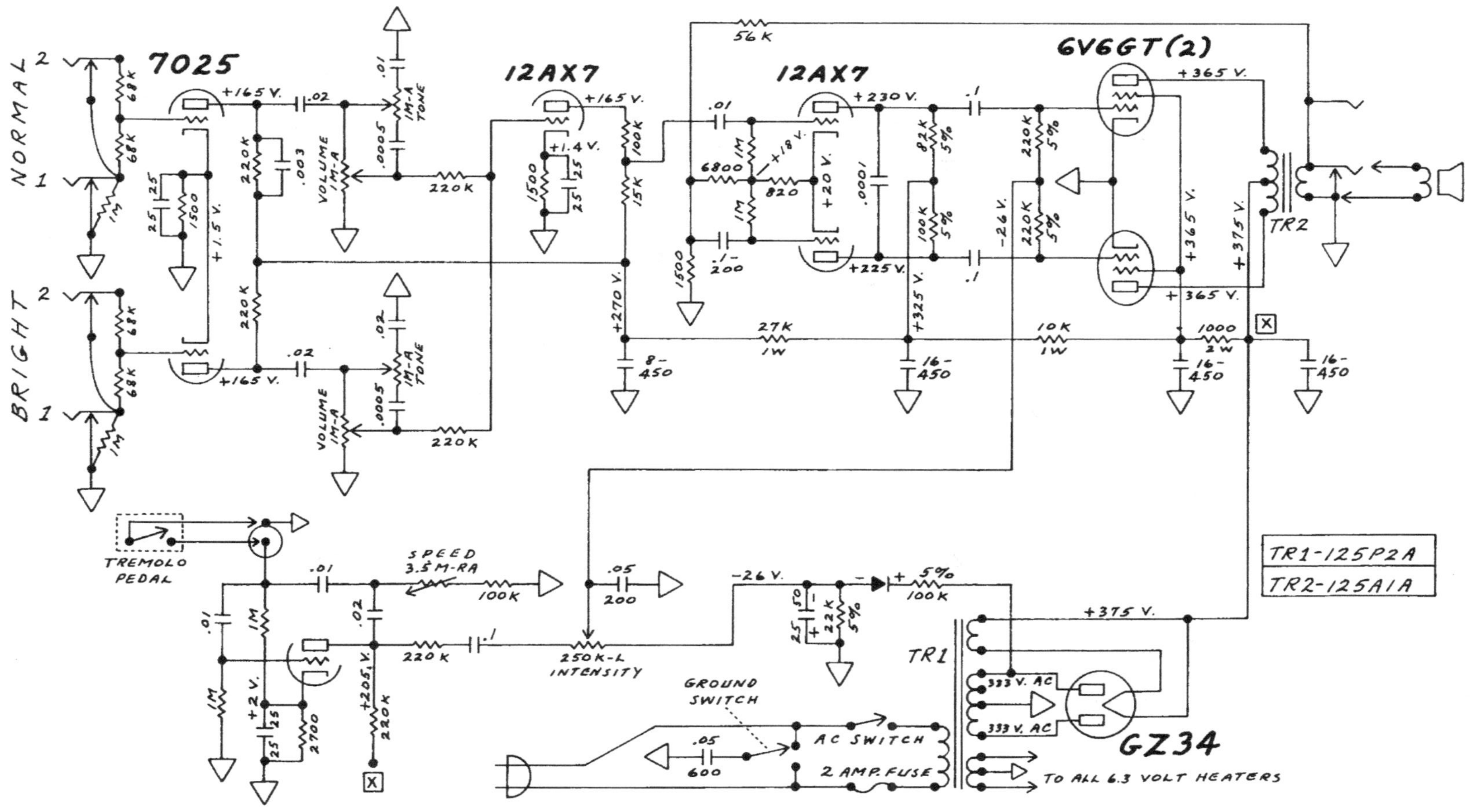

TR1-125P2A
TR2-125A1A

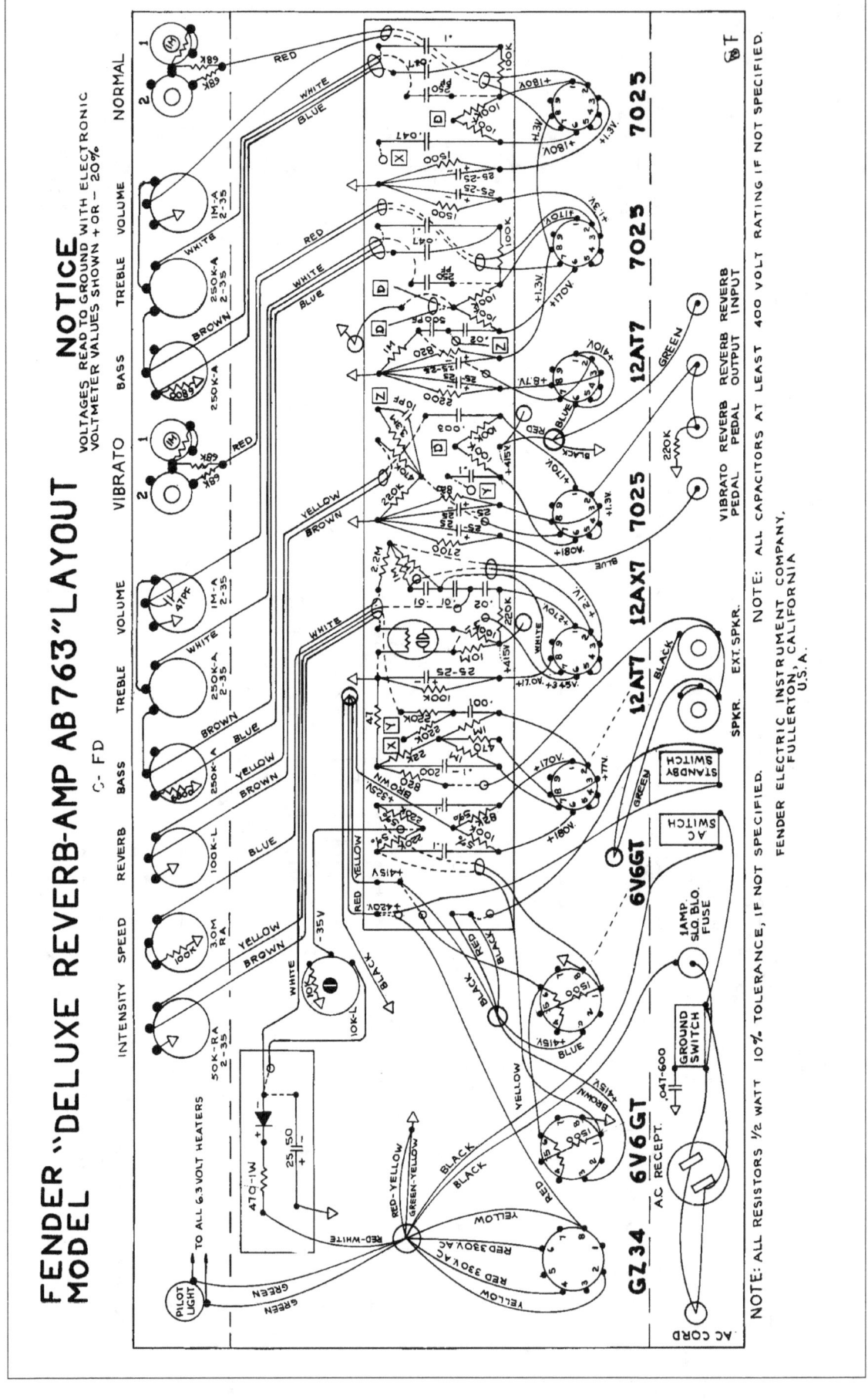

FENDER MODEL "DELUXE REVERB-AMP AB763" LAYOUT

C- FD

NOTICE
VOLTAGES READ TO GROUND WITH ELECTRONIC VOLTMETER VALUES SHOWN + OR − 20%

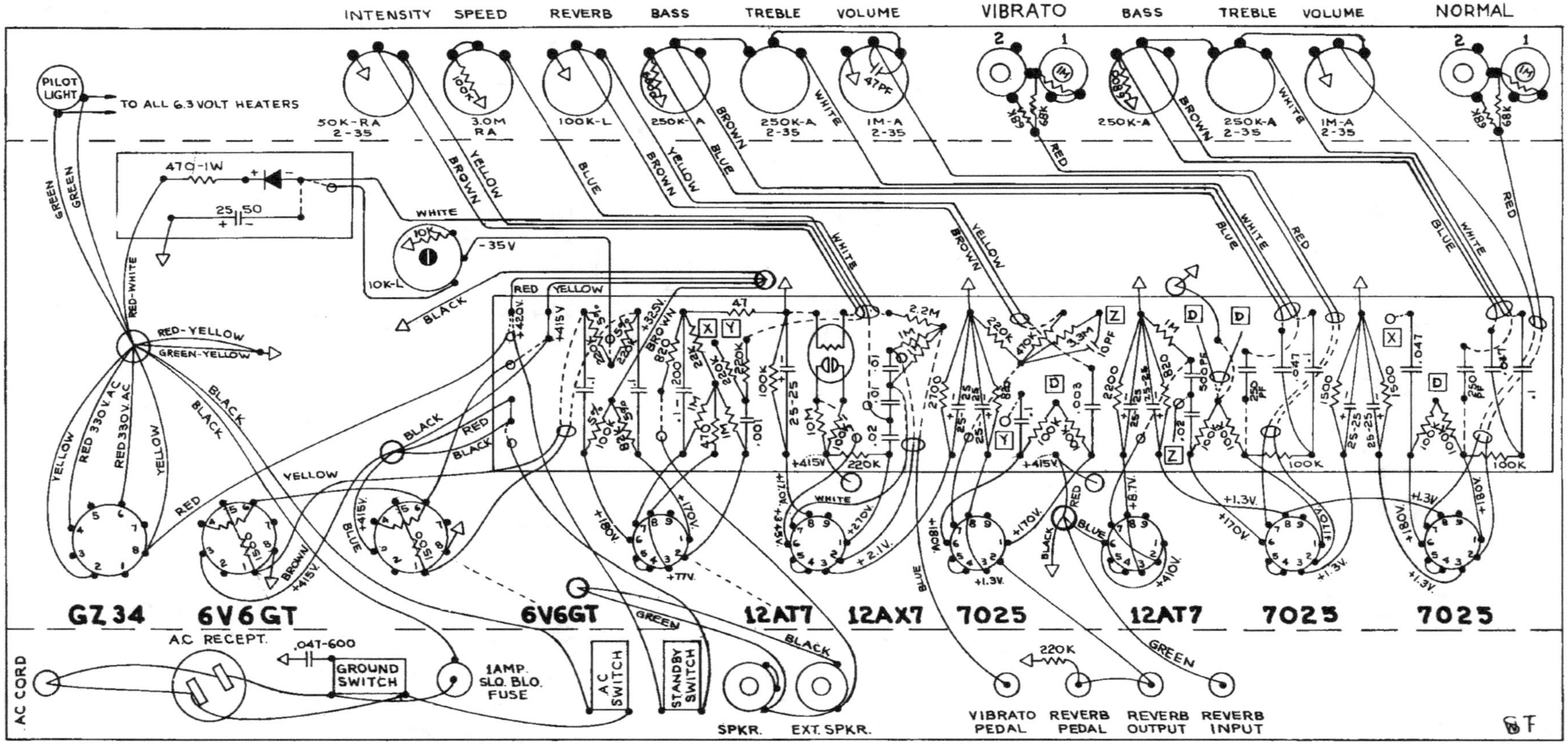

NOTE: ALL RESISTORS ½ WATT 10% TOLERANCE, IF NOT SPECIFIED.

NOTE: ALL CAPACITORS AT LEAST 400 VOLT RATING IF NOT SPECIFIED.

FENDER ELECTRIC INSTRUMENT COMPANY,
FULLERTON, CALIFORNIA
U.S.A.

DELUXE REVERB-AMP **AB763**

Fender

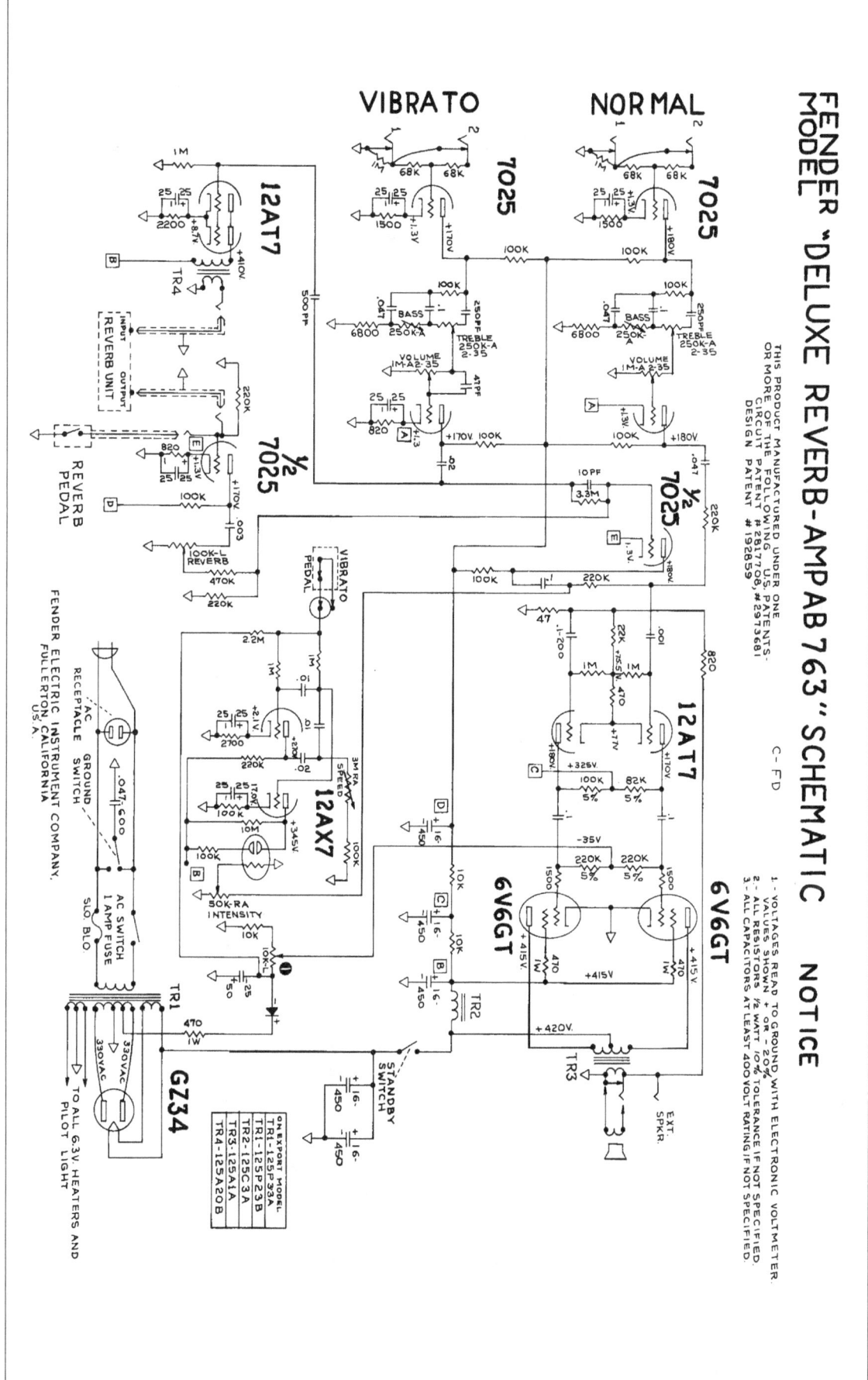

FENDER MODEL "DELUXE REVERB-AMP AB763" SCHEMATIC

THIS PRODUCT MANUFACTURED UNDER ONE OR MORE OF THE FOLLOWING U.S. PATENTS-
CIRCUIT PATENT #2817708, #2973681
DESIGN PATENT #192859

C-FD

NOTICE

1.- VOLTAGES READ TO GROUND WITH ELECTRONIC VOLTMETER. VALUES SHOWN + OR - 20%
2.- ALL RESISTORS ½ WATT 10% TOLERANCE IF NOT SPECIFIED.
3.- ALL CAPACITORS AT LEAST 400 VOLT RATING IF NOT SPECIFIED.

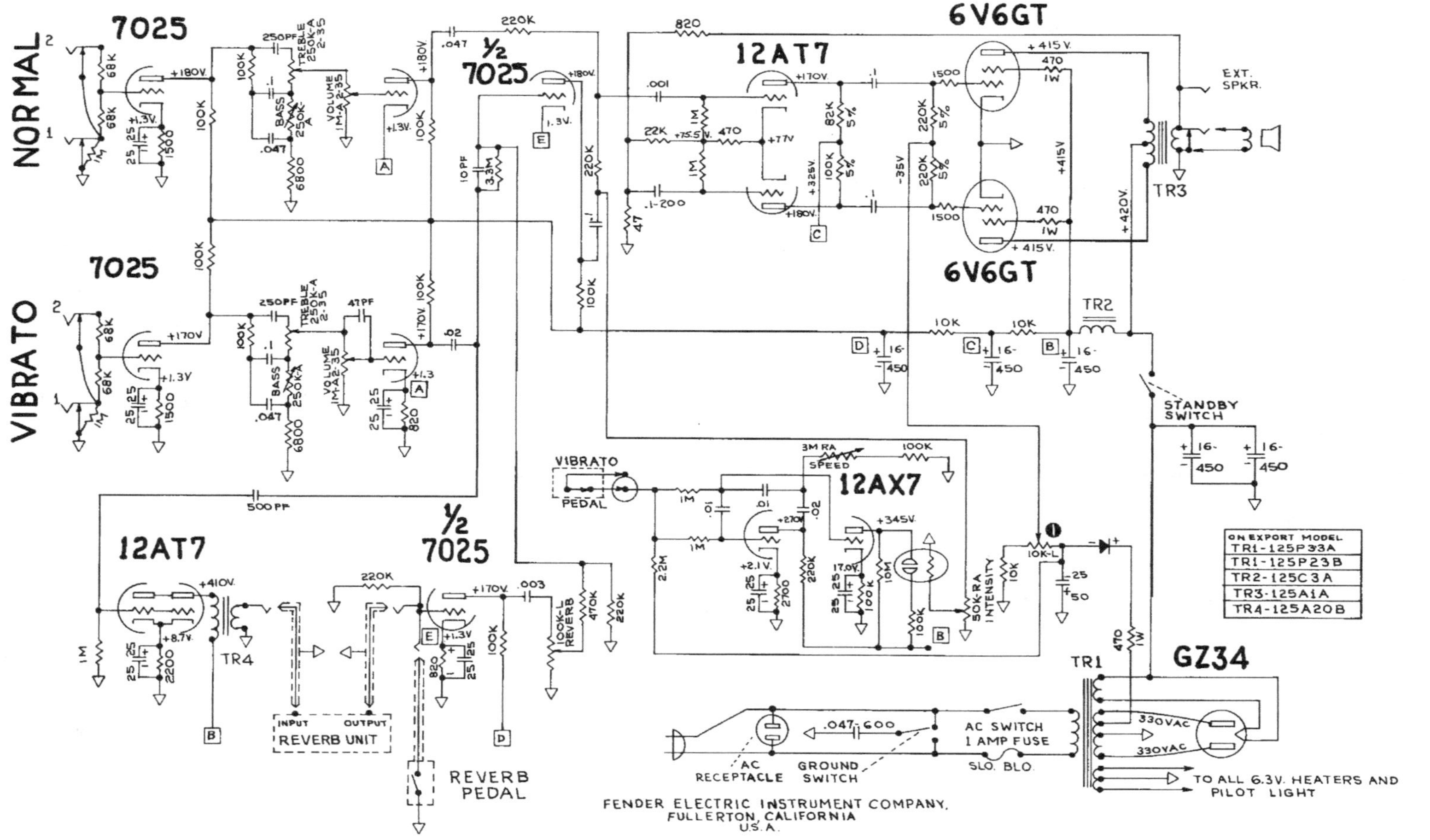

FENDER ELECTRIC INSTRUMENT COMPANY,
FULLERTON, CALIFORNIA
U.S.A.

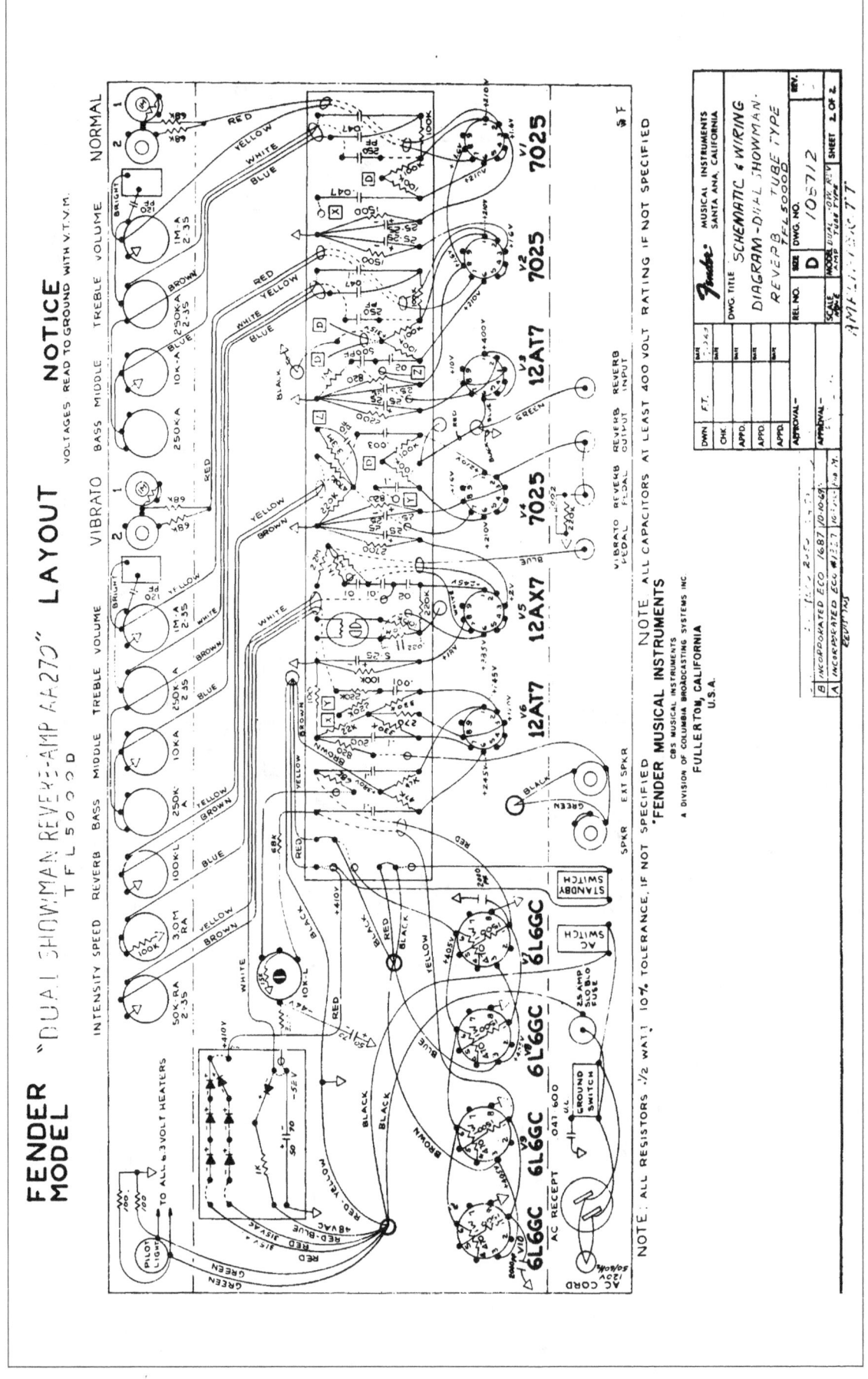

FENDER MODEL "DUAL SHOWMAN REVERB-AMP AA270" **LAYOUT**

TFL5000D

NOTICE

VOLTAGES READ TO GROUND WITH V.T.V.M.

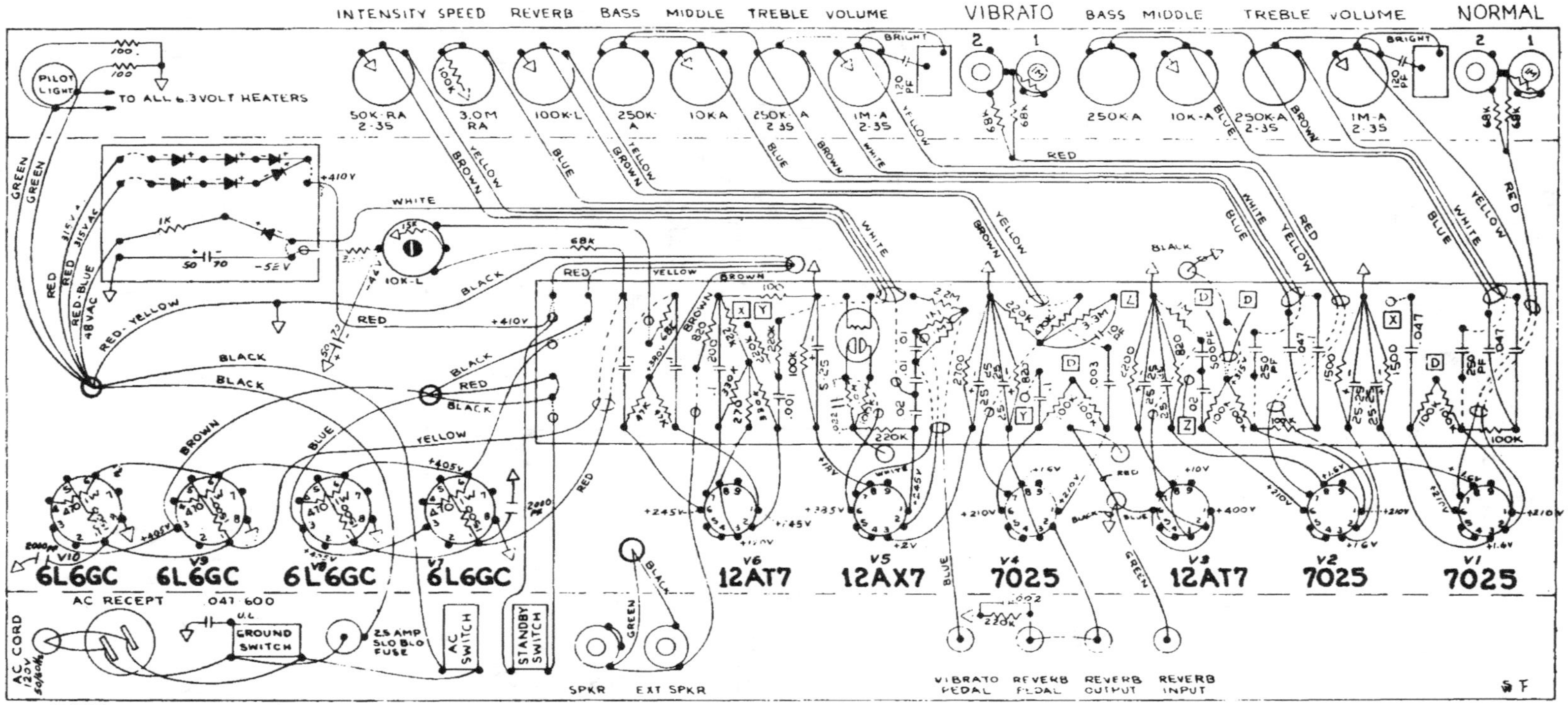

NOTE: ALL RESISTORS 1/2 WATT 10% TOLERANCE, IF NOT SPECIFIED

NOTE: ALL CAPACITORS AT LEAST 400 VOLT RATING IF NOT SPECIFIED

FENDER MUSICAL INSTRUMENTS
CBS MUSICAL INSTRUMENTS
A DIVISION OF COLUMBIA BROADCASTING SYSTEMS INC.
FULLERTON, CALIFORNIA
U.S.A.

DWN	F.T.
CHK	
APPD.	
APPD.	
APPD.	
APPROVAL–	
APPROVAL–	

Fender MUSICAL INSTRUMENTS SANTA ANA, CALIFORNIA

DWG. TITLE SCHEMATIC & WIRING DIAGRAM – DUAL SHOWMAN REVERB TUBE TYPE TFL5000D

REL. NO.	SIZE	DWG. NO.	REV.
	D	105712	

SCALE NONE · MODEL DUAL SHOWMAN REV AMP TUBE TYPE · SHEET 2 OF 2

B	INCORPORATED ECO 1687	10-10-69	
A	INCORPORATED ECO #1227		

REVISIONS

AMPLIFIER

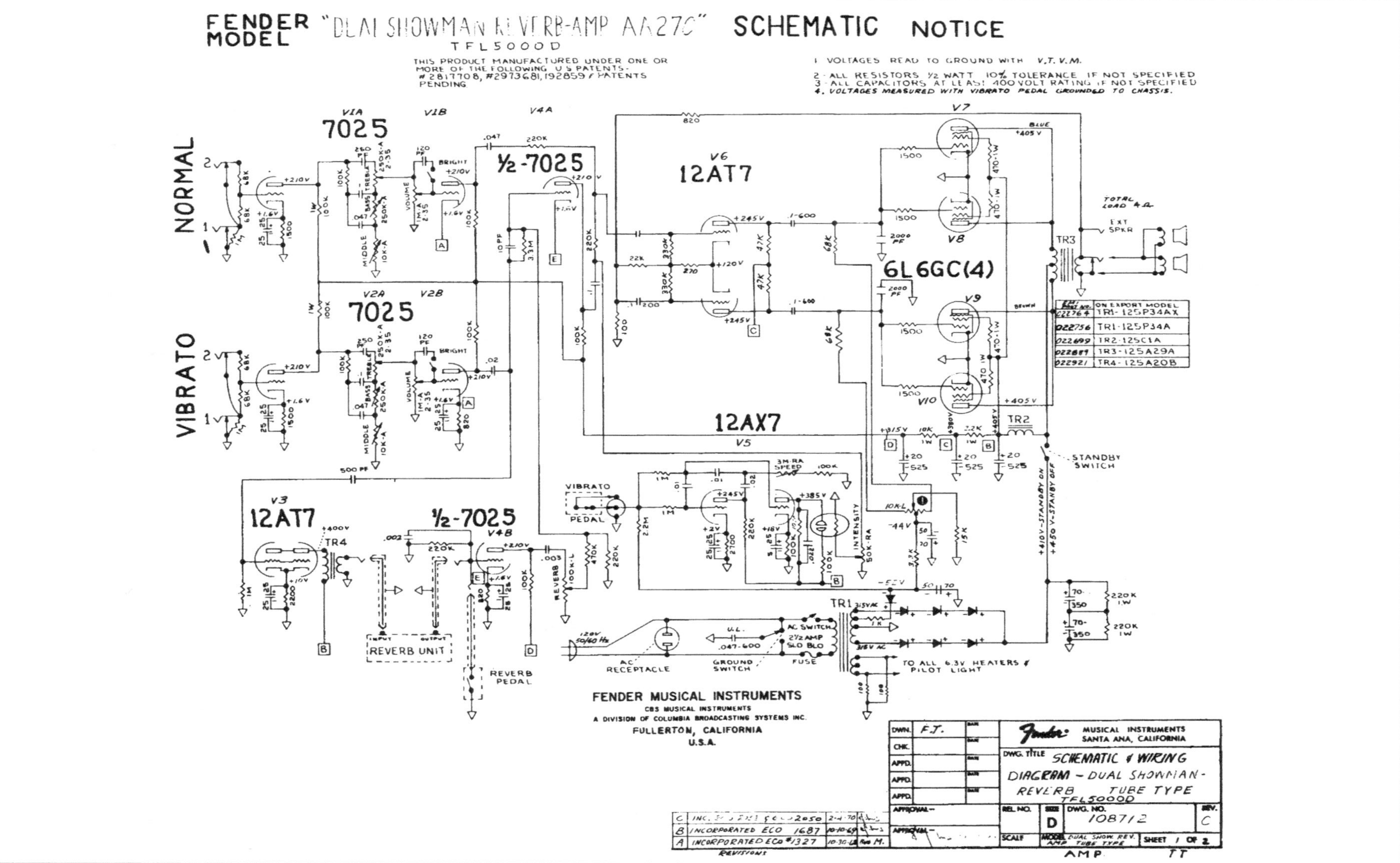
FENDER MODEL "DUAL SHOWMAN REVERB-AMP AA270" SCHEMATIC
TFL5000D
THIS PRODUCT MANUFACTURED UNDER ONE OR MORE OF THE FOLLOWING U.S PATENTS. #2817708, #2973681, 192859 & PATENTS PENDING
NOTICE
1 VOLTAGES READ TO GROUND WITH V.T.V.M.
2 ALL RESISTORS ½ WATT 10% TOLERANCE IF NOT SPECIFIED
3 ALL CAPACITORS AT LEAST 400 VOLT RATING IF NOT SPECIFIED
4. VOLTAGES MEASURED WITH VIBRATO PEDAL GROUNDED TO CHASSIS.
NORMAL
VIBRATO
V1A 7025
V1B
V2A 7025
V2B
V4A ½-7025
V6 12AT7
V5 12AX7
V3 12AT7
V4B ½-7025
6L6GC(4)
V7
V8
V9
V10
TR1
TR2
TR3
TR4
REVERB UNIT
REVERB PEDAL
VIBRATO PEDAL
INTENSITY
SPEED
TREBLE
BASS
MIDDLE
VOLUME
BRIGHT
REVERB
STANDBY SWITCH
+410V-STANDBY ON
+450V-STANDBY OFF
EXT SPKR
TOTAL LOAD 4Ω
AC RECEPTACLE
GROUND SWITCH
AC SWITCH
2½ AMP SLO BLO
FUSE
TO ALL 6.3V HEATERS & PILOT LIGHT
ON EXPORT MODEL
022764 TR1-125P34AX
022756 TR1-125P34A
022699 TR2-125C1A
022889 TR3-125A29A
022921 TR4-125A20B
FENDER MUSICAL INSTRUMENTS
CBS MUSICAL INSTRUMENTS
A DIVISION OF COLUMBIA BROADCASTING SYSTEMS INC.
FULLERTON, CALIFORNIA
U.S.A.
Fender MUSICAL INSTRUMENTS SANTA ANA, CALIFORNIA
DWN. F.J.
DWG. TITLE SCHEMATIC & WIRING DIAGRAM - DUAL SHOWMAN-REVERB TUBE TYPE TFL5000D
SIZE D
DWG. NO. 108712
REV. C
SHEET 1 OF 2
AMP
TT
C INC. ECO 2050
B INCORPORATED ECO 1687
A INCORPORATED ECO #1327
REVISIONS

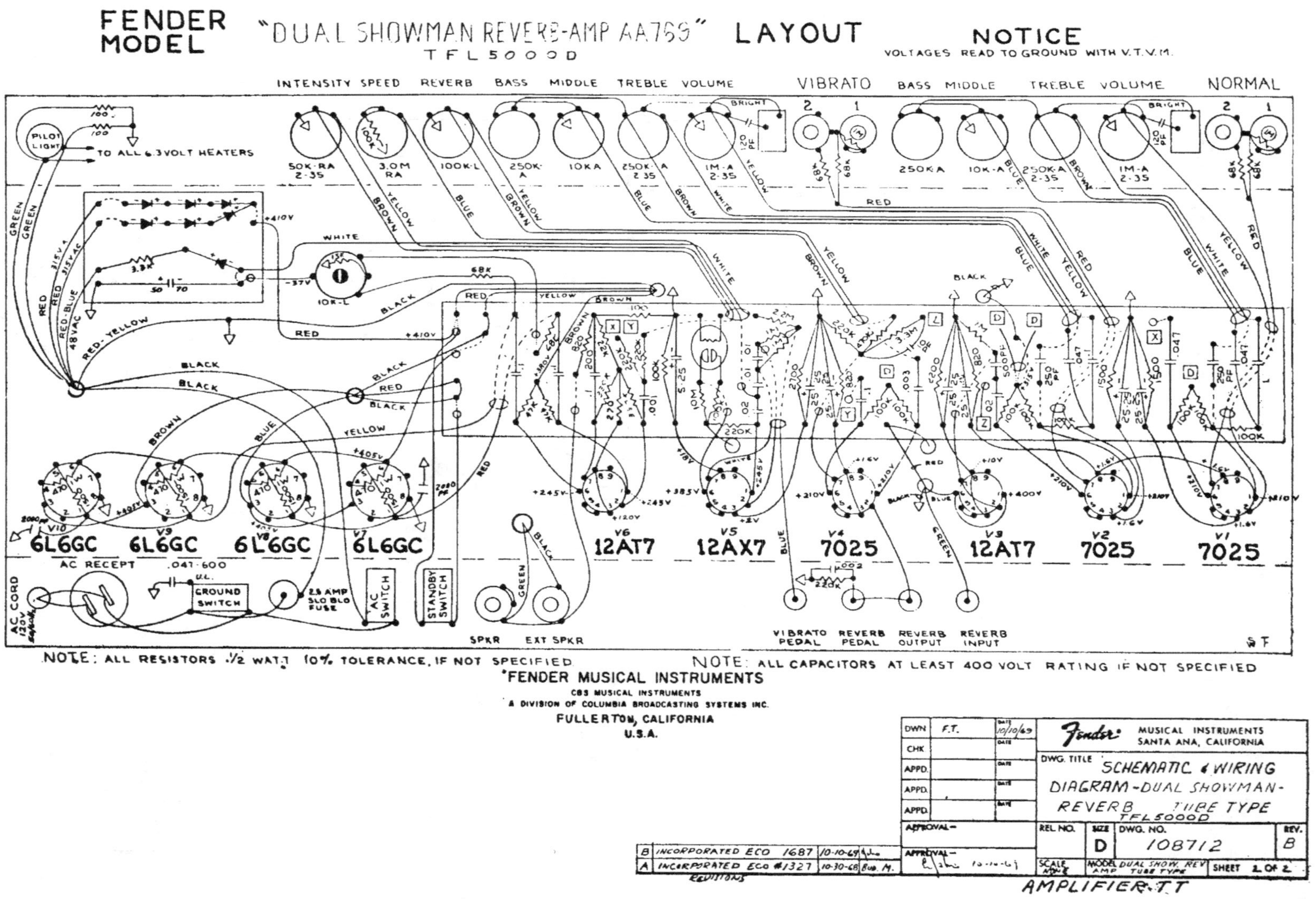
FENDER MODEL "DUAL SHOWMAN REVERB-AMP AA769" LAYOUT
TFL5000D
NOTICE
VOLTAGES READ TO GROUND WITH V.T.V.M.
INTENSITY SPEED REVERB BASS MIDDLE TREBLE VOLUME
VIBRATO
BASS MIDDLE TREBLE VOLUME
NORMAL
PILOT LIGHT
TO ALL 6.3 VOLT HEATERS
50K-RA 2-35
3.0M RA
100K-L
250K-A
10K A
250K-A 2-35
1M-A 2-35
BRIGHT
120 PF
10K-L
6L6GC
6L6GC
6L6GC
6L6GC
12AT7
12AX7
7025
12AT7
7025
7025
V10
V9
V8
V7
V6
V5
V4
V3
V2
V1
AC RECEPT
.047-600
GROUND SWITCH
2.5 AMP SLO BLO FUSE
AC SWITCH
STANDBY SWITCH
AC CORD 120V
SPKR
EXT SPKR
VIBRATO PEDAL
REVERB PEDAL
REVERB OUTPUT
REVERB INPUT
NOTE: ALL RESISTORS 1/2 WATT 10% TOLERANCE, IF NOT SPECIFIED
NOTE: ALL CAPACITORS AT LEAST 400 VOLT RATING IF NOT SPECIFIED
FENDER MUSICAL INSTRUMENTS
CBS MUSICAL INSTRUMENTS
A DIVISION OF COLUMBIA BROADCASTING SYSTEMS INC.
FULLERTON, CALIFORNIA
U.S.A.
Fender MUSICAL INSTRUMENTS SANTA ANA, CALIFORNIA
DWG. TITLE SCHEMATIC & WIRING DIAGRAM - DUAL SHOWMAN-REVERB TUBE TYPE TFL5000D
DWN F.T. 10/10/69
DWG. NO. 108712
SIZE D
REV. B
SHEET 2 OF 2
B INCORPORATED ECO 1687 10-10-69
A INCORPORATED ECO #1327 10-30-68
REVISIONS
AMPLIFIER-T.T

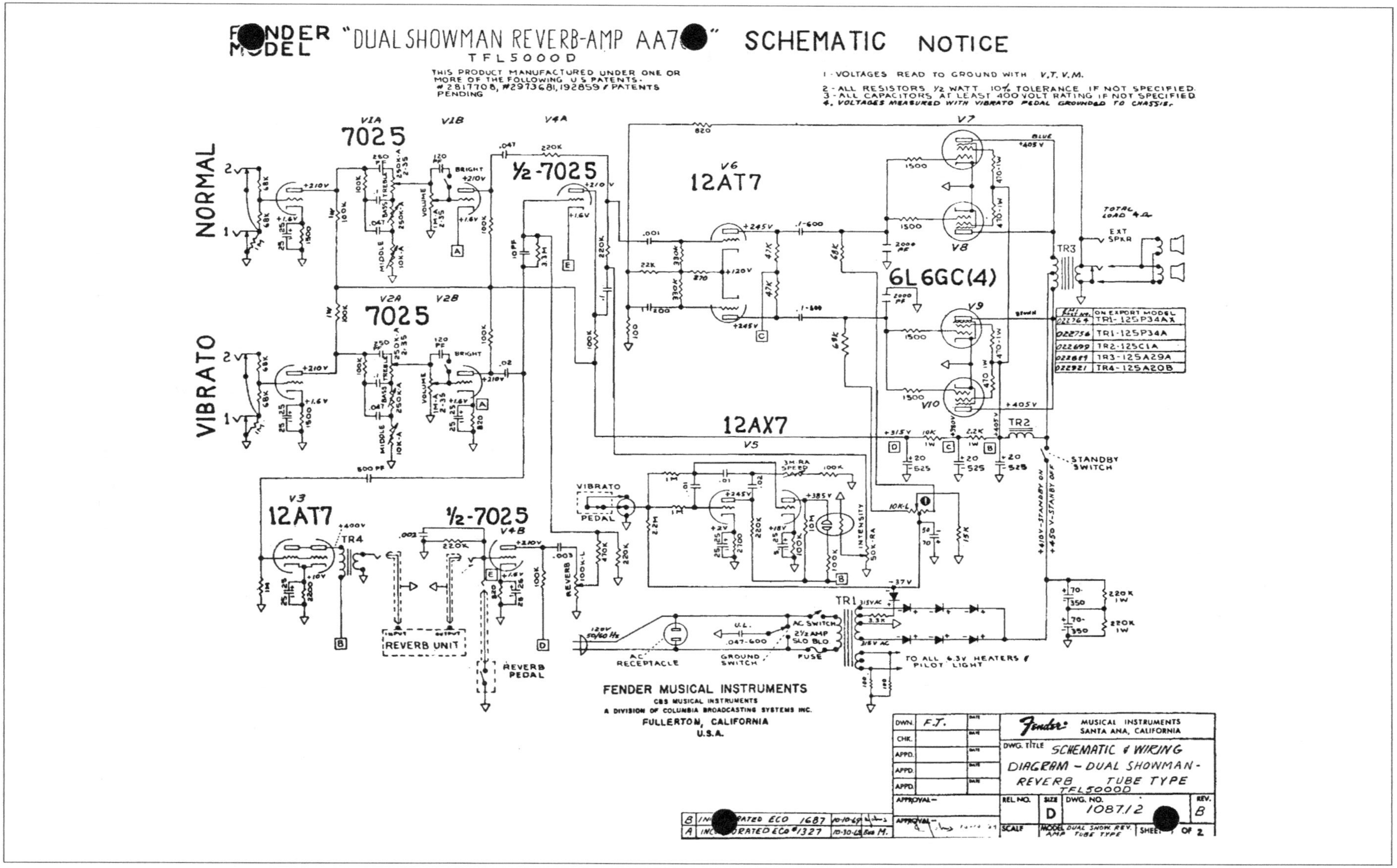
FENDER MODEL "DUAL SHOWMAN REVERB-AMP AA7 " SCHEMATIC
TFL5000D
THIS PRODUCT MANUFACTURED UNDER ONE OR MORE OF THE FOLLOWING U.S. PATENTS- #2817708, #2973681, 192859 & PATENTS PENDING
NOTICE
1 - VOLTAGES READ TO GROUND WITH V.T.V.M.
2 - ALL RESISTORS 1/2 WATT 10% TOLERANCE IF NOT SPECIFIED.
3 - ALL CAPACITORS AT LEAST 400 VOLT RATING IF NOT SPECIFIED
4. VOLTAGES MEASURED WITH VIBRATO PEDAL GROUNDED TO CHASSIS.
NORMAL
VIBRATO
V1A 7025
V1B
V2A V2B 7025
V4A 1/2-7025
V6 12AT7
V7
V8
V9
V10
6L6GC(4)
12AX7
V5
V3 12AT7
TR4
1/2-7025 V4B
REVERB UNIT
REVERB PEDAL
VIBRATO PEDAL
INTENSITY
SPEED
TR1
TR2
TR3
STANDBY SWITCH
+610V-STANDBY ON
+450V-STANDBY OFF
TOTAL LOAD 4Ω
EXT SPKR
AC RECEPTACLE
GROUND SWITCH
FUSE
AC SWITCH
2 1/2 AMP SLO BLO
TO ALL 6.3V HEATERS & PILOT LIGHT
ON EXPORT MODEL
022764 TR1-125P34AX
022756 TR1-125P34A
022699 TR2-125C1A
022689 TR3-125A29A
022921 TR4-125A20B
FENDER MUSICAL INSTRUMENTS
CBS MUSICAL INSTRUMENTS
A DIVISION OF COLUMBIA BROADCASTING SYSTEMS INC.
FULLERTON, CALIFORNIA
U.S.A.
Fender MUSICAL INSTRUMENTS SANTA ANA, CALIFORNIA
DWN. F.T.
DWG. TITLE SCHEMATIC & WIRING DIAGRAM - DUAL SHOWMAN-REVERB TUBE TYPE TFL5000D
SIZE D
DWG. NO. 108712
REV. B
OF 2
INCORPORATED ECO 1687 10-10-69
INCORPORATED ECO #1327 10-30-68

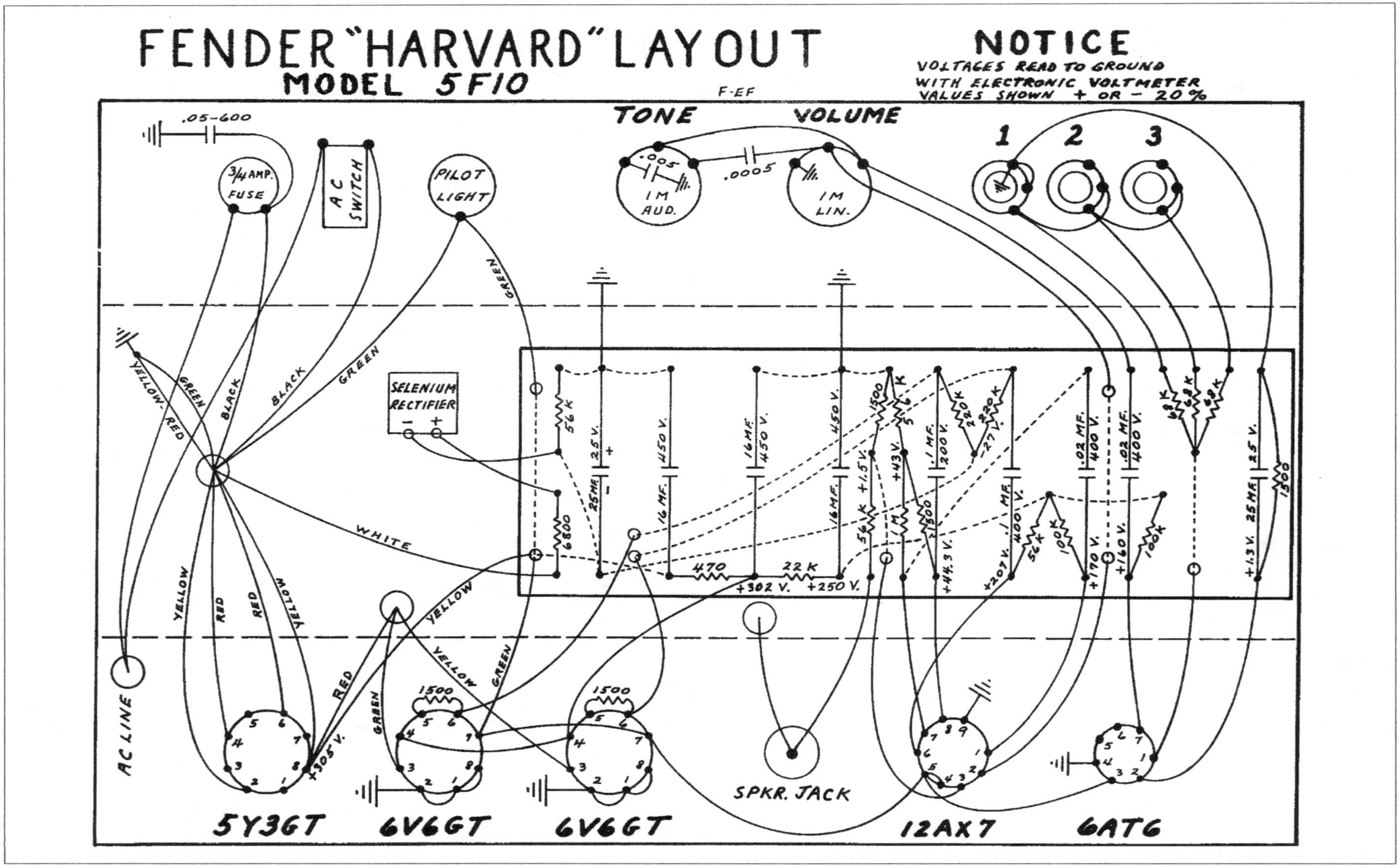
FENDER "HARVARD" LAYOUT
MODEL 5F10
F-EF
NOTICE
VOLTAGES READ TO GROUND
WITH ELECTRONIC VOLTMETER
VALUES SHOWN + OR - 20%
TONE
VOLUME
1
2
3
.05-600
3/4 AMP. FUSE
AC SWITCH
PILOT LIGHT
.005
1M AUD.
.0005
1M LIN.
SELENIUM RECTIFIER
WHITE
GREEN
BLACK
YELLOW-RED
YELLOW
RED
470
22K
+302 V.
+250 V.
AC LINE
SPKR. JACK
1500
+305 V.
5Y3GT
6V6GT
6V6GT
12AX7
6AT6

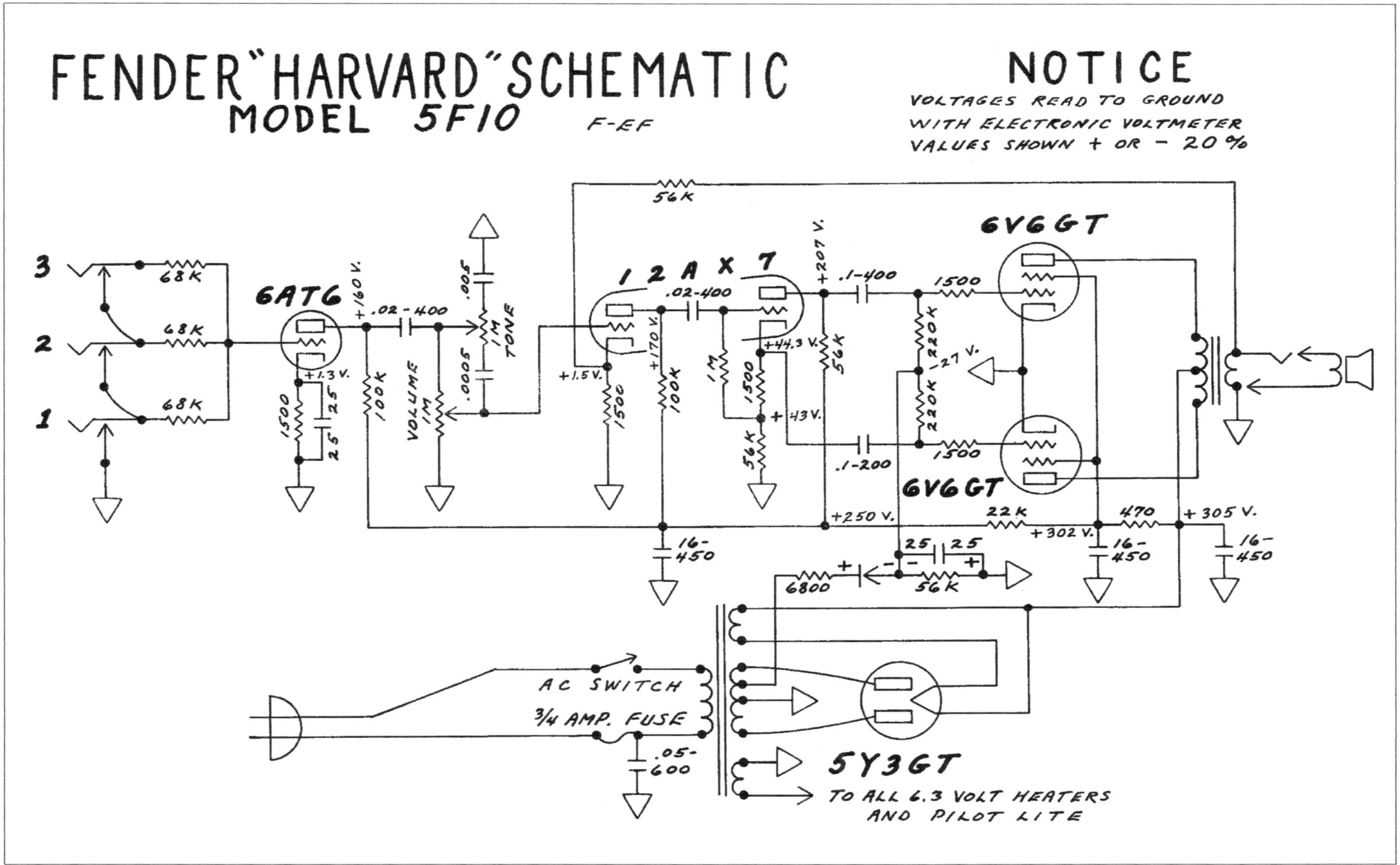
FENDER "HARVARD" SCHEMATIC
MODEL 5F10
F-EF
NOTICE
VOLTAGES READ TO GROUND
WITH ELECTRONIC VOLTMETER
VALUES SHOWN + OR − 20%
3
2
1
68K
68K
68K
6AT6
+1.3 V.
1500
25
25
+160 V.
100K
.02-400
VOLUME
1M
.005
.0005
1M
TONE
56K
12AX7
+1.5 V.
1500
.02-400
+170 V.
100K
1M
+44.3 V.
1500
+43 V.
56K
+207 V.
56K
.1-400
.1-200
1500
220K
-27 V.
220K
1500
6V6GT
6V6GT
+250 V.
22K
+302 V.
470
+305 V.
16-
450
16-
450
16-
450
25
25
6800
56K
AC SWITCH
3/4 AMP. FUSE
.05-
600
5Y3GT
TO ALL 6.3 VOLT HEATERS
AND PILOT LITE

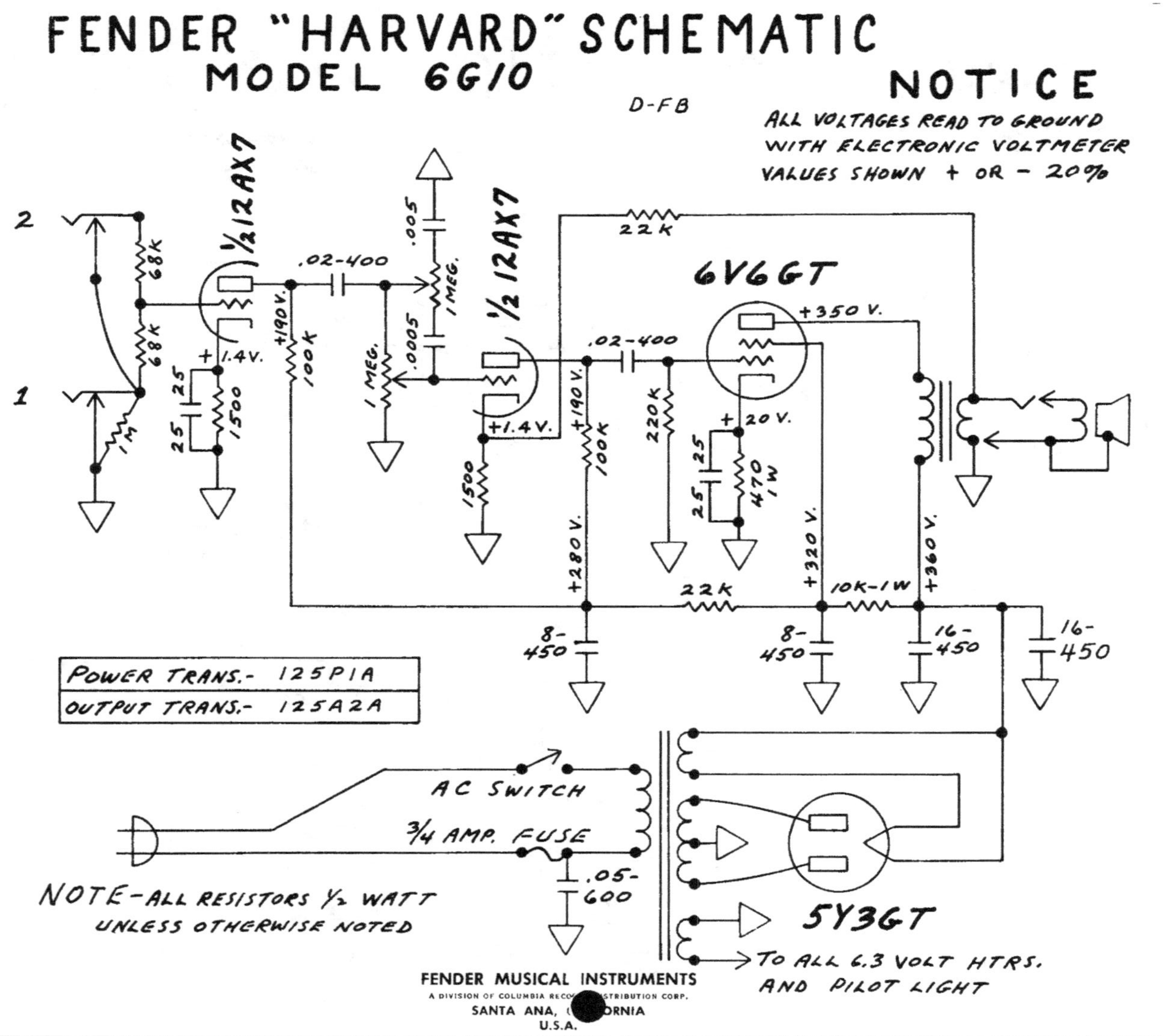
FENDER "HARVARD" SCHEMATIC
MODEL 6G10
D-FB
NOTICE
ALL VOLTAGES READ TO GROUND
WITH ELECTRONIC VOLTMETER
VALUES SHOWN + OR – 20%
2
1
68K
68K
1M
½ 12AX7
+1.4V.
25
25
1500
.02-400
+190V.
100K
.005
1 MEG.
.0005
1 MEG.
½ 12AX7
+1.4V.
1500
22K
6V6GT
+350 V.
.02-400
+190V.
100K
220K
+20 V.
25
25
470
1W
+280 V.
22K
+320 V.
10K-1W
+360 V.
8-
450
8-
450
16-
450
16-
450
POWER TRANS.- 125P1A
OUTPUT TRANS.- 125A2A
AC SWITCH
¾ AMP. FUSE
.05-
600
NOTE-ALL RESISTORS ½ WATT
UNLESS OTHERWISE NOTED
5Y3GT
TO ALL 6.3 VOLT HTRS.
AND PILOT LIGHT
FENDER MUSICAL INSTRUMENTS
SANTA ANA,
U.S.A.

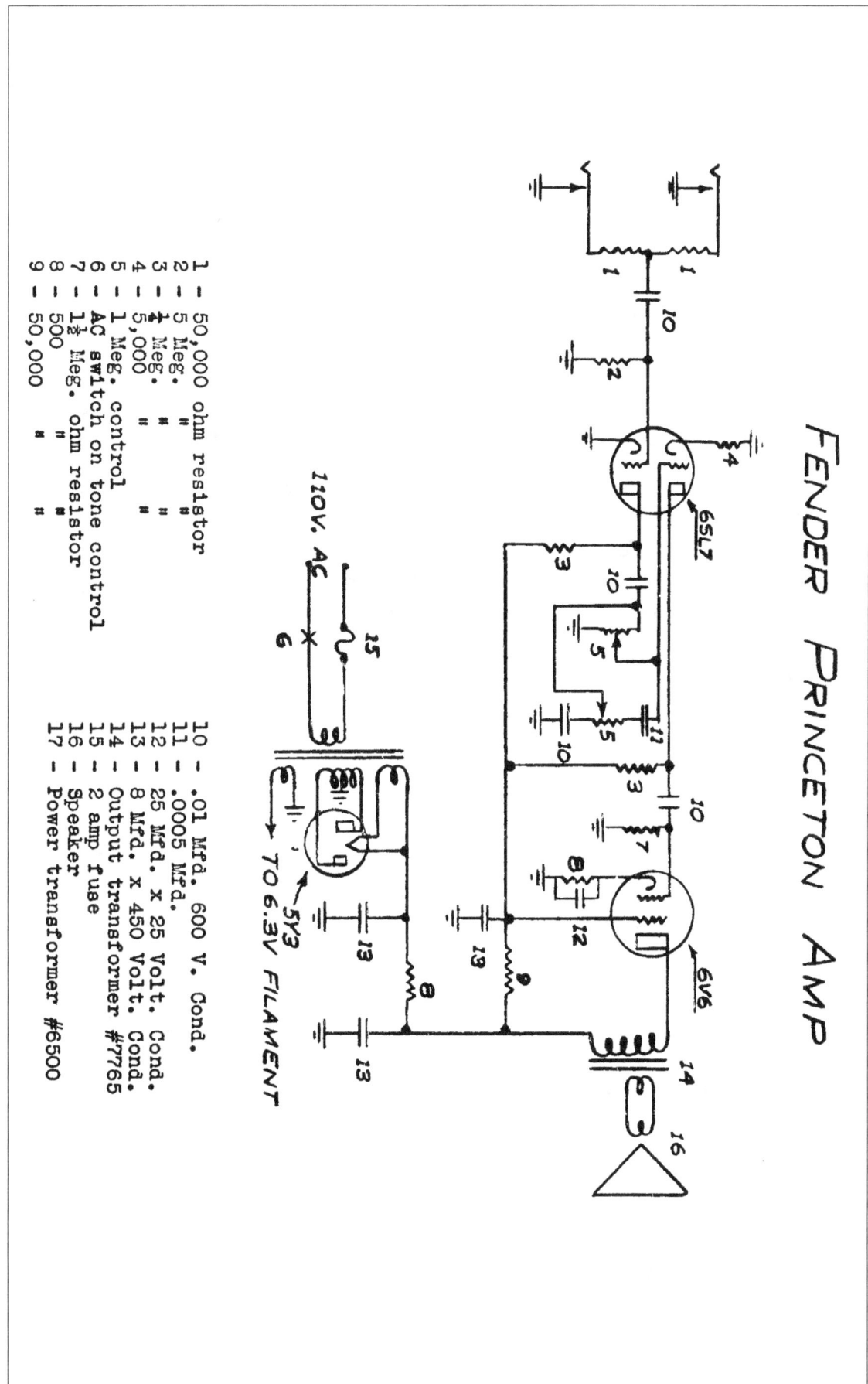

FENDER PRINCETON AMP

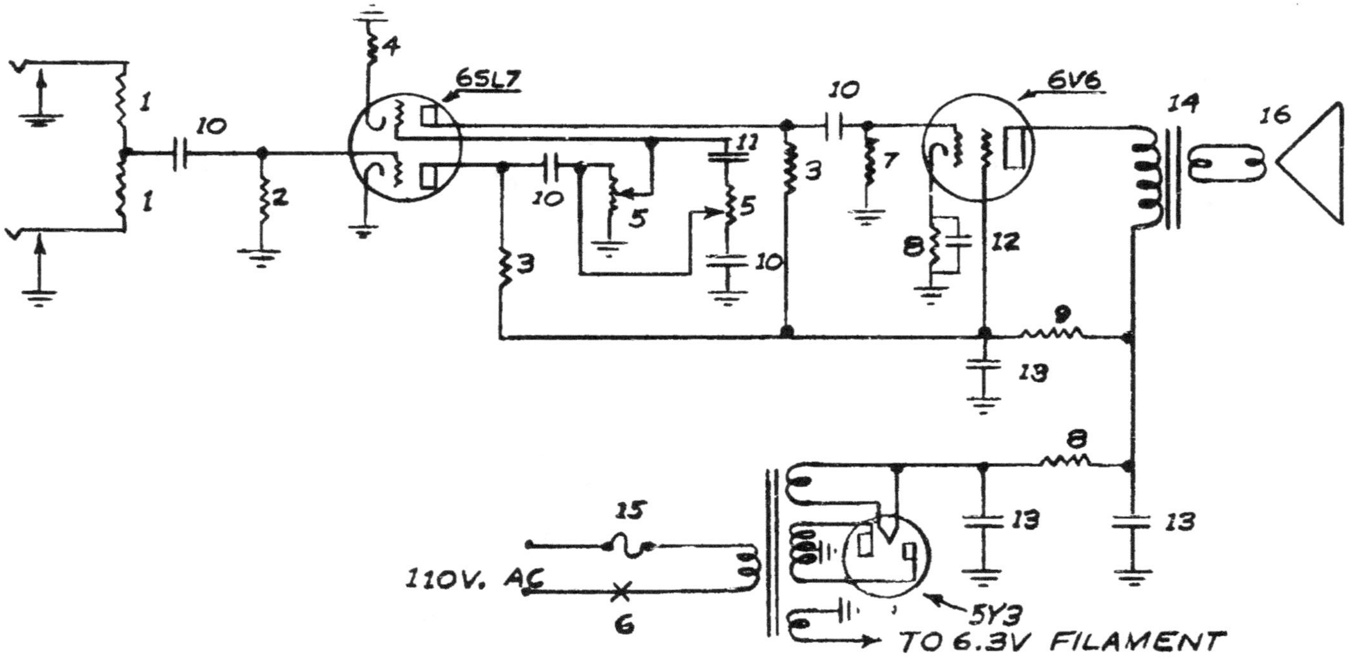

1 - 50,000 ohm resistor
2 - 5 Meg. " "
3 - ¼ Meg. " "
4 - 5,000 " "
5 - 1 Meg. control
6 - AC switch on tone control
7 - 1½ Meg. ohm resistor
8 - 500 " "
9 - 50,000 " "

10 - .01 Mfd. 600 V. Cond.
11 - .0005 Mfd.
12 - 25 Mfd. x 25 Volt. Cond.
13 - 8 Mfd. x 450 Volt. Cond.
14 - Output transformer #7765
15 - 2 amp fuse
16 - Speaker
17 - Power transformer #6500

5B2

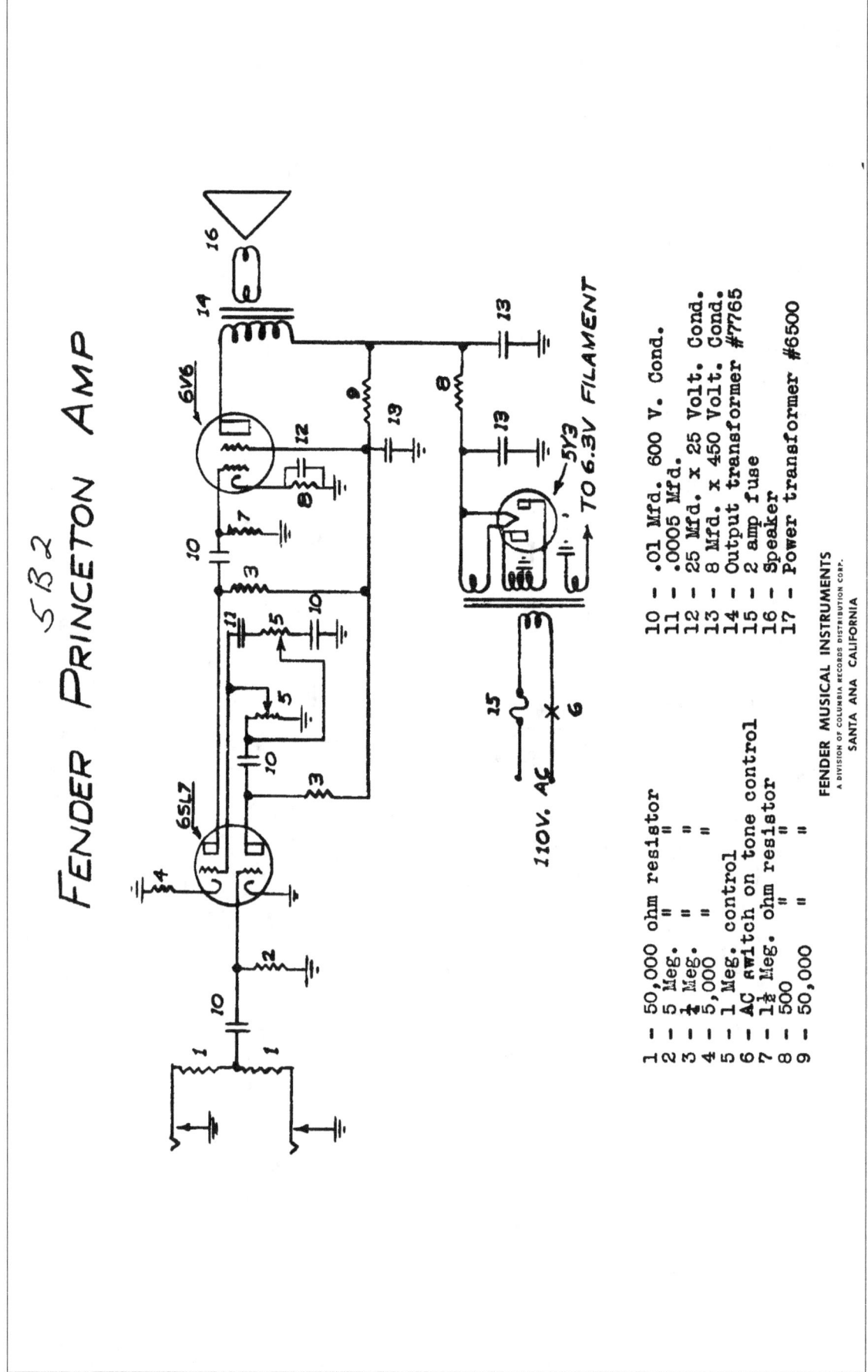

FENDER PRINCETON AMP

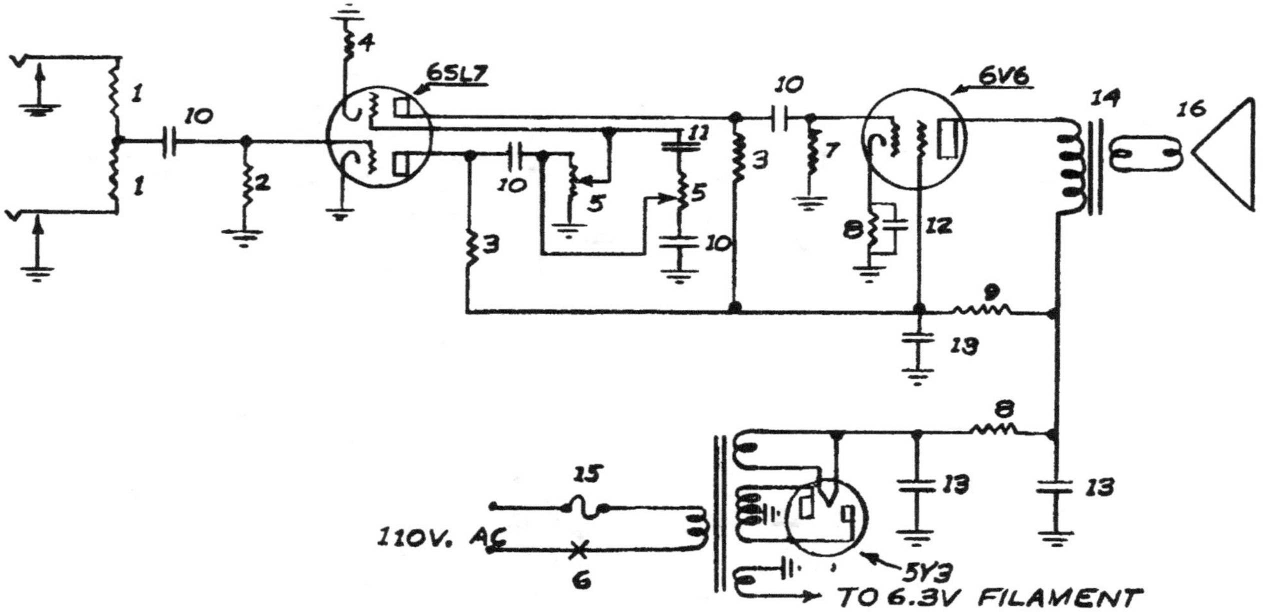

1 - 50,000 ohm resistor
2 - 5 Meg. " "
3 - ¼ Meg. " "
4 - 5,000 " "
5 - 1 Meg. control
6 - AC switch on tone control
7 - 1½ Meg. ohm resistor
8 - 500 " "
9 - 50,000 " "

10 - .01 Mfd. 600 V. Cond.
11 - .0005 Mfd.
12 - 25 Mfd. x 25 Volt. Cond.
13 - 8 Mfd. x 450 Volt. Cond.
14 - Output transformer #7765
15 - 2 amp fuse
16 - Speaker
17 - Power transformer #6500

FENDER MUSICAL INSTRUMENTS
A DIVISION OF COLUMBIA RECORDS DISTRIBUTION CORP.
SANTA ANA CALIFORNIA

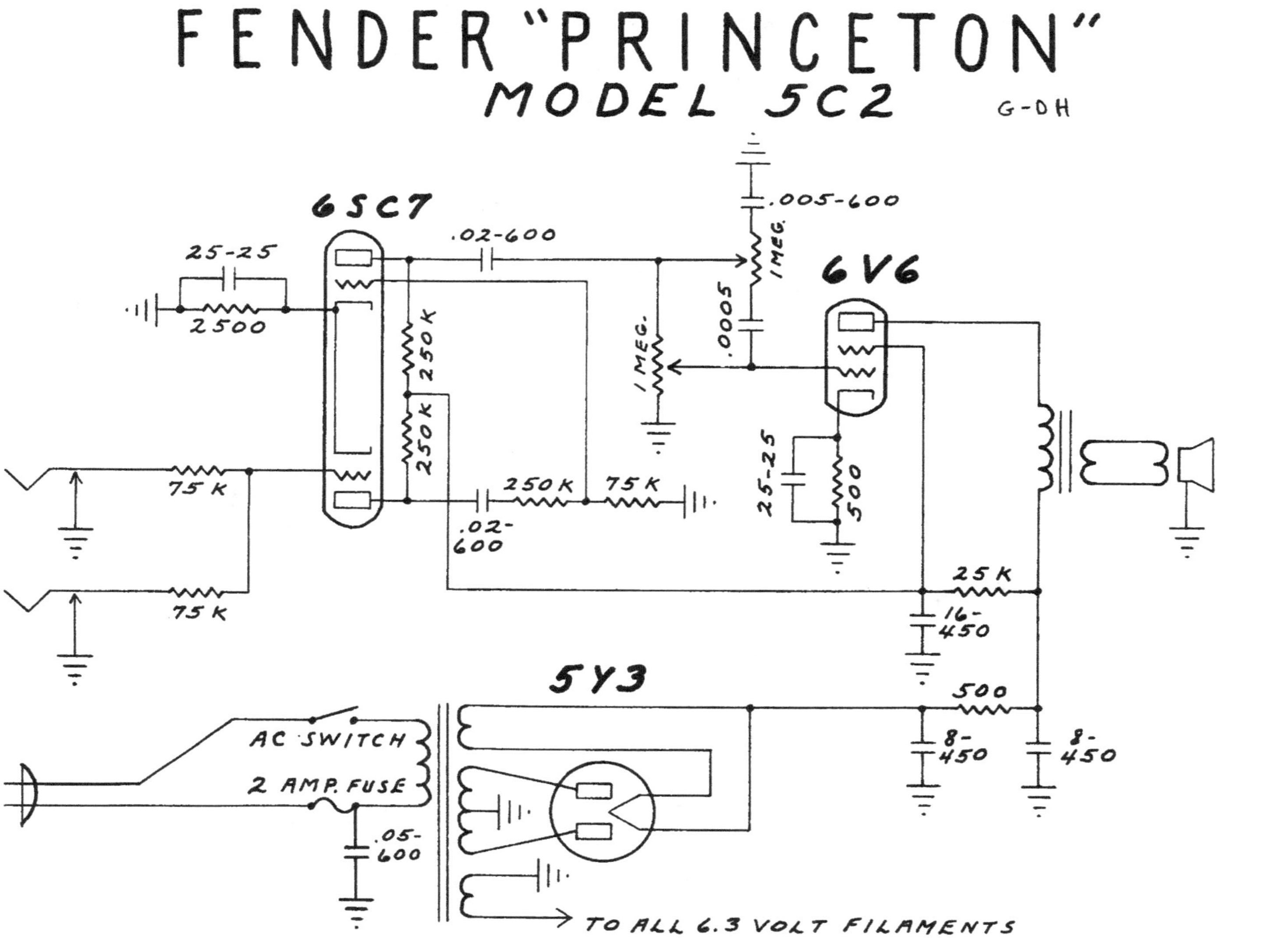
FENDER "PRINCETON"
MODEL 5C2
G-DH
6SC7
6V6
5Y3
25-25
2500
.02-600
250K
250K
250K
75K
75K
75K
.02-600
1 MEG.
1 MEG.
.005-600
.0005
25-25
500
25K
16-450
500
8-450
8-450
AC SWITCH
2 AMP. FUSE
.05-600
TO ALL 6.3 VOLT FILAMENTS

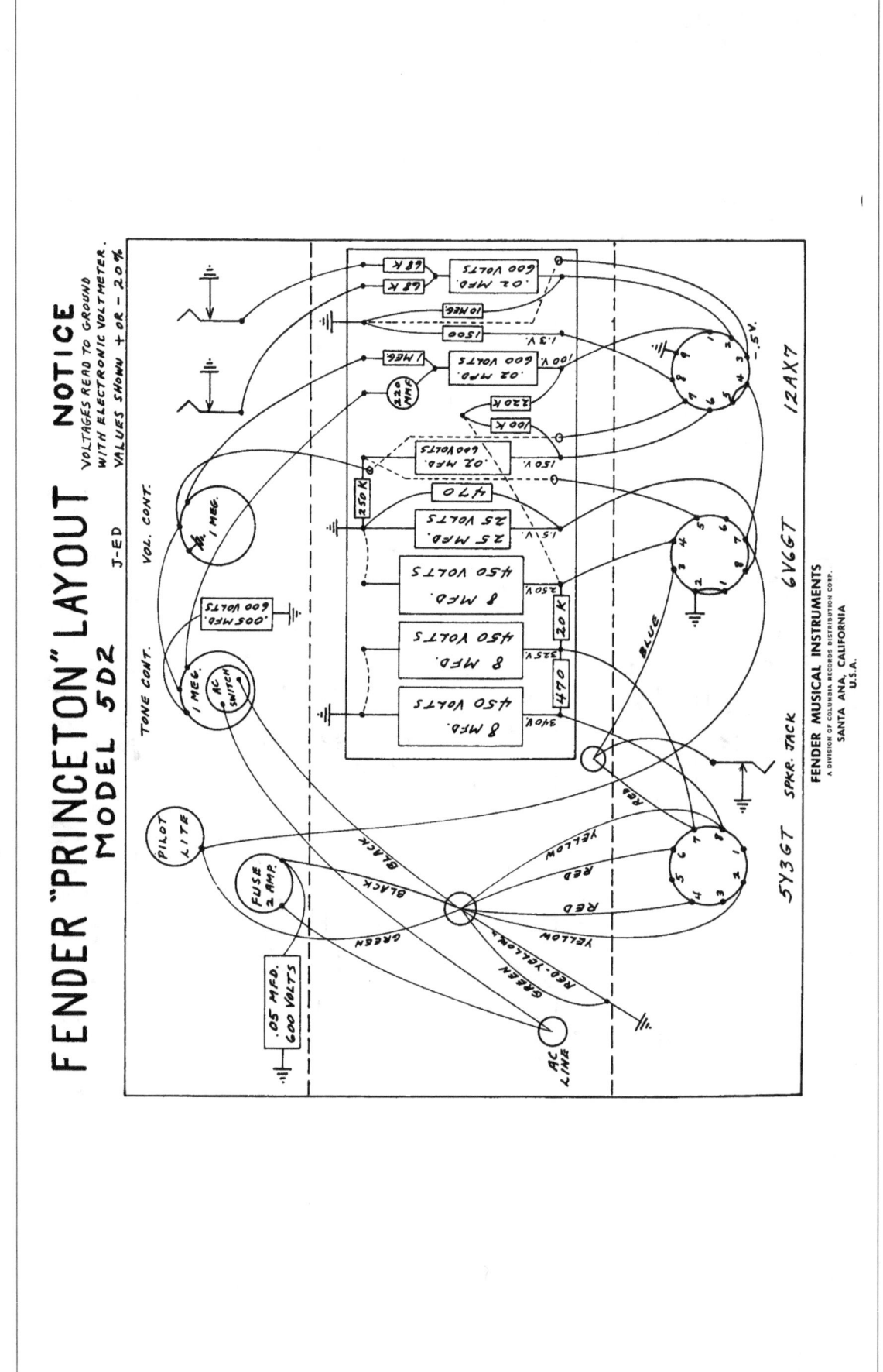

FENDER "PRINCETON" LAYOUT
MODEL 5D2

NOTICE
VOLTAGES READ TO GROUND WITH ELECTRONIC VOLTMETER.
VALUES SHOWN + OR - 20%

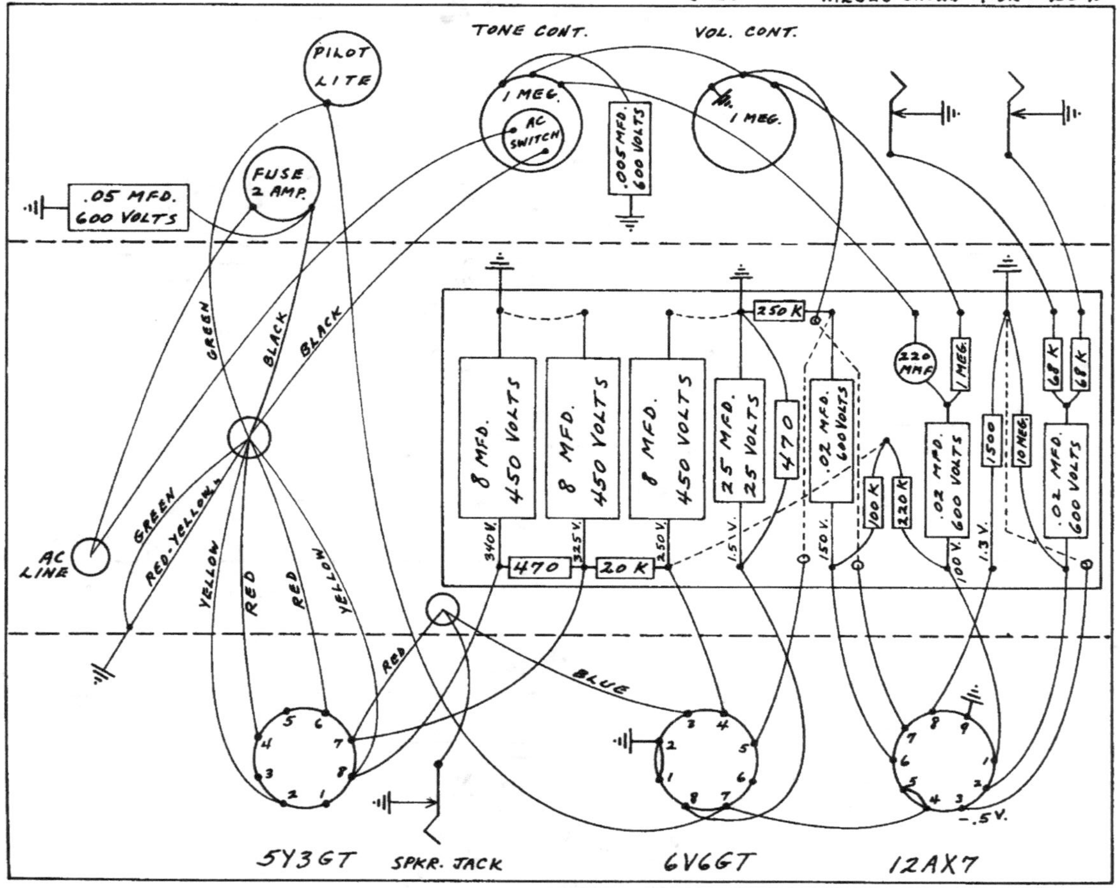

FENDER MUSICAL INSTRUMENTS
A DIVISION OF COLUMBIA RECORDS DISTRIBUTION CORP.
SANTA ANA, CALIFORNIA
U.S.A.

FENDER "PRINCETON" SCHEMATIC

MODEL 5D2

J-ED

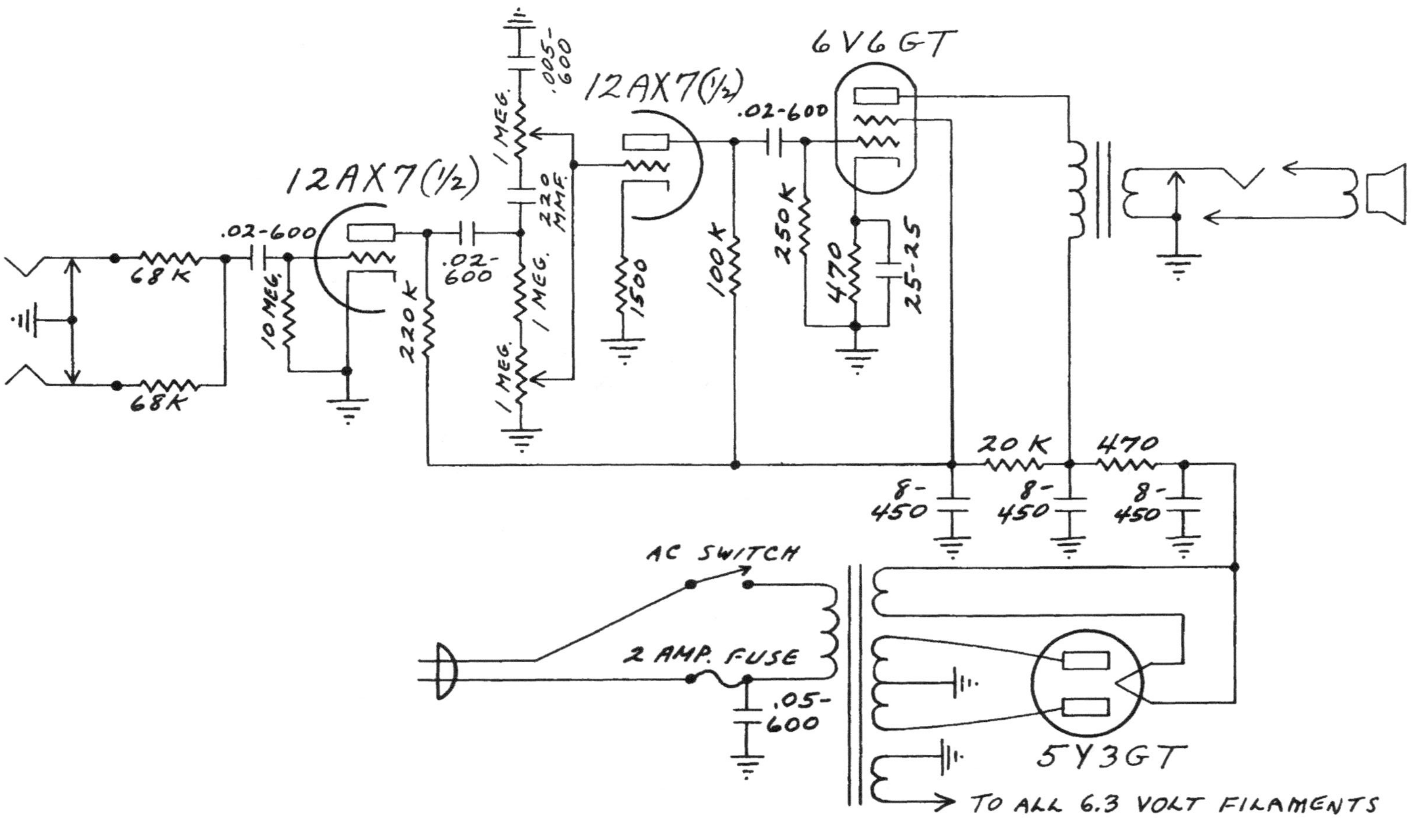

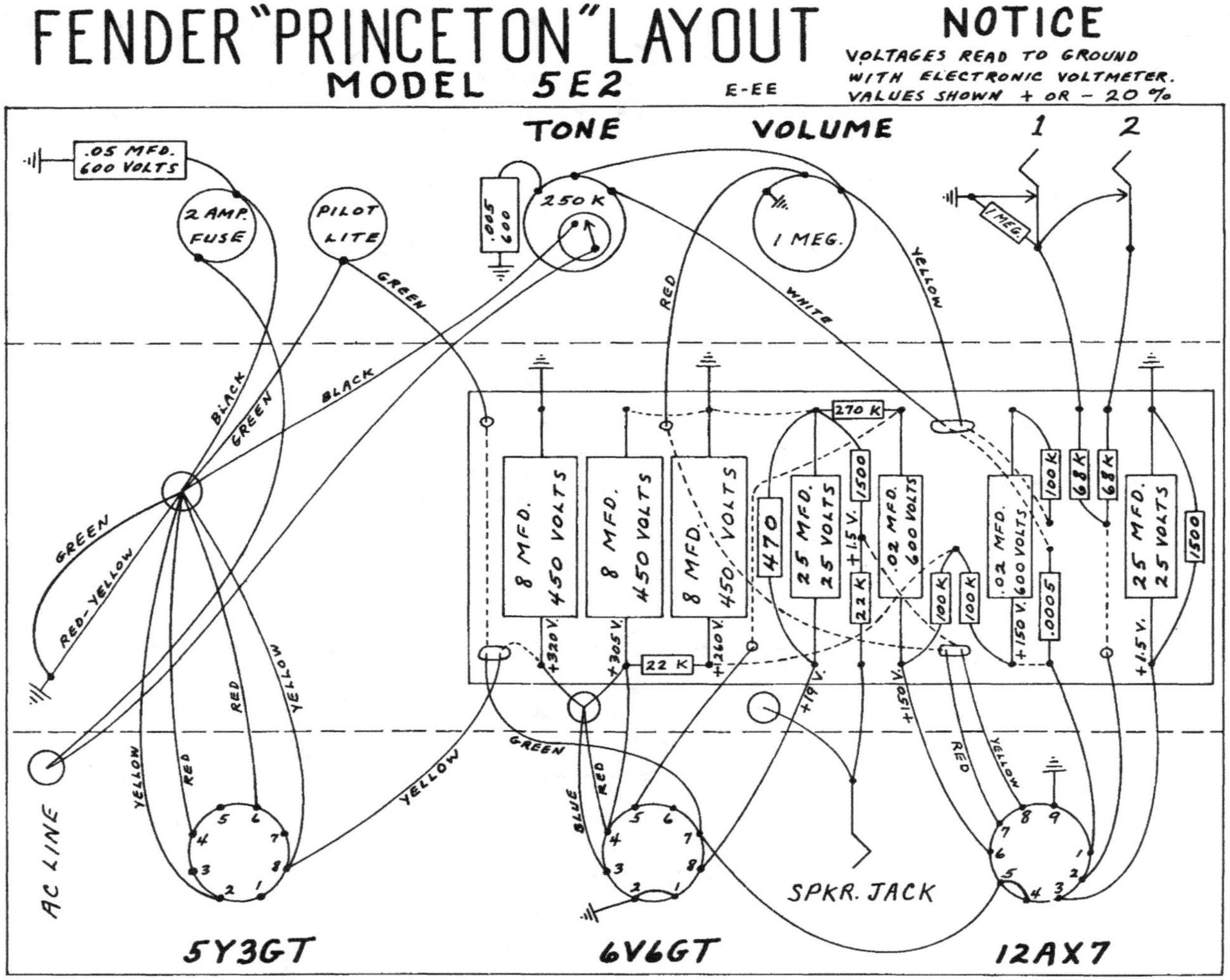

FENDER MUSICAL INSTRUMENTS
A DIVISION OF COLUMBIA RECORDS DISTRIBUTION CORP.
SANTA ANA, CALIFORNIA
U.S.A.

FENDER "PRINCETON" SCHEMATIC

MODEL 5E2

E-EE

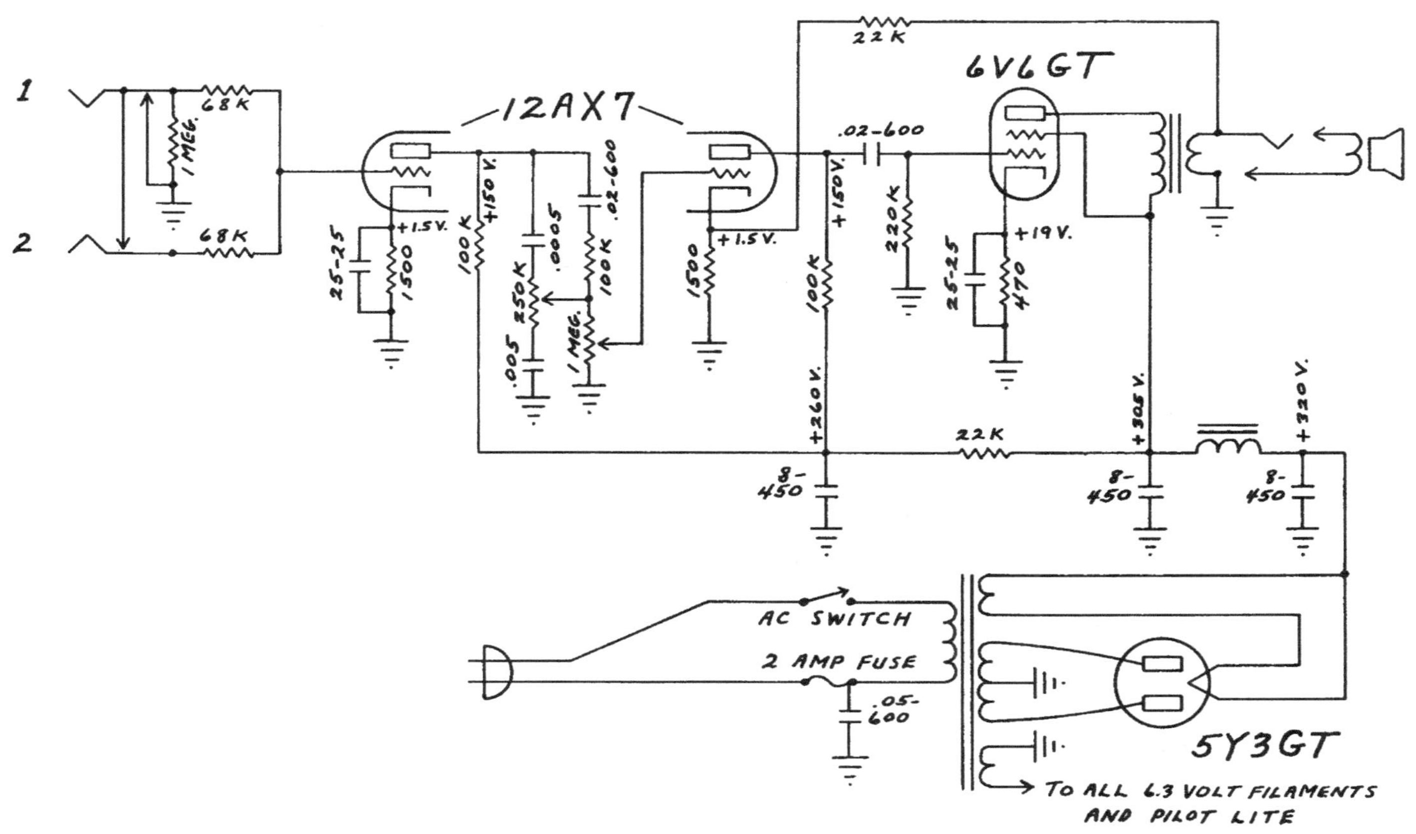

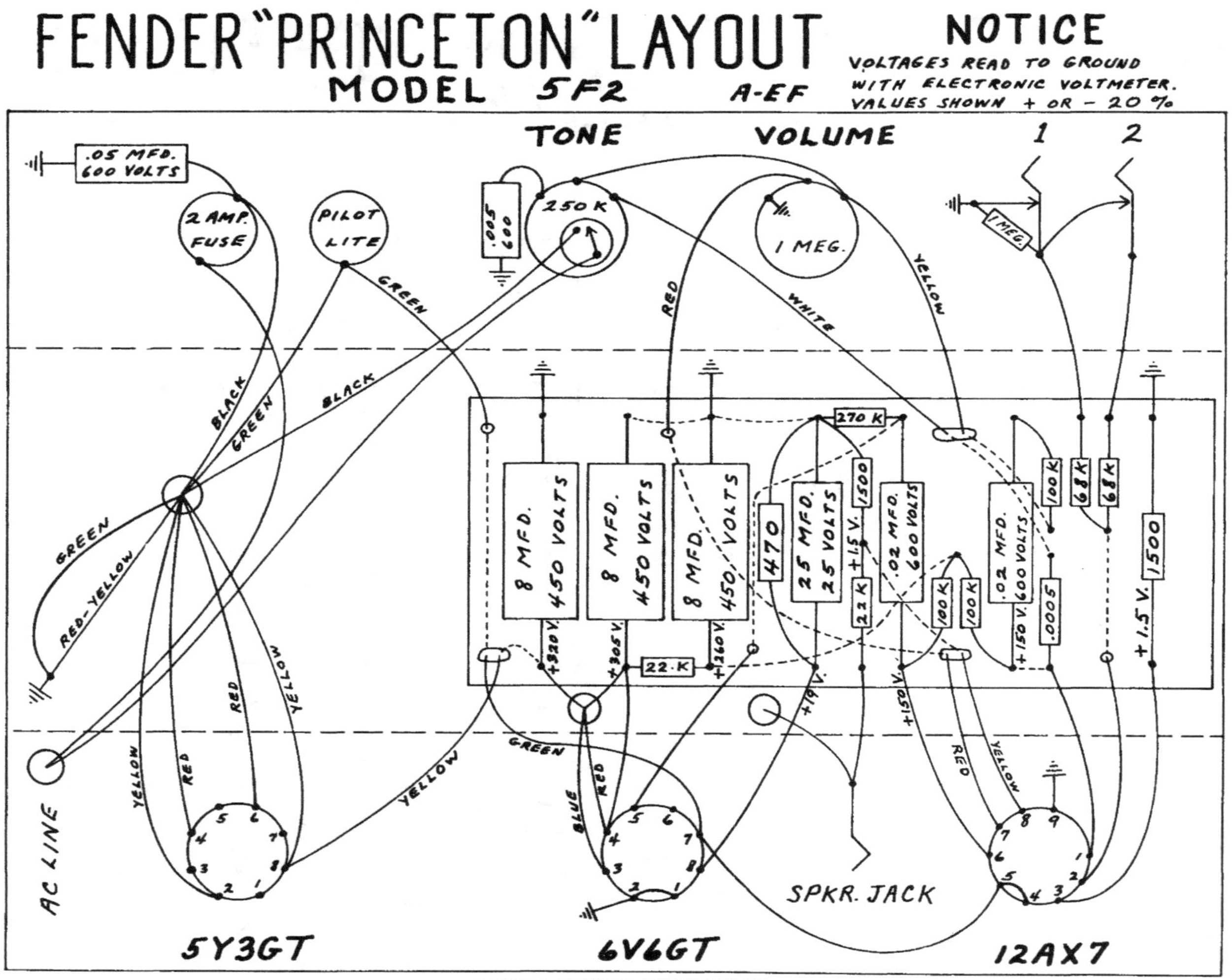

FENDER MUSICAL INSTRUMENTS
A DIVISION OF COLUMBIA RECORDS DISTRIBUTION CORP.
SANTA ANA, CALIFORNIA
U.S.A.

FENDER "PRINCETON" SCHEMATIC

MODEL 5F2

A-EF

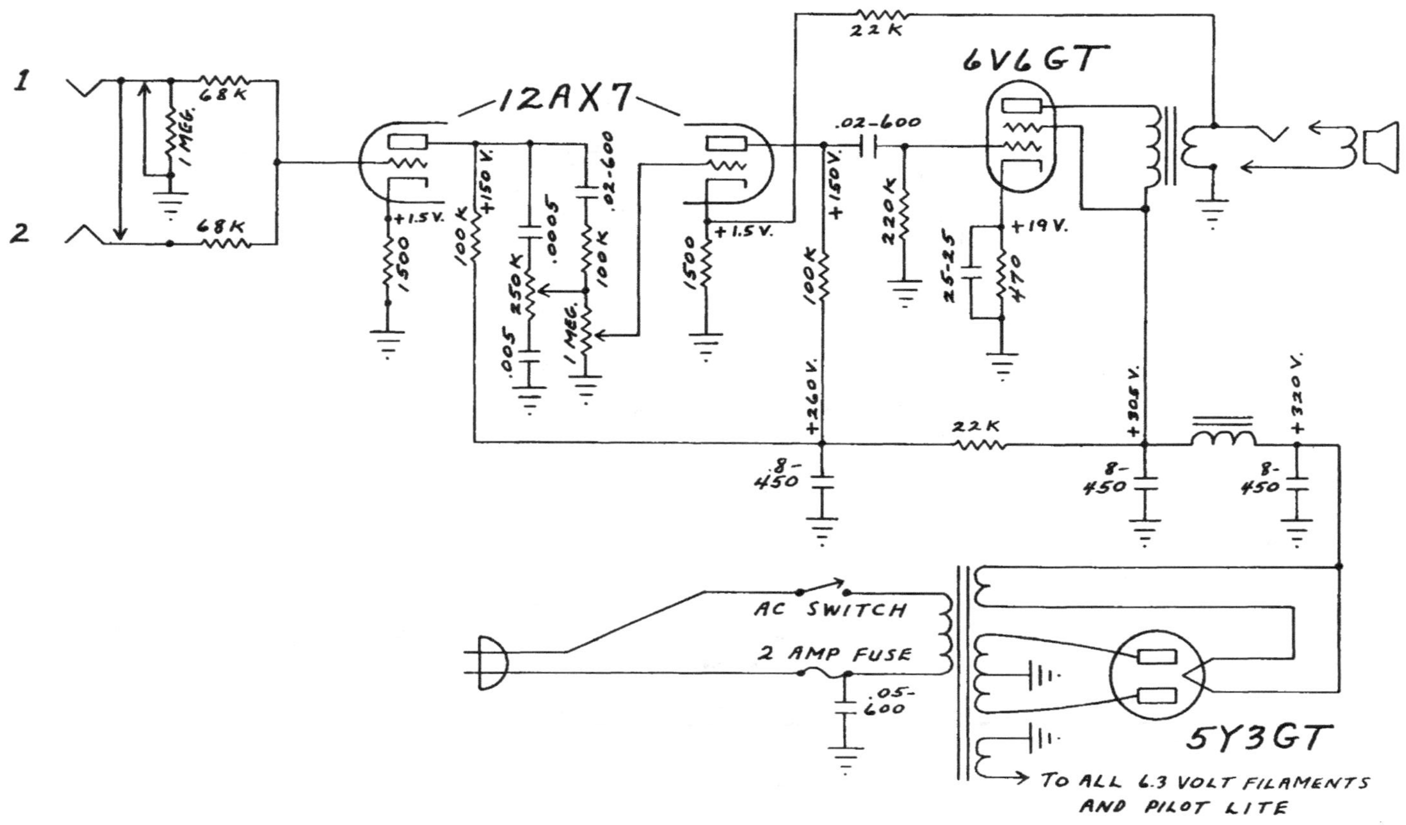

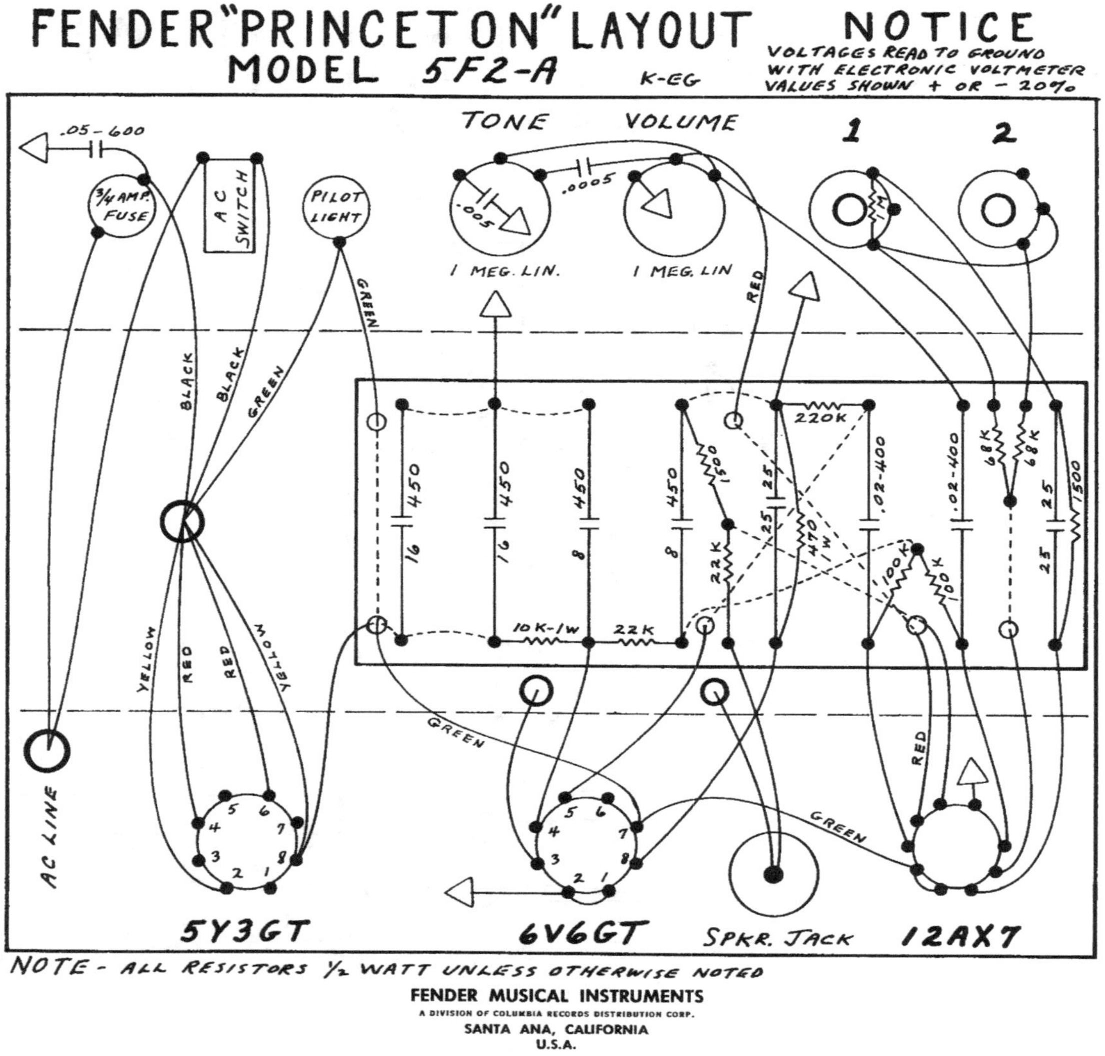
FENDER "PRINCETON" LAYOUT
MODEL 5F2-A
K-EG
NOTICE
VOLTAGES READ TO GROUND
WITH ELECTRONIC VOLTMETER
VALUES SHOWN + OR - 20%
.05-600
3/4 AMP. FUSE
AC SWITCH
PILOT LIGHT
TONE
VOLUME
1
2
.005
.0005
1 MEG. LIN.
1 MEG. LIN
RED
GREEN
BLACK
BLACK
GREEN
220K
1500
25
25
470 W
.02-400
.02-400
68K
68K
25
25
1500
450
16
450
16
450
8
450
8
22K
100K
100K
10K-1W
22K
YELLOW
RED
RED
YELLOW
GREEN
RED
GREEN
AC LINE
5Y3GT
6V6GT
SPKR. JACK
12AX7
NOTE - ALL RESISTORS 1/2 WATT UNLESS OTHERWISE NOTED
FENDER MUSICAL INSTRUMENTS
A DIVISION OF COLUMBIA RECORDS DISTRIBUTION CORP.
SANTA ANA, CALIFORNIA
U.S.A.

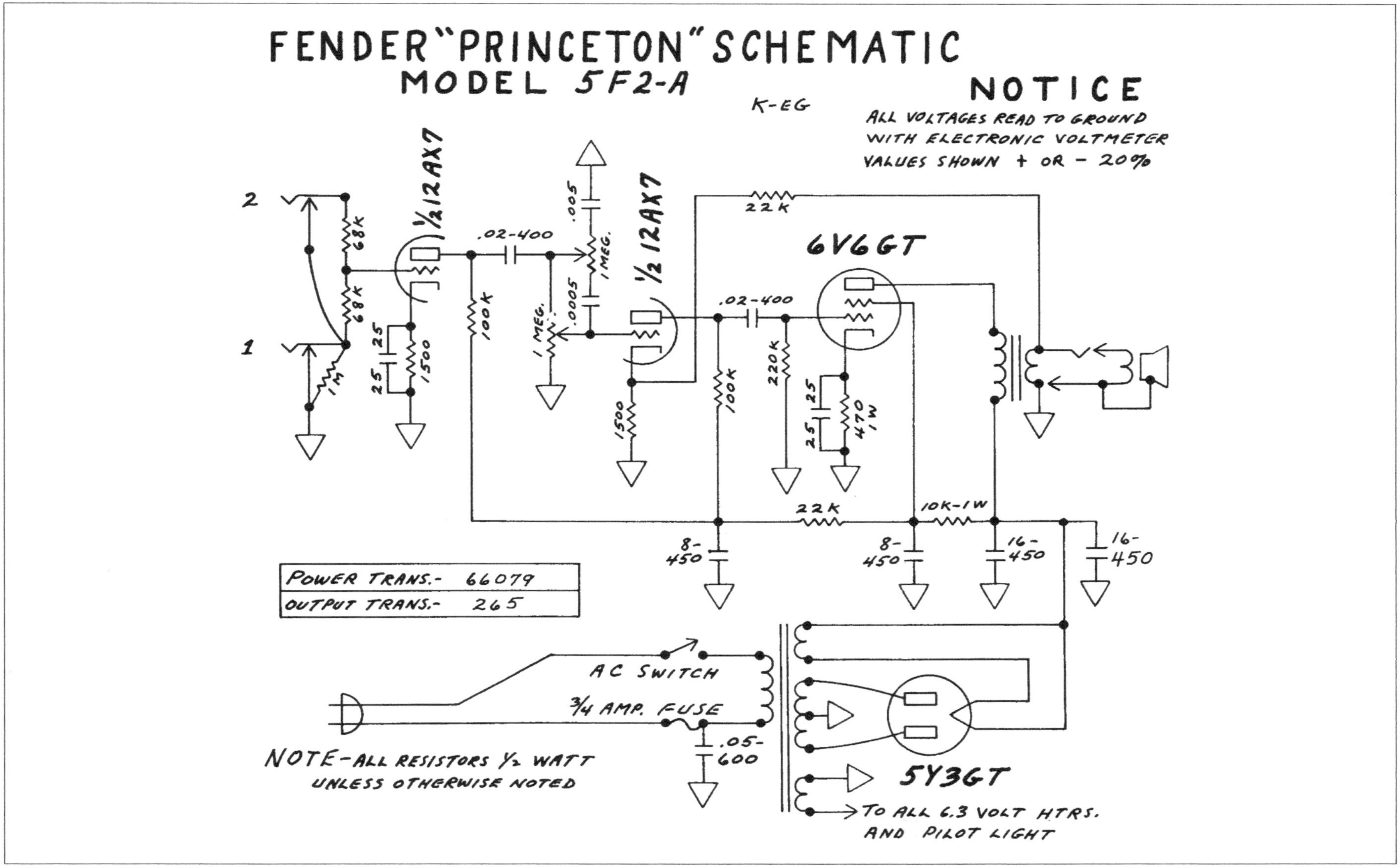
FENDER "PRINCETON" SCHEMATIC
MODEL 5F2-A
K-EG
NOTICE
ALL VOLTAGES READ TO GROUND
WITH ELECTRONIC VOLTMETER
VALUES SHOWN + OR - 20%
2
1
68K
68K
1M
½12AX7
25
25
1500
.02-400
100K
.005
1 MEG.
.0005
1 MEG.
½ 12AX7
1500
22K
6V6GT
.02-400
100K
220K
25
25
470
1W
22K
10K-1W
8-
450
8-
450
16-
450
16-
450
POWER TRANS.- 66079
OUTPUT TRANS.- 265
AC SWITCH
¾ AMP. FUSE
.05-
600
NOTE-ALL RESISTORS ½ WATT
UNLESS OTHERWISE NOTED
5Y3GT
TO ALL 6.3 VOLT HTRS.
AND PILOT LIGHT

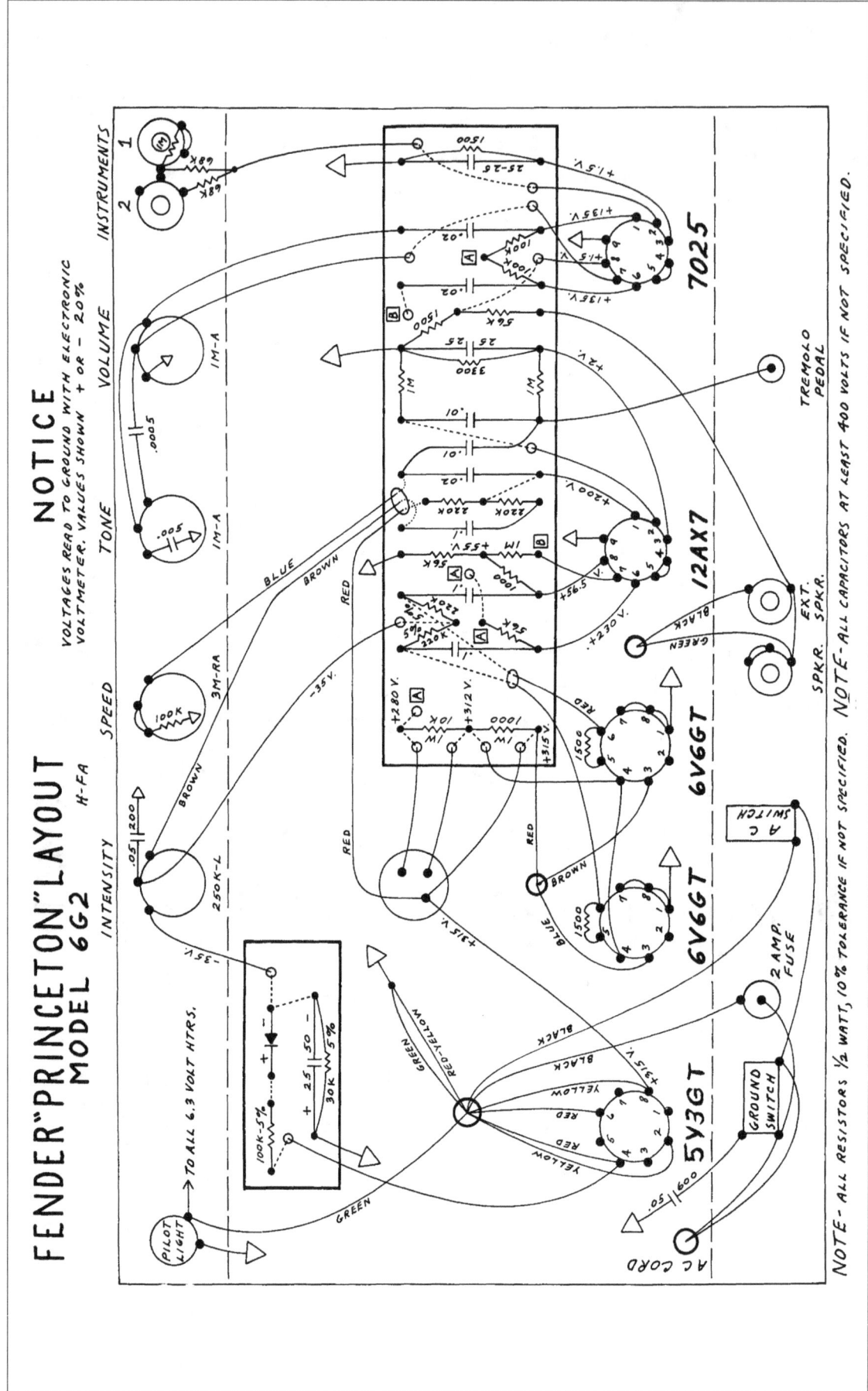

FENDER "PRINCETON" LAYOUT
MODEL 6G2

H-FA

NOTICE

VOLTAGES READ TO GROUND WITH ELECTRONIC VOLTMETER. VALUES SHOWN + OR − 20%

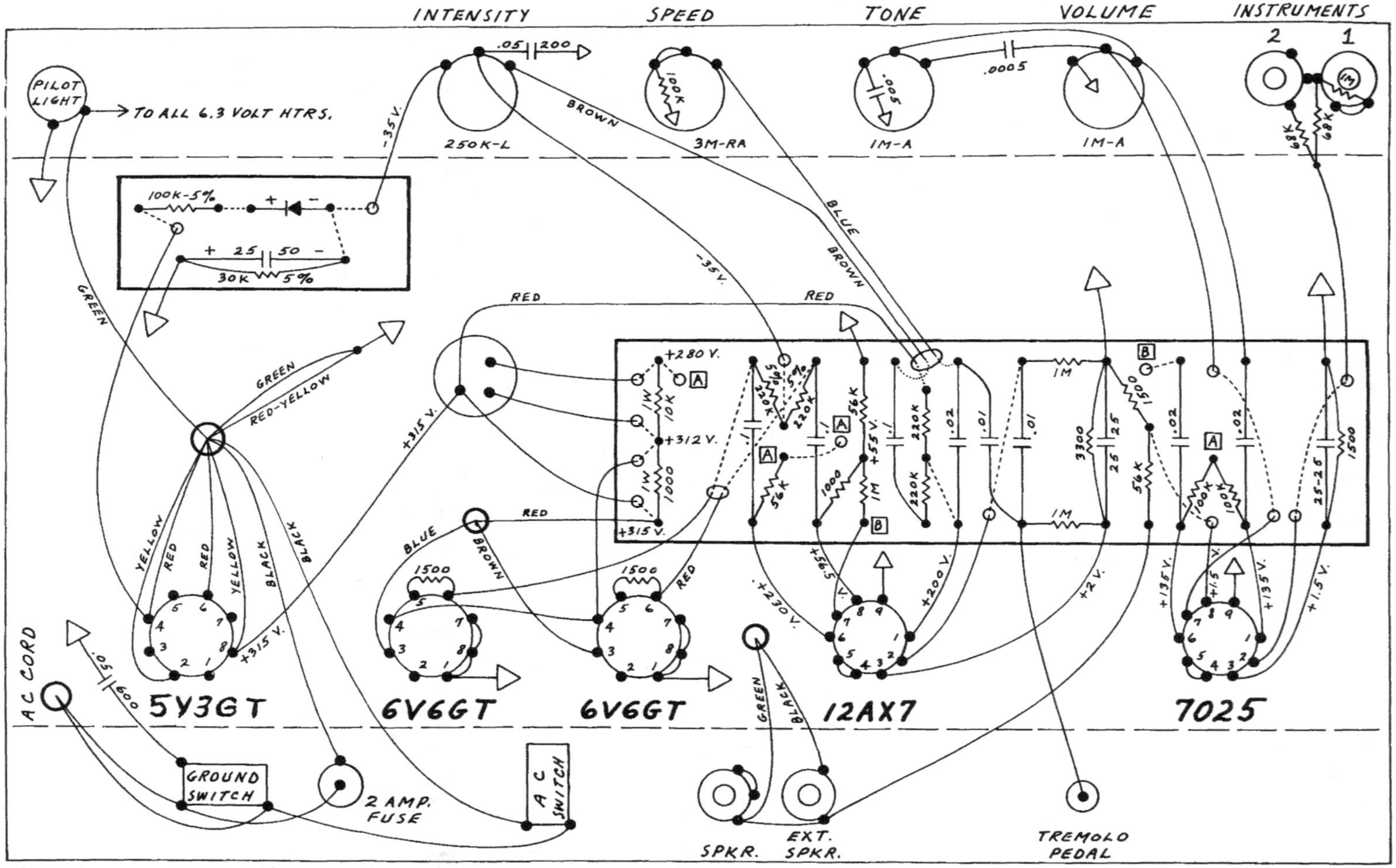

NOTE - ALL RESISTORS ½ WATT, 10% TOLERANCE IF NOT SPECIFIED. NOTE - ALL CAPACITORS AT LEAST 400 VOLTS IF NOT SPECIFIED.

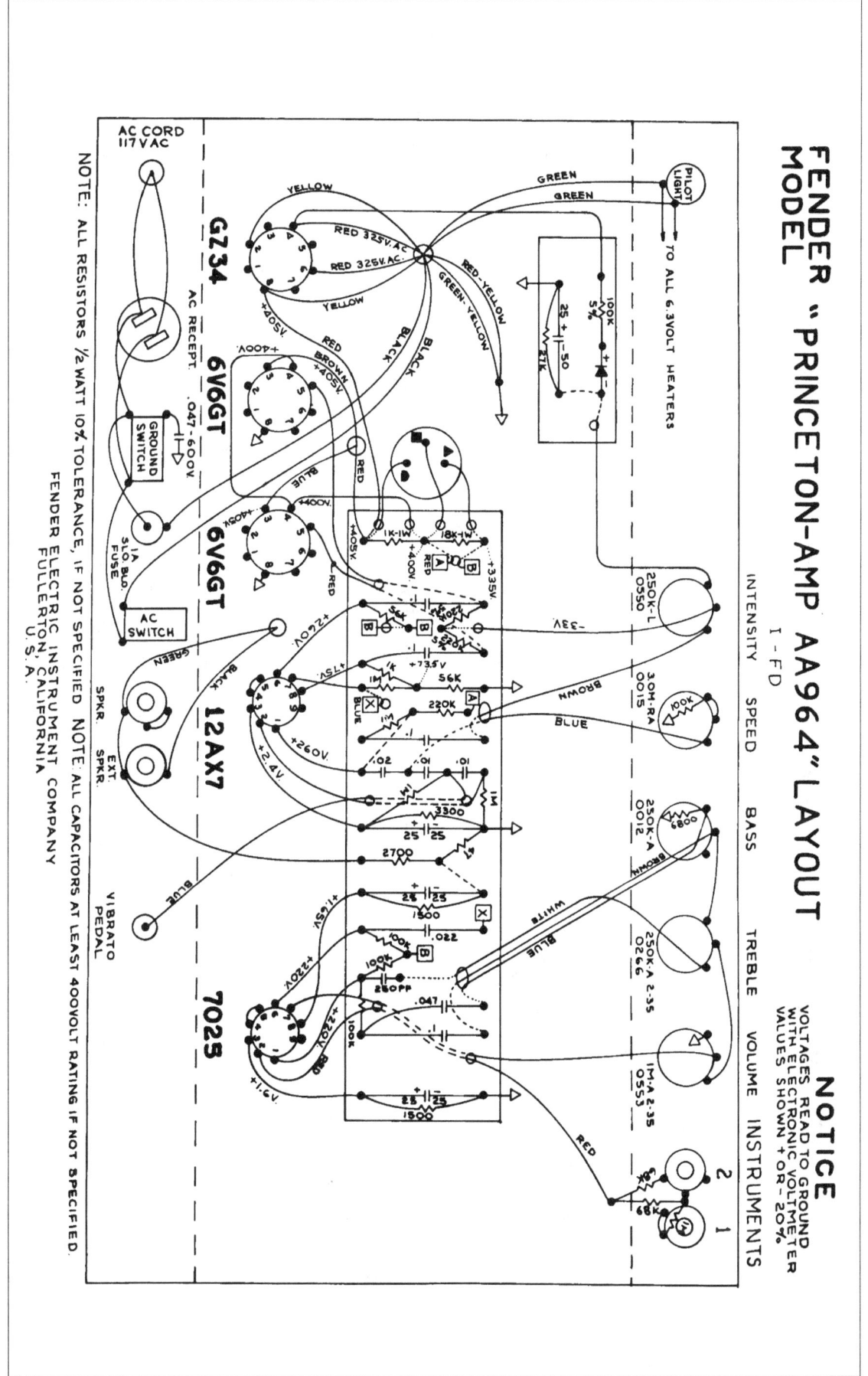

FENDER MODEL "PRINCETON-AMP AA964" LAYOUT

I - FD

NOTICE
VOLTAGES READ TO GROUND WITH ELECTRONIC VOLTMETER VALUES SHOWN + OR - 20%

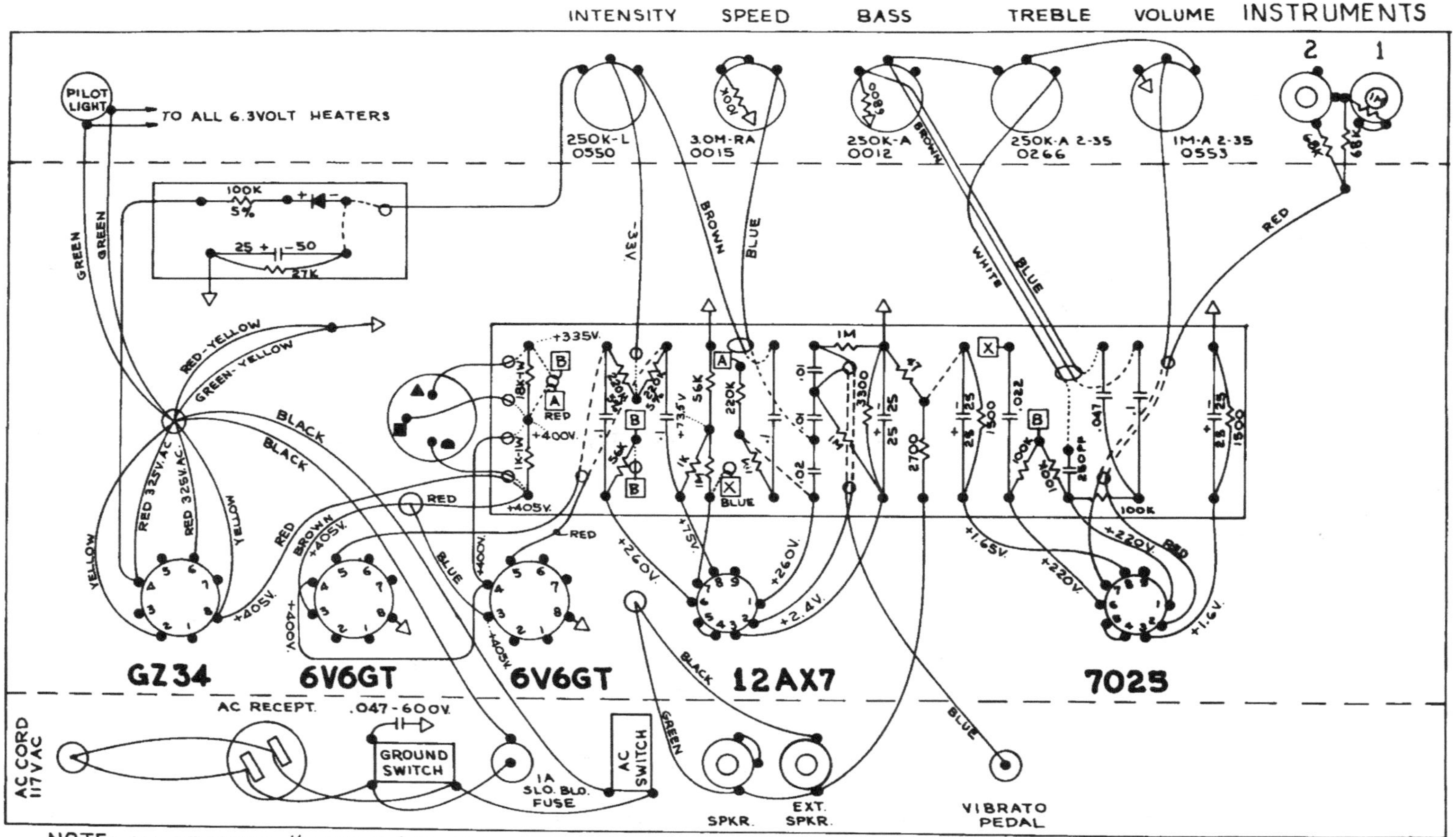

NOTE: ALL RESISTORS ½ WATT 10% TOLERANCE, IF NOT SPECIFIED NOTE: ALL CAPACITORS AT LEAST 400VOLT RATING IF NOT SPECIFIED.

FENDER ELECTRIC INSTRUMENT COMPANY
FULLERTON, CALIFORNIA
U.S.A.

FENDER MODEL **"PRINCETON-AMP AA964" SCHEMATIC**

THIS PRODUCT MANUFACTURED UNDER ONE OR MORE OF THE FOLLOWING U.S. PATENTS - #2817708, 2973681, 192859 & PATENTS PENDING.

NOTICE

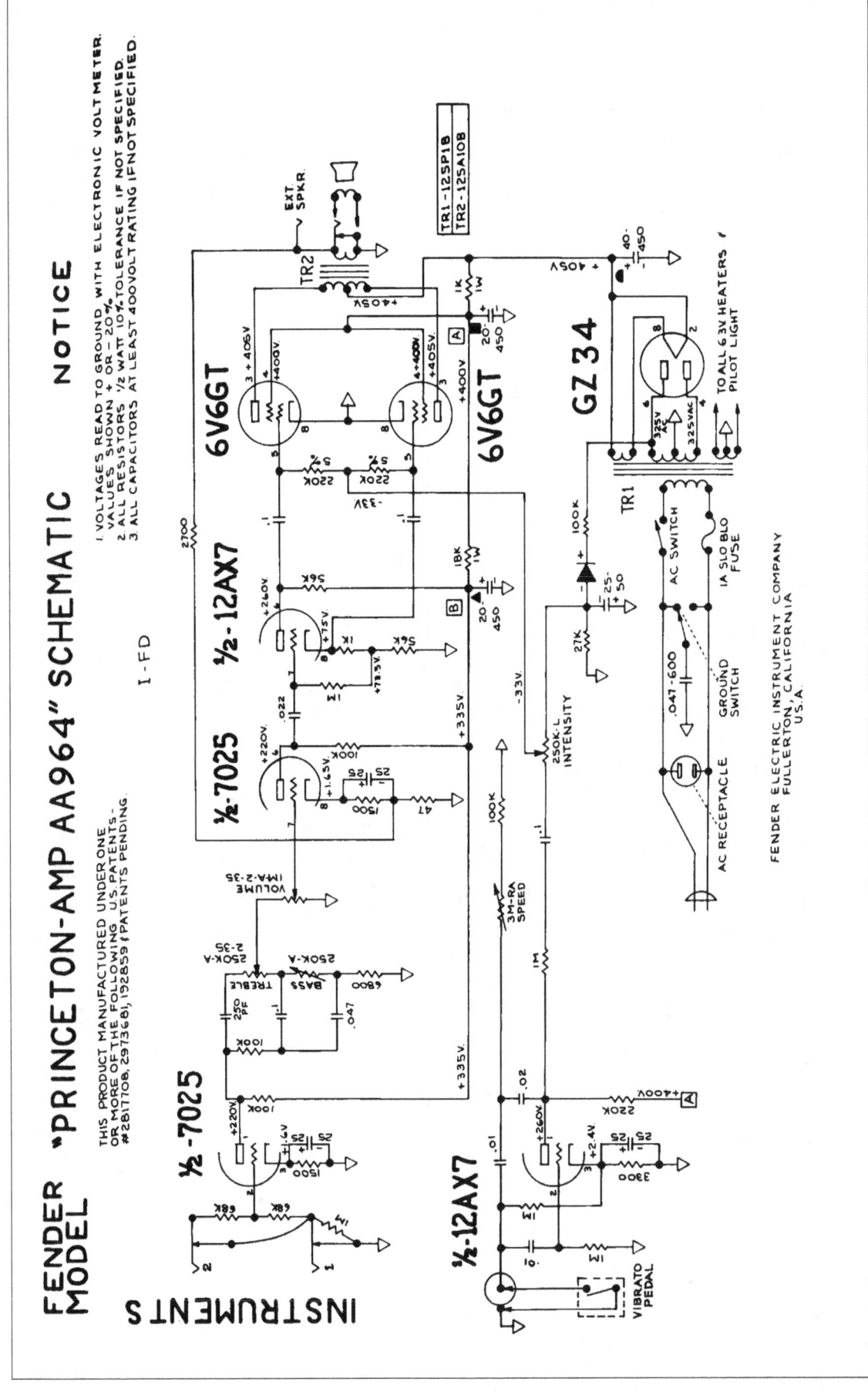

1. VOLTAGES READ TO GROUND WITH ELECTRONIC VOLTMETER. VALUES SHOWN + OR − 20%
2. ALL RESISTORS 1/2 WATT 10% TOLERANCE IF NOT SPECIFIED.
3. ALL CAPACITORS AT LEAST 400 VOLT RATING IF NOT SPECIFIED.

I-FD

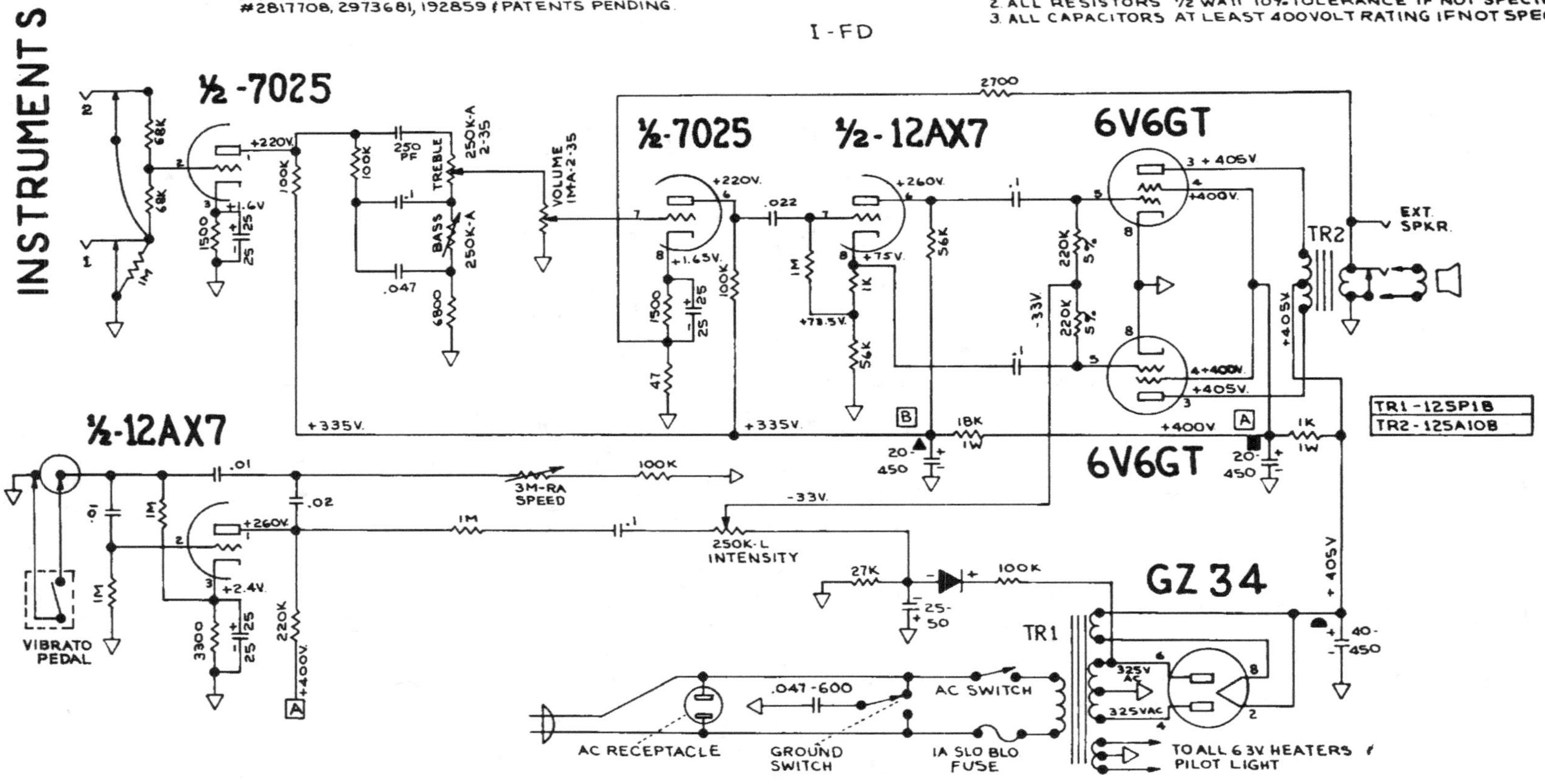

FENDER ELECTRIC INSTRUMENT COMPANY
FULLERTON, CALIFORNIA
U.S.A.

FENDER MODEL "PRINCETON REVERB-AMP AA1164" LAYOUT NOTICE

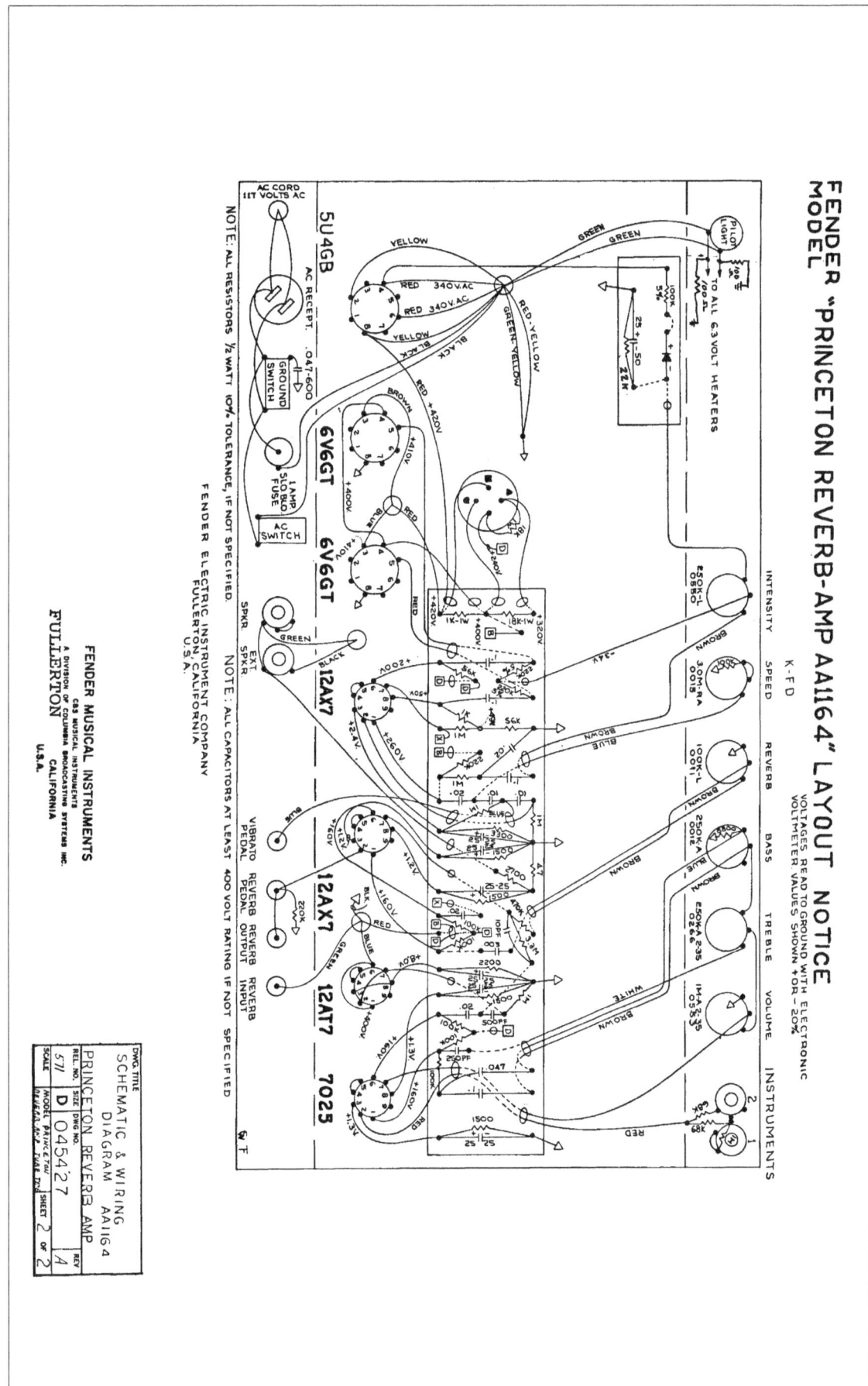

VOLTAGES READ TO GROUND WITH ELECTRONIC VOLTMETER. VALUES SHOWN +OR − 20%

K-FD

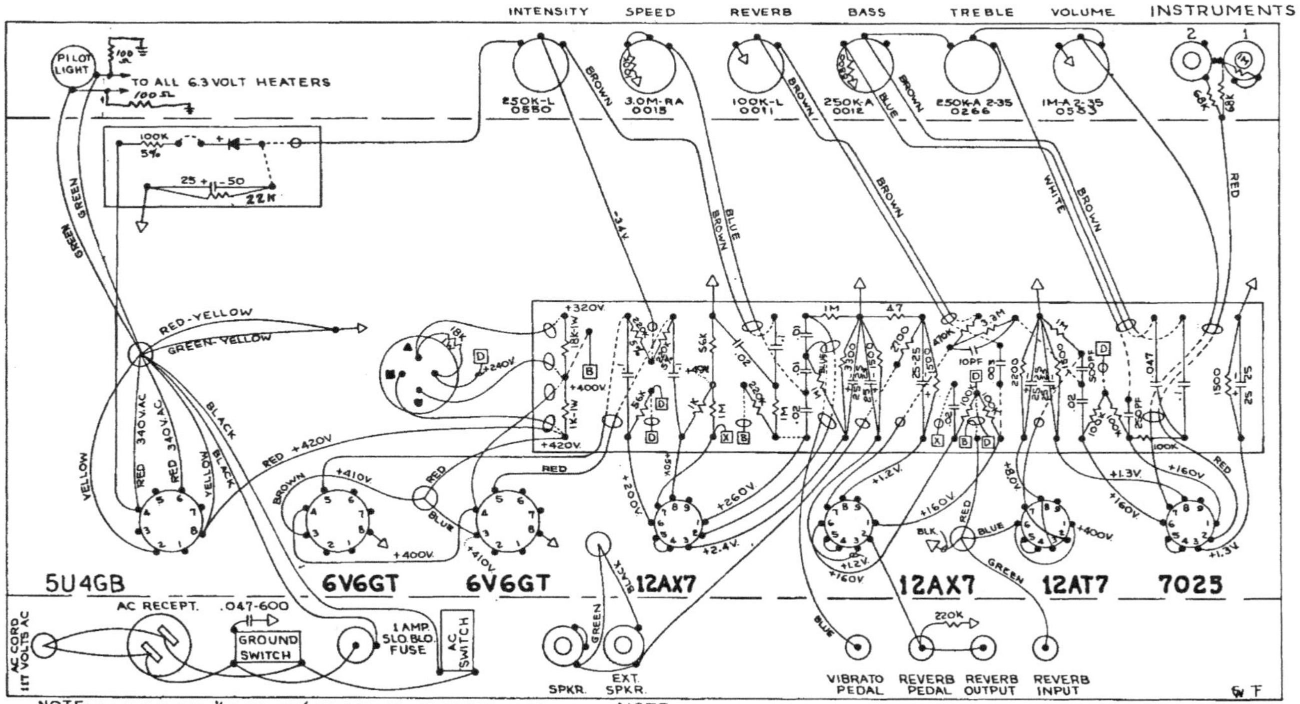

NOTE: ALL RESISTORS ½ WATT 10% TOLERANCE, IF NOT SPECIFIED.

NOTE: ALL CAPACITORS AT LEAST 400 VOLT RATING IF NOT SPECIFIED

FENDER ELECTRIC INSTRUMENT COMPANY
FULLERTON, CALIFORNIA
U.S.A.

FENDER MUSICAL INSTRUMENTS
CBS MUSICAL INSTRUMENTS
A DIVISION OF COLUMBIA BROADCASTING SYSTEMS INC.
FULLERTON CALIFORNIA
U.S.A.

DWG. TITLE			
SCHEMATIC & WIRING DIAGRAM AA1164 PRINCETON REVERB AMP			
REL. NO. 571	SIZE D	DWG NO. 045427	REV A
SCALE	MODEL PRINCETON REVERB-AMP TUBE TCB	SHEET 2 OF 2	

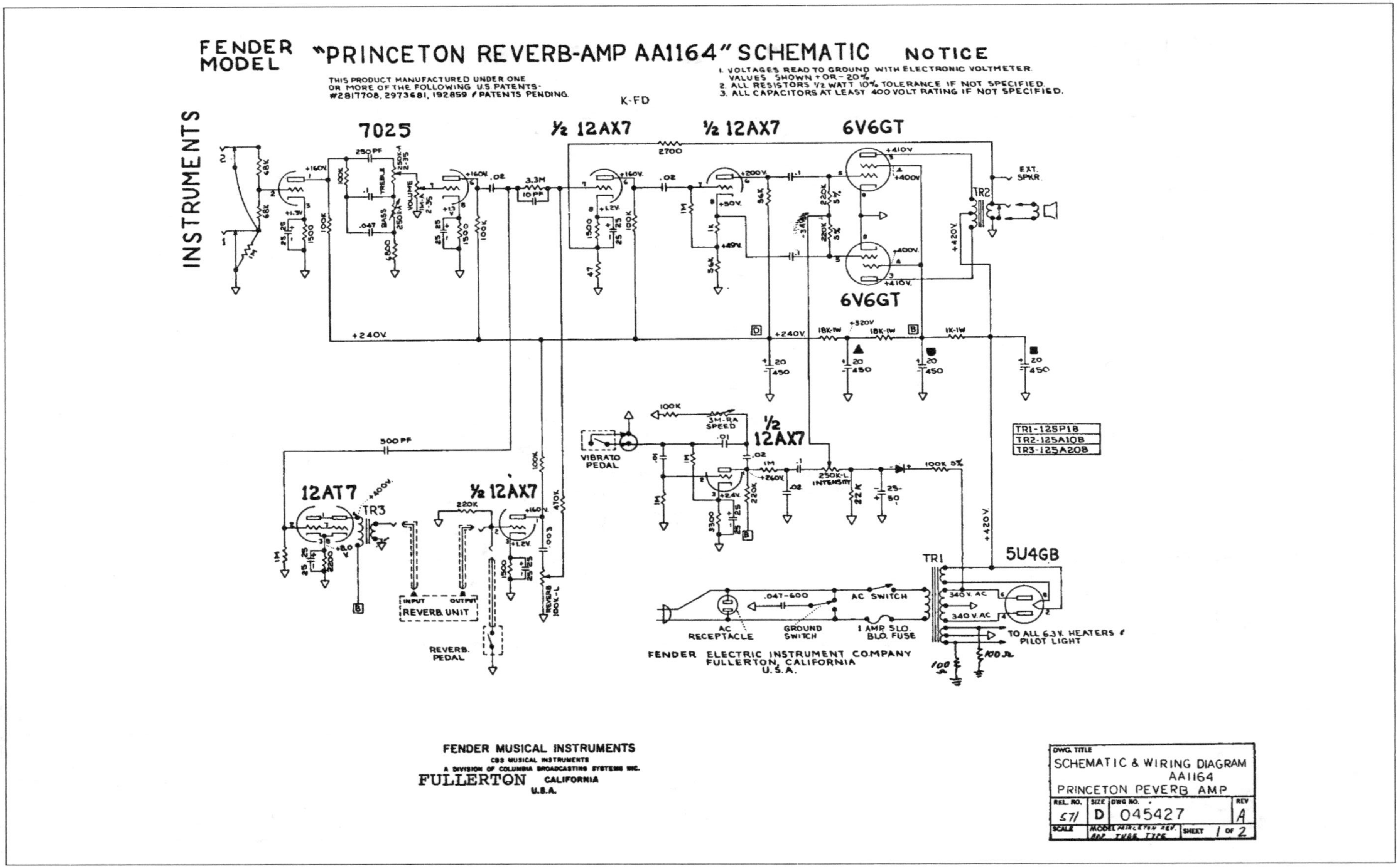
FENDER MODEL "PRINCETON REVERB-AMP AA1164" SCHEMATIC
THIS PRODUCT MANUFACTURED UNDER ONE OR MORE OF THE FOLLOWING U.S. PATENTS- #2817708, 2973681, 192859 / PATENTS PENDING.
K-FD
NOTICE
1. VOLTAGES READ TO GROUND WITH ELECTRONIC VOLTMETER. VALUES SHOWN + OR - 20%
2. ALL RESISTORS 1/2 WATT 10% TOLERANCE IF NOT SPECIFIED.
3. ALL CAPACITORS AT LEAST 400 VOLT RATING IF NOT SPECIFIED.
INSTRUMENTS
7025
½ 12AX7
½ 12AX7
6V6GT
6V6GT
EXT. SPKR.
TR2
TR1-125P1B
TR2-125A10B
TR3-125A20B
VIBRATO PEDAL
SPEED
INTENSITY
½ 12AX7
12AT7
TR3
½ 12AX7
REVERB UNIT
INPUT
OUTPUT
REVERB. PEDAL
TR1
5U4GB
AC SWITCH
AC RECEPTACLE
GROUND SWITCH
1 AMP SLO BLO. FUSE
TO ALL 6.3V. HEATERS & PILOT LIGHT
FENDER ELECTRIC INSTRUMENT COMPANY FULLERTON, CALIFORNIA U.S.A.
FENDER MUSICAL INSTRUMENTS
CBS MUSICAL INSTRUMENTS
A DIVISION OF COLUMBIA BROADCASTING SYSTEMS INC.
FULLERTON CALIFORNIA U.S.A.
DWG. TITLE
SCHEMATIC & WIRING DIAGRAM AA1164
PRINCETON PEVERB AMP
REL. NO. 571
SIZE D
DWG NO. 045427
REV A
SHEET 1 OF 2

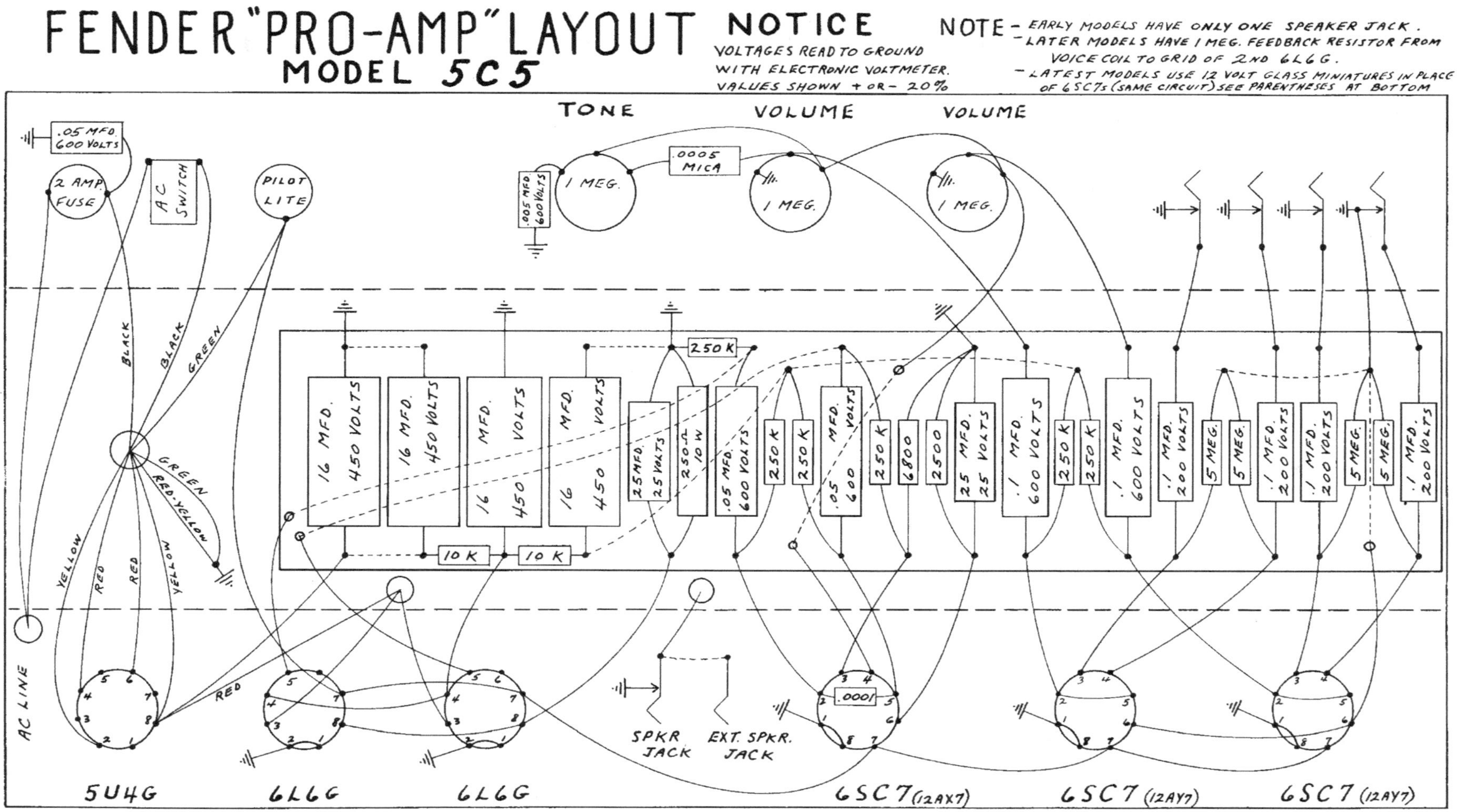
FENDER "PRO-AMP" LAYOUT
MODEL 5C5
NOTICE
VOLTAGES READ TO GROUND WITH ELECTRONIC VOLTMETER. VALUES SHOWN + OR - 20%
NOTE - EARLY MODELS HAVE ONLY ONE SPEAKER JACK.
- LATER MODELS HAVE 1 MEG. FEEDBACK RESISTOR FROM VOICE COIL TO GRID OF 2ND 6L6G.
- LATEST MODELS USE 12 VOLT GLASS MINIATURES IN PLACE OF 6SC7s (SAME CIRCUIT) SEE PARENTHESES AT BOTTOM
TONE
VOLUME
VOLUME
.05 MFD. 600 VOLTS
2 AMP. FUSE
AC SWITCH
PILOT LITE
.005 MFD. 600 VOLTS
1 MEG.
.0005 MICA
1 MEG.
1 MEG.
BLACK
BLACK
GREEN
GREEN
RED-YELLOW
YELLOW
RED
RED
YELLOW
RED
AC LINE
16 MFD. 450 VOLTS
16 MFD. 450 VOLTS
16 MFD. 450 VOLTS
16 MFD. 450 VOLTS
10 K
10 K
250 K
25 MFD. 25 VOLTS
250Ω 10 W
.05 MFD. 600 VOLTS
250 K
250 K
.05 MFD. 600 VOLTS
250 K
6800
2500
25 MFD. 25 VOLTS
.1 MFD. 600 VOLTS
250 K
250 K
.1 MFD. 600 VOLTS
.1 MFD. 200 VOLTS
5 MEG.
5 MEG.
.1 MFD. 200 VOLTS
.1 MFD. 200 VOLTS
5 MEG.
5 MEG.
.1 MFD. 200 VOLTS
.0001
SPKR JACK
EXT. SPKR. JACK
5U4G
6L6G
6L6G
6SC7 (12AX7)
6SC7 (12AY7)
6SC7 (12AY7)

FENDER "PRO-AMP" SCHEMATIC
MODEL 5C5

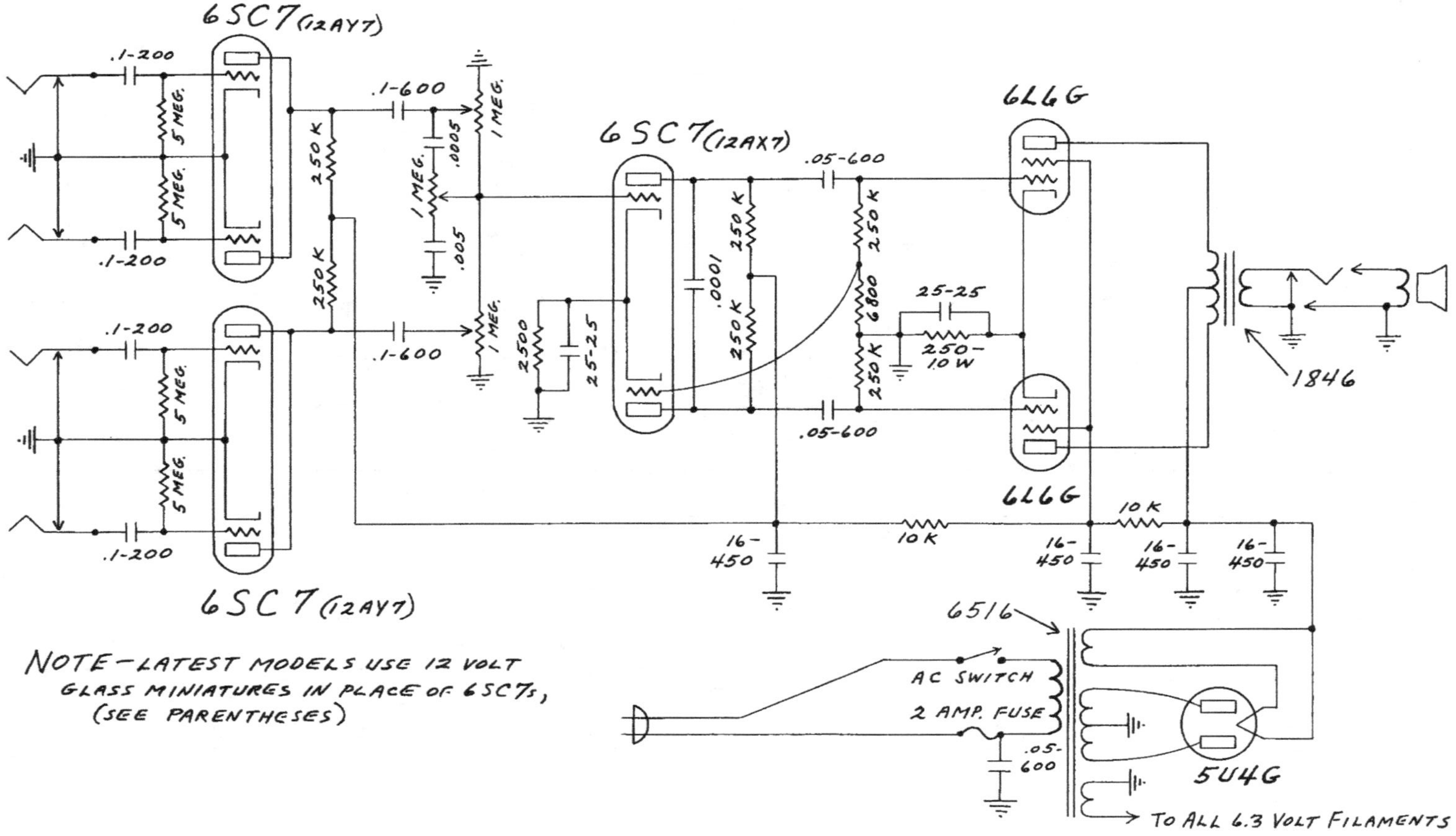

FENDER "PRO-AMP" SCHEMATIC
MODEL 5D5

I-ED

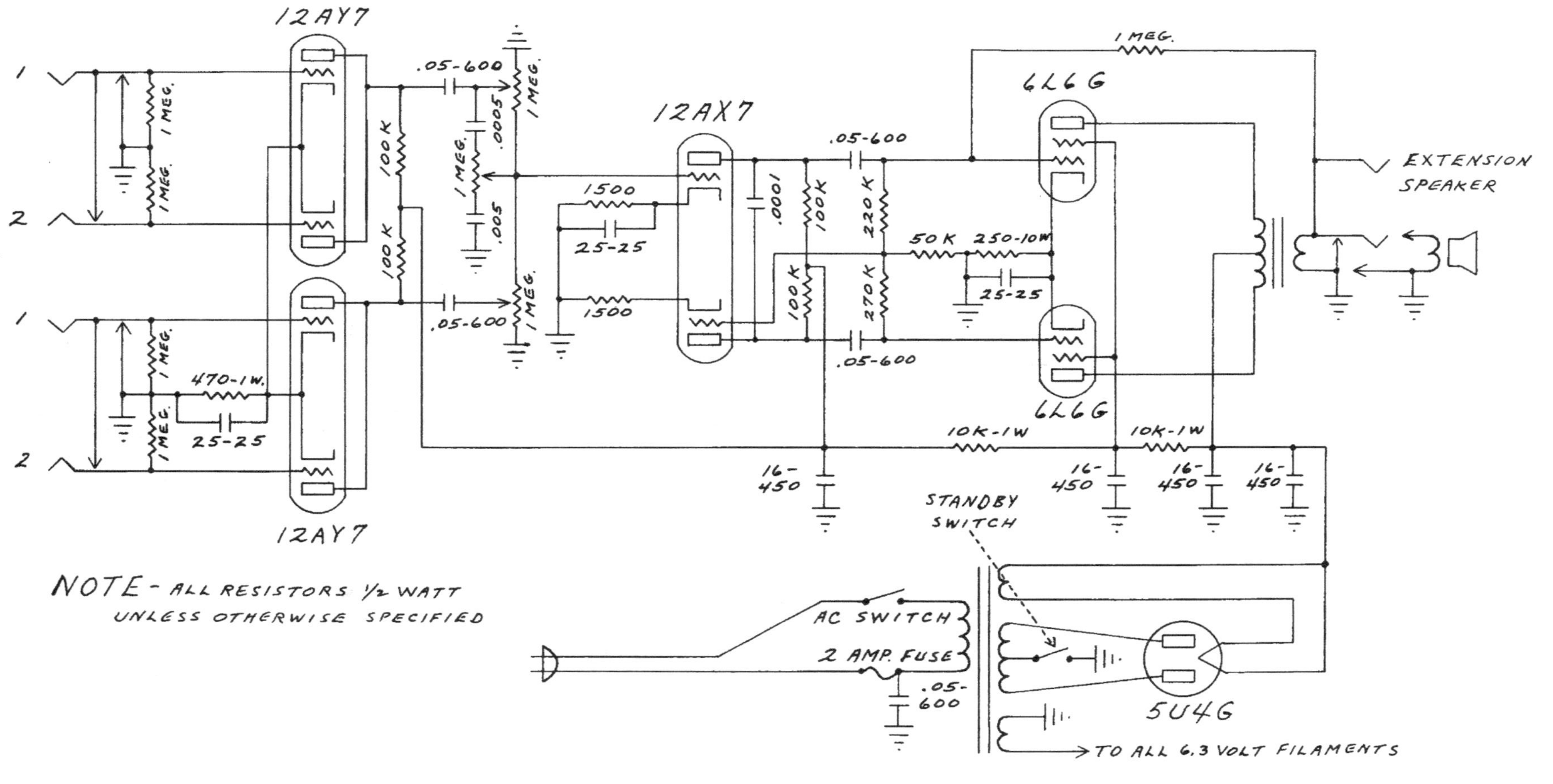

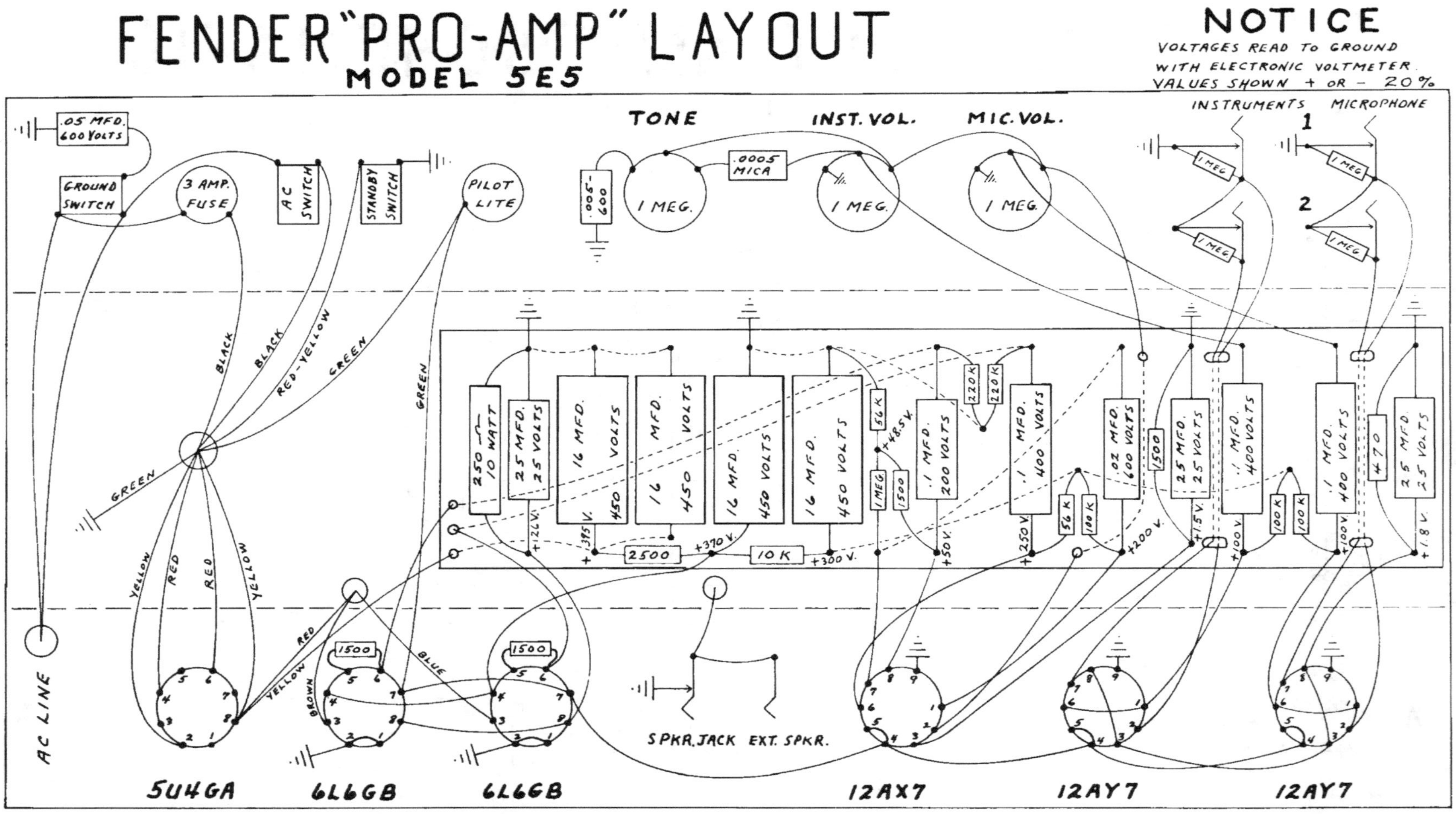
FENDER "PRO-AMP" LAYOUT
MODEL 5E5
NOTICE
VOLTAGES READ TO GROUND
WITH ELECTRONIC VOLTMETER.
VALUES SHOWN + OR - 20%
INSTRUMENTS
MICROPHONE
1
2
TONE
INST. VOL.
MIC. VOL.
.05 MFD. 600 VOLTS
GROUND SWITCH
3 AMP. FUSE
AC SWITCH
STANDBY SWITCH
PILOT LITE
.005-600
1 MEG.
.0005 MICA
1 MEG.
1 MEG.
1 MEG
BLACK
BLACK
RED-YELLOW
GREEN
GREEN
GREEN
YELLOW
RED
RED
YELLOW
RED
YELLOW
BROWN
BLUE
250 Ω 10 WATT
25 MFD. 25 VOLTS
16 MFD. 450 VOLTS
16 MFD. 450 VOLTS
16 MFD. 450 VOLTS
16 MFD. 450 VOLTS
56 K
1 MEG
1500
+48.5 V.
220 K
220 K
.1 MFD. 200 VOLTS
.1 MFD. 400 VOLTS
56 K
100 K
.02 MFD. 600 VOLTS
1500
25 MFD. 25 VOLTS
.1 MFD. 400 VOLTS
100 K
100 K
.1 MFD. 400 VOLTS
470
25 MFD. 25 VOLTS
+12 V.
+395 V.
2500
+370 V.
10 K
+300 V.
+50 V.
+250 V
+200 V.
+15 V.
+100 V
+100 V
+1.8 V.
1500
1500
AC LINE
SPKR. JACK
EXT. SPKR.
5U4GA
6L6GB
6L6GB
12AX7
12AY7
12AY7

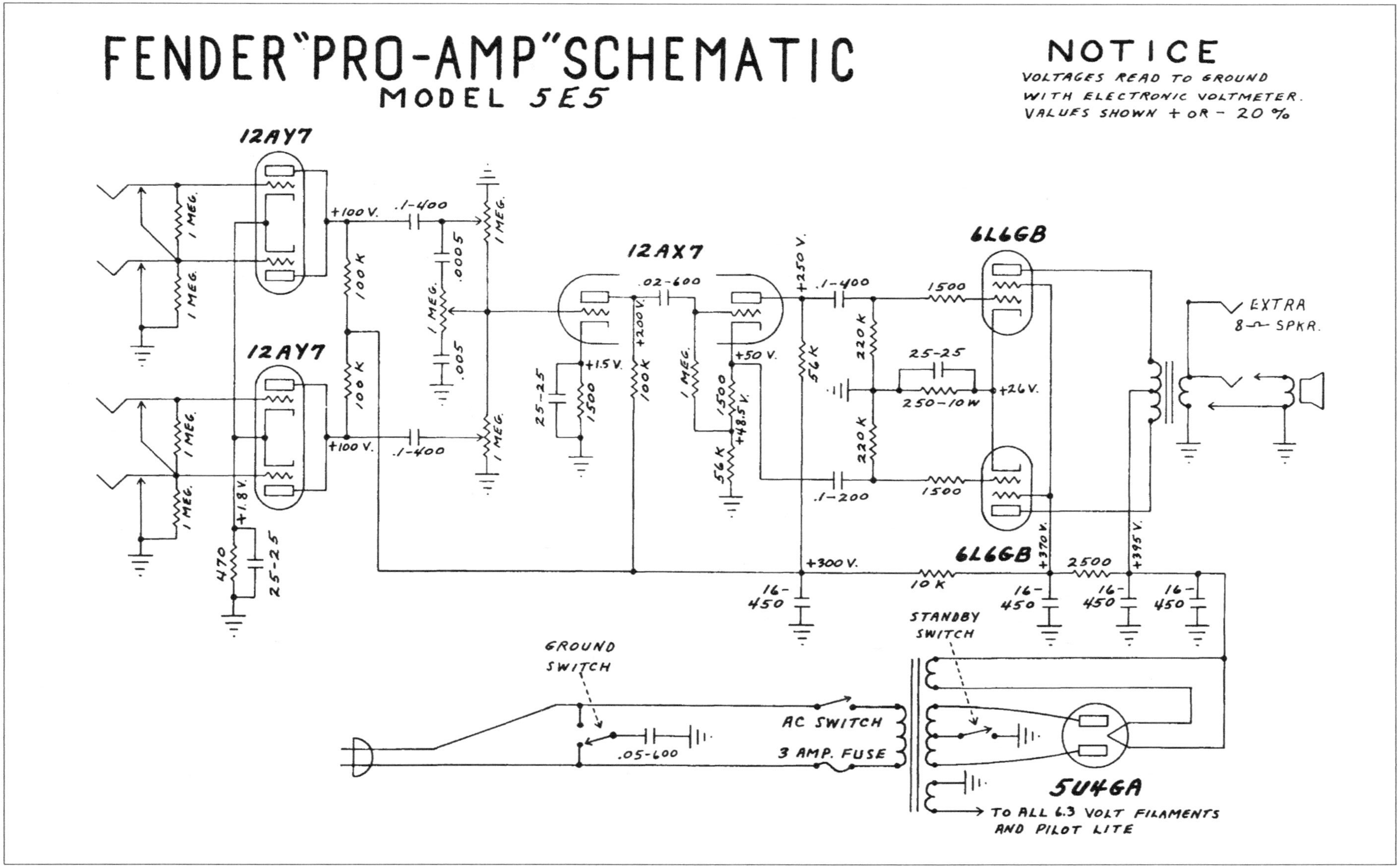
FENDER "PRO-AMP" SCHEMATIC
MODEL 5E5
NOTICE
VOLTAGES READ TO GROUND
WITH ELECTRONIC VOLTMETER.
VALUES SHOWN + OR - 20 %
12AY7
12AY7
12AX7
6L6GB
6L6GB
5U4GA
1 MEG.
470
25-25
+1.8 V.
+100 V.
.1-400
100 K
.0005
.005
+1.5 V.
1500
.02-600
+200 V.
+50 V.
+48.5 V.
56 K
+250 V.
220 K
250-10W
+26 V.
.1-200
+300 V.
10 K
16-450
+370 V.
2500
+395 V.
EXTRA
8-Ω SPKR.
GROUND SWITCH
.05-600
AC SWITCH
3 AMP. FUSE
STANDBY SWITCH
TO ALL 6.3 VOLT FILAMENTS AND PILOT LITE

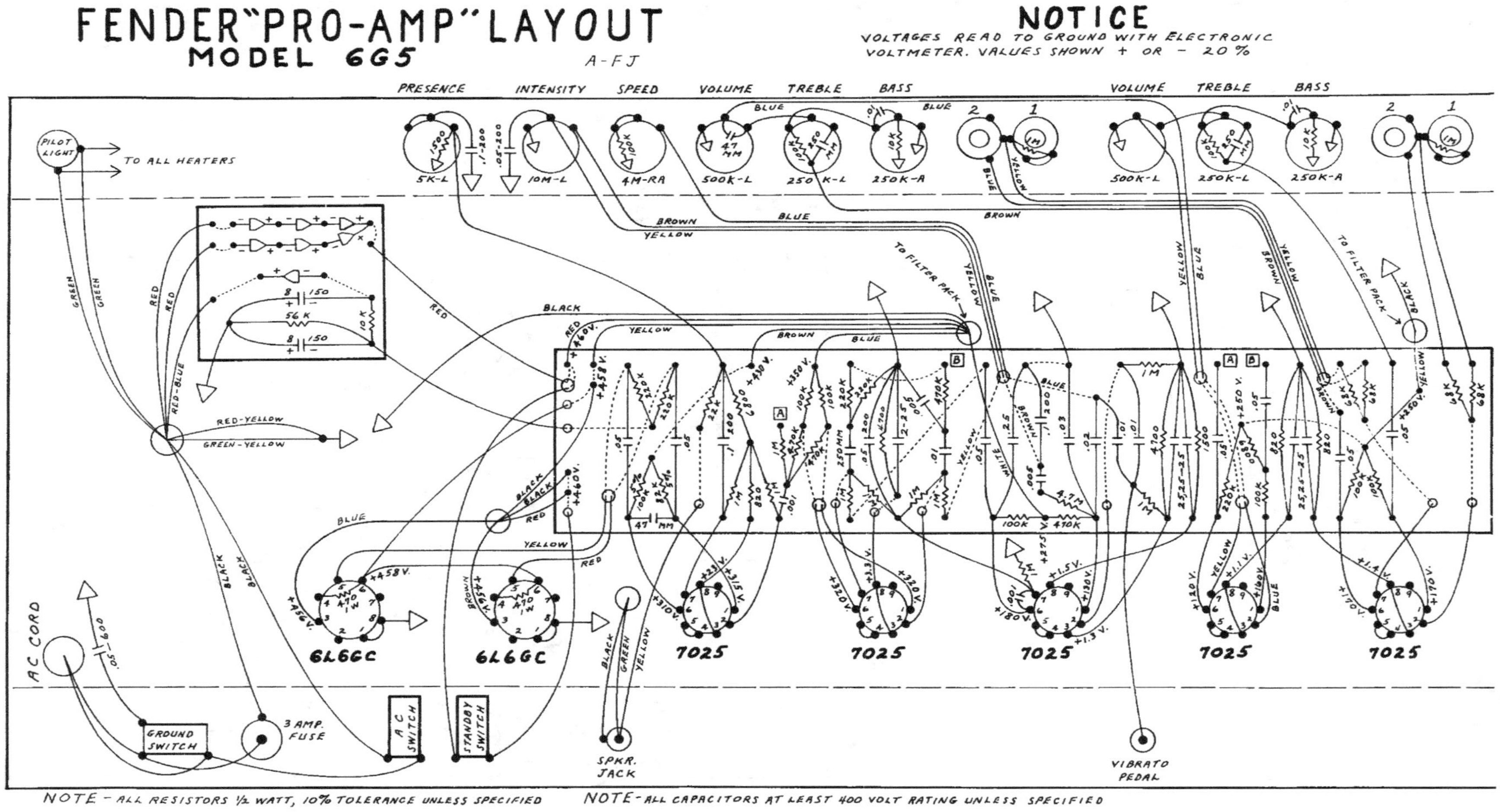
FENDER "PRO-AMP" LAYOUT
MODEL 6G5
A-FJ
NOTICE
VOLTAGES READ TO GROUND WITH ELECTRONIC VOLTMETER. VALUES SHOWN + OR - 20%
PRESENCE
INTENSITY
SPEED
VOLUME
TREBLE
BASS
VOLUME
TREBLE
BASS
PILOT LIGHT
TO ALL HEATERS
TO FILTER PACK
TO FILTER PACK
6L6GC
6L6GC
7025
7025
7025
7025
7025
AC CORD
GROUND SWITCH
3 AMP. FUSE
AC SWITCH
STANDBY SWITCH
SPKR. JACK
VIBRATO PEDAL
NOTE - ALL RESISTORS 1/2 WATT, 10% TOLERANCE UNLESS SPECIFIED
NOTE - ALL CAPACITORS AT LEAST 400 VOLT RATING UNLESS SPECIFIED

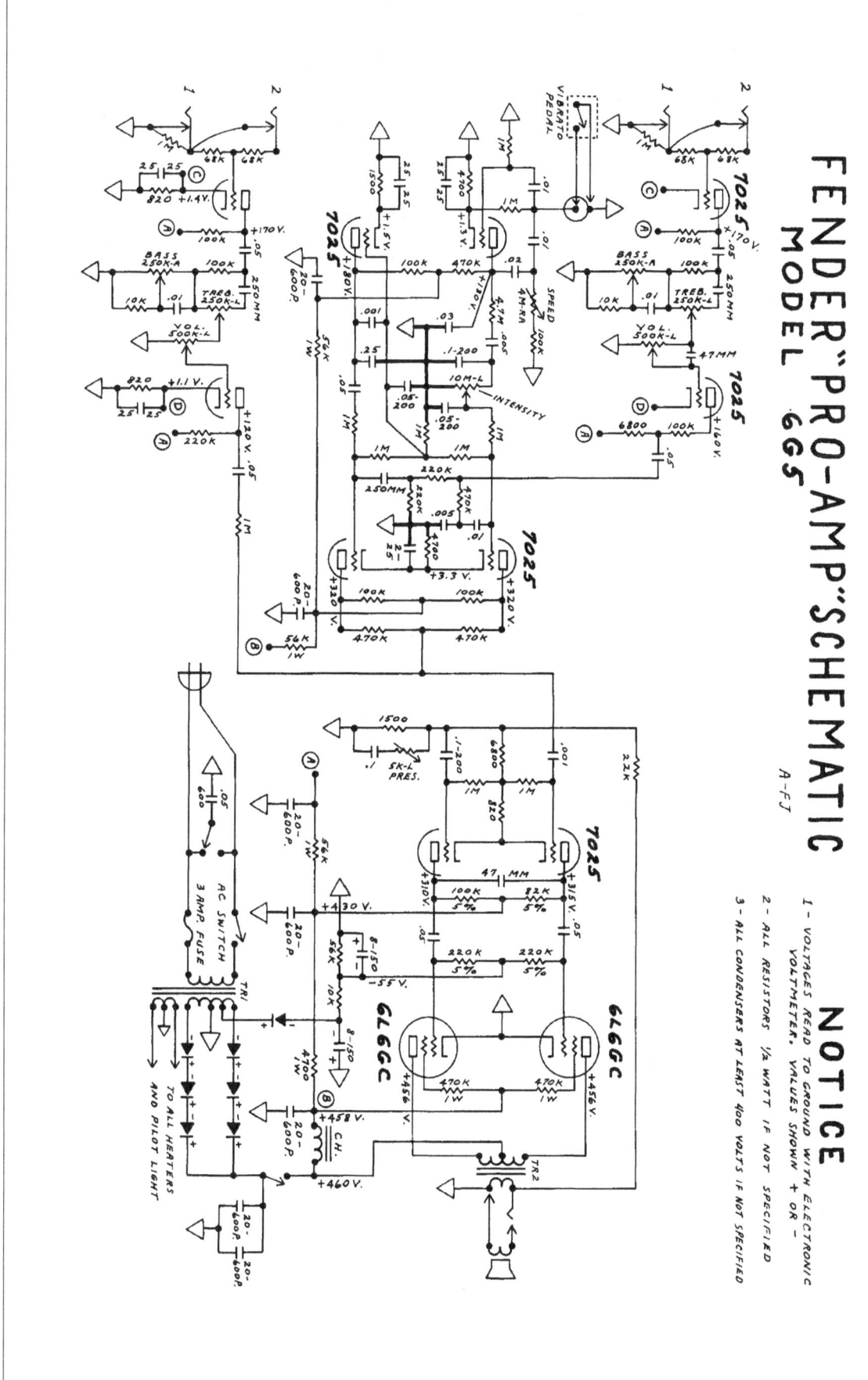

FENDER "PRO-AMP" SCHEMATIC
MODEL 6G5

A-FJ

NOTICE

1 - VOLTAGES READ TO GROUND WITH ELECTRONIC VOLTMETER. VALUES SHOWN + OR -

2 - ALL RESISTORS ½ WATT IF NOT SPECIFIED

3 - ALL CONDENSERS AT LEAST 400 VOLTS IF NOT SPECIFIED

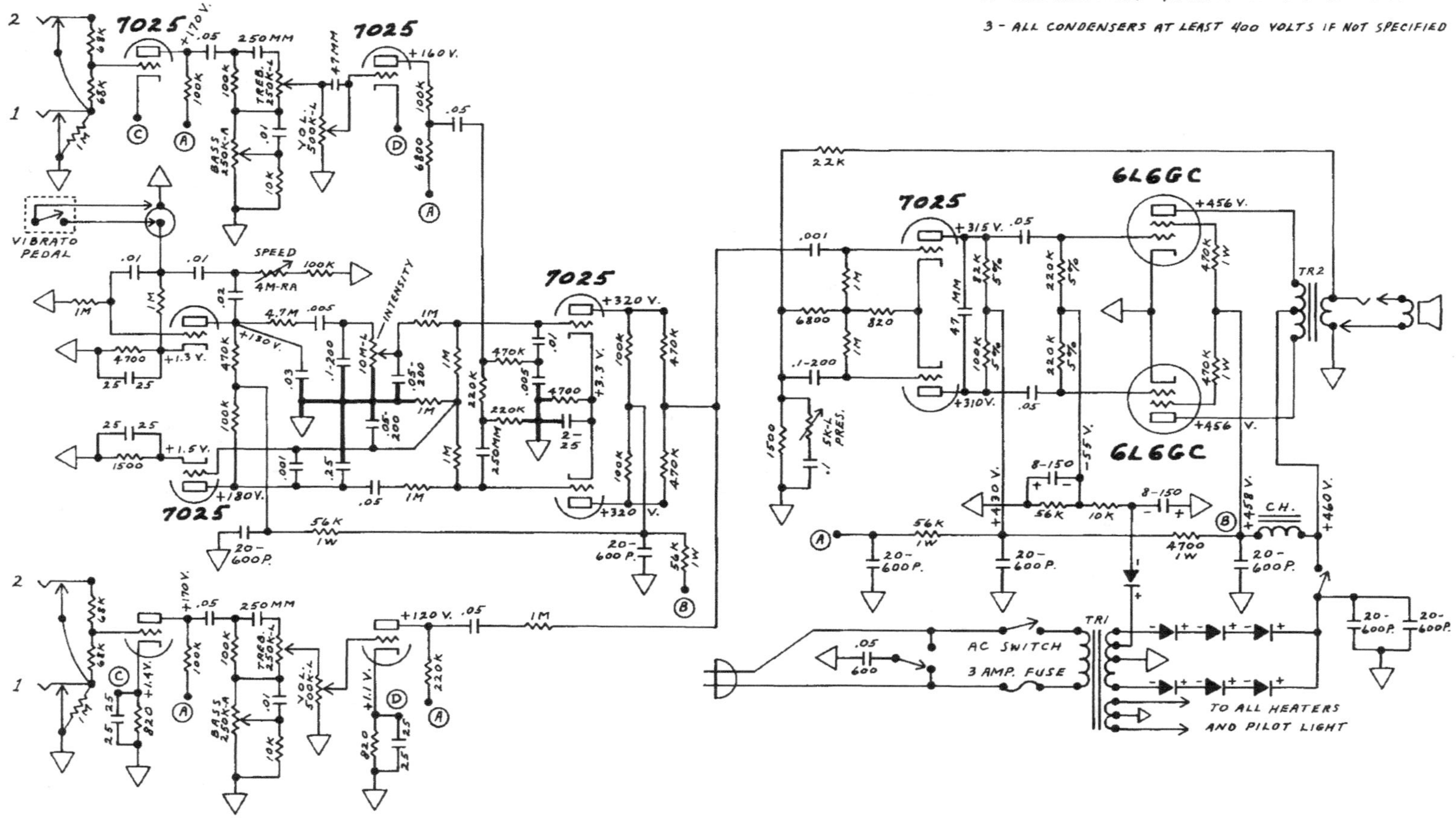

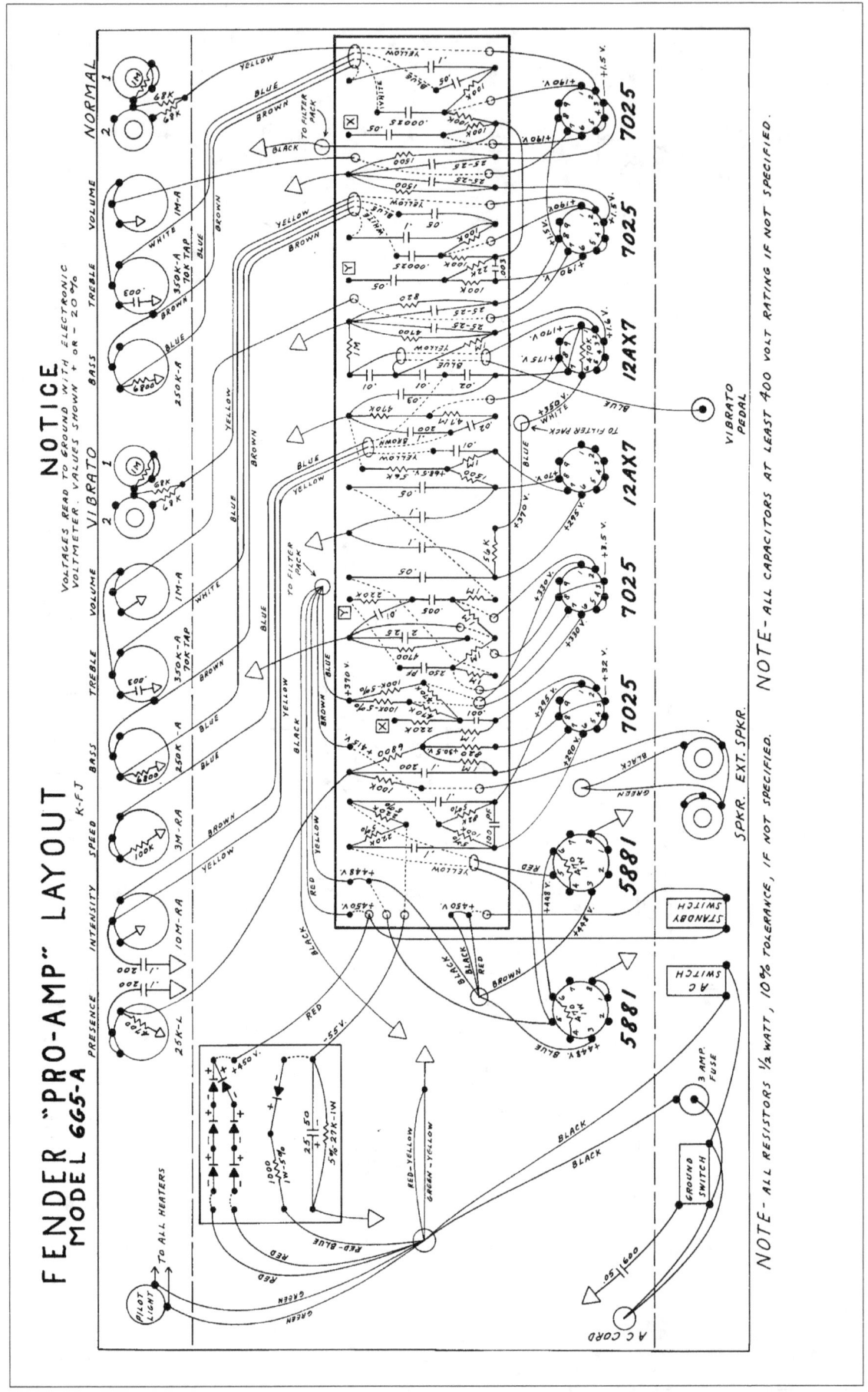

FENDER "PRO-AMP" LAYOUT
MODEL 6G5-A

K-FJ

NOTICE
VOLTAGES READ TO GROUND WITH ELECTRONIC VOLTMETER. VALUES SHOWN + OR − 20%

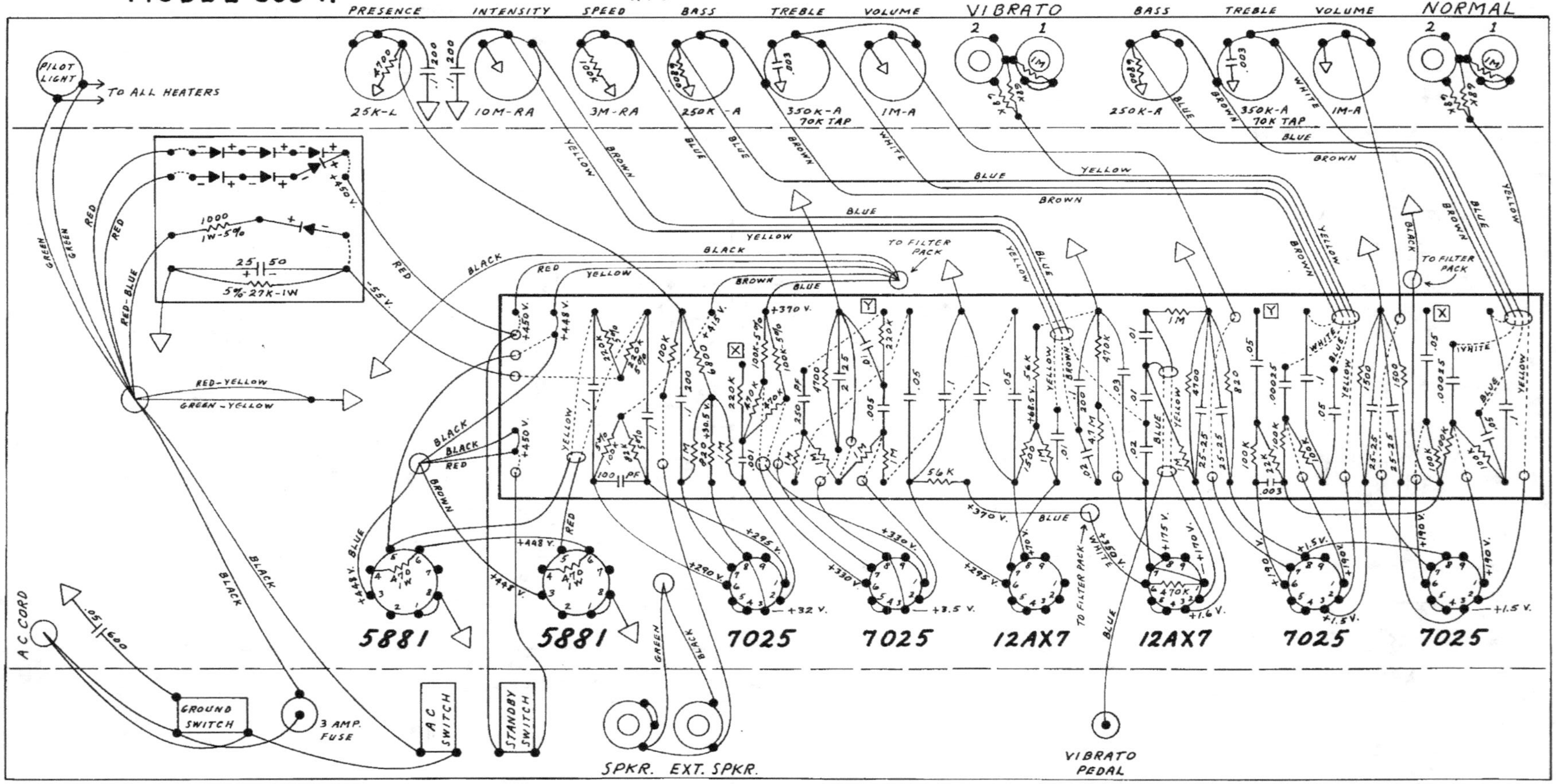

NOTE- ALL RESISTORS ½ WATT, 10% TOLERANCE, IF NOT SPECIFIED.

NOTE- ALL CAPACITORS AT LEAST 400 VOLT RATING IF NOT SPECIFIED.

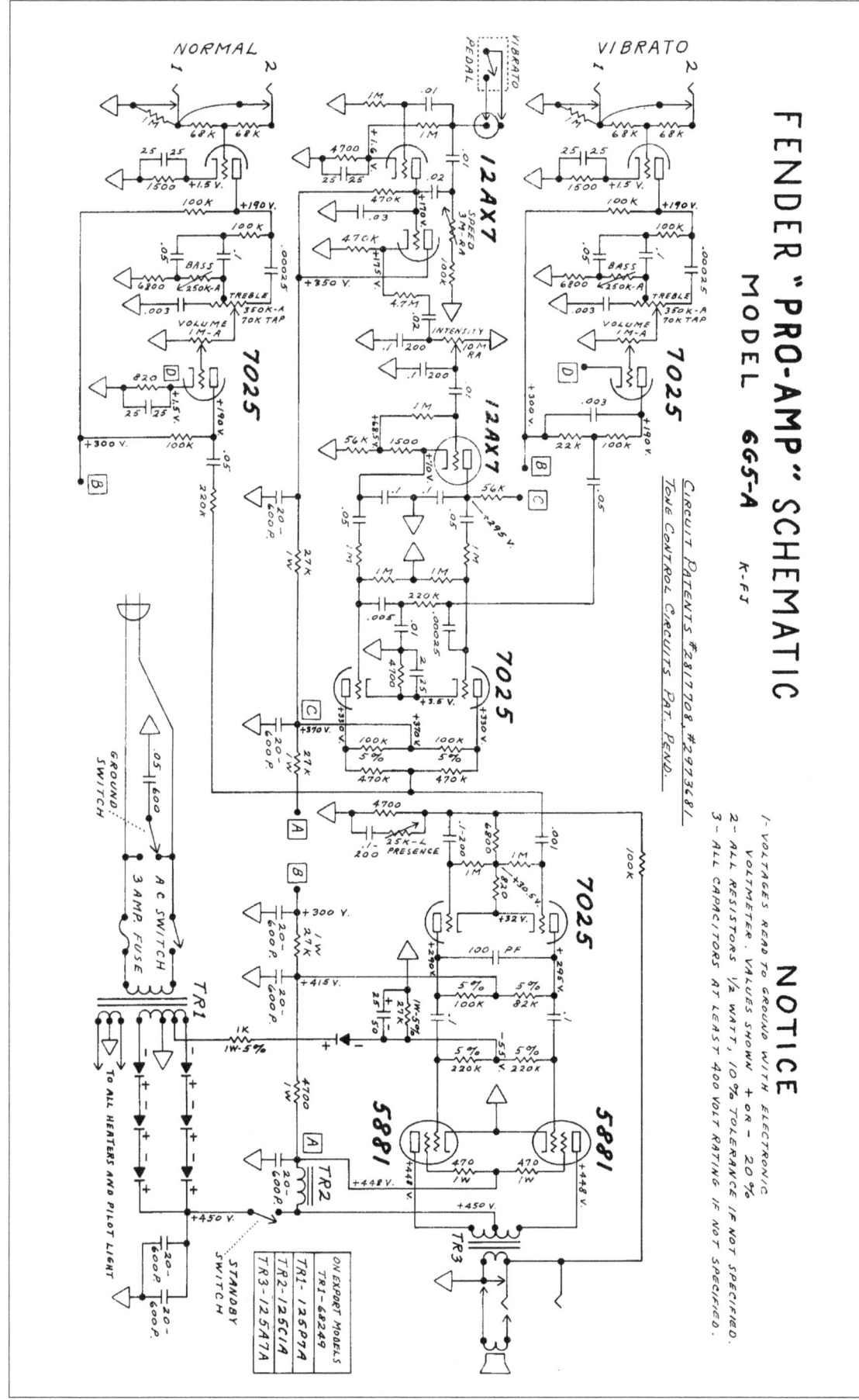

FENDER "PRO-AMP" SCHEMATIC

MODEL 6G5-A

K-FJ

NOTICE

1- VOLTAGES READ TO GROUND WITH ELECTRONIC VOLTMETER. VALUES SHOWN + OR − 20%
2- ALL RESISTORS ½ WATT, 10% TOLERANCE IF NOT SPECIFIED.
3- ALL CAPACITORS AT LEAST 400 VOLT RATING IF NOT SPECIFIED.

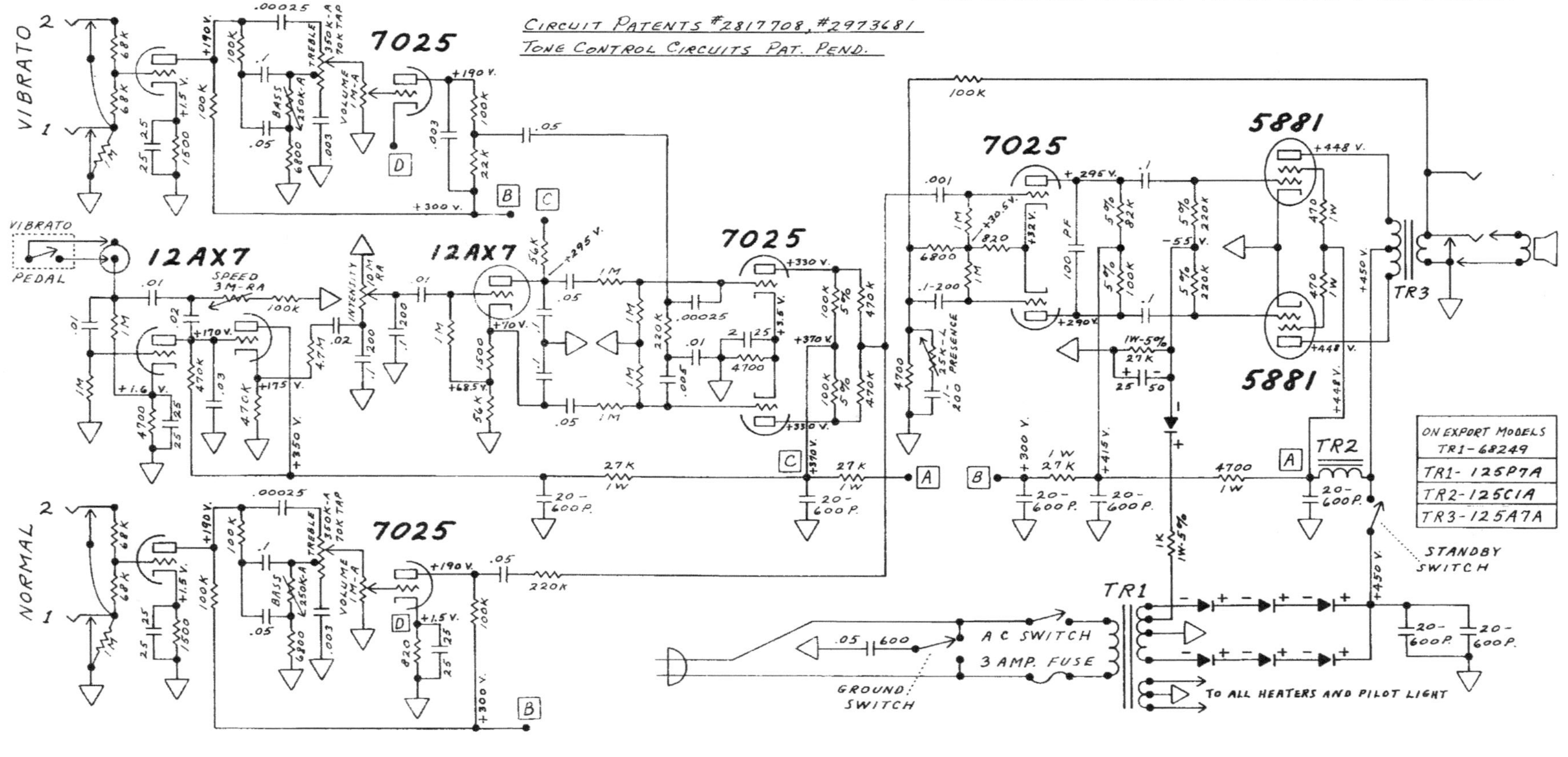

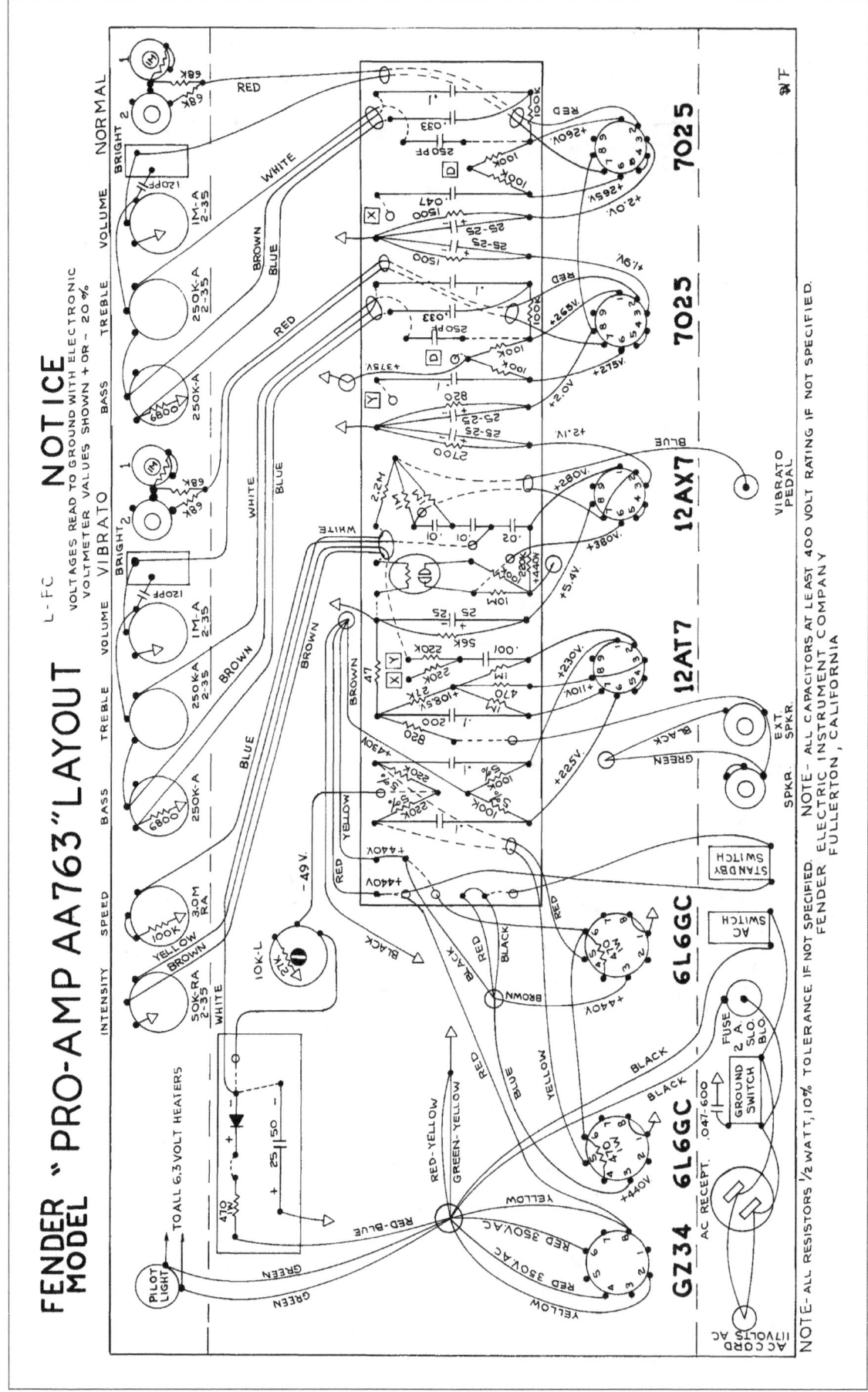

FENDER MODEL "PRO-AMP AA763" LAYOUT

L-FC

NOTICE

VOLTAGES READ TO GROUND WITH ELECTRONIC VOLTMETER VALUES SHOWN + OR − 20%

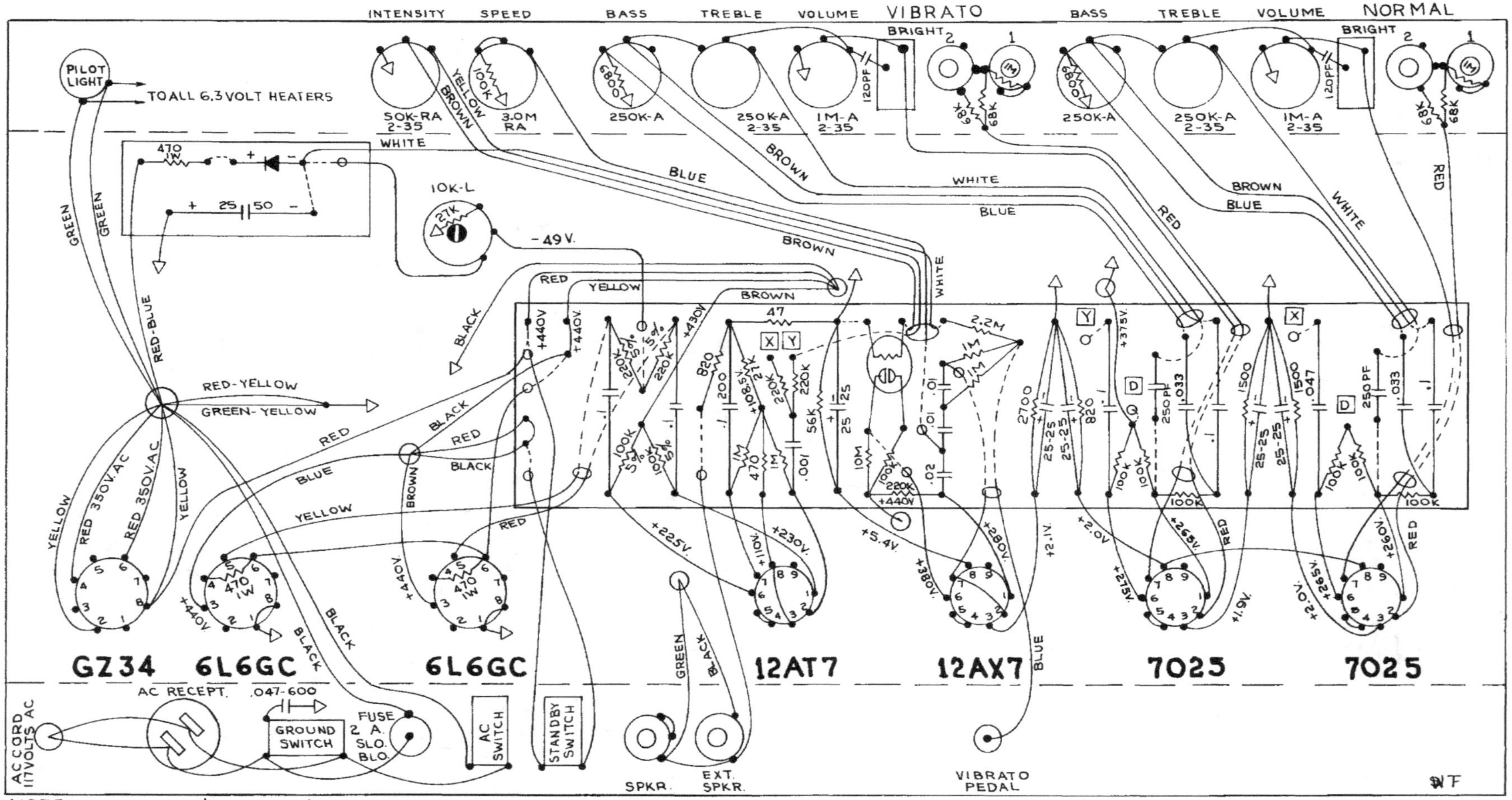

NOTE- ALL RESISTORS 1/2 WATT, 10% TOLERANCE IF NOT SPECIFIED. NOTE- ALL CAPACITORS AT LEAST 400 VOLT RATING IF NOT SPECIFIED.

FENDER ELECTRIC INSTRUMENT COMPANY
FULLERTON, CALIFORNIA

PRO-AMP **AA763** | Fender

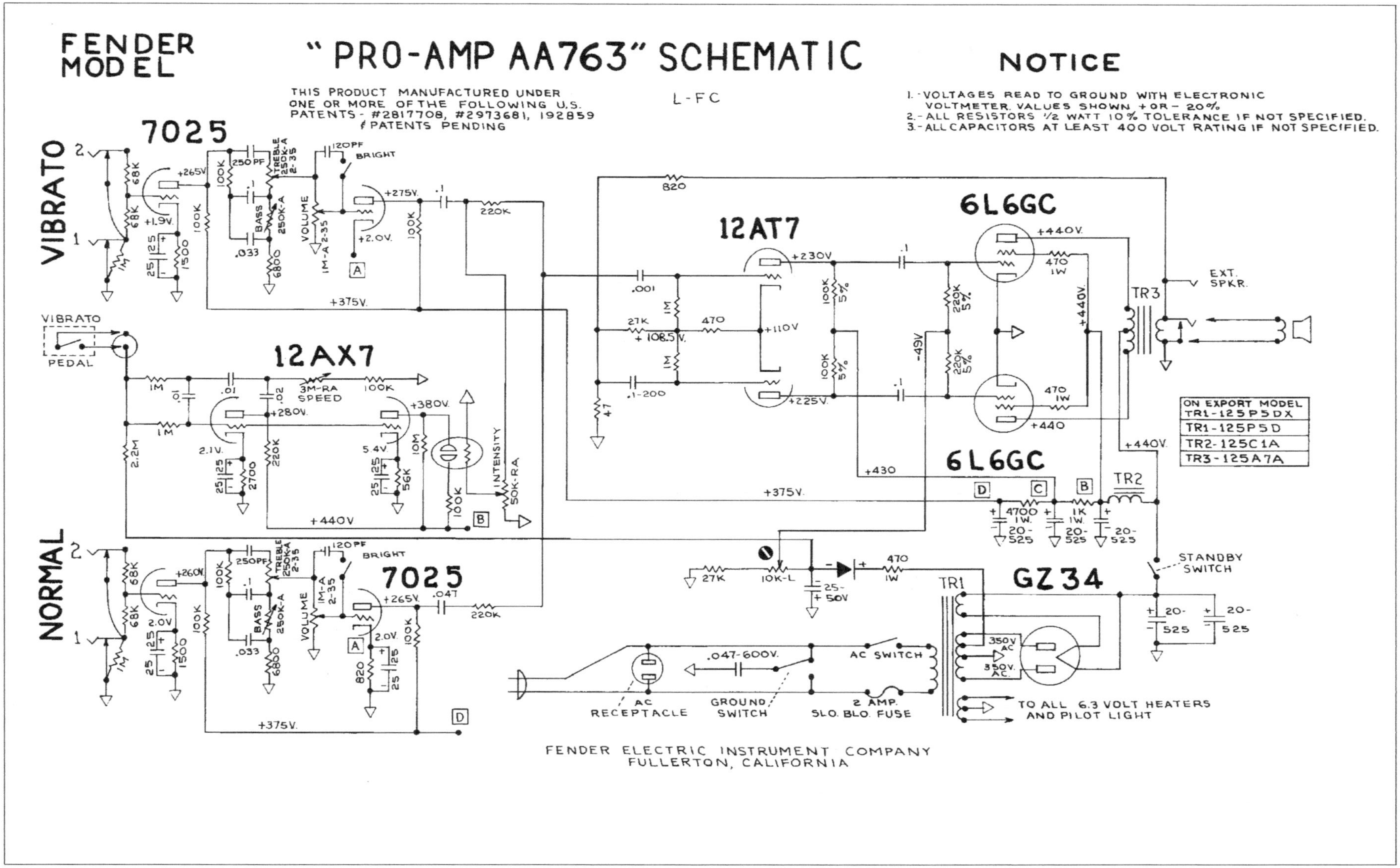

FENDER MODEL "PRO-AMP AB763" LAYOUT

C-FD

NOTICE

VOLTAGES READ TO GROUND WITH ELECTRONIC VOLTMETER VALUES SHOWN + OR − 20%

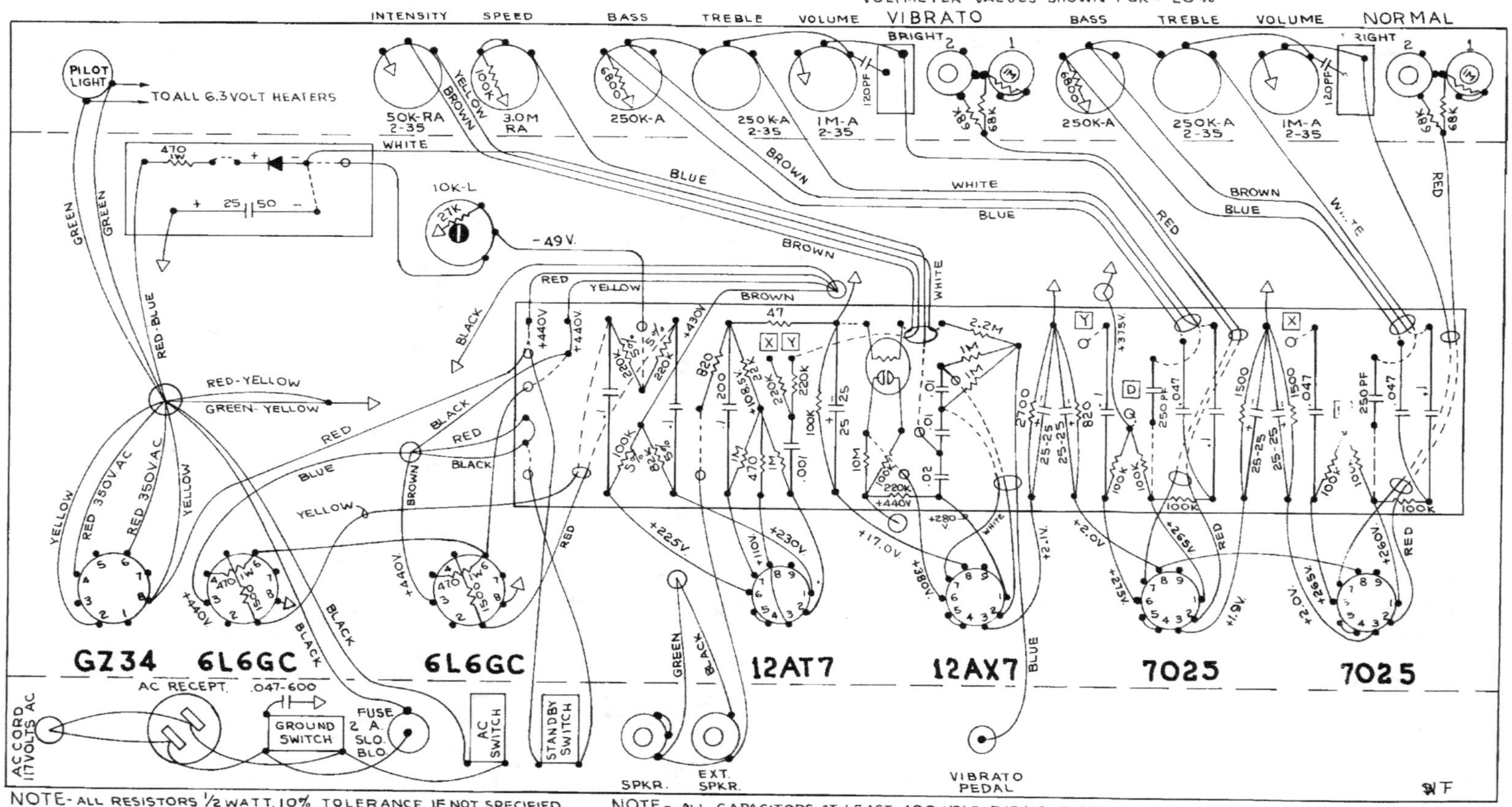

NOTE- ALL RESISTORS ½ WATT, 10% TOLERANCE IF NOT SPECIFIED. NOTE- ALL CAPACITORS AT LEAST 400 VOLT RATING IF NOT SPECIFIED.

FENDER ELECTRIC INSTRUMENT COMPANY
FULLERTON, CALIFORNIA
U.S.A.

FENDER MUSICAL INSTRUMENTS
A DIVISION OF COLUMBIA RECORDS DISTRIBUTION CORP.
SANTA ANA, CALIFORNIA
U.S.A.

FENDER
MODEL

"PRO-AMP AB763" SCHEMATIC

C-FD

THIS PRODUCT MANUFACTURED UNDER ONE OR MORE OF THE FOLLOWING U.S. PATENTS - #2817708, #2973681, 192859
PATENTS PENDING

NOTICE

1. VOLTAGES READ TO GROUND WITH ELECTRONIC VOLTMETER. VALUES SHOWN + OR − 20%
2. ALL RESISTORS 1/2 WATT 10% TOLERANCE IF NOT SPECIFIED.
3. ALL CAPACITORS AT LEAST 400 VOLT RATING IF NOT SPECIFIED.

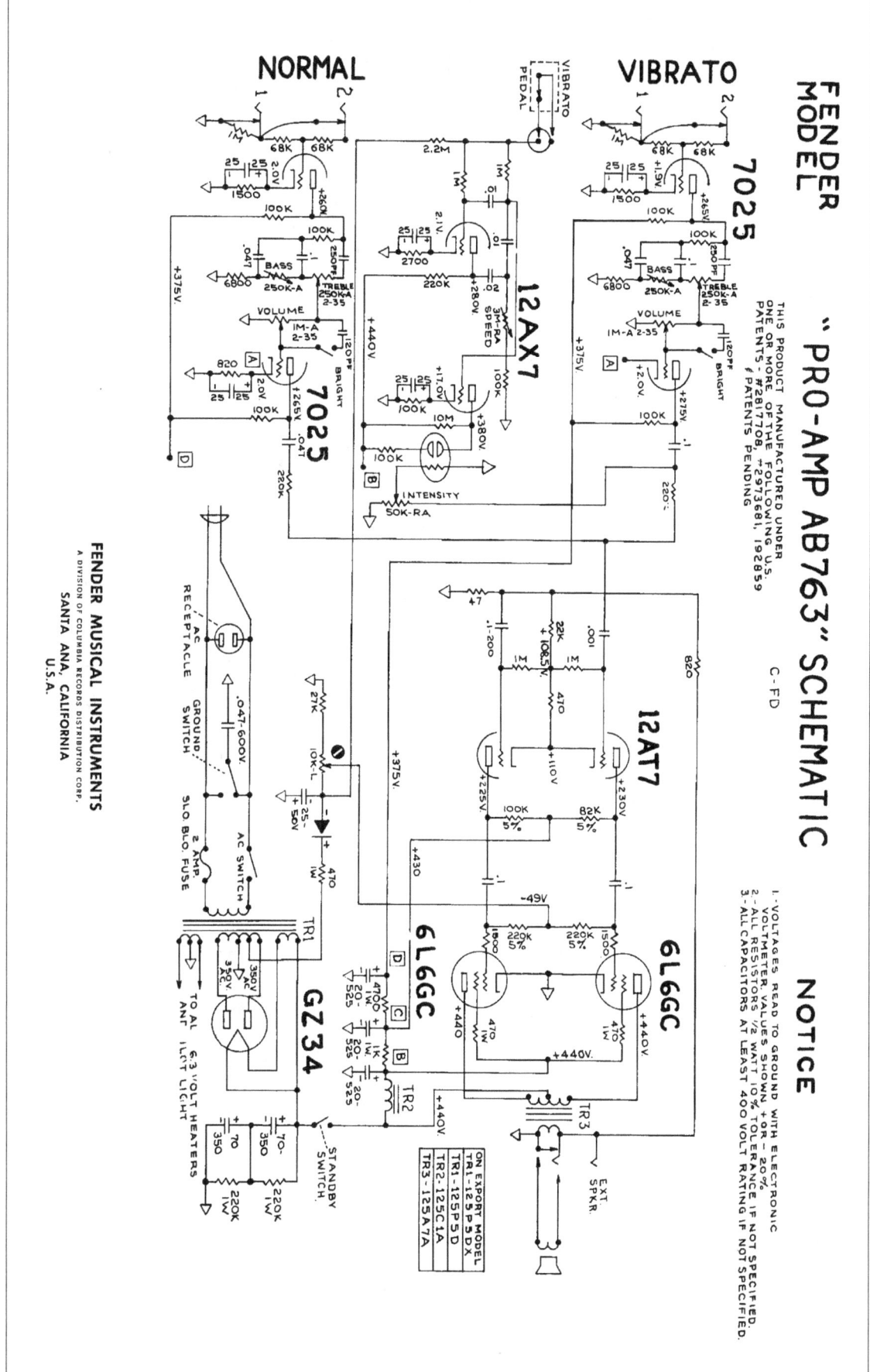

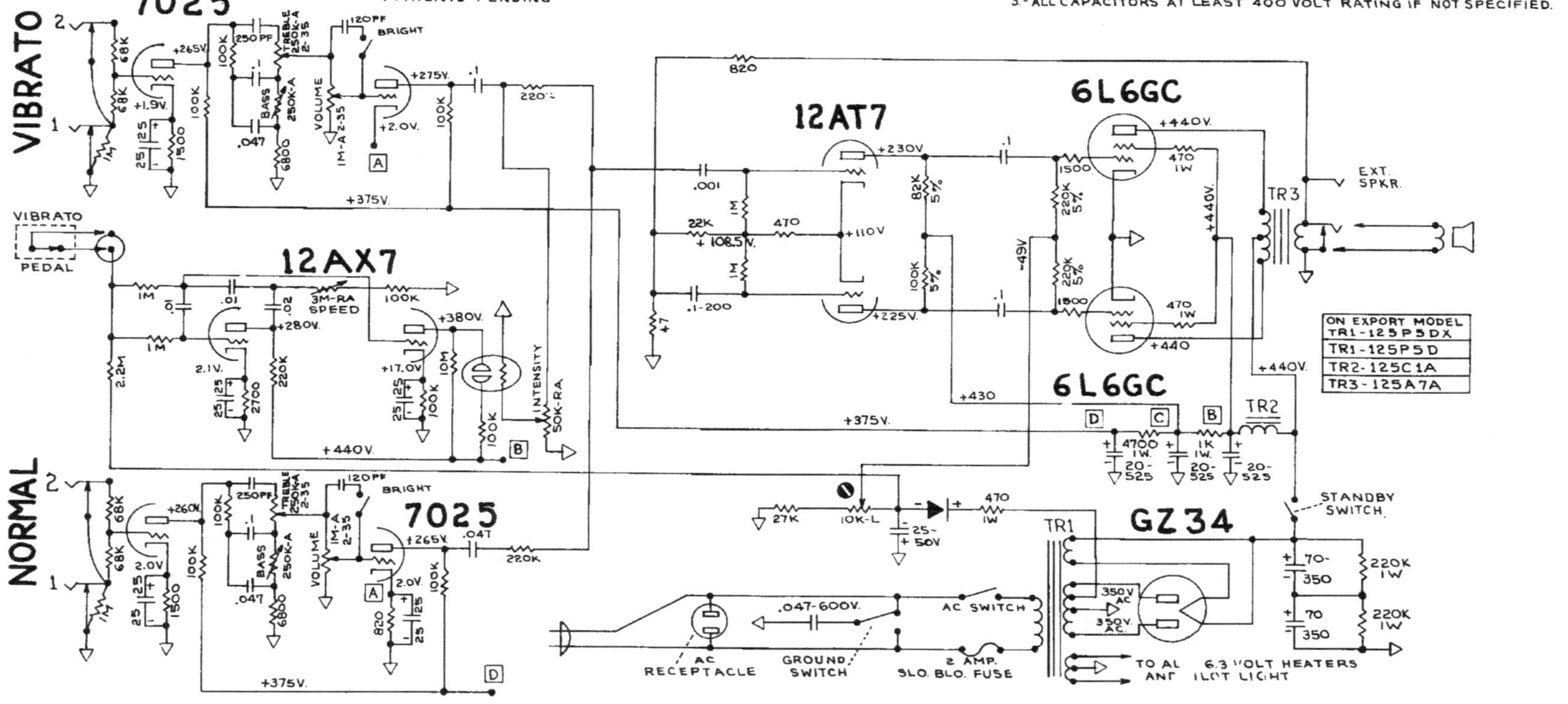

FENDER MUSICAL INSTRUMENTS
A DIVISION OF COLUMBIA RECORDS DISTRIBUTION CORP.
SANTA ANA, CALIFORNIA
U.S.A.

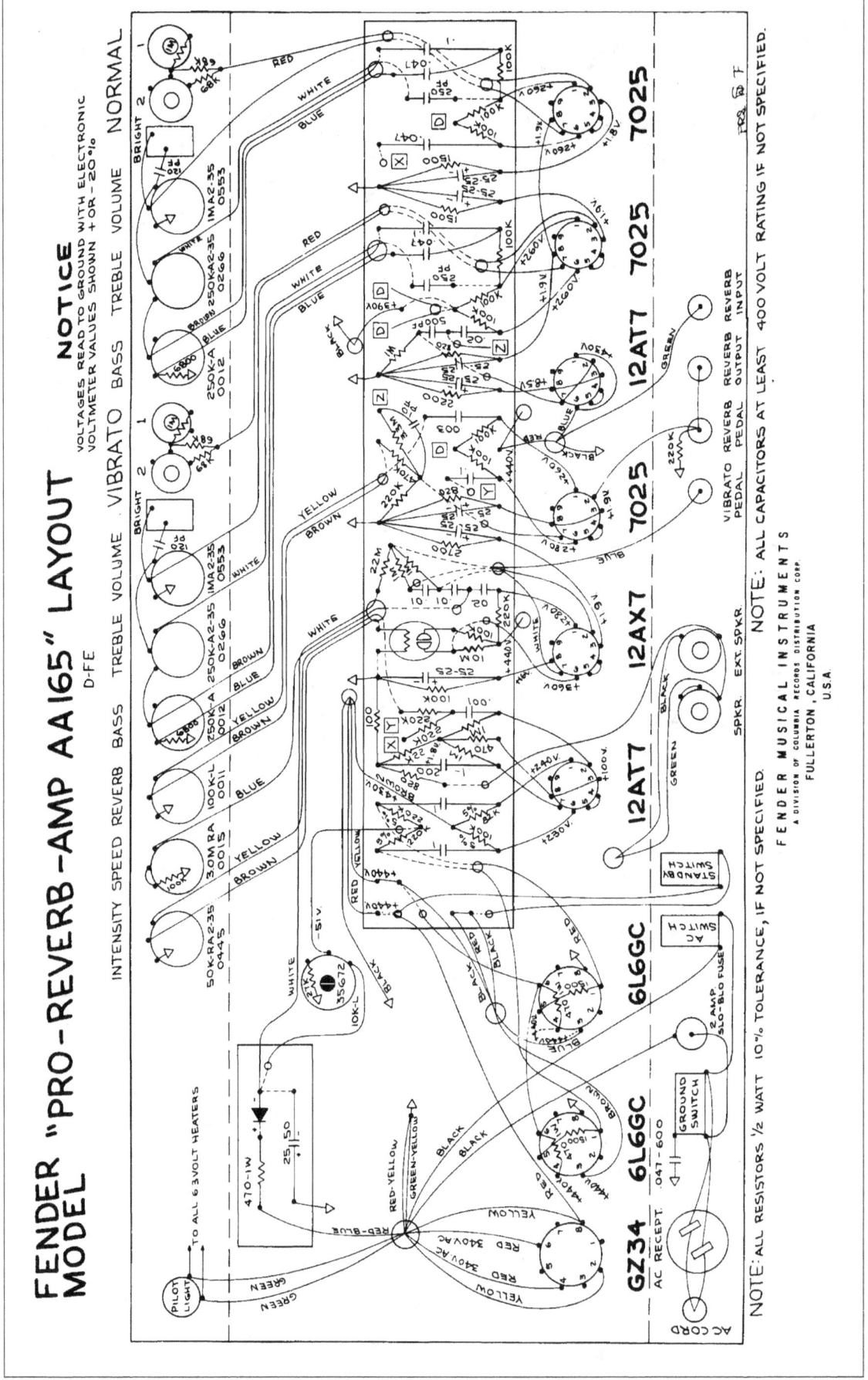

FENDER MODEL "PRO-REVERB-AMP AA165" LAYOUT

D-FE

NOTICE

VOLTAGES READ TO GROUND WITH ELECTRONIC VOLTMETER VALUES SHOWN + OR - 20%

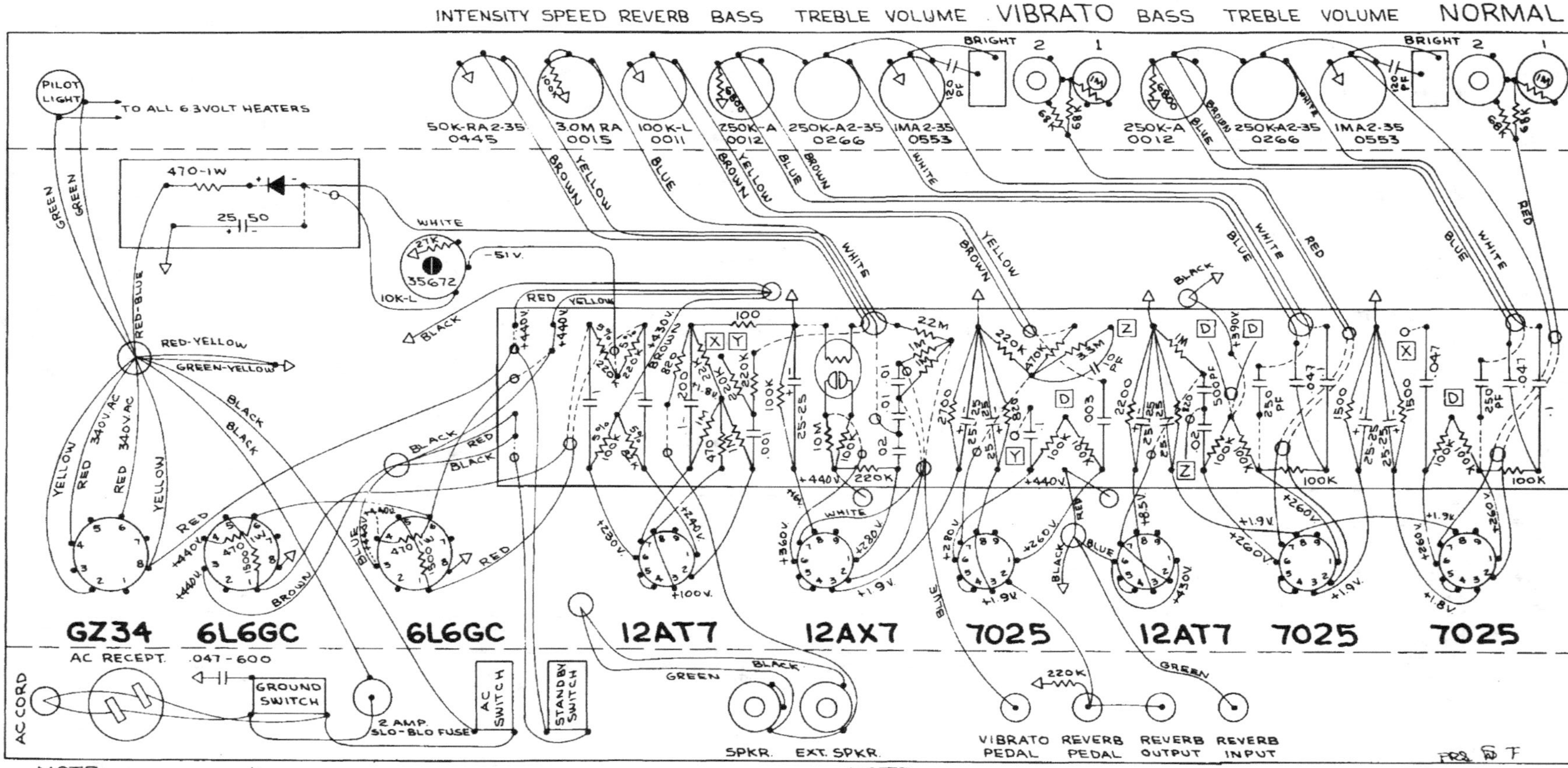

NOTE: ALL RESISTORS ½ WATT 10% TOLERANCE, IF NOT SPECIFIED.

NOTE: ALL CAPACITORS AT LEAST 400 VOLT RATING IF NOT SPECIFIED.

FENDER MUSICAL INSTRUMENTS
A DIVISION OF COLUMBIA RECORDS DISTRIBUTION CORP.
FULLERTON, CALIFORNIA
U.S.A.

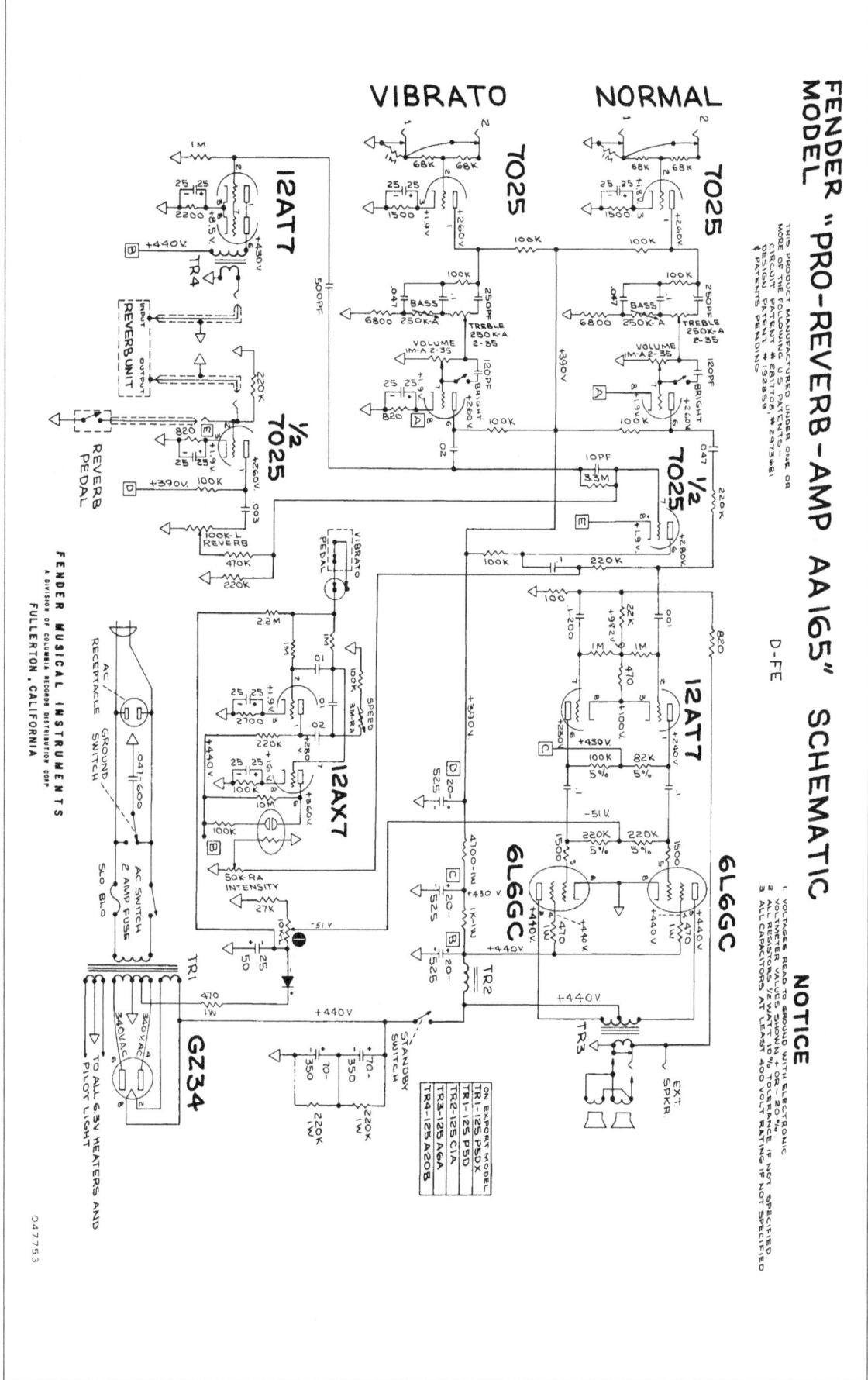

FENDER MODEL "PRO-REVERB-AMP AA165" SCHEMATIC

THIS PRODUCT MANUFACTURED UNDER ONE OR MORE OF THE FOLLOWING U.S. PATENTS –
CIRCUIT PATENT # 2817708, # 2973681
DESIGN PATENT # 192859
& PATENTS PENDING

D-FE

NOTICE

1. VOLTAGES READ TO GROUND WITH ELECTRONIC VOLTMETER VALUES SHOWN + OR – 20 %
2. ALL RESISTORS 1/2 WATT 10% TOLERANCE IF NOT SPECIFIED
3. ALL CAPACITORS AT LEAST 400 VOLT RATING IF NOT SPECIFIED

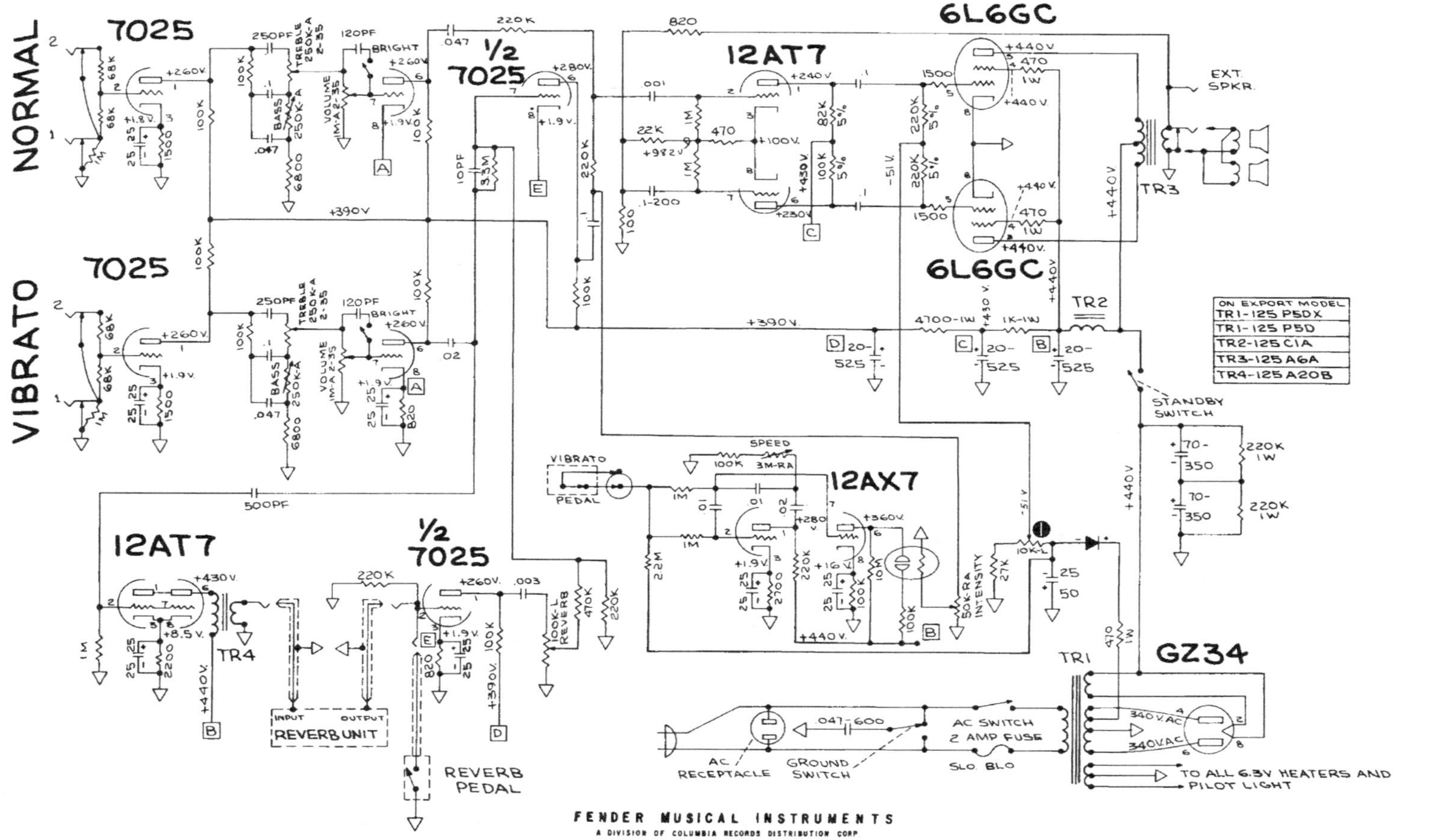

FENDER MUSICAL INSTRUMENTS
A DIVISION OF COLUMBIA RECORDS DISTRIBUTION CORP.
FULLERTON, CALIFORNIA

047753

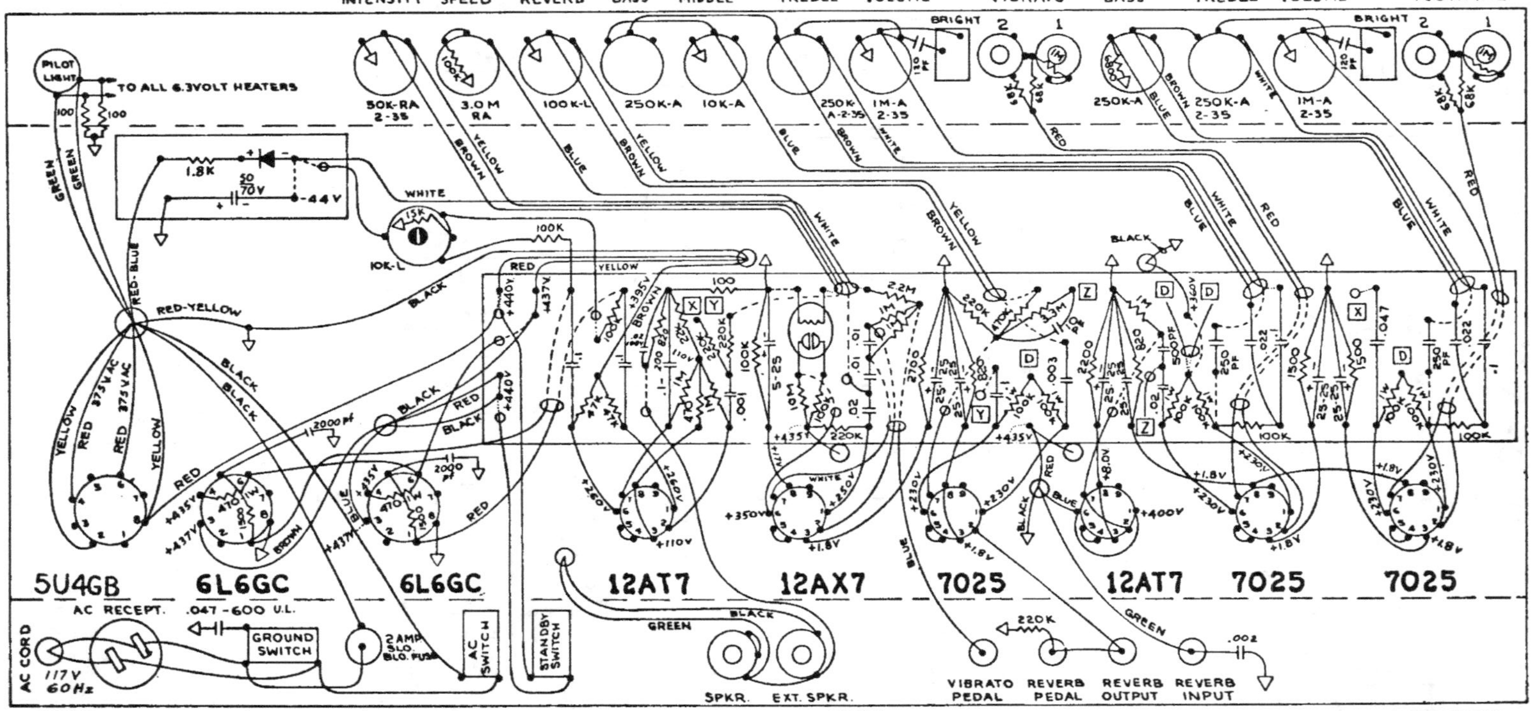

NOTE: ALL RESISTORS ½ WATT 10% TOLERANCE, IF NOT SPECIFIED.

NOTE: ALL CAPACITORS AT LEAST 600 VOLT RATING IF NOT SPECIFIED.

FENDER MUSICAL INSTRUMENTS
CBS MUSICAL INSTRUMENTS
A DIVISION OF COLUMBIA BROADCASTING SYSTEMS INC.
FULLERTON, CALIFORNIA
U.S.A.

DRAWN	F	10-16-69	
APPROVED			

DWG. TITLE: SCHEMATIC & WIRING DIAGRAM PRO REVERB-AMP

SIZE	DWG. NO.	REV
D	047753	D

SCALE NONE | MODEL TUBE TYPE PRO REVERB-AMP | SHEET 2 OF 2

D	INCORPORATED ECO #1689	10-23-69	
C	INCORPORATED ECO #1428	1-6-69	
B	INCORPORATED ECO #1081	6-10-68	E. James

REVISIONS

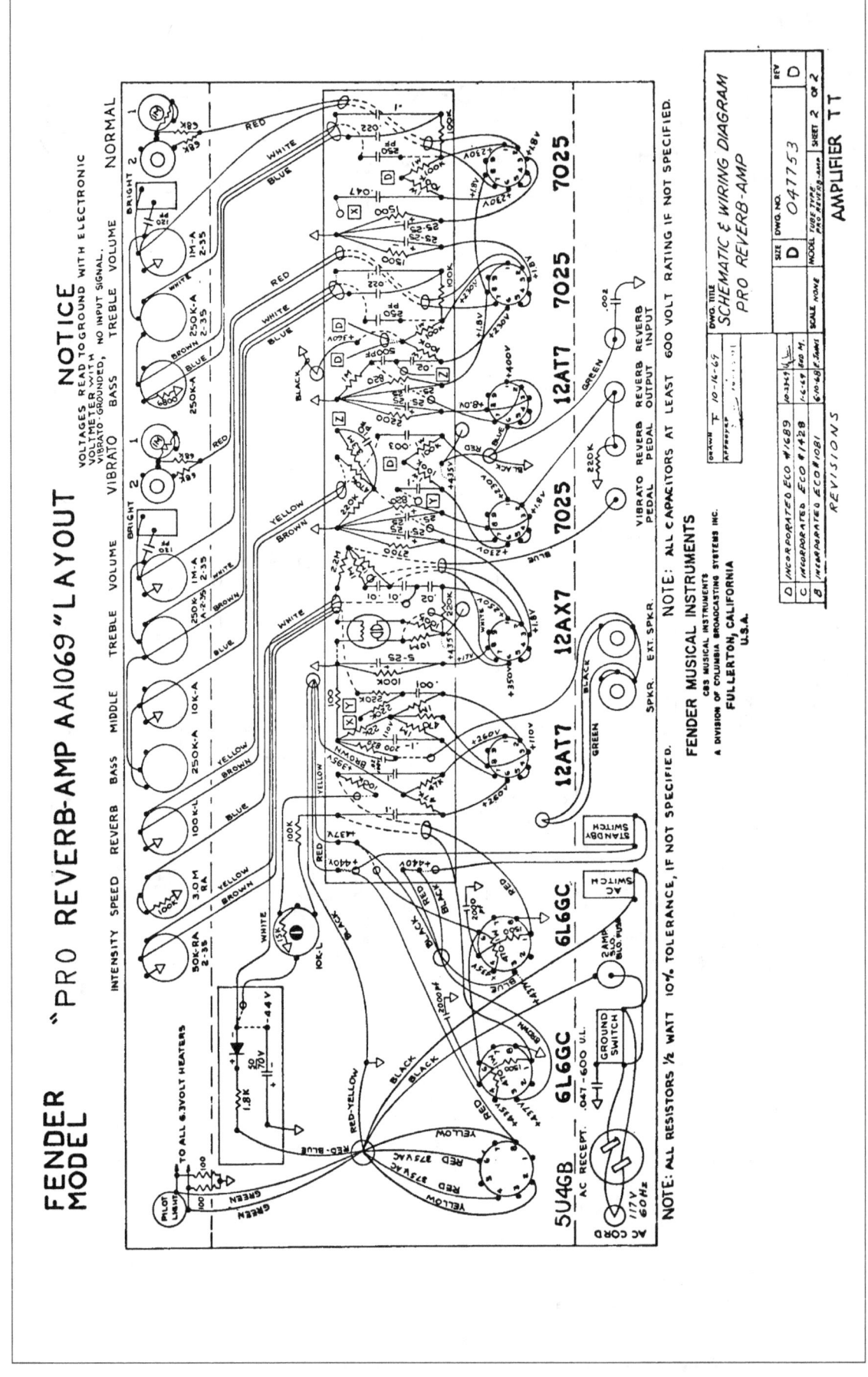

AMPLIFIER TT

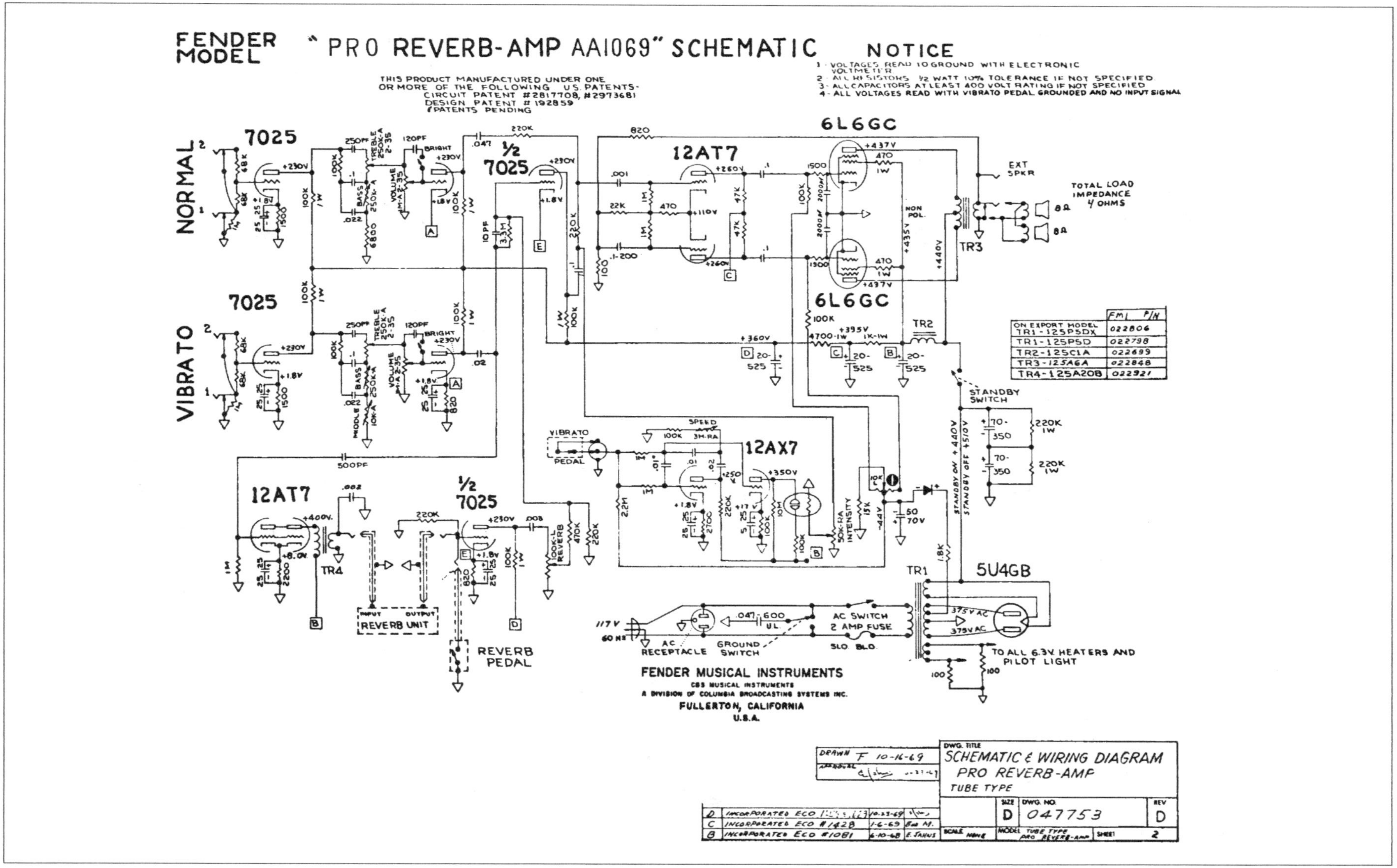

FENDER MODEL "PRO REVERB-AMP AA1069" SCHEMATIC
THIS PRODUCT MANUFACTURED UNDER ONE OR MORE OF THE FOLLOWING US PATENTS- CIRCUIT PATENT #2817708, #2973681 DESIGN PATENT # 192859 PATENTS PENDING
NOTICE
1 - VOLTAGES READ TO GROUND WITH ELECTRONIC VOLTMETER
2 - ALL RESISTORS 1/2 WATT 10% TOLERANCE IF NOT SPECIFIED
3 - ALL CAPACITORS AT LEAST 400 VOLT RATING IF NOT SPECIFIED
4 - ALL VOLTAGES READ WITH VIBRATO PEDAL GROUNDED AND NO INPUT SIGNAL
NORMAL
VIBRATO
7025
1/2 7025
12AT7
6L6GC
12AX7
5U4GB
TR1
TR2
TR3
TR4
EXT SPKR
TOTAL LOAD IMPEDANCE 4 OHMS
REVERB UNIT
REVERB PEDAL
VIBRATO PEDAL
STANDBY SWITCH
AC RECEPTACLE
GROUND SWITCH
AC SWITCH 2 AMP FUSE
TO ALL 6.3V HEATERS AND PILOT LIGHT
ON EXPORT MODEL TR1-125P5DX 022806
TR1-125P5D 022798
TR2-125C1A 022699
TR3-125A6A 022848
TR4-125A20B 022921
FENDER MUSICAL INSTRUMENTS
CBS MUSICAL INSTRUMENTS
A DIVISION OF COLUMBIA BROADCASTING SYSTEMS INC.
FULLERTON, CALIFORNIA U.S.A.
SCHEMATIC & WIRING DIAGRAM PRO REVERB-AMP TUBE TYPE
D 047753 D
DRAWN F 10-16-69
INCORPORATED ECO #1428 1-6-69
INCORPORATED ECO #1081 6-10-68

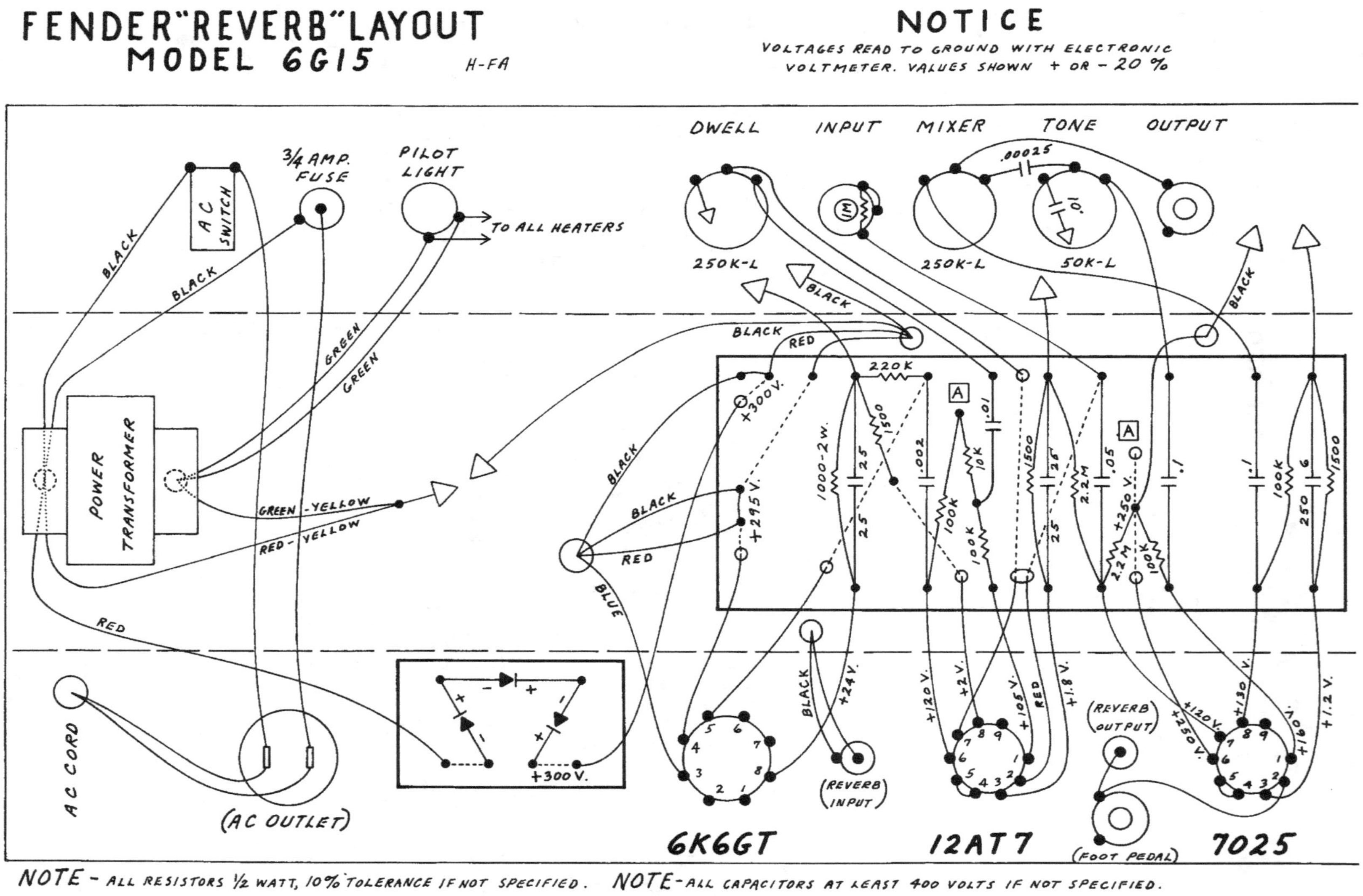
FENDER "REVERB" LAYOUT
MODEL 6G15
H-FA
NOTICE
VOLTAGES READ TO GROUND WITH ELECTRONIC VOLTMETER. VALUES SHOWN + OR − 20 %
DWELL
INPUT
MIXER
TONE
OUTPUT
250K-L
250K-L
50K-L
.00025
3/4 AMP. FUSE
PILOT LIGHT
TO ALL HEATERS
AC SWITCH
POWER TRANSFORMER
BLACK
GREEN
GREEN-YELLOW
RED-YELLOW
RED
BLUE
220K
1000-2W.
+300 V.
+295 V.
+250 V.
+24 V.
+120 V.
+2 V.
+105 V.
+1.8 V.
+130 V.
+160 V.
+1.2 V.
(REVERB INPUT)
(REVERB OUTPUT)
(FOOT PEDAL)
AC CORD
(AC OUTLET)
6K6GT
12AT7
7025
NOTE - ALL RESISTORS 1/2 WATT, 10% TOLERANCE IF NOT SPECIFIED.
NOTE - ALL CAPACITORS AT LEAST 400 VOLTS IF NOT SPECIFIED.

FENDER "REVERB" SCHEMATIC
MODEL 6G15

H-FA

NOTICE

1- VOLTAGES READ TO GROUND WITH ELECTRONIC VOLTMETER. VALUES SHOWN + OR - 20%.

2- ALL RESISTORS ½ WATT IF NOT SPECIFIED.

3- ALL CAPACITORS AT LEAST 400 VOLTS IF NOT SPECIFIED.

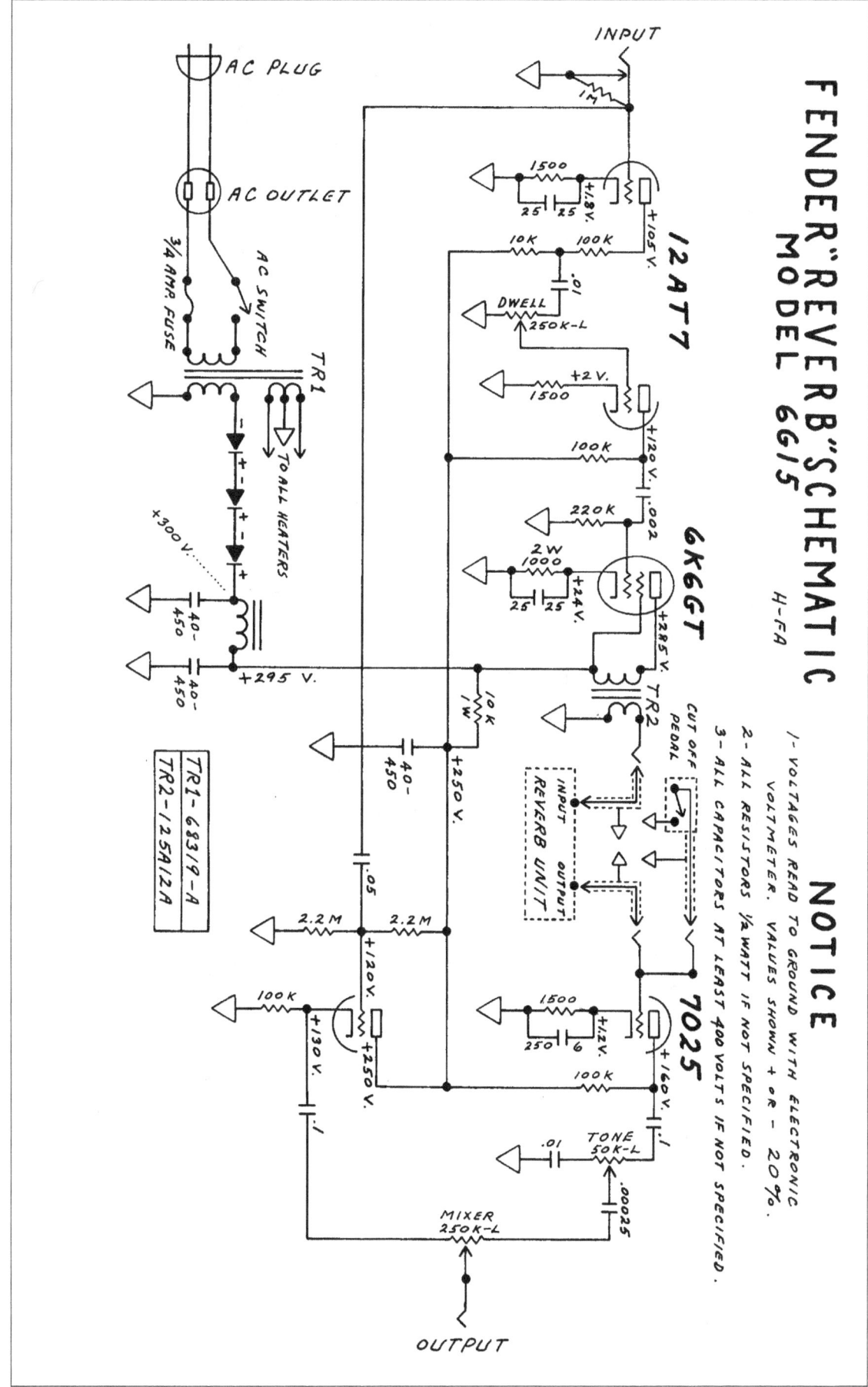

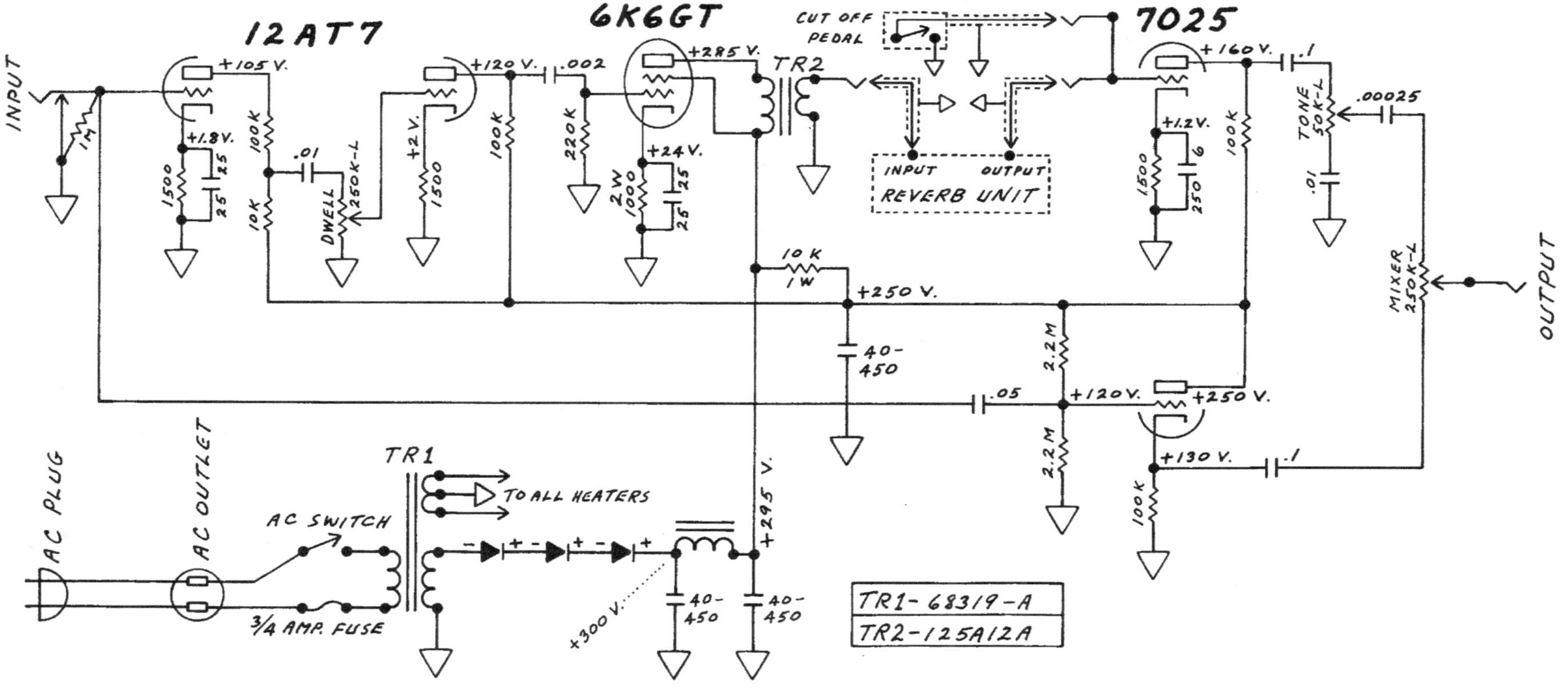

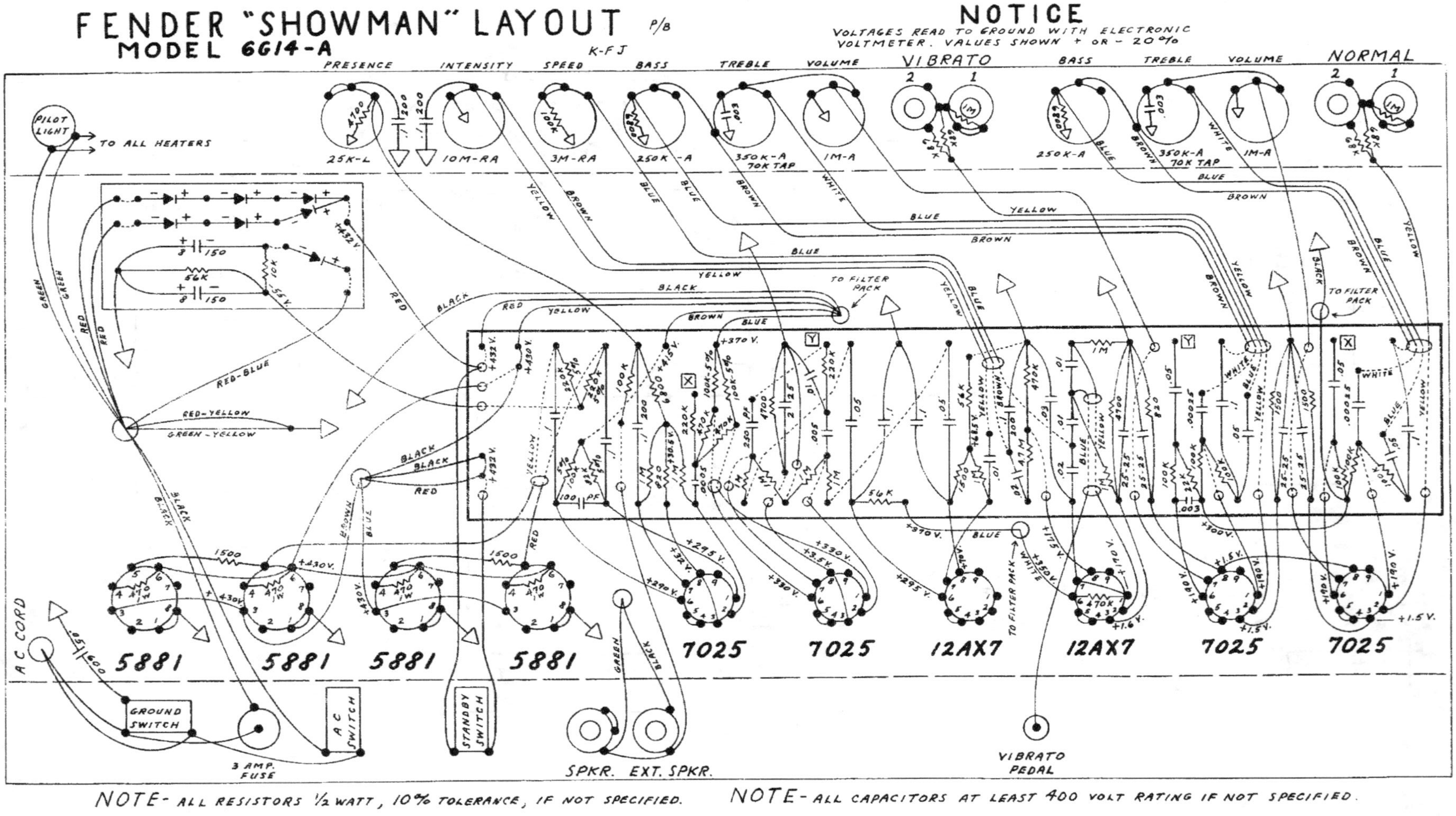
FENDER "SHOWMAN" LAYOUT
MODEL 6G14-A
NOTICE
VOLTAGES READ TO GROUND WITH ELECTRONIC VOLTMETER. VALUES SHOWN + OR - 20%
PRESENCE
INTENSITY
SPEED
BASS
TREBLE
VOLUME
VIBRATO
NORMAL
PILOT LIGHT
TO ALL HEATERS
TO FILTER PACK
AC CORD
GROUND SWITCH
3 AMP. FUSE
AC SWITCH
STANDBY SWITCH
SPKR.
EXT. SPKR.
VIBRATO PEDAL
5881
5881
5881
5881
7025
7025
12AX7
12AX7
7025
7025
NOTE- ALL RESISTORS ½ WATT, 10% TOLERANCE, IF NOT SPECIFIED.
NOTE- ALL CAPACITORS AT LEAST 400 VOLT RATING IF NOT SPECIFIED.

SHOWMAN **6G14-A**

Fender

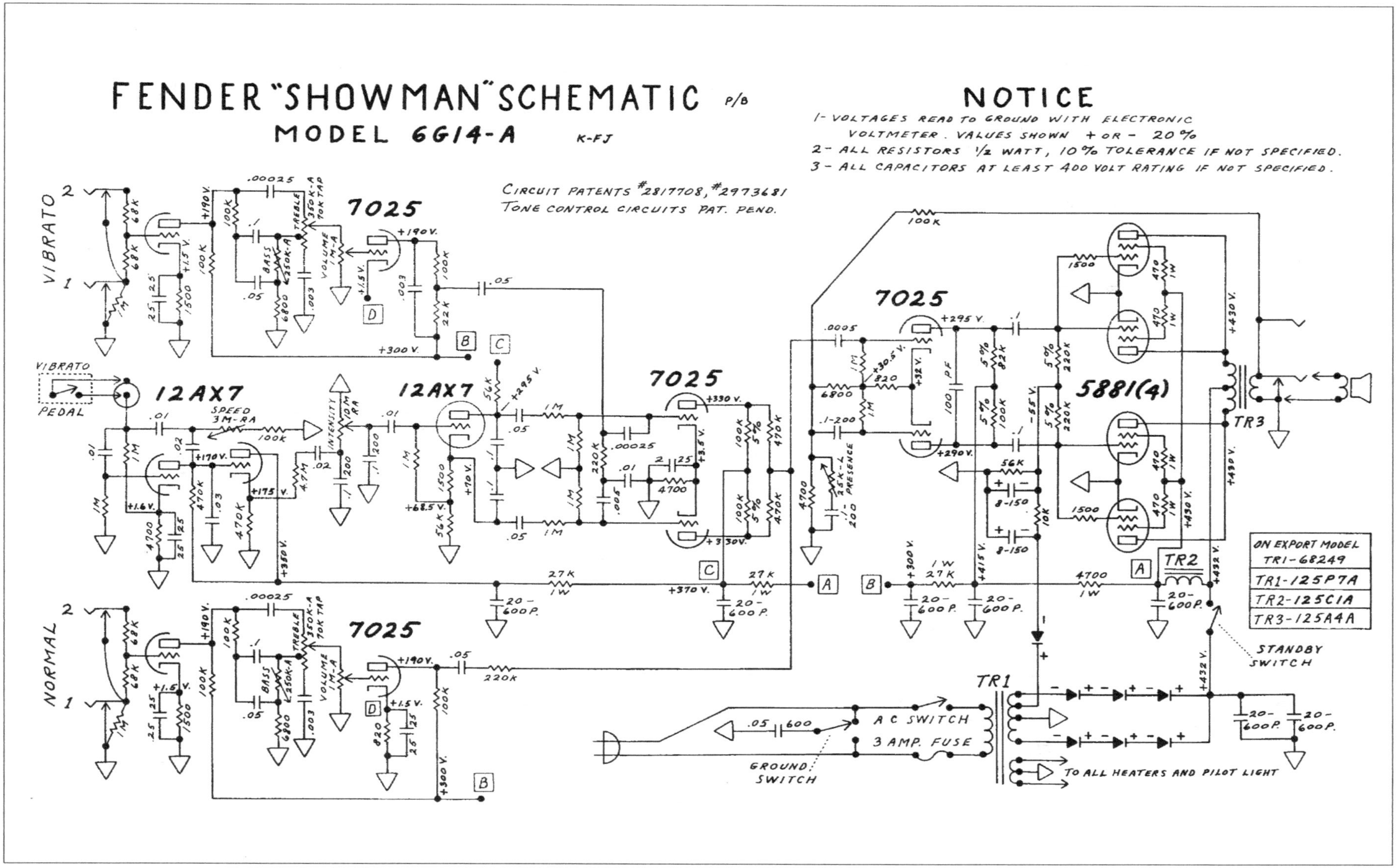

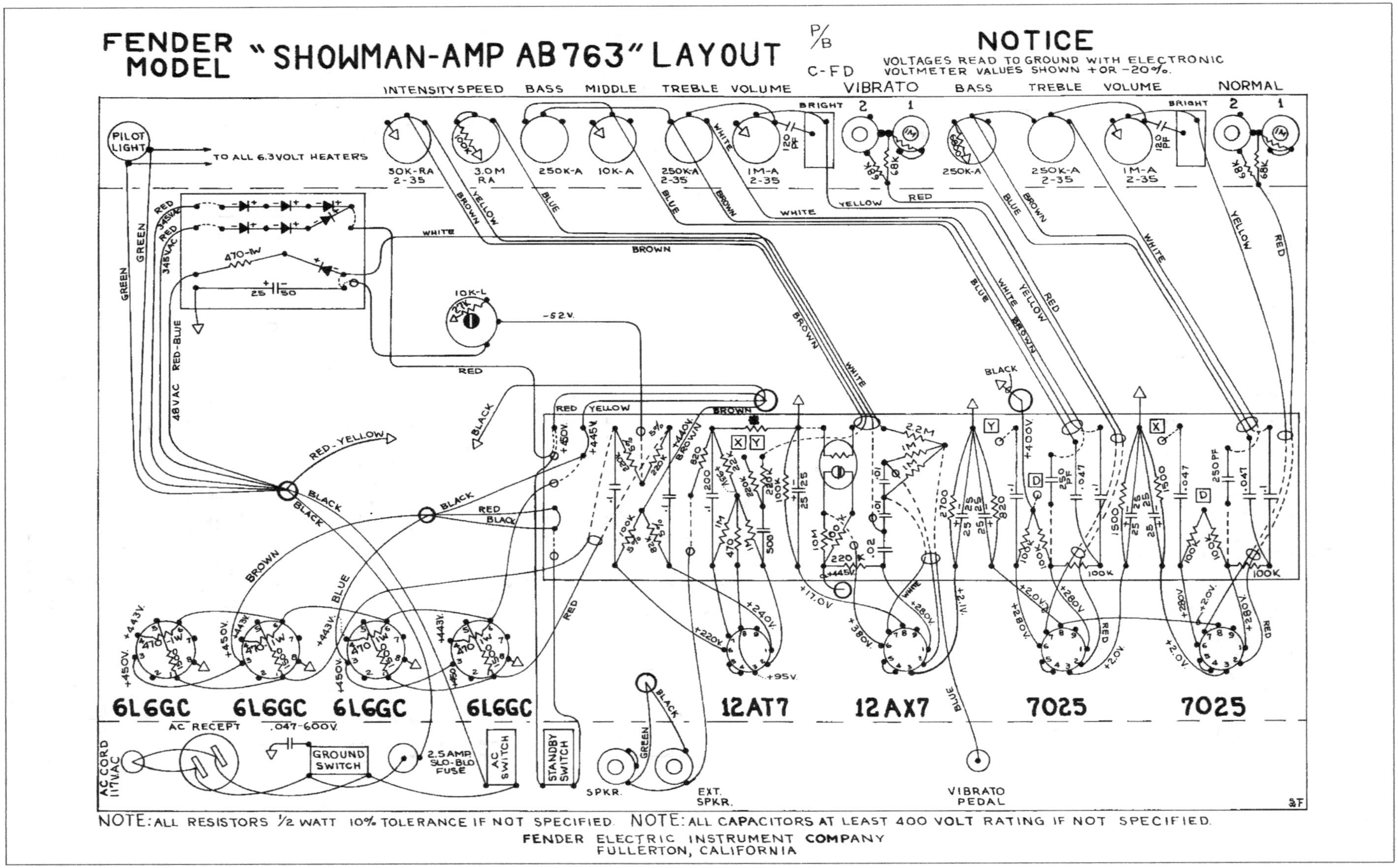
FENDER MODEL "SHOWMAN-AMP AB763" LAYOUT
P/B C-FD
NOTICE
VOLTAGES READ TO GROUND WITH ELECTRONIC VOLTMETER VALUES SHOWN +OR -20%.
INTENSITY SPEED BASS MIDDLE TREBLE VOLUME
VIBRATO 2 1
BASS TREBLE VOLUME
NORMAL 2 1
BRIGHT
PILOT LIGHT
TO ALL 6.3VOLT HEATERS
6L6GC
6L6GC
6L6GC
6L6GC
12AT7
12AX7
7025
7025
AC CORD 117VAC
AC RECEPT
.047-600V.
GROUND SWITCH
2.5AMP. SLO-BLO FUSE
AC SWITCH
STANDBY SWITCH
SPKR.
EXT. SPKR.
VIBRATO PEDAL
NOTE: ALL RESISTORS 1/2 WATT 10% TOLERANCE IF NOT SPECIFIED.
NOTE: ALL CAPACITORS AT LEAST 400 VOLT RATING IF NOT SPECIFIED.
FENDER ELECTRIC INSTRUMENT COMPANY
FULLERTON, CALIFORNIA

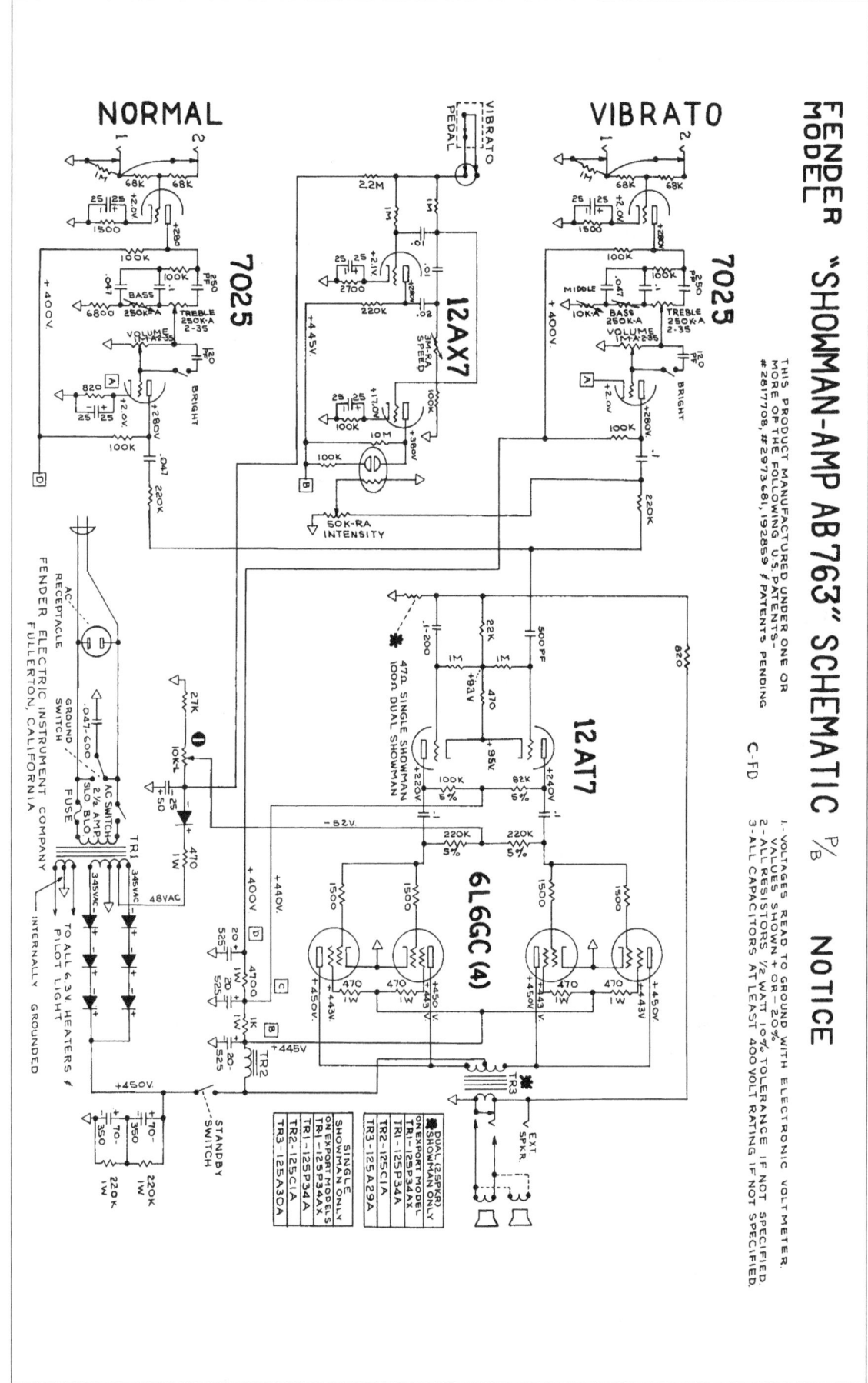

FENDER MODEL "SHOWMAN-AMP AB763" SCHEMATIC P/B

THIS PRODUCT MANUFACTURED UNDER ONE OR MORE OF THE FOLLOWING U.S. PATENTS- #2817708, #2973681, 192859 ≠ PATENTS PENDING

C-FD

NOTICE

1.- VOLTAGES READ TO GROUND WITH ELECTRONIC VOLTMETER. VALUES SHOWN + OR − 20%
2 - ALL RESISTORS ½ WATT 10% TOLERANCE IF NOT SPECIFIED.
3 - ALL CAPACITORS AT LEAST 400 VOLT RATING IF NOT SPECIFIED.

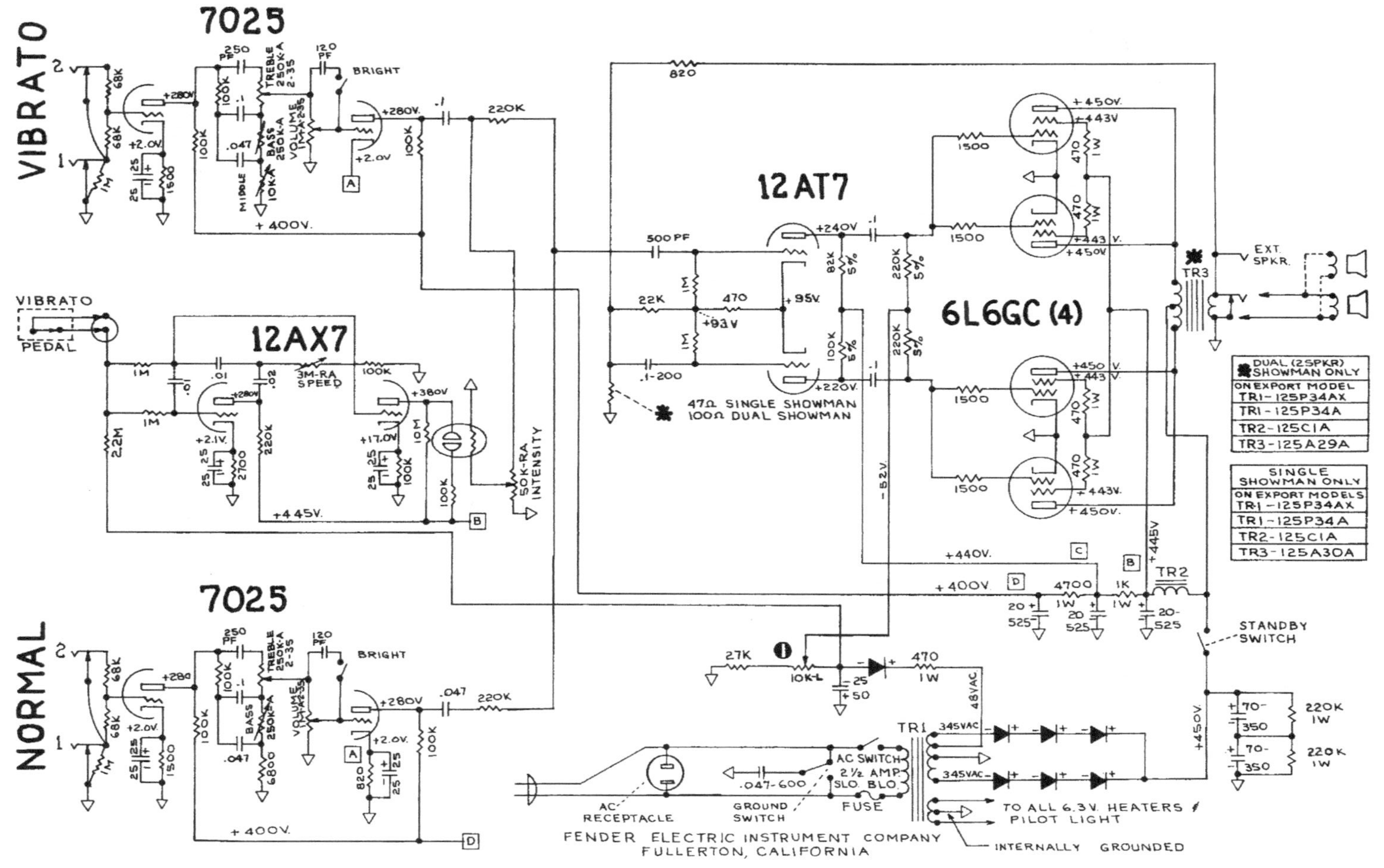

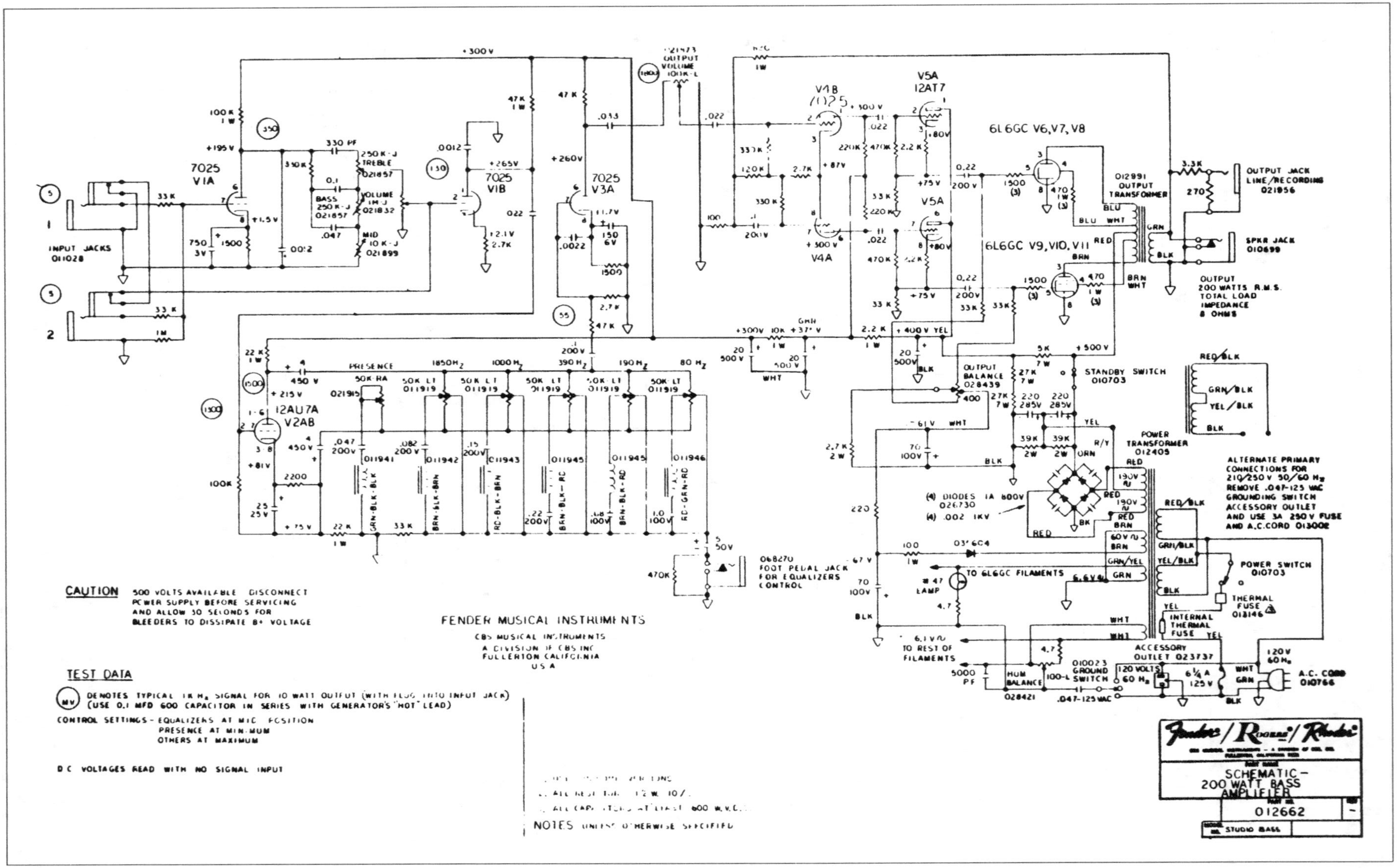
+300V
INPUT JACKS 011028
7025 V1A
7025 V1B
7025 V3A
V1B 7025
V4A
V5A 12AT7
V5A
6L6GC V6, V7, V8
6L6GC V9, V10, V11
12AU7A V2AB
PRESENCE
012991 OUTPUT TRANSFORMER
OUTPUT JACK LINE/RECORDING 021956
SPKR JACK 010699
OUTPUT 200 WATTS R.M.S. TOTAL LOAD IMPEDANCE 8 OHMS
STANDBY SWITCH 010703
POWER TRANSFORMER 012405
OUTPUT BALANCE 028439
(4) DIODES 1A 800V 026730
(4) .002 1KV
ALTERNATE PRIMARY CONNECTIONS FOR 210/250V 50/60 Hz REMOVE .047-125 VAC GROUNDING SWITCH ACCESSORY OUTLET AND USE 3A 250V FUSE AND A.C. CORD 013002
POWER SWITCH 010703
THERMAL FUSE 013146
INTERNAL THERMAL FUSE
ACCESSORY OUTLET 023737
A.C. CORD 010766
010023 GROUND SWITCH
HUM BALANCE 028421
TO 6L6GC FILAMENTS
#47 LAMP
6.1V TO REST OF FILAMENTS
068270 FOOT PEDAL JACK FOR EQUALIZERS CONTROL
CAUTION 500 VOLTS AVAILABLE DISCONNECT POWER SUPPLY BEFORE SERVICING AND ALLOW 30 SECONDS FOR BLEEDERS TO DISSIPATE B+ VOLTAGE
TEST DATA
DENOTES TYPICAL 1K Hz SIGNAL FOR 10 WATT OUTPUT (WITH PLUG INTO INPUT JACK) (USE 0.1 MFD 600 CAPACITOR IN SERIES WITH GENERATOR'S "HOT" LEAD)
CONTROL SETTINGS - EQUALIZERS AT MID POSITION PRESENCE AT MINIMUM OTHERS AT MAXIMUM
D C VOLTAGES READ WITH NO SIGNAL INPUT
FENDER MUSICAL INSTRUMENTS
CBS MUSICAL INSTRUMENTS
A DIVISION OF CBS INC
FULLERTON CALIFORNIA
U S A
NOTES UNLESS OTHERWISE SPECIFIED
Fender / Rhodes
SCHEMATIC - 200 WATT BASS AMPLIFIER
012662
STUDIO BASS

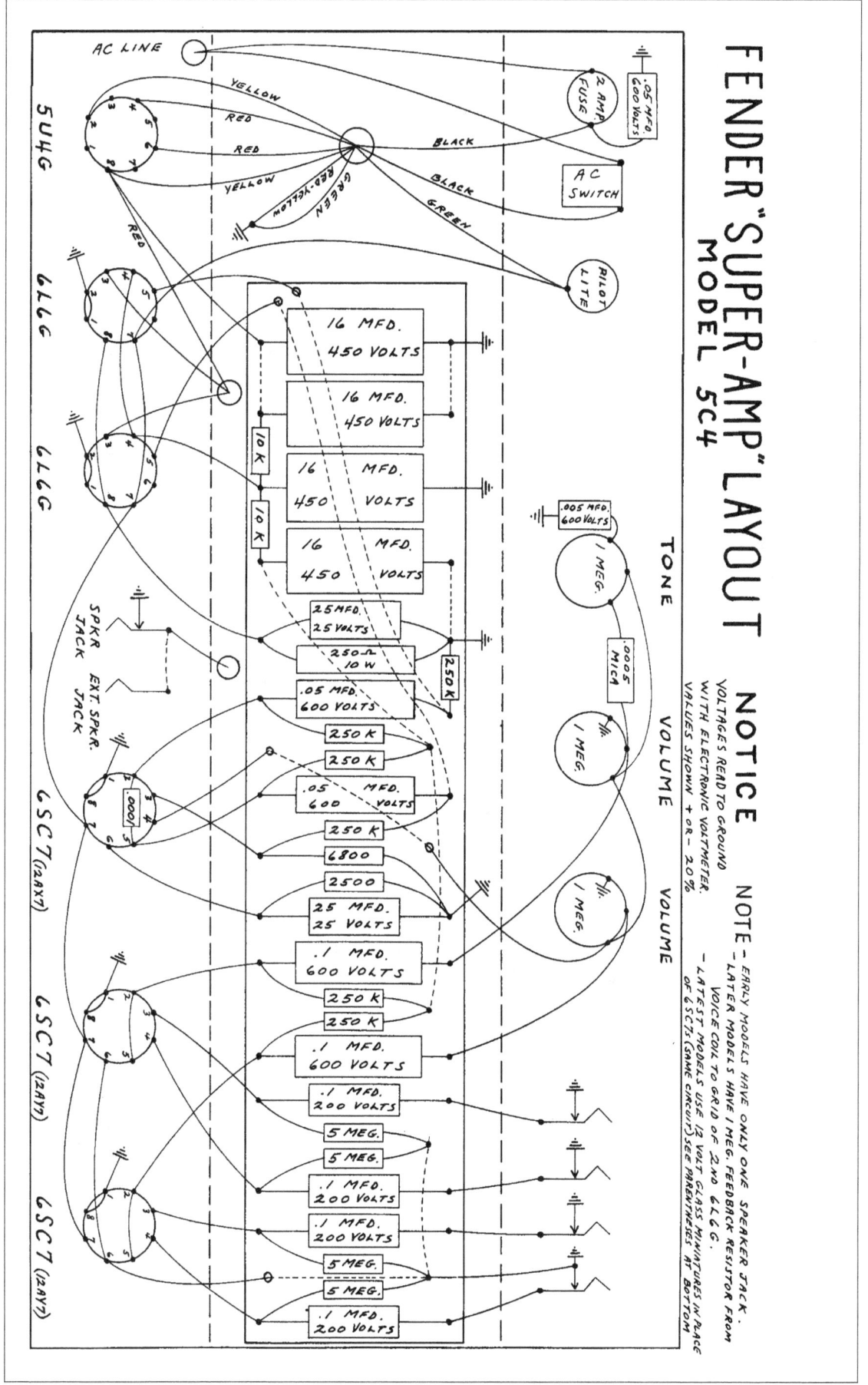

FENDER "SUPER-AMP" LAYOUT
MODEL 5C4

NOTICE
VOLTAGES READ TO GROUND
WITH ELECTRONIC VOLTMETER.
VALUES SHOWN + OR − 20%

NOTE – EARLY MODELS HAVE ONLY ONE SPEAKER JACK.
– LATER MODELS HAVE 1 MEG. FEEDBACK RESISTOR FROM VOICE COIL TO GRID OF 2ND 6L6G.
– LATEST MODELS USE 12 VOLT GLASS MINIATURES IN PLACE OF 6SC7s (SAME CIRCUIT) SEE PARENTHESES AT BOTTOM

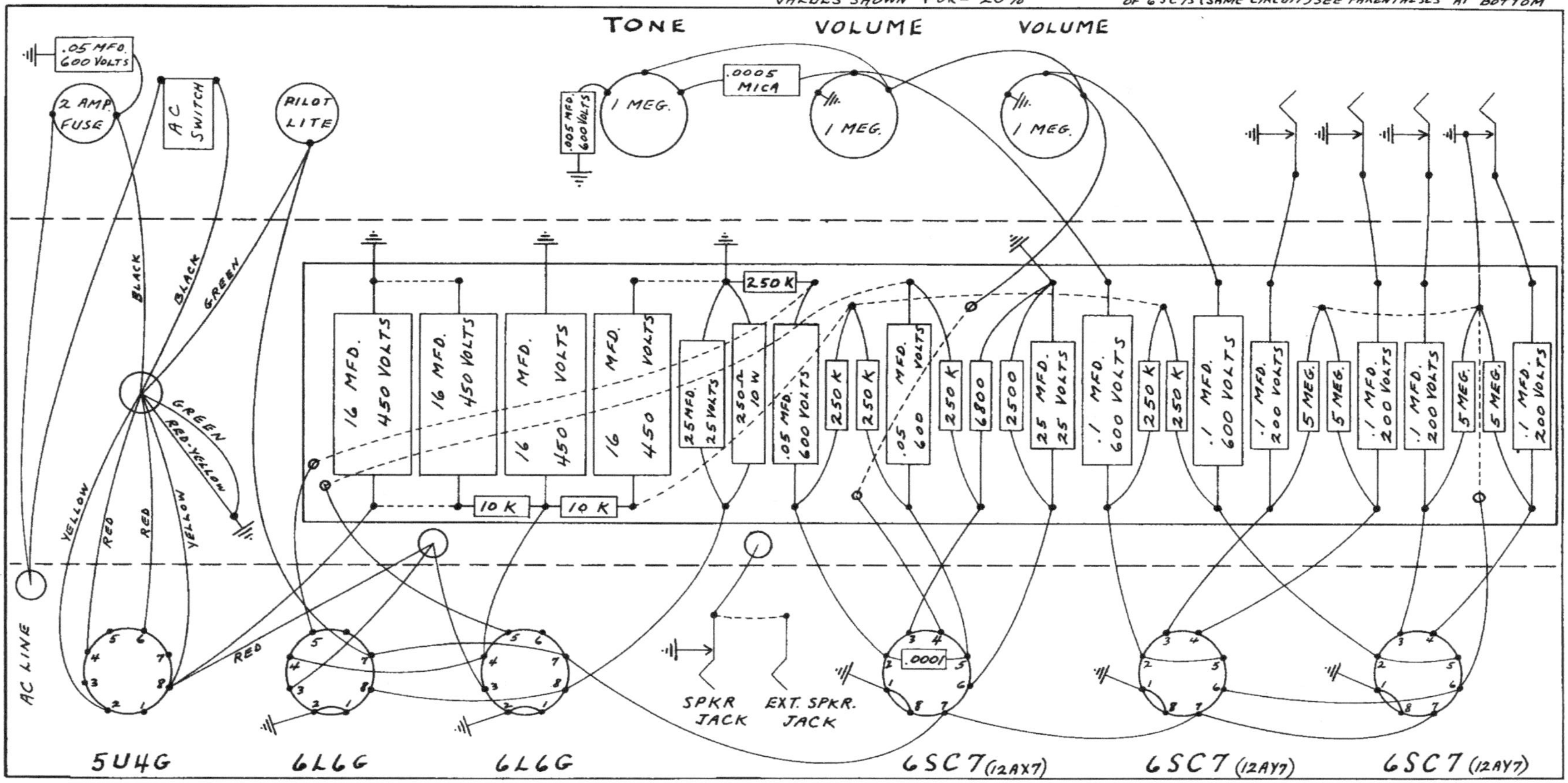

FENDER "SUPER-AMP" SCHEMATIC

MODEL 5C4

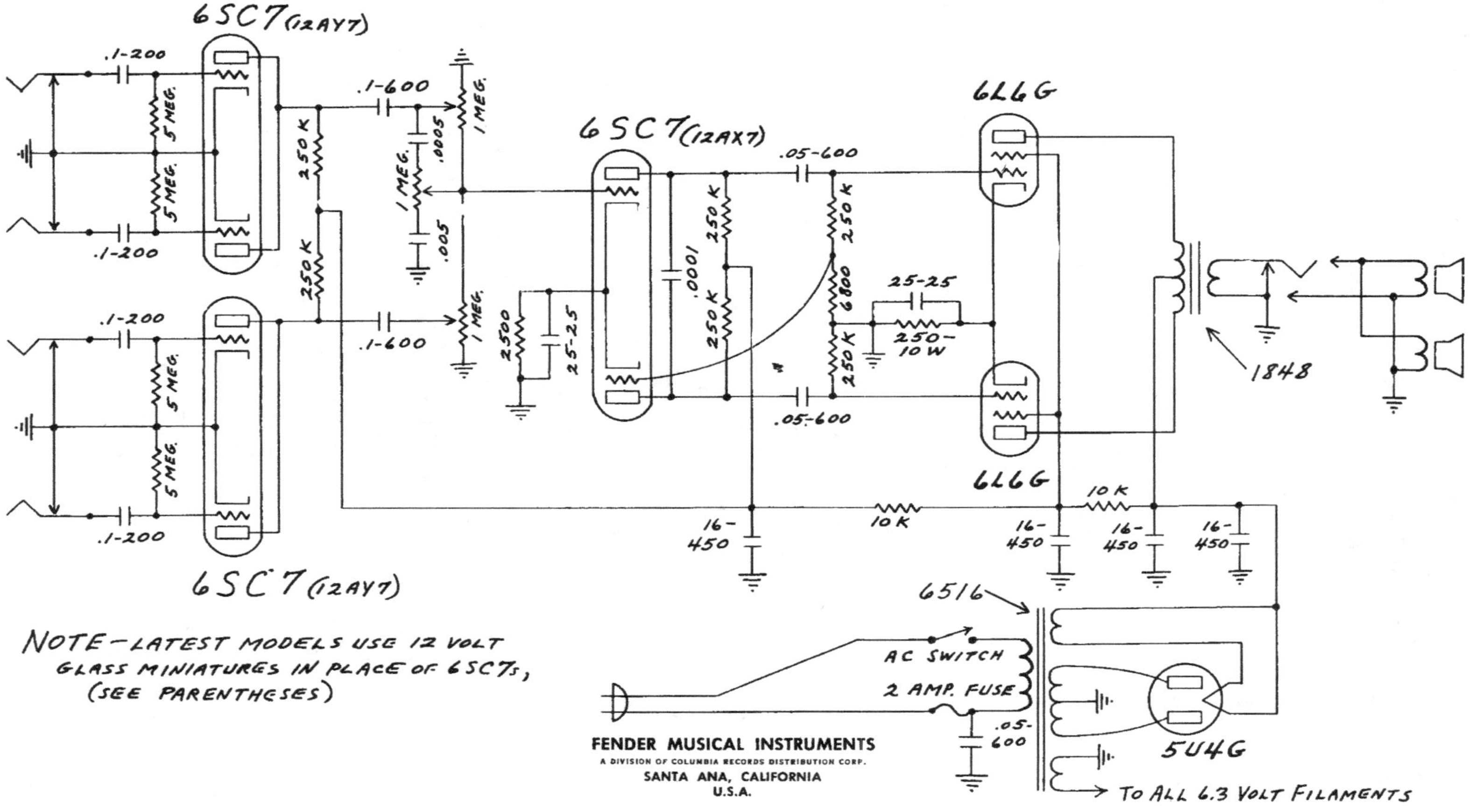

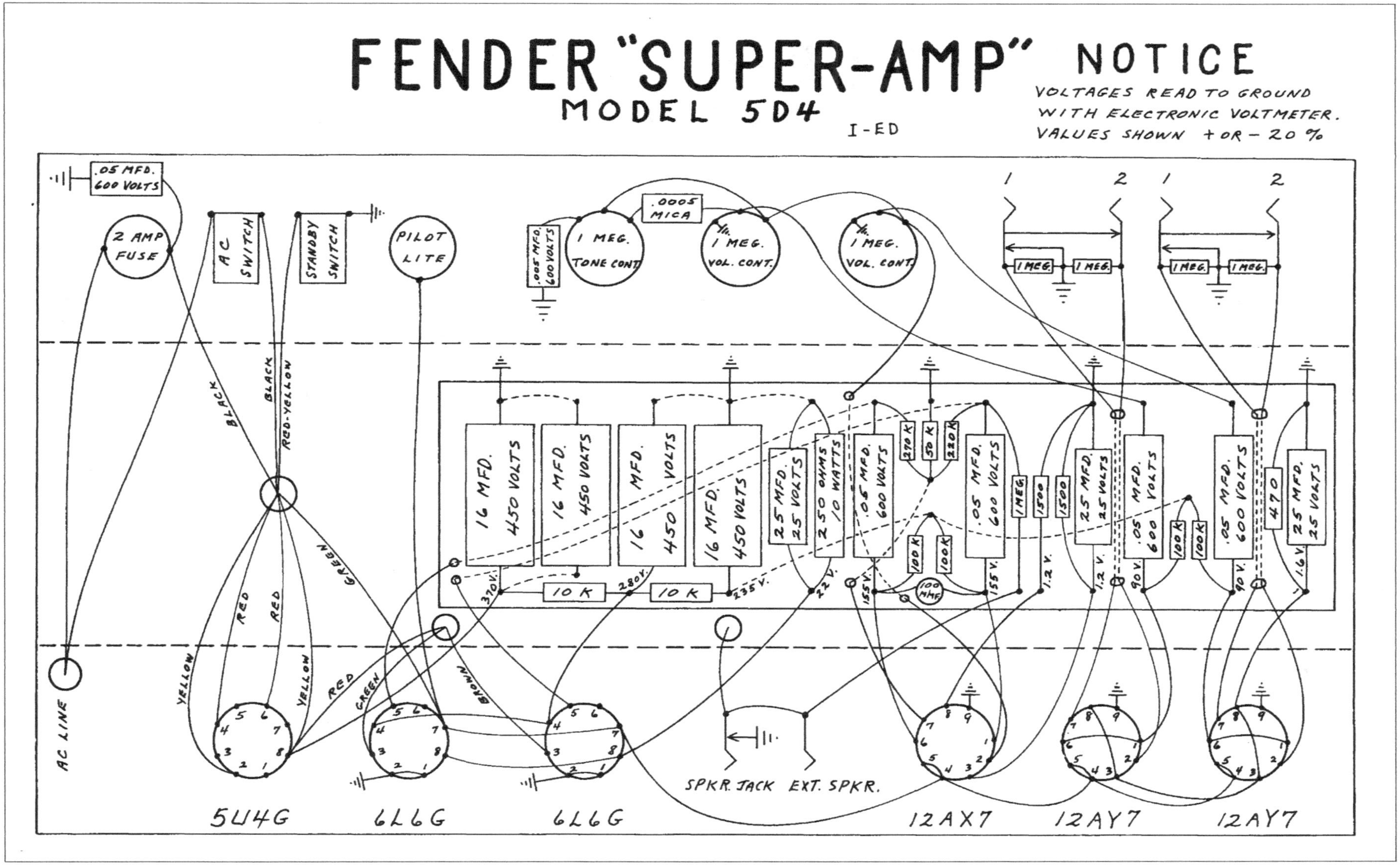
FENDER "SUPER-AMP"
MODEL 5D4
I-ED
NOTICE
VOLTAGES READ TO GROUND
WITH ELECTRONIC VOLTMETER.
VALUES SHOWN + OR − 20 %
.05 MFD. 600 VOLTS
2 AMP FUSE
AC SWITCH
STANDBY SWITCH
PILOT LITE
.005 MFD. 600 VOLTS
1 MEG. TONE CONT.
.0005 MICA
1 MEG. VOL. CONT.
1 MEG. VOL. CONT.
1 MEG.
1 MEG.
1 MEG.
1 MEG.
1
2
1
2
BLACK
BLACK
RED-YELLOW
GREEN
RED
RED
YELLOW
YELLOW
RED
GREEN
BROWN
AC LINE
16 MFD. 450 VOLTS
16 MFD. 450 VOLTS
16 MFD. 450 VOLTS
16 MFD. 450 VOLTS
10 K
10 K
370V.
280V.
235V.
25 MFD. 25 VOLTS
250 OHMS 10 WATTS
22 V.
.05 MFD. 600 VOLTS
270 K
50 K
220 K
100 K
100 K
100 MMF
155 V.
155 V.
.05 MFD. 600 VOLTS
1 MEG.
1500
1500
1.2 V.
1.2 V.
25 MFD. 25 VOLTS
90 V.
.05 MFD. 600 VOLTS
100 K
100 K
90 V.
.05 MFD. 600 VOLTS
470
1.6 V.
25 MFD. 25 VOLTS
SPKR. JACK
EXT. SPKR.
5U4G
6L6G
6L6G
12AX7
12AY7
12AY7

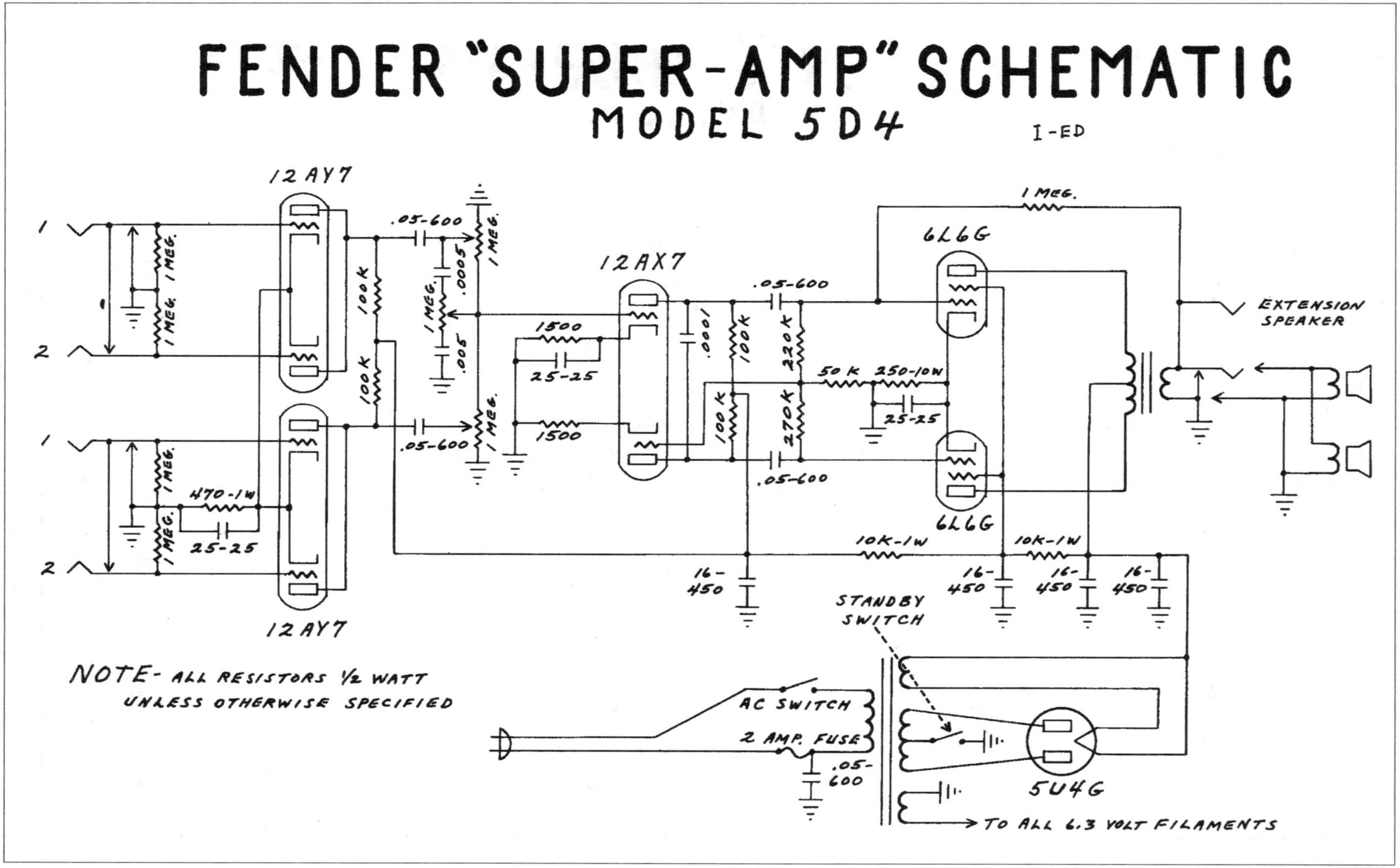
FENDER "SUPER-AMP" SCHEMATIC
MODEL 5D4
I-ED
12AY7
12AY7
12AX7
6L6G
6L6G
5U4G
1 MEG.
.05-600
.0005
.005
100 K
1500
25-25
.001
220 K
270 K
50 K
250-10W
470-1W
10K-1W
16-450
EXTENSION SPEAKER
STANDBY SWITCH
AC SWITCH
2 AMP. FUSE
.05-600
TO ALL 6.3 VOLT FILAMENTS
NOTE- ALL RESISTORS ½ WATT UNLESS OTHERWISE SPECIFIED

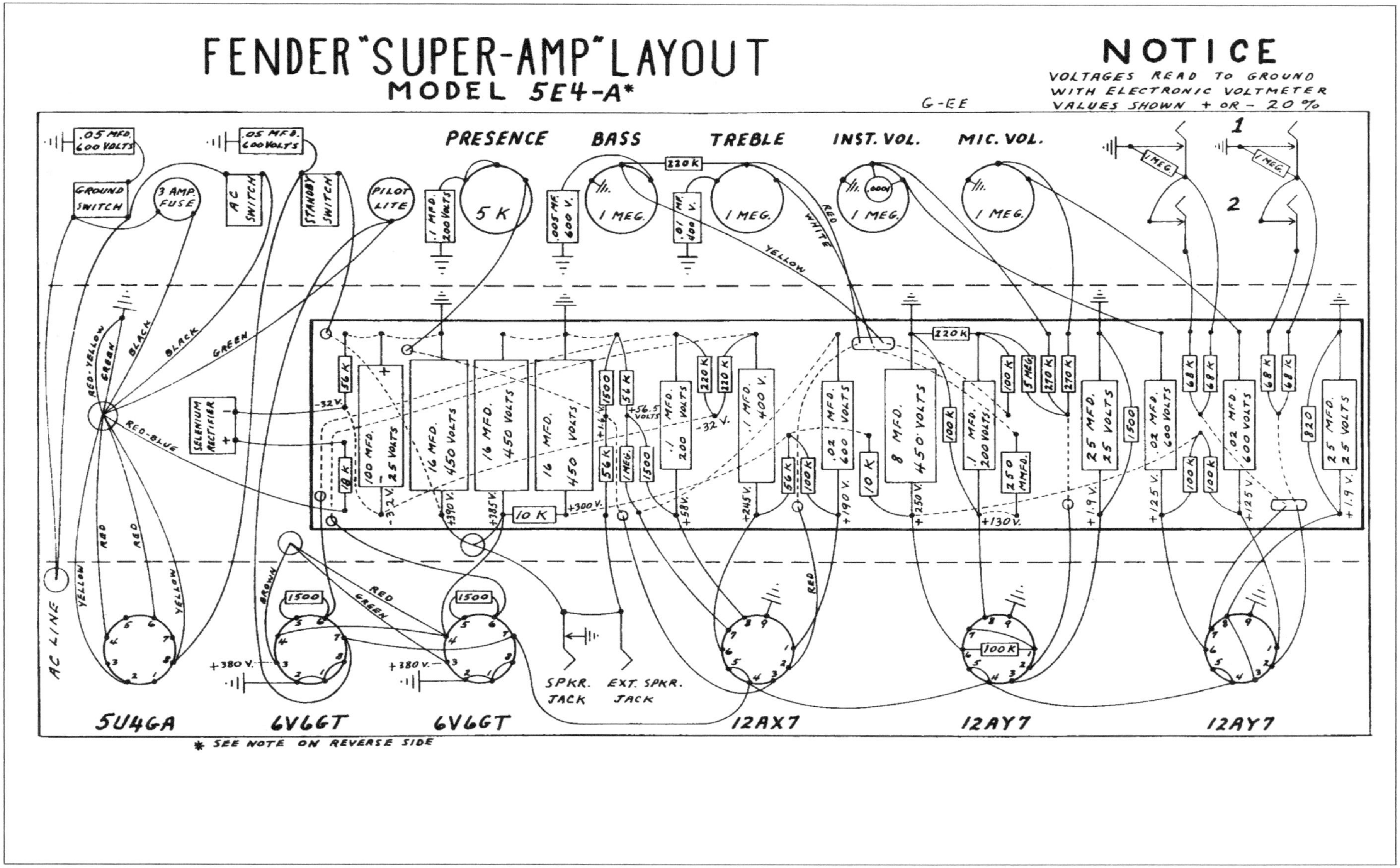
FENDER "SUPER-AMP" LAYOUT
MODEL 5E4-A*
NOTICE
VOLTAGES READ TO GROUND
WITH ELECTRONIC VOLTMETER
VALUES SHOWN + OR – 20 %
G-EE
PRESENCE
BASS
TREBLE
INST. VOL.
MIC. VOL.
.05 MFD. 600 VOLTS
GROUND SWITCH
3 AMP. FUSE
AC SWITCH
STANDBY SWITCH
PILOT LITE
SELENIUM RECTIFIER
AC LINE
RED-YELLOW
GREEN
BLACK
RED-BLUE
YELLOW
RED
BROWN
WHITE
SPKR. JACK
EXT. SPKR. JACK
5U4GA
6V6GT
12AX7
12AY7
* SEE NOTE ON REVERSE SIDE

FENDER "SUPER-AMP" SCHEMATIC

MODEL 5E4-A*

G-EE

NOTICE

VOLTAGES READ TO GROUND
WITH ELECTRONIC VOLTMETER
VALUES SHOWN + OR − 20%

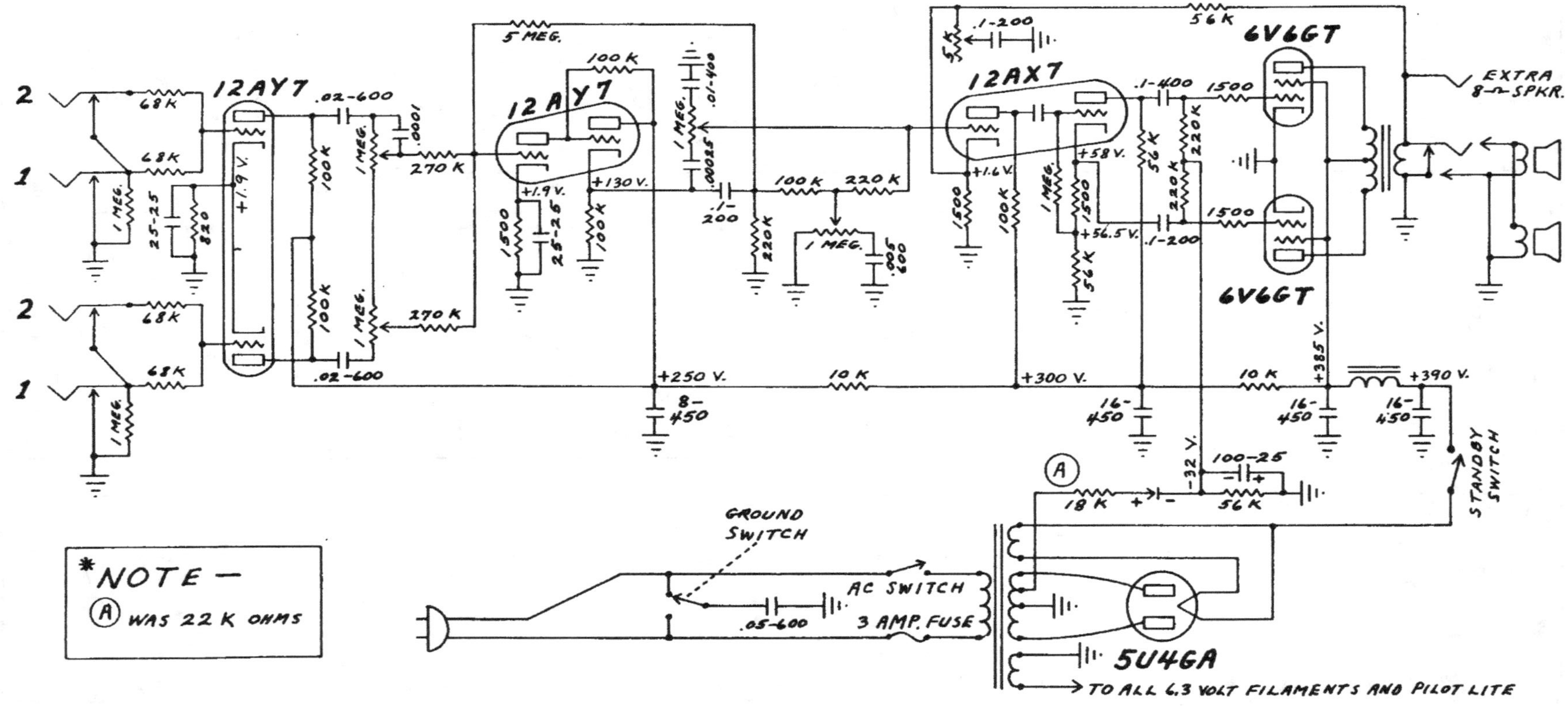

*NOTE —
(A) WAS 22 K OHMS

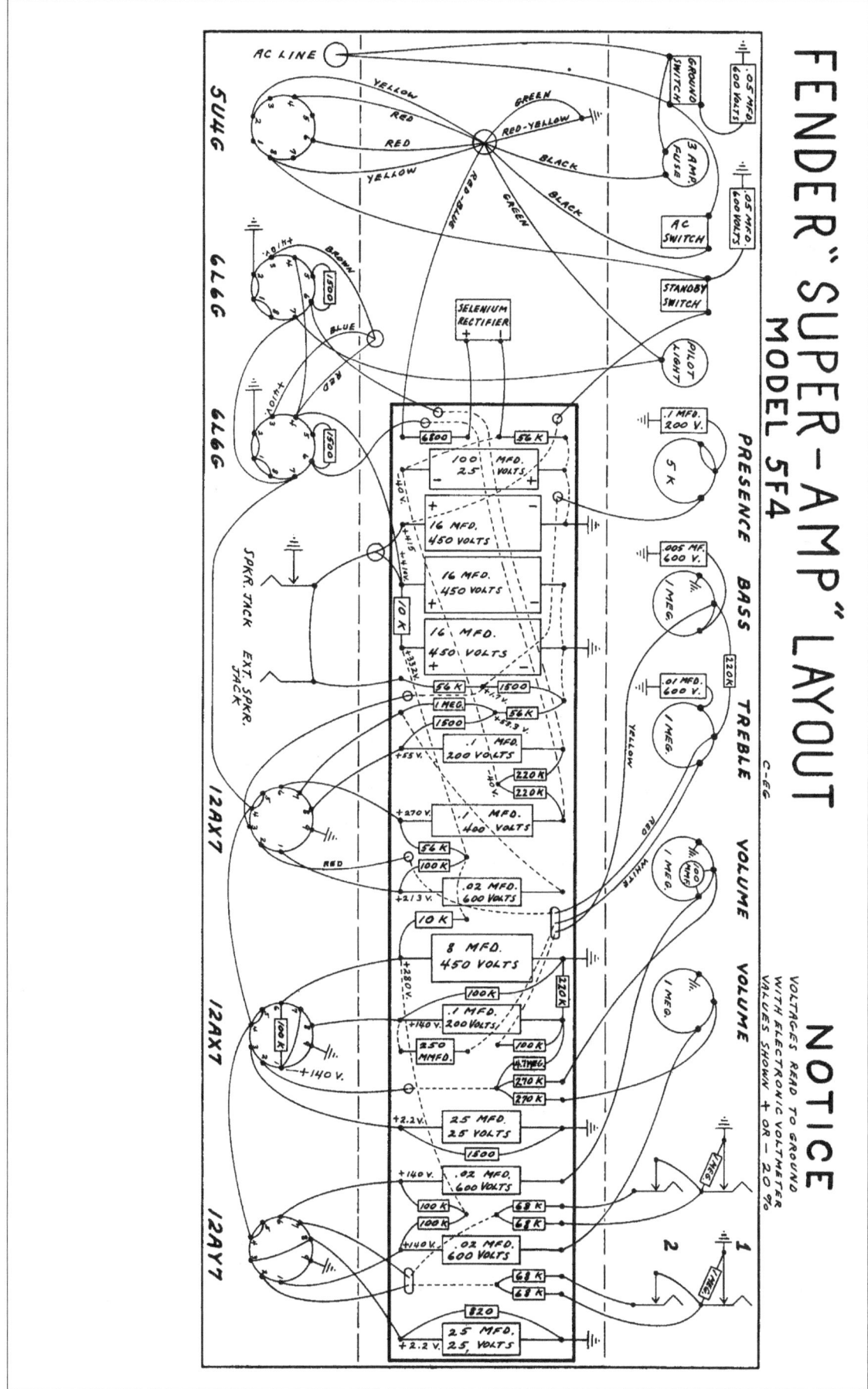

FENDER "SUPER-AMP" LAYOUT

MODEL 5F4

C-EG

NOTICE

VOLTAGES READ TO GROUND
WITH ELECTRONIC VOLTMETER
VALUES SHOWN + OR − 20 %

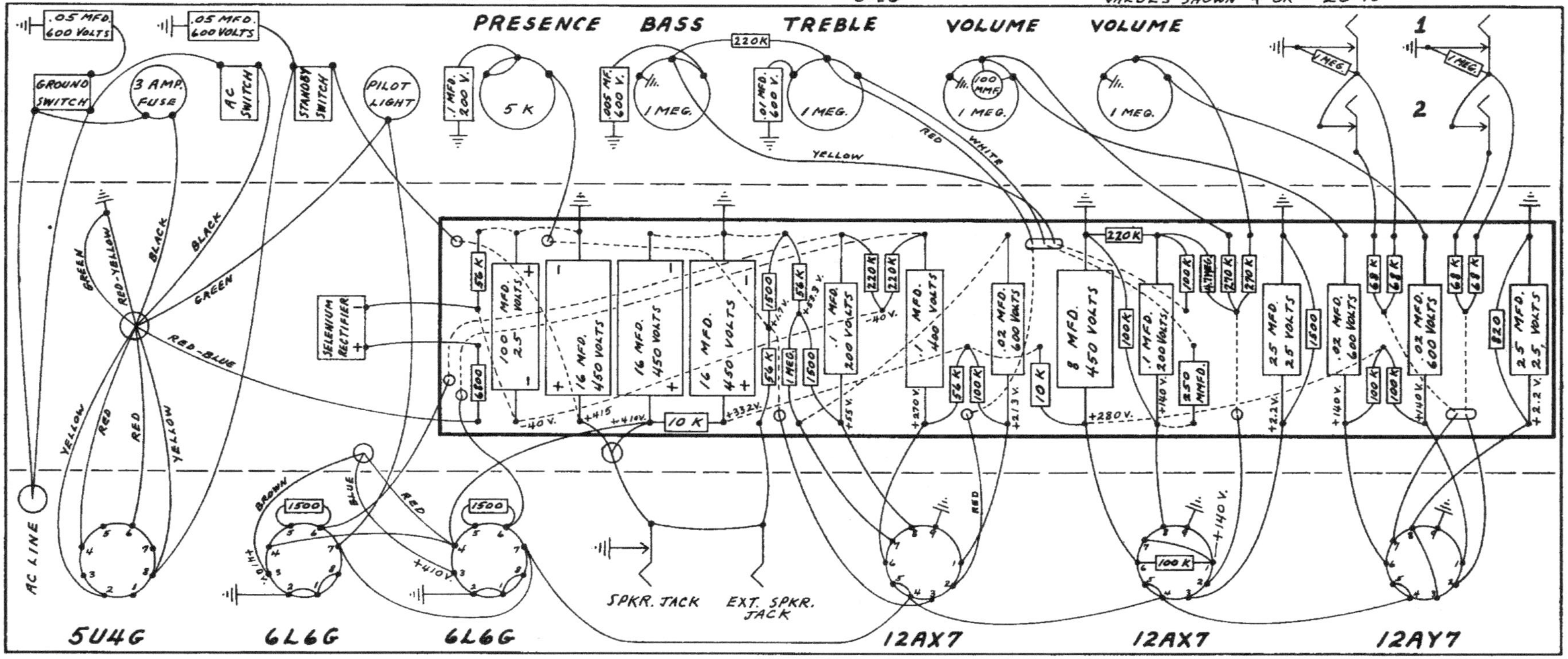

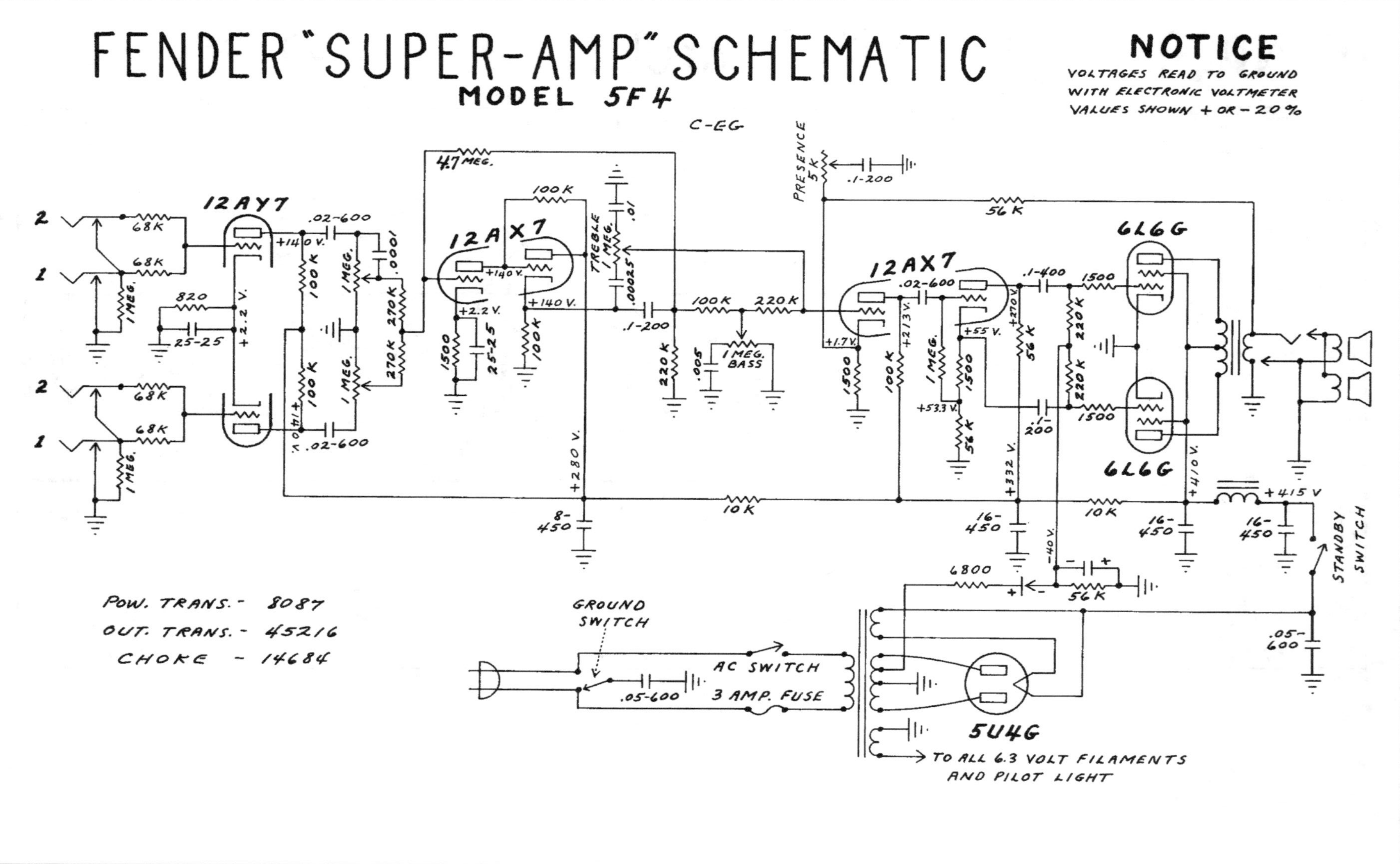
FENDER "SUPER-AMP" SCHEMATIC
MODEL 5F4
C-EG
NOTICE
VOLTAGES READ TO GROUND
WITH ELECTRONIC VOLTMETER
VALUES SHOWN + OR - 20 %
12AY7
12AX7
12AX7
6L6G
6L6G
5U4G
PRESENCE 5K
TREBLE 1 MEG.
1 MEG. BASS
GROUND SWITCH
AC SWITCH
3 AMP. FUSE
STANDBY SWITCH
TO ALL 6.3 VOLT FILAMENTS AND PILOT LIGHT
POW. TRANS. - 8087
OUT. TRANS. - 45216
CHOKE - 14684

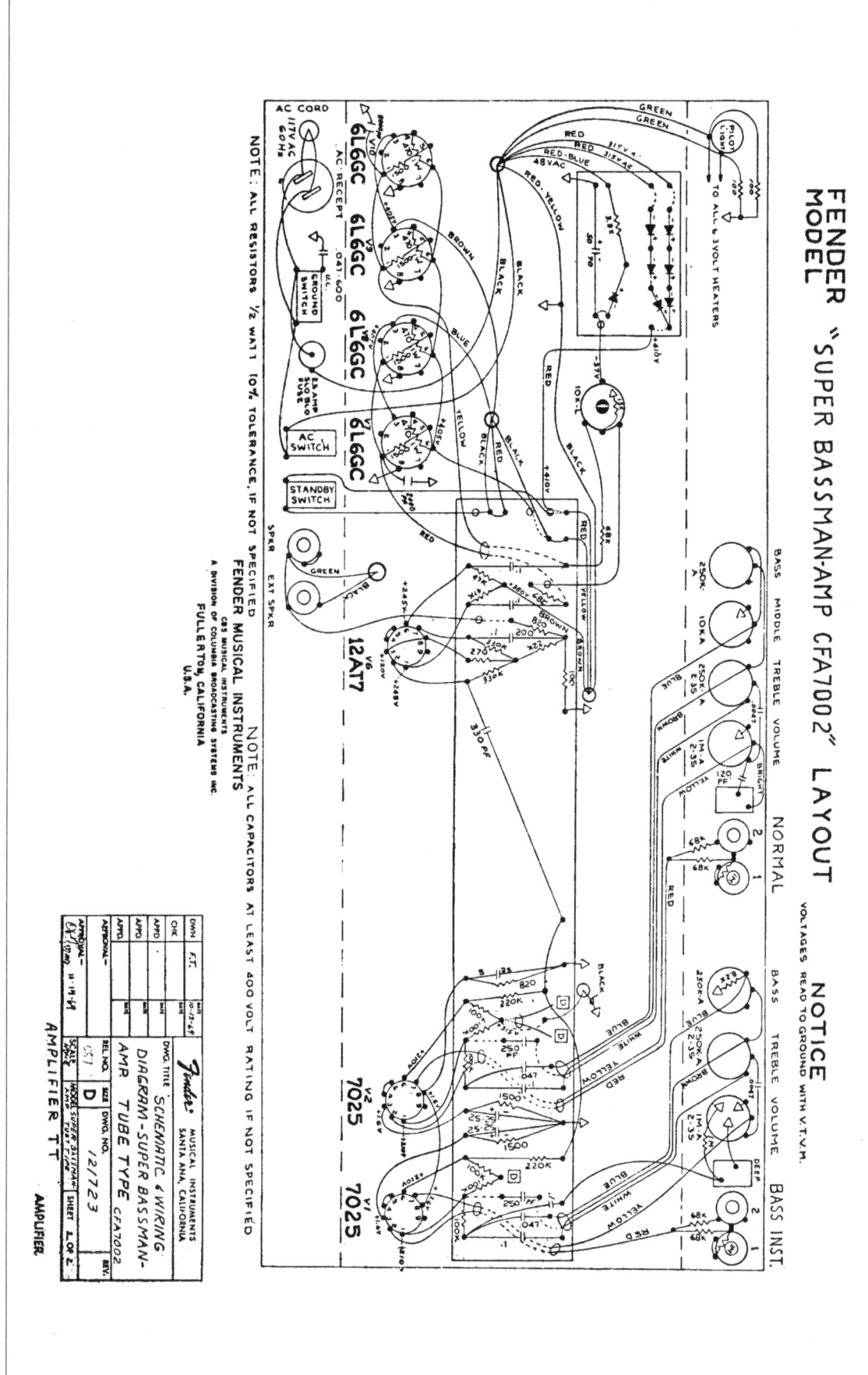

FENDER MODEL "SUPER BASSMAN-AMP CFA7002" LAYOUT

NOTICE
VOLTAGES READ TO GROUND WITH V.T.V.M.

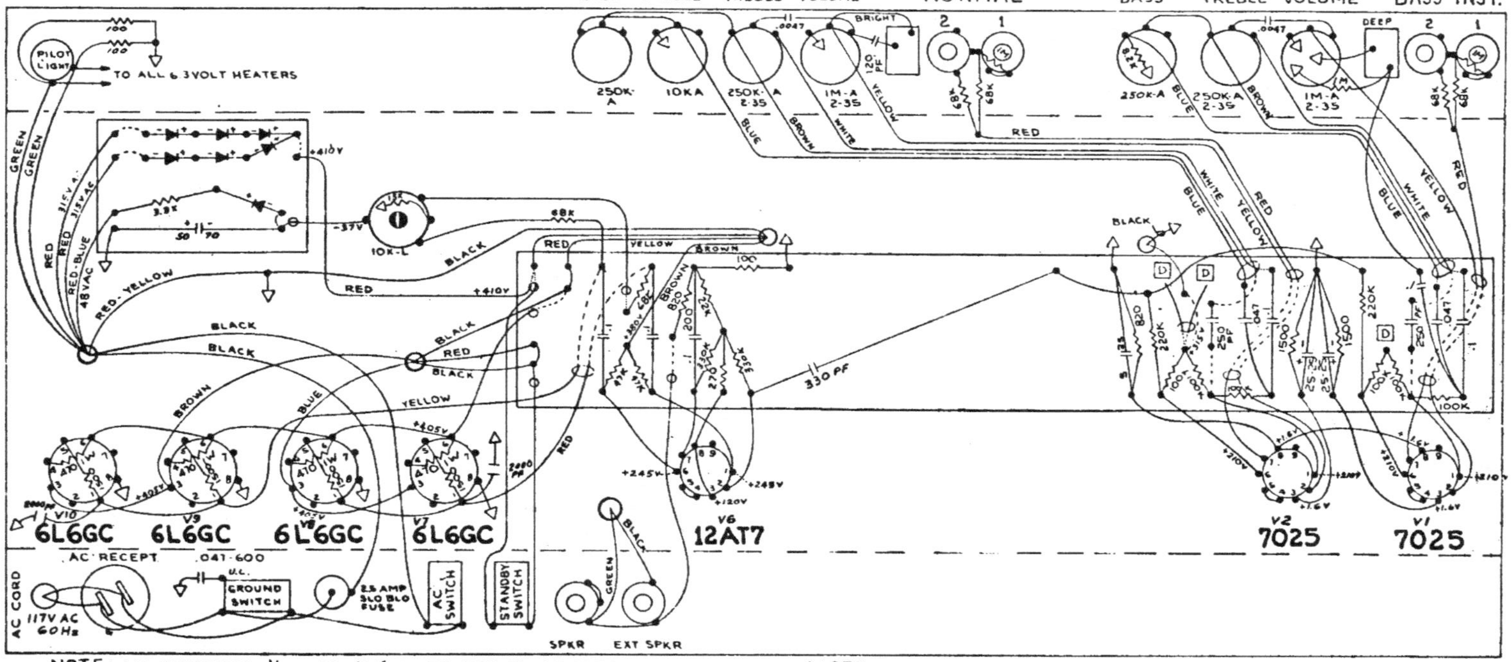

NOTE: ALL RESISTORS ½ WATT 10% TOLERANCE, IF NOT SPECIFIED

NOTE: ALL CAPACITORS AT LEAST 400 VOLT RATING IF NOT SPECIFIED

FENDER MUSICAL INSTRUMENTS
CBS MUSICAL INSTRUMENTS
A DIVISION OF COLUMBIA BROADCASTING SYSTEMS INC.
FULLERTON, CALIFORNIA
U.S.A.

DWN	F.T.	DATE 10-13-69	Fender MUSICAL INSTRUMENTS SANTA ANA, CALIFORNIA
CHK		DATE	DWG. TITLE SCHEMATIC & WIRING DIAGRAM-SUPER BASSMAN-AMP TUBE TYPE CFA7002
APPD		DATE	
APPD		DATE	
APPD		DATE	
APPROVAL–			REL. NO. — SIZE D — DWG. NO. 121723 — REV.
APPROVAL–		11-14-69	SCALE NONE — MODEL SUPER BASSMAN-AMP TUBE TYPE — SHEET 2 OF 2

AMPLIFIER TT

AMPLIFIER

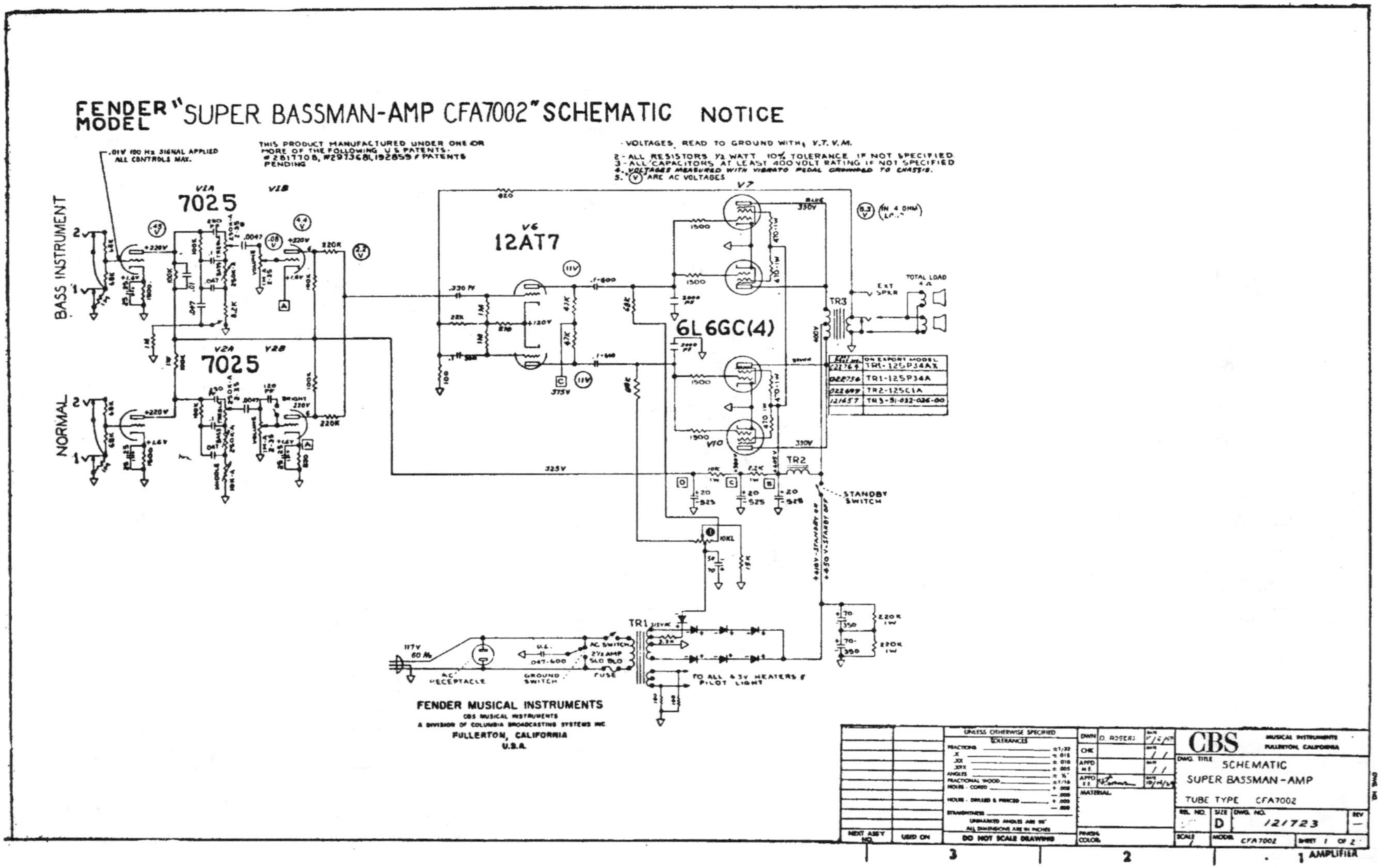
FENDER MODEL "SUPER BASSMAN-AMP CFA7002" SCHEMATIC
NOTICE
THIS PRODUCT MANUFACTURED UNDER ONE OR MORE OF THE FOLLOWING U.S. PATENTS: #2817708, #2973681, 192859 & PATENTS PENDING
.01V 100 Hz SIGNAL APPLIED ALL CONTROLS MAX.
VOLTAGES READ TO GROUND WITH V.T.V.M.
2 - ALL RESISTORS ½ WATT 10% TOLERANCE IF NOT SPECIFIED
3 - ALL CAPACITORS AT LEAST 400 VOLT RATING IF NOT SPECIFIED
BASS INSTRUMENT
NORMAL
V1A
7025
V1B
V2A
V2B
7025
V6
12AT7
V7
6L6GC(4)
V10
TR1
TR2
TR3
EXT SPKR
TOTAL LOAD 4 Ω
STANDBY SWITCH
117V 60 Hz
AC RECEPTACLE
GROUND SWITCH
FUSE
TO ALL 6.3V HEATERS & PILOT LIGHT
ON EXPORT MODEL
TR1-125P34AX
TR1-125P34A
TR2-125C1A
TR3-51-032-026-00
FENDER MUSICAL INSTRUMENTS
CBS MUSICAL INSTRUMENTS
A DIVISION OF COLUMBIA BROADCASTING SYSTEMS INC
FULLERTON, CALIFORNIA
U.S.A.
CBS
MUSICAL INSTRUMENTS
FULLERTON, CALIFORNIA
SCHEMATIC
SUPER BASSMAN-AMP
TUBE TYPE CFA7002
D
121723
DO NOT SCALE DRAWING
AMPLIFIER

FENDER MODEL "SUPER REVERB-AMP AA1069" LAYOUT

NOTICE
VOLTAGES READ TO GROUND WITH ELECTRONIC VOLTMETER WITH VIBRATO-GROUNDED, NO INPUT SIGNAL.

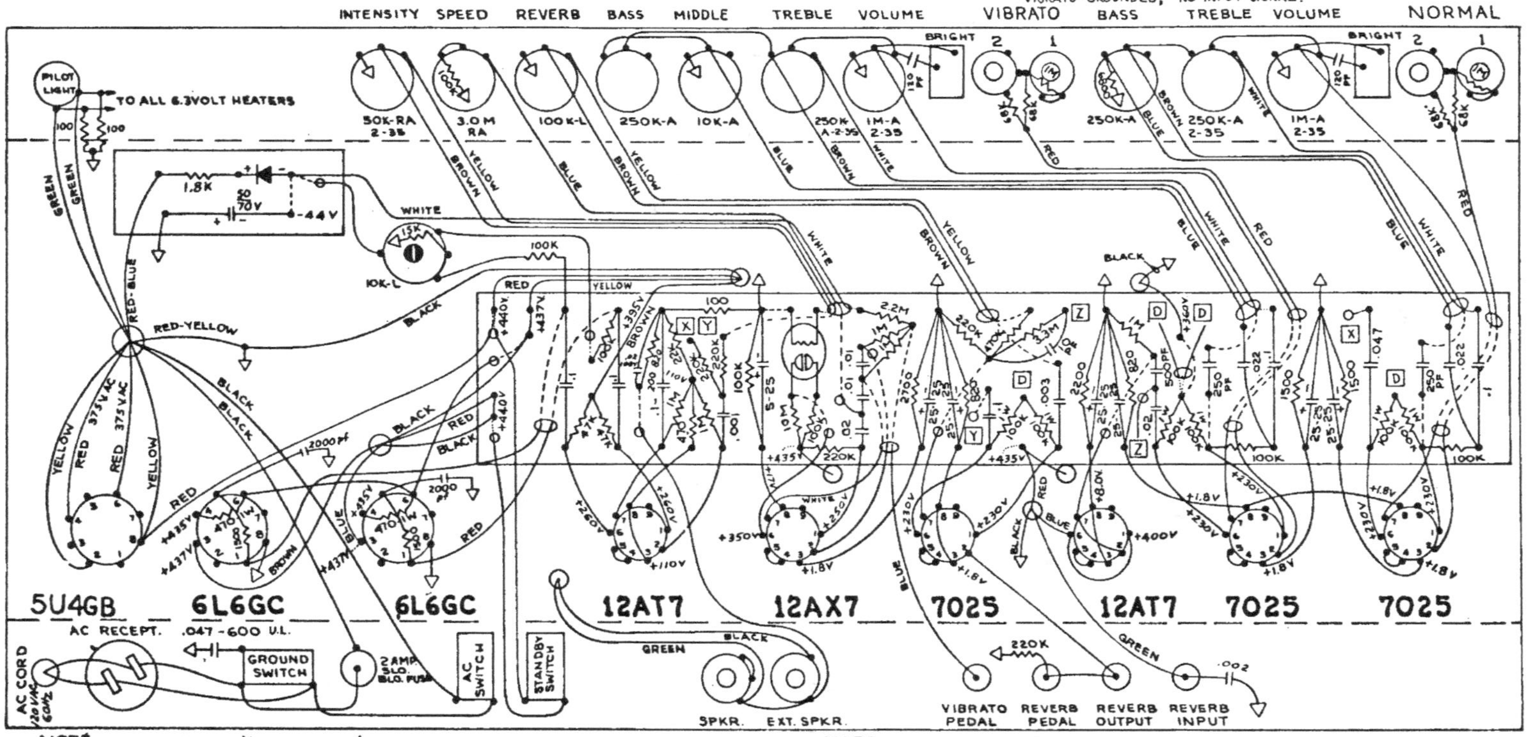

NOTE: ALL RESISTORS ½ WATT 10% TOLERANCE, IF NOT SPECIFIED.

NOTE: ALL CAPACITORS AT LEAST 600 VOLT RATING IF NOT SPECIFIED.

FENDER MUSICAL INSTRUMENTS
CBS MUSICAL INSTRUMENTS
A DIVISION OF COLUMBIA BROADCASTING SYSTEMS INC.
FULLERTON, CALIFORNIA
U.S.A.

DRAWN	F	10-13-69	
APPROVED		10-14-69	
DWG. TITLE	SCHEMATIC & WIRING DIAGRAM SUPER REVERB-AMP		
SIZE D	DWG. NO. 047746	REV D	
SCALE NONE	MODEL TUBE TYPE SUPER REVERB AMP	SHEET 2 OF 2	

D	INCORPORATED ECO 1689		
C	INCORPORATED ECO 1438	4/6-9	
B	INCORPORATED ECO 1044		

REVISIONS

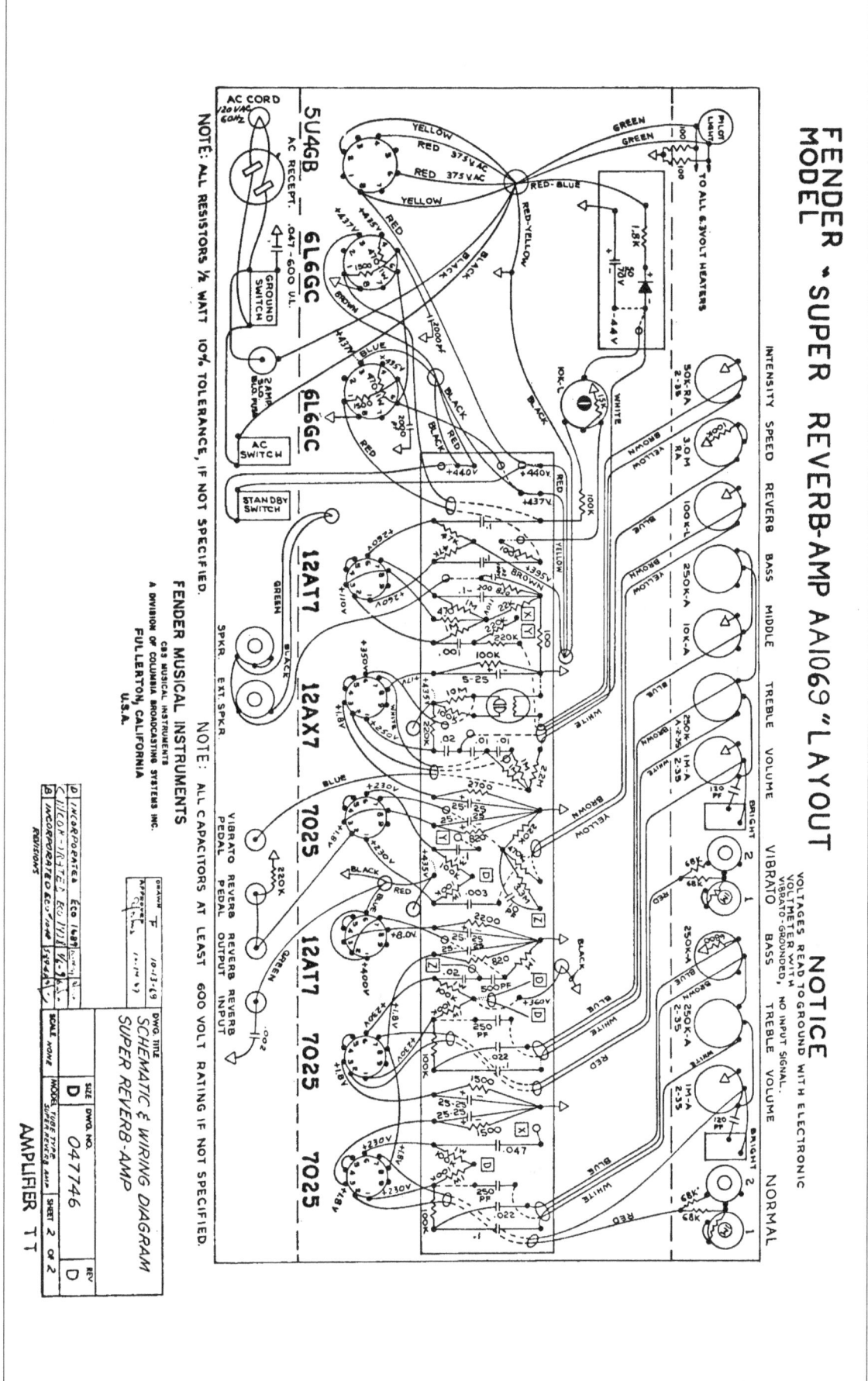

AMPLIFIER TT

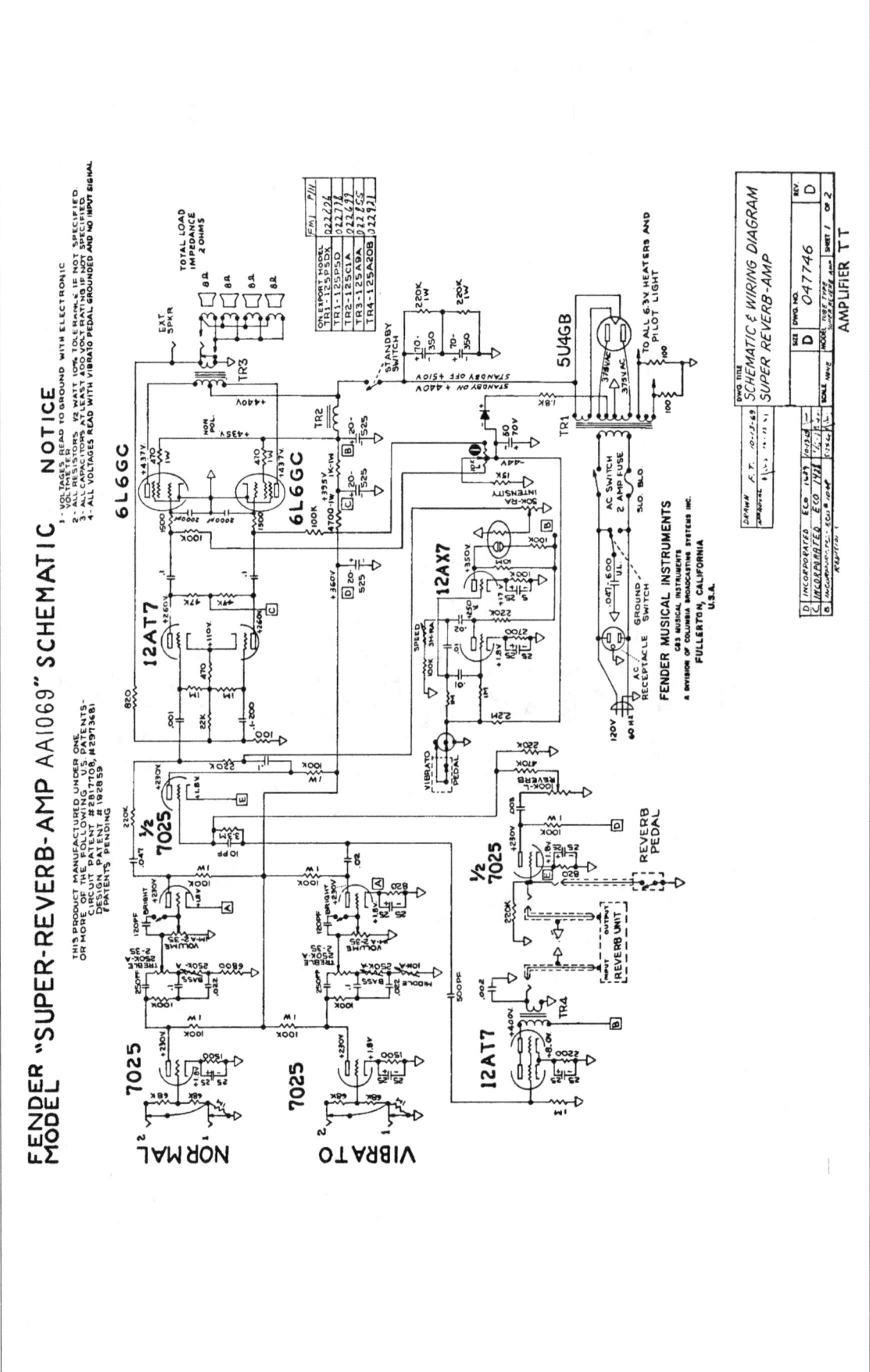

FENDER MODEL "SUPER-REVERB-AMP AA1069" SCHEMATIC

THIS PRODUCT MANUFACTURED UNDER ONE OR MORE OF THE FOLLOWING U.S. PATENTS-
CIRCUIT PATENT #2817708, #2973681
DESIGN PATENT # 192859
& PATENTS PENDING

NOTICE

1 - VOLTAGES READ TO GROUND WITH ELECTRONIC VOLTMETER
2 - ALL RESISTORS 1/2 WATT 10% TOLERANCE IF NOT SPECIFIED.
3 - ALL CAPACITORS AT LEAST 400 VOLT RATING IF NOT SPECIFIED.
4 - ALL VOLTAGES READ WITH VIBRATO PEDAL GROUNDED AND NO INPUT SIGNAL

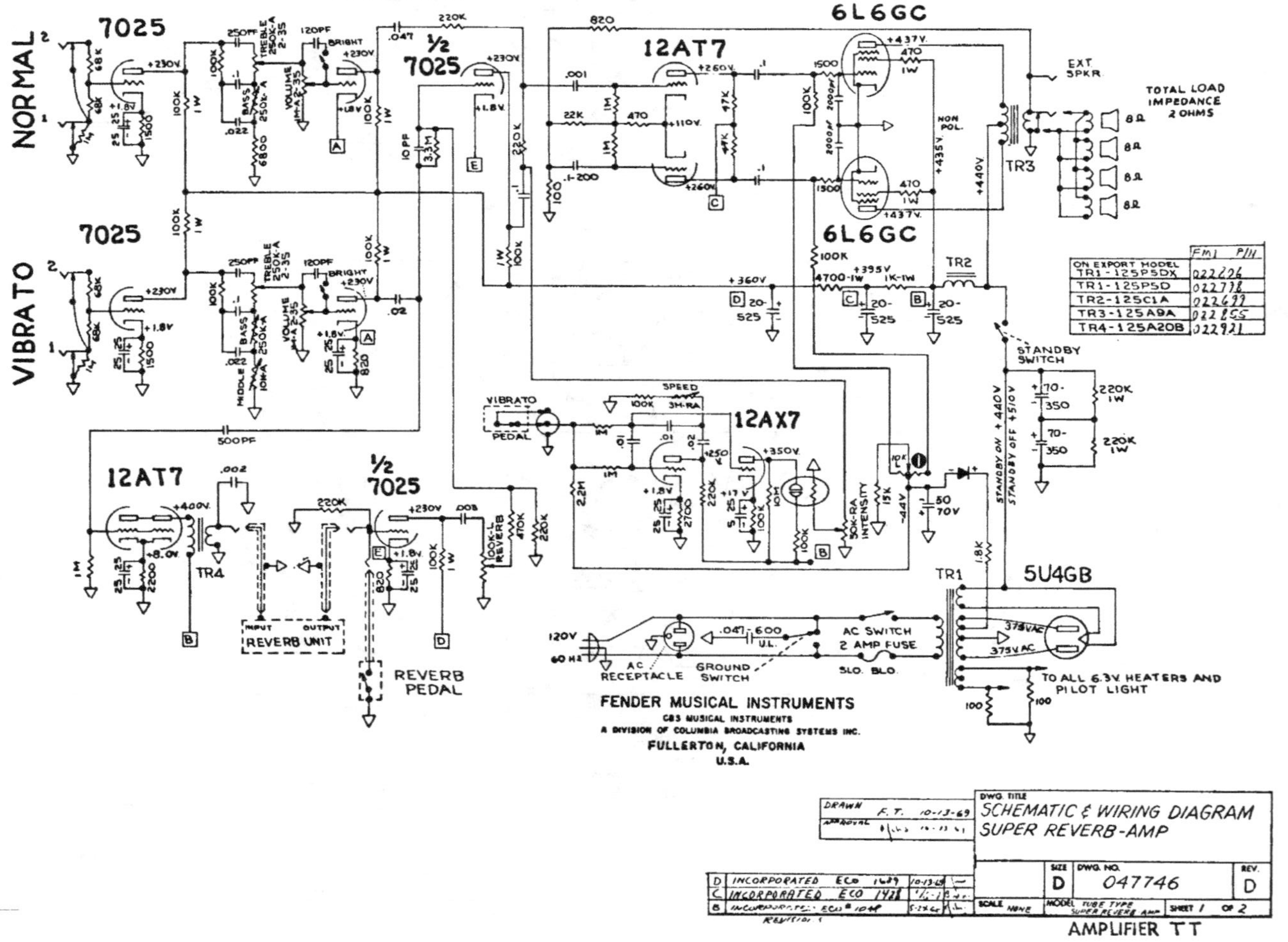

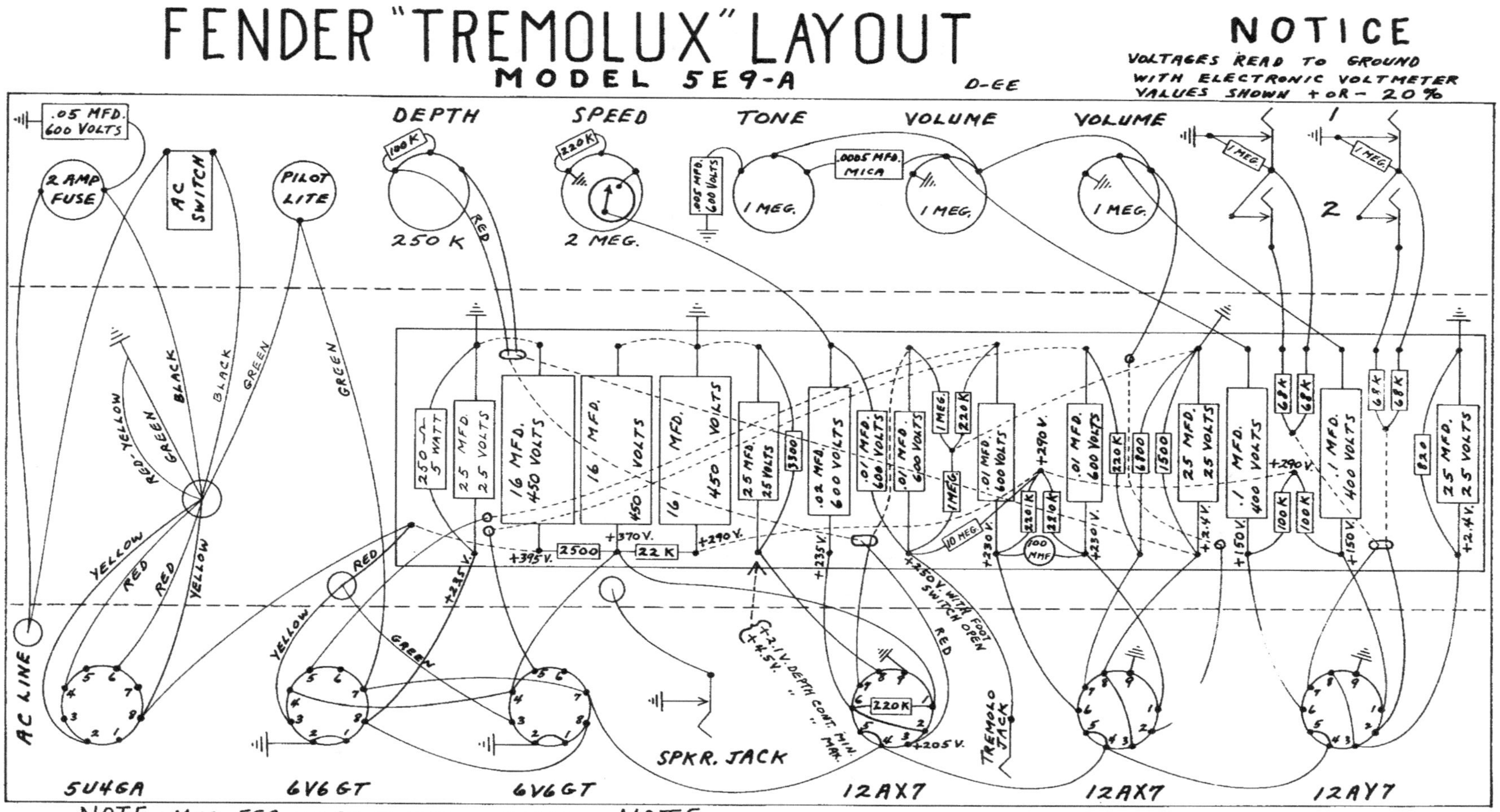

NOTE – MODEL 5E9 USES 5Y3GT INSTEAD OF 5U4GA

NOTE – EARLY MODELS HAVE SWITCH ON DEPTH CONTROL INSTEAD OF SPEED CONTROL. CIRCUIT IS SAME.

FENDER "TREMOLUX" SCHEMATIC
MODEL 5E9-A

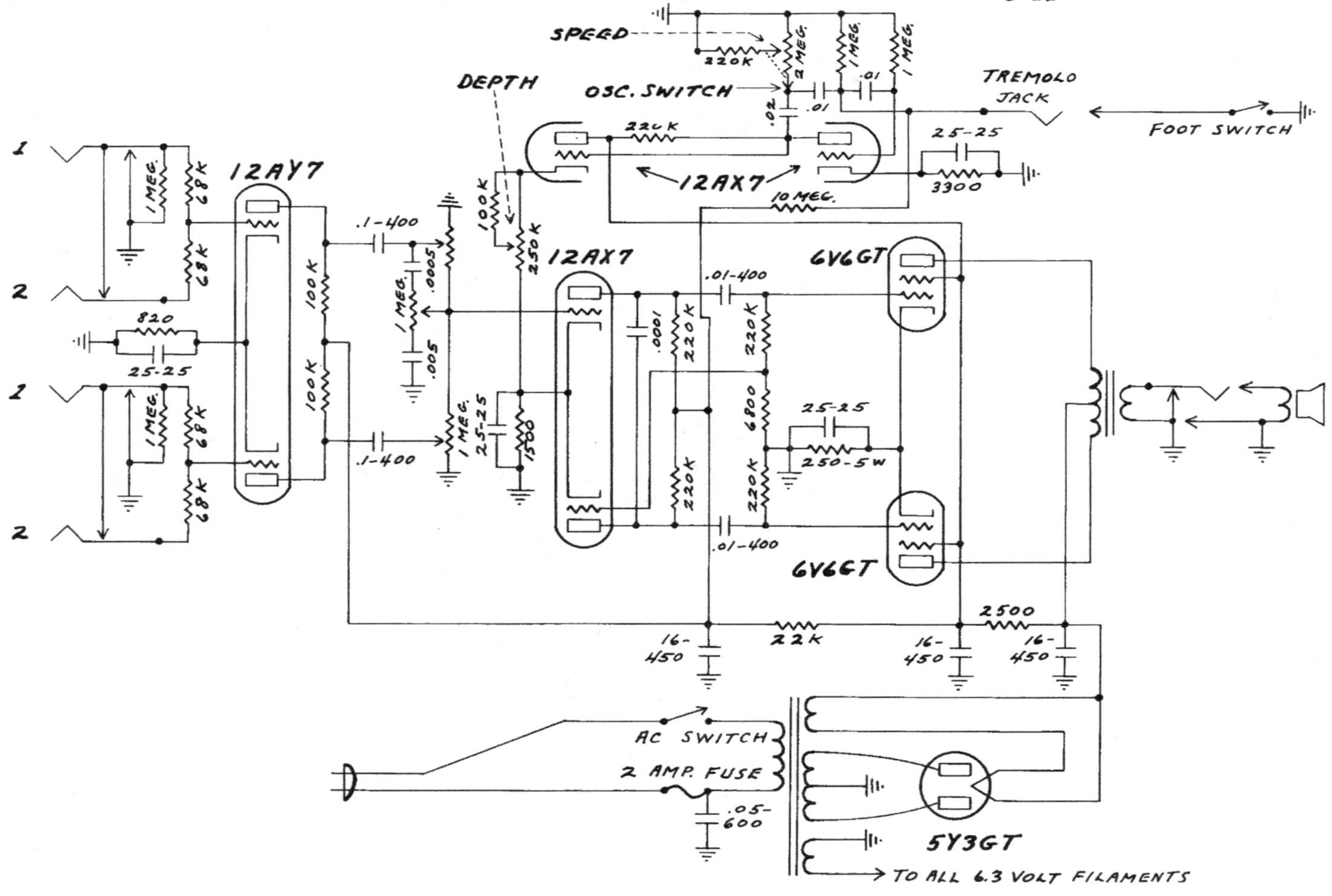

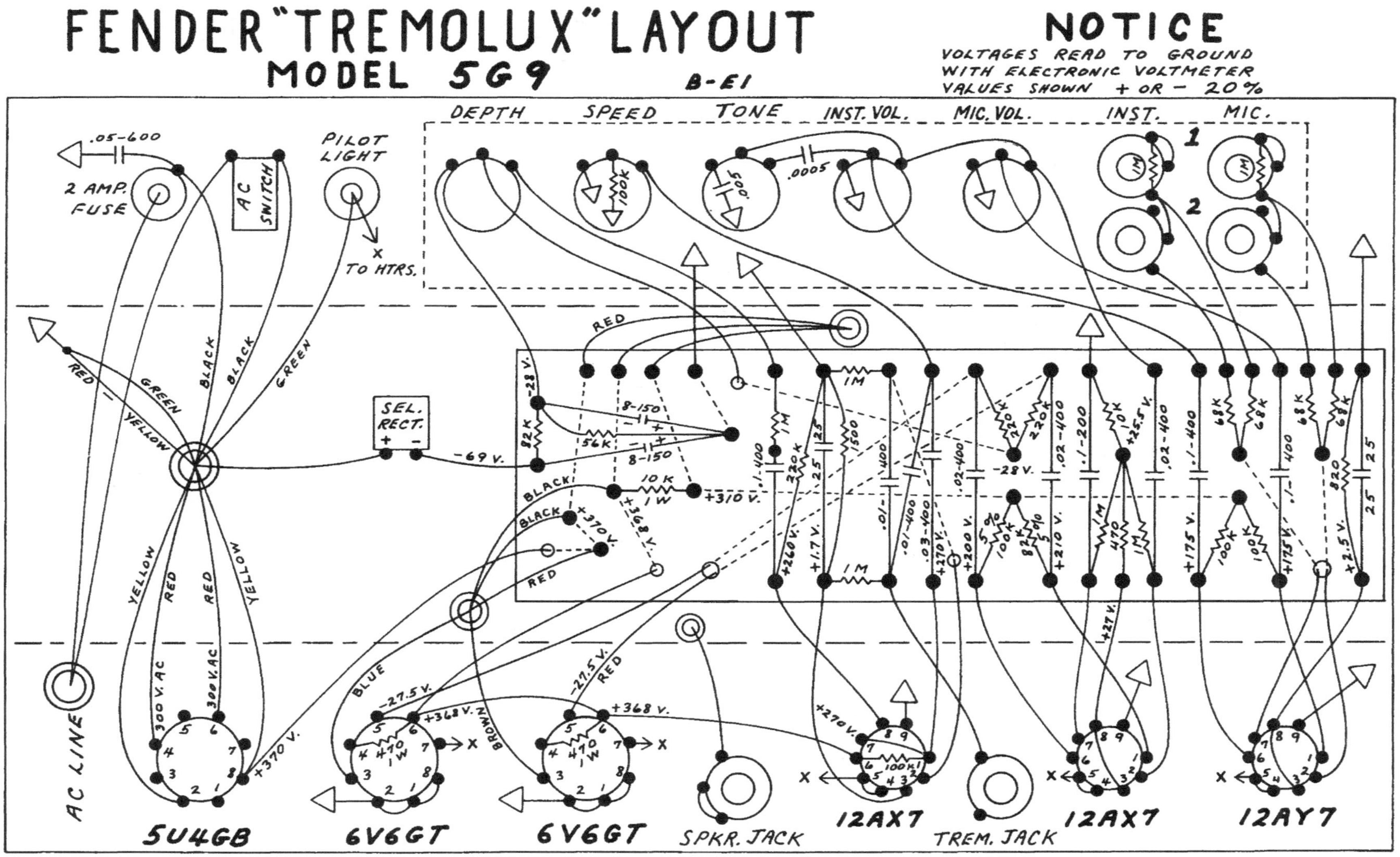
FENDER "TREMOLUX" LAYOUT
MODEL 5G9
B-E1
NOTICE
VOLTAGES READ TO GROUND
WITH ELECTRONIC VOLTMETER
VALUES SHOWN + OR − 20%
DEPTH
SPEED
TONE
INST. VOL.
MIC. VOL.
INST.
MIC.
1
2
.05-600
2 AMP. FUSE
AC SWITCH
PILOT LIGHT
TO HTRS.
SEL. RECT.
−69 V.
−28 V.
82 K
56 K
8-150
10 K 1 W
+310 V.
+368 V.
+370 V.
RED
GREEN
YELLOW
BLACK
BLUE
BROWN
300 V. AC
+260 V.
+1.7 V.
+270 V.
+200 V.
+210 V.
+27 V.
+25.5 V.
+175 V.
+2.5 V.
−27.5 V.
1M
220 K
1500
.1-400
.01-400
.03-400
.02-400
.1-200
68 K
100 K
470
820
25
AC LINE
5U4GB
6V6GT
SPKR. JACK
12AX7
TREM. JACK
12AY7

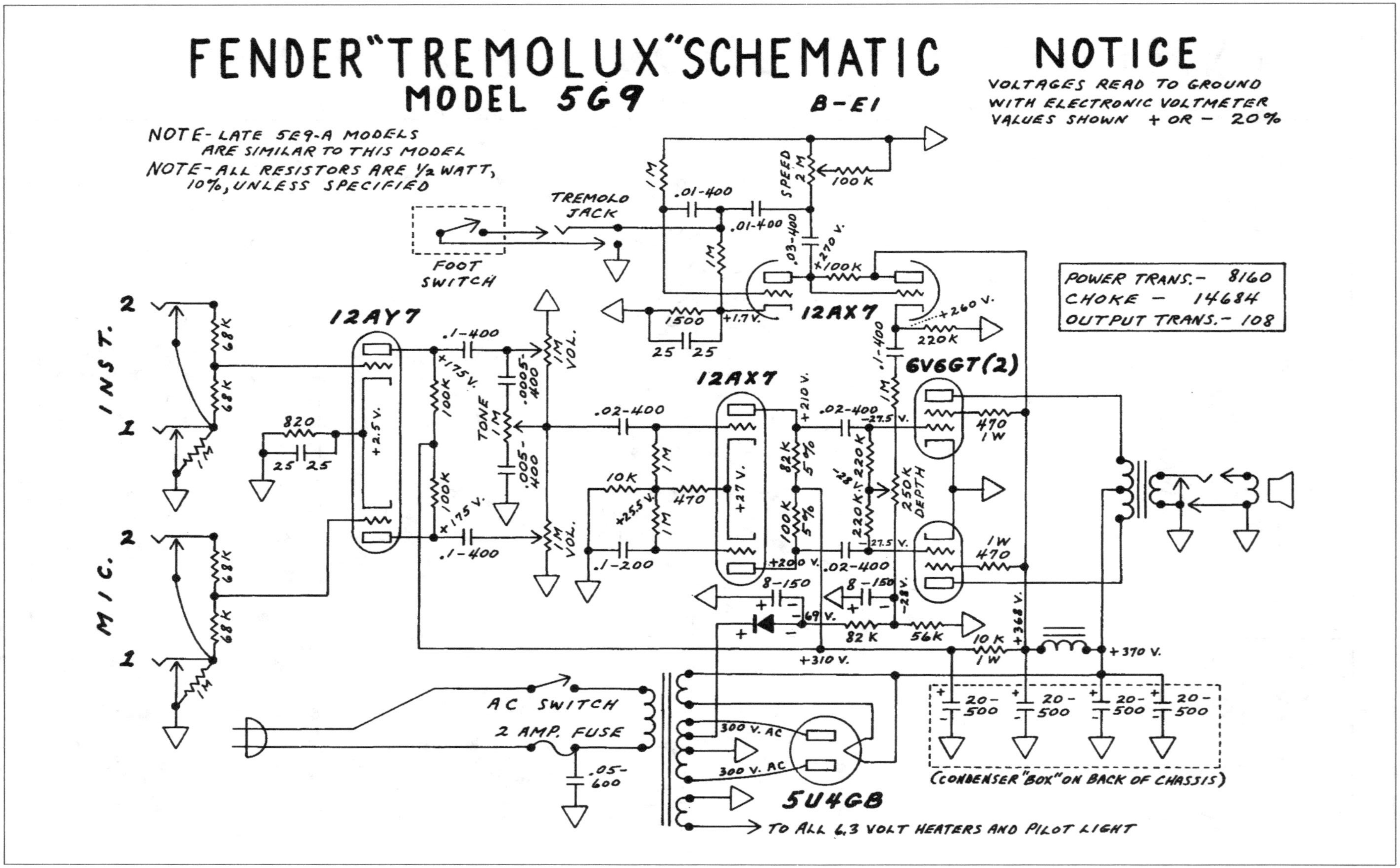
FENDER "TREMOLUX" SCHEMATIC
MODEL 5G9
B-E1
NOTICE
VOLTAGES READ TO GROUND WITH ELECTRONIC VOLTMETER VALUES SHOWN + OR − 20%
NOTE- LATE 5E9-A MODELS ARE SIMILAR TO THIS MODEL
NOTE- ALL RESISTORS ARE ½ WATT, 10%, UNLESS SPECIFIED
POWER TRANS.- 8160
CHOKE - 14684
OUTPUT TRANS.- 108
TREMOLO JACK
FOOT SWITCH
SPEED
DEPTH
TONE
VOL.
INST.
MIC.
12AY7
12AX7
6V6GT(2)
5U4GB
AC SWITCH
2 AMP. FUSE
(CONDENSER "BOX" ON BACK OF CHASSIS)
TO ALL 6.3 VOLT HEATERS AND PILOT LIGHT

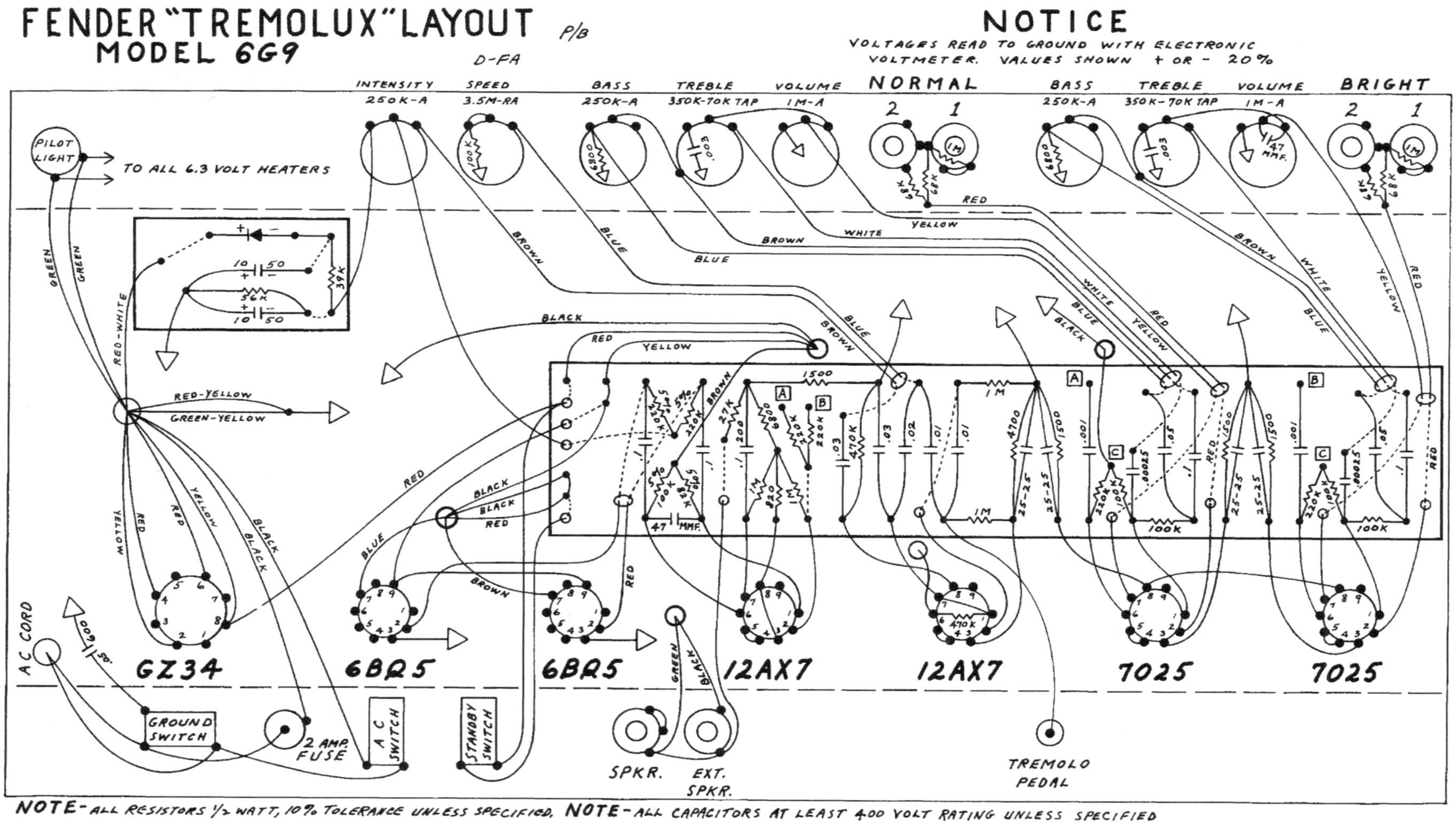
FENDER "TREMOLUX" LAYOUT
MODEL 6G9
P/B
D-FA
NOTICE
VOLTAGES READ TO GROUND WITH ELECTRONIC VOLTMETER. VALUES SHOWN + OR - 20%
INTENSITY 250K-A
SPEED 3.5M-RA
BASS 250K-A
TREBLE 350K-70K TAP
VOLUME 1M-A
NORMAL
BASS 250K-A
TREBLE 350K-70K TAP
VOLUME 1M-A
BRIGHT
PILOT LIGHT
TO ALL 6.3 VOLT HEATERS
GREEN
GREEN
RED-WHITE
RED-YELLOW
GREEN-YELLOW
YELLOW
RED
RED
YELLOW
BLACK
BLACK
BROWN
BLUE
BLUE
BROWN
WHITE
YELLOW
RED
WHITE
RED
YELLOW
BLACK
BLUE
39K
56K
GZ34
6BQ5
6BQ5
12AX7
12AX7
7025
7025
AC CORD
GROUND SWITCH
2 AMP. FUSE
A C SWITCH
STANDBY SWITCH
SPKR.
EXT. SPKR.
TREMOLO PEDAL
47 MMF.
NOTE- ALL RESISTORS 1/2 WATT, 10% TOLERANCE UNLESS SPECIFIED. NOTE- ALL CAPACITORS AT LEAST 400 VOLT RATING UNLESS SPECIFIED

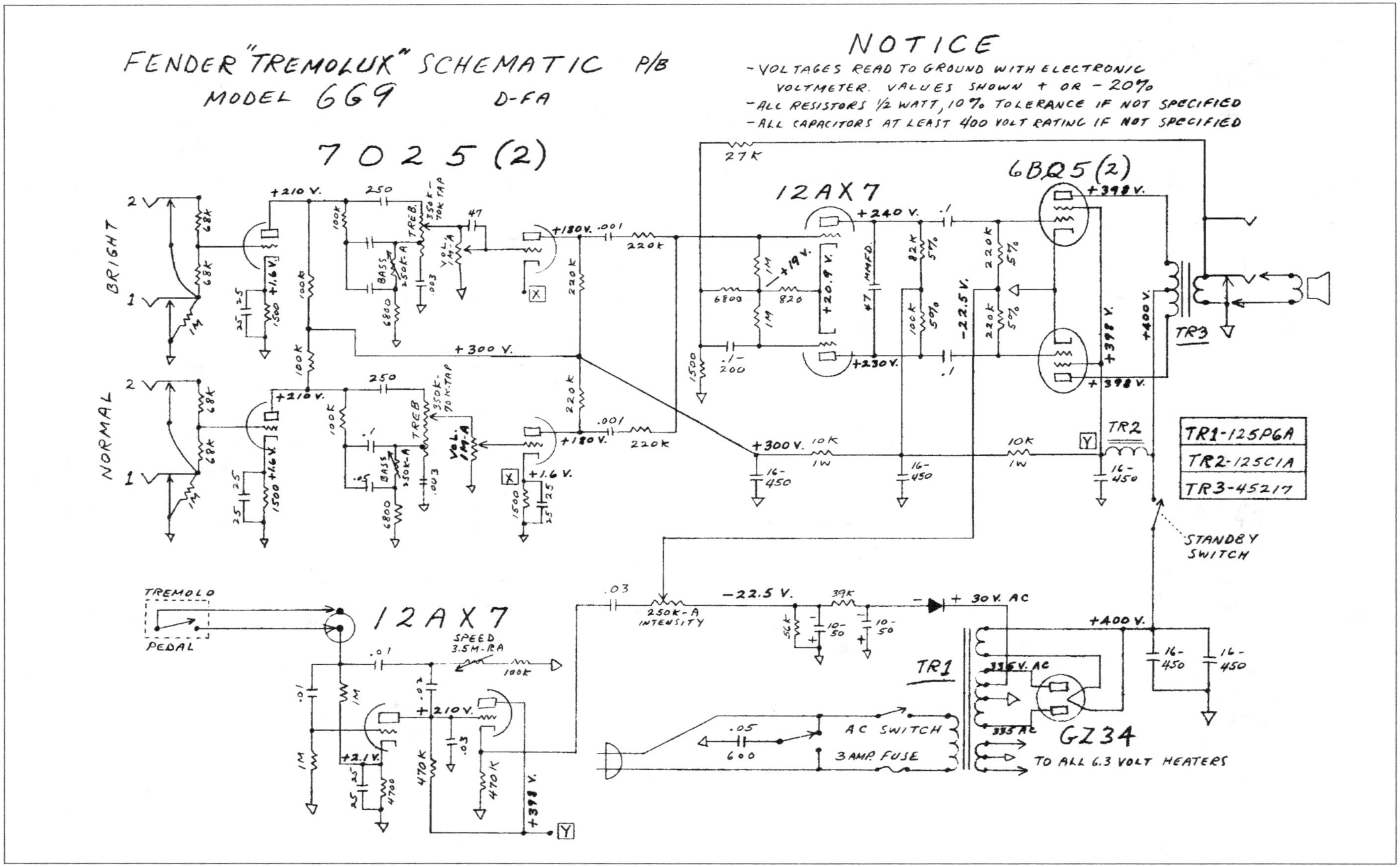
FENDER "TREMOLUX" SCHEMATIC P/B
MODEL 6G9 D-FA
7025 (2)
NOTICE
- VOLTAGES READ TO GROUND WITH ELECTRONIC VOLTMETER. VALUES SHOWN + OR - 20%
- ALL RESISTORS ½ WATT, 10% TOLERANCE IF NOT SPECIFIED
- ALL CAPACITORS AT LEAST 400 VOLT RATING IF NOT SPECIFIED
12AX7
6BQ5 (2)
BRIGHT
NORMAL
TREMOLO
PEDAL
12AX7
SPEED 3.5M-RA
250K-A INTENSITY
TR1
TR2
TR3
TR1-125P6A
TR2-125C1A
TR3-45217
STANDBY SWITCH
GZ34
AC SWITCH
3 AMP. FUSE
TO ALL 6.3 VOLT HEATERS

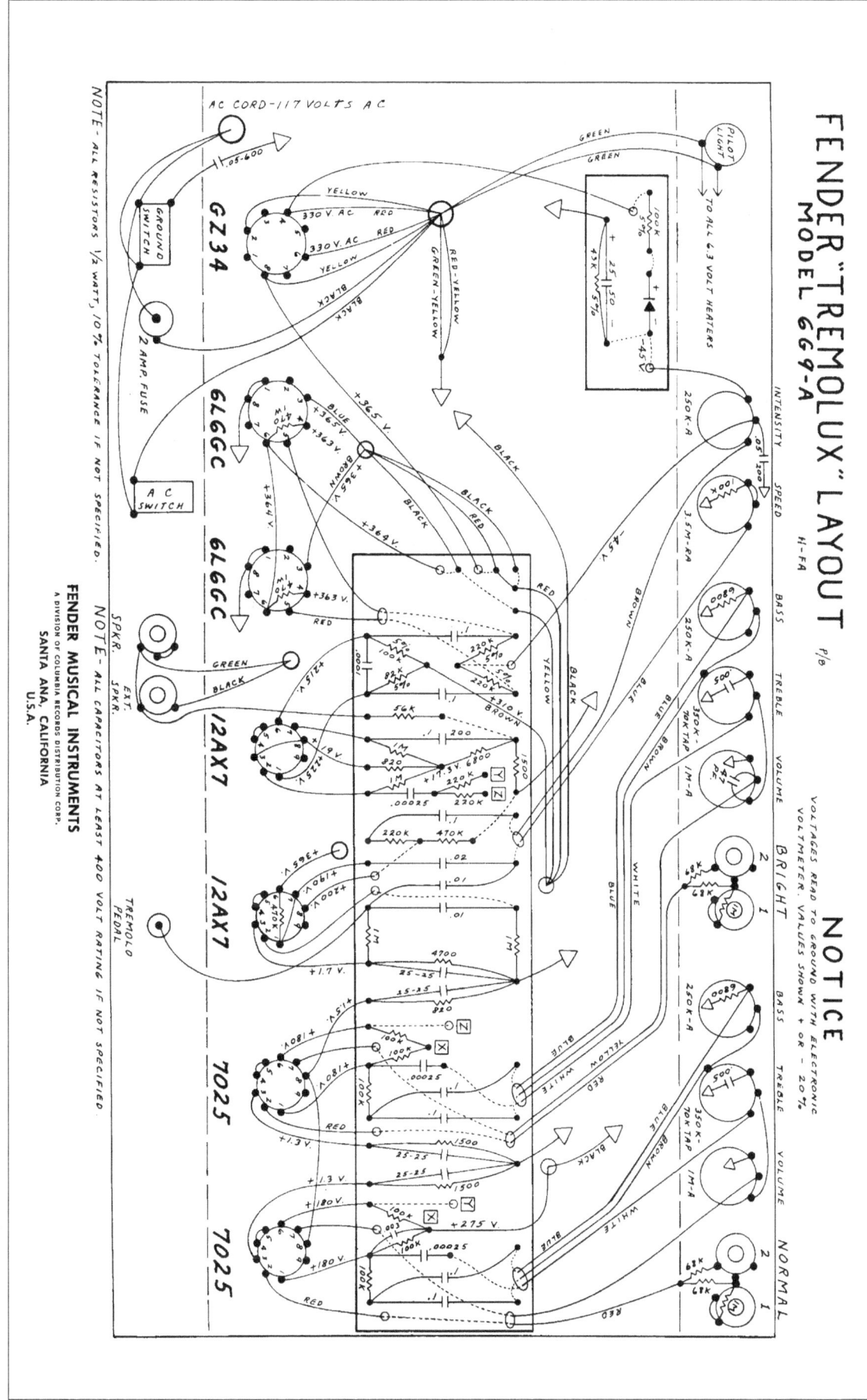

FENDER "TREMOLUX" LAYOUT

P/B

MODEL 6G9-A

H-FA

NOTICE

VOLTAGES READ TO GROUND WITH ELECTRONIC VOLTMETER. VALUES SHOWN + OR − 20%

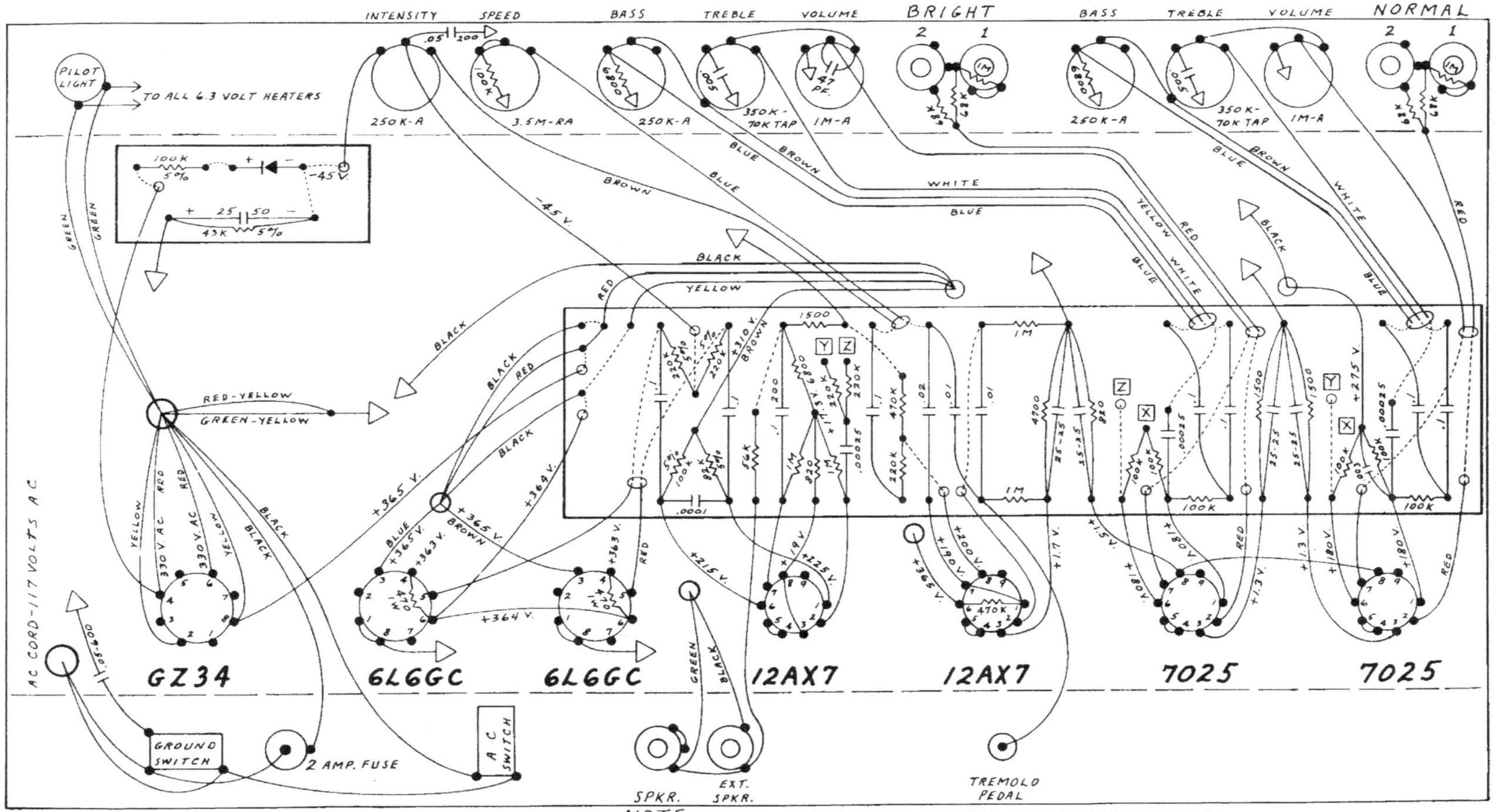

NOTE- ALL RESISTORS 1/2 WATT, 10% TOLERANCE IF NOT SPECIFIED.

NOTE- ALL CAPACITORS AT LEAST 400 VOLT RATING IF NOT SPECIFIED

FENDER MUSICAL INSTRUMENTS

A DIVISION OF COLUMBIA RECORDS DISTRIBUTION CORP.

SANTA ANA, CALIFORNIA

U.S.A.

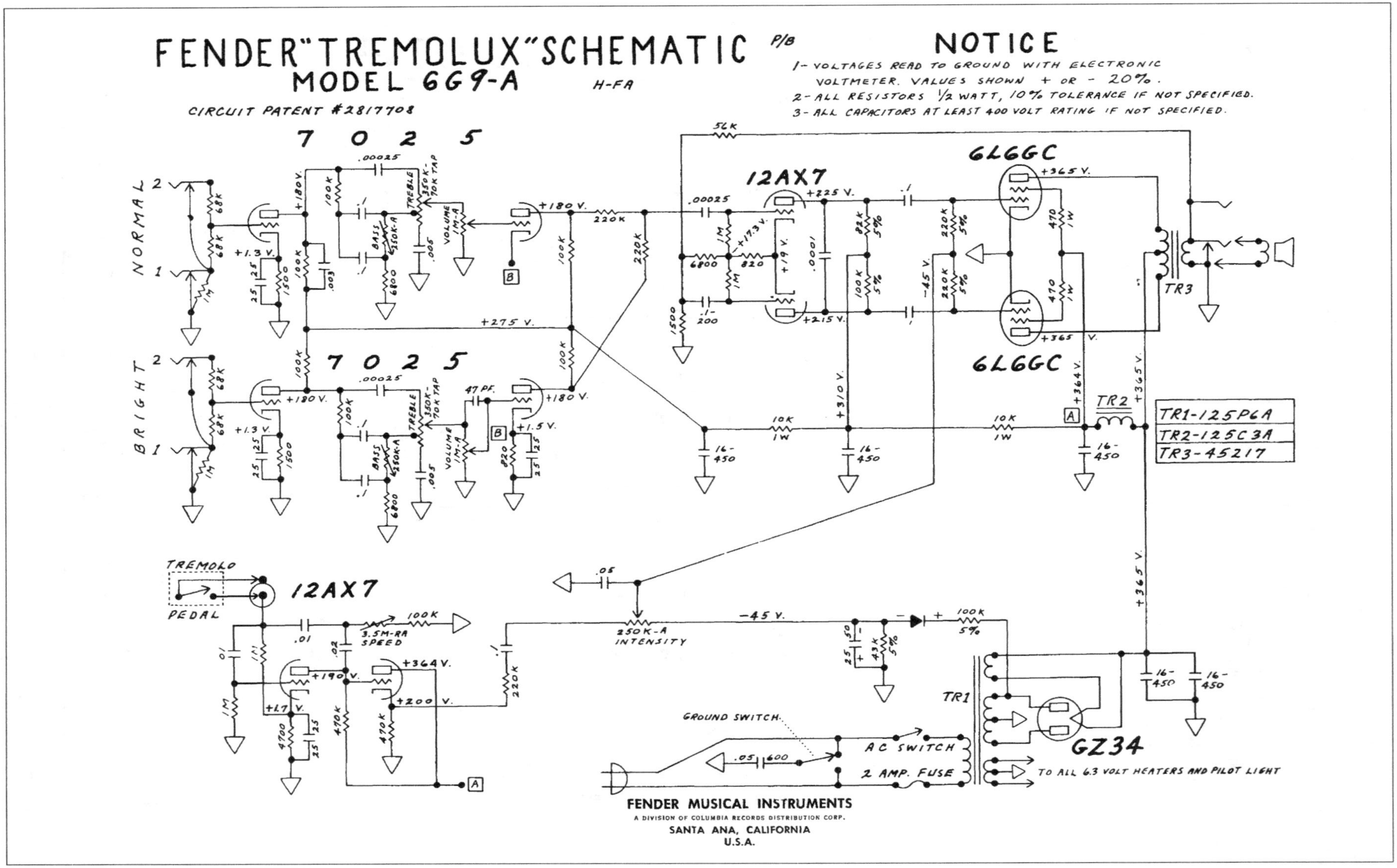
FENDER "TREMOLUX" SCHEMATIC
MODEL 6G9-A
CIRCUIT PATENT #2817708
NOTICE
1- VOLTAGES READ TO GROUND WITH ELECTRONIC VOLTMETER. VALUES SHOWN + OR - 20%.
2- ALL RESISTORS 1/2 WATT, 10% TOLERANCE IF NOT SPECIFIED.
3- ALL CAPACITORS AT LEAST 400 VOLT RATING IF NOT SPECIFIED.
7025
12AX7
6L6GC
GZ34
NORMAL
BRIGHT
TREMOLO PEDAL
TR1-125P6A
TR2-125C3A
TR3-45217
250K-A INTENSITY
3.5M-RA SPEED
GROUND SWITCH
AC SWITCH
2 AMP. FUSE
TO ALL 6.3 VOLT HEATERS AND PILOT LIGHT
FENDER MUSICAL INSTRUMENTS
A DIVISION OF COLUMBIA RECORDS DISTRIBUTION CORP.
SANTA ANA, CALIFORNIA
U.S.A.

FENDER MODEL "TREMOLUX-AMP AA763" LAYOUT

I-FC

NOTICE

VOLTAGES READ TO GROUND WITH ELECTRONIC VOLTMETER VALUES SHOWN + OR – 20 %

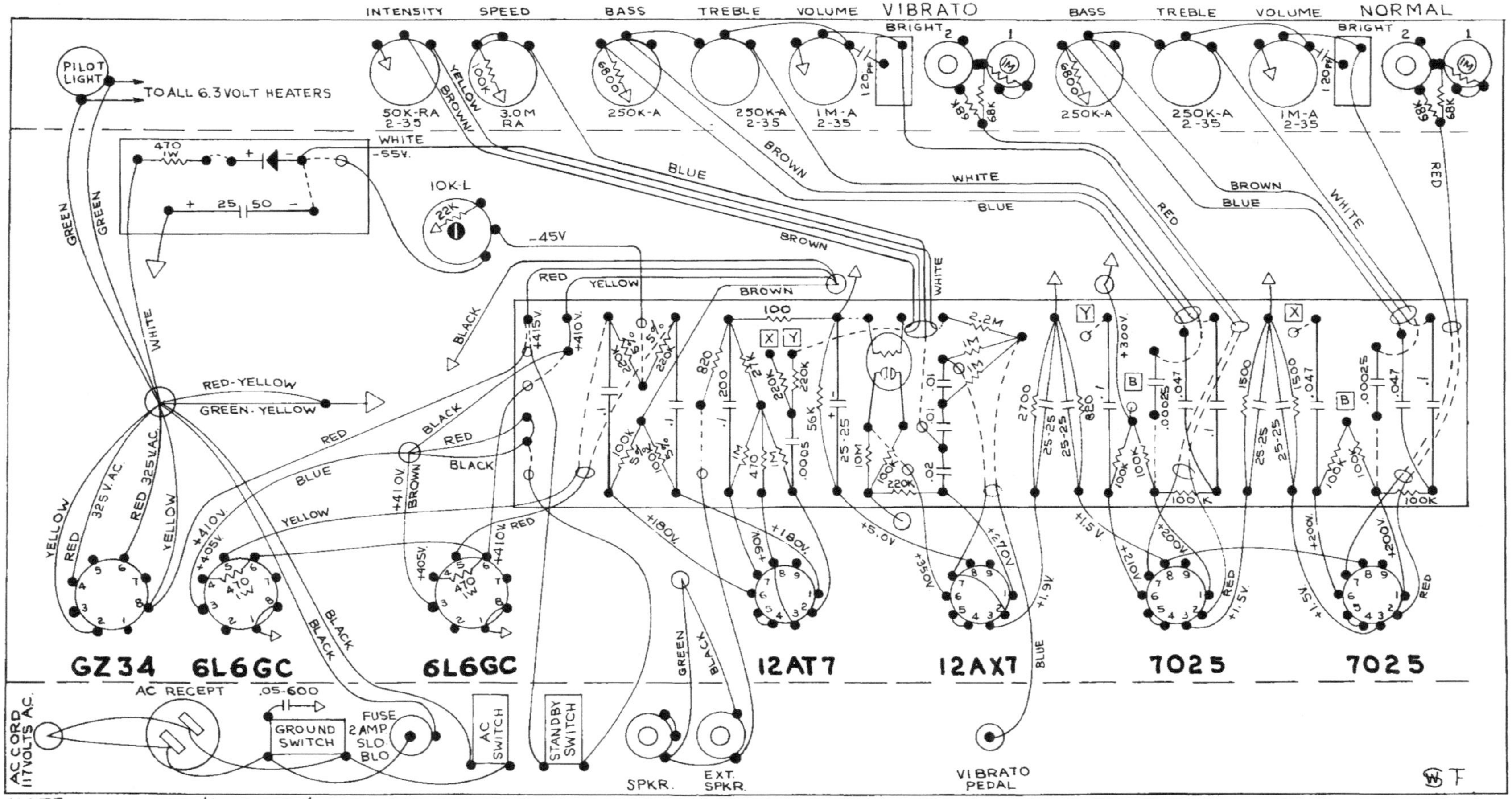

NOTE- ALL RESISTORS ½ WATT, 10% TOLERANCE IF NOT SPECIFIED.

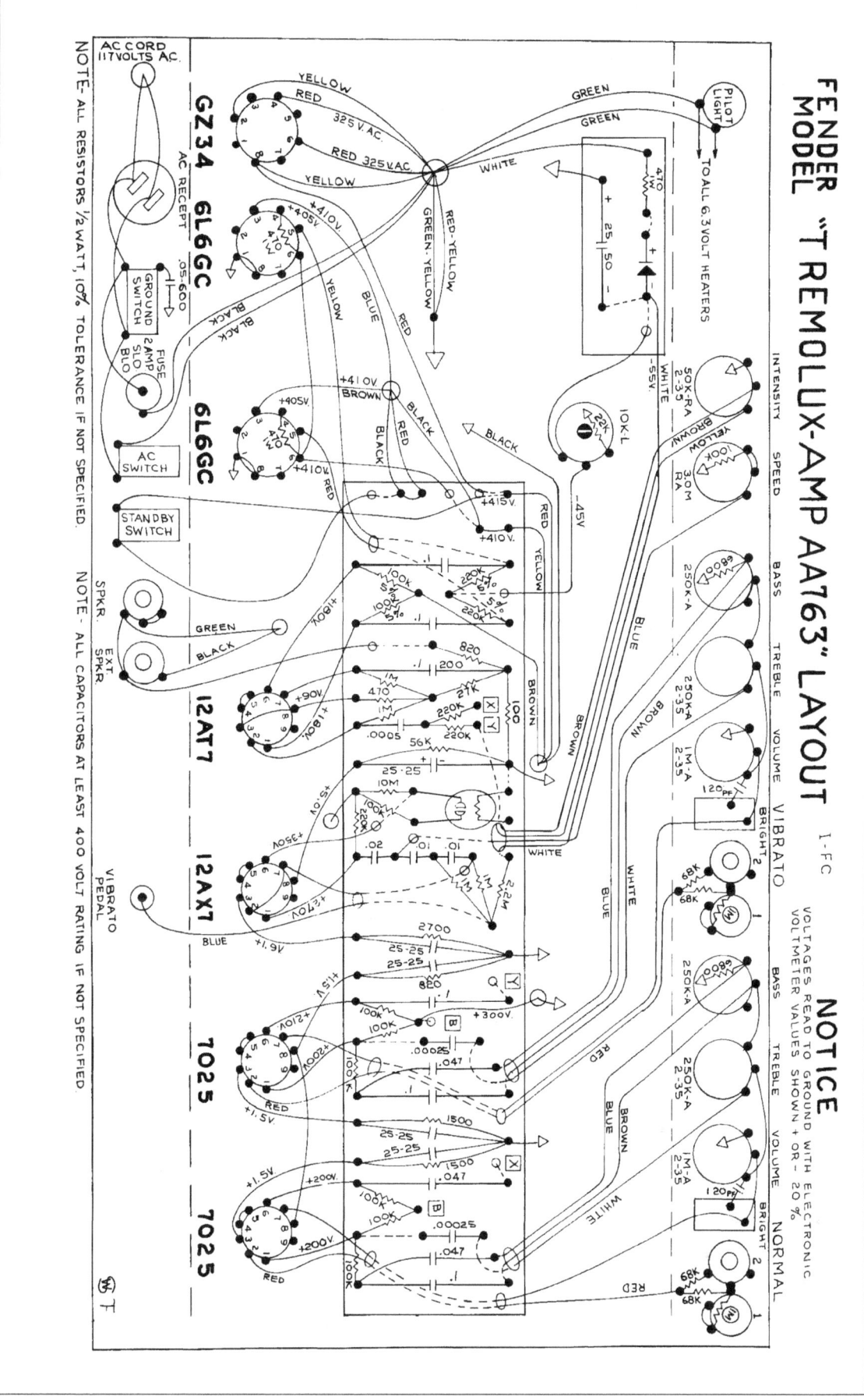

NOTE- ALL CAPACITORS AT LEAST 400 VOLT RATING IF NOT SPECIFIED.

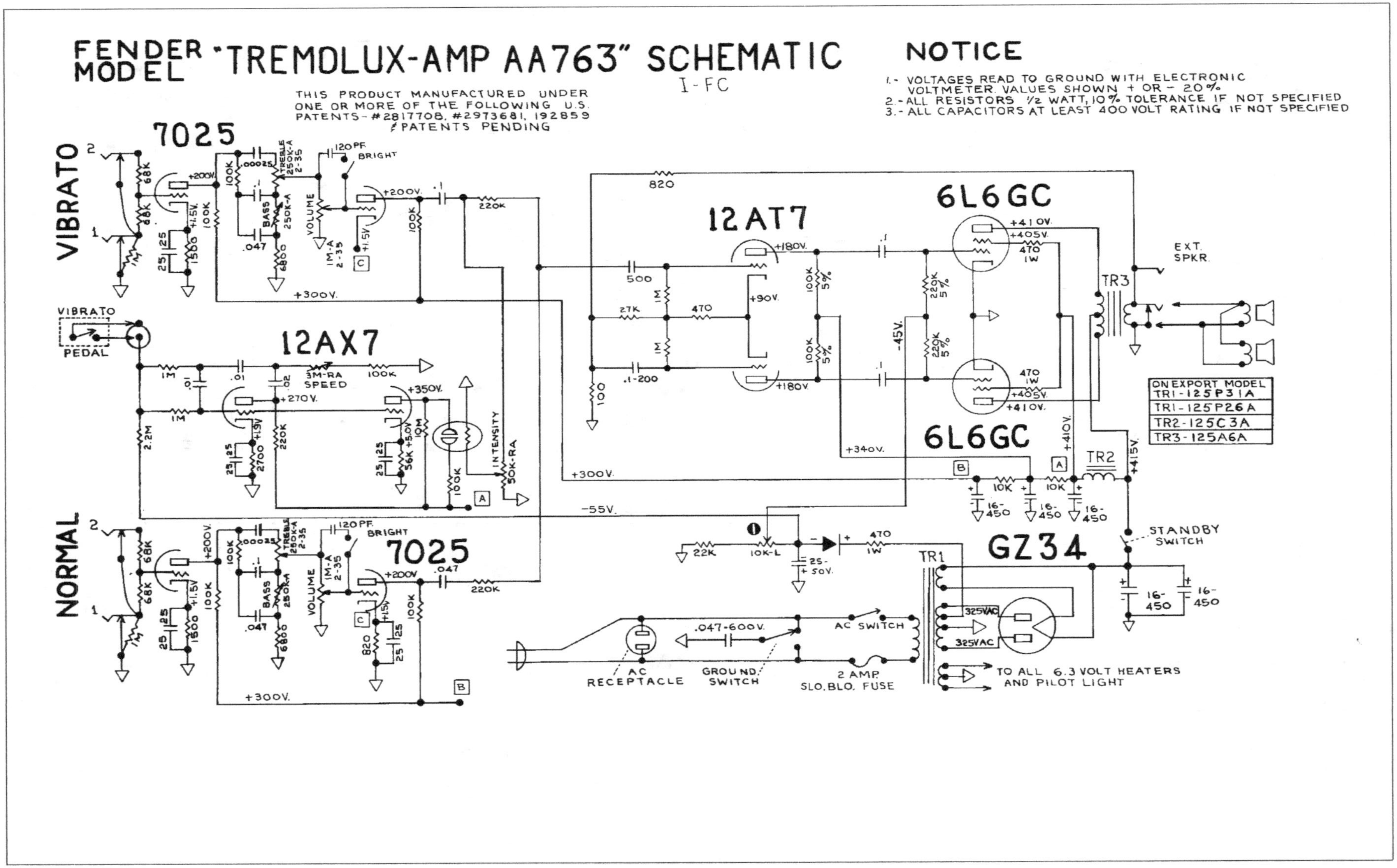
FENDER MODEL "TREMOLUX-AMP AA763" SCHEMATIC
I-FC
THIS PRODUCT MANUFACTURED UNDER ONE OR MORE OF THE FOLLOWING U.S. PATENTS-#2817708, #2973681, 192859 ≠PATENTS PENDING
NOTICE
1.- VOLTAGES READ TO GROUND WITH ELECTRONIC VOLTMETER. VALUES SHOWN + OR - 20%
2.- ALL RESISTORS 1/2 WATT, 10% TOLERANCE IF NOT SPECIFIED
3.- ALL CAPACITORS AT LEAST 400 VOLT RATING IF NOT SPECIFIED
VIBRATO
NORMAL
7025
12AX7
12AT7
6L6GC
6L6GC
GZ34
VIBRATO PEDAL
BRIGHT
VOLUME
TREBLE
BASS
3M-RA SPEED
INTENSITY 50K-RA
EXT. SPKR.
TR1
TR2
TR3
ON EXPORT MODEL
TR1-125P31A
TR1-125P26A
TR2-125C3A
TR3-125A6A
STANDBY SWITCH
AC SWITCH
2 AMP. SLO.BLO. FUSE
GROUND SWITCH
AC RECEPTACLE
.047-600V.
325VAC
TO ALL 6.3 VOLT HEATERS AND PILOT LIGHT
+300V.
-55V.
-45V.
+340V.
+410V.
+415V.

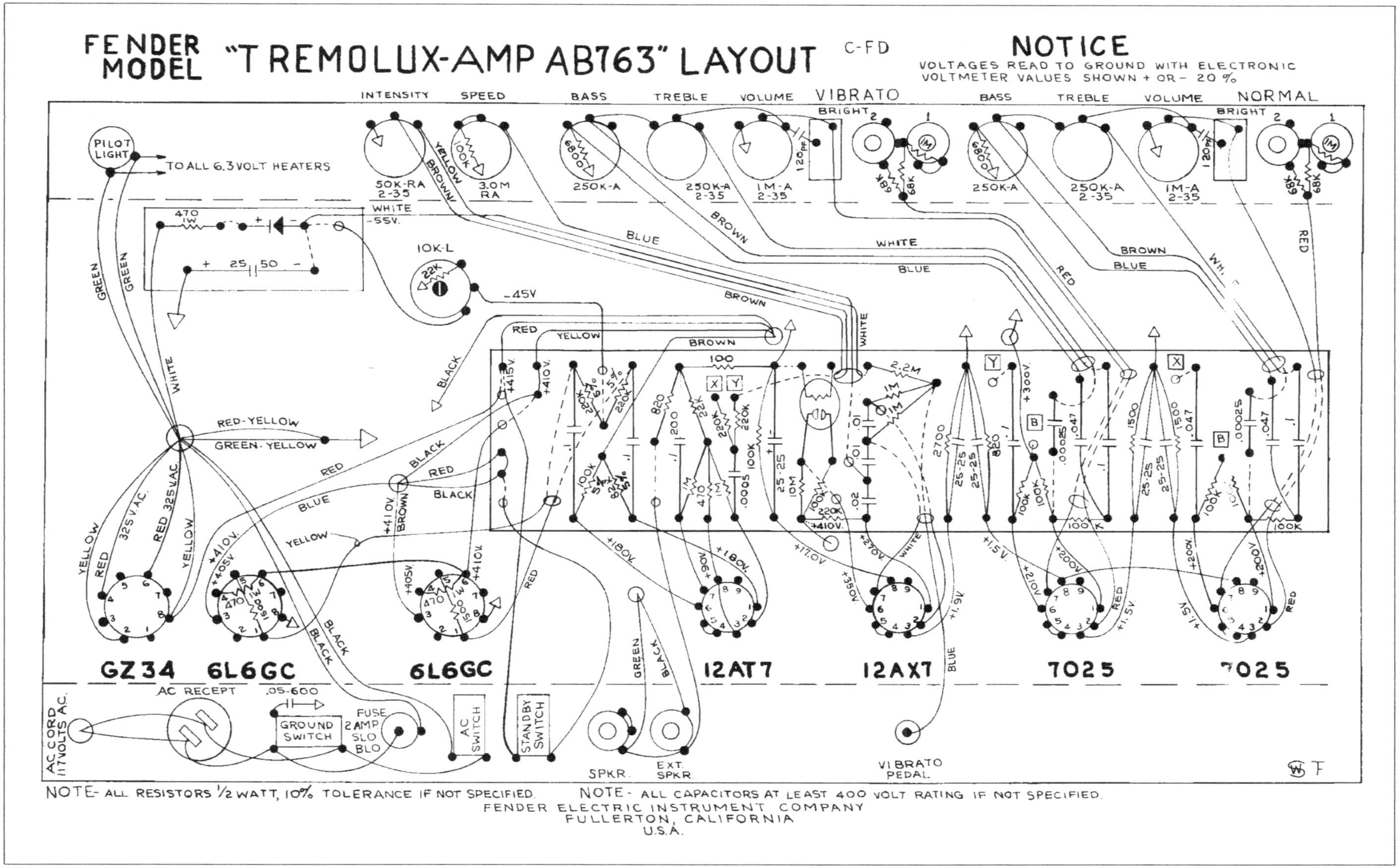
FENDER MODEL "TREMOLUX-AMP AB763" LAYOUT
C-FD
NOTICE
VOLTAGES READ TO GROUND WITH ELECTRONIC VOLTMETER VALUES SHOWN + OR − 20 %
INTENSITY
SPEED
BASS
TREBLE
VOLUME
VIBRATO
BRIGHT
NORMAL
PILOT LIGHT
TO ALL 6.3 VOLT HEATERS
GZ34
6L6GC
6L6GC
12AT7
12AX7
7025
7025
AC CORD 117 VOLTS AC.
AC RECEPT
GROUND SWITCH
FUSE 2 AMP SLO BLO
AC SWITCH
STANDBY SWITCH
SPKR.
EXT. SPKR.
VIBRATO PEDAL
NOTE- ALL RESISTORS 1/2 WATT, 10% TOLERANCE IF NOT SPECIFIED.
NOTE- ALL CAPACITORS AT LEAST 400 VOLT RATING IF NOT SPECIFIED.
FENDER ELECTRIC INSTRUMENT COMPANY
FULLERTON, CALIFORNIA
U.S.A.

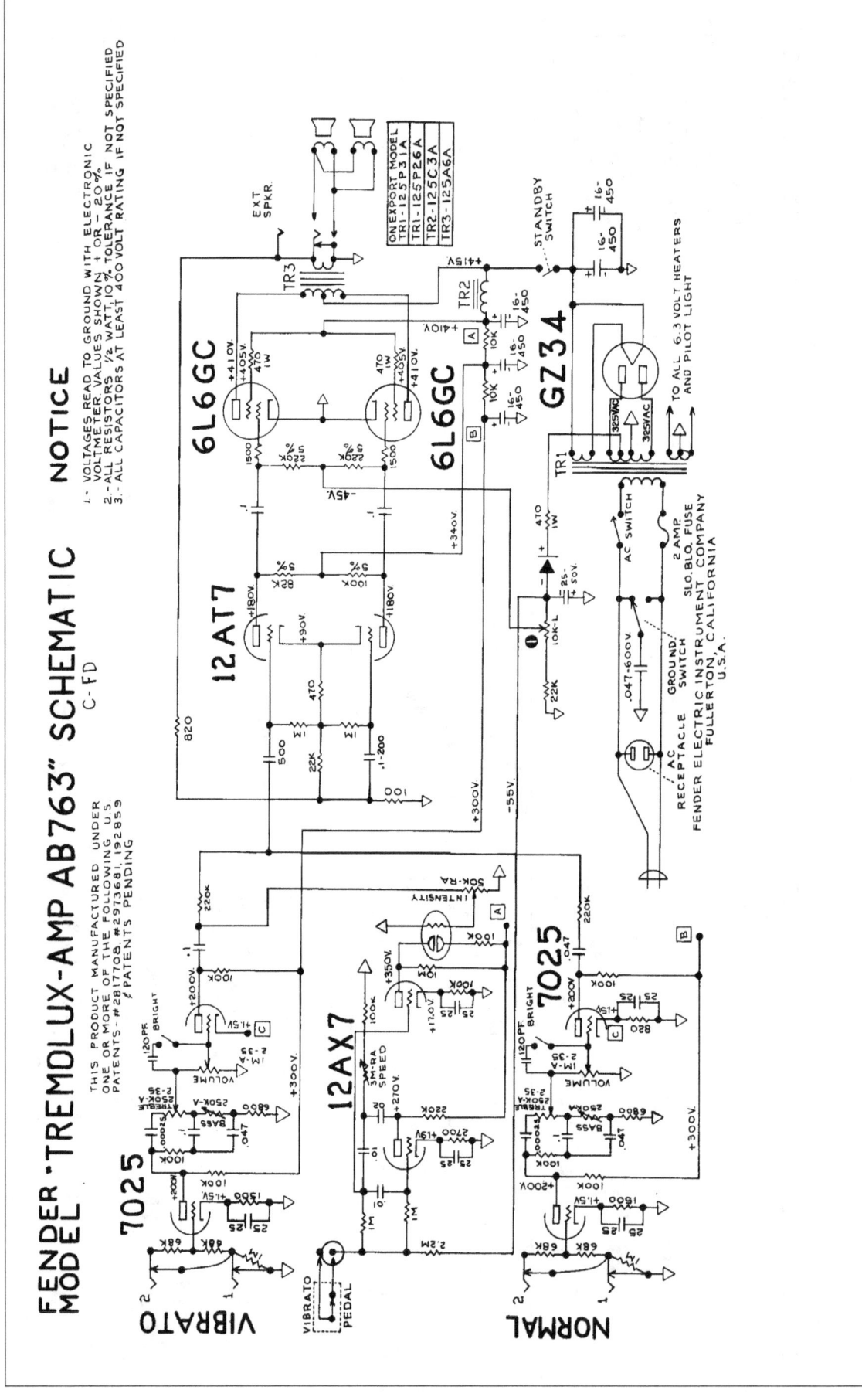

FENDER MODEL "TREMOLUX-AMP AB763" SCHEMATIC

C-FD

THIS PRODUCT MANUFACTURED UNDER ONE OR MORE OF THE FOLLOWING U.S. PATENTS - #2817708, #2973681, 192859 & PATENTS PENDING

NOTICE

1.- VOLTAGES READ TO GROUND WITH ELECTRONIC VOLTMETER. VALUES SHOWN + OR - 20%
2.- ALL RESISTORS 1/2 WATT, 10% TOLERANCE IF NOT SPECIFIED
3.- ALL CAPACITORS AT LEAST 400 VOLT RATING IF NOT SPECIFIED

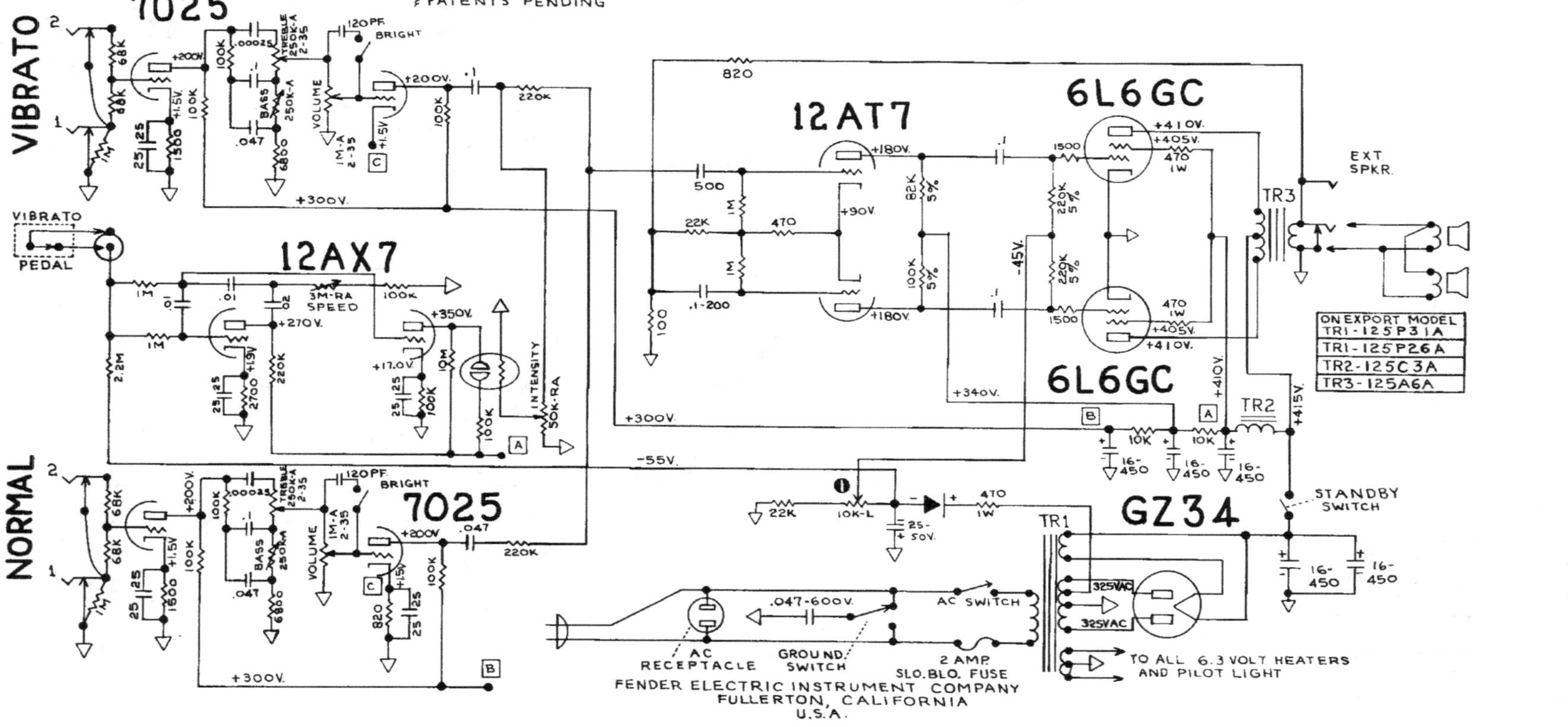

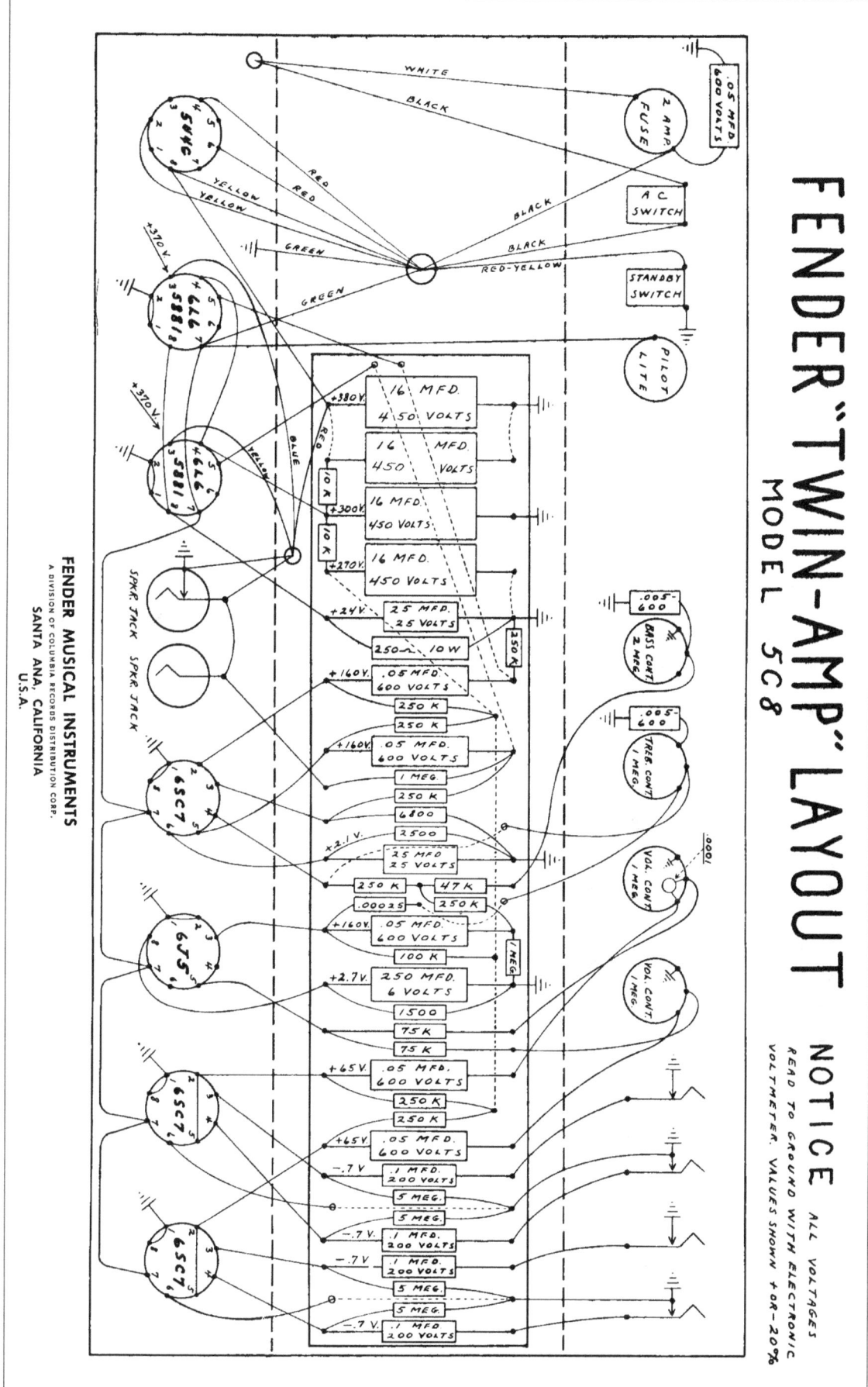

FENDER "TWIN-AMP" LAYOUT

MODEL 5C8

NOTICE ALL VOLTAGES READ TO GROUND WITH ELECTRONIC VOLTMETER. VALUES SHOWN + OR − 20%

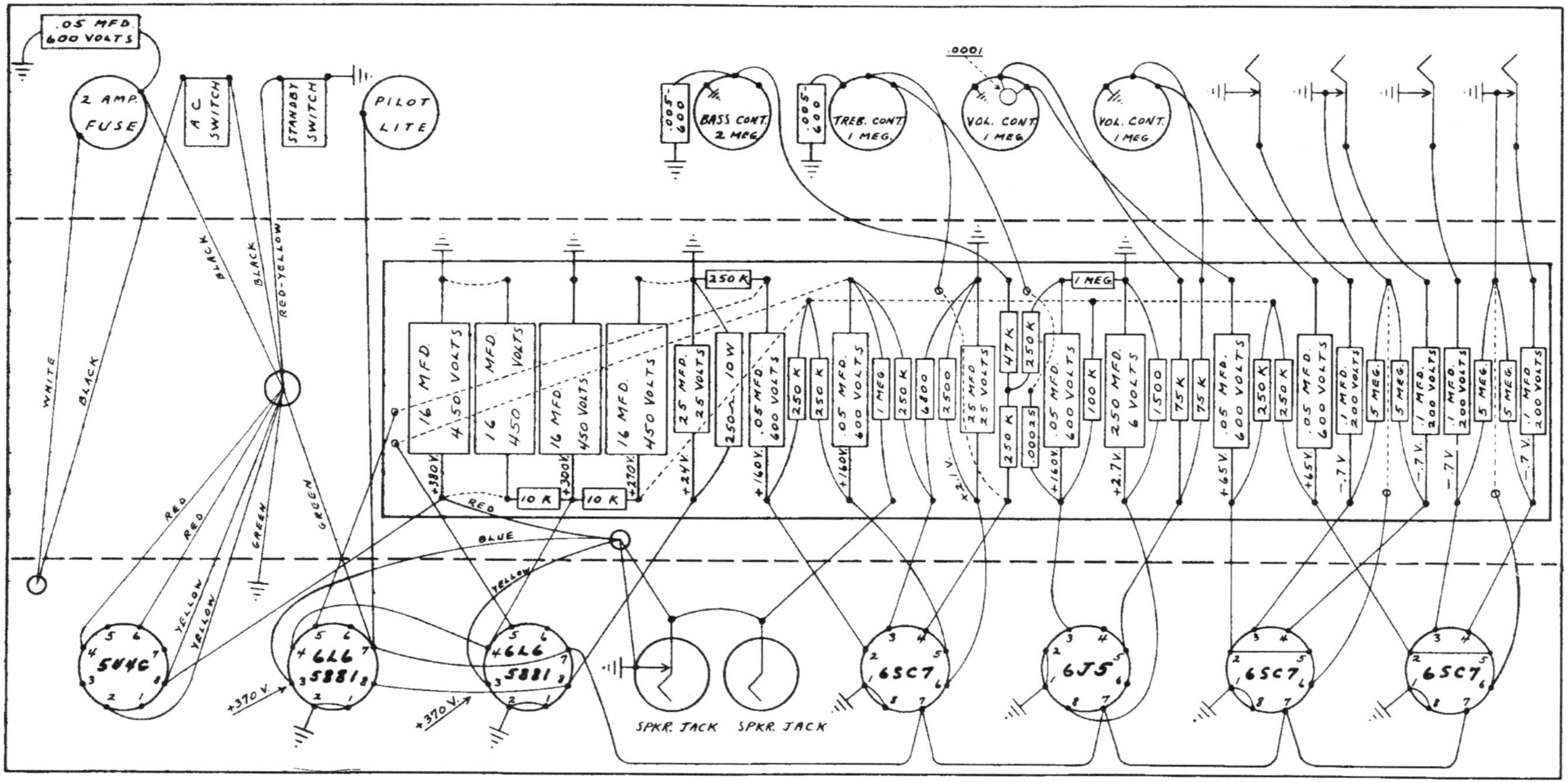

FENDER MUSICAL INSTRUMENTS

A DIVISION OF COLUMBIA RECORDS DISTRIBUTION CORP.

SANTA ANA, CALIFORNIA

U.S.A.

FENDER "TWIN-AMP" SCHEMATIC
MODEL 5C8

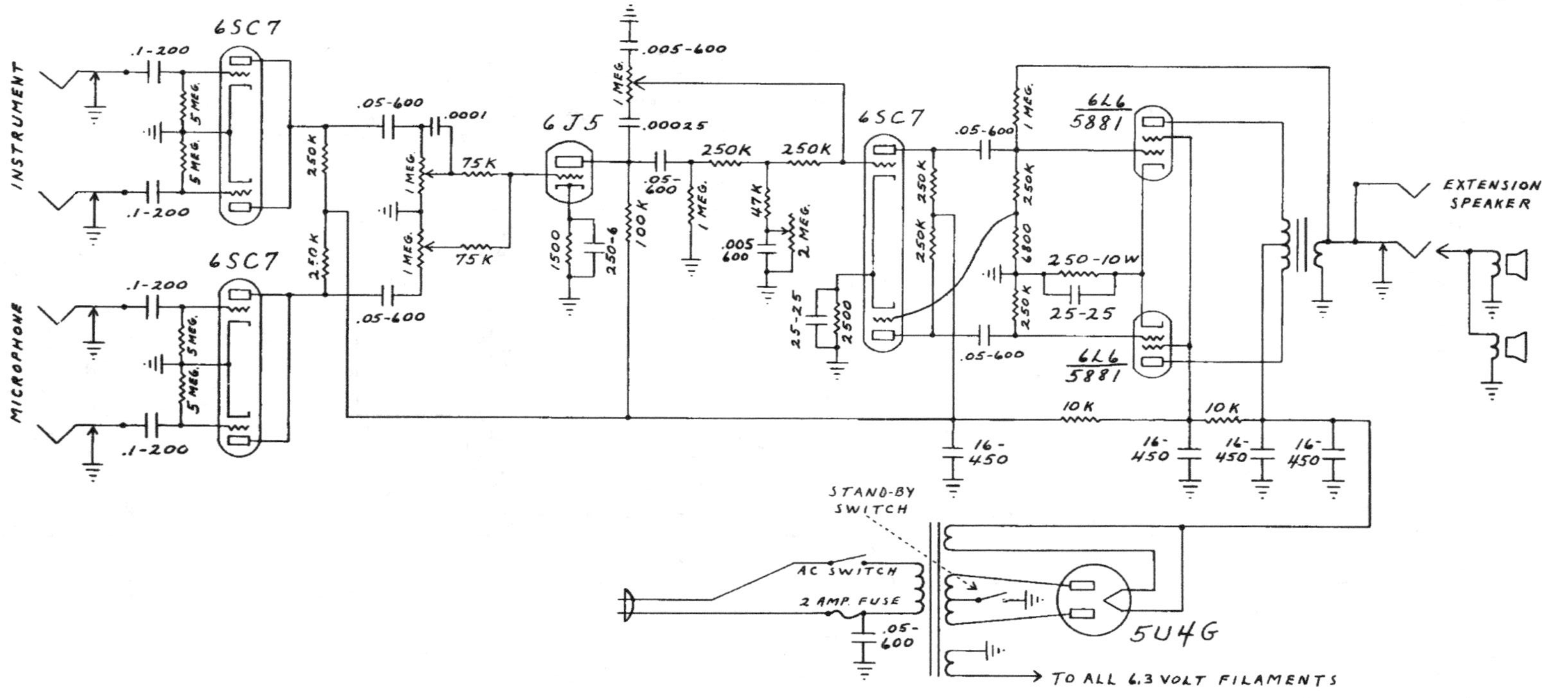

FENDER MUSICAL INSTRUMENTS
A DIVISION OF COLUMBIA RECORDS DISTRIBUTION CORP.
SANTA ANA, CALIFORNIA
U.S.A.

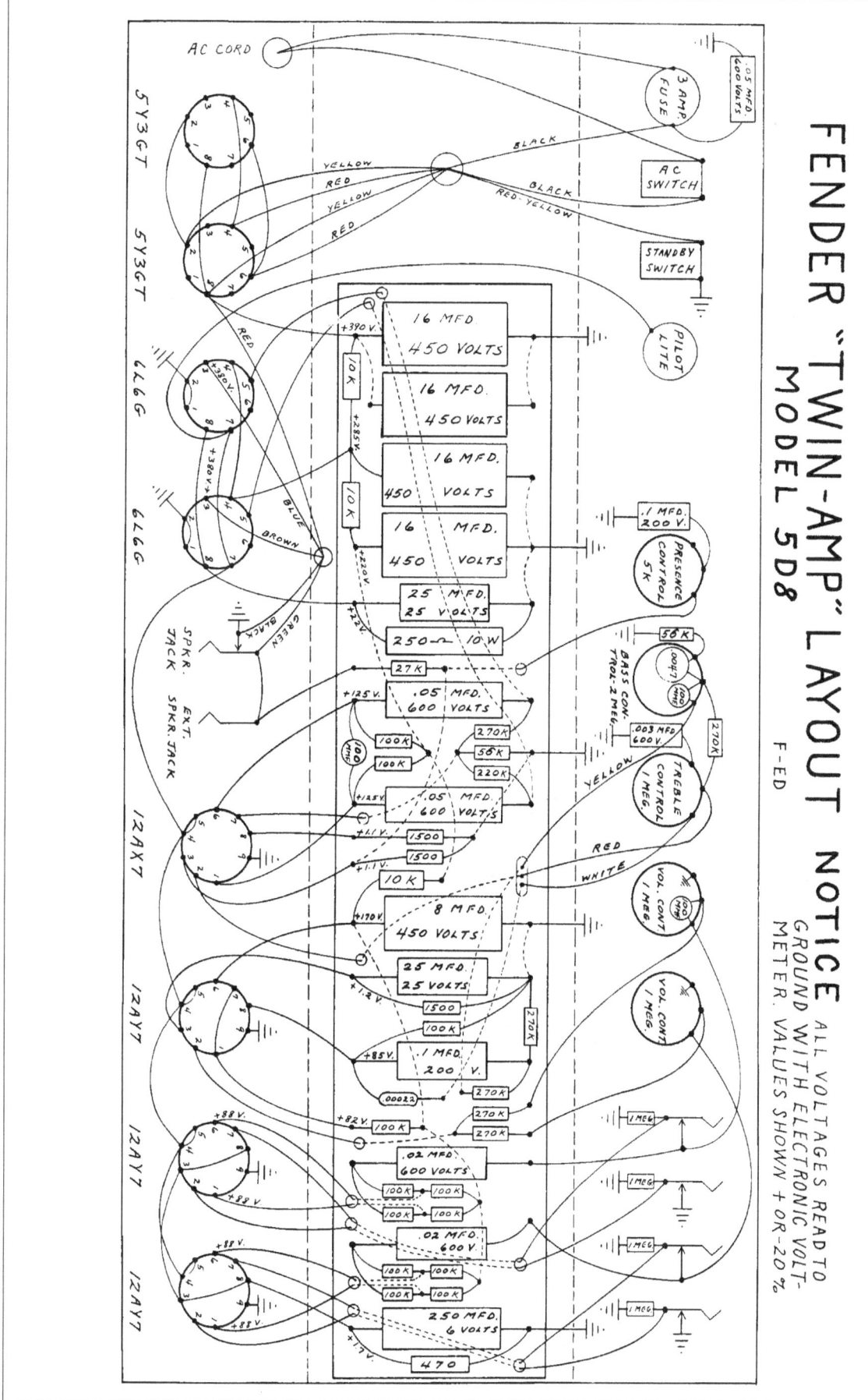

FENDER "TWIN-AMP" LAYOUT
MODEL 5D8

F-ED

NOTICE ALL VOLTAGES READ TO GROUND WITH ELECTRONIC VOLT-METER. VALUES SHOWN + OR -20%

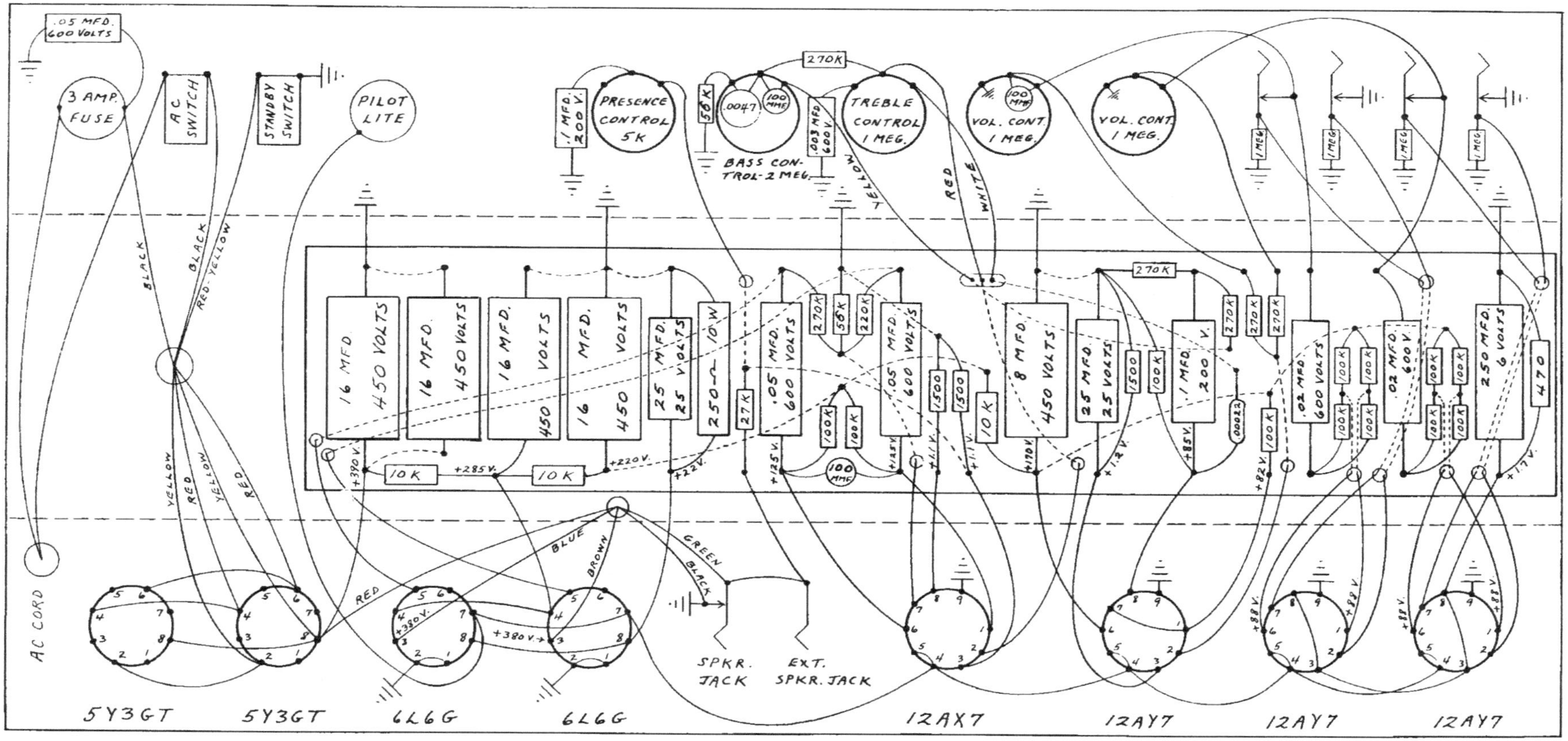

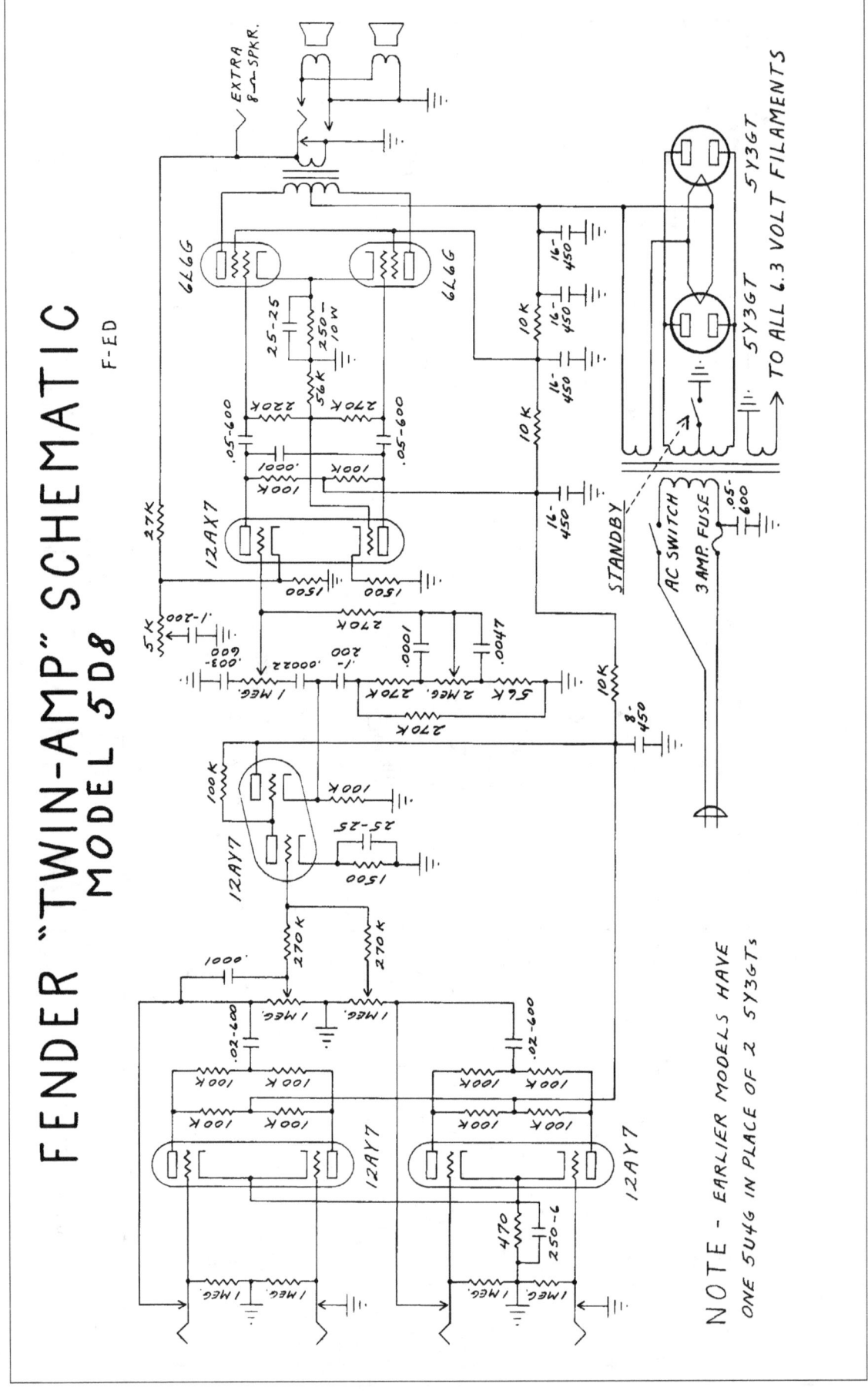

FENDER "TWIN-AMP" SCHEMATIC
MODEL 5D8

F-ED

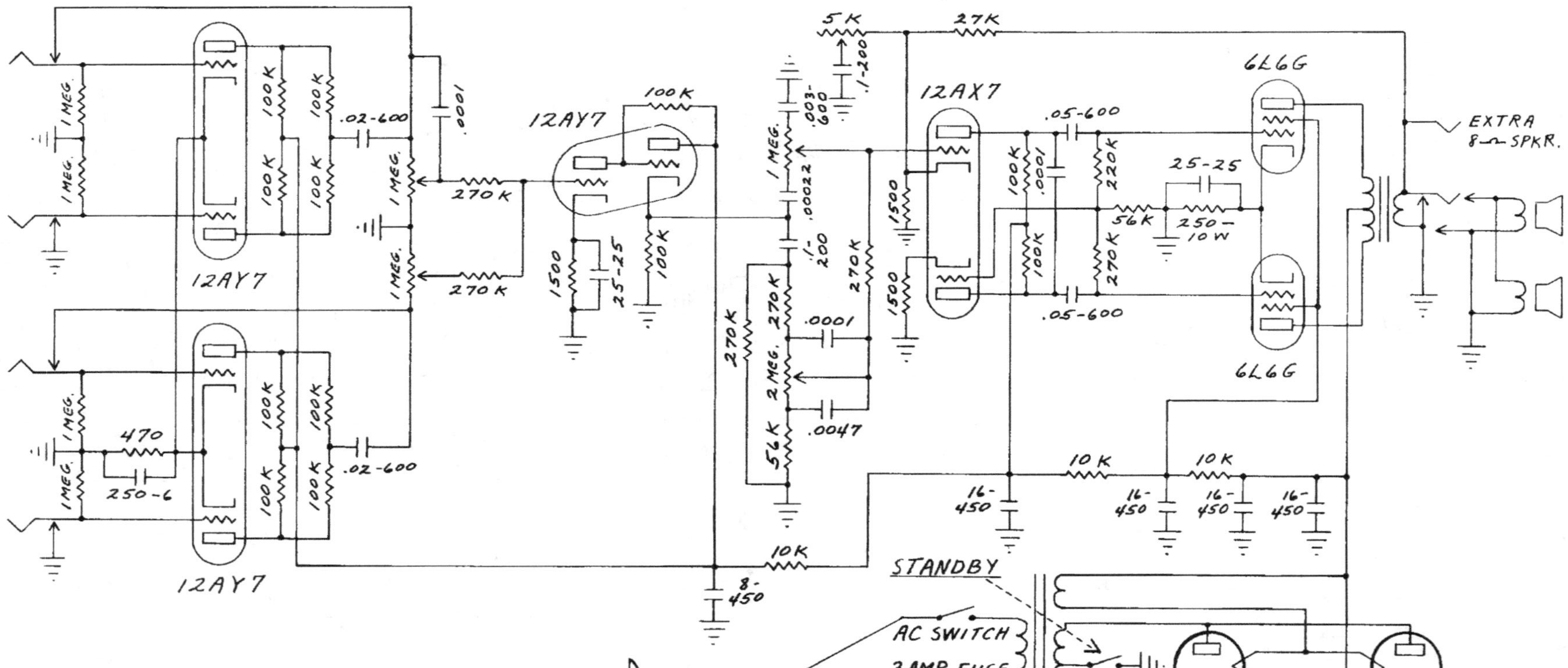

NOTE - EARLIER MODELS HAVE ONE 5U4G IN PLACE OF 2 5Y3GTs

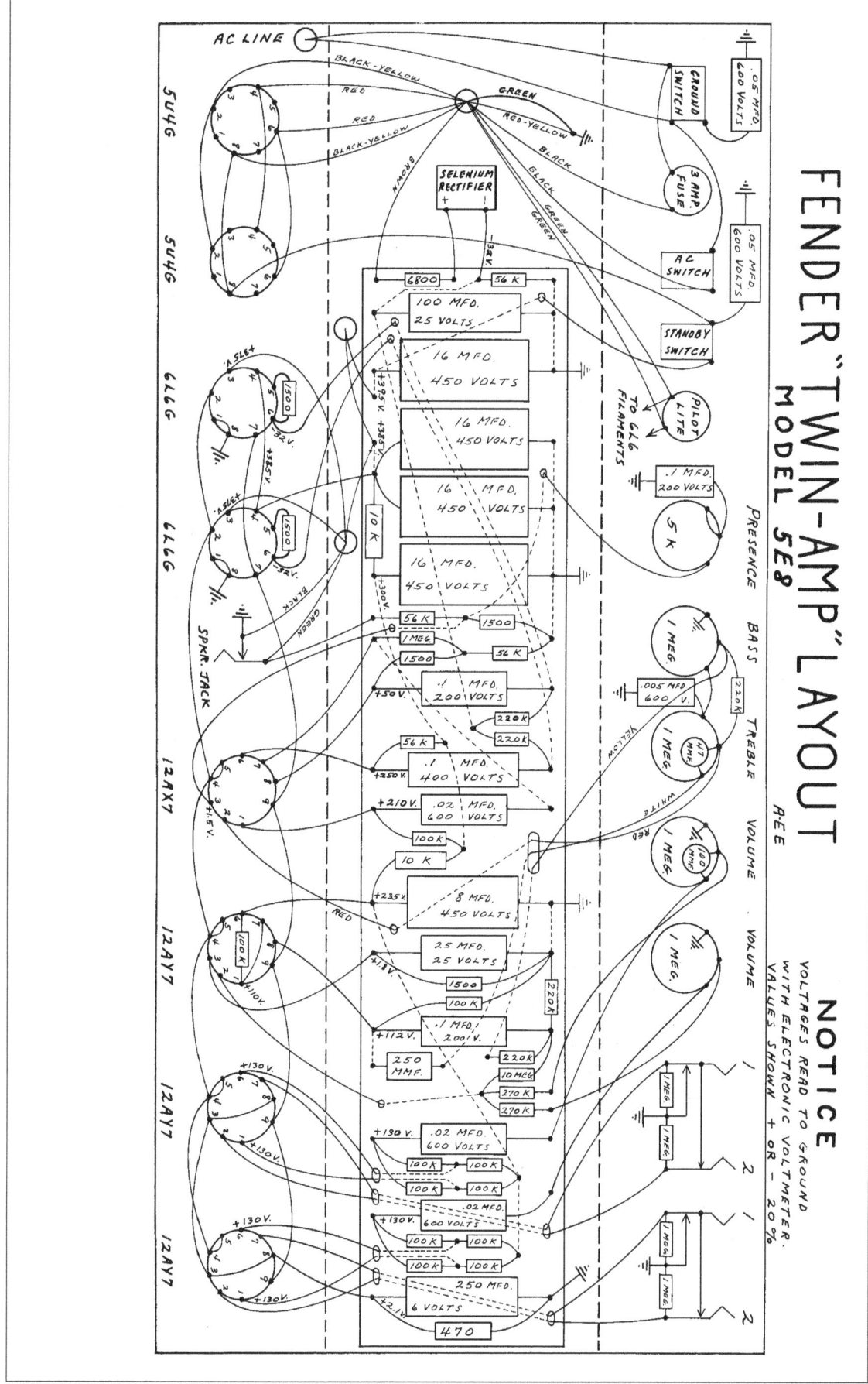

FENDER "TWIN-AMP" LAYOUT

MODEL 5E8

A·EE

NOTICE

VOLTAGES READ TO GROUND
WITH ELECTRONIC VOLTMETER.
VALUES SHOWN + OR − 20%

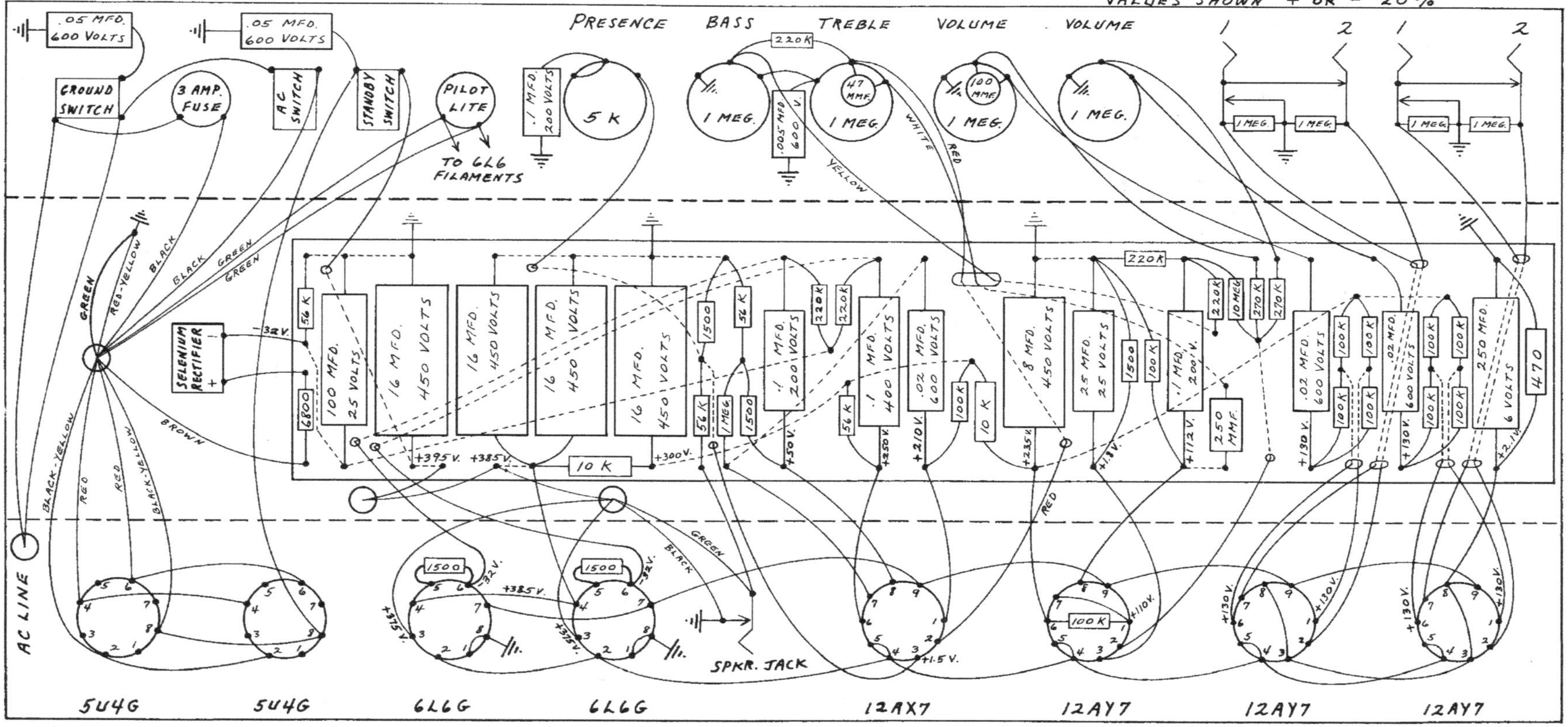

FENDER "TWIN-AMP" SCHEMATIC

MODEL 5E8

A-EE

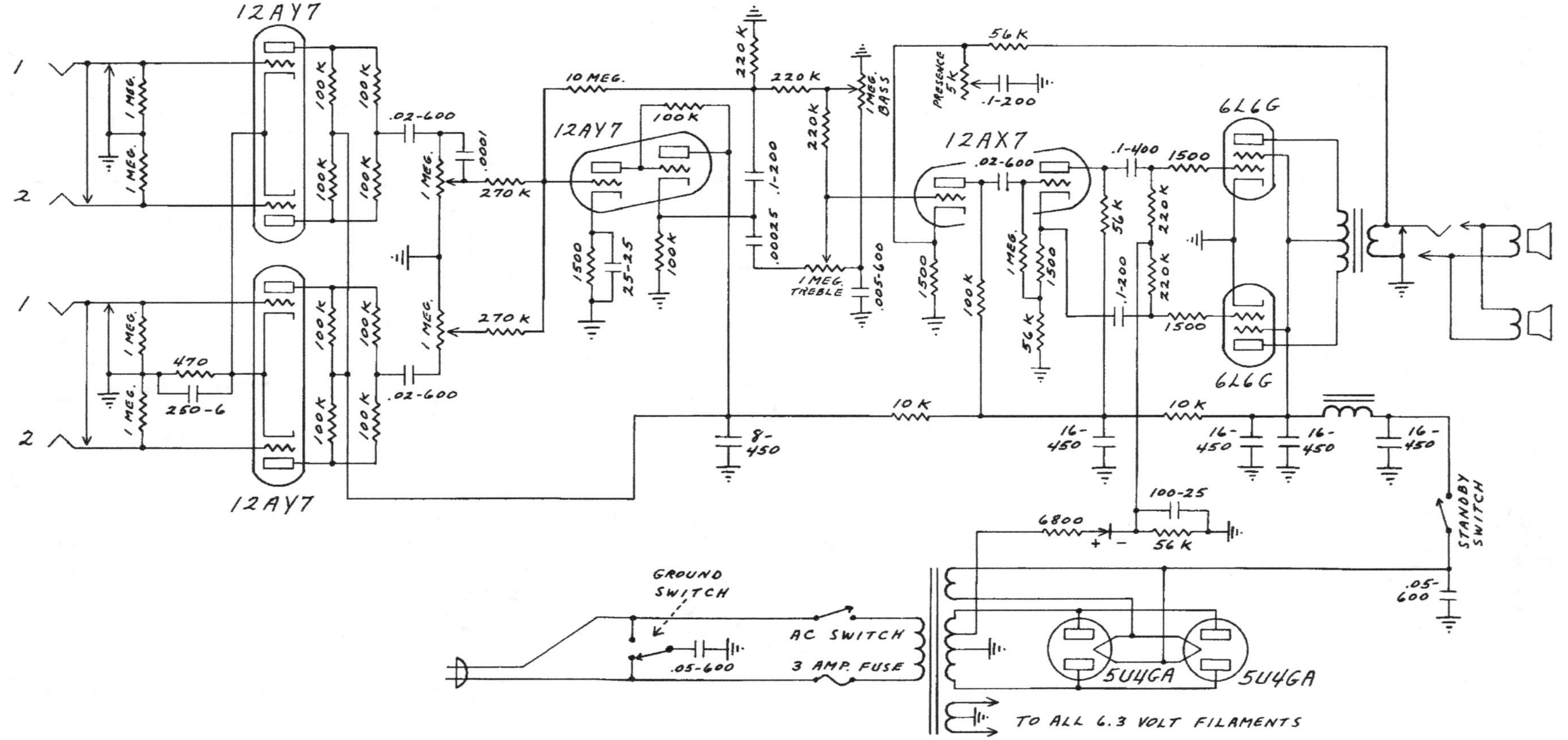

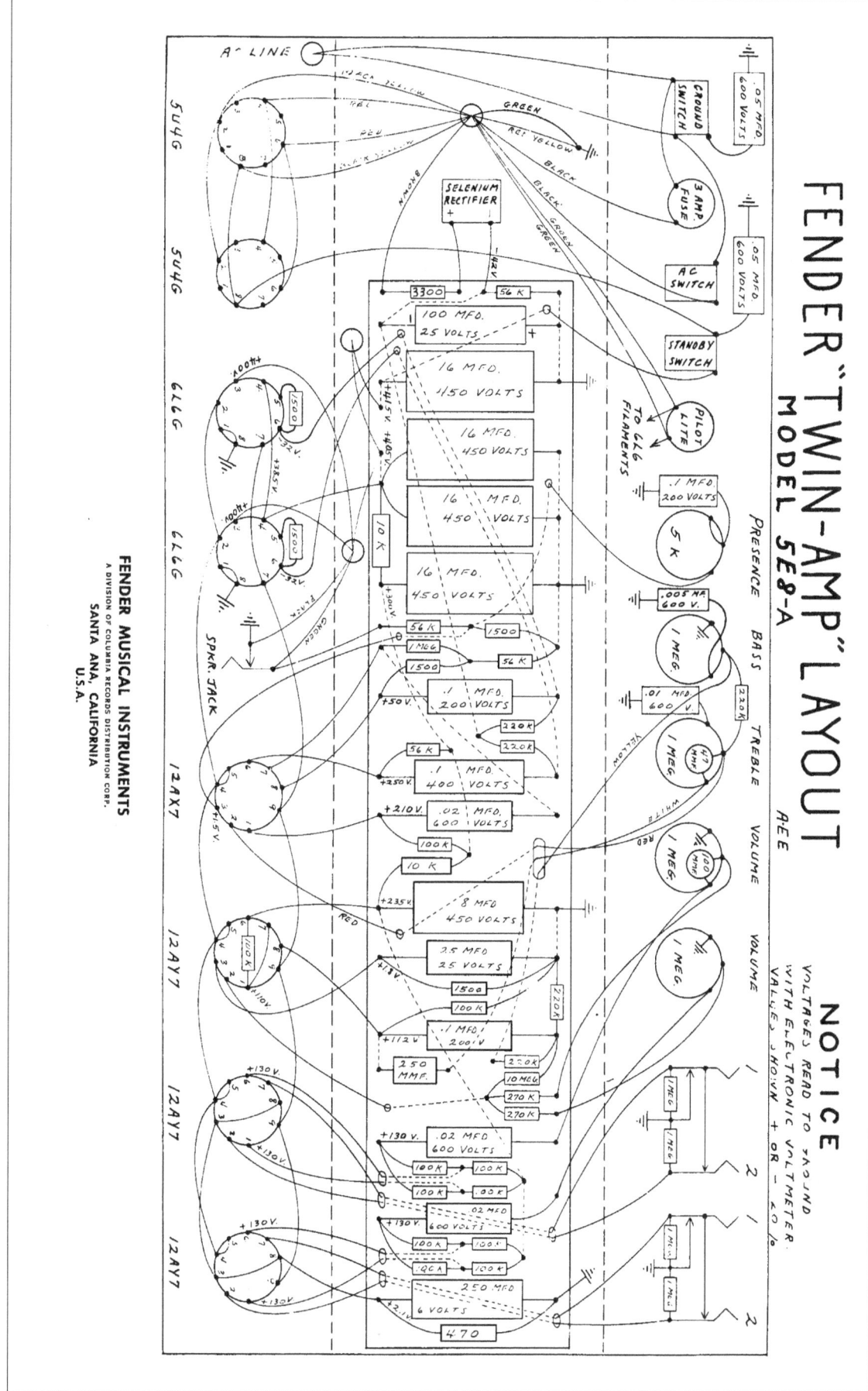

FENDER "TWIN-AMP" LAYOUT

MODEL 5E8-A

A·E·E

NOTICE

VOLTAGES READ TO GROUND
WITH ELECTRONIC VOLTMETER.
VALUES SHOWN + OR – 20 %

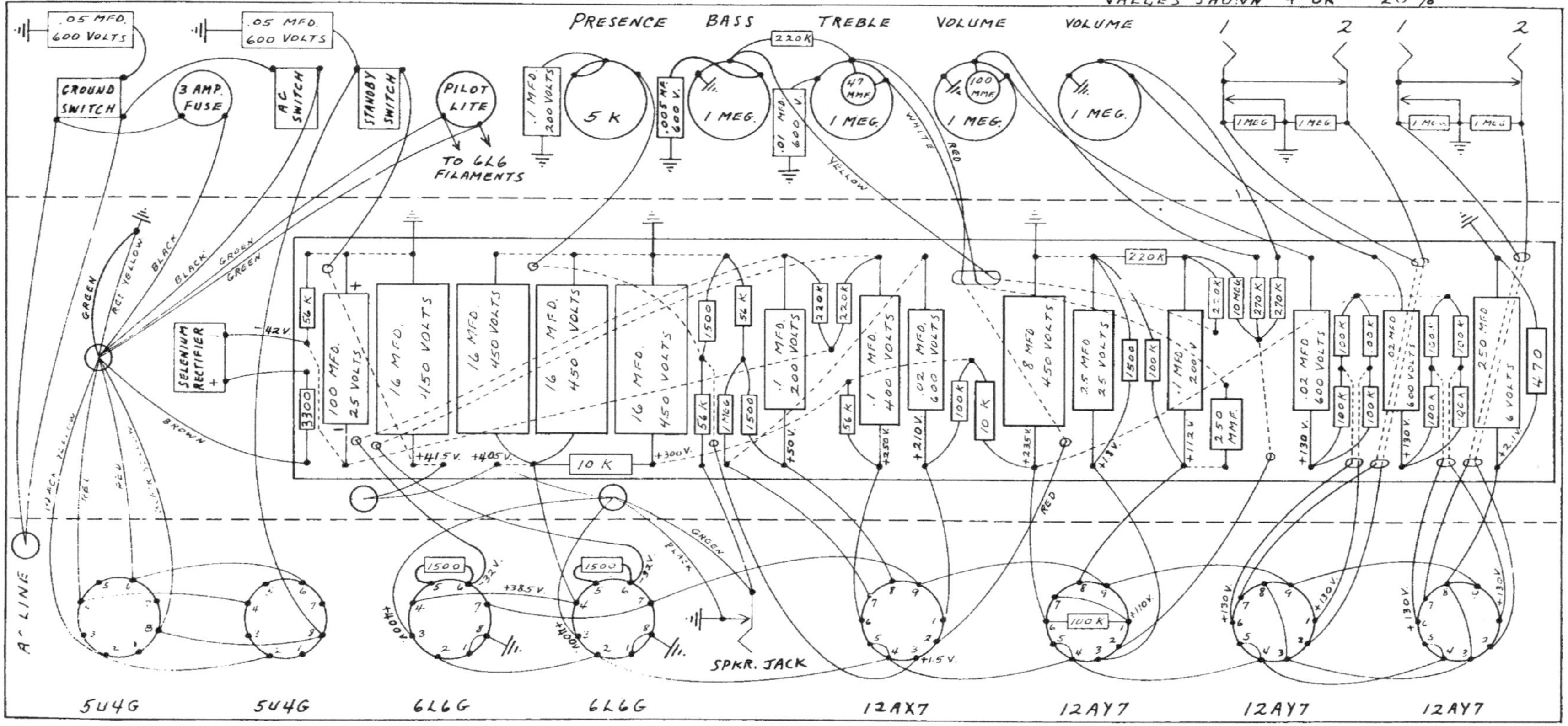

FENDER MUSICAL INSTRUMENTS
A DIVISION OF COLUMBIA RECORDS DISTRIBUTION CORP.
SANTA ANA, CALIFORNIA
U.S.A.

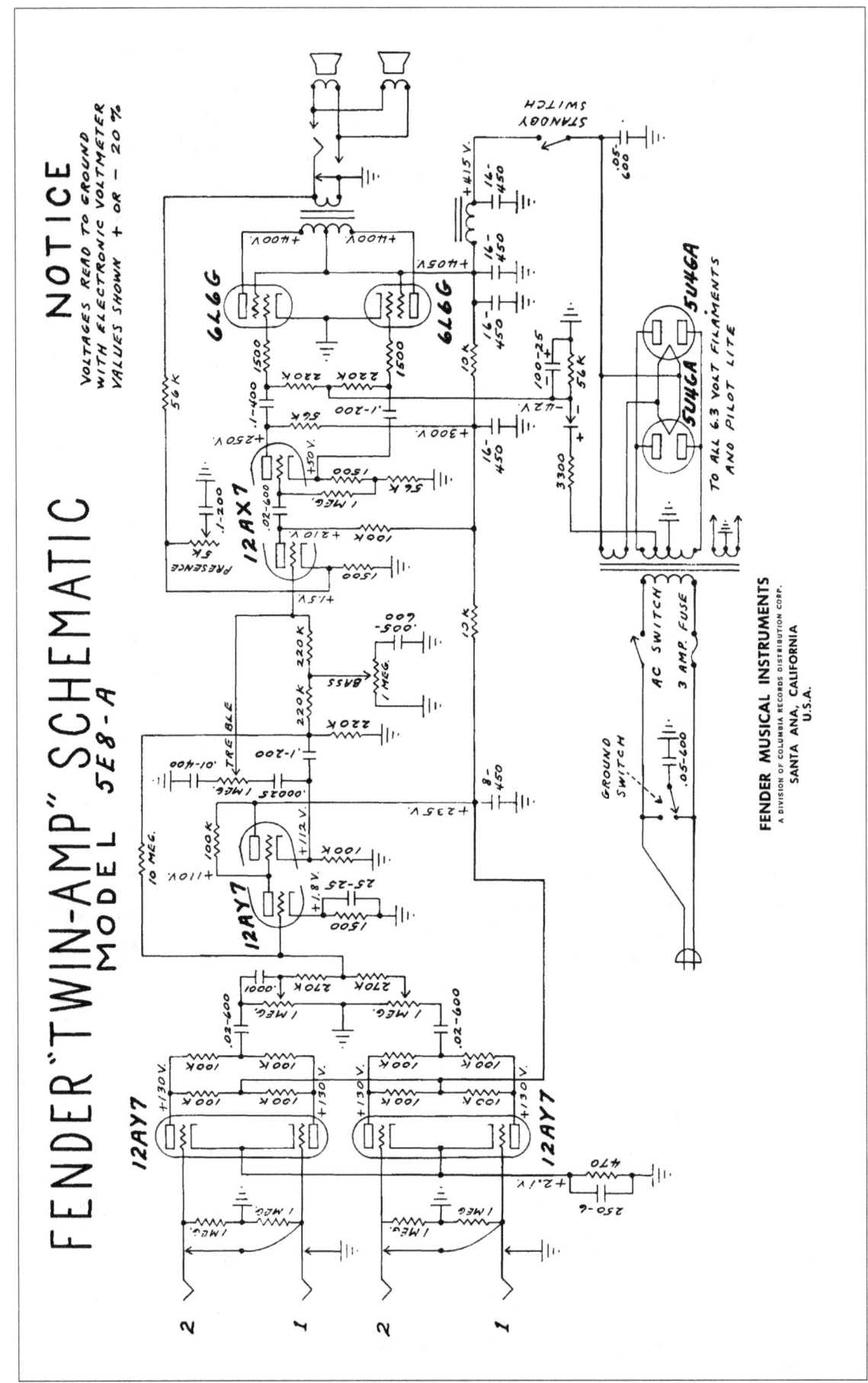

FENDER "TWIN-AMP" SCHEMATIC
MODEL 5E8-A

NOTICE

VOLTAGES READ TO GROUND WITH ELECTRONIC VOLTMETER VALUES SHOWN + OR − 20 %

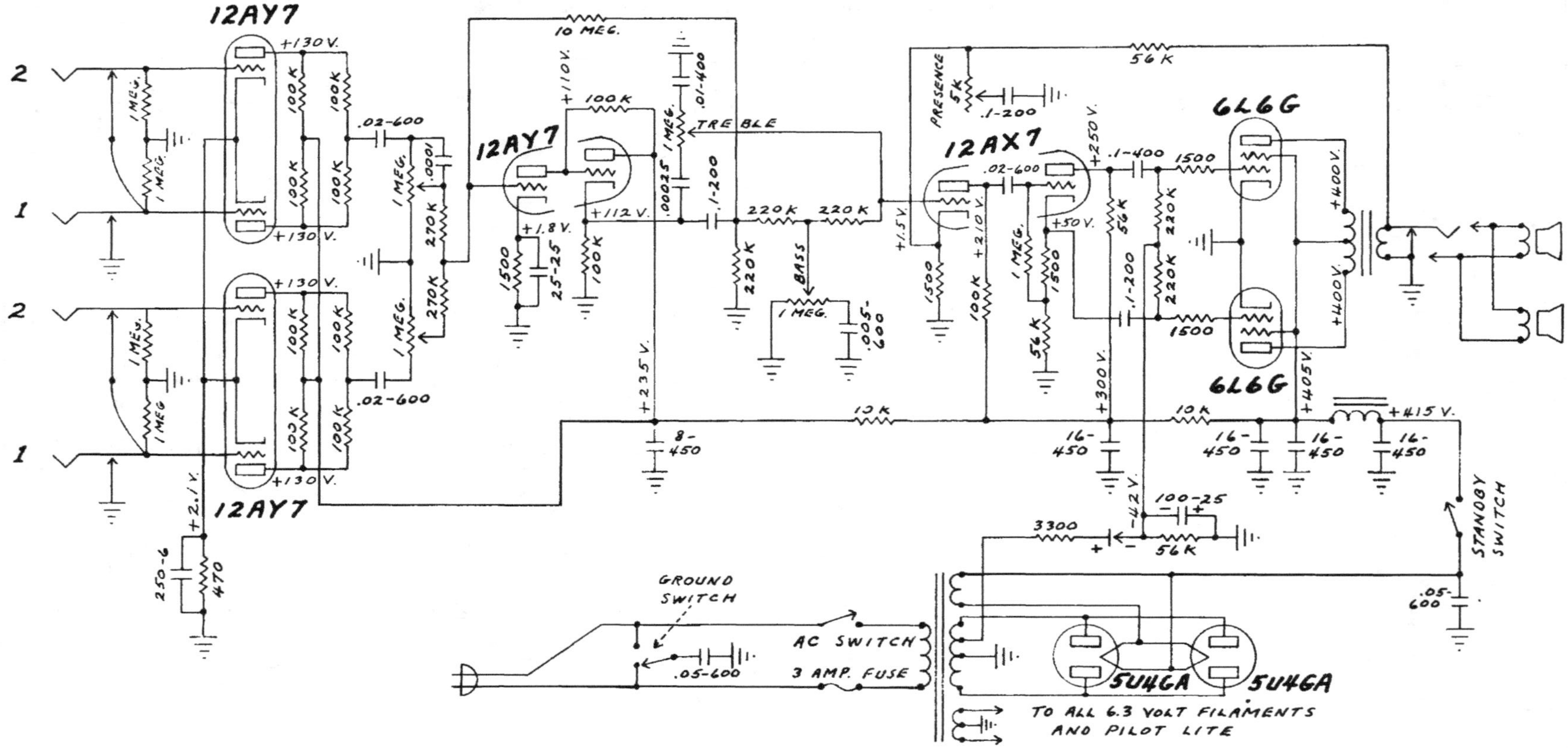

FENDER MUSICAL INSTRUMENTS
A DIVISION OF COLUMBIA RECORDS DISTRIBUTION CORP.
SANTA ANA, CALIFORNIA
U.S.A.

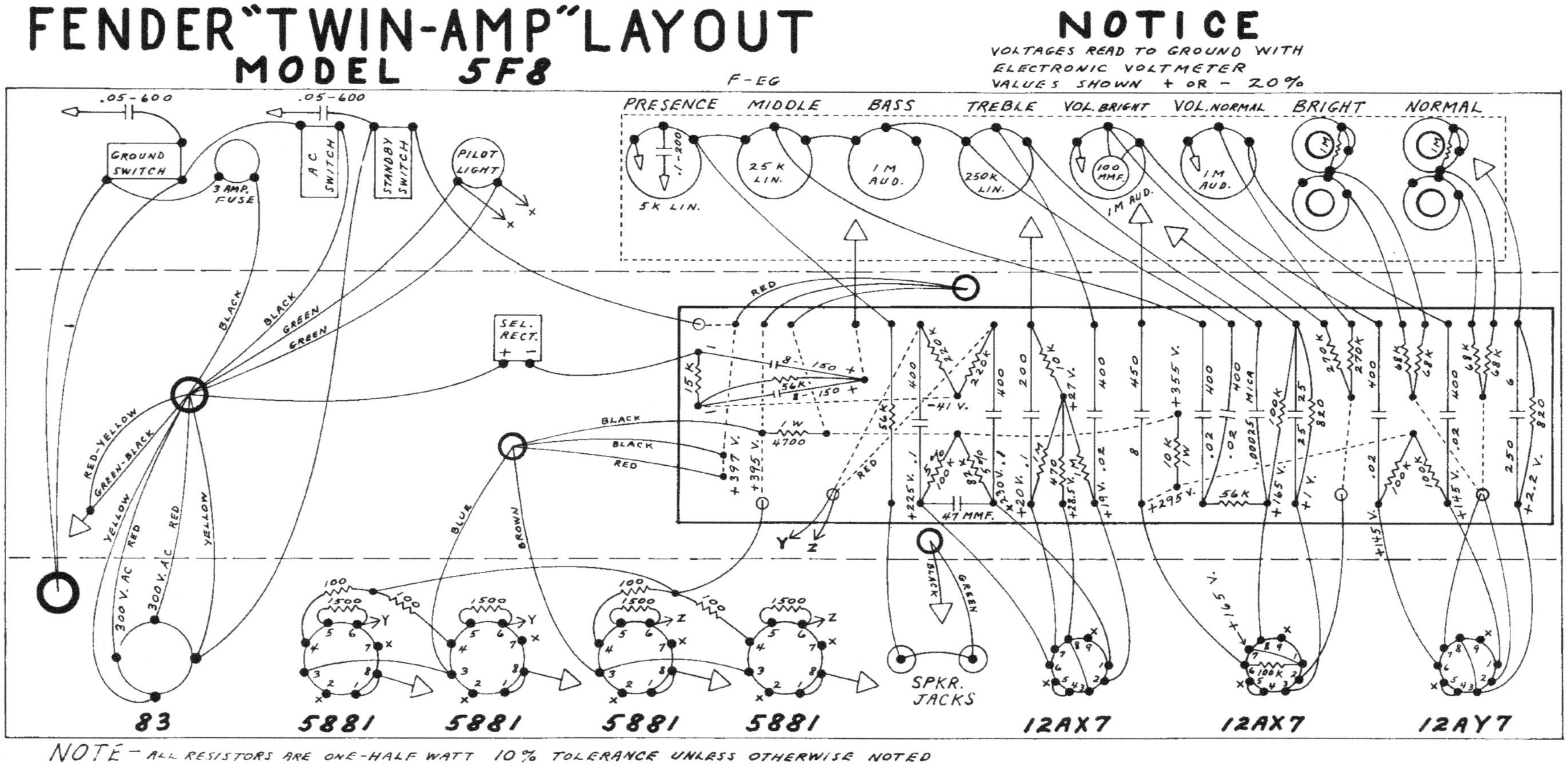
FENDER "TWIN-AMP" LAYOUT
MODEL 5F8
NOTICE
VOLTAGES READ TO GROUND WITH ELECTRONIC VOLTMETER
VALUES SHOWN + OR - 20%
F-EG
PRESENCE
MIDDLE
BASS
TREBLE
VOL. BRIGHT
VOL. NORMAL
BRIGHT
NORMAL
5K LIN.
25K LIN.
1M AUD.
250K LIN.
100 MMF.
1M AUD.
.05-600
GROUND SWITCH
3 AMP. FUSE
AC SWITCH
STANDBY SWITCH
PILOT LIGHT
SEL. RECT.
RED-YELLOW
GREEN-BLACK
YELLOW
300 V. AC RED
300 V. AC RED
YELLOW
BLACK
GREEN
BLUE
BROWN
RED
+397 V.
+395 V.
4700
1W
15K
56K
8-150
+225 V.
-41 V.
220K
47 MMF.
+230 V.
+20 V.
+27 V.
+28.5 V.
+19 V.
+355 V.
10K 1W
+295 V.
+165 V.
.00025 MICA
+1 V.
270K
+145 V.
68K
+2.2 V.
820
SPKR. JACKS
83
5881
5881
5881
5881
12AX7
12AX7
12AY7
NOTE - ALL RESISTORS ARE ONE-HALF WATT 10% TOLERANCE UNLESS OTHERWISE NOTED

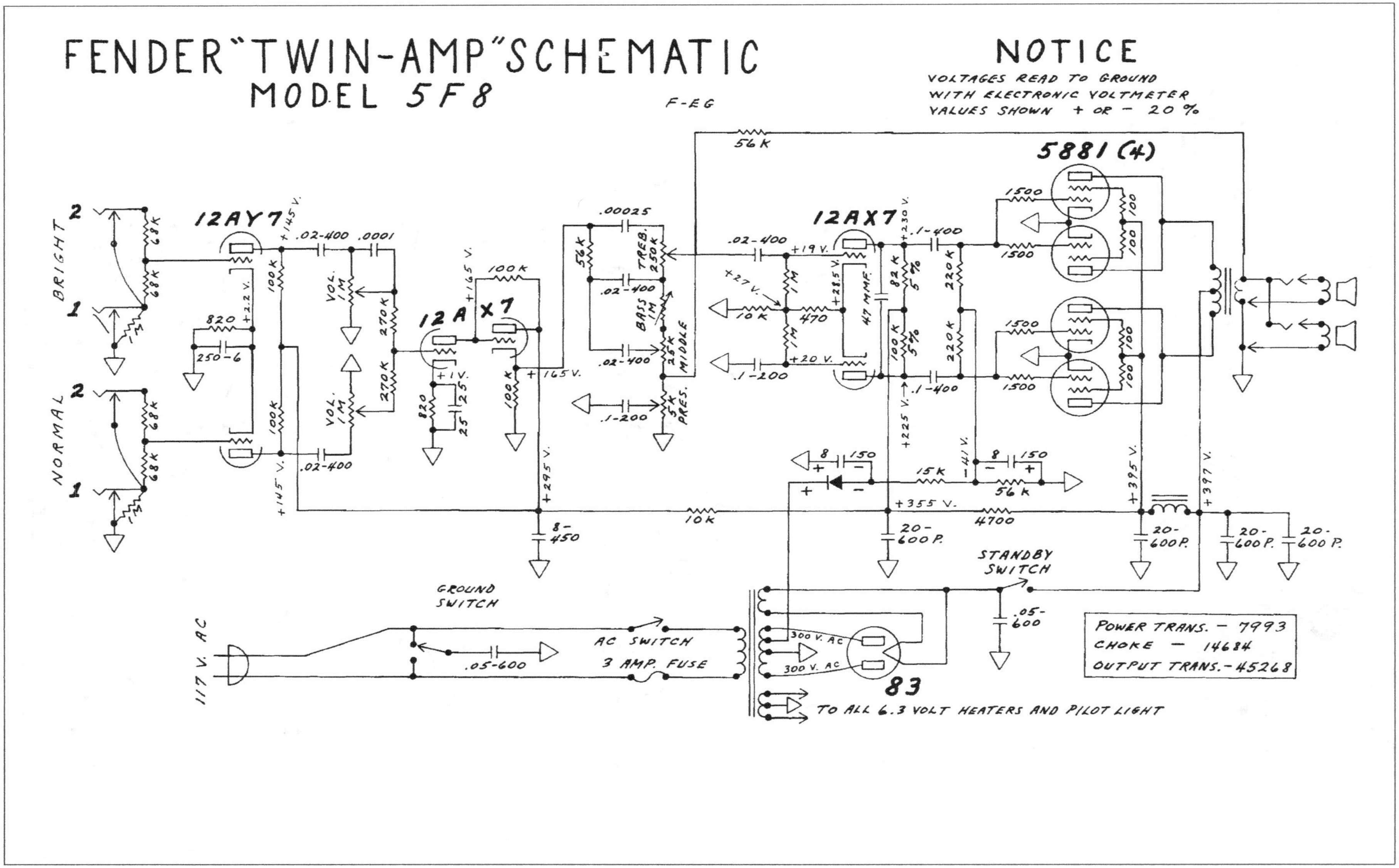
FENDER "TWIN-AMP" SCHEMATIC
MODEL 5F8
F-EG
NOTICE
VOLTAGES READ TO GROUND
WITH ELECTRONIC VOLTMETER
VALUES SHOWN + OR − 20 %
5881 (4)
12AY7
12AX7
12A X7
83
BRIGHT
NORMAL
VOL. 1M
TREB. 250K
BASS 1M
25K MIDDLE
5K PRES.
STANDBY SWITCH
GROUND SWITCH
AC SWITCH
3 AMP. FUSE
117 V. AC
300 V. AC
TO ALL 6.3 VOLT HEATERS AND PILOT LIGHT
POWER TRANS. − 7993
CHOKE − 14684
OUTPUT TRANS. − 45268

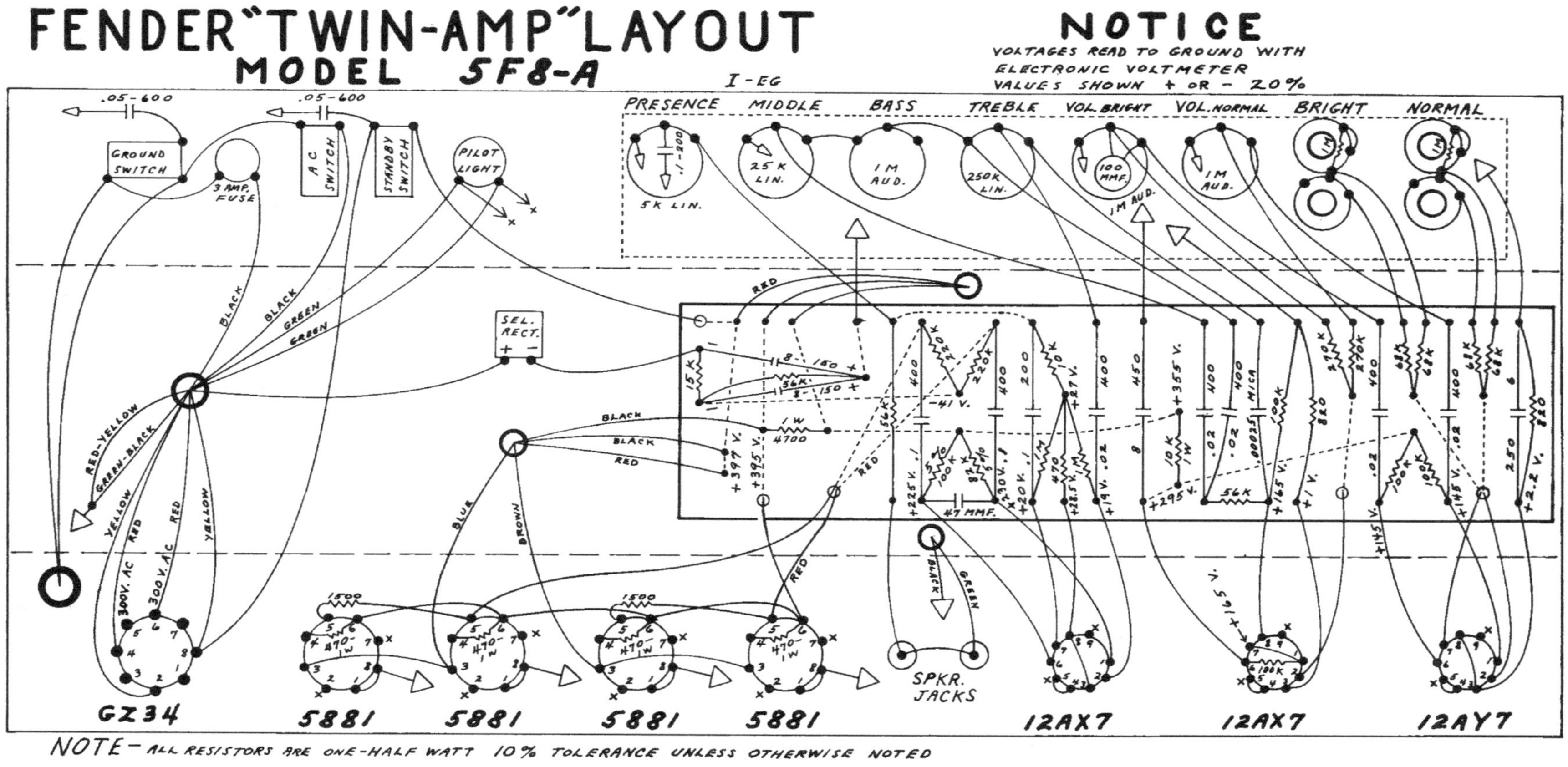
FENDER "TWIN-AMP" LAYOUT
MODEL 5F8-A
I-EG
NOTICE
VOLTAGES READ TO GROUND WITH ELECTRONIC VOLTMETER
VALUES SHOWN + OR - 20%
PRESENCE
MIDDLE
BASS
TREBLE
VOL. BRIGHT
VOL. NORMAL
BRIGHT
NORMAL
GROUND SWITCH
3 AMP. FUSE
AC SWITCH
STANDBY SWITCH
PILOT LIGHT
SEL. RECT.
SPKR. JACKS
GZ34
5881
5881
5881
5881
12AX7
12AX7
12AY7
NOTE - ALL RESISTORS ARE ONE-HALF WATT 10% TOLERANCE UNLESS OTHERWISE NOTED

FENDER "TWIN-AMP" SCHEMATIC
MODEL 5F8-A

I-EG

NOTICE

VOLTAGES READ TO GROUND
WITH ELECTRONIC VOLTMETER
VALUES SHOWN + OR − 20 %

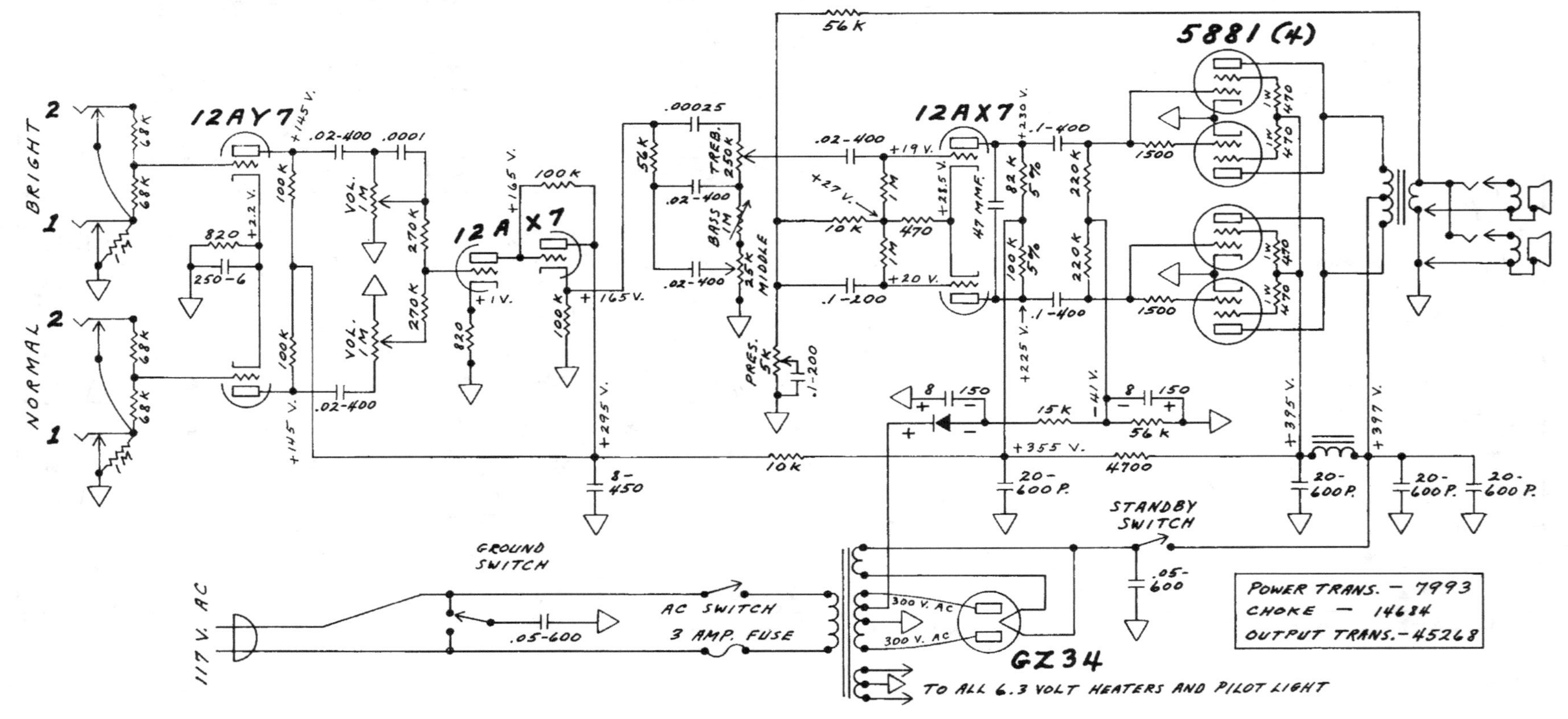

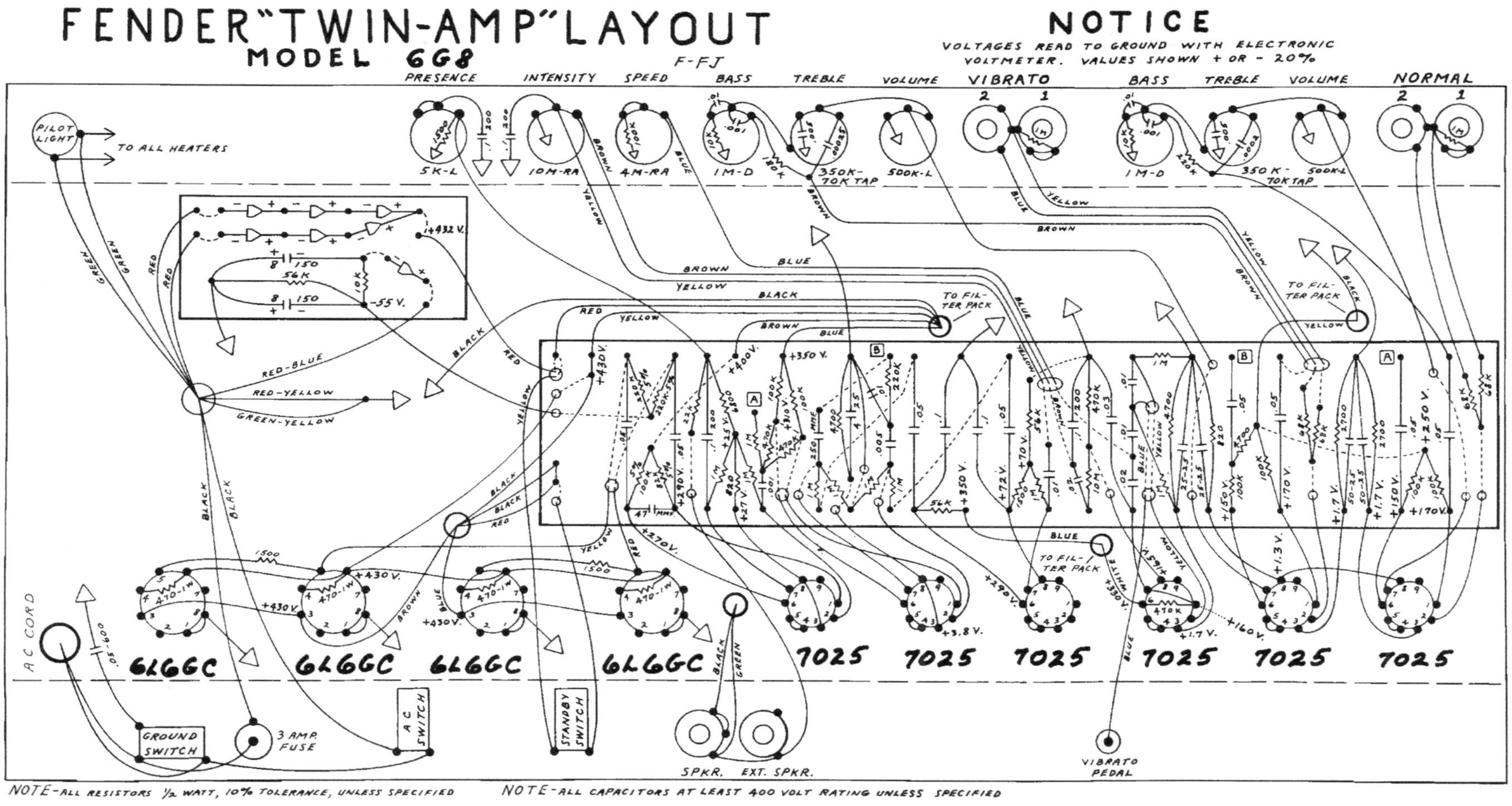
FENDER "TWIN-AMP" LAYOUT
MODEL 6G8
F-FJ
NOTICE
VOLTAGES READ TO GROUND WITH ELECTRONIC VOLTMETER. VALUES SHOWN + OR - 20%
PRESENCE
INTENSITY
SPEED
BASS
TREBLE
VOLUME
VIBRATO
BASS
TREBLE
VOLUME
NORMAL
PILOT LIGHT
TO ALL HEATERS
AC CORD
GROUND SWITCH
3 AMP. FUSE
AC SWITCH
STANDBY SWITCH
SPKR.
EXT. SPKR.
VIBRATO PEDAL
TO FILTER PACK
6L6GC
6L6GC
6L6GC
6L6GC
7025
7025
7025
7025
7025
7025
NOTE-ALL RESISTORS 1/2 WATT, 10% TOLERANCE, UNLESS SPECIFIED
NOTE-ALL CAPACITORS AT LEAST 400 VOLT RATING UNLESS SPECIFIED

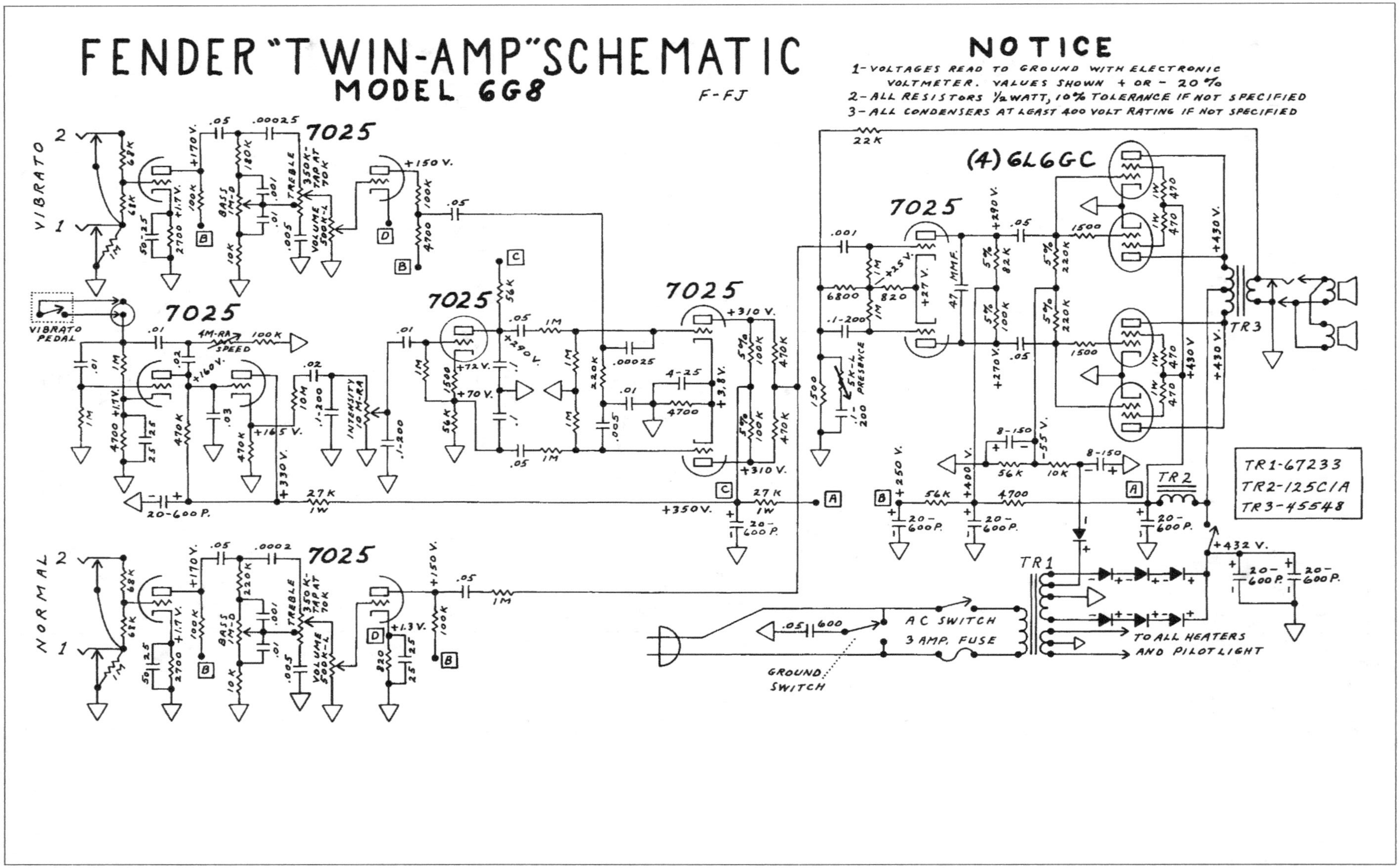
FENDER "TWIN-AMP" SCHEMATIC
MODEL 6G8
F-FJ
NOTICE
1-VOLTAGES READ TO GROUND WITH ELECTRONIC VOLTMETER. VALUES SHOWN + OR - 20%
2-ALL RESISTORS ½ WATT, 10% TOLERANCE IF NOT SPECIFIED
3-ALL CONDENSERS AT LEAST 400 VOLT RATING IF NOT SPECIFIED
(4) 6L6GC
7025
VIBRATO
NORMAL
VIBRATO PEDAL
AC SWITCH
3 AMP. FUSE
GROUND SWITCH
TO ALL HEATERS AND PILOT LIGHT
TR1-67233
TR2-125C1A
TR3-45548

TWIN-AMP **6G8-A**

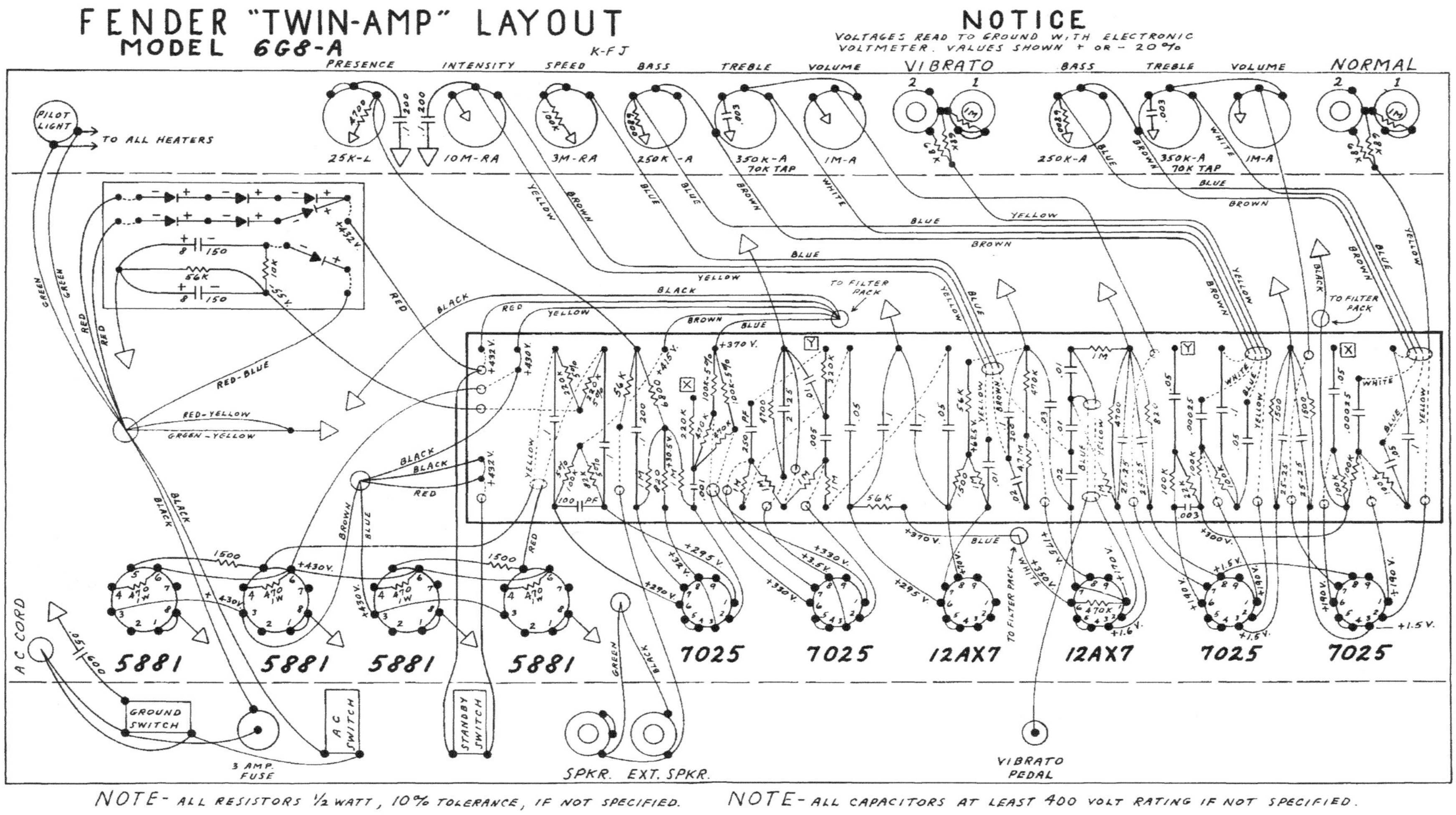

FENDER "TWIN-AMP" SCHEMATIC

MODEL 6G8-A

K-FJ

NOTICE

1- VOLTAGES READ TO GROUND WITH ELECTRONIC VOLTMETER. VALUES SHOWN + OR - 20%

2- ALL RESISTORS ½ WATT, 10% TOLERANCE IF NOT SPECIFIED.

3- ALL CAPACITORS AT LEAST 400 VOLT RATING IF NOT SPECIFIED.

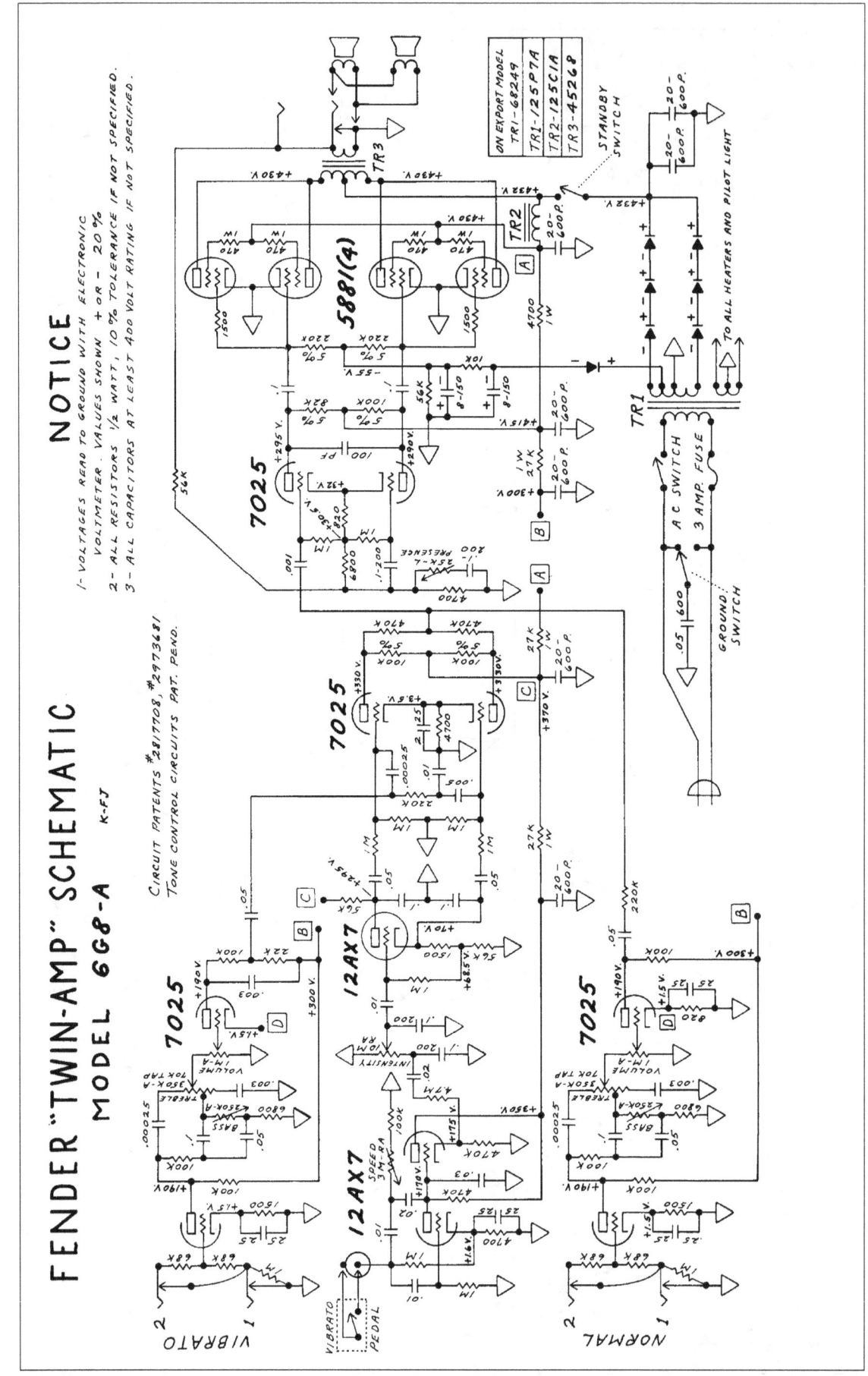

CIRCUIT PATENTS #2817708, #2973681
TONE CONTROL CIRCUITS PAT. PEND.

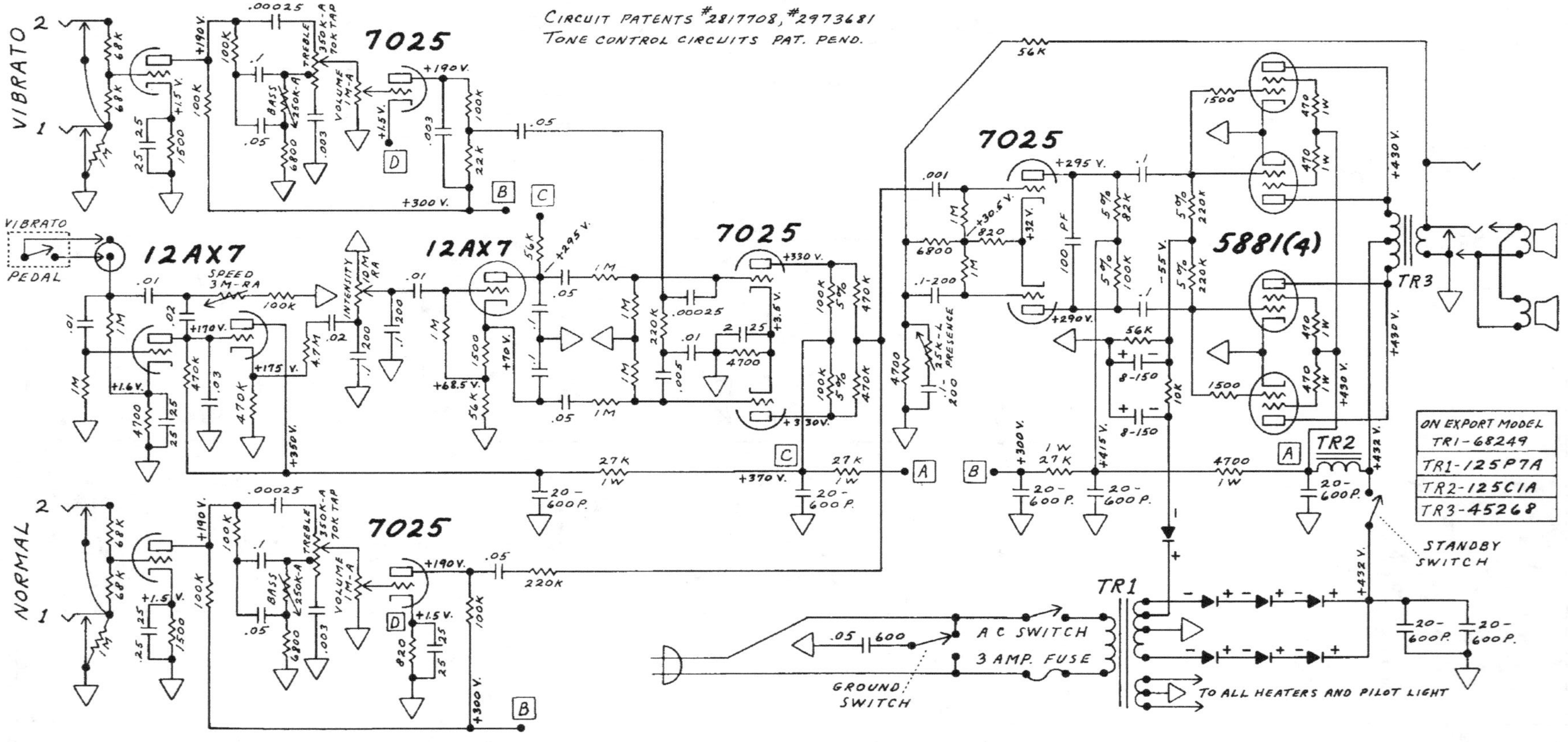

TWIN REVERB **AA270**

Fender

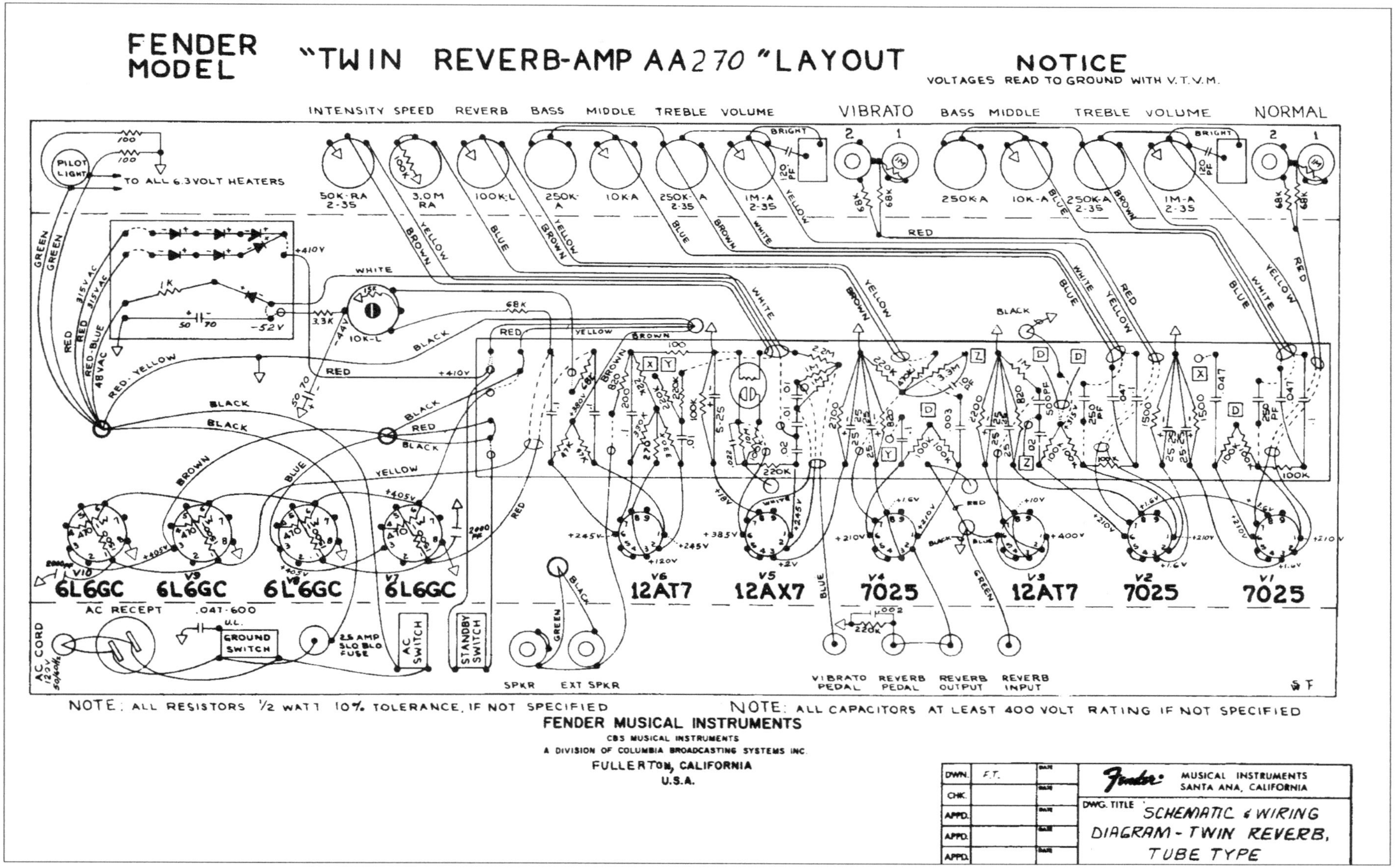

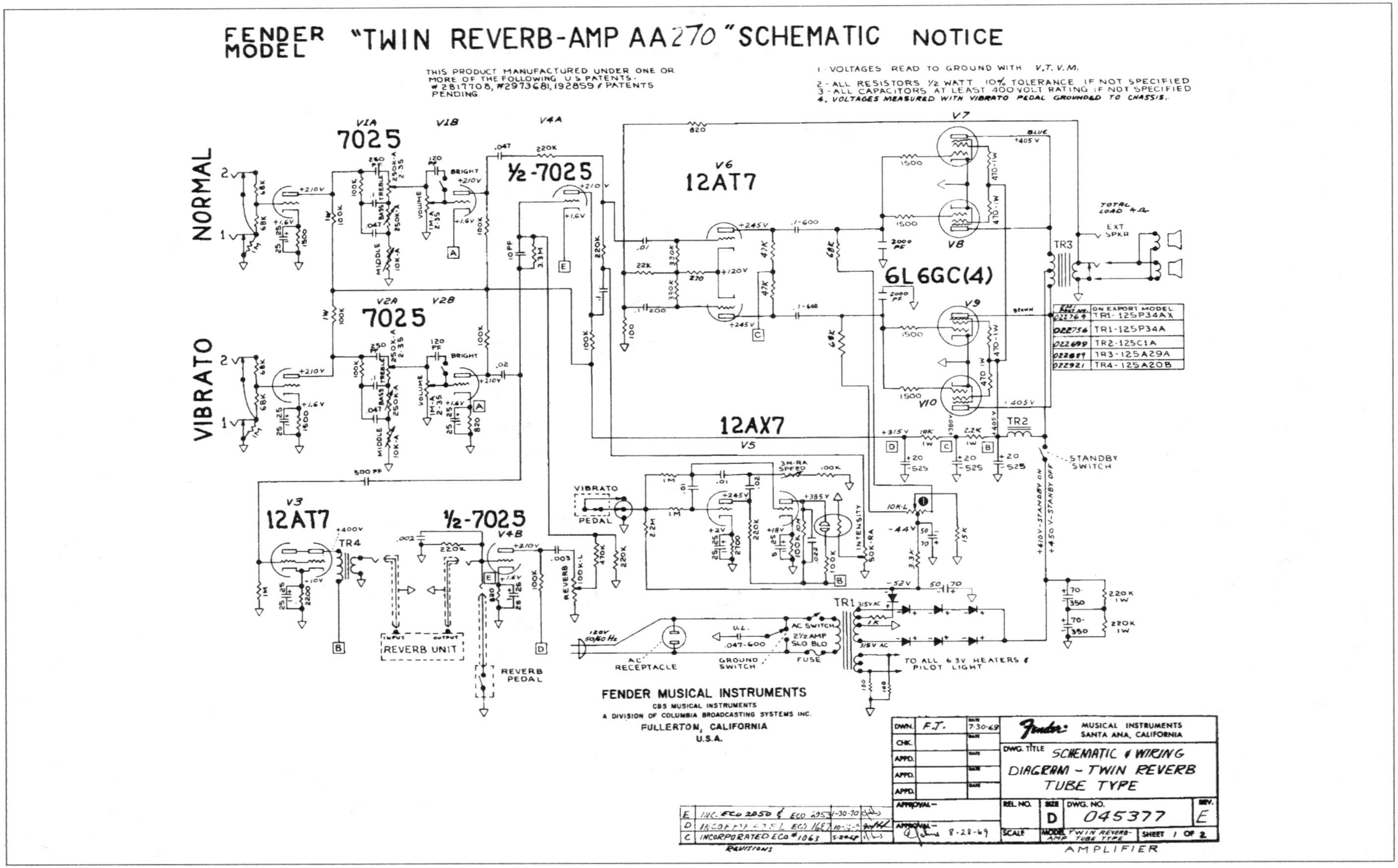
FENDER MODEL "TWIN REVERB-AMP AA270" SCHEMATIC NOTICE
THIS PRODUCT MANUFACTURED UNDER ONE OR MORE OF THE FOLLOWING U.S. PATENTS #2817708, #2973681, 192859 & PATENTS PENDING
1 - VOLTAGES READ TO GROUND WITH V.T.V.M.
2 - ALL RESISTORS 1/2 WATT 10% TOLERANCE IF NOT SPECIFIED
3 - ALL CAPACITORS AT LEAST 400 VOLT RATING IF NOT SPECIFIED
4. VOLTAGES MEASURED WITH VIBRATO PEDAL GROUNDED TO CHASSIS.
NORMAL
VIBRATO
V1A 7025
V1B
V2A 7025
V2B
V4A 1/2-7025
V6 12AT7
6L6GC(4)
V7
V8
V9
V10
12AX7
V5
V3 12AT7
TR4
1/2-7025 V4B
REVERB UNIT
REVERB PEDAL
VIBRATO PEDAL
INTENSITY
SPEED
TOTAL LOAD 4Ω
EXT SPKR
TR3
TR2
TR1
STANDBY SWITCH
+410V - STANDBY ON
+450 V - STANDBY OFF
AC RECEPTACLE
GROUND SWITCH
FUSE
AC SWITCH
2 1/2 AMP SLO BLO
TO ALL 6.3V HEATERS & PILOT LIGHT
ON EXPORT MODEL
022764 TR1-125P34AX
022756 TR1-125P34A
022699 TR2-125C1A
022689 TR3-125A29A
022921 TR4-125A20B
FENDER MUSICAL INSTRUMENTS
CBS MUSICAL INSTRUMENTS
A DIVISION OF COLUMBIA BROADCASTING SYSTEMS INC.
FULLERTON, CALIFORNIA
U.S.A.
DWN. F.J. 7-30-69
MUSICAL INSTRUMENTS SANTA ANA, CALIFORNIA
DWG. TITLE SCHEMATIC & WIRING DIAGRAM - TWIN REVERB TUBE TYPE
SIZE D
DWG. NO. 045377
REV. E
SHEET 1 OF 2
AMPLIFIER
8-28-69
E INC. ECO 2050 & ECO 2053
C INCORPORATED ECO #1063
REVISIONS

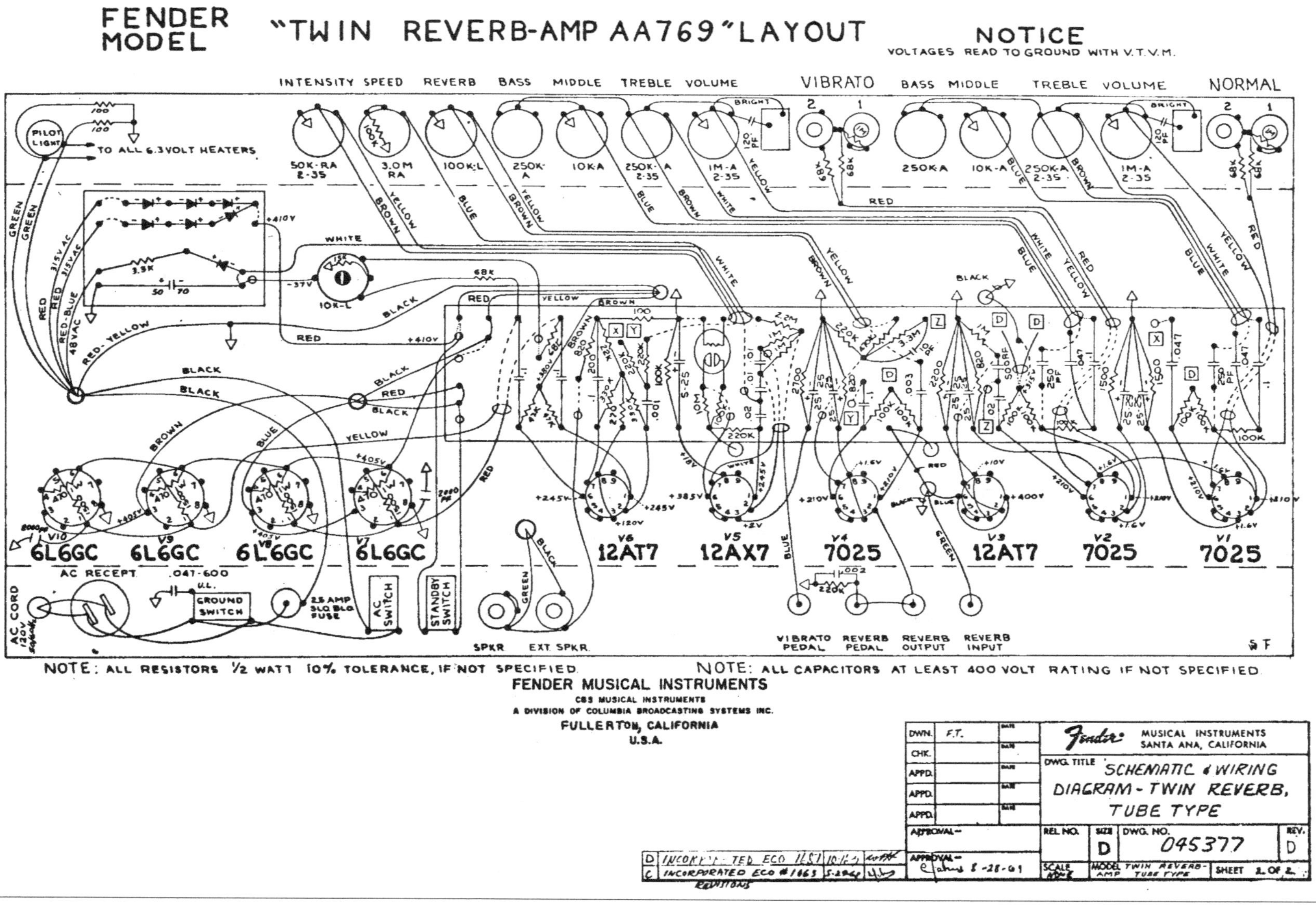
FENDER MODEL "TWIN REVERB-AMP AA769" LAYOUT
NOTICE
VOLTAGES READ TO GROUND WITH V.T.V.M.
INTENSITY SPEED REVERB BASS MIDDLE TREBLE VOLUME
VIBRATO
BASS MIDDLE TREBLE VOLUME
NORMAL
PILOT LIGHT
TO ALL 6.3 VOLT HEATERS
6L6GC
6L6GC
6L6GC
6L6GC
12AT7
12AX7
7025
12AT7
7025
7025
AC CORD
AC RECEPT
GROUND SWITCH
AC SWITCH
STANDBY SWITCH
SPKR
EXT. SPKR.
VIBRATO PEDAL
REVERB PEDAL
REVERB OUTPUT
REVERB INPUT
NOTE: ALL RESISTORS ½ WATT 10% TOLERANCE, IF NOT SPECIFIED.
NOTE: ALL CAPACITORS AT LEAST 400 VOLT RATING IF NOT SPECIFIED
FENDER MUSICAL INSTRUMENTS
CBS MUSICAL INSTRUMENTS
A DIVISION OF COLUMBIA BROADCASTING SYSTEMS INC.
FULLERTON, CALIFORNIA
U.S.A.
MUSICAL INSTRUMENTS SANTA ANA, CALIFORNIA
DWG. TITLE SCHEMATIC & WIRING DIAGRAM - TWIN REVERB, TUBE TYPE
DWN. F.T.
SIZE D
DWG. NO. 045377
REV. D
SHEET 2 OF 2
C INCORPORATED ECO #1863
REVISIONS

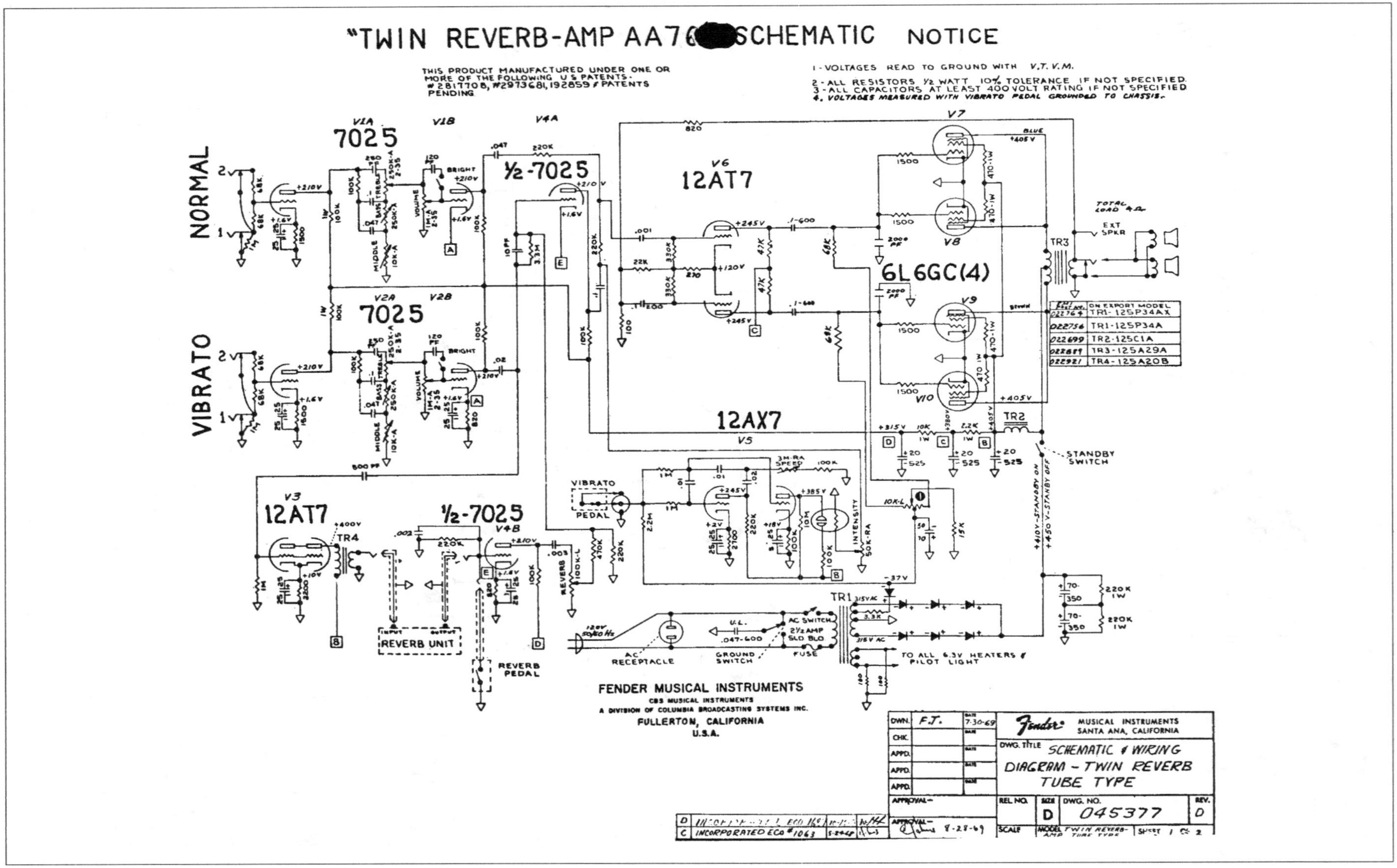
"TWIN REVERB-AMP AA76[redacted]SCHEMATIC
NOTICE
THIS PRODUCT MANUFACTURED UNDER ONE OR MORE OF THE FOLLOWING U.S. PATENTS. #2817708, #2973681, 192859 & PATENTS PENDING
1 - VOLTAGES READ TO GROUND WITH V.T.V.M.
2 - ALL RESISTORS ½ WATT 10% TOLERANCE IF NOT SPECIFIED
3 - ALL CAPACITORS AT LEAST 400 VOLT RATING IF NOT SPECIFIED
4. VOLTAGES MEASURED WITH VIBRATO PEDAL GROUNDED TO CHASSIS.
NORMAL
VIBRATO
V1A 7025
V1B
V2A 7025
V2B
V4A ½-7025
V6 12AT7
V7
V8
6L6GC(4)
V9
V10
12AX7
V5
V3 12AT7
TR4
½-7025
V4B
REVERB UNIT
REVERB PEDAL
VIBRATO PEDAL
TR1
TR2
TR3
TOTAL LOAD 4Ω
EXT SPKR
STANDBY SWITCH
AC RECEPTACLE
GROUND SWITCH
AC SWITCH
2½ AMP SLO BLO
FUSE
TO ALL 6.3V HEATERS & PILOT LIGHT
INTENSITY
SPEED
FENDER MUSICAL INSTRUMENTS
CBS MUSICAL INSTRUMENTS
A DIVISION OF COLUMBIA BROADCASTING SYSTEMS INC.
FULLERTON, CALIFORNIA
U.S.A.
MUSICAL INSTRUMENTS SANTA ANA, CALIFORNIA
DWG. TITLE SCHEMATIC & WIRING DIAGRAM - TWIN REVERB TUBE TYPE
DWN. F.J.
DATE 7-30-69
SIZE D
DWG. NO. 045377
REV. D
INCORPORATED ECO #1063

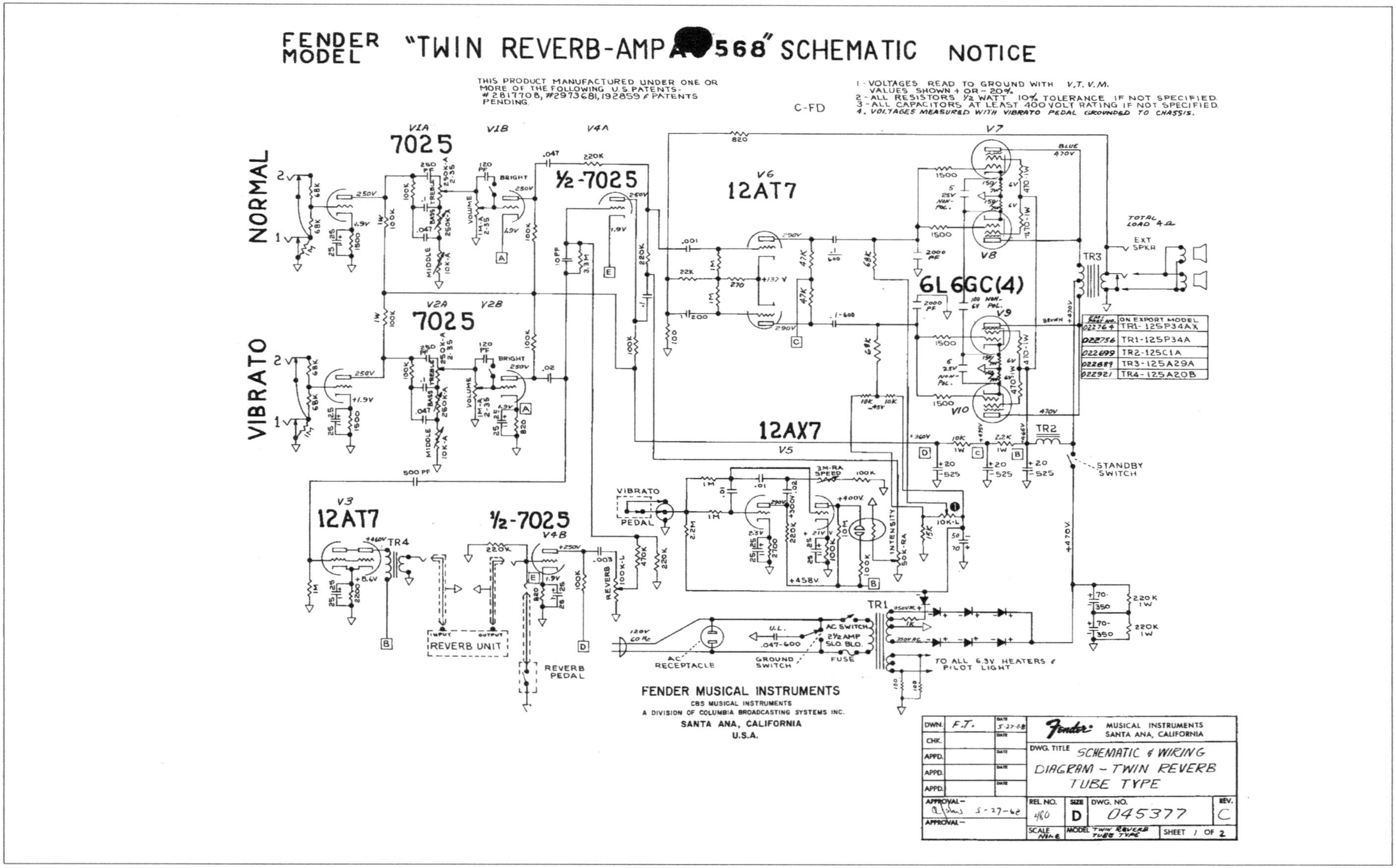
FENDER MODEL "TWIN REVERB-AMP 568" SCHEMATIC NOTICE
THIS PRODUCT MANUFACTURED UNDER ONE OR MORE OF THE FOLLOWING U.S. PATENTS- #2817708, #2973681, 192859 & PATENTS PENDING.
C-FD
1- VOLTAGES READ TO GROUND WITH V.T.V.M. VALUES SHOWN + OR - 20%
2- ALL RESISTORS ½ WATT 10% TOLERANCE IF NOT SPECIFIED
3- ALL CAPACITORS AT LEAST 400 VOLT RATING IF NOT SPECIFIED
4. VOLTAGES MEASURED WITH VIBRATO PEDAL GROUNDED TO CHASSIS.
NORMAL
VIBRATO
V1A
7025
V1B
V4A
½-7025
V2A
V2B
7025
V6
12AT7
V7
V8
6L6GC(4)
V9
V10
12AX7
V5
V3
12AT7
½-7025
V4B
TR4
REVERB UNIT
REVERB PEDAL
VIBRATO PEDAL
INTENSITY
SPEED
REVERB
TR1
TR2
TR3
STANDBY SWITCH
EXT SPKR
TOTAL LOAD 4 Ω
AC RECEPTACLE
GROUND SWITCH
FUSE
AC SWITCH
TO ALL 6.3V HEATERS & PILOT LIGHT
ON EXPORT MODEL TR1-125P34AX
022756 TR1-125P34A
022699 TR2-125C1A
022689 TR3-125A29A
022921 TR4-125A20B
FENDER MUSICAL INSTRUMENTS
CBS MUSICAL INSTRUMENTS
A DIVISION OF COLUMBIA BROADCASTING SYSTEMS INC.
SANTA ANA, CALIFORNIA
U.S.A.
Fender MUSICAL INSTRUMENTS SANTA ANA, CALIFORNIA
DWG. TITLE SCHEMATIC & WIRING DIAGRAM - TWIN REVERB TUBE TYPE
DWN. F.T. 5-27-68
REL. NO. 480
SIZE D
DWG. NO. 045377
REV. C
SHEET 1 OF 2

FENDER MODEL "TWIN REVERB-AMP AB763" SCHEMATIC

THIS PRODUCT MANUFACTURED UNDER ONE OR MORE OF THE FOLLOWING U.S. PATENTS - #2817708, #2973681, 192859 & PATENTS PENDING.

C-FD

NOTICE

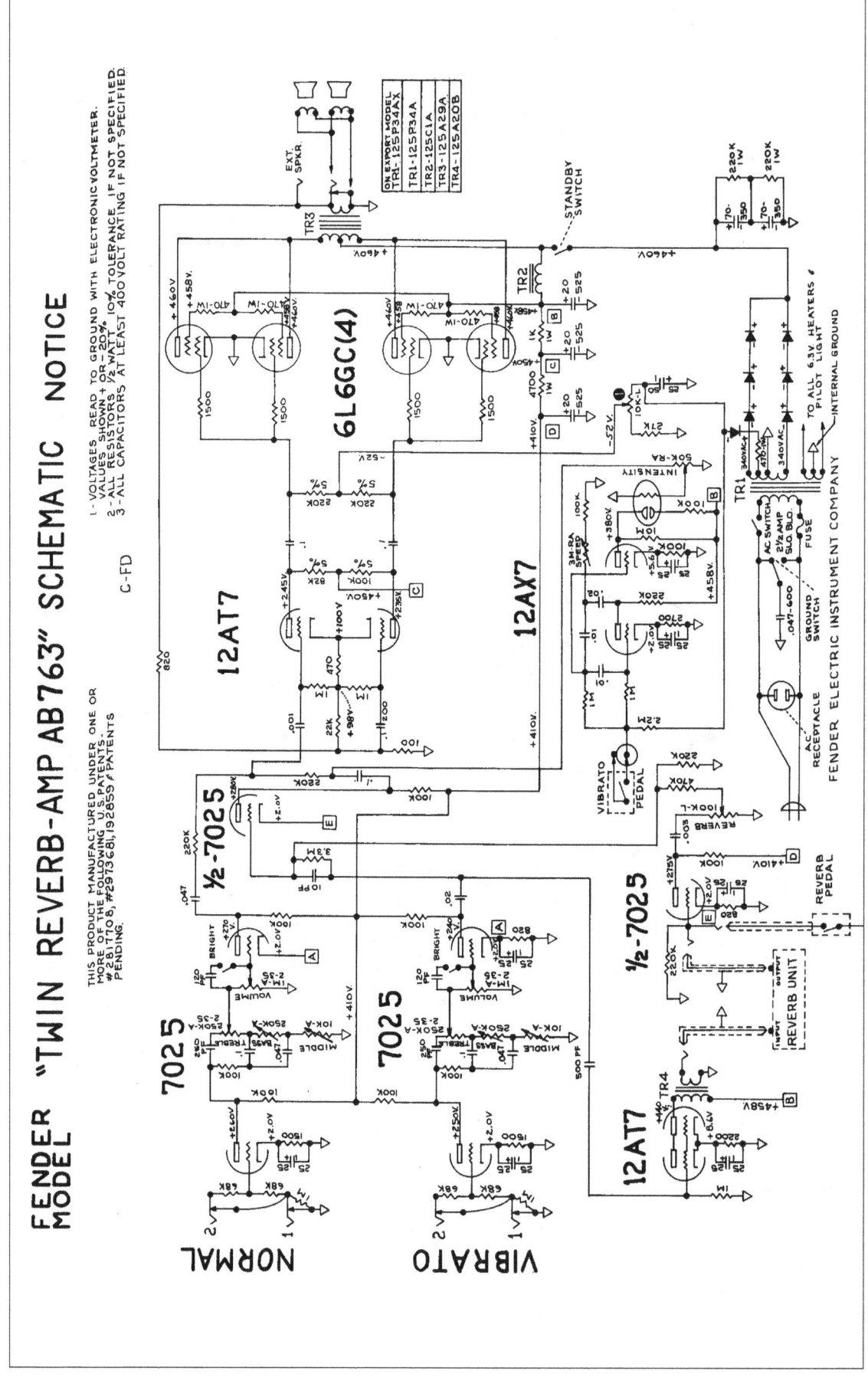

1 - VOLTAGES READ TO GROUND WITH ELECTRONIC VOLTMETER. VALUES SHOWN + OR - 20%.
2 - ALL RESISTORS ½ WATT 10% TOLERANCE IF NOT SPECIFIED.
3 - ALL CAPACITORS AT LEAST 400 VOLT RATING IF NOT SPECIFIED.

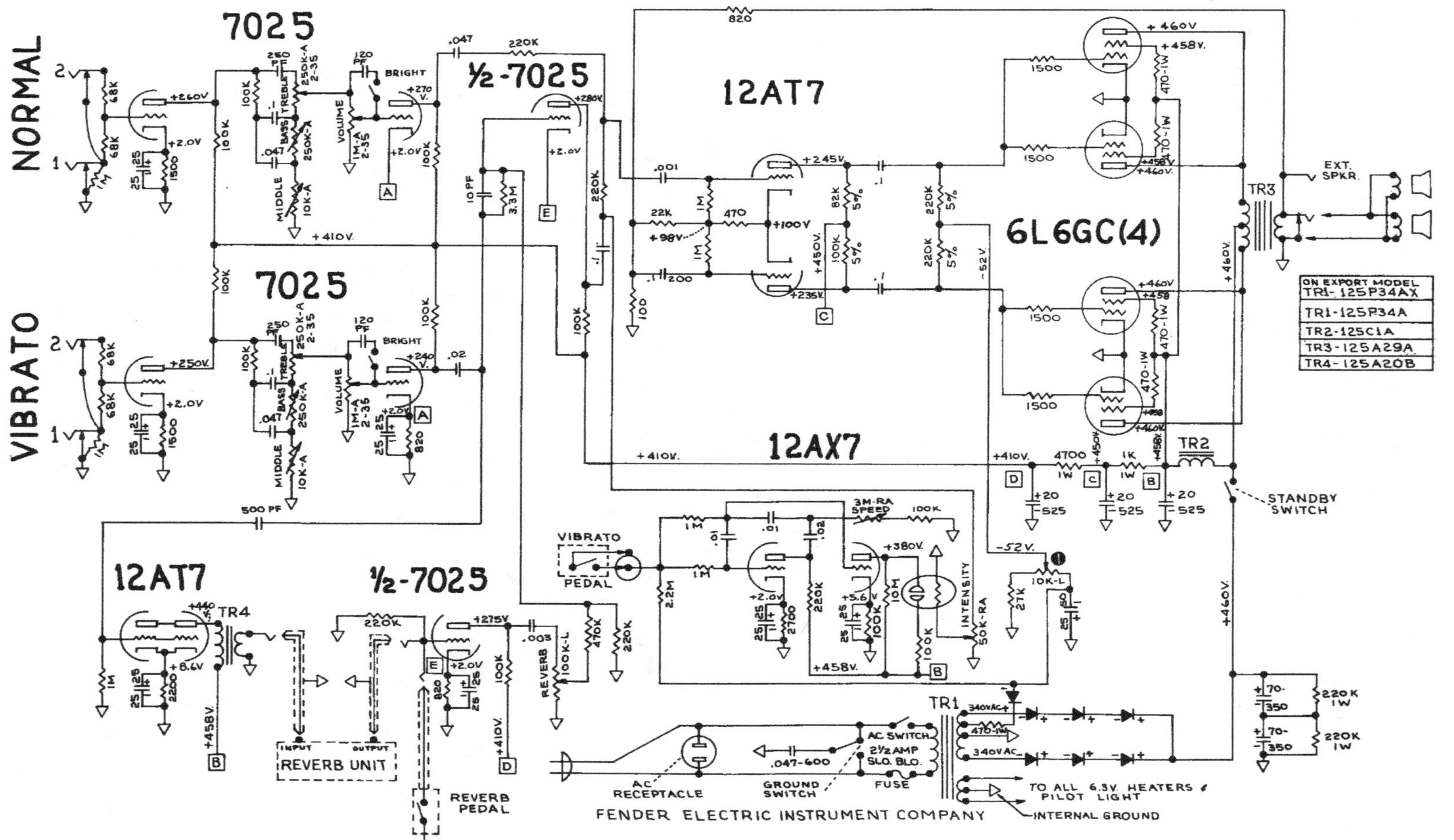

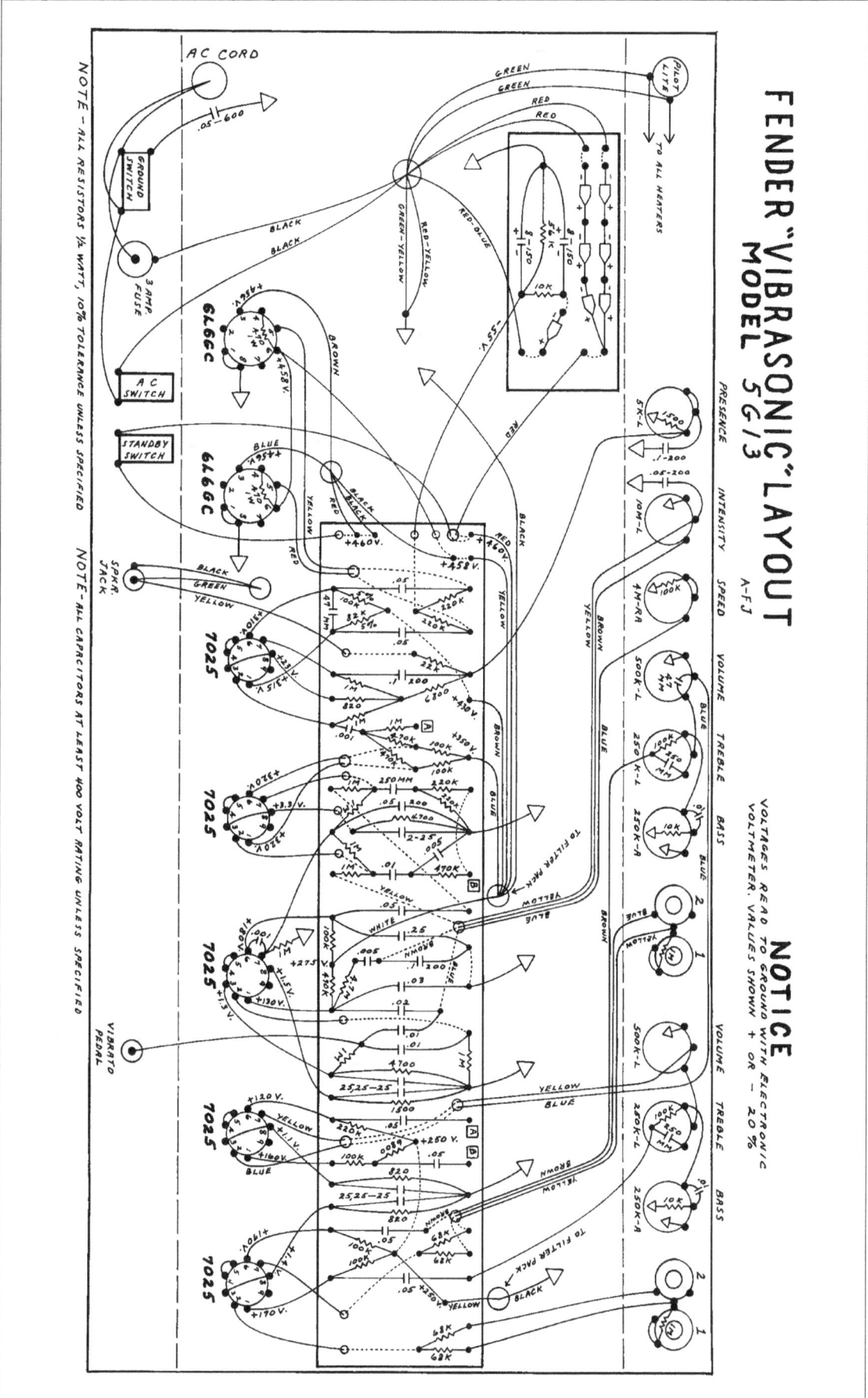

FENDER "VIBRASONIC" LAYOUT
MODEL 5G13

A-FJ

NOTICE

VOLTAGES READ TO GROUND WITH ELECTRONIC VOLTMETER. VALUES SHOWN + OR − 20%

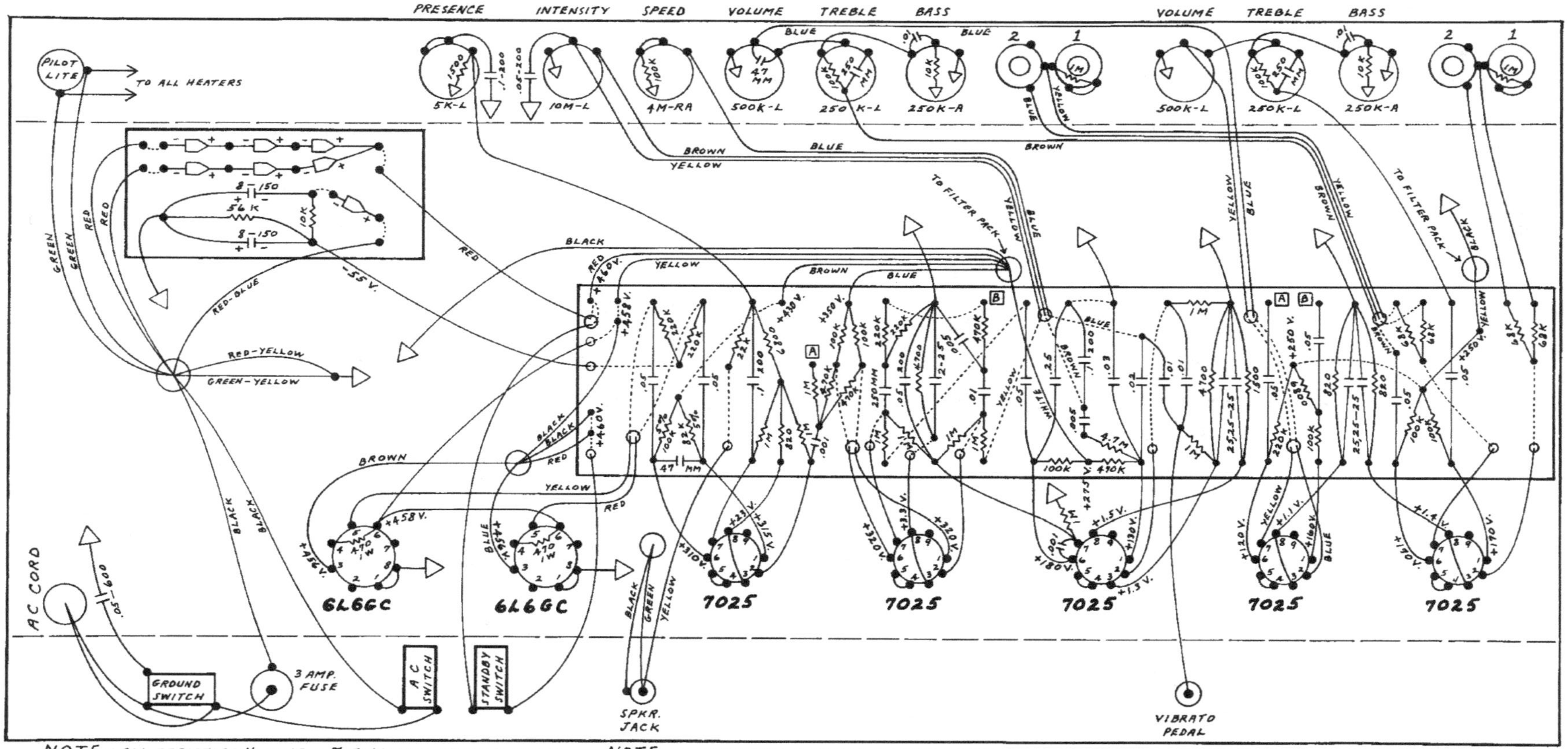

NOTE – ALL RESISTORS ½ WATT, 10% TOLERANCE UNLESS SPECIFIED

NOTE – ALL CAPACITORS AT LEAST 400 VOLT RATING UNLESS SPECIFIED

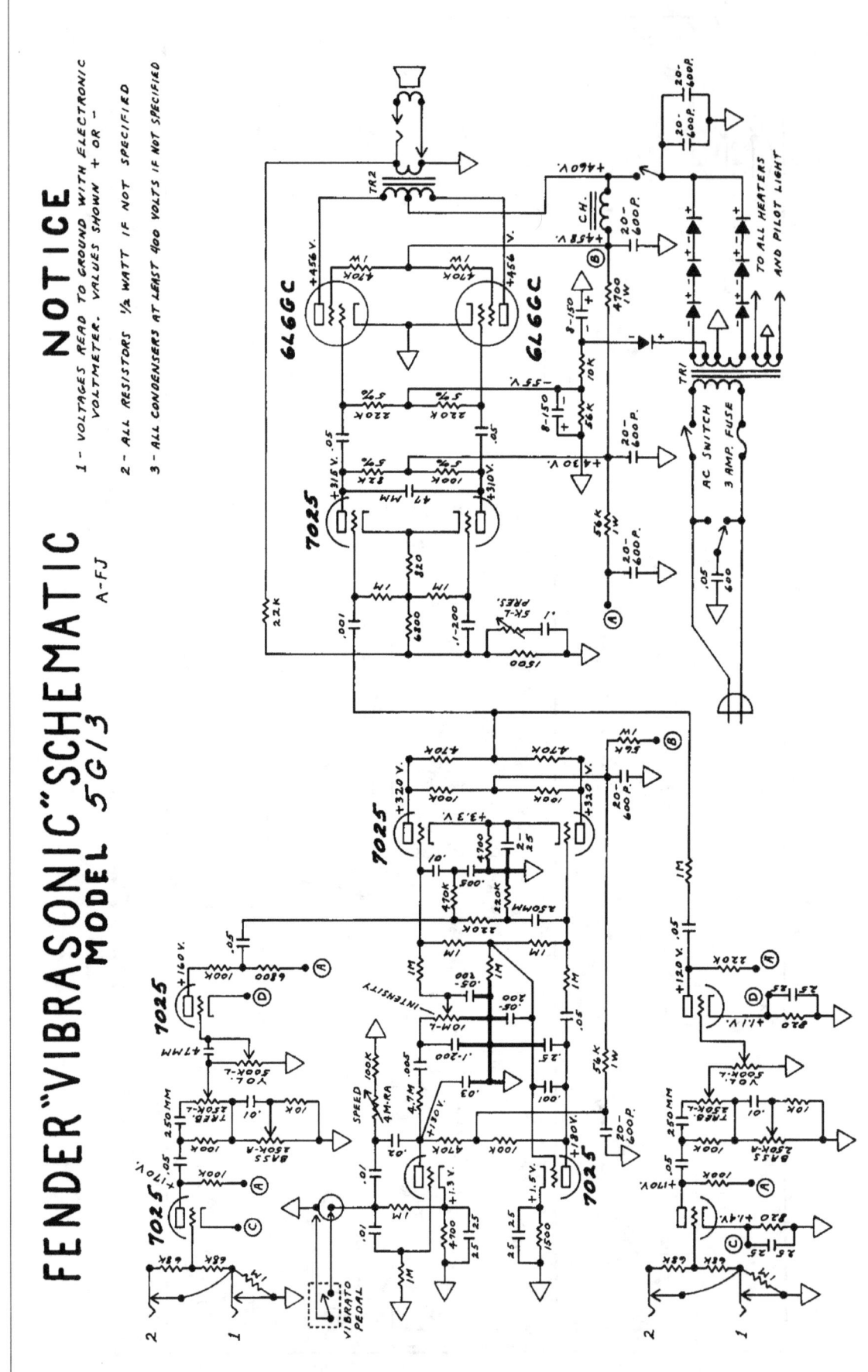

FENDER "VIBRASONIC" SCHEMATIC
MODEL 5G13

A-FJ

NOTICE

1 - VOLTAGES READ TO GROUND WITH ELECTRONIC VOLTMETER. VALUES SHOWN + OR -

2 - ALL RESISTORS ½ WATT IF NOT SPECIFIED

3 - ALL CONDENSERS AT LEAST 400 VOLTS IF NOT SPECIFIED

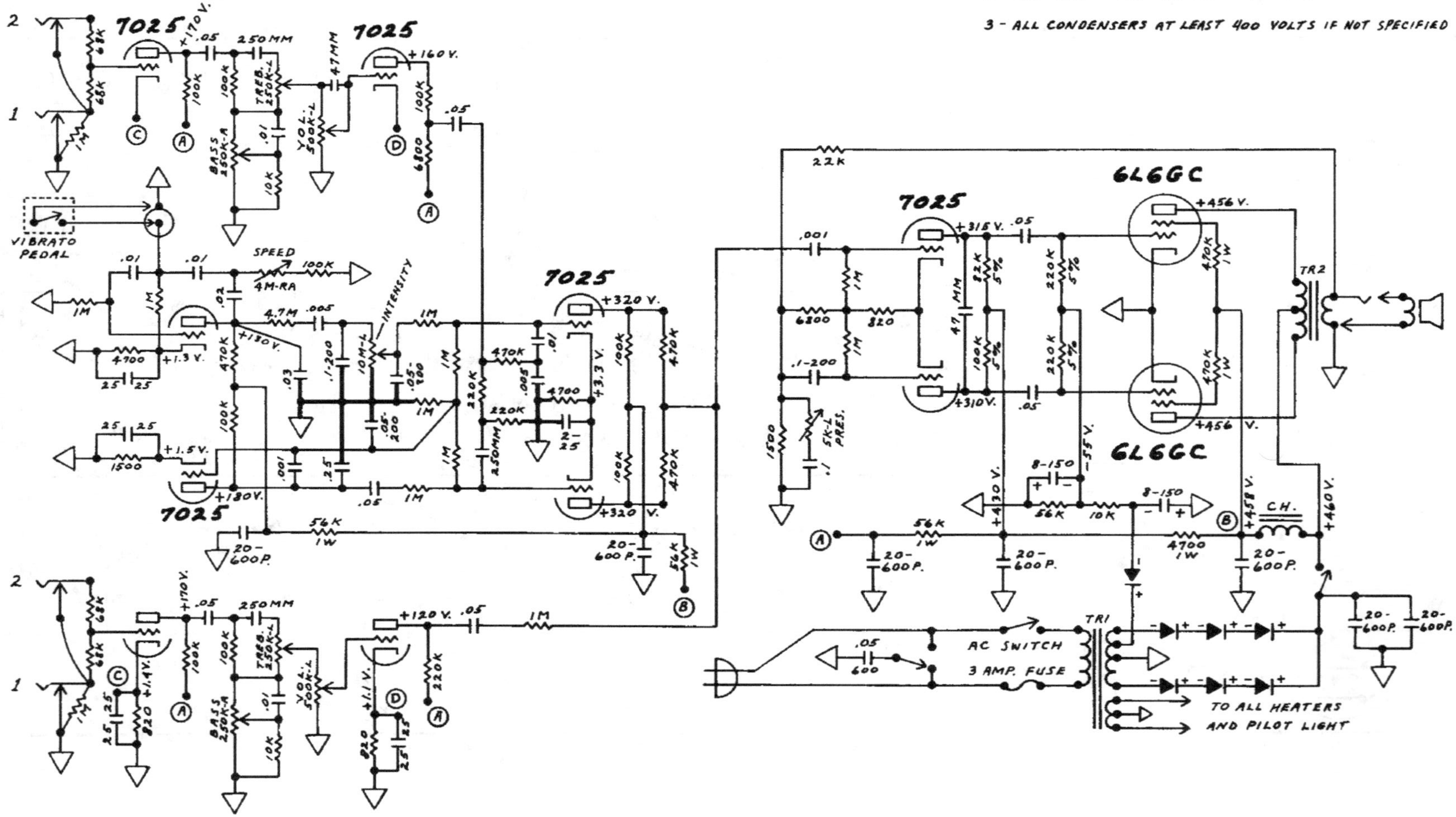

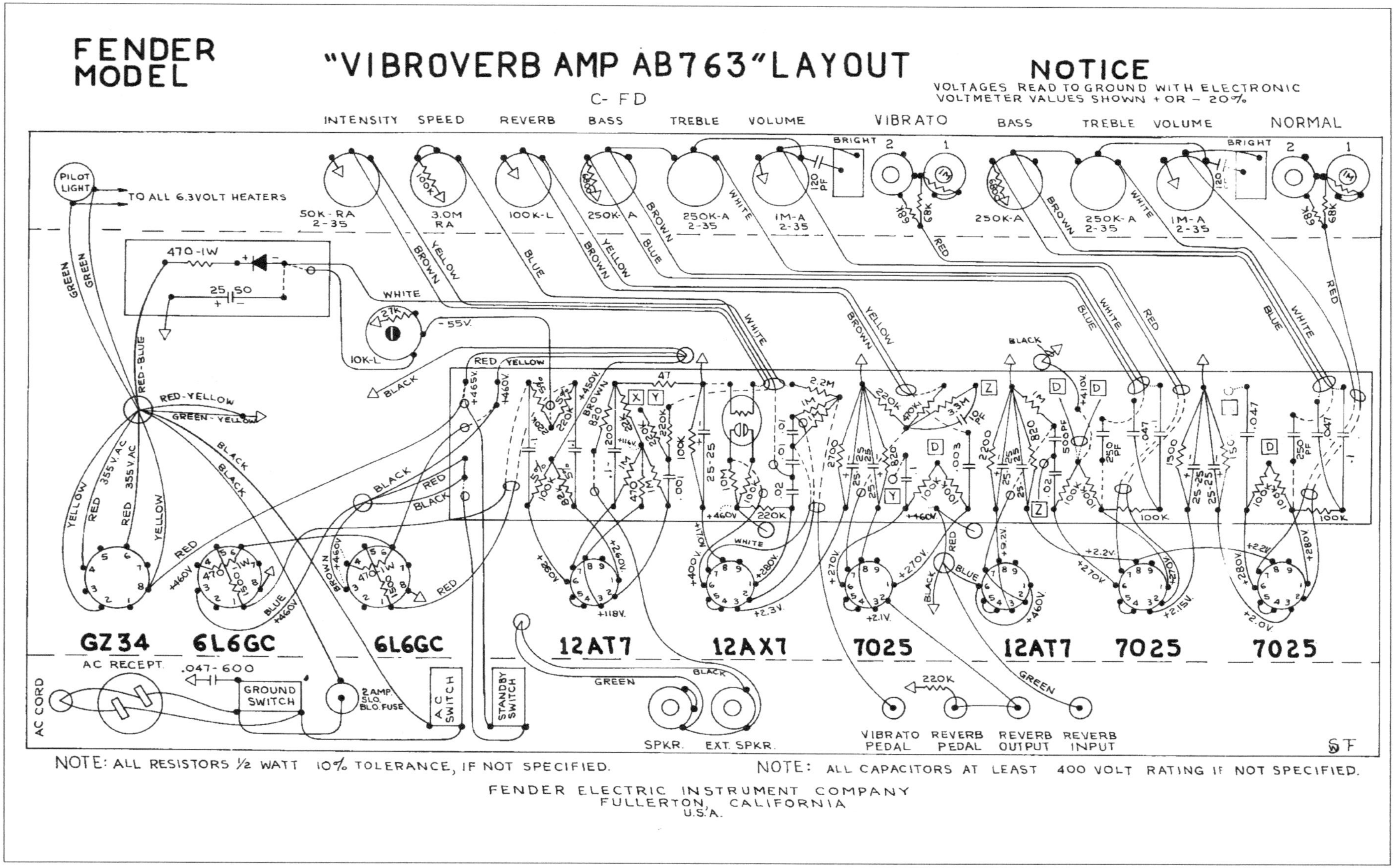
FENDER MODEL
"VIBROVERB AMP AB763" LAYOUT
C-FD
NOTICE
VOLTAGES READ TO GROUND WITH ELECTRONIC VOLTMETER VALUES SHOWN + OR − 20%
INTENSITY
SPEED
REVERB
BASS
TREBLE
VOLUME
VIBRATO
BASS
TREBLE
VOLUME
NORMAL
BRIGHT
PILOT LIGHT
TO ALL 6.3VOLT HEATERS
GZ34
6L6GC
6L6GC
12AT7
12AX7
7025
12AT7
7025
7025
AC RECEPT.
AC CORD
GROUND SWITCH
2AMP. SLO. BLO. FUSE
A C SWITCH
STANDBY SWITCH
SPKR.
EXT. SPKR.
VIBRATO PEDAL
REVERB PEDAL
REVERB OUTPUT
REVERB INPUT
NOTE: ALL RESISTORS ½ WATT 10% TOLERANCE, IF NOT SPECIFIED.
NOTE: ALL CAPACITORS AT LEAST 400 VOLT RATING IF NOT SPECIFIED.
FENDER ELECTRIC INSTRUMENT COMPANY
FULLERTON, CALIFORNIA
U.S.A.

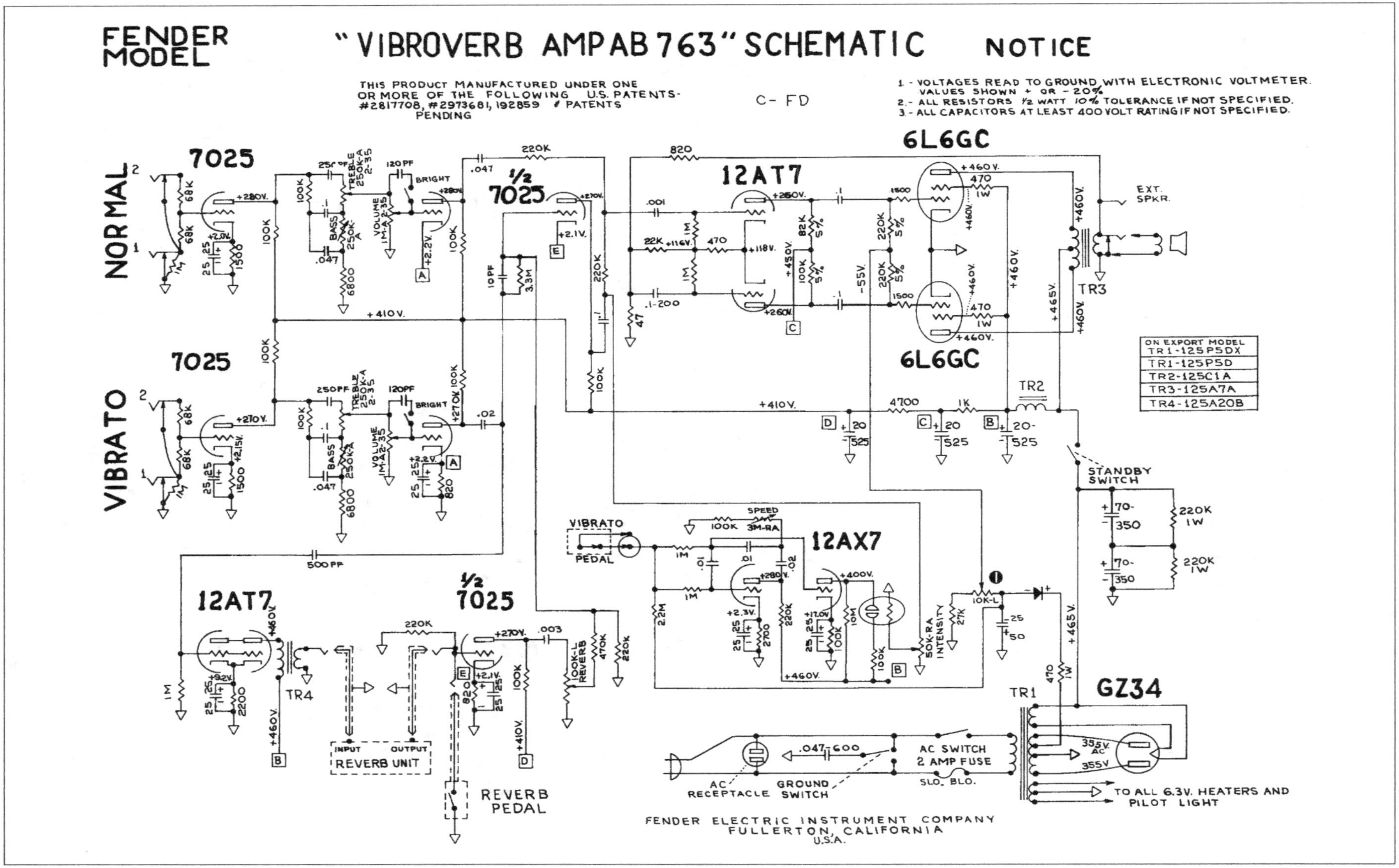
FENDER MODEL
"VIBROVERB AMPAB763" SCHEMATIC
NOTICE
THIS PRODUCT MANUFACTURED UNDER ONE OR MORE OF THE FOLLOWING U.S. PATENTS- #2817708, #2973681, 192859 & PATENTS PENDING
C- FD
1. - VOLTAGES READ TO GROUND WITH ELECTRONIC VOLTMETER. VALUES SHOWN + OR - 20%
2. - ALL RESISTORS ½ WATT 10% TOLERANCE IF NOT SPECIFIED.
3. - ALL CAPACITORS AT LEAST 400 VOLT RATING IF NOT SPECIFIED.
NORMAL
VIBRATO
7025
7025
½ 7025
12AT7
6L6GC
6L6GC
12AX7
12AT7
½ 7025
GZ34
EXT. SPKR.
TR1
TR2
TR3
TR4
ON EXPORT MODEL TR1-125P5DX
TR1-125P5D
TR2-125C1A
TR3-125A7A
TR4-125A20B
STANDBY SWITCH
VIBRATO PEDAL
SPEED 3M-RA
50K-RA INTENSITY
INPUT
OUTPUT
REVERB UNIT
REVERB PEDAL
100K-L REVERB
AC RECEPTACLE
GROUND SWITCH
AC SWITCH 2 AMP FUSE
SLO. BLO.
TO ALL 6.3V. HEATERS AND PILOT LIGHT
FENDER ELECTRIC INSTRUMENT COMPANY FULLERTON, CALIFORNIA U.S.A.

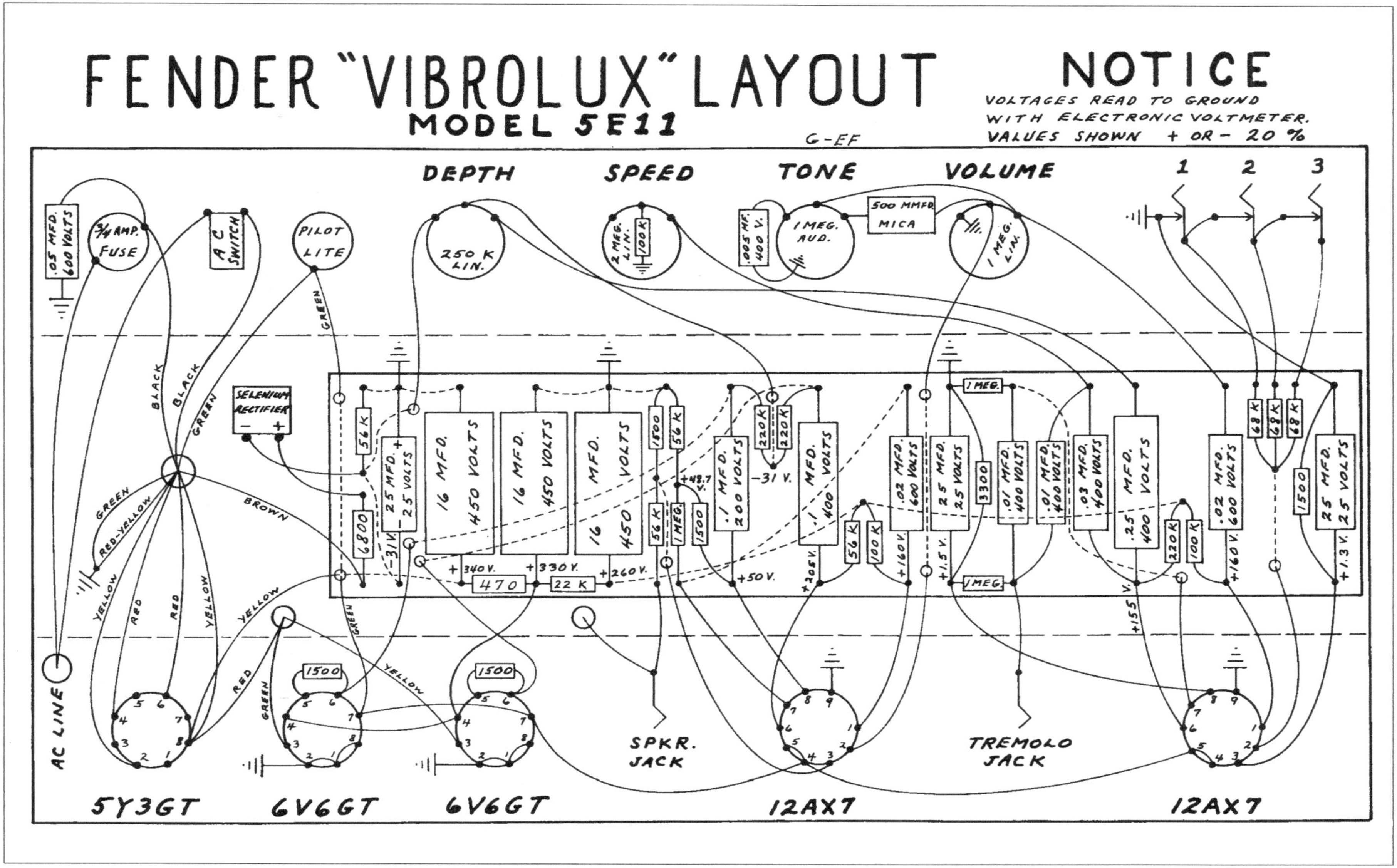
FENDER "VIBROLUX" LAYOUT
MODEL 5E11
NOTICE
VOLTAGES READ TO GROUND
WITH ELECTRONIC VOLTMETER.
VALUES SHOWN + OR - 20 %
G-EF
DEPTH
SPEED
TONE
VOLUME
1
2
3
.05 MFD. 600 VOLTS
3/4 AMP. FUSE
A C SWITCH
PILOT LITE
250 K LIN.
2 MEG. LIN.
100 K
.005 MF. 400 V.
1 MEG. AUD.
500 MMFD MICA
1 MEG. LIN.
SELENIUM RECTIFIER
BLACK
BLACK
GREEN
GREEN
RED-YELLOW
BROWN
YELLOW
RED
RED
YELLOW
YELLOW
RED
GREEN
YELLOW
56 K
6800
-31V.
25 MFD. 25 VOLTS
16 MFD. 450 VOLTS
16 MFD. 450 VOLTS
16 MFD. 450 VOLTS
+340V.
+330V.
+260V.
470
22 K
1500
56 K
56 K
1 MEG.
1500
+48.7 V.
.1 MFD. 200 VOLTS
220 K
220 K
-31 V.
+50 V.
.1 MFD. 400 VOLTS
+205 V.
56 K
100 K
.02 MFD. 600 VOLTS
+160 V.
25 MFD. 25 VOLTS
+1.5 V.
1 MEG.
1 MEG.
3300
.01 MFD. 400 VOLTS
.01 MFD. 400 VOLTS
.03 MFD. 400 VOLTS
.25 MFD. 400 VOLTS
+155 V.
220 K
100 K
.02 MFD. 600 VOLTS
+160 V.
68 K
68 K
68 K
1500
25 MFD. 25 VOLTS
+1.3 V.
1500
1500
AC LINE
SPKR. JACK
TREMOLO JACK
5Y3GT
6V6GT
6V6GT
12AX7
12AX7

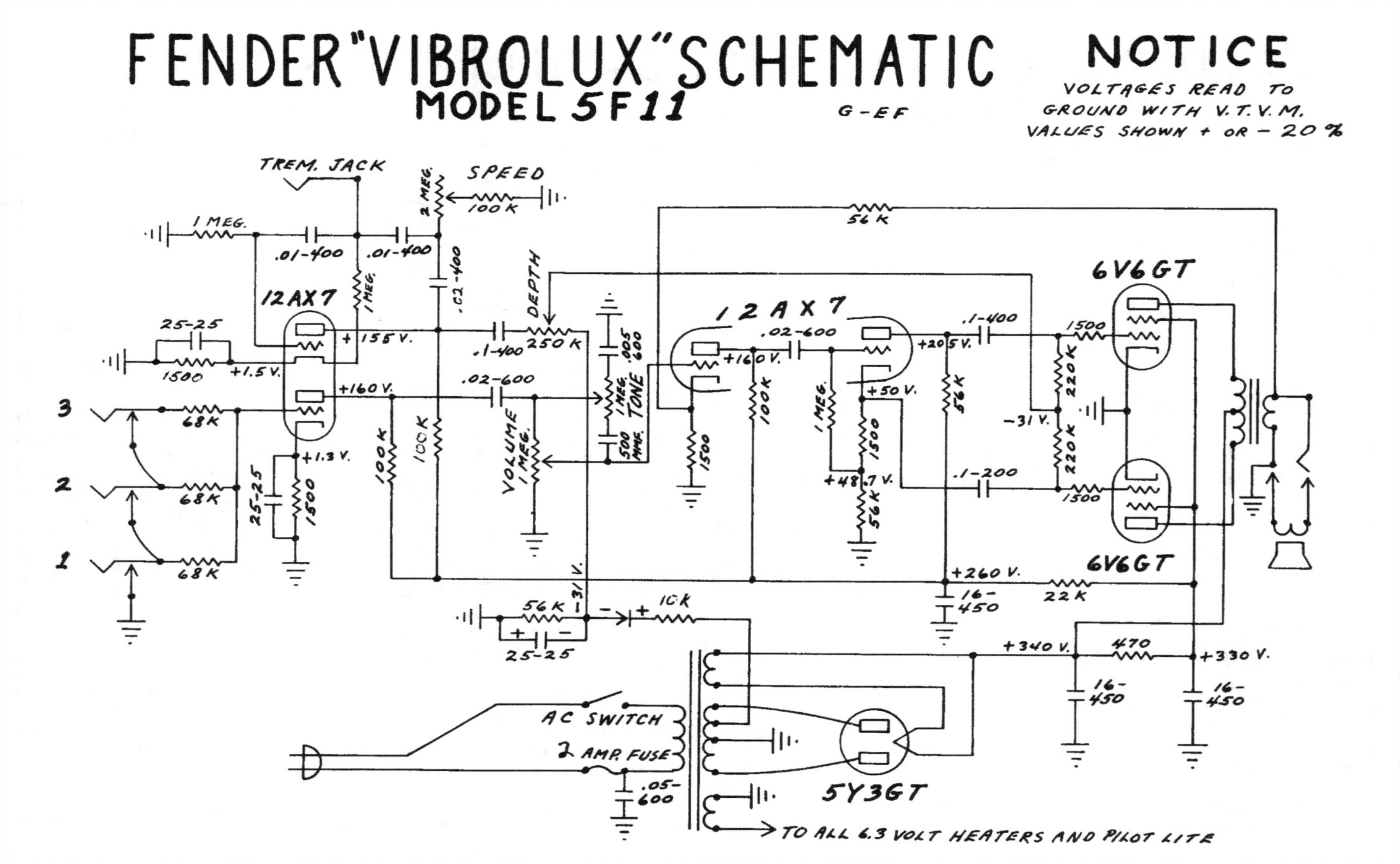
FENDER "VIBROLUX" SCHEMATIC
MODEL 5F11
G-EF
NOTICE
VOLTAGES READ TO GROUND WITH V.T.V.M.
VALUES SHOWN + OR − 20%
TREM. JACK
SPEED
100 K
2 MEG.
1 MEG.
.01-400
.01-400
.02-400
12AX7
1 MEG.
25-25
1500
+1.5 V.
+155 V.
+160 V.
+1.3 V.
3
2
1
68 K
68 K
68 K
25-25
1500
DEPTH
.1-400
250 K
.02-600
100 K
100 K
VOLUME 1 MEG.
.005 600
1 MEG. TONE
500 MMF.
12AX7
.02-600
+160 V.
1500
100 K
1 MEG.
+50 V.
1500
+48.7 V.
56 K
56 K
+205 V.
.1-400
56 K
6V6GT
1500
220 K
-31 V.
220 K
.1-200
1500
6V6GT
+260 V.
16-450
22 K
-31 V.
56 K
25-25
10 K
+340 V.
470
+330 V.
16-450
16-450
AC SWITCH
2 AMP. FUSE
.05-600
5Y3GT
TO ALL 6.3 VOLT HEATERS AND PILOT LITE

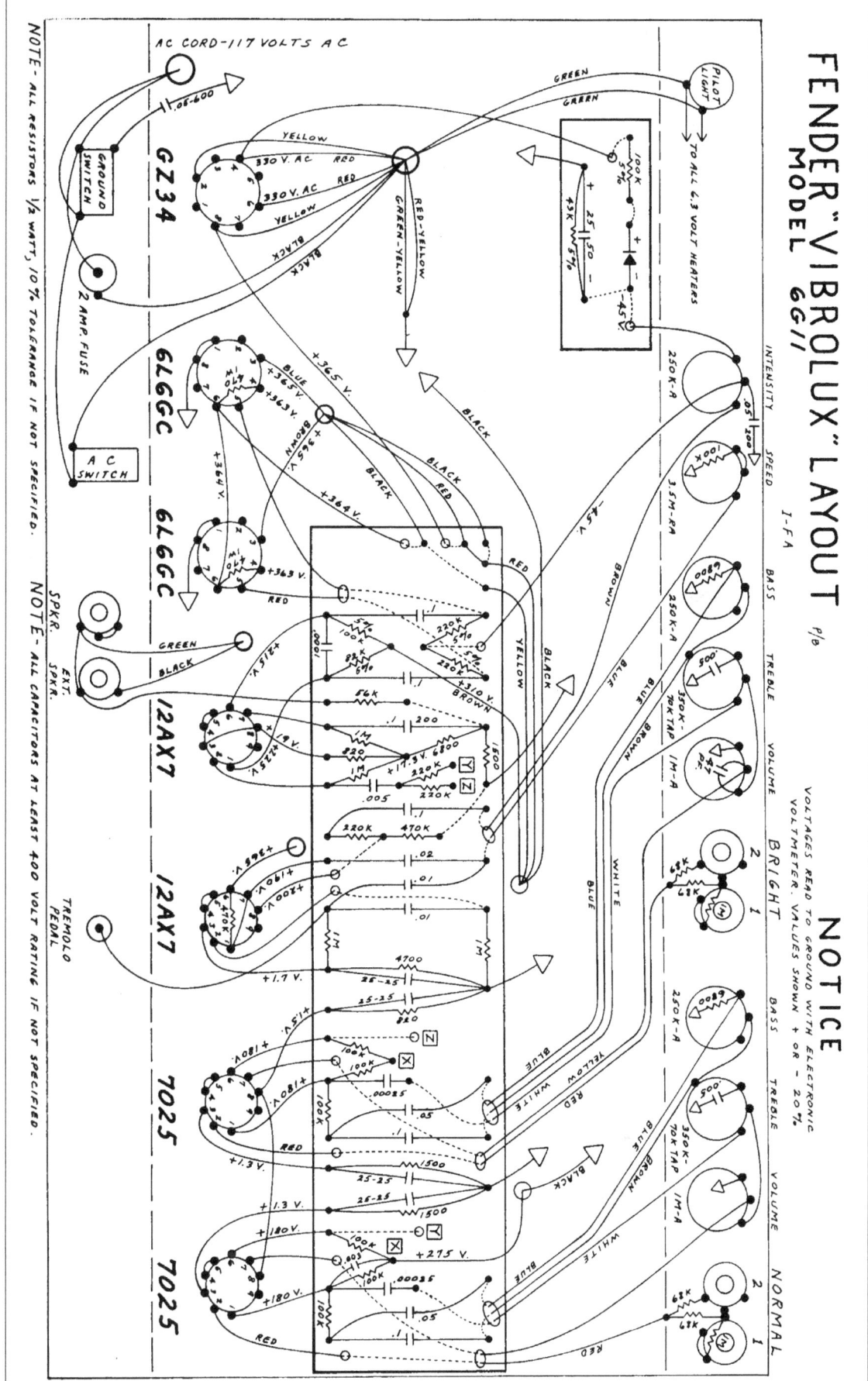

FENDER "VIBROLUX" LAYOUT P/B
MODEL 6G11

I-FA

NOTICE
VOLTAGES READ TO GROUND WITH ELECTRONIC VOLTMETER. VALUES SHOWN + OR − 20%

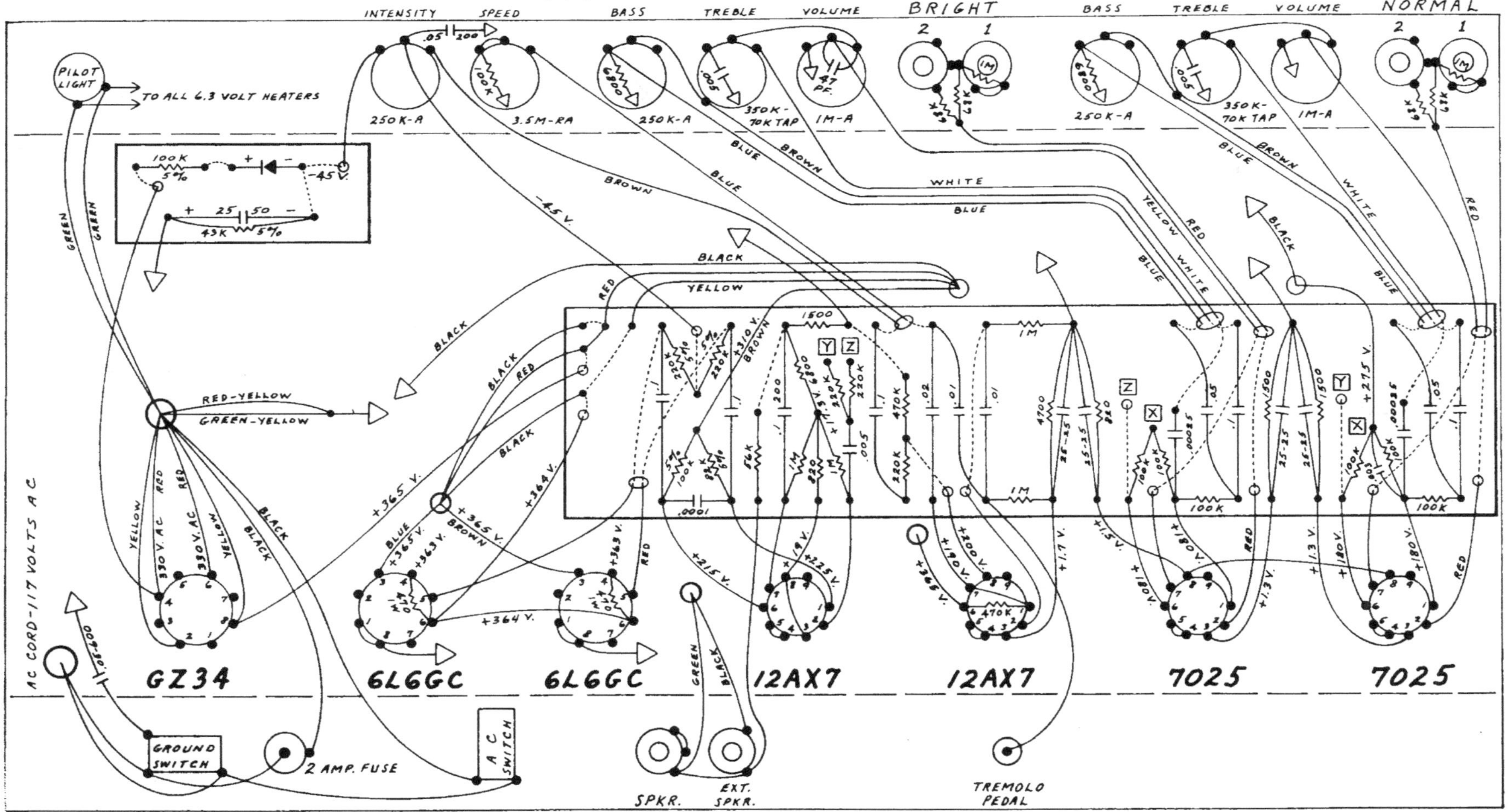

NOTE- ALL RESISTORS ½ WATT, 10% TOLERANCE IF NOT SPECIFIED.

NOTE- ALL CAPACITORS AT LEAST 400 VOLT RATING IF NOT SPECIFIED.

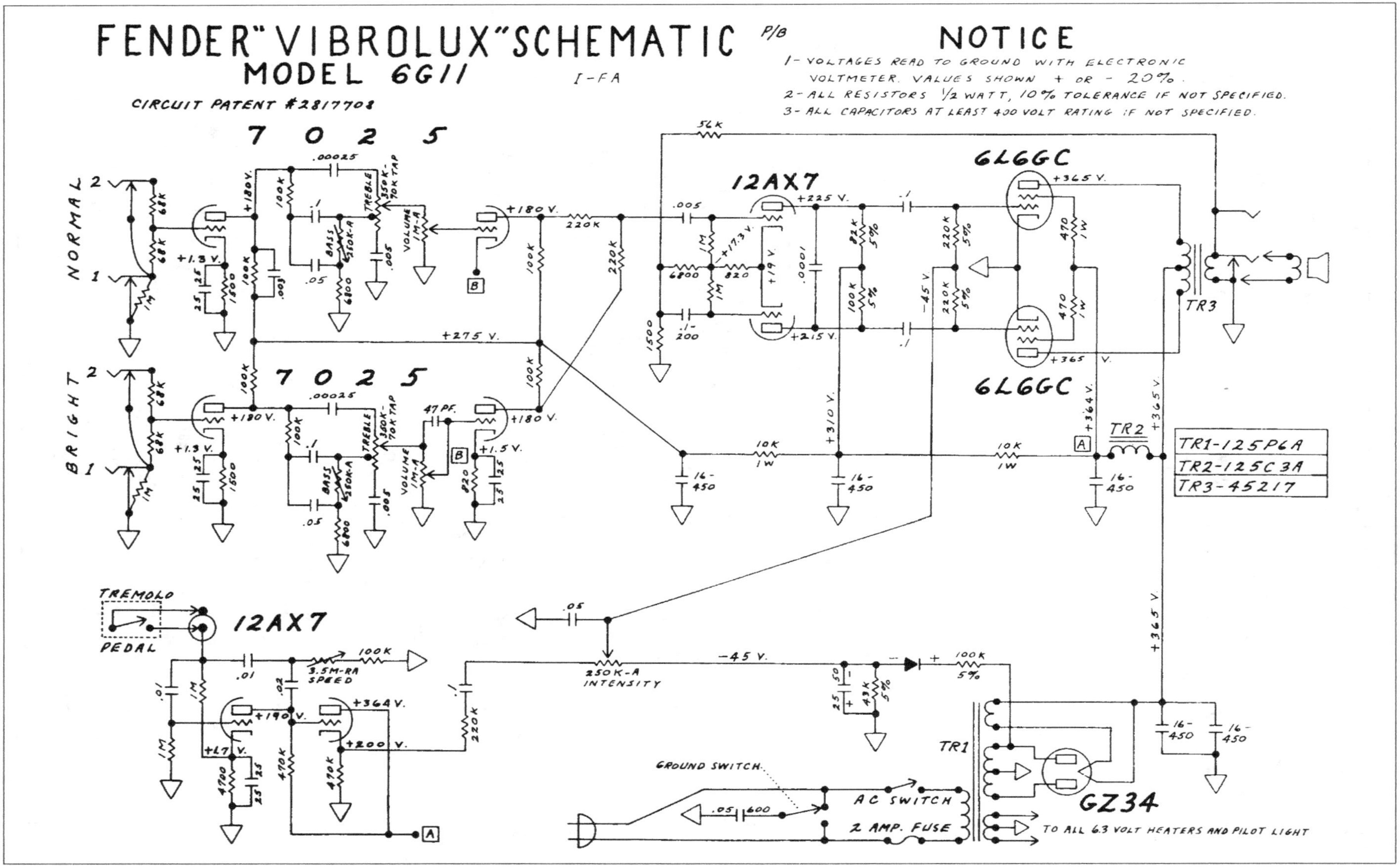
FENDER "VIBROLUX" SCHEMATIC
P/B
MODEL 6G11
I-FA
CIRCUIT PATENT #2817708
NOTICE
1- VOLTAGES READ TO GROUND WITH ELECTRONIC VOLTMETER. VALUES SHOWN + OR - 20%.
2- ALL RESISTORS 1/2 WATT, 10% TOLERANCE IF NOT SPECIFIED.
3- ALL CAPACITORS AT LEAST 400 VOLT RATING IF NOT SPECIFIED.
7025
12AX7
6L6GC
GZ34
NORMAL
BRIGHT
TREMOLO
PEDAL
TREBLE
BASS
VOLUME
3.5M-RA SPEED
250K-A INTENSITY
GROUND SWITCH
AC SWITCH
2 AMP. FUSE
TO ALL 6.3 VOLT HEATERS AND PILOT LIGHT
TR1-125P6A
TR2-125C3A
TR3-45217

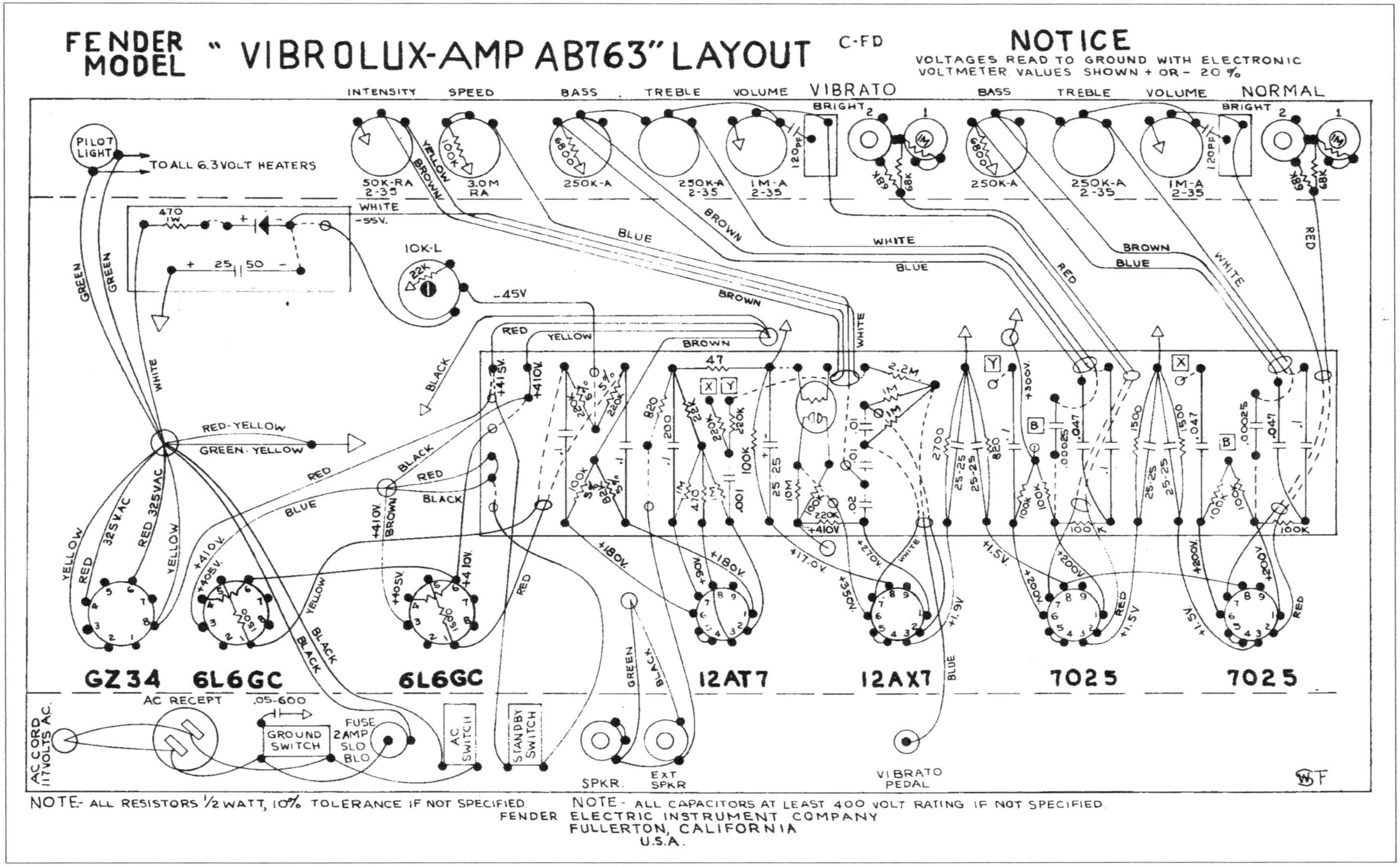
FENDER MODEL " VIBROLUX-AMP AB763" LAYOUT
C-FD
NOTICE
VOLTAGES READ TO GROUND WITH ELECTRONIC VOLTMETER VALUES SHOWN + OR - 20 %
INTENSITY
SPEED
BASS
TREBLE
VOLUME
VIBRATO
BRIGHT
NORMAL
PILOT LIGHT
TO ALL 6.3 VOLT HEATERS
50K-RA 2-35
3.0M RA
250K-A
250K-A 2-35
1M-A 2-35
120PF
68K
10K-L
22K
-45V
-55V.
470 1W
25 50
GREEN
WHITE
BLUE
BROWN
RED
YELLOW
BLACK
RED-YELLOW
GREEN-YELLOW
325V.AC
RED 325V.AC
+410V.
+405V.
+415V.
+180V.
+90V.
+17.0V.
+270V.
+350V.
+1.9V.
+1.5V.
+200V.
+300V.
GZ34
6L6GC
6L6GC
12AT7
12AX7
7025
7025
AC RECEPT
.05-600
GROUND SWITCH
FUSE 2AMP SLO BLO
AC SWITCH
STANDBY SWITCH
AC CORD 117VOLTS A.C.
SPKR.
EXT. SPKR
VIBRATO PEDAL
NOTE- ALL RESISTORS ½ WATT, 10% TOLERANCE IF NOT SPECIFIED
NOTE- ALL CAPACITORS AT LEAST 400 VOLT RATING IF NOT SPECIFIED
FENDER ELECTRIC INSTRUMENT COMPANY
FULLERTON, CALIFORNIA
U.S.A.

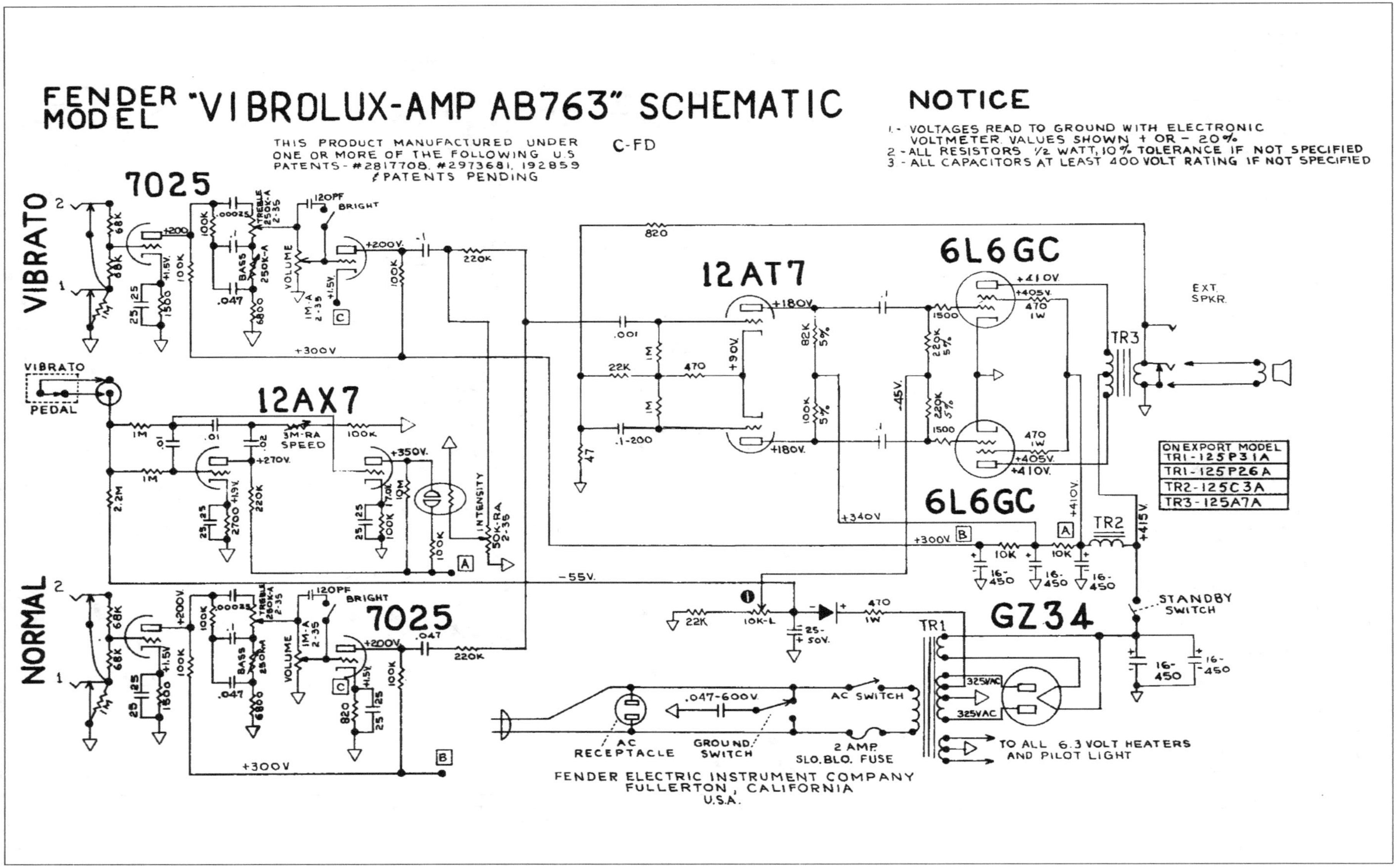
FENDER MODEL "VIBROLUX-AMP AB763" SCHEMATIC
THIS PRODUCT MANUFACTURED UNDER ONE OR MORE OF THE FOLLOWING U.S. PATENTS - #2817708, #2973681, 192859 & PATENTS PENDING
C-FD
NOTICE
1 - VOLTAGES READ TO GROUND WITH ELECTRONIC VOLTMETER. VALUES SHOWN + OR - 20%
2 - ALL RESISTORS 1/2 WATT, 10% TOLERANCE IF NOT SPECIFIED
3 - ALL CAPACITORS AT LEAST 400 VOLT RATING IF NOT SPECIFIED
VIBRATO
NORMAL
7025
12AX7
12AT7
6L6GC
GZ34
VIBRATO PEDAL
EXT. SPKR.
ON EXPORT MODEL TR1-125P31A
TR1-125P26A
TR2-125C3A
TR3-125A7A
STANDBY SWITCH
TO ALL 6.3 VOLT HEATERS AND PILOT LIGHT
AC RECEPTACLE
GROUND SWITCH
AC SWITCH
2 AMP SLO.BLO. FUSE
FENDER ELECTRIC INSTRUMENT COMPANY
FULLERTON, CALIFORNIA
U.S.A.

FENDER MODEL "VIBROLUX REVERB-AMP AA270" SCHEMATIC

THIS PRODUCT MANUFACTURED UNDER ONE OR MORE OF THE FOLLOWING U.S. PATENTS- #2817708, #2973681, 192859 & PATENTS PENDING.

NOTICE

1- VOLTAGES READ TO GROUND WITH ELECTRONIC VOLTMETER VALUES SHOWN + OR - 20%.
2- ALL RESISTORS ½ WATT 10% TOLERANCE IF NOT SPECIFIED.
3- ALL CAPACITORS AT LEAST 400 VOLT RATING IF NOT SPECIFIED.
4.- VIBRATO GROUNDED

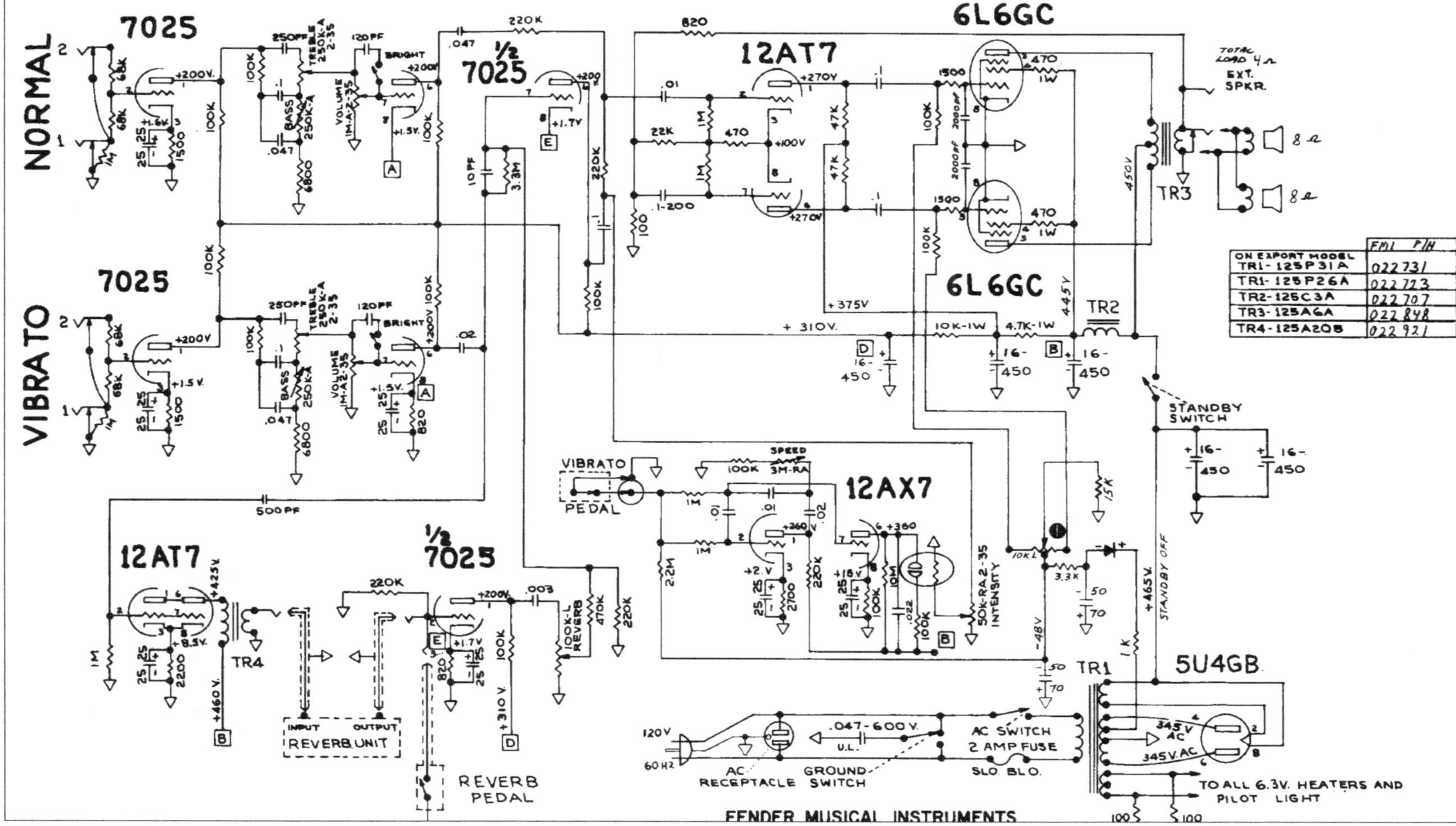

ON EXPORT MODEL	FMI P/N
TR1- 125P31A	022731
TR1- 125P26A	022723
TR2- 125C3A	022707
TR3- 125A6A	022848
TR4- 125A20B	022921

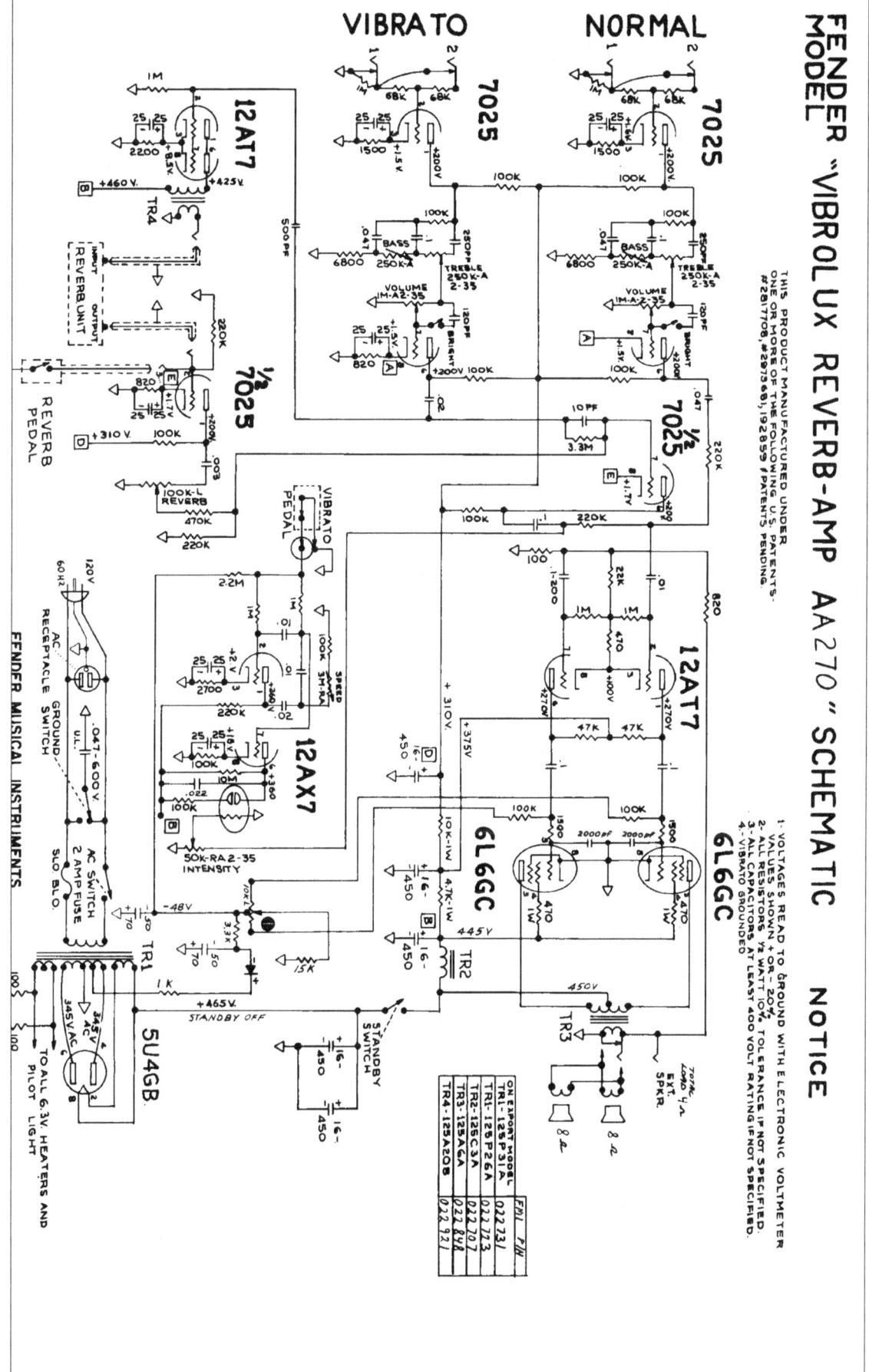

CBS 100w **TWIN REV/SUPER**

Fender

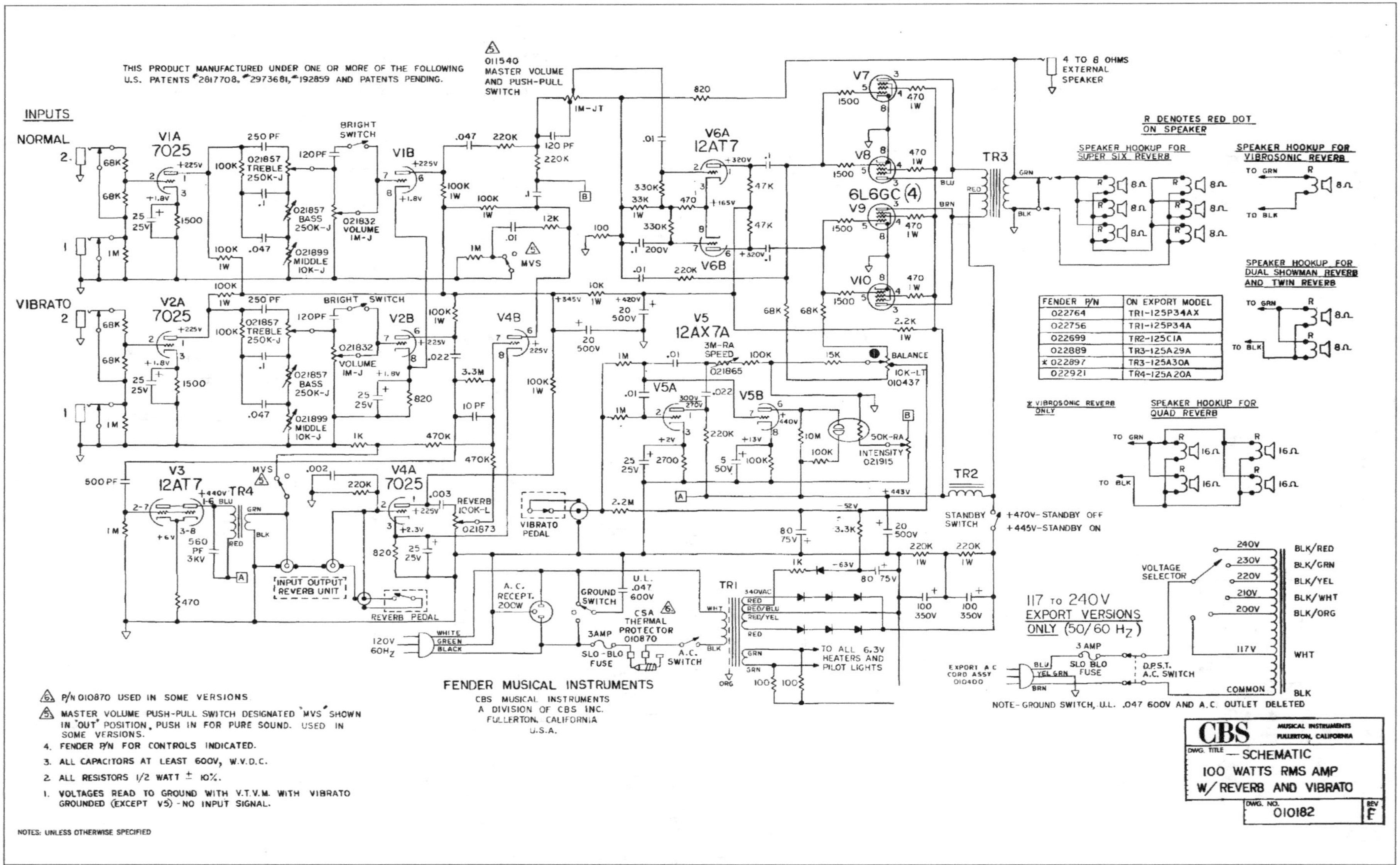

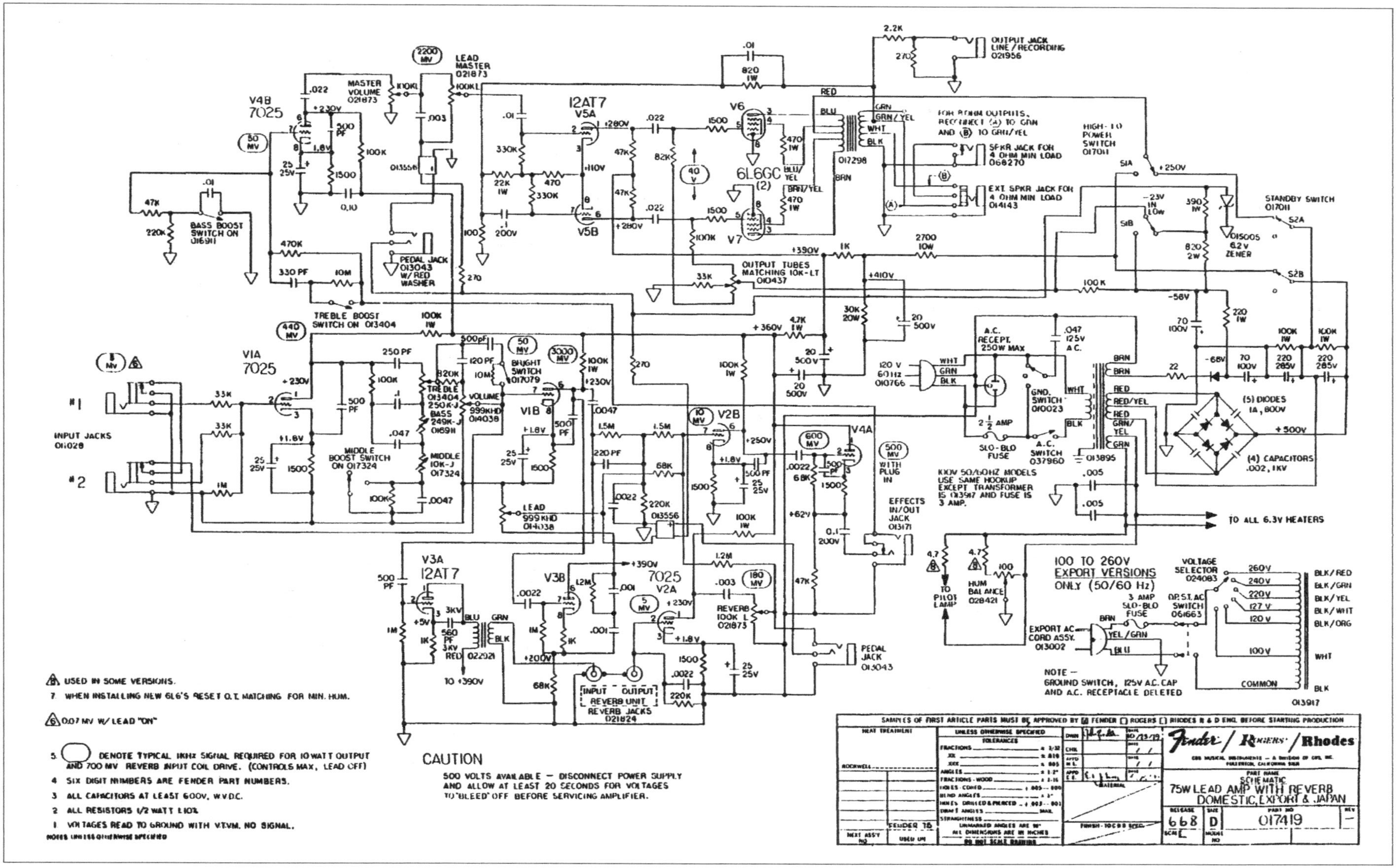

INPUT JACKS 011028
V1A 7025
V1B
V2A 7025
V2B
V3A 12AT7
V3B
V4A
V4B 7025
V5A 12AT7
V5B
V6
V7
6L6GC (2)
OUTPUT TUBES MATCHING 10K-LT 010437
MASTER VOLUME 021873
LEAD MASTER 021873
BASS BOOST SWITCH ON 016911
TREBLE BOOST SWITCH ON 013404
MIDDLE BOOST SWITCH ON 017324
BRIGHT SWITCH 017079
PEDAL JACK 013043 W/RED WASHER
INPUT OUTPUT REVERB UNIT REVERB JACKS 021824
REVERB 100K L 021873
EFFECTS IN/OUT JACK 013171
OUTPUT JACK LINE/RECORDING 021956
FOR 8 OHM OUTPUTS, RECONNECT (A) TO GRN AND (B) TO GRN/YEL
SPKR JACK FOR 4 OHM MIN LOAD 068270
EXT SPKR JACK FOR 4 OHM MIN LOAD 014143
HIGH-LO POWER SWITCH 017011
STANDBY SWITCH 017011
A.C. RECEPT. 250W MAX
GND. SWITCH 010023
A.C. SWITCH 037960
2 1/2 AMP SLO-BLO FUSE
100V 50/60HZ MODELS USE SAME HOOKUP EXCEPT TRANSFORMER IS 013917 AND FUSE IS 3 AMP.
(5) DIODES 1A, 800V
(4) CAPACITORS .002, 1KV
TO ALL 6.3V HEATERS
HUM BALANCE 028421
TO PILOT LAMP
100 TO 260V EXPORT VERSIONS ONLY (50/60 Hz)
3 AMP SLO-BLO FUSE
OP.ST.AC SWITCH 061663
VOLTAGE SELECTOR 024083
EXPORT AC CORD ASSY. 013002
NOTE – GROUND SWITCH, 125V A.C. CAP AND A.C. RECEPTACLE DELETED
8 USED IN SOME VERSIONS.
7 WHEN INSTALLING NEW 6L6'S RESET O.T. MATCHING FOR MIN. HUM.
6 0.07 MV W/ LEAD "ON"
5 DENOTE TYPICAL 1KHz SIGNAL REQUIRED FOR 10 WATT OUTPUT AND 700 MV REVERB INPUT COIL DRIVE. (CONTROLS MAX, LEAD OFF)
4 SIX DIGIT NUMBERS ARE FENDER PART NUMBERS.
3 ALL CAPACITORS AT LEAST 600V, W.V.D.C.
2 ALL RESISTORS 1/2 WATT ±10%
1 VOLTAGES READ TO GROUND WITH V.T.V.M. NO SIGNAL.
NOTES UNLESS OTHERWISE SPECIFIED
CAUTION
500 VOLTS AVAILABLE – DISCONNECT POWER SUPPLY AND ALLOW AT LEAST 20 SECONDS FOR VOLTAGES TO "BLEED" OFF BEFORE SERVICING AMPLIFIER.
SAMPLES OF FIRST ARTICLE PARTS MUST BE APPROVED BY FENDER ROGERS RHODES R & D ENG. BEFORE STARTING PRODUCTION
Fender / Rogers / Rhodes
SCHEMATIC
75W LEAD AMP WITH REVERB DOMESTIC, EXPORT & JAPAN
668 D 017419

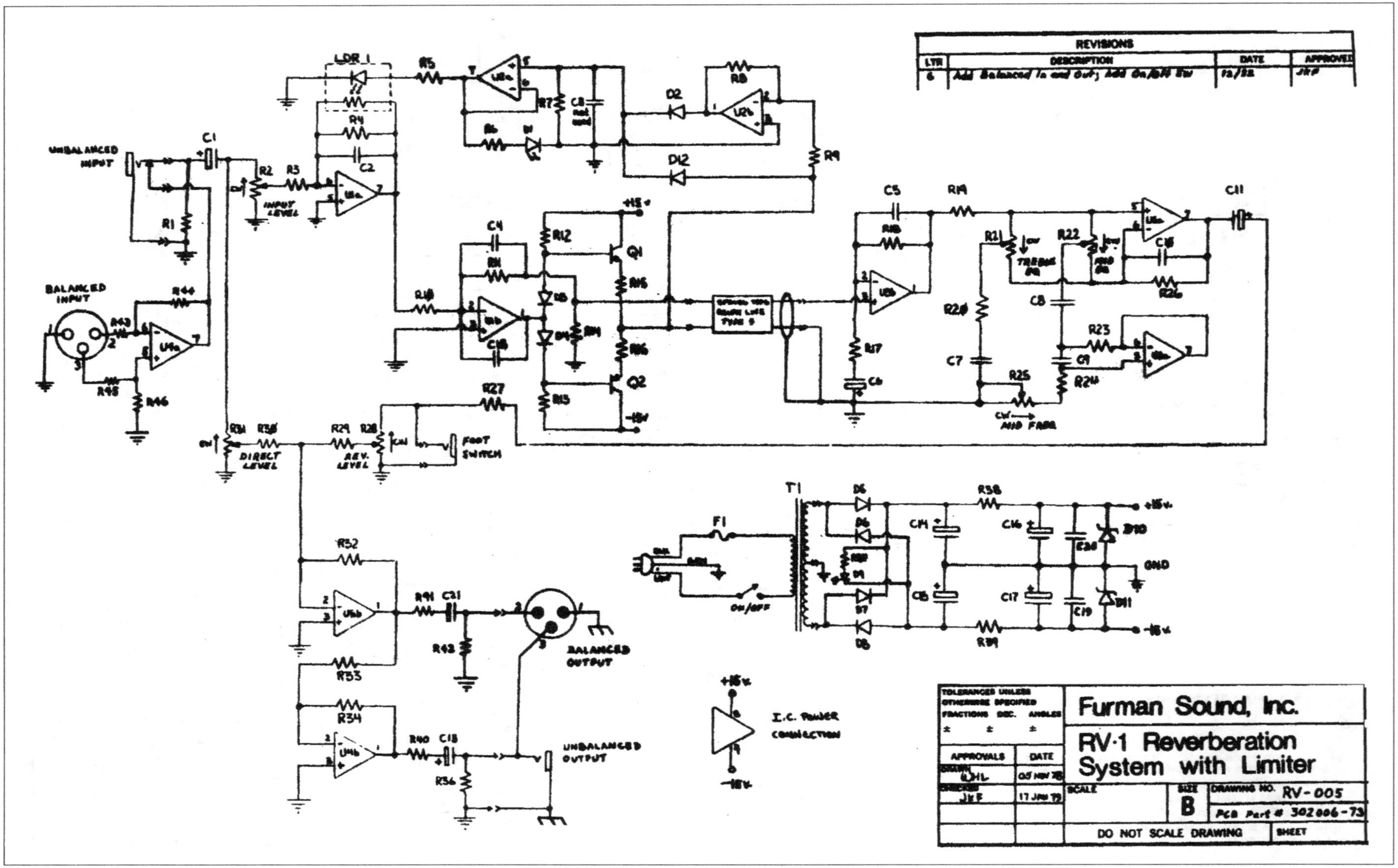
REVISIONS
LTR
DESCRIPTION
DATE
APPROVED
UNBALANCED INPUT
BALANCED INPUT
INPUT LEVEL
DIRECT LEVEL
REV. LEVEL
FOOT SWITCH
BALANCED OUTPUT
UNBALANCED OUTPUT
ON/OFF
GND
+15v
-15v
I.C. POWER CONNECTION
LDR 1
R5
R2
R3
R1
R44
R43
R45
R46
R4
C1
C2
R8
R9
D2
D12
C4
R12
R15
Q1
Q2
R27
R29
R28
C5
R19
R17
C6
C7
R20
R21
R22
C8
C9
R23
R24
R25
C11
R26
T1
F1
D6
D8
R38
R39
C14
C16
C17
R32
R33
R34
R40
R41
C21
C18
R42
R36
Furman Sound, Inc.
RV·1 Reverberation System with Limiter
TOLERANCES UNLESS OTHERWISE SPECIFIED
FRACTIONS
DEC.
ANGLES
APPROVALS
DATE
DRAWN
CHECKED
05 NOV 78
17 JAN 79
SCALE
SIZE
B
DRAWING NO. RV-005
PCB Part # 302006-73
DO NOT SCALE DRAWING
SHEET

www.ingramcontent.com/pod-product-compliance
Lightning Source LLC
Chambersburg PA
CBHW081432170526
45166CB00008B/2183
* 9 7 8 1 5 0 0 4 4 7 3 6 6 *